modern
soil
microbiology

BOOKS IN SOILS, PLANTS, AND THE ENVIRONMENT

Soil Biochemistry, Volume 1, edited by A. D. McLaren and G. H. Peterson
Soil Biochemistry, Volume 2, edited by A. D. McLaren and J. Skujiňš
Soil Biochemistry, Volume 3, edited by E. A. Paul and A. D. McLaren
Soil Biochemistry, Volume 4, edited by E. A. Paul and A. D. McLaren
Soil Biochemistry, Volume 5, edited by E. A. Paul and J. N. Ladd
Soil Biochemistry, Volume 6, edited by Jean-Marc Bollag and G. Stotzky
Soil Biochemistry, Volume 7, edited by G. Stotzky and Jean-Marc Bollag
Soil Biochemistry, Volume 8, edited by Jean-Marc Bollag and G. Stotzky
Soil Biochemistry, Volume 9, edited by G. Stotzky and Jean-Marc Bollag

Organic Chemicals in the Soil Environment, Volumes 1 and 2, edited by C. A. I. Goring and J. W. Hamaker
Humic Substances in the Environment, M. Schnitzer and S. U. Khan
Microbial Life in the Soil: An Introduction, T. Hattori
Principles of Soil Chemistry, Kim H. Tan
Soil Analysis: Instrumental Techniques and Related Procedures, edited by Keith A. Smith
Soil Reclamation Processes: Microbiological Analyses and Applications, edited by Robert L. Tate III and Donald A. Klein
Symbiotic Nitrogen Fixation Technology, edited by Gerald H. Elkan
Soil/Water Interactions: Mechanisms and Applications, Shingo Iwata and Toshio Tabuchi with Benno P. Warkentin
Soil Analysis: Modern Instrumental Techniques, Second Edition, edited by Keith A. Smith
Soil Analysis: Physical Methods, edited by Keith A. Smith and Chris E. Mullins
Growth and Mineral Nutrition of Field Crops, N. K. Fageria, V. C. Baligar, and Charles Allan Jones
Semiarid Lands and Deserts: Soil Resource and Reclamation, edited by J. Skujiňš
Plant Roots: The Hidden Half, edited by Yoav Waisel, Amram Eshel, and Uzi Kafkafi
Plant Biochemical Regulators, edited by Harold W. Gausman
Maximizing Crop Yields, N. K. Fageria

Modern Soil Microbiology, edited by J. D. van Elsas, J. T. Trevors, and E. M. H. Wellington

Growth and Mineral Nutrition of Field Crops: Second Edition, N. K. Fageria, V. C. Baligar, and Charles Allan Jones

Fungal Pathogenesis in Plants and Crops: Molecular Biology and Host Defense Mechansims, P. Vidhyasekaran

Additional Volumes in Preparation

Plant Pathogen Detection and Disease Diagnosis, P. Narayanasamy

Agricultural Systems Modeling and Simulation, edited by Robert M. Peart and R. Bruce Curry

Agricultural Biotechnology, edited by Arie Altman

Plant–Microbe Interactions and Biological Control, edited by Gregory J. Boland and L. David Kuykendall

Handbook of Soil Conditioners, edited by Arthur Wallace and Richard E. Terry

modern
soil
microbiology

edited by

Jan Dirk van Elsas
Soil Biotechnology Group
Research Institute for Plant Protection (IPO-DLO)
Wageningen, The Netherlands

Jack T. Trevors
Department of Environmental Biology
University of Guelph
Guelph, Ontario, Canada

Elizabeth M. H. Wellington
Department of Biological Sciences
University of Warwick
Coventry, United Kingdom

 MARCEL DEKKER, INC.   NEW YORK · BASEL · HONG KONG

Library of Congress Cataloging-in-Publication Data

Modern soil microbiology / edited by Jan Dirk van Elsas, Jack T. Trevors, Elizabeth M. H. Wellington
    p. cm. — (Books in soils, plants, and the environment: v. 56)
    Includes bibliographical references and index
    ISBN 0-8247-9436-2 (alk. paper)
    1. Soil microbiology. 2. Molecular microbiology. I. Elsas, J. D. van (Jan D.) . II. Trevors, Jack T. III. Wellington, E. M. H. (Elizabeth M. H.). IV. Series.
QR 111.M58 1997
597'.1757—dc21

97-12600
CIP

The publisher offers discounts on this book when ordered in bulk quantities. For more information, write to Special Sales/Professional Marketing at the address below.

This book is printed on acid-free paper.

MARCEL DEKKER, INC.
270 Madison Avenue, New York, New York 10016
http://www.dekker.com

Current printing (last digit):
10 9 8 7 6 5 4 3 2 1

PRINTED IN THE UNITED STATES OF AMERICA

Preface

Soil microbiology is a discipline that describes the numbers, fate, activity, and interactions of microorganisms present in soil, and how they are affected by their environment. The soil microbial community is largely responsible for the cycling of carbonaceous, nitrogenous, and phosphorous compounds. Soil microorganisms can be involved in plant–pathogenic reactions, as well as in biological transformations of xenobiotic compounds added to soil. Soil also represents a natural reservoir of genetic information and soil organisms, many of which are unculturable and hence unknown.

There are good reasons to study and understand the principles underlying soil microbiological phenomena. Understanding the function of the soil ecosystem in relation to ever changing soil conditions is key to understanding the basic mechanisms of soil productivity. This is important in light of the urgency to change agricultural practices and also because of current problems of xenobiotic compounds in soils. The capacity of soil microbial populations to cope with xenobiotic and other stresses—the possible perturbations caused by pollution, intense agricultural practice, or changing land use—are of major interest. In addition, the possibility of involvement of nonculturable or minute cell fractions in these and other important soil processes presents an intriguing topic for innovative research.

Recent developments in the application of molecular biological techniques to ecological questions have revolutionized concepts in soil microbial ecology. In particular, soil DNA and RNA extraction, nucleic acid reassociation, determination of G + C content of whole microbial populations, specific hybridization and polymerase chain reaction (PCR) techniques, cloning and sequencing, denaturing gradient gel electrophoresis (DGGE) of 16S rDNA based soil DNA amplicons, as well as reporter gene technology, have proven to be powerful tools for the assessment of the ecology of microorganisms in their natural environment. These as well as techniques based on the detection of other cellular macromolecules, such as fatty acid lipids, have also provided new insight into elucidation of the nonculturable fraction (the silent majority) in soil microbial populations and have shed light on the quantity and activity of the culturable bacterial fractions.

In addition, the molecular biology of the starvation and stress survival programs in certain nondifferentiating soil bacteria is beginning to be elucidated, as is the response of bacteria to environmental triggers such as those occurring in bulk soil or in the rhizosphere. Knowledge of the effect of the soil environment on the physiological state of bacteria in soil is an important developing area of research, and interesting information on the effects of different root parts versus bulk soil have been obtained.

The issue of putative biosafety problems associated with the release of genetically modified microorganisms to soil has spurred research on the extent to which gene transfer between microorganisms takes place in soil as affected by soil factors. These studies have also provided insight into the role gene transfer plays in the evolution of microorganisms in their natural environment.

In light of the intrinsic complexity of soil and the potential offered by several recent exciting methodological developments, soil microbiology can be considered one of the last frontiers in science. There has been great progress lately, and even greater progress will be made in the near future in addressing and unraveling some of the great unknowns of soil microbiology.

The novel techniques applied recently to soil and new insights into the diversity as well as physiology of bacteria in this habitat represent knowledge not presently covered in any other textbook. This text provides a description of currently accepted principles of soil microbiology, updated and combined with insights obtained via the use of molecular techniques. The intended readership are graduate and undergraduate students as well as professionals who require a quick entry into this fascinating intersection of classical soil microbiology and rapidly evolving molecular biology applied to soil microbial ecology.

The first section describes accepted concepts of soil microbiology, picturing the soil and rhizosphere as habitats for microbes to live in, discussing different soil microbial groups (bacteria, fungi, protozoa, and nematodes), and microbial processes and interactions. The second section is devoted to the composition and activity (gene expression) of soil microbial populations, as determined by molecular techniques. This new information is integrated with current knowledge on soil microbiology. These chapters are unique in that no other current soil microbiology volume has managed to cover all the different new developments. The final section addresses a series of applied aspects of soil microbiology. Here, basic microbial ecology in areas such as the ecology of plant pathogenesis, bioremediation of soil, perturbations by heavy metal pollution, and soil manipulation via plowing or manuring is discussed.

We would like to thank all the authors who contributed to this text for their expertise and patience. Special thanks are also due to the editors at Marcel Dekker, Inc. for their excellent work.

Jan Dirk van Elsas
Jack T. Trevors
Elizabeth M. H. Wellington

Contents

Contents

Section II.   Modern Methodology and Approaches

Section III. Soil Microbial Processes and Soil Perturbance

Contributors

Bernhard Aßmus, Ph.D. Institute for Soil Ecology, GSF-National Research Center for Environment and Health, Neuherberg, Germany

Hans-Jürgen Bach, Ph.D. Institute for Soil Ecology, GSF-National Research Center for Environment and Health, Neuherberg, Germany

Mark J. Bailey, Ph.D. Institute of Virology and Environmental Microbiology, Laboratory of Molecular Microbial Ecology, Oxford, United Kingdom

Lars Reier Bakken, Ph.D. Department of Soil and Water Sciences, Agricultural University of Norway, Aas, Norway

Richard D. Bardgett, Ph.D. School of Biological Sciences, University of Manchester, Manchester, United Kingdom

Eric G. Beauchamp, Ph.D. Department of Land Resource Science, University of Guelph, Guelph, Ontario, Canada

Jaap Bloem, Ph.D. DLO Research Institute for Agrobiology and Soil Fertility (AB-DLO), Haren, The Netherlands

Lucas Bouwman, Ph.D. DLO Research Institute for Agrobiology and Soil Fertility (AB-DLO), Haren, The Netherlands

Penny A. Bramwell, Ph.D. Institute of Virology and Environmental Microbiology, Laboratory for Molecular Microbial Ecology, Oxford, United Kingdom

John Burden, Ph.D. Department of Biological Sciences, University of Warwick, Coventry, United Kingdom

Mike Cassidy, B.Sc. Department of Environmental Biology, University of Guelph, Guelph, Ontario, Canada

Peter de Ruiter, Ph.D. DLO Research Institute for Agrobiology and Soil Fertility (AB-DLO), Haren, The Netherlands

Richard J. Ellis, Ph.D. Institute of Virology and Environmental Microbiology, Laboratory for Molecular Microbial Ecology, Oxford, United Kingdom

Deena Errampalli, Ph.D. Department of Environmental Biology, University of Guelph, Guelph, Ontario, Canada

Bryan S. Griffiths, Ph.D. Unit of Integrative Bioscience, Cellular and Environmental Physiology Department, Scottish Crop Research Institute, Dundee, United Kingdom

Anton Hartmann, Ph.D. Institute for Soil Ecology, GSF-National Research Center for Environment and Health, Neuherberg, Germany

B. Hall Department of Industrial Microbiology, University College Dublin, Dublin, Ireland

Holger Heuer Institute for Biochemistry and Plant Virology, Federal Biological Research Centre for Agriculture and Forestry, Braunschweig, Germany

David J. Hume, Ph.D. Crop Science Department, University of Guelph, Guelph, Ontario, Canada

Peter H. Janssen, D.Phil. Department of Microbiology, University of Melbourne, Parkville, Victoria, Australia

Gundrun Kirchhof, Ph.D. Institute for Soil Ecology, GSF-National Research Center for Environment and Health, Neuherberg, Germany

Hung Lee, Ph.D. Department of Environmental Biology, University of Guelph, Guelph, Ontario, Canada

Kam Tin Leung, Ph.D. Center for Environmental Biotechnology, University of Tennessee, Knoxville, Tennessee

Werner Liesack, Ph.D. Department of Biogeochemistry, Max-Planck-Institut für terrestrische Mikrobiologie, Marburg, Germany

Andrew K. Lilley, Ph.D. Institute of Virology and Environmental Microbiology, Laboratory for Molecular Microbial Ecology, Oxford, United Kingdom

Peter Marsh, B.Sc. (Hons), Ph.D. Department of Biological Sciences, University of Warwick, Coventry, United Kingdom

Brian B. McSpadden Gardener Department of Botany and Plant Pathology, Michigan State University, East Lansing, Michigan

Max Mergeay, Ph.D. Department of Environmental Technology, Flemish Institute for Technological Research (VITO), Mol, Belgium

J. Alun W. Morgan, Ph.D. Department of Plant Pathology and Microbiology, Horticulture Research International, Wellesbourne, United Kingdom

Hideo Okamura, Ph.D. Research Institute for Bioresources, Okayama University, Kurashiki, Japan

James I. Prosser, Ph.D. Department of Molecular and Cell Biology, Institute of Medical Sciences, University of Aberdeen, Aberdeen, United Kingdom

Frederick A. Rainey, D.Phil. Deutsche Sammlung von Mikroorganismen und Zellkulturen GmbH, Braunschweig, Germany

Michael Schloter, Ph.D. Institute for Soil Ecology, GSF-National Research Institute for Environment and Health, Neuherberg, Germany

Kornelia Smalla, Ph.D. Institute for Biochemistry and Plant Virology, Federal Biological Research Centre for Agriculture and Forestry, Braunschweig, Germany

Jan Sørensen, Ph.D. Section of Genetics and Microbiology, Department of Ecology and Molecular Biology, Royal Veterinary and Agricultural University, Copenhagen, Denmark

Guy Soulas, Ph.D. Department of Soil Science, National Institute of Agricultural Research (INRA), Dijon, France

Erko Stackebrandt, Ph.D. Department of Molecular Ecology, Deutsche Sammlung von Mikroorganismen und Zellkulturen GmbH, Braunschweig, Germany

G. Stotzky, Ph.D. Department of Biology, New York University, New York, New York

Ian P. Thompson, Ph.D. Microbial Diversity Group, Institute of Virology and Environmental Microbiology, Oxford, United Kingdom

Greg Thorn, Ph.D. Department of Botany, University of Wyoming, Laramie, Wyoming

Edward Topp, Ph.D. Pest Management Research Centre, Agriculture and Agri-Food Canada, London, Ontario, Canada

Jack T. Trevors, Ph.D. Department of Environmental Biology, University of Guelph, Guelph, Ontario, Canada

Tatiana Vallaeys, Ph.D. Laboratoire de Microbiologie des Sols, Centre de Microbiologie des Sols et de l'Environnement, National Institute of Agricultural Research (INRA), Dijon, France

Jan Dirk van Elsas, Ph.D. Soil Biotechnology Group, Research Institute for Plant Protection (IPO-DLO), Wageningen, The Netherlands

Leo van Overbeek, Ph.D. Soil Biotechnology Group, Research Institute for Plant Protection (IPO-DLO), Wageningen, The Netherlands

Naomi L. Ward-Rainey, Ph.D. Deutsche Sammlung von Mikroorganismen und Zellkulturen GmbH, Braunschweig, Germany

Joy Elizabeth Margaret Watts, B.Sc. Department of Biological Sciences, University of Warwick, Coventry, United Kingdom

Elizabeth M. H. Wellington, Ph.D. Department of Biological Sciences, University of Warwick, Coventry, United Kingdom

John M. Whipps, Ph.D. Plant Pathology and Microbiology Department, Horticulture Research International, Wellesbourne, Warwick, United Kingdom

Craig Winstanley, B.Sc. Ph.D. School of Natural and Environmental Sciences, Coventry University, Coventry, United Kingdom

Stefan Wuertz,* Ph.D. Department of Environmental Technology, Flemish Institute for Technological Research (VITO), Mol, Belgium

---

*Current affiliation: Technical University of Munich, Garching, Germany

modern
soil
microbiology

1

Soil as an Environment for Microbial Life

G. STOTZKY New York University, New York, New York

1. INTRODUCTION

The concept of soil as an environment for microbial life is based on a number of truisms and on a number of traditional but unsupported assumptions. Among the truisms are that soil is a complex habitat and that it has a high solid/liquid ratio, which distinguishes it from most other natural habitats. Among the unsupported assumptions are that soil is an unfavorable environment for microbes and that the ecology and species diversity of microbes in soil cannot be defined. The assumptions are, for the most part, nonsensical and based on the failure of investigators in this area to ask the correct and pertinent questions and to divorce themselves from anthropocentric concepts of what is unfavorable ("bad") or favorable ("good") for microbes. On the basis of a cursory and superficial overview, soil should be a "bad" habitat for microbes, as it is generally poor in available nutrients (especially in available carbon and energy sources, although limitations in other nutrients may be as or more restrictive), and it is constantly exposed to the vagaries and extremes of the environmental conditions that are conducive to microbial growth (e.g., available water, temperature, radiation, nutrients, osmotic pressure). However, soil contains more genera and species of microorganisms than other microbial habitats, as soil is exposed to and eventually receives essentially all microbes present on Earth. Some of these species are present in low numbers, probably because the conditions for their survival and growth are restricted to discrete sites in which the nutritional and other physicochemical environmental factors necessary for their establishment, growth, and survival are located. This is one of the results of a structured environment with a high solid/liquid ratio. Consequently, in terms of the numbers and diversity of microbes in soil, it appears to be a "good" habitat for microorganisms, despite the apparent limitations for their survival and growth, probably because microbes indigenous to soil are well adapted to this austere environment. However, many exogenous microbes are not naturally adapted to and do not survive in soil. Most microbes indigenous (autochthonous) to soil are oligotrophic or zymogenous, which may explain their higher numbers and greater diversity in soil than in other habitats that are more nutrient-rich (e.g., skin, oral and vaginal cavities, gastrointestinal and urinary tracts). It may also explain why many copiotrophic species from these habitats do not apparently establish, grow, and survive in soil.

Microbes, perhaps more than any other organisms, are highly adaptable, both physiologically and genetically, to varying conditions. This has been necessary, as they have been around for more than 3.5×10^9 years. In addition, soil obviously selects for and enriches some microbes, and there is a homeostasis that has probably existed for millions, if not billions, of years. It is this homeostasis that human beings are constantly attempting to alter for the presumed benefit of mankind (e.g., enhanced production of food and fiber). These alterations have involved primarily inoculation of soil (e.g., with *Rhizobium* sp. to enhance nitrogen fixation by legumes; with mycorrhizal fungi to enhance uptake of phosphorus by plants; with *Phanerochaete* sp. to enhance degradation of toxic and persistent organic pollutants; with a variety of rhizobacteria and other microbial antagonists of soil-borne root-infecting pathogens). Hundreds, probably thousands, of papers have been published on the results of such inoculations. Most of the data in these papers have been empirical, and the "bottom line" is that, for the most part, these inoculations have been unsuccessful in situ, especially in the long term, although some have been successful and appeared promising in the laboratory and even in the greenhouse.

2. INOCULATION OF SOIL

Why have soil inoculations usually not been successful? It is probably because the emphasis has been directed to altering, presumably improving, the physiological characteristics of the microbes. This emphasis has been ill directed, as it incorrectly assumes that the physiological characteristics of microbes, which have evolved for billions of years and have adapted to and have been selected by the soil environment, can be changed significantly to improve their establishment, growth, survival, and desired activities in soil. Even when microbes have been genetically modified by the techniques of molecular biology, which presumably simulates the millennia of evolution, their survival in soil is hardly improved. In the case of genetically modified organisms, the introduced genes usually code for catabolic functions (e.g., degradation of xenobiotics), for markers (e.g., antibiotic resistance, *lux* genes) that allow monitoring their persistence in soil, or for some product (e.g., the insecticidal toxins produced by subspecies of *Bacillus thuringiensis*). However, even in the presence of the specific substrates on which, for example, the products of catabolic genes function, which would presumably provide an ecological advantage to the modified organisms, the introduced genetically modified organisms seldom survive longer in soil than the unmodified parental strains (e.g., 8,21,25,26). Does this suggest that there are "survival" genes? If so, then genetic engineers who want to develop strains that will perform the functions for which they have been genetically modified and will persist long enough in soil to perform these functions should look for and clone such genes. However, it is doubtful that such specific survival genes exist, as survival is probably a function of the synergistic action of numerous "normal" genes that regulate the overall metabolism and acquisition of nutrients from soil or any other environment.

Consequently, the major emphasis should not be on the physiological features of potential inoculants but rather on the physicochemical characteristics of soil that affect the establishment, growth, survival, and activity of the inoculants, as well as of the indigenous microbiota. The importance of the indigenous microbiota in achieving the aims of mankind should not be minimized. For example, control of pathogens (the term "pathogens" is another anthropocentric concept) occurs naturally in soil as part of the normal competitive, antagonistic, parasitic, and predatory interplay between microbes, although often at levels below those desired by mankind. Most organisms that have been introduced to soil as biocontrol agents were originally isolated from soil (e.g., nematode-trapping fungi, antibiotic-producing actinomycetes, antagonistic rhizobacteria, *B. thuringiensis*) where they were exerting some level of control. The large-scale in vitro cultivations of these organisms and their subsequent reintroductions to soil have been, for the most part, unsuccessful, probably because the physicochemical and biological factors that kept these indigenous organisms at some low level of activity also reduced their numbers to the same low levels shortly after reintroduction. Many novices when first beginning their studies will isolate a microorganism from soil (usually an actinomycete or a fungus) that will show significant inhibition of a plant pathogen on an agar plate (remember Sir A. Fleming!) and become excited. However, when this antagonistic organism, after in vitro cultivation, is added to natural soil, either with or without the target pathogen, its population density quickly declines, as does its biocontrol activity (e.g., 19,20).

The same phenomena have been observed for the introduction of microbes that, for example, degrade xenobiotics, fix nitrogen as either nodule formers or free-living microorganisms, or establish mycorrhizal associations. Again, the colonization and subsequent activity of these inoculants are not restricted by their physiological characteristics but by the physicochemical characteristics of the recipient soil. However, the possibility that in vitro cultivation of these microorganisms changes their physiological traits so that they lose their ability to survive and grow in soil after reintroduction needs to be studied.

3. SOIL PHYSIOCHEMICAL CHARACTERISTICS THAT AFFECT BACTERIAL GROWTH AND SURVIVAL

The physicochemical characteristics of soil that have been demonstrated, to varying degrees, to affect the activity, ecology, and population dynamics of microorganisms in soil are listed in Table 1 (the first 13 characteristics) and are briefly discussed later. If the specific characteristics that control the growth and survival of individual species were known, it might be possible to manipulate

Table 1. Factors Affecting the Activity, Ecology, and Population Dynamics of Microorganisms in Natural Habitats

Carbon and energy sources
Mineral nutrients
Growth factors
Ionic composition
Available water
Temperature
Pressure
Atmospheric composition
Electromagnetic radiation
pH
Oxidation–reduction potential
Surfaces
Spatial relationships
Genetics of the microorganisms
Interactions between microorganisms

From Ref. 21.

those characteristics to encourage the growth of desirable microorganisms (e.g., nitrogen fixers, biocontrol agents, degraders of xenobiotics) and discourage the growth of undesirable ones (e.g., pathogens of plants and animals). Different physicochemical factors obviously differentially affect different groups of organisms (e.g., bacteria vs. fungi) and even different species. Control of organisms in soil by the appropriate manipulation of these factors would decrease the use of pesticides, fertilizers, and other xenobiotics, which have numerous deleterious effects on the biosphere. Unfortunately, there have been few concerted research efforts to determine the influence of individual factors, either alone or in various permutations, on organisms in soil. A major reason for this paucity of studies appears to be the reluctance of funding agencies to support such basic research, which would both reduce nonproductive expenditures for trial-and-error (empirical) studies and produce important, unique, and focused information on the factors that control microbial activities not only in soil but also in other microbial habitats, including animal systems.

The importance and effects of the physicochemical factors have been discussed in detail, with respect to both the soil microbiota in general (e.g., 19,20) and, more specifically, the transfer of genetic information among bacteria in soil (e.g., 21,25). Consequently, they will be discussed only in general terms here.

Soil differs from most other microbial habitats in that it is dominated by a solid phase consisting of particles of different sizes and which is surrounded by aqueous and gaseous phases, the amount and composition of which fluctuate markedly in time and space. The solid phase is a tripartite system composed of finely divided minerals (both primary and secondary); residues of plants, animals, and microorganisms in various stages of decay; and a living microbiota. These particles exist as independent entities and mixed conglomerates. The aqueous phase, in which inorganic and organic substances that serve as nutrients for or inhibitors of the microbiota are dissolved, is normally discontinuous, except when soil is saturated (e.g., after a heavy rain, snow melt, flood, or extensive irrigation). This discontinuity restricts the movement of microbes, especially of bacteria and other nonfilamentous forms, and results in local accumulations of nutrients and toxicants, escape of cells from grazing predators, a low probability of genetic transfer, and other phenomena that do not occur in habitats with a continuous water phase. The pore space not filled with the aqueous phase is filled with air and other gases and volatiles. Because of the low diffusion coefficients of most gases and the quiescent nature of soil, in contrast to the continuous movement and wave action of aquatic environments, some gases (e.g., CO_2, CH_4, CO, NO_x) and volatiles (e.g., short carbon-chain fatty acids, alcohols, and aldehydes; aromatic compounds) may accumulate and other gases (e.g., O_2) may be depleted, thus affecting the composition and activity of the microbiota (e.g., aerobes vs. facultatives, anaerobes, microaerophiles).

4. MICROHABITATS IN SOIL

All microorganisms can be considered to be aquatic creatures, and their metabolism in soil is restricted to those sites where there is a continual supply of available water. Hence, their distribution in soil is restricted essentially to sites that contain clay minerals, as sand and silt do not retain water against gravitational pull. Clay minerals, because of their surface activity (which results from their unique crystalline structure, as they are essentially secondary minerals), retain water against this pull, as the water adjacent to their active surfaces and coordinated with charge-compensating ions on the clays becomes sufficiently ordered to form a quasi-crystalline structure (i.e., the strong attraction of water molecules to the negatively and positively charged surfaces of clay minerals and to their charge-compensating ions enhances the hydrogen bonding of adjacent water molecules). The ordering of this clay-associated water decreases with distance from the clay surface until a distance is reached at which water is no longer under the attraction of the clay and is susceptible to gravity (20).

Clay minerals do not generally exist free in soil but as coatings, or cutans, on larger sand and silt particles or as oriented clusters, or domains, between these particles. The clay-coated particles cluster together, primarily as the result of electrostatic attraction between the net negatively charged faces and the net positively charged edges of clays, into microaggregates, which, in turn, cluster together to form aggregates that can range from 0.5 to 5 mm in diameter and that are stabilized by organic matter and precipitated inorganic materials. These aggregates retain water, the thickness and permanence of which depend on the type and amount of clay and organic matter within the aggregates, and this water may form "bridges" with the water of adjacent aggregates. These aggregates or clusters of aggregates, with their adjacent water, constitute the microhabitats in soil wherein microbes function. A schematic two-dimensional representation of microhabitats in soil, as envisioned by this author, is shown in Fig. 1.

As the result of the discreteness of microhabitats in soil, interactions between microbes, both positive (e.g., commensalistic, protocooperative, mutualistic, synergistic) and negative (e.g., competitive, amensalistic, parasitic, predatory), probably occur less frequently and more sporadically in soil than in habitats wherein water is continuous. Even when the pore space is saturated with the aqueous phase or where water bridges between adjacent microhabitats occur, movement of bacteria between microhabitats may be restricted, as the surface tension of the ordered water around aggregates may be too great to allow passive movement of cells or even active movement by flagellated cells. There is no convincing evidence that bacteria are flagellated in soil, even though they may have the genetic capability to produce flagella when isolated from soil and cultured in liquid media or on agar. However, filamentous fungi are able to cross pore spaces between microhabitats, even when the pore spaces are not filled with the

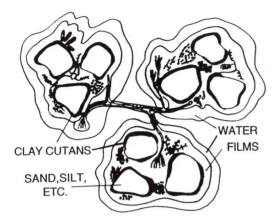

Figure 1. Schematic representation of three microhabitats in soil. The space between the microhabitats is the pore space that is filled with the aqueous and gas phases. Note the hyphae growing through the pore space, even when devoid of the aqueous phase, from one microhabitat to the other two. The hyphae are surrounded by water, which may contain bacteria, bacteriophages, and dissolved substances. (Not to scale.) (From Ref. 21.)

aqueous phase. Filamentous fungi grow by apical and lateral extension of hyphae from a food and water base in a microhabitat, and as they are able to translocate nutrients and water internally, they are independent of the ambient nutritional and aqueous conditions of the ramifying mycelia. Moreover, the hyphae are surrounded by water films in which bacteria, bacteriophages, nutrients, etc., can be transported from one microhabitat to another.

The conditions within a microhabitat, or a portion of it, can also affect microbial activities. A simple two-dimensional representation of presumed conditions within a microhabitat, as envisioned also by this author, is shown in Fig. 2. A bacterium that sticks, if it sticks (see later discussion), on a clay cutan or domain after its chemotropic attraction to adsorbed organic substances may be nutritionally deprived after it consumes the adsorbed substances, if it can, in fact, utilize the adsorbed substances (see later discussion), as it will be dependent on the rate of diffusion of new substances for its nutrition (i.e., $K_{replenishment}$). If the adsorbed substances to which it was attracted are toxic, the bacterium may die.

The mechanisms by which organic molecules are bound on clays will determine the tenacity with which they are held and the ability of microbes, usually with the aid of extracellular enzymes, to utilize these molecules as substrates. The presence of chaotropic ions (which decrease the structure of water and tend to disrupt hydrophobic interactions by increasing the accommodation of nonpo-

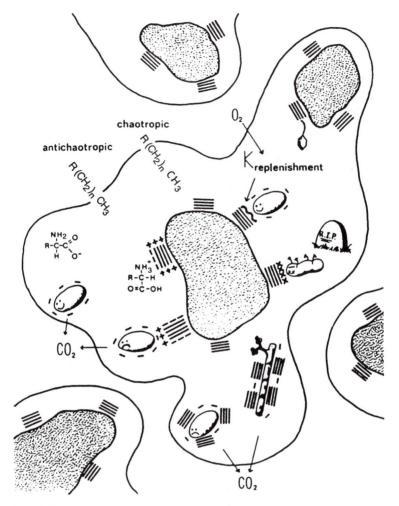

Figure 2. Schematic representation within a microhabitat in soil. The stippled areas represent sand or silt particles, and the families of adjacent short lines represent packets of clay minerals, with the locations of positive and negative charges indicated. The wavy line on one clay packet indicates bound substrates, and the series of x's on another packet indicates bound substances toxic to microbes. The adsorption of a positively charged protein (i.e., at a pH below its pI) on negative sites on clay, of a negatively charged bacterium on positive sites on clay, and of a bacteriophage by its negatively charged tail on positive sites on clay is indicated. The physiological state of bacteria in different associations with particles in the microhabitat is indicated. (Not to scale.) (From Ref. 21.)

lar compounds in aqueous solutions) and of antichaotropic ions (which increase the structure of water and, thereby, increase hydrophobic interactions by reducing the ability of aqueous solutions to accommodate nonpolar groups) will affect the nutritional status of the microhabitat. Even though clay cutans and domains are relatively stable, some clay particles may become dislodged and attach to the surface of microbes in the microhabitat. Such attachment reduces the effective surface area of microbes for transmembrane transfer of nutrients and waste products (12).

These conditions in soil differ markedly from those in sediments of aquatic systems. Although clay minerals in sediments also occur as cutans on larger particles and as domains in aggregates, water-dependent microhabitats do not occur as they do in soil, as the water in sediments is continuous from one aggregate to the next. Moreover, microbes appear to colonize primarily sand and silt particles, rather than clays, in sediments, as water surrounds these particles, and the need to overcome the electrokinetic repulsion between net negatively charged clay minerals and microbial cells is reduced or eliminated (20).

5. CLAY MINERALS AND SOIL MICROBES

The type of clay minerals present in soil has been shown to have a profound influence on numerous microbial activities in soil (e.g., growth; heterotrophic, autotrophic, and mixotrophic metabolism; organic and inorganic nutrition; spore germination; competition, amensalism, predation, and parasitism; transfer of genetic information; pathogenesis), as well as providing protection against the toxicity of acid precipitation, heavy metals, organics, gases, volatiles, hypertonic osmotic pressures, elevated temperatures, desiccation, ultraviolet light, and x-rays (20). How clay minerals affect these activities is not always clear. In many cases, the effects of clays appear to be *indirect* by modifying the physicochemical characteristics of the microhabitats (e.g., the pH, water potential, nutritional status, activity of toxicants), thereby either enhancing or attenuating the growth and metabolic activities of individual populations, which, in turn, influence the growth and activities of other populations. The indirect nature of these effects has been demonstrated in studies in which the same results were obtained when the clays and the microbes were separated or not separated by a dialysis membrane (20).

The demonstration of *direct* effects, which involve surface interactions (e.g., adhesion) between clays and microbes, is more difficult. There is empirical evidence to suggest that such surface interactions occur in situ: e.g., lack of movement of large numbers of microbes from surface to underlying soil layers and then to groundwater during heavy rains, snow melts, floods, or irrigation; failure to wash substantial numbers of microbes from soil columns in perfusion or leaching experiments; partial removal of microbes from wastewater in percolation beds; increased release of microbes from soil by sonication, surfactants, and other methods to enhance the numbers of microbes enumerated. However, there is limited evidence for such surface interactions from carefully controlled studies (e.g., using electron microscopy and changes in particle-size distributions). Moreover, on theoretical grounds, such surface interactions are difficult to understand, as the high electrokinetic potentials, of the same net negative charge, on both clays and cells at the pH of most soils mitigate against such interactions. In fact, such surface interactions can be demonstrated only (1) when the pH is lowered to levels at which most microbes cannot grow (e.g., pH 2 to 3) and the net charge on the microbes (and of some clays) becomes positive, whereas the net charge on most clays, because of their pH-independent charge that results from isomorphous substitution within the crystal structure, remains negative; (2) when the electrokinetic potentials of the clays and microbes are reduced by polyvalent cations; or (3) when the net charge is reversed to positive at higher pH values (e.g., pH 6 to 8) by the first hydrolysis product of some heavy metal ions (6,7,20). These conditions seldom occur in soil. Moreover, as indicated earlier, it may be detrimental to a microbe to stick to a clay-containing aggregate. Consequently, microbes probably reside in the relatively permanent water films associated with such aggregates rather than on the surface of the aggregates. The apparent lack of movement of microbes through soil may be the result of hysteresis rather than of direct surface interactions.

Similar questions must be asked about the binding of organic molecules on clay. Because of their small size relative to that of microbial cells, organic molecules bind on clay, not only by ionic interactions, which are also dependent on the ambient pH and the isoelectric point of amphoteric molecules and the pK of nonampholytes, but especially by multiple hydrogen bonds. Although such binding serves to concentrate these molecules at the solid–liquid interface, it also reduces their availability to microbes. Numerous studies with proteins, peptides, amino acids, polysaccharides, nucleic acids, nucleotides, and other organic molecules bound on clay minerals have demonstrated this resistance to biodegradation (20). Despite this resistance, the activity of some of these molecules (e.g., enzymes [20], transforming deoxyribonucleic acid [DNA] [e.g., 10,11], insecticidal proteins from subspecies of *B. thuringiensis* [27]) is not eliminated, although it is usually reduced. Similar relations between binding, degradation, and activity have been reported for antibiotics and other inhibitors (20).

In contrast to organic molecules, inorganic ions on clays are readily exchangeable and can be used as nutrients by microbes. In addition, clays can scavenge protons and toxic heavy metal cations by cation exchange and, thereby, reduce their toxicity (2,4).

Many of the effects attributed to clay minerals are probably also caused by organic particles (e.g., humic substances) in soil, as they share some of the characteristics of clays (e.g., high surface area and activity, ion-exchange capacity).

However, because these substances are difficult to extract, purify, and character-ize, these effects have been insufficiently studied. Moreover, because of their or-ganic nature, the permanence of these particles is less and their composition more variable than that of clay minerals.

6. HETEROGENEITY OF MICROHABITATS IN SOIL

In addition to their discreteness, microhabitats are highly variable and heteroge-neous, primarily as the result of the heterogeneity of the surfaces (which are coated, partially or completely, with clay minerals, hydrous metal oxides, and or-ganic matter with pH-dependent charges) and of fluctuations in the types and con-centrations of organic and inorganic solutes in the soil solution. Even over small distances, the composition and size of the particles, the amounts and types of solu-tion, nutrients, and gases, and the pH, E_h, ionic strength, and other physicochemi-cal characteristics can vary. This variability in abiotic factors is reflected in the heterogeneity of the microbiota, which is demonstrated by the simultaneous occur-rence in the same soil sample of autotrophs and heterotrophs (both oligotrophs and copiotrophs), aerobes and anaerobes, vegetative cells and spores, prokaryotes and eukaryotes, and cells with different requirements for and tolerances to ambient and extreme conditions of pH, E_h, temperature, osmotic pressure, etc. It is the task of the soil microbiologist to determine which abiotic factors control the activity and population dynamics of specific components of this heterogeneous microbiota and then to devise methods for manipulating the responsible factors, to enhance or at-tenuate the activities of specific components for some desired benefit.

7. MANIPULATION OF PHYSICOCHEMICAL AND BIOLOGICAL FACTORS IN SOIL

A brief discussion of each physicochemical and biological factor (Table 1) and speculations on how it may be manipulated follow. These discussions and specu-lations are not intended to be exhaustive (19,20).

1. Carbon and energy sources. Although most soils contain considerable amounts of organic matter, this material is relatively unavailable as a source of energy and carbon, as it has been humified. The addition of a source of readily available carbon, such as fresh plant residues (green manures), bagasses, or sug-ars, will enhance the growth of the microbiota until the source has been mineral-ized and the activity and density of the microbiota decline to the original state. Although this stimulation of the microbiota is generally nonspecific, it may re-sult in some biocontrol of soil-borne plant pathogens and in the degradation of some recalcitrant materials (e.g., some xenobiotics), as the result of a general

"priming action." The addition of specific compounds that only desired members of the microbiota can utilize could enhance biocontrol, degradation, and other activities beneficial to mankind. Unfortunately, compounds with such specificity have not been identified, and many indigenous microbes share the same enzymatic capabilities. Nevertheless, the ability to enhance, even if only temporarily, the growth and survival in soil of genetically modified bacteria by amending the soil with the specific substrates on which the novel genes function suggests that this use of substrates for the enrichment of specific indigenous microbes should be further explored.

2. Mineral nutrients. As with carbon and energy sources, most soils contain sufficient total amounts of mineral nutrients, but most are present in unavailable forms. The sequence of requirements of the major inorganic nutrients for optimal microbial growth is $N > P > S$. Fertilization of soil with these nutrients will enhance growth, but also generally nonspecifically. In some cases, too much of a mineral nutrient will suppress the growth or activity of a desired population. For example, the presence of elevated amounts of inorganic N will inhibit the ligninolytic activity of *Phanerochaete chrysosporium* (5). The addition of materials with a high C:N ratio (e.g., saw dust, corn cobs) to immobilize the inorganic nitrogen may be an effective and inexpensive method to enhance the activity of this fungus and its degradation of various recalcitrant xenobiotics (23).

3. Growth factors. Because the requirements for specific growth factors differ for different species, the addition of such factors may enhance the growth of desired species better than the addition of nonspecific carbon and energy sources. Unfortunately, too little is known about the specific requirements for growth factors of specific species, although vitamin B_{12} (cyanocobalamin) and various siderophores (which are here considered to be growth factors) are good candidates. It is unfortunate that the early studies by Lochhead and colleagues (14,15) on the effects of growth factors on microbial events in soil have not been extended. However, even if specific growth factors for specific components of the soil microbiota can be identified, their persistence and activity in soil may be limited; e.g., they may be rapidly mineralized as are other added organic molecules, or, in the case of siderophores, they may bind on surfaces, such as clay minerals, and not perform their function (13).

4. Ionic composition. The soil solution is essentially a weak electrolyte composed of a variety of organic and inorganic cations and anions. Because the interface between cells and the soil solution is essentially ionic, even small changes in ionic composition and strength probably have a significant effect on the growth and activity of the microorganisms. Unfortunately, not enough is known about how the ionic composition of soil affects specific organisms to suggest how the ionic composition can be manipulated beneficially. However, care must be taken in any such manipulation, as large changes in ionic composition may alter the osmotic pressure and the availability of water.

5. Available water. Perhaps the most important physicochemical factor that affects microbes in soil is an adequate supply of available water. As stated earlier, microbes are essentially aquatic organisms and even in soil require a sufficiently high water activity (a_w) for growth. Except for short periods after rain, snow melt, flood, or irrigation, most of the soil contains insufficient available water to support microbial activity, as the water is rapidly drained by gravity. Primarily the clay fraction, because of its high surface charge, retains water against gravity and bestows a degree of permanence to microbes present in this water. Consequently, the apparent correlation between microbial activity and the amount and types of clay in soil may be the result primarily of the fact that this is where the water is, and the role of clays in concentrating nutrients, removing inhibitors, direct surface interactions, etc. may be only secondary in importance (20). However, some of the water retained by clays is so tightly bound that its a_w is too low for use by microbes (e.g., temperatures in excess of 160°C are necessary to remove this water). The amount of such tightly bound water depends on the structure of the clay minerals: 2:1, Si:Al clays (e.g., smectites) bind water much more tightly than 1:1 clays (e.g., kaolinite) (i.e., the matric potential of smectites is higher). Consequently, the total amount of water present in a soil (as determined by drying) or the amount that a soil can hold (the water-holding capacity) does not indicate accurately the amount of water available to microbes. Microbes function best at a water potential of −33 kPa (which is very close to the field capacity, which is defined as the amount of water retained by a soil 48 hours after its saturation), where no more water will be removed by gravity, and the balance between the amount of available water and O_2 is optimal for growth. This water potential is most easily determined with a pressure membrane.

Particulate organic matter also retains ordered water, primarily as the result of the polar groups on the organic matter, and this water is also tenaciously held and not all is available to microbes. For example, the water-holding capacity of an organic soil (85% organic matter) was 262% (w/w), but only when the water content was at the −33-kPa water tension of the soil (157% w/w) was the a_w sufficiently high for significant microbial activity (19). Moreover, drying of soil can render the organic matter difficult to rewet, as a result of its structural reorientation and exposure of hydrophobic regions. This development of hydrophobicity can also cause clay–organic aggregates to become difficult to rewet (18).

Different groups of microbes have different tolerances to a_w (e.g., fungi can metabolize at a significantly lower a_w than bacteria). The water content of soils is easily controlled by irrigation, but "fine-tuning" of the manipulation of this physicochemical factor for the differential control of microbes will require considerably more study. In many cases, the effects on microbes of changes in the water content of soil are the result of changes in the O_2 content.

6. Temperature. Temperature is another primary physicochemical factor that affects microbes, especially in vivo. Extremes in temperature can cause the

majority of the indigenous microbial populations to be psychrophilic or thermophilic, depending on the geographical characteristics of the soils. However, the majority of soils worldwide contain predominantly mesophilic microbes. Regardless of the geographic location of the soils, the temperature of the top few centimeters of most soils, wherein most of the microbial activity usually occurs, does not fluctuate more than a few degrees Celsius in any season of the year. Consequently, the temperature of soils in situ cannot be conveniently manipulated for extended periods. Although heating of soils for the control of some plant pathogens (e.g., solar heating or steam after covering the soil with black plastic) can be accomplished relatively easily, this is generally a nonspecific treatment, as essentially all indigenous microbes are inhibited and then return to their original population levels after the soils revert to their normal temperature range and are naturally inoculated with the same microbes, both pathogenic and nonpathogenic, from adjacent areas. To achieve long-term control (i.e., elimination or reductions) of specific populations, other critical changes in the physicochemical characteristics of the soil need to be made before heating.

7.   Pressure. Changes in atmospheric pressure are probably too small to affect the soil microbiota significantly, and the overall solute concentration of normal soil solutions is usually not high enough to suggest that osmotic pressure inhibits the microbiota. However, within the microhabitats, where the solute levels are probably higher because of their concentration at solid–liquid interfaces, especially at charged surfaces, and during periods of drying, the osmotic pressure may increase sufficiently to affect some microbes. This is especially true in saline and alkaline soils. Most microbes in soil are probably stenohaline and, therefore, susceptible to hypertonic osmotic pressures. Moreover, marked increases in solutes will affect the osmotic potential of water and decrease its availability. Altering the osmotic pressure of soil by the addition of organic or inorganic solutes is relatively easy. Some years ago, control of nematodes in soil was achieved by the addition of copious quantities of molasses, presumably because of the increase in osmotic pressure (W. Feder, personal communication). However, as large changes in solute concentrations can affect numerous other physicochemical characteristics (e.g., available water, pH, O_2, and mineral nutrients if the solute can be used as a carbon source), caution is recommended when the manipulation of this factor for the control of microbes in soil is considered, especially as the effects may be nonspecific.

8.   Atmospheric composition. Because air and water share the same pore space between microhabitats, the atmospheric composition in soil can be easily manipulated, at least on an overall gross basis, by altering the water content. Flooding has been used in an attempt to control some soil-borne fungal pathogens of plants (e.g., Fusarium wilt of bananas), as most filamentous fungi are obligate aerobes. However, the various spores of fungi, especially chlamydospores, appear to be able to resist long periods of anaerobiosis, and when the

soil subsequently becomes aerobic again, the spores germinate and reestablish the fungal colonies. Moreover, many fungi—especially, but not exclusively, pathogens of animals—are dimorphic and can persist in the yeast form under anaerobic conditions.

The addition of large quantities of available substrates will also reduce the amount of O_2 and increase that of CO_2 in soil. However, even under "normal" conditions, the content of CO_2 is higher and that of O_2 lower in soil than in the overlying atmosphere, as the result of microbial metabolism and the slow diffusion rate of these gases. Even in well-aerated soils, anaerobic sites are present, as indicated by the ability to isolate obligate anaerobes from such soils. In addition to O_2 and CO_2, other gases (e.g., CH_4, NO_x) and volatiles (e.g., short-chain organic acids, aldehydes, alcohols, esters, hydrocarbons, ethylene) are present in the atmosphere of soil and can serve as either substrates for or inhibitors of microbes (3,24).

Manipulation of the soil atmosphere for the control of specific components of the microbiota is the basis for the fumigation of soil with a variety of xenobiotics. In many cases, however, the effects of these pesticides are transient, and they often also affect nontarget organisms. Nevertheless, manipulation of the soil atmosphere appears to have a good potential for controlling specific portions of the microbiota in soil, especially if it can be fine-tuned and coupled with the manipulation of other physicochemical factors to enhance the duration of its effects.

9. Electromagnetic radiation. Although light probably affects only microbes residing on the surface of soils, it can be important in arid and semiarid soils, where photosynthesis in algal crusts, both prokaryotic and eukaryotic, is probably the major source of carbon and energy (17). In such environments, resistance to ultraviolet radiation (e.g., pigments) is probably necessary. Microbes residing below the soil surface do not generally have such resistance, and manipulation of the soil that periodically exposes these microbes to direct solar radiation (e.g., cultivation) can exert some control, albeit relatively nonspecific, on the microbiota. Although in aquatic environments, radiation is a physicochemical factor of major importance (as is hydrostatic pressure), it is generally of minor importance in soil and, hence, difficult to manipulate for the control of specific species.

10. pH. The hydrogen ion concentration is a major physicochemical characteristic of soil that is amenable to manipulation in situ. In general, fungi predominate in acidic soils (below pH 5.5), whereas eubacteria, including actinomycetes, predominate in near-neutral or moderately alkaline soils. The apparent lower numbers of fungi in the latter soils does not occur because fungi are intolerant of these pH values but because bacteria are efficient competitors at these pH values and prevent the establishment and proliferation of fungi. In contrast, the lower numbers of bacteria in acidic soils is the result of their intolerance to the elevated concentrations of H^+ that fungi can tolerate, and, therefore, the fungi can proliferate in the absence of competition. In addition to numerous effects of

pH on a spectrum of physiological, morphological, and metabolic responses of the microbiota, the pH also affects the solubility, availability, and toxicity of mineral nutrients; the speciation of heavy metals (4,6); the sign of the net surface charge of amphoteric materials and the negative charge of ionizable materials, which will affect the adsorption and subsequent binding of cells and organics on surfaces and, hence, in their availability as nutrients and their bioactivity; and numerous other phenomena. Manipulation of the pH of soil is routinely done, although primarily to enhance the growth of plants, by the addition of lime to raise the pH or of gypsum or some other relatively cheap source of sulfur to reduce the pH. A prime example of the manipulation of the pH of soil for biological control is the reduction in pH to control a disease ("scab") of potato caused by *Streptomyces scabies.*

11.   Oxidation–reduction potential (E_h). The E_h of soil is related primarily to the O_2 content of soil and, therefore, can be manipulated, to some extent, by cultural practices (e.g., cultivation, improvement of drainage) (9). The E_h is also influenced by pH, temperature, pressure, and nutrients. The ability to isolate aerobes, facultatives, and anaerobes (including fermenters, methanogens, and sulfate reducers) from the same soil sample indicates that microhabitats that differ in E_h by approximately 1000 mV (e.g., from +700 to –300 mV) can coexist in the same soil. Consequently, fine-tuned manipulation of the E_h of soil to control specific components of the microbiota would appear to be difficult.

12.   Surfaces. The great importance of surfaces to microorganisms and their activities in soil, especially those of clay minerals, was discussed earlier. The incorporation into soil of clay minerals, especially of montmorillonite, can affect the establishment, proliferation, and activity of fungal pathogens of plants and animals (i.e., help in converting "conducive" to "nonconducive" soils [1]) (20), probably by affecting the growth of and competition by bacteria, as well as by binding siderophores necessary for the iron nutrition of the fungi (13). Although such changes may initially appear to be positive, the numerous direct and indirect effects that clay minerals have on many other physicochemical characteristics of soil and, therefore, on the soil microbiota, indicate that all possible permutations of altering the clay mineralogical features of a soil—a relatively permanent change—should be carefully evaluated before such additions of clay are made. Less concern is necessary before the amount of organic surfaces is increased by the incorporation of plant and other organic residues into soil, as the permanence of such surfaces is more transient, because of their susceptibility to biodegradation. However, as stated earlier, the relative importance of surfaces of organic matter to microbial events is not as well understood as that of clay minerals. Nevertheless, because of the profound and multifaceted effects that surfaces have on microbes in soil, their manipulation for the control of specific portions of the soil microbiota should be extensively investigated.

13. Spatial relationships. The spatial relationships between microbes in soil are influenced primarily by the structure of the microhabitats, as discussed earlier. Although some cultural practices (e.g., cultivation, irrigation) will affect these relationships, these effects are probably transitory. Hence, manipulation of spatial relationships may not be an easy or effective way to control specific microbes in soil.

14. Genetic characteristics of the microorganisms. As mentioned earlier, autochthonous microbes in soil are highly adaptable, both physiologically and genetically, and they have established a homeostasis in soil. Until recently, attempts to alter this homeostasis have generally been unsuccessful. However, with the increasing sophistication in genetic engineering, it is now possible to alter the genetics of microorganisms, especially of bacteria, and introduce these genetically altered organisms to soil for the purpose of performing specific functions. However, introduction is not always accompanied by establishment (colonization), growth, and survival of the altered organisms. Consequently, until the physicochemical characteristics of soil can be manipulated to ensure a significant expression of the introduced genes, any effects of these genes will be transient. If and when such manipulation becomes possible, other problems, such as the transfer of the genes to the indigenous microbiota, which is probably more adapted to the soil environment than the introduced organisms, and potential ecological and health hazards of the products of the novel genes, will need to be considered. Inasmuch as these and other aspects of the release of genetically altered organisms to the environment have been extensively discussed (e.g., 8,21,22,25,26), they will not be addressed here.

15. Interactions between microorganisms. Although the physicochemical factors briefly discussed will influence the growth and survival of microbes in soil, interactions, primarily negative ones, between microbes within a microhabitat have a marked effect. For example, there are numerous studies that show the survival and growth in sterile soil of a spectrum of microorganisms, both genetically engineered and nonengineered, as well as microorganisms that are not normal residents of soil, whereas the same organisms do not survive or grow when introduced to the same soil when not sterile (25). Although various positive, negative, and presumably neutral interactions between and among organisms have been individually demonstrated, most of these studies have been done with model systems consisting of only a few species of interacting organisms. Inasmuch as the microhabitats in soil are heterogeneous and their inhabitants also display a high degree of heterogeneity, numerous interactions between the inhabitants are undoubtedly occurring simultaneously and are in constant flux. By understanding the specific physicochemical factors that affect positive and negative interactions, it might be possible to manipulate these factors to enhance natural biocontrol of desired and undesired microbes.

8. CONCLUSIONS

Increased knowledge of how individual physicochemical factors affect microbes in soil may provide some clues as to how to manipulate such factors. However, it is obvious that a change in one factor will result in changes in numerous other factors and that this cascade of changes may either augment or cancel the initial effect. Most of the factors, which are constantly fluctuating, are interrelated and, therefore, must be studied, understood, and manipulated in concert. In addition, the presumed effects of such changes must be carefully and critically evaluated (16). For example, the failure to recover an introduced microorganism from soil, especially from nonsterile soil, may only reflect a "viable but nonculturable" stage of the organism (see Chapter 3). Microorganisms in soil, especially introduced organisms, sometimes become so debilitated or otherwise altered that they cannot be recovered, especially on selective media, even though they could be surviving, and possibly even growing, in the soil (21). Consequently, it must be clearly and unequivocally established that the apparent lack of survival and growth is not an artifact of the experimental procedures.

There is still much to learn about soil as an environment for microbial life. Inasmuch as the soil is so basic an environment, as entire food webs ultimately depend on it, it is difficult to understand why some soil microbiologists sometimes appear to express feelings of inferiority (e.g., many prefer to be called "microbial ecologists") and believe that they have no valid questions to investigate. There is, obviously, a plethora of questions, many of which are old but still unresolved and complex. As are other biologists, soil microbiologists are clambering onto the "bandwagon" of molecular biology and are becoming obsessed with new, especially molecular, techniques, even though some of these techniques may provide little new information. Hence, although some concepts and techniques of molecular biology will undoubtedly be very helpful in answering some of these questions, it would be imprudent to forget and not also use older tested methods (e.g., respirometry, assays for enzymes and adenosine triphosphate [ATP], plating on selective media, x-ray diffractometry, Fourier–transform infrared spectrometry, electron microscopy, measurement of surface charges). In many cases, selecting the methodology is secondary to asking the pertinent questions. Moreover, it must be remembered that because of the complexity of soil, the concepts and techniques of microbiology, chemistry (ranging from biochemistry to inorganic and physical chemistry), physics, mathematics, and agronomy, in addition to those of molecular biology, must be invoked.

Although the use of molecular techniques has provided some new information on the microbiological features of soil, it is not always certain that the amount and value of this information are commensurate with its cost. Moreover, there are many paradoxes that need to be resolved between these and more classical techniques. For example, the use of DNA probes and immunological techniques is extremely helpful in monitoring the fate and ecological aspects of individual species in soil. However, these techniques require that the species be cultured to obtain their DNA or antigens. The application of the polymerase chain reaction (PCR), DNA–DNA hybridization, flow cytometry or other methods to measure antigen–antibody reactions, etc. is secondary to the ability to culture the organisms in vitro. Inasmuch as only 1% to 10% of the soil microbiota can presumably be cultured currently (see Chapter 3), the number of species that can be monitored is limited. Similarly, although 16S ribosomal ribonucleic acid (rRNA) analyses can provide valuable information on changes in the apparent species diversity of microbes in soil over time, cultivation of these species will eventually be necessary to identify them, if for nothing else than to establish data bases relating the RNA patterns to identifiable species, as well as to study their physiological features, functions, production of potential valuable (anthropocentric) metabolites, etc. Despite these current limitations of molecular techniques to answer some relevant questions relating to microbial life in soil, the application of molecular biology, in conjunction with other techniques and with the correct and relevant questions, should provide much valuable information in the future. Moreover, the use of molecular techniques should facilitate obtaining funds for basic studies in soil microbiology, as well as make soil microbiologists feel that they are in the mainstream of current biology.

Consequently, let us get on with it! We know the questions (or, at least, we should know many of them). Let us get the answers! In a time when tremendous advances are being made in many biomedical areas, we need to make similar advances in the sustained production of food and fiber, especially in many parts of the world where production is too low to sustain the high and burgeoning human populations. Many of the current advances in biomedical research have limited impact in these parts of the world, as an adequate source of nutrients is the primary limiting factor to human life.

REFERENCES

1. C. Alabouvette, H. Höper, P. Lemanceau, and C. Steinberg, *Soil Biochemistry* (G. Stotzky and J.-M. Bollag, eds.), Marcel Dekker, New York, p. 371 (1996).
2. H. Babich and G. Stotzky, *Appl. Environ. Microbiol. 33*:696 (1977).
3. H. Babich and G. Stotzky, *Experimental Microbial Ecology* (R. G. Burns and J. H. Slater, eds.), Blackwell Scientific, Oxford, p. 631 (1992).
4. H. Babich and G. Stotzky, *Environ. Res. 36*:111 (1985).
5. J. A. Bumpus, *Soil Biochemistry* (J.-M. Bollag and G. Stotzky, eds.), Marcel Dekker, New York, p. 65 (1993).
6. Y. E. Collins and G. Stotzky, *Metal Ions and Bacteria* (T. J. Beveridge and R. J. Doyle, eds.), Wiley & Sons, New York, p. 31 (1989).

7. Y. E. Collins and G. Stotzky, *Appl. Environ. Microbiol. 58*:1592 (1992).

8. J. D. Doyle, G. Stotzky, G. McClung, and C. W. Hendricks, *Adv. Appl. Microbiol. 40*:237 (1995).

9. L. F. Elliott, J.-M. Lynch, and R. I. Papendick, *Soil Biochemistry* (G. Stotzky and J.-M. Bollag, eds.), Marcel Dekker, New York, p. 1 (1996).

10. E. Gallori, M. Bazzicalupo, L. Dal Canto, R. Fani, P. Nannipieri, C. Vettori, and G. Stotzky, *FEMS Microbiol. Ecol. 15*:119 (1994).

11. M. Khanna and G. Stotzky, *Appl. Environ. Microbiol. 58*:1930 (1992).

12. S. Lavie and G. Stotzky, *Appl. Environ. Microbiol. 51*:65 (1986).

13. S. Lavie and G. Stotzky, *Appl. Environ. Microbiol. 51*:74 (1986).

14. A. G. Lochhead and F. E. Chase, *Soil Sci. 55*:185 (1943).

15. A. G. Lochhead and M. O. Burton, *Can. J. Microbiol. 3*:35 (1957).

16. E. Madsen, *Soil Biochemistry* (G. Stotzky and J.-M. Bollag, eds.), Marcel Dekker, New York, p. 287 (1996).

17. J. Skujins (ed.), *Semiarid Lands and Deserts: Soil Resource and Reclamation*, Marcel Dekker, New York (1991).

18. J. T. Staley, J. B. Adams, and F. E. Palmer, *Soil Biochemistry* (G. Stotzky and J.-M. Bollag, eds.), Marcel Dekker, New York, p. 173 (1992).

19. G. Stotzky, *Microbial Ecology* (A. I. Laskin and H. Lechevalier, eds.), CRC Press, Boca Raton, Fla., p. 57 (1974).

20. G. Stotzky, *Interactions of Soil Minerals with Natural Organics and Microbes* (P. M. Huang and M. Schnitzer, eds.), Soil Science Society of America, Madison, Wisc., p. 305 (1986).

21. G. Stotzky, *Gene Transfer in the Environment* (S. B. Levy and R. V. Miller, eds.), McGraw-Hill, New York, p. 165 (1989).

22. G. Stotzky, *Proceedings of the 2nd International Symposium on the Biosafety Results of Field Tests of Genetically Modified Plants and Microorganisms* (R. Casper and J. Landsmann, eds.), Biologische Bundesanstalt für Land und Fortswirtschaft, Braunschweig, p. 122 (1992).

23. G. Stotzky, *The International Workshop on Establishment of Microbial Inocula in Soil* (L. F. Elliott and J. M. Lynch, eds.), *Am. J. Altern. Agric. 10*:54 (1995).

24. G. Stotzky and S. Schenck, *Crit. Rev. Microbiol. 4*:333 (1976).

25. G. Stotzky, L. R. Zeph, and M. A. Devanas, *Assessing Ecological Risks of Biotechnology* (L. R. Ginzburg, ed.), Butterworth-Heinemann, Stoneham, Mass., p. 95 (1991).

26. G. Stotzky, M. W. Broder, J. D. Doyle, and R. A. Jones, *Adv. Appl. Microbiol. 38*:1 (1993).

27. H. Tapp and G. Stotzky, *Appl. Environ. Microbiol. 61*:1786 (1995).

2

The Rhizosphere as a Habitat for Soil Microorganisms

JAN SØRENSEN   Royal Veterinary and Agricultural University, Copenhagen, Denmark

1. INTRODUCTION

Hiltner (1) first introduced the term "rhizosphere," defined as that volume of soil surrounding roots in which bacterial growth is stimulated. Over the years, however, the term has been redefined several times, mostly to incorporate parts of the root tissue. Microorganisms often invade the surface tissue of roots, where they may cause a number of plant diseases. Other microorganisms on the roots provide an important link for nutrient transport between the plant and soil. The term "endorhizosphere" (2) may now appropriately be used to describe the multilayered microenvironment, which includes a mucoid layer of plant- or microbe-derived polysaccharide, the epidermal layer including the root hairs, and the cortical layer. The rhizoplane (root surface) should be defined as the epidermal layer, including its associated polysaccharide matrix.

By comparison, the ectorhizosphere comprises the rhizosphere soil, which usually extends a few millimeters from the root surface. A useful separation of rhizosphere and bulk soil may be obtained by shaking the root system manually. Soil adhering to the root is defined as the rhizosphere. For some purposes, further separation of rhizosphere soil compartments may be obtained during a subsequent washing procedure, whereby the loosely adhering soil is loosened from the root. However, the compartments are often difficult to define, because the rhizosphere and bulk soil components have been mixed by diffusion and motility of microorganisms and soil fauna. Also, plants may develop a dense "rhizosheath," which is a strongly adhering layer of root hairs, mucoid material, microorganisms, and soil particles. The rhizosheath develops at specific segments of the root, as shown in Fig. 1, but may also extend to the whole root system (3).

As the rhizosphere is a dynamic environment, depending on root elongation, development of root hairs and adventitious roots, and different stages of maturation and senescence, the microorganisms meet a large number of challenges and risks of extinction. Not surprisingly, the microbial community is composed of microorganisms with different types of metabolism and adaptive responses to the variable supply of water, oxygen, organic carbon sources, and nutrients. We shall discuss a number of selected microorganisms in the endo- and ectorhizospheres in relation to root colonization, diversity, and responses to plant exudation, starvation, stress, and predation by protozoa in the rhizosphere environment.

2. STRUCTURE AND CHEMISTRY OF THE RHIZOSPHERE

2.1 The Endorhizosphere

The mucigel

Because roots can advance quickly through soil, or because root cap cells are sloughed off by mechanical shear, the apex is typically devoid of microorganisms. However, an important role of the root cap is its production of mucilage (polysaccharide), which covers epidermal cells and acts as a lubricant while the root advances through soil. The mucilage is also a site for microbial attachment and protects against desiccation, by forming a matrix for absorption and transport of water between epidermal cells and soil particles.

The mucilage layer on the epidermis is best developed in young roots and may later become very thin except over cell junctions. The composition of root mucilage may vary. For example, in maize (*Zea mays*), the mucus is rich in the carbohydrates galactose and fucose, with lesser amounts of xylose, arabinose, and galacturonic acid (4). It is interesting that the mucilage layer on the young root may contain the chemical components of importance in host–pathogen or host–symbiont recognition systems. The fucose component in maize has been shown to be involved in early root attachment by *Azospirillum lipoferum* (5). During subsequent root elongation, the mucilage layer may be supplemented by slime secretion from the epidermis. This mucilage may

Figure 1. Barley (*Hordeum vulgare*) plant showing rhizosheaths (dense matrix of mucigel, root hairs, and soil particles) at the oldest, upper 1–3 cm, of the roots.

consist of polygalacturonic acid, a pectinlike mucus (6). The combined layer of mucilages of different origins has been termed mucigel by Jenny and Grossenbacher (7).

Exterior to the surface-associated inner mucigel, a more tenuous exterior layer of amorphous slime of microbial origin is often present. The exopolysaccharide compounds surrounding bacterial cells or semidegraded mucigel of sloughed-off root cells form this diffuse outer mucigel. In the mature root, the mucigel layers bind clay particles, organic matter, and microorganisms into a coherent matrix around the root (8).

Epidermal cells and root hairs

In the young plant root, the uptake of nutrients and water occurs through an intact epidermal layer including the root hairs. Normally, this layer is quite short-lived (9) as a result of the mechanical tension or desiccation of the rhizosphere. However, the dense matrix of clay particles, organic matter, and microorganisms held together by extensive mucigel formation protects the epidermal layer, sometimes thoughout the whole growing season (9,10).

Living epidermal cells and root hairs become densely colonized by microorganisms, notably bacteria, which depend completely on simple organic molecules exuded from the plant cells (see later discussion). This group of early-colonizing bacteria includes primarily the so-called rhizobacteria of aerobic, gram-negative heterotrophs, e.g., *Pseudomonas* and *Azospirillum* spp. The close association that may develop between rhizobacteria and the root surface is illustrated in Fig. 2, which shows a *Pseudomonas fluorescens* strain located between epidermal cells on the surface of a barley (*Hordeum vulgare*) root. The bacteria grow poorly in the bulk soil because of the lack of short-lived organic substrates. As a consequence, they compete strongly with each other and with symbiotic and plant pathogenic microorganisms during the early phase of root colonization. Bacterial invasion of living epidermal cells and root hairs may also occur, e.g., by the *Agrobacterium tumefaciens* pathogen or by *Rhizobium* spp. symbionts.

The early root colonization phase sometimes involves a chemotactic response by bacteria to seed exudate (11), active movement toward and along the root sur-

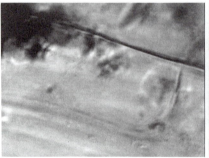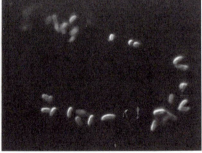

Figure 2. *Pseudomonas fluorescens* cells on the surface of a barley (*Hordeum vulgare*) root, detected by fluorescent antibody labeling and confocal laser scanning microscopy. Bacterial colonization (right) is observed to occur at the edges of the root epidermal cells (left). The bacteria are approximately 2 µm long.

face (12,13), and specific molecular recognition during attachment (14). The first attachment is a reversible phase, in which bacteria overcome charged, repulsive energy barriers. Some bacteria are initially held by specific plant glycoproteins (agglutinins), which are exposed from the root cells. Bacterial adhesion polymers involved in this early phase may be fibrillar proteins (fimbriae) (15) or lipopolysaccharide (LPS) components (16). Subsequently, a firmer and irreversible anchoring of bacteria may involve exopolysaccharides (EPS) such as those produced in *Agrobacterium* (17) and *Rhizobium* spp. (18).

The cortex layer

Disrupted or dead plant cells in the epidermal layer are a common site of entry of pathogenic as well as saprophytic microorganisms, which utilize a range of complex organic molecules (proteins, glycoproteins, lignins, cellulose, or other polysaccharides). Small openings in the plant epidermis provide rapid access to other thin-walled cells in the cortical layer. From here, the pathogens can proliferate farther into the cortical cells.

Many pathogens are completely dependent on the plant for extensive proliferation. Their relationship to the environment and to other rhizosphere-inhabiting microorganisms is therefore particularly complex, because the life cycle occurs partly outside and partly inside the root. During root colonization, they must compete strongly with other rhizobacteria proliferating on short-lived organic substrates (carbohydrates and amino acids) in root exudate. The pathogens represent a number of bacterial genera, including *Pseudomonas* and *Erwinia* spp. In addition, soil fungi such as *Pythium*, *Fusarium*, and *Rhizoctonia* spp. are of major significance as pathogens in agriculture and forestry.

Mycorrhizae and saprophytic fungi such as *Trichoderma* sp. (19) are known as early invaders of the root and quickly occupy an ecological niche on the root. As a consequence of their ability to utilize complex substrates, these microorganisms do not completely depend on the plant in their life cycle. Also, they may not compete for substrates with rhizobacteria to any significant extent. However, both mycorrhizae and *Trichoderma* spp. produce antibiotics such as peptides and may thus protect the root from subsequent invasion of pathogens; such inhibition may also explain why some rhizobacteria, e.g., *Pseudomonas* spp., occur in lower numbers on plant roots with mycorrhizae (*Glomus* spp.) (20,21).

2.2 The Ectorhizosphere

Root exudates and rhizodeposition

The production of mucilaginous material and the exudation of soluble organic compounds from the epidermal plant cells play an important role in root colonization and maintenance of microbial growth in the rhizosphere. The greatest quantity of carbon loss from the root actually occurs at the root tip (corresponding to root cap sloughing), but a considerable amount is lost as diffusible substrates from the zone of elongation (22). The exudates are typically carbohydrate monomers (sugars), amino acids, and organic acids, which are suitable substrates for a wide range of rhizobacteria.

Different plant species or cultivars vary in their exudation of organic carbon compounds to the rhizosphere. Materials released from maize roots were incorporated into soil microbial biomass to a higher degree than materials from wheat (*Triticum aestivum*) roots (23). In a particular cultivar, the exudation may further depend on both moisture (24) and temperature (25) in the soil. Finally, the N or P nutrient levels may also influence the composition and quantity of root exudates (see later discussion).

Microbial activity is expected to be high in the rhizosphere, where readily degradable substrates are exuded from the plant. Unfortunately, it has so far been difficult to measure the actual heterotrophic activity in the narrow zones of the rhizosphere. As shown in Fig. 3, an oxygen microsensor approach has been used to determine microprofiles of O_2 concentrations and microbial O_2 consumption (heterotrophic activity) in a gel-stabilized barley rhizosphere system (26). The study indicated that exudate-supported microbial activity varied significantly along the young barley roots.

Nutrient availability

Application of N fertilizer stimulates root exudation in agricultural plants and may indirectly affect microbial growth in the rhizosphere (27). However, in wheat cultivars stimulation of microbial growth seemed to be due to increased utilization of the root exudates rather than increased exudation rates (28). Liljeroth et al. (29) observed that N application resulted in higher bacterial abundance on seminal roots of young barley plants (10-day-old seedlings). It was further noted that NO_3^- application, because the latter resulted in a low rhizosphere pH from the proton release associated with root NH_4^+ uptake. Alternatively, N fertilizer may sometimes stimulate root growth at the expense of exudation, and cultivars may respond differently in terms of stimulated exudation or increased root growth (29). In conclusion, N fertilization may not lead to predictable effects on higher microbial activity and growth in the rhizosphere.

In nonfertilized soil lower nutrient availability may limit utilization of the root-release carbon compounds (30). Plants may have dual and counteracting effects on microbial activity: Stimulation may occur as a result of the organic substrate, but plants may also inhibit microorganisms by scavenging inorganic nutrients such as ammonium (NH_4^+), nitrate (NO_3^-), and phosphate (PO_4^{3-}). Microscale techniques to study the turnover of these compounds in the rhizosphere

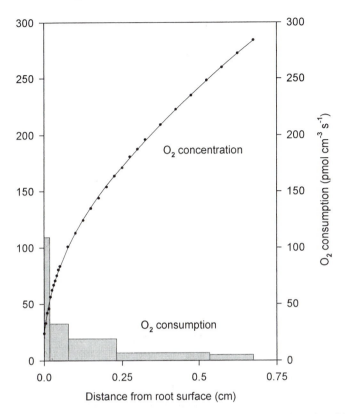

Figure 3.   Microscale determination of O_2 profiles in gel-stabilized barley (*Hordeum vulgare*) rhizosphere, using an O_2 microelectrode. The O_2 consumption rates were calculated from the profiles, using two different diffusion-reaction models. (From Ref. 26.)

are also in great demand. Fig. 4 shows an example of microdistributions of NO_3^- concentrations and microbial biomass in the vicinity of 2-month-old barley roots. The samples at the root surface show higher NO_3^- and biomass content than the bulk soil. Binnerup and Sørensen (31) subsequently developed a microscale assay to analyze small (10-mg) rhizosphere samples and demonstrated accumulations of both NO_3^- and nitrite (NO_2^-) at the root surface, presumably due to enhanced nitrification activity.

Finally, phosphate limitation due to uptake by the root may also occur in the rhizosphere, and a distinct P-depleted zone is formed around the root. One example of the symbiotic interactions between root and microorganisms is the development of mycorrhizae, which are efficient P scavengers in the soil. The ability

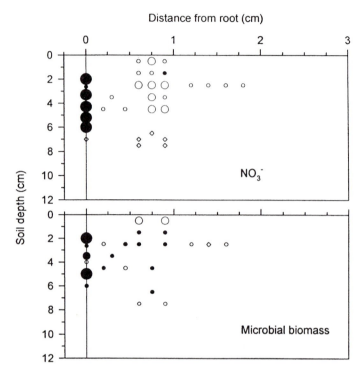

Figure 4. Microscale dissection sampling of soil in barley (*Hordeum vulgare*) rhizosphere for determination of NO_3^- concentrations and microbial biomass. The symbols indicate relative differences from values measured in the bulk soil (∘, 0%–50% smaller; ○, 50%–100% smaller; •, 0%–100% larger; •, 100%–200% larger; ●, >200% larger; ◊, not significantly different).

of vesicular–arbuscular (VA) mycorrhizae to stimulate plant growth can thus, in most cases, be attributed to increased P transport to the plant. The external hyphae extend several centimeters into the soil and provide an efficient P carrier to the plant (32,33).

Aggregate formation

As the root develops there is typically a significant increase of soil microbial biomass in the ectorhizosphere soil, supported by continued exudation or leakage of organic substrates from damaged root cells. This increase may be observed in barley roots within 1–2 weeks of growth and may be localized at particular segments of the root (34). During the same period, the rhizosphere shows increasing aggregation of soil, presumably a result of extensive mucigel formation. The microbial exopolysaccharides bind clay minerals and humic components into microaggregates (approximately 50 μm), while fungal hyphae may bind these into (larger) macroaggregates (approximately 1–2 mm) (35). The interconnecting roots hairs and the fungal hyphae, including those of mycorrhizae, maintain efficient supply lines for substrate transport between the microaggregated bacteria and the root tissue.

An additional role of the microaggregates is to serve as protective microhabitats for bacteria against soil predators, such as the amoebae. On the other hand, some bacteria occurring in the small voids of the microaggregates lose access to external resources. Inactive or dormant populations of soil bacteria may therefore be formed as the rhizosphere soil aggregates.

3. MICROBIAL POPULATIONS IN THE RHIZOSPHERE
3.1 Enumeration and Isolation
Total populations

Numerous studies of rhizosphere samples using fluorescence microscopy or electron microscopy (EM) have documented microbial colonization of mucigel, epidermis, and cortex layers. Such direct observations have provided useful information on microdistributions and relative abundances of bacteria and fungi on roots. Rovira (36,37) thus presented early evidence that bacteria are distributed in clumps or patches on the roots, interspersed by areas where they are sparse or absent. Even the smallest microcolonies of bacteria seem to occur in patchy patterns, indicating that individual root segments provide different conditions for microbial growth (38).

Total counts of whole microbial populations provide information on preferred habitats of microorganisms, e.g., in root-colonization experiments. Preparation of root specimens for staining and microscopy is easier than preparation of those of soil-containing samples. The latter may be obtained by thin sectioning of rhizosphere soil after embedding in a resin (39). Unfortunately, fluorescence microscopy is difficult because of the autofluorescence from plant tissue, humic compounds, and light absorbance in the resin. It is promising at this time, though, that the new technology of confocal laser scanning microscopy (CLSM) allows relatively thick (ca. 1-mm) root specimens to be analyzed in fluorescent light, and that this technique eliminates the problem associated with autofluorescence (40) (discussed later).

Because of a lack of suitable techniques for easy determination of total microbial numbers or biomass in the rhizosphere, a number of indirect measures have long been used. By mistake, the microbial biomass has often been assumed to reflect soil fertility or plant productivity; the result has been unfortunate attention to biomass determinations in soil. The relevance of microbial biomass to soil fertility is related more to the amounts of plant nutrients and the timing of their becoming available through microbial turnover. Hence, there is now a greater need for turnover data (mineralization of nutrients) and for localization and estimation of in situ rates of activity (mineralization and other nutrient transformations) related to the active fraction of microbial biomass. Using a fumigation–extraction technique to determine microbial biomass (41) and a microscale-dissection method to obtain small soil samples (100 mg), Stoumann Jensen and Sørensen (34) recently demonstrated a buildup of microbial biomass in specific barley root segments. By using the substrate-induced respiration method of West and Sparling (42) to determine the actively respiring fraction of the biomass, the root basis (immediately below the seed) and the root tip proximal segment in particular were found to support an active respiring biomass of microorganisms.

Viable and culturable subpopulations

Culturability of microorganisms on laboratory media has been the basis for the majority of studies on microbial populations in the rhizosphere. Even cell viability has often been determined from estimates of colony-forming units (CFUs), as percentage of total counts. This figure is variable, but estimates of 10% viability for bacterial populations in rhizosphere soil are common (43,44). By comparison, much lower apparent viabilities, e.g., 0.1% to 2%, are found in bulk soil (45,46).

It must be remembered that formation of visible CFU on agar plates never offers a complete count of the viable population of bacteria. It is well established that plate counts inherently underestimate the actual number of viable bacteria for several reasons. Physical attachment of bacteria to soil particles will lead to underestimations. In addition, some bacteria may be killed in the dilution medium and others may fail to grow on various plating media. Many unknown microorganisms may require specific media or growth conditions, and some bacteria will not grow on nutritionally rich laboratory media, especially after a period of nutrient starvation in soil (47). Studies using CFU counts are therefore hampered by the fact that bacteria frequently enter a dormant state in which they are viable, but nonculturable (VBNC).

Several studies have taken the challenge to improve culture techniques for determinations of bacterial numbers in rhizosphere and soil samples. One method, the microcolony technique, is based on detecting microcolonies formed after a few initial cell divisions (micro-CFU) rather than the full-size, visible colonies (CFU) on laboratory media (48). Using a rhizosphere isolate, *Pseudomonas fluorescens* strain DF57, it was shown that microcolony formation gave a much higher estimate of surviving cells in soil, and that the numbers apparently included a subpopulation of viable, but nonculturable cells of the strain. It appears that improvements in the enumeration and detection techniques can be obtained if attention is paid to physiological stress and recovery from nongrowth conditions. For example, energy-starved *Pseudomonas aeruginosa* (lacking on electron acceptor for respiration) showed better recovery in an anaerobic medium with NO_3 as the final electron acceptor than in an aerobic medium (49).

A special application of the plating technique is the study of colony-formation curves, which are progress curves for CFU formation over extended incubation periods (50). The lambda value (λ) is calculated from the equation $\ln(N_\infty - N) = \ln N_\infty - \lambda(t - t_x)$, where N_∞ is the estimated number of cells capable of forming visible colonies, N is the number of colonies observed at time t, and t_x is the time interval between plating and the initiation of colony appearance. Lambda is the probability of a cell forming a visible colony per unit time. Bååth et al. (44) used this approach to show that some bacteria had better growth conditions on the root surface than in the rhizosphere soil. Fig. 5 shows an application of the assay to distinguish between two different subpopulations of bacteria in the rhizosphere of barley plants. One population with high λ value (0.41) was missing on one of

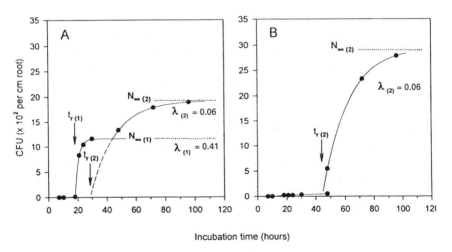

Incubation time (hours)

Figure 5.   Colony-forming units (CFU on Kings B agar at 30°C) in bacterial cell suspension extracted from rhizosphere of two barley (*Hordeum vulgare*) plants (A and B). Two physiologically different subpopulations of bacteria are seen from the progress curves; the rapid-growing subpopulation is missing on plant B. The *t* values indicate calculated incubation time for first appearance of subpopulation CFUs. The *N* values indicate calculated size of total subpopulation.

the two plants, while another population with low λ value (0.06) was present on both plants (M. Neiendam Nielsen and J. Sørensen, unpublished results).

3.2 Diversity

Germinating seeds and young roots

Rhizosphere microorganisms may arise from seed-borne populations which survive seed storage and germination or form soil-borne populations. The microbiological characteristics of the germinating seed are not often addressed but could be important for establishment of some saprophytic microorganisms in the endorhizosphere. It is thought, however, that most recruitment of microorganisms for subsequent colonization of plant root and rhizosphere soil takes place after growth stimulation of the microorganisms by the advancing root. The root tip is thus a slowly advancing point source of carbon substrate and acts as a stimulus for resumed activity and growth by the dormant microorganisms in the bulk soil. One evidence of this mechanism is the fungal germination gradient, which develops perpendicularly to the roots immediately after the root tip has passed (51). However, although chemical attractants in root exudates may be involved in fungal spore activation or growth in the rhizosphere, this mechanism is rarely known.

Considerable attention has been paid to early microbial successions in the rhizosphere. Van Vuurde and Schippers (52) reported that an invasion sequence of (in order of appearance) various rhizobacteria (excluding coryneforms and true actinomycetes), coryneforms, true actinomycetes, and microfungi was associated with onset of epidermal and cortical cell senescence in young roots (1 to 2 weeks old).

Liljeroth et al. (53) washed 1- to 2-day-old root tips and 8- to 9-day-old root bases of wheat seedlings before maceration and plating on 1/10 strength tryptic soy agar at 20°C. The early colonizers of the root surface (endorhizosphere) were characterized in terms of carbon substrate utilization and clustered into 11 groups of both gram-negative and gram-positive bacteria. Interestingly, utilization of simple sugars (lactose, galactose, mannose, xylose, and mannitol) and carboxylic acids (citrate and succinate) was found significantly more often in populations from the root tips than in those from the root bases. The advancing root tip is considered an important site for exudation of simple sugars and organic acids, which may have stimulated the growth of rhizobacteria utilizing these compounds. It must be remembered that growth of these bacteria on the agar plates may also occur because exudates provide a stimulus to relieve dormancy and obtain culturability.

To isolate bacteria from the endorhizosphere of wheat and barley, Kleeberger et al. (54) washed and surface-sterilized root segments before samples were macerated and dilution-plated on plate count agar (casein peptone, glucose, and yeast extract). Among the gram-negative, fluorescent *Pseudomonas* spp., belonging to the *P. fluorescens–P. putida* complex, formed the largest group (Table 1). It was further observed that the coryneform group of bacteria was predominant among the gram-positive rhizosphere colonizers. However, the low abundance of coryneforms in the surface-sterilized root samples indicated that they were primarily located in the external part of the wheat and barley endorhizosphere, i.e., the mucilage layer.

Mature roots

Several studies have indicated that the species diversity of indigenous soil populations will influence the composition of the ectorhizosphere populations. As for the endorhizosphere studies (see previous discussion), however, it must be remembered that observations of dominant groups in the rhizosphere soil may be biased by the media used for their isolation.

Seasonal successions may be observed, as the soil microbial activity varies with temperature, water content, and nutrition. Lambert et al. (55,56) introduced a rapid screening of microbial diversity, based on comparisons of the cellular protein compositions in the culturable bacteria. Such protein profiles, characterizing the protein composition in each clone of the culturable bacteria, were used to distinguish specific strain clusters among fast-growing rhizobacteria in the endorhizosphere of maize and sugar beets. Growth took place on

Table 1. Relative Abundance (Percentage of Total Counts) of Different Groups of Rhizobacteria on Barley [*Hordeum vulgare*] and Wheat [*Triticum aestivum*] Roots.

	Barley		Wheat	
	Untreated ($n = 61$)	Sterilized ($n = 45$)	Untreated ($n = 28$)	Sterilized ($n = 62$)
Pseudomonads	34	53	46	52
Enterobacteriaceae	—	16	4	23
Other gram-negative bacteria	—	18	—	8
Coryneform bacteria	61	13	50	16
Spore formers	5	—	—	2

[a]The data are based on CFU counts on plate count agar, using either untreated roots representing whole rhizosphere or surface-sterilized roots representing endorhizosphere. The *n* value indicates the total number of isolates analyzed.
From Ref. 54.

1/10-strength tryptic soy agar and the protein profiles in whole-cell digests were compared after polyacrylamide gel electrophoresis (PAGE) (56). *Pseudomonas fluorescens* or *Sphingomonas paucimobilis* was predominant on the root surface of young sugar beets until June, after which *Xanthomonas maltophilia* (formerly *P. maltophilia*) and *Phyllobacterium* sp. (*Rhizobiaceae*) were found at increasing densities. The three former species had considerable genetic variability; the protein profiles of identical or different species replaced each other during plant growth, however, suggesting that the variability had an ecological significance.

During the plant growth season, microorganisms become increasingly dependent on mobilization of autochtonous organic matter. Gram-positive microorganisms, including coryneforms (e.g., *Arthrobacter* sp.), and true actinomycetes (e.g., *Streptomyces* sp.) become increasingly abundant in the rhizosphere of maturing plants. Miller et al. (57) used 1/10-strength tryptic soy agar to obtain bacterial total counts in rhizosphere soil of 1- to 2-month-old wheat cultivars. Using two selective media for the fluorescent *Pseudomonas* spp. (*P. fluorescens–P. putida*) it was concluded that this group accounted for only a minor (approximately 1%) fraction of the total population. This may reflect that pseudomonads are only abundant among rhizobacteria during the initial growth phases of wheat (58).

Coryneforms, enumerated on the complex medium nutrient agar enriched with casein and glucose, constituted a significant fraction of the total population. In two different wheat cultivars, the coryneforms were 4% to 6% and 6% to 15% of the total CFU counts, respectively. By comparison, the root-free soil contained up to 30%, and the soil-borne coryneforms clearly invaded the rhizosphere soil from the bulk soil as the plants became older (57).

In a subsequent study, Miller et al. (59) also tested the relative abundance of true actinomycetes in the rhizosphere soil of two 3- to 10-week-old wheat cultivars. A selective medium, chitin oatmeal agar, gave very stable numbers of these bacteria over the 3- to 10-week period of plant growth. As for the coryneforms, root exudation seemed not to control the establishment of true actinomycetes in the rhizosphere soil. It was plausible that their high capacity for polymer degradation (sugar polymers and phenols) or resistance to desiccation, toxic compounds, etc., provides these microorganisms with a selective advantage when the root exudates become sparser in mature plants.

As the root becomes increasingly leaky, wounds can be infected by pathogens, mycorrhizae, and other saprophytic microfungi. Whereas some fungi are most abundant on the root surface and are only active during the early phases of root senescence (e.g., the saprophytic *Trichoderma viride*), others occur in all layers of the mature root and are also active at later stages of decay (e.g., the pathogenic *Fusarium oxysporium*) (60).

3.3 Activity and Growth

Root colonization

Many studies of root colonization are based on seed inoculation with a specific bacterial strain and subsequent determinations of surviving numbers on the root. To follow growth on selective media, a particular microorganism is typically equipped with one or more antibiotic resistance markers, e.g., kanamycin, rifampicin, or streptomycin resistance. Another approach is to equip an organism with a marker gene for catabolism of specific substrates, made detectable by cleavage of dyes. The *lac*Z (β-galactosidase), *gus*A (β-glucuronidase), and *xy*/E (2,3-catechol dioxygenase) genes have all been used as chromogenic marker genes to study the fate of bacteria in the rhizosphere (61). It is noteworthy, however, that most of these studies still rely on additional antibiotic resitance genes in the organisms, to obtain a primary distinction of colonies of the engineered strains from those of the indigenous rhizosphere bacteria.

Loper et al. (62) used rifampicin-resistant mutants of *Pseudomonas fluorescens–P. putida* to follow colonization of potato (*Solanum tuberosum*) roots (Fig. 6). Seed pieces were dipped into bacterial inoculum (suspension of 2-day-old culture grown on Kings B agar) and planted in soil. After 2–3 weeks of plant growth, the bacterial CFUs on different root segments indicated that only some strains were good root colonizers, as judged from their presence on the root tip–proximal segments. A study by Thompson et al. (63) of wheat root colonization and survival of an *Arthrobacter* sp. (streptomycin-resistant strain A109) illustrates that these coryneforms are well adapted to long-term survival in both rhizosphere and bulk soil. Wheat seeds inoculated with *Arthrobacter* A109 were germinated and plants were grown for 3 months. Endorhizosphere CFUs were followed after washing and macerating the root and plating onto nutrient agar with streptomycin. The constant numbers over time (expressed per gram of root) indicated that the *Arthrobacter* A109 population was profilerating or constantly supplemented by cells invading from the soil.

Survival (stress)

Since the soil water content at the root surface of an actively transpiring plant may reach a low level, below that of the nonrhizosphere soil, the resistance of rhizosphere bacteria to low water content may be critical to their survival. In the study by Loper et al. (62) on root-colonizing *Pseudomonas* strains, there was an apparent relationship between colonization potential and in vitro osmotolerance among the strains, which indicated that drought resistance could be critical to survival after root colonization (Fig. 7). Among the major gram-positive groups in the rhizosphere, drought resistance is obtained by endospore and actinospore for-

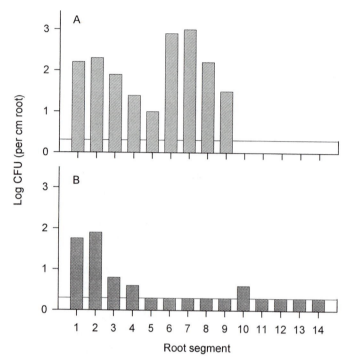

Figure 6. Spatial distribution of root-colonizing *Pseudomonas fluorescens–P. putida* strains B4 (A) and B10 (B) on 1-cm longitudinal segments of a potato (*Solanum tuberosum*) root. Root segment 1 is that closest to the seed piece. Detection limit is indicated by horizontal line. (From Ref. 62.)

mations in *Bacillus* spp. and true actinomycetes, respectively, while arthrospores in *Arthrobacter* spp. may serve as an example of drought-resistant forms among the coryneforms. These spore forms are fully viable and spontaneously proliferate into normal rod-formed cells when growth conditions are suitable.

In addition to the tolerance against low matric potential (including osmotolerance and drought resistance), the ability of bacteria to protect themselves against oxidative stress (see previous discussion) may be important during colonization. Plant root cells respond quickly to approaching microorganisms by releasing activated oxygen species (superoxide and hydrogen peroxide) (64). Protection against the activated oxygen species, the oxidative-stress response, has been shown in gram-negative bacteria to include a coupled synthesis of several new enzymes involved in uptake, energy metabolism, and DNA repair, which provide cross-protection against low water content, activated oxygen species, and certain antibiotics (65). Although the present knowledge of stress responses in root-col-

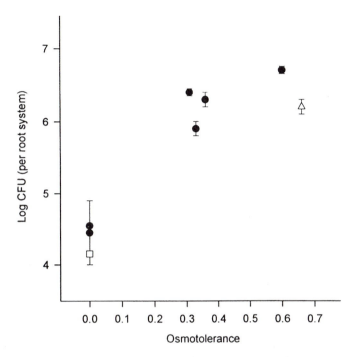

Figure 7. Correlation between potato (*Solanum tuberosum*) root colonization and osmotolerance, expressed as relative growth under specific osmotic potentials, in eight *P. fluorescens–P. putida* strains, including B4 (△) showing good colonization and high osmotolerance and B10 (□) showing poor colonization and low osmotolerance. See also Fig. 6. (From Ref. 62.)

onizing *Pseudomonas* spp. is limited, such cross-protection to adverse soil factors has recently been found in phytopathogenic *Pseudomonas syringae* (66) in culture experiments.

In general, resistance to abiotic soil factors such as low water availability, oxidative stress, or starvation may strongly influence the ability of gram-negative rhizobacteria to compete with other microorganisms and match predation factors in the rhizosphere. Attempts to introduce plant-beneficial *Pseudomonas* spp. into rhizosphere soil illustrate that bacteria face a high risk of extinction if cells are not preadapted to growth in the rhizosphere or if a protective microhabitat is not available. Vandenhove et al. (67) showed that the physiological state of *Pseudomonas fluorescens* (strain 88W1) cells introduced into soil had a marked influence on its survival. Cells from the late exponential phase of the inoculum culture survived better than those from the early exponential or stationary phase.

Much effort has therefore been put into design of inoculation protocols, by which plant-benefial rhizobacteria survive their initial exposure to adverse stress factors in the soil. Improved root colonization and survival were shown when *Pseudomonas fluorescens* cells were encapsulated in protecting polymer (alginate) beads (68). The extended protection combined with a slow release of cells from the alginate beads resulted in successful growth in the rhizosphere environment.

Extinction (predation and lysis)

While decreasing CFUs in time-dependent studies of colonization and survival may be attributed to formation of viable, but nonculturable cell forms (discussed previously), there are undoubtedly significant death and turnover of the bacterial populations due to autolysis, lysis by bacteriophages, and predation by protozoa. The latter, including amoebae, flagellates, and ciliates, are common components of the rhizosphere soil (69,70), and selective predation by grazing amoebae may influence patterns of dominating bacteria and fungi. Stout and Heal (71) thus showed that some soil amoebae prefer to graze on gram-negative bacteria such as *Pseudomonas* spp.

Although protozoans have been shown to limit the numbers of rhizobacteria (72,73), the predators would not be expected to result in selective effects on the strain level as do the bacteriophages. Stephens et al. (74) reported that two different *Pseudomonas* strains, M11/4 and B2/6, declined at different rates under colonization of sugar beet roots in nonsterile soil. This was explained by the presence of large numbers of a bacteriophage capable of lysing one of the strains, B2/6 (Fig. 8). Unfortunately, the soil phage reservoir for interaction with *Pseudomonas* spp. and other root-colonizing bacteria is yet poorly investigated, although several new species of large *Pseudomonas* phages were recently isolated from barley rhizosphere (75). All phages were lytic and hosted by several *Pseudomonas* spp., but their slow multiplication rates suggested a possible mechanism of balanced phage–host coexistence in the rhizosphere.

4. PROSPECTS IN RHIZOSPHERE MICROBIOLOGY: REVIVAL OF THE MICROSCOPE

Scanning and transmission electron microscopes continue to be important instruments for research in rhizosphere microbiology. One example is a fascinating study by Achouak et al. (76), in which a N_2-fixing isolate of *Enterobacter agglomerans* was shown to form cell aggregates on rice, but not on wheat roots (Fig. 9). The aggregates or symplasmata, already observed by Beijerinck in 1888 (77), are tightly packed cells within a common extracellular sheath. It was suggested that the sheath surrounding the bacteria may act as a selective barrier for

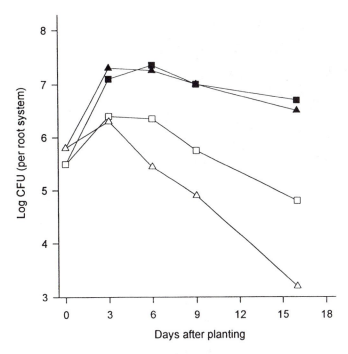

Figure 8. Survival of root-colonizing *Pseudomonas fluorescens* strains M11/4 (■, □) and B2/6 (▲, △) on sugarbeet (*Beta vulgaris*) roots in sterile (■, ▲) and nonsterile (□, △) soil. Soil bacteriophages may cause differential survival of bacterial strains in nonsterile soil. (From Ref. 74.)

oxygen diffusion into the bacteria, allowing optimal (microaerophilic or low-oxygen) conditions for nitrogenase activity (enzymatic reduction of N_2 to NH_4^+). It will be interesting to determine whether such adaptations also occur among other N_2-fixing rhizobacteria, e.g., the closely related plant-pathogenic *Erwinia herbicola*, which also contain N_2-fixing strains (78). The study by Kleeberger et al. (54) further indicated that two subgroups of *Enterobacter agglomerans* appeared almost exclusively in samples of surface-sterilized roots and thus seemed to be genuine colonizers of the endorhizosphere. One of the groups represented facultative anaerobes, among which about one-third were N_2-fixing.

Immunodetection of bacteria by the fluorescent antibody technique was introduced in soil microbiology more than 20 years ago (79). This technique can be most specific, if the antibody recognizes a stable surface component of the target bacterium, usually a membrane protein or a lipopolysaccharide (LPS) molecule. Compared to the use of plate counts requiring a culturable organism, the immunotechniques also recognize dead bacteria and may therefore overestimate vi-

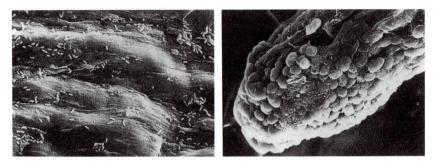

Figure 9. Scanning electron micrograph of *Enterobacter agglomerans* strain NO30 incubated for 24 hr with wheat (*Triticum sativum*) roots (left) and rice (*Oryza sativa*) roots (right). Note the presence of polysaccharide-encapsulated aggregates of bacteria (symplasmata, approximately 6 μm long) on the rice roots, whereas the wheat roots only contain normal bacterial cells (approximately 1.5 μm long). (From Ref. 76.)

able cell numbers. It has been argued, though, that dead bacteria disappear within several weeks in soil and thus cannot account for major enumeration differences between immunofluorescence (IF) and plate counts (80).

Within the last few years, the rapid development of dual- and multilabeling techniques using several fluorochromes such as fluorescein and rhodamine, as well as advanced-microscopes, have provided a new stimulus to rhizosphere microbiologists. Multiple labeling experiments to follow several interacting microorganisms at the same time are only one example of future directions of this research. Immunofluorescence detection has further gained a new renaissance by the development of very specific monoclonal antibody techniques to detect microbial products such as enzymes and antibiotics in complex microbial communities. The rapid development of new detection techniques using fluorescent oligonucleotide probes has also been breath-taking, as these probes allow for both highly specific (e.g., strain) and less specific (e.g., species or higher taxon) detection of microorganisms or their genes in the environment. Most importantly, the large number of sophisticated techniques for determining localization and activity of microorganisms in situ makes it possible to pose new questions concerning their life strategies in natural environments. One example of new microscopic techniques was demonstrated in Fig. 2; this photo was made from a three-dimensional image of *Pseudomonas fluorescens* strain DF57 on barley roots accomplished by fluorescence-labeled monoclonal antibody staining and CLSM. Schloter et al. (40) and Assmus et al. (81) used immunodetection and ribosomal ribonucleic acid (rRNA)-targeted probes, respectively, and CLSM to study an *Azospirillum* strain colonizing the surface of wheat roots. The CLSM reduces the background fluorescence (autofluorescence) from plant tissue to a minimum, and, supported by advanced image analysis software, the spatial distribution of bacteria on plant surfaces can be observed. In the next step, in situ three-dimensional localization of the specific bacterial cell and demonstration of one or more of its functioning, cell-associated proteins (enzymes) are certainly within reach. They wil provide details, not only on the microscale distribution of cells, but also on their physiological state and activity.

Finally, the molecular technology has now made it possible to detect in situ physiological activity in specific cells labeled with reporter genes such as *lux*. Insertion of *lux* genes, which code for enzymes involved in bioluminescence (82), into particular bacterial strains is promising for studies of single cells in their rhizosphere microhabitats. To date, no bioluminescent bacteria have been reported as natural constituents of rhizosphere microbial communities on the basis of culturing techniques, suggesting that *lux* may be a useful marker gene for studies of rhizosphere bacteria (83). Bacteria expressing these genes emit photons that can be detected and quantified by microscopy. De Weger et al. (84) first used autophotography and sensitive light detectors to visualize *lux*-marked bacteria on roots of soybean (*Glycine max*). Fig. 10 demonstrates the localization of *lux*-equipped *Pseudomonas fluorescens* strain DF57 on barley roots. The strain is a good colonizer and seems evenly distributed along the roots. Finally, insertion of *lux* in environmentally regulated genes will facilitate in situ analysis of the expression of specific functions, such as the production of selected enzymes or antibiotics. Eventually, reporter gene technology will allow researchers to

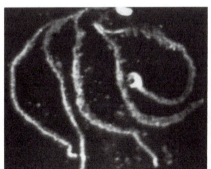

Figure 10. Use of bioluminescence as a marker for visualizing root colonization by a *Pseudomonas fluorescens* strain DF57 (*lux*AB transformant) in barley (*Hordeum vulgare*) rhizosphere. Root colonization is evident by comparing bioluminescence (right) and root structure seen with incident light (left). The seed is located at the top and the roots are approximately 10 cm long.

determine both the in situ microlocalization and physiological state of specific microorganisms in situ in the rhizosphere.

ACKNOWLEDGMENTS

I am grateful to Drs. Svend Binnerup, Mikael Hansen, Ole Højberg, Lene Kragelund, Mette Neiendam Nielsen, and Ole Nybroe at the Section of Genetics and Microbiology, Royal Veterinary and Agricultural University, for the illustrations and constructive comments on the manuscript.

REFERENCES

1. L. Hiltner, *Arb. Dtsch. Landwirt. Ges. 98*:59 (1904).
2. J. Belandreau and R. Knowles, *Interactions Between Non-Pathogenic Soil Microorganisms and Plants* (Y. R. Dommergues and S. V. Krupa, eds.), Elsevier, Amsterdam, p. 243 (1978).
3. R. W. Duell and G. R. Peacock, *Crop Sci. 25*:880 (1985).
4. M. Rougier, *Plant Carbohydrates II, Extracellular Carbohydrates* (W. Tanner and F. A. Leowus, eds.), Springer-Verlag, Berlin, p. 542 (1981).
5. J. Vermeer and M. E. McCully, *Protoplasma 109*:233 (1981).
6. G. G. Leppard and J. R. Colvin, *J. Polymer Sci. C36*:321 (1971).
7. H. Jenny and K. Grossenbacher, *Soil Sci. Soc. Am. Proc. 27*:273 (1963).
8. R. Campbell, *Can. J. Microbiol. 29*:39 (1983).
9. M. E. McCully, *Root Development and Function* (P. J. Gregory, J. V. Lake, and D. A. Rose, eds.), p. 53 (1987).
10. J. Vermeer and M. E. McCully, *Planta 156*:45 (1982).
11. R. van Peer, H. L. M. Punte, L. A. de Weger, and B. Schippers, *Appl. Environ. Microbiol. 56*:2462 (1990).
12. Y. Bashan, *J. Gen. Microbiol. 132*:3407 (1986).
13. I. J. Misaghi, M. W. Olsen, J. M. Billotte, and R. M. Sonoda, *Soil Biol. Biochem. 24*:287 (1992).
14. A. J. Anderson, P. Habibzadegah, and C. S. Tepper, *Appl. Environ. Microbiol. 54*:375 (1988).
15. S. J. Vesper, *Appl. Environ. Microbiol. 53*:1397 (1987).
16. L. A. de Weger, B. Jann, K. Jann, and B. Lugtenberg, *J. Bacteriol. 169*:1441 (1987).
17. C. J. Douglas, W. Halperin, and E. W. Nester, *J. Bacteriol. 152*:1265 (1982).
18. F. B. Dazzo, G. L. Truchet, J. E. Sherwood, E. M. Hrabak, M. Abe, and S. H. Pankratz, *Appl. Environ. Microbiol. 48*:1140 (1984).
19. D. Parkinson, G. S. Taylor, and R. Pearson, *Plant Soil 19*:332 (1963).

20. R. N. Ames, C. P. P. Reid, and E. R. Ingram, *New Phytol.* 96:555 (1984).
21. J. R. Meyer and R. G. Lindermann, *Soil Biol. Biochem.* 18:191 (1986).
22. B. M. McDougall, "The exudation of 14C-labeled substrates from roots of wheat seedlings," Transactions of the Ninth International Congress of Soil Science, Adelaide, pp. 647–655 (1968).
23. R. Merckx, A. Dijkstra, A. den Hartog, and J. A. van Veen, *Biol. Fertil. Soils* 5:126 (1987).
24. J. K. Martin, *Soil Biol. Biochem.* 9:303 (1977).
25. J. K. Martin and J. R. Kemp, *Soil Biol. Biochem.* 12:551 (1980).
26. O. Højberg and J. Sørensen, *Appl. Environ. Microbiol.* 59:431 (1993).
27. W. Kolb and P. Martin, *Soil Biol. Biochem.* 20:221 (1988).
28. E. Liljeroth, J. A. van Veen, and H. J. Miller, *Soil Biol. Biochem.* 22:1015 (1990a).
29. E. Liljeroth, E. Bååth, I. Mathiason, and T. Lundborg, *Plant Soil* 127:81 (1990b).
30. J. A. van Veen, R. Merckx, and S. C. van de Geijn, *Plant Soil* 115:179 (1989).
31. S. J. Binnerup and J. Sørensen, *Appl. Environ. Microbiol.* 58:2375 (1992).
32. F. E. Sanders and P. B. Tinker, *Nature* 233:278 (1971).
33. I. Jakobsen, L. K. Abbott, and A. D. Robson, *New Phytol.* 120:371 (1992).
34. L. Stoumann Jensen and J. Sørensen, *Plant Soil* 162:151 (1994).
35. J. M. Tisdall and J. M. Oades, *J. Soil Sci.* 33:141 (1982).
36. A. D. Rovira, *J. Appl. Bacteriol.* 19:72 (1956).
37. A. D. Rovira, *Ecology of Soil-Borne Plant Pathogens* (K. F. Baker and W. C. Snyder, eds.), Murray, London, p. 170 (1965).
38. E. I. Newman and H. J. Bowen, *Soil Biol. Biochem.* 6:205 (1974).
39. J. Postma and H. -J. Altemüller, *Soil Biol. Biochem.* 22:89 (1990).
40. M. Schloter, R. Borlinghaus, W. Bode, and A. Hartmann, *J. Microscopy* 171:173 (1993).
41. P. C. Brookes, J. F. Kragt, D. S. Powlson, and D. S. Jenkinson, *Soil Biol. Biochem.* 17:837 (1985).
42. A. W. West and G. P. Sparling, *J. Microbiol. Meth.* 5:177 (1986).
43. A. D. Rovira, E. I. Newman, H. J. Bowen, and R. Campbell, *Soil Biol. Biochem.* 6:211 (1974).
44. E. Bååth, S. Olsson, and A. Tunlid, *FEMS Microbiol. Ecol.* 53:355 (1988).
45. A. Faegri, V. Torsvik, and J. Goksoyr, *Soil Biol. Biochem.* 9:105 (1977).
46. R. A. Olsen and L. R. Bakken, *Microb. Ecol.* 13:59 (1987).
47. D. B. Roszak and R. R. Colwell, *Microbiol. Rev.* 51:365 (1987).
48. S. J. Binnerup, D. F. Jensen, H. Thordahl-Christensen, and J. Sørensen, *FEMS Microbiol. Ecol.* 12:97 (1993).
49. S. J. Binnerup and J. Sørensen, *FEMS Microbiol. Ecol.* 13:79 (1993).

50. T. Hattori, *J. Gen. Appl. Microbiol. 28*:13 (1982).

51. S. Olsson and B. Nordbring-Hertz, *FEMS Microbiol. Ecol. 31*:293 (1985).

52. J. W. L. van Vuurde and B. Schippers, *Soil Biol. Biochem. 12*:559 (1980).

53. E. Liljeroth, S. L. G. E. Burgers, and J. A. van Veen, *Biol. Fertil. Soils 10*:276 (1991).

54. A. Kleeberger, H. Castorph, and W. Klingmüller, *Arch. Microbiol. 136*:306 (1983).

55. B. Lambert, F. Leyns, L. van Rooyen, F. Gossele, Y. Papon, and J. Swings, *Appl. Environ. Microbiol. 53*:1866 (1987).

56. B. Lambert, P. Meire, H. Joos, P. Lens, and J. Swings, *Appl. Environ. Microbiol. 56*:3375 (1990).

57. H. J. Miller, G. Henken, and J. A. van Veen, *Can. J. Microbiol. 35*:656 (1989).

58. K. Vagnerova, J. Macura, and V. Catska, *Folia Microbiol. 5*:298 (1960).

59. H. J. Miller, E. Liljeroth, G. Henken, and J. A. van Veen, *Can. J. Microbiol. 36*:254 (1990).

60. G. S. Taylor and D. Parkinson, *Plant Soil 22*:1 (1965).

61. J. W. Kloepper and C. J. Beauchamps, *Can. J. Microbiol. 38*:1219 (1992).

62. J. E. Loper, C. Haack, and M. N. Schroth, *Appl. Environ. Microbiol. 49*:416 (1985).

63. I. P. Thompson, K. A. Cook, G. Lethbridge, and R. G. Burns, *Soil Biol. Biochem. 8*:1029 (1990).

64. J. Katsuwon and A. J. Anderson, *Appl. Environ. Microbiol. 55*:2985 (1989).

65. D. A. Siegele and R. Kolter, *J. Bacteriol. 174*:345 (1992).

66. M. G. Klotz, *Can. J. Microbiol. 39*:948 (1993).

67. H. Vandenhove, R. Merckx, H. Wilmots, and K. Vlassak, *Soil Biol. Biochem. 23*:1133 (1991).

68. J. D. van Elsas, J. T. Trevors, D. Jain, A. C. Wolters, C. E. Heijnen, and L. S. van Overbeek, *Biol. Fertil. Soils 14*:14 (1992).

69. J. F. Darbyshire and M. P. Greaves, *Can. J. Microbiol. 13*:1057 (1967).

70. R. C. Foster and J. F. Dormaar, *Biol. Fertil. Soils 11*:83 (1991).

71. J. D. Stout and O. W. Heal, *Soil Biology* (A. Burges and F. Raw, eds.), Academic Press, New York, p. 149 (1967).

72. M. Habte and M. Alexander, *Appl. Microbiol. 29*:159 (1975).

73. C. Ramirez and M. Alexander, *Appl. Environ. Microbiol. 40*:492(1980).

74. P. M. Stephens, M. O'Sullivan, and F. O'Gara, *Appl. Environ. Microbiol. 53*:1164 (1987).

75. J. I. A. Campbell, M. Albrechtsen, and J. Sørensen, *FEMS Microbiol. Ecol. 18*:63 (1995).

76. W. Achouak, T. Heulin, G. Villemin, and J. Balandreau, *FEMS Microbiol. Ecol. 13*:287 (1994).

77. M. W. Beijerinck, *Bot. Ztg. 46*:740 (1888).

78. H. Papen and D. Werner, *Arch. Microbiol. 120*:25 (1979).
79. E. L. Schmidt and E. A. Paul, *Methods of Soil Analysis.* Part 2. *Chemical and Microbiological Properties* (A. L. Page, R. H. Miller, and D. R. Keeney, eds.), American Society for Agronomy, Inc., and Soil Science Society of American, Inc., Madison, Wisc., p. 803. (1982).
80. J. Postma, J. D. van Elsas, J. M. Govaert, and J. A. van Veen, *FEMS Microbiol. Ecol. 53*251 (1988).
81. B. Assmus, P. Hutzler, G. Kirchhoh, R. Amann, J. R. Lawrence, and A. Hartmann, *Appl. Environ. Microbiol. 61*:1013 (1995).
82. J. J. Shaw and C. I. Kado, *Bio/Technology 4*:560 (1986).
83. C. J. Beauchamp, J. W. Kloepper, and P. A. Lemke, *Plant Growth-Promoting Rhizobacteria—Progress and Prospects* (C. Keel, B. Koller, and G. Defago, eds.), IOBC/WPRS Bulletin XIV, 8, p. 243 (1991).
84. L. A. de Weger, P. Dunbar, W. F. Mahaffee, and B. J. J. Lugtenberg, *Appl. Environ. Microbiol. 57*:3641 (1991).

3

Culturable and Nonculturable Bacteria in Soil

LARS REIER BAKKEN Agricultural University of Norway, Aas, Norway

1. INTRODUCTION

For at least half a century, it has been known that the major part of the structurally intact bacterial cells in soil appear to lack the capacity to grow on standard laboratory media. This consensus is based on numerous attempts to compare total microscopic counts with viable counts on laboratory media (1–3). Heterotrophic bacteria have been assumed to be the dominant types among soil bacteria, since the consumption and mineralization of organic materials represent most of the energy flux through the soil biota. For this reason, all attempts to culture the majority of soil bacteria have concentrated on the heterotrophic bacteria. For the same reason, the majority of apparently nonculturable bacteria in soil are assumed to be heterotrophic.

This is important to keep in mind when discussing observations of nonculturable bacteria in soils. If we were not biased as to their metabolism, the absence of growth on a "nonselective" organic medium would hardly be significant. But there are few reasons to throw this bias overboard, and it remains a "fact" that the majority of the structurally intact heterotrophic bacteria in soil are unable to grow on nutrient media tested so far (1–3). This casts some doubt as to the relevance of classic studies of soil bacteria by culturing and isolation. The bias of any cultural method appears to be so serious that only a minor fraction of the bacterial flora will be detected and available for further characterization in pure cultures.

A question of utmost importance is therefore whether the cultured bacteria represent important indigenous bacterial species in soil. The low frequency of culturable cells strongly suggests a negative answer, but the situation may not be as bleak if we take other factors such as metabolic activity and cell dimensions into account. The purpose of this chapter is to clarify some facts, probabilities, hypotheses, and new research strategies regarding the cultured and noncultured heterotrophic soil bacteria. The sources of information broadly fall into two groups. One is a number of experiments in which indigenous soil bacteria have been extracted from the soil and screened with various methods to elucidate their metabolic integrity and activity. The other is the recent progress in the studies of physiology under starvation of cultured heterotrophic bacteria.

2. THE CULTURED BACTERIA

Taxonomic characterization of whole soil bacterial communities has not been a hot topic in soil biology for some time, for various reasons. One is the general lack of confidence of soil microbiologists in culturing/isolation methodology, regarding the representativeness of the isolated organisms. Second, the task is a herculean one, considering the large diversity and variability of the populations (4). Third, the classic taxonomical methodology is difficult to apply to microorganisms isolated from soil. Taken together with a fundamental lack of confidence in classic taxonomy, these factors explain the general lack of interest in the taxonomy of soil bacteria.

Given this background, only brief comments will be provided on the taxonomy of cultured soil bacteria. Alexander (5) has made a brave attempt to summarize a great number of classic taxonomical studies of bacteria prevalent in soil and presents the following ranges (percentage of total viable counts): *Arthrobacter*: 5%–60%, *Bacillus*: 7%–67%, *Pseudomonas*: 3%–15%, *Agrobacterium*: up to 20%, *Alcaligenes*: 2%–12%, *Flavobacterium*: 2%–10%. The list suggests variable composition of the culturable bacterial flora of soil. The variability is likely to reflect differences in culture conditions applied as well as true differences between soils. Sørheim et al. (3) compared the microflora (in beech forest soil) growing on a number of different nutrient media and found only partial overlap between the populations growing on fairly similar media. All media indicated a prevalence of gram-negative bacteria, however. The dominance of gram-negatives is commonly assumed to be the rule (particularly in the rhizosphere), but this view is challenged by those who claim that coryneform bacteria are often more numerous (6).

3. DIRECT OBSERVATIONS OF WHOLE BACTERIAL COMMUNITIES

Early direct observations of soil bacteria by microscopy (light, fluorescence, or electron microscopy) revealed vast numbers of extremely small and apparently structurally intact bacteria. The total numbers were two to three orders of magnitude higher than the numbers of "viable" cells as counted by any cultural methods for heterotrophic microorganisms tested.

The criteria for recognizing a cell are critical, and methodology-dependent. The (surface) morphological features are the only available criteria in scanning electron microscopy (SEM), whereas transmission electron microscopy (TEM) will reveal cell walls and membranes as recognizable cellular structures. In a pioneering study of indigenous soil bacteria by TEM, Bae and Casida (7) demonstrated that the majority of recognizable cells (cell walls, membranes, and various other structures revealed) had diameters below 0.3 μm and that the relative abundance of such dwarf cells was transiently lowered by nutrient (glucose) additions. The authors tentatively hypothesized that this could be due to a transient cell enlargement in response to nutrients, followed by shrinkage as nutrients were exhausted.

Fluorescence microscopy involves staining with fluorochromes which bind specifically to cellular components; deoxyribonucleic acid (DNA) is the most common target molecule. Thus, fluorescence microscopic counting of cells is based on a combination of specific staining and recognizable morphological features, provided the dimensions of the cells are above the resolution limits of the microscope. The specificity of the staining is disputable (8), and the determination of cell dimensions in light microscopy is not precise, at least when based on the judgment of the microscopist (9). Image analysis and new fluorescence microscopic technology (confocal laser scanning) represent a significant improvement with respect to reproducible and objective enumeration and size determinations, although calibration of the size determination still represents a problem when analyzing mixtures of cells (10). The results with this new technology seem to corroborate earlier findings regarding the dominance of small cells in unamended soils. At least 50% of the cells were below 0.1 μm^3, and less than 5% had volumes above 1 μm^3 (10–12).

Aquatic systems are similarly dominated by small cells (13,14), the average cell volumes appear to be even lower (0.04–0.07 μm^3 [14]) than those in soil. Torrella and Morita (15) observed slow growth of extremely small cells when transferred to microslide cultures and used the term "ultramicrobacteria" to denote such cells. For the sake of convenience, I shall use the term "dwarfs," coined by Bae and Casida (7), to denote cells with diameters below 0.3 μm, arbitrarily defined in our studies as cells with volumes below 0.07 μm³ (16).

Direct observations of whole bacterial communities in soil indicated a connection between energy starvation and the presence of dwarfs. Long-term starvation of the heterotrophic soil community by eliminating plant growth for several years resulted in a higher frequency of dwarfs (17). In response to short pulses of nutrients, the proportion of larger bacteria increase significantly (7,18–20). Initial bacterial colonization of the rhizosphere of barley and ryegrass was dominated by larger bacterial cells, but the number of dwarfs increased toward the end of the plant growth period. The estimated average generation time for the dwarf bacterial population under these conditions was 10–20 days (20).

These observations could be taken to indicate that dwarfs represent miniature cells of otherwise more normal-sized bacteria, with the size reduction a result of asymmetric cell division or cell shrinkage due to energy starvation. Cell shrinkage in response to starvation is observed in pure cultures of gram-negative bacteria. Gram-positive bacteria have a variable cell size, depending on the growth rate, and asymmetric cell division during extremely slow growth has been demonstrated (21,22).

An alternative and more hypothetical explanation would be that dwarfs are the normal form of obligate oligotrophic bacteria, and that the small size is one of the traits of a successful "model" oligotrophic microorganisms (23). Although the existence of obligate oligotrophs is questionable (14), the term "oligotroph" can be a useful collector of traits that are assumed to be essential for survival at low substrate levels: high surface to volume ratios (24), high-affinity uptake systems for substrates (13), and efficient shutdown of metabolism upon starvation (25).

4. FILTERABILITY, CELL CONSTITUENTS, AND CULTURABILITY

In experiments with indigenous soil bacteria, we designed an experimental protocol that allowed inspection of the relationship between cell dimensions and culturability of the indigenous soil bacteria, and the relationship between original cell dimensions and the dimensions after onset of growth. The steps in the protocol are illustrated in Fig. 1. First, cells were dislodged from soil particles by dispersion and centrifugation [1]. Second, the cells were purified by high-speed centrifugation on a cushion of colloidal silica (11). The final cell suspensions were relatively free of colloids other than bacterial cells, thus allowing reasonable amounts of cells to be screened according to cell dimen-

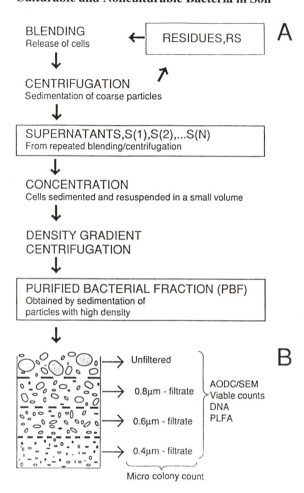

Figure 1. Protocol for separating soil bacteria according to cell size. Cells were first extracted and purified by density gradient centrifugation (A). The cells were then filtered (B) through polycarbonate membranes with successively smaller pore dimensions. The filtrates were then subjected to various analyses, as indicated: AODC, direct fluorescence microscopic counts of acridine orange–stained cells; SEM, scanning electron microscopy; DNA, determination of average DNA content of cells; PLFA, phospholipid fatty acid analysis.

sions by filtration through polycarbonate membranes with successively smaller pore diameters. The suspensions obtained allowed investigation of the relationship between cell dimensions and the viability, cell size changes prior to growth, lag phases, growth rates, and DNA content of the cells. The method of cell extraction was later simplified (26,27), and the same filtration procedure has been used for analysis of the microbial phospholipid fatty acid contents and profiles (28), which can be used as signatures of the microbial biomass.

The information gained from these filtration experiments can be summarized as follows:

Of the cells with diameters below 0.4 μm 0.2%–0.5% were able to grow on complex low-strength nutrient media (10 mg l^{-1} of five organic substances). In contrast, the percentage viabilities on the same medium were 9% and 38% for cells with diameters 0.6–0.8 μm and >0.8 μm, respectively (29). Rich nutrient media restricted the growth of small cells, but generally not that of larger cells (2).

For the majority of the dwarf cells that were able to multiply on agar or in liquid nutrient medium, the small cell diameter was retained during growth (29). Thus, the experiments offered little evidence for the hypothesis that dwarfs are smaller forms of normally sized bacteria. However, this absence of evidence may be far from evidence of absence, considering the extremely low percentage culturability (<0.1%) of the cells that passed the 0.4-μm-pore-diameter polycarbonate filters.

The number of microcolonies observed by direct microscopy corresponded well to the number of visible colonies observed after a 3- to 4-week incubation period on the same medium, which again corresponded well with the estimated most probable number (MPN) based on growth in liquid media (29).

The microcolony observations allowed investigation of lag phase and generation times of growing cells among dwarf cells. Lag phases up to 13 hours were observed, and the specific growth rate ranged from 0.05 to 0.75 h^{-1} (generation times from 1 to 40 hr).

Measurement of DNA in dwarfs revealed that their average DNA content (around 2 fg cell^{-1}) was similar to that of larger cells in the same soil (30). The level is high enough to suggest that a majority of the dwarfs contain intact genomes. Flow cytometric determination of DNA in soil bacteria from the same soil (9) corroborated the DNA levels but demonstrated a positive relationship between cell size and DNA content. The occurrence of "ghost cells" containing little or no DNA (as indicated for marine bacteria [8]) did not appear to be important, neither among dwarfs nor among the larger cells. The DNA levels are similar to those measured in marine bacteria by flow cytometry of DAPI stained cells (14).

The amounts of phospholipid fatty acids per cell surface area were the same for the dwarfs as that for the larger cells (28), indicating that a majority of the dwarfs may have intact membranes.

Although dwarfs outnumber other cells and contain a similarly large share of the total bacterial DNA, their biovolume is only 10% of the total bacterial biovolume. This suggests that their metabolism accounts for a minor fraction of the total energy flux through the microbial community in soil.

These observations clearly suggest that the major part of the nonculturable bacteria in soil are dwarf cells. The dwarfs appear to be structurally intact and metabolically competent at least in having an intact genome. Thus, they carry most of the genetic information (DNA) present in soil. However, their share of the metabolic activity and energy flux may be moderate (12,16) because of their small size and apparently low activity.

A similar filtration protocol was used by Bottomley and Dughri (31), in combination with immunofluorescence microscopy, to enumerate and characterize cells of *Rhizobium leguminosarum* in soil. Small cells (passing through a 0.4-μm polycarbonate membrane) represented 20%–30% of the total numbers within each of the indigenous populations of four serogroups found in soil. However, attempts to induce growth or nodulation by these small cells were not successful. The fraction of dwarfs within each population increased with soil depth (rather abruptly at 15–20 cm), indicating a relationship between energy availability and the presence of dwarfs within each population. The observations represent a strong case for the hypothesis that dwarf cells are miniature cells of otherwise more normal-sized bacteria.

5. FILTERABILITY AND METABOLIC ACTIVITY

Bååth (12) filtered indigenous soil bacteria through polycarbonate membranes and measured the incorporation rates of ^{3}H-thymidine and ^{14}C-leucine in the different size fractions. The turnover times of cellular thymidine and leucine (total amounts per cell divided by incorporation rate per cell) were estimated. Such turnover times can be tentatively used as inverse values of the average growth rate of the populations. The results showed a negative relationship between cell size and turnover times. The calculated turnover times, based on thymidine incorporation, were 2.5–4.5 days for cells with diameters >0.6 μm, 9.6 days for cells with diameters 0.4–0.6 μm, and 32 days for cells with diameter <0.4 μm. The turnover times based on leucine incorporation were 2.4–5.9, 6.1, and 13.6 days for the diameter groups >0.6, 0.4–0.6, and <0.4 μm, respectively. The turnover times for dwarfs agree surprisingly well with the apparent growth rates of the dwarf cell populations (0.02–0.03/day) during colonization of the rhizosphere of barley and ryegrass (20).

Further, Bååth (12) found an almost perfect linear correlation ($r^2 = 0.995$) between the thymidine incorporation rate per cell and the percentage of colony-forming cells. Taken to the extreme, this would indicate that the culturable bacteria are responsible for practically all the growth (thymidine incorporation) and de novo protein synthesis (leucine incorporation) activity in the soil; hence the noncultured soil bacteria would be of minimal metabolic importance.

This relationship between cell size and growth rate of indigenous soil bacteria has also been demonstrated by a combination of fluorescence microscopic and flow cytometric determinations of cell volumes and DNA during culturing of suspensions of soil bacteria (9). In this experiment, indigenous soil bacteria were extracted and suspended in a buffered medium supplemented with moderate amounts (10 mg l^{-1}) of glucose, fructose, xylose, peptone, and yeast extract. During the 60-hr incubation period, the numbers of dwarf cells did not increase significantly, nor did their DNA content. In contrast, significant growth (and increased amounts of DNA per cell) occurred within the other volume groups, and the apparent growth rates were strongly correlated with the cell size. The apparent growth rate of the three volume groups 0.07–0.18, 0.18–0.32, and >0.32 μm^3 were 0.02, 0.1, and 0.15 hr^{-1} respectively (equivalent to doubling times of 32, 6.6, and 4.5 hours, respectively).

The calculated growth rates for the different volume groups are disputable, since a shift in cell size prior to growth would obscure the result (and in fact tend to produce exactly the same correlation between cell size and apparent growth rate). However, the difference in apparent growth rates between the volume groups persisted throughout the whole experiment, whereas a shift in cell size would mainly occur during initiation of growth.

The absence of detectable growth among the dwarf cells in such short experiments is not surprising. Assuming that the dwarf cell population grows at a rate similar to that found by Bååth (12) and Breland and Bakken (20), i.e., with doubling times of 10–20 days, their numbers would not increase more than 15%–25% during the 60-hr experimental period, which was within the range of error of the determinations.

6. THE STARVATION RESPONSE OF BACTERIA

Successful bacterial colonization of an environment is bound to result in starvation, unless heavy predation or other "catastrophes" keeps the population far below the carrying capacity of the system (13,32,33). Soil is a chemically relatively stable environment, and protection against predation is offered by small pores (34). Most bacteria in soil appear to withstand catastrophes such as drought, although their metabolism and culturability may be transiently reduced (35). The relatively stable bacterial biomass at high levels compared to the total energy flux through the system probably reflects reasonable stability of the system. It seems likely, therefore, that most soil bacteria are starved most of the time. Hence, knowledge of starvation physiological mechanisms in cultured organisms is an important source of information to obtain an understanding of the physiological characteristics of the indigenous soil bacteria.

For more extensive reviews of the starvation response of bacteria, the reader is referred to a number of recent publications (13,25,33,36). Starvation almost invariably results in a reduction of the amounts of ribosomal RNA, followed by a degradation of proteins (25). The degradation of RNA is "rational" since the cellular demand for ribosomes is proportional to the rate of protein synthesis. Protein degradation may be a necessity to provide substrates to support mainte- nance metabolism as well as de novo synthesis of starvation proteins (see later discussion).

Most bacteria seem to respond actively to starvation by expression of starva- tion-specific genes (25,33). Such "starvation proteins" overlap partly with other stress-induced proteins; that may be one reason why starved cells are generally more stress-tolerant than actively growing cells (heat, hyperosmosis, H_2O_2, acid, and disinfectants [25]). De novo synthesis of stress proteins does not seem to be the only mechanism responsible for the stress tolerance of starved cells. Jouper- Jaan et al. (37) claimed that the initial starvation-induced heat tolerance in *Vibrio* sp. and *Escherichia coli* depends on de novo protein synthesis, whereas a further increase in heat resistance observed during long-term starvation appeared *not* to depend on such protein synthesis. Whatever the mechanisms are, an ecologically important rule emerges from these studies: growth is a risky affair compared to dormancy, since growing cells are more vulnerable than dormant ones to numer- ous stress factors. This characteristic may add to the ecological fitness of species which are able to produce frequent or "deep" dormancy stages. This is equiva- lent to stating that a risk-averse strategy for a species population would be to produce resting stages even under optimal conditions, and to secure that only a fraction of the resting stages would become active in response to onset of favor- able conditions.

There appears to be a striking diversity in morphological responses to starva- tion. A number of gram-negative bacteria will "fragment" into very small cells as a first stage after onset of starvation, whereas others appear to remain intact but with a gradual cell shrinkage during prolonged starvation (38). Gram-positive bacteria appear to react differently with respect to morphological mechanisms. *Arthrobacter* spp, which form rod-shaped cells during rapid growth and spheri- cal cells during slow (substrate-limited) growth, appear not to change their mor- phological characteristics if growth is completely restricted by starvation (39). Similar observations have been made for a number of other bacteria within the coryneform group (40). This finding was taken to indicate that gram-positive cells do not shrink during starvation, but miniature cells may be produced by asymmetric cell division (16). The observed miniature cells in a starving culture of *Micrococcus luteus* (21) may reflect such asymmetric division.

Starvation of bacteria rarely results in significant cell lysis, and most cells seem to remain physically intact within the time frames (days to months) applicable for such experiments. On the other hand, a loss of culturability occurs frequently, as revealed by a gradual reduction in the number of cells able to initiate cell division. There appears to exist a great diversity in the apparent "success" of starvation survival, when measured by viable counting of starving cultures. Morita (33) recognized four different patterns in viability response of laboratory cultures when exposed to starvation (1) a rapid decline in viability, (2) an initial increase in viability followed by a decline below original numbers, (3) an initial increase followed by relatively stable number of viable cells, (4) a constant viability. However, the culturability of the starved cells is not necessarily a good measure of the success of the organism under natural conditions. It may rather reflect the success of the microbiologists, with respect to their ability to find the right resuscitation and growth conditions for starved cells, as will be evident from the following discussion.

It has repeatedly been observed that starvation, together with many other sublethal stress factors, may turn a fraction of the population into a "nonculturable" state in the sense that they do not readily grow on otherwise suitable media. But the cells generally remain structurally intact and appear to retain a minimum of metabolic integrity and competence as measured by direct observation of metabolic activities in single cells (41). Such metabolically competent but nonculturable cells have been called "viable, but nonculturable" (42). Thus, starvation appears to take the cells beyond a "point of no return," where they lose the capability of returning to a growing stage. The cells in this stage appear to be moribund, but the criteria for culturable, viable, and moribund/dead cells are as yet conditional (operational). It is difficult to refute the argument "His life could have been saved if adequate resuscitation had been performed," unless there exists a consensus on some positive criteria for death.

One criterion defining cell death would be the loss of membrane integrity. Maintaining the membrane functions is considered to be a vital aspect of starvation survival of bacteria (38). This criterion for cell death (or viability) has been used to distinguish between dead and viable cells by flow cytometry; "dead" cells are those which are stained with DNA-specific dyes for which an intact membrane is an efficient barrier (43). This definition of cell death was challenged by Votyakova et al. (44) in an experiment with the gram-positive bacterium *Micrococcus luteus*. They found that cells which had become nonculturable as a result of prolonged starvation could be efficiently resuscitated by exposure to culture filtrates. They further demonstrated that an increasing fraction of a starving population would become permeable to the "death stain" PO-PRO-3 (45), a fluorescent DNA stain to which intact membranes are normally impermeable (hence the term). This implies that the membrane functions were degraded in starving cells. Finally, they were able to demonstrate that such "dead" cells were able to reestablish the membrane function upon adequate resuscitation and finally start growing again. Resuscitation by temperature upshifts (room temperature) has also been observed (46). The resuscitation was inhibited by chloramphenicol, indicating that de novo protein synthesis was involved.

The foregoing discussion illustrates that both culturability and membrane integrity are too stringent criteria for "metabolic competence" and "viability" of bacteria. On the other hand, the absence of an intact genome is clearly a too stringent criterion for cell death. The chloroform fumigation method to measure microbial biomass implies an alternative clearly operational criterion for living and dead cells. Biomass is defined as organic structures which are rendered decomposable by a killing agent (chloroform); in other words, a living organism is an organism that can be killed. The definition of killing is that it involves a destruction of a metabolic and structural integrity that protects the organism against degradation.

As mentioned earlier, the metabolic integrity of "viable, but nonculturable" cells has traditionally been assessed by fluorescence-based detection of metabolic activity in single cells, and nonculturability has been recognized as the lack of cell division. The latter criterion has been challenged by Binnerup et al. (47), who claim that a fraction of a population of starved cells may enter a stage where they are able to perform only two to three cell divisions, after which growth is arrested (see Chapter 2). The strain used was a kanamycin-resistant *Pseudomonas fluorescens* which was introduced into soil. The cells rapidly lost their culturability, defined as the ability to form visible colonies, but 20% of the initial inoculum appeared to be able to go through two to three cell divisions (revealed by fluorescence microscopic observations of microcolonies developing on a polycarbonate membrane which was placed on an agar surface). The authors defined this fraction as viable, but nonculturable forms of bacteria. At first glance, the observed growth restriction after two to three cell divisions is hard to understand. If the cells were able to divide at all, what mechanism is responsible for such delayed growth restriction? The authors offer no explanation, but one is tempted to speculate. Flow cytometric determinations of DNA in cultured organisms have revealed that a fraction of the cells in a population will carry more than one genome, even in a stationary phase culture (48). Such cells would be able to go through one or more cell divisions (depending on the number of genomes) without initiating replication of their genome. If starvation for some reason results in a blocking of genome replication, then we would get the type of phenomenon observed by Binnerup et al. (49).

In conclusion, dormancy induced by starvation seems to be common among nonsporulating bacteria. Bacteria show great diversity with respect to the mechanism responsible for the dormancy and their requirements for resuscitation. Since the frequency of culturable, viable but nonculturable, and dead cells within each species population seems to vary over a wide range, the question of the representativeness of the cultured organisms may not have a straightforward answer.

7. NOVEL METHODS

Molecular biology potentially represents a breakthrough in the study of indigenous populations of bacteria in soil. It is beyond the scope of this chapter to discuss all possible applications, and the discussion will be limited to those most relevant to the topic.

The analysis of 16S ribosomal RNA (rRNA) genes extracted from whole communities is an interesting strategy for circumventing problems with culturability. The 16S rRNA genes from whole communities can be analyzed after polymerase chain reaction (PCR) amplification with general primers. Sequencing can be performed after cloning, but a more promising method is first to separate the PCR products (according to sequence) by denaturing gradient gel electrophoresis (DGGE) (50). As further outlined in Chapter 12b, this allows analysis of the phylogenetic diversity of whole communities and further analysis of particular elements (bands) by reamplification and sequencing (or cloning). Hybridization with oligonucleotides specific for functional or phylogenetic groups is another interesting possibility offered. If applied successfully to soil bacteria, this opens a new era in the study of the uncultured ones.

As with all novel developments, there are numerous methodological problems (49). One which needs to be stressed in this context is the possible bias introduced by cell lysis (prior to DNA extraction). Large cells lyse readily when sonicated, but dwarfs have been found to be extremely sonication-resistant (30). Similar problems were encountered when bead beating (slurry of 0.1 mm zirconia/silica beads in 4% sodiumdodecyl sulfate (SDS), shaken in a bead mill homogenizer) was used to lyse cells from sediments (51). This is a bias that hits where it hurts most, since the dwarf cells (which do not release their DNA) represent the majority of the nonculturable cells in soil. The sonication energies that we found necessary to disrupt cells and extract genomic DNA (30) are probably inadequate for genetic analyses, since fragmentation of DNA to very short pieces would occur. In conclusion, there is a need for thorough methodological efforts to lyse dwarf cells adequately.

Further attempts to resuscitate and cultivate a larger fraction of the organisms (as done for marine bacteria by Schut et al. [14]) are highly relevant and should be pursued. Such studies should be guided by starvation experiments with cultured organisms. The breaking of dormancy and the resuscitation of "almost dead" cells provide a field for much trial and error (44), but are probably worth the effort. Enhanced culturability of dwarf cells has been obtained by dilution culturing (MPN) using low-nutrient media incubated for extended periods (Kari Aa, AUN Norway, personal communication). Comparisons of viabilities in bulk soil and in the rhizosphere have indicated a much higher viability of dwarf cells in the rhizosphere (16).

The use of fluorescence-labeled 16S rRNA oligonucleotide probes is a powerful tool that allows resolution of communities at different phylogenetic levels (52). The application of the method in combination with a high-sensitivity light detection system may allow the detection of cells with low numbers of ribosomes per cell. However, background fluorescence is a problem when looking at intact soils (Henrik Christensen, pers comm). Separation of cells from soil and size discrimination by filtration are an obvious strategy to increase the signal to noise ratio.

The metabolic activity of dwarf cells needs to be further elucidated. The data obtained so far indicate that their share of the total energy flux in soil is very small. Refinement of techniques for measuring specific metabolic activities in dwarf cells is clearly desirable.

REFERENCES

1. A. Fægri, V. L. Torsvik and J. Goksøyr, *Soil Biol. Biochem. 9*:105 (1977).
2. R. A. Olsen and L. R. Bakken, *Microbial Ecol. 13*:59 (1987).
3. R. Sørheim, V. L. Torsvik, and J. Goksøyr, *Microbial Ecol.* 17:181 (1989).
4. V. L. Torsvik, J. Goksøyr, and F. L. Daae, *Appl. Environ. Microbiol. 56*:782–787.
5. M. Alexander. *Introduction to Soil Microbiology*, Wiley & Sons, New York (1977).
6. J. D. van Elsas and L. S. van Overbeek. *Starvation in Bacteria* (S. Kjelleberg, ed), Plenum Press, New York, p. 9 (1993).
7. H. C. Bae and L. E. Casida, *J. Bacteriol., 113*:1462 (1973).
8. U. L. Zweifel and å. Hagström, *Appl. Environ. Microbiol., 61*:2180 (1995).
9. H. Christensen, R. A. Olsen, and L. R. Bakken, *Microbial. Ecol. 29*: (1995).
10. J. Bloem, M. Veninga, and J. Shepherd, *Appl. Environ. Microbiol. 61*:926 (1995).
11. L. R. Bakken, *Appl. Environ. Microbiol. 49*:1482 (1985).
12. E. Bååth, *Microbial Ecol. 27*:267 (1994).
13. J. C. Gottschal, *J. Appl. Bact. Symp. Suppl. 73*:39S (1992).
14. F. Schut, J. V. Egbert, J. C. Gottschal, B. R. Robertson, W. Harder, R. A. Prins, and D. K. Button, *Appl. Environ. Microbiol. 59*:2150 (1993).
15. F. Torrella and R. Y. Morita, *Appl. Environ. Microbiol. 41*:518 (1981).
16. L. R. Bakken and R. A. Olsen, "Perspectives of Microbial Ecology," Proceedings of 4th. Int. Symp. Microbial Ecol. Ljubljana, Yugoslavia, pp. 563–566 (1986).
17. J. Schnürer, M. Clarholm, and T. Rosswall, *Soil Biol. Biochem. 17*:611 (1985).
18. M. Clarholm and T. Rosswall, *Soil Biol. Biochem. 12*:49 (1980).
19. L. R. Bakken, *Sci. Rep. Agric. Univ. Norway, 65*:14 (1986).

20. T. A. Breland and L. R. Bakken, *Biol. Fert. Soils 12*:154 (1991).
21. A. S. Kaprelyants and D. B. Kell, *Appl. Environ. Microbiol. 59*:3187 (1993).
22. A. S. Kaprelyants, J. C. Gottschal, and D. B. Kell, *FEMS Microbiol. Rev. 104*:271 (1994).
23. N. S. Panikov, *Microbial Growth Kinetics*, Chapman & Hall, London (1994).
24. P. Hirsch, M. Berhard, S. S. Cohen, J. C. Ensign, H. W. Jannasch, A. L. Koch, K. C. Marshall, A. Matin, J. S. Pointdexter, S. C. Rittenberg, D. C. Smith and H. Veldkamp, *Strategies of Microbial Life in Extreme Environments* (M. Shilo, ed.), Verlag Chemie (1989).
25. A. Matin (1992), *J. Appl. Bact. Symp. Suppl. 73*:49S (1992).
26. L. R. Bakken and V. Lindahl, In: *Nucleic Acids in the Environment: Methods and Applications* (J. T. Trevors and J. D. vanElsas, eds), Springer Verlag, Berlin, p. 9 (1995).
27. V. Lindahl and L. R. Bakken, *FEMS Microbiol. Ecol., 16*:135 (1995).
28. V. Lindahl, Å. Frostegård, L. R. Bakken and E. Bååth, *Soil Biol. Biochem.*, in press (1997).
29. L. R. Bakken and R. A. Olsen, *Microbial Ecol., 13*:103 (1987).
30. L. R. Bakken and R. A. Olsen, *Soil Biol. Biochem. 21*:789 (1989).
31. P. J. Bottomley and M. H. Dughri, *Appl. Environ. Microbiol. 55*:959 (1989).
32. R. Y. Morita, *Can. J. Microbiol. 34*:436 (1988).
33. R. Y. Morita, *Experientia* (Birchhäuser Verlag, Basel, Switzerland) *46*: 813 (1990).
34. J. Postma, C. H. Hok-A-Hin and J. A. van Veen. Appl. Environ. *Microbiol. 56*:495 (1990).
35. A. Winding, S. J. Binnerup and J. Sørensen, *Appl. Environ. Microbiol. 60*:2869 (1994).
36. S. Kjelleberg, *Starvation in Bacteria* Plenum Press, New York, 227 p. (1993).
37. Å. Jouper-Jaan, A. E. Goodman and S. Kjelleberg, *FEMS Microbiol. Ecol. 101*:229 (1992).
38. S. Kjelleberg, M. Hermansson, P. Mården and G. W. Jones, *Annual Rev Microbiol, 41*:25–49 (1987).
39. J. C. Ensign, *J. Bact. 103*:569 (1970).
40. C. W. Boylen and M. H. Mulks, *J. Gen. Microbiol. 105*:323 (1978).
41. J. I. Prosser, *Microbiology 140*:5 (1994).
42. R. R. Colwell, B. R. Brayton, D. J. Grimes, D. B. Roszak, S. A. Huq and L. M. Palmer (1985), *Biotechnology, 3*:817 (1985).
43. H. M. Shapiro, *Practical Flow Cytometry*. Allan Liss Inc, New York. 295p (1985).
44. T. V. Votyakova, A. S. Kaprelyants and D. B. Kell, *Appl. Environ. Microbiol. 60*:3284 (1994).

45. R. P. Haugland *Molecular Probes Handbook of Fluorescent Probes and Research Chemicals*, 5th ed., Molecular Probes Inc, Eugene, Oregon.
46. L. Nilsson, J. D. Oliver and S. Kjelleberg, *J. Bacteriol. 173*:5054 (1991).
47. S. J. Binnerup, D. F. Jensen, H. T. Thordal, and J. Sørensen, *FEMS Microbiol. Ecol. 12*:97 (1993).
48. H. Christensen, L. R. Bakken, and R. A. Olsen, *FEMS Microbiol. Ecol. 102*:128 (1993).
49. J. T. Trevors and J. D. van Elsas (eds.) *Nucleic Acids in the Environment*, Springer Verlag, Berlin (1995).
50. G. Muyzer, E. C. de Waal, and A. G. Uitterlinden, *Appl. Environ. Microbiol. 59*:695 (1993).
51. M. I. Moré, J. B. Herrick, M. C. Silva, W. C. Ghiorse, and E. L. Madsen, *Appl. Environ. Microbiol. 60*:1572 (1994).
52. R. I. Amann, B. Zarda, D. A. Stahl and K. H. Schleifer, *Appl. Environ. Microbiol. 58*:3001 (1992).

4

The Fungi in Soil

GREG THORN   University of Wyoming, Laramie, Wyoming

1. INTRODUCTION

Soil is a complex ecosystem composed of multiple minute habitats and harbors almost all major taxonomic groups of fungi (Table 1). We walk unknowingly on a great quantity and morphological and physiological diversity of soil fungi. These vary from gliding, bacterium-engulfing slime molds to wood decay, mycorrhizal and root-pathogenic basidiomycetes with differentiated trophic mycelia, and foraging, rootlike rhizomorphs. Soil fungi range from the microscopic, in which the whole organism consists of a single cell with dry weight of less than 1 $\times$ 10^{-12} g, to the immense, in which filaments of a single clonal microorganism can occur over an area of 15 ha, and the individual has been estimated to have a weight of 1×10^7 g (1).

The numbers of species indicated in Table 1 are conservative estimates of the numbers of taxonomically distinct described species of fungi. Many species among the groups indicated as having some species occurring in soil are not truly soil fungi: they may be associated with above-ground parts of living plants, or plant litter, or may be lichenized (form stable associations with algae or

Table 1. Major Taxonomic Groups of Fungi

Hawksworth et al. (4)	Hawksworth et al. (5)	This chapter
Kingdom FUNGI	Kingdom PROTOZOA	
Division Myxomycota (625)[a]	Division Myxomycota (719)	"Motile fungi"
Class Myxomycetes (500)[b]	Class Myxomycetes (690)[b]	
Class Protosteliomycetes (21)[b]	Class Protosteliomycetes (29)[b]	
Class Ceratiomyxomycetes (3)[b]	[Including former Ceratiomyxomycetes]	
Class Acrasiomycetes (13)[b]	Division Acrasiomycota (12)[b]	
Class Dictyosteliomycetes (23)[b]	Division Dictyosteliomycota (46)[b]	
Class Plasmodiophoromycetes (45)[b]	Division Plasmodiophoromycota (46)[b]	
Class Labyrinthulomycetes (21)	Kingdom CHROMISTA	
Division Eumycota (64000)	Division Labyrinthulomycota (42)	
Subdivision Mastigomycotina (1170)	Division Hyphochytriomycota (24)[b]	
Class Hyphochytriomycetes (16)[b]	Division Oomycota (694)[b]	
Class Oomycetes (580)[b]	Kingdom FUNGI	
Class Chytridiomycetes (575)[b]	Division Chytridiomycota (793)[b]	
Subdivision Zygomycotina (800)	Division Zygomycota (1056)	"Zygomycetes"
Class Zygomycetes (665)[b]	Class Zygomycetes (867)[b]	
Class Trichomycetes (136)	Class Trichomycetes (189)	

Subdivision Ascomycotina (28650)[b] [including lichenized species]	Division Ascomycota (32267)[b] [including lichenized species] "Mitosporic fungi" [including former Deuteromycotina] (14104)[b]	"Ascomycetes" [including lichens and asexual forms]
Subdivision Deuteromycotina (17000) Class Coelomycetes (8000)[b] Class Hyphomycetes (9000)[b]		
Subdivision Basidiomycotina (16000) Class Hymenomycetes (8000)[b] Class Gasteromycetes (1060)[b] Class Urediniomycetes (6000) Class Ustilaginomycetes (980)	Division Basidiomycota (22244) Class Basidiomycetes (13857)[b] [including former Gasteromycetes] Class Teliomycetes (7134) Class Ustomycetes (1064)	"Basidiomycetes" [including asexual forms]

[a]Number of species.
[b]Present in soil.
Data from Refs. 4, 5.

cyanobacteria; see discussion below. Fewer than 15,000 species can be considered true soil fungi. This number, which may seem shocking in contrast to the approximately 450 species listed in the *Compendium of Soil Fungi* (2), is swollen in part by the ranks of the soil basidiomycetes, which are often overlooked in discussions of soil fungi. A more recent and widely cited estimate of the number of species of fungi is approximately 1.5 million (3). There are no realistic estimates based on this number for the number of fungi that may be encountered in soil. In any case, on the basis of their species diversity, fungi may outnumber all other groups of soil organisms.

Lichens are stable symbiotic associations between fungi and green algae or cyanobacteria, in which the fungal partner (mycobiont) forms a characteristic structure (thallus) that encloses and protects the alga or cyanobacterium (photobiont) (6). Lichens play vital roles in soil biogenesis and stabilization (7–8). Except for brief mention of these roles, they will be excluded from consideration in this chapter. Ascomycete lichens are especially diverse and account for about one-fifth of the 72,000 described species of fungi (5).

Fungi are also often dominant in soils in terms of their biomass. In some soils, fungal biomass may exceed that of all other soil microbes, plants, and animals combined (9–11). As such, their biomass may represent a significant portion of the nutrient pool, and their activities are key to providing or limiting access to nutrients for plant growth. In contrast, in a tropical rain forest soil, fungal biomass was estimated to be one-third that of the bacteria (12), and in certain agricultural soils in the Netherlands, only 1%–2% that of the bacteria (13). Fungi dominate in temperate or polar soils that are well aerated and oligotrophic or carbon-rich and nitrogen-poor (10).

Fungi play dominant roles in soil processes of nutrient cycling and interactions with all other soil-inhabiting organisms, including plants. In most ecosystems, fungi are key to the recycling of nutrients from above-ground and below-ground plant litter (14–15). Certain fungi form mutualistic associations called mycorrhizae with the roots of vascular and nonvascular plants, supplying them with nutrients and protection from drought and root pathogens (16–17). Root pathogens such as species of *Armillaria, Phellinus, Fusarium,* and *Rhizoctonia* may act as agents of natural disturbance in native plant communities (e.g., 18) and cause major losses in agricultural crops such as wheat, cotton, canola, and oil and date palms (see Chapter 17). All soil fungi are involved in the complex food webs in soil through interactions with soil fauna and microbes: as food for various nematodes, mites, colembolans, and tardigrades (see Chapter 6); as parasites and predators of these and other soil fauna and microbes; and as recyclers of waste products and other chemicals secreted and excreted by plant roots, animals, and microbes (19–25). Various soil fungi are the source of a number of important pharmaceutical compounds, such as penicillin, the cyclosporins, and lovastatin, and others are being investigated as potential agents of biological control in agricultural systems because of their roles as parasites, predators, and antagonists of plant-pathogenic nematodes or fungi (26–32). A mycocentric, rather than phytocentric, viewpoint indicates that the fungi not only determine the functioning of terrestrial ecosystems now, but have done so since earliest evolution, making possible the colonization of land by plants in the late Silurian, over 400 million years ago (33–35).

This chapter can only provide a brief introduction to the quantity, diversity, and importance of fungi in soil and will necessarily have the biased emphasis of its writer's interests and experience. Even for the topics covered, the references listed represent only a selection but should serve to direct readers into the literature. Several reviews provide details of methods for analysis of fungal communities in soil (36–45); discussions of fungal occurrence, activities, or community structure in soil and litter (10,46–59); and guides for identification of fungi isolated from soil (2,60–64). Several additional sources of literature for the identification of soil fungi are available (5,65–66). Soil fungi as plant pathogens will be discussed in Chapter 17, and soil fungi as food for soil invertebrates are discussed in Chapter 6. In this chapter, certain names, such as ascomycetes, basidiomycetes, and zygomycetes (uncapitalized), are used informally, without reference to taxonomic rank, and the term "fungi" (in double quotation marks) is used to include both fungi and funguslike organisms usually studied by mycologists. Likewise, the traditional terms saprophyte and saprophytic (rather than sapromycete or saprobic) are used to refer to fungi that derive their nutrition from dead plant or animal remains.

2. TAXONOMIC DIVERSITY

2.1 Motile Fungi

The motile fungi of soil include organisms related to the protozoa, the chromistan algae, and true fungi. Among the protozoan fungi are the cellular and plasmodial slime molds (dictyostelids, myxomycetes, and relatives), with amoebalike stages that engulf microbes and fungal spores, and also plant pathogens such as *Plasmodiophora* and *Spongospora* (67–71). The recent discovery that the feeding plasmodia of the slime molds *Physarum* and *Stemonitis* produce extracellular amylase that breaks down starches to simple sugars (72) indicates that myxomycetes may have a combination of absorptive nutrition (characteristic of the fungi) and phagotrophic nutrition (characteristic of protozoa and animals). Myxomycetes and dictyostelids may represent a significant proportion of the amoebae in agricultural soil (73). The water molds (oomycetes) and hyphochytrids (both related to chromistan algae) and the chytrids (true fungi) have stages that swim through soil water films and aquatic habitats elsewhere propelled by one or more flagellae. Among the water molds, some members of *Pythium* and *Phytophthora* are notable plant pathogens; others are myco-parasites and thus potential biological control agents of plant pathogenic fungi (74–76). The chytrids and hyphochytrids are microscopic parasites of soil (and freshwater or marine) algae and animals, or saprobes of organic debris (77–79). Most require special techniques for recovery from soil and study, and mycologists specializing in these groups are few.

2.2 Zygomycota

The Zygomycota include three highly divergent groups with remarkably different morphology, life-styles, and nutrition: (1) the symbiotic, arbuscular mycorrhizal (endomycorrhizal) Glomales (Fig. 1 A), with spores up to 1 mm in diameter and containing thousands of genetically different nuclei (80,81); (2) the Trichomycetes, mostly microscopic gut symbionts of insects (and possibly nematodes [82]); and (3) the Zygomycetes, which include the bread mold (*Rhizopus*) and other filamentous saprobes and mycoparasites (62,83–85). Of the groups in soil, the endomycorrhizal fungi and mycoparasitic members of the Zygomycetes cannot be grown in culture without their hosts, whereas many of the saprophytic Zygomycetes grow so readily and so quickly in laboratory culture that they can overwhelm less readily isolated soil fungi.

2.3 Ascomycetes, Yeasts, and Molds

Fungal nomenclature is unusual (and confusing to nonmycologists) in allowing duplicate names and classification systems for sexually reproducing and asexually reproducing forms.[1] Mycological tradition has recognized the Ascomycotina and Basidiomycotina for fungi reproducing sexually by means of an ascus (pl. asci) or basidium (pl. basidia), containing or bearing asco- or basidiospores, respectively, and the Deuteromycotina, for these same fungi reproducing asexually (e.g., [87]; see Table 1). Thus, there are often two correct names for a species of fungus: an anamorph name based on the morphological features of its asexual reproductive state (e.g., *Aspergillus glaucus*) and a teleomorph name based on the morphological features of its sexual reproductive state (e.g., *Eurotium herbariorum*, the same species). Anamorph names are also referred to as form names (e.g., form species, form genera), recognizing that morphological characters, inadequate for a natural classification, are often all that has been available. The name of the teleomorph, if known, is considered the correct name for the total organism, or holomorph. For example, when only the anamorph is evident, one can speak of the *Aspergillus* state of *Eurotium herbariorum*. However, many deuteromycetes have no known sexual

---

[1]Thurston (86) addresses this problem in a footnote that begins, "The fungi . . . are at the mercy of taxonomists whose capacity to wreak havoc in the literature is untold."

Figure 1. A. A group of spores and hyphae of *Glomus aggregatum*, a zygomycete that forms endomycorrhizae with a variety of vascular plants on sandy soils (scale bar approximately 100 μm). B. A broomlike cluster of asexual spores produced by *Penicillium*, one of the most familiar ascomycetous molds of soil (scale bar 10 μm). C. The fruiting body (mushroom) of *Coprinus comatus*, a basidiomycete that decays plant litter in soil and also attacks and consumes living bacteria (scale bar 2.5 cm). D. A 100 mm Petri dish showing a diversity of ascomycetous molds such as *Aspergillus, Penicillium*, and *Cladosporium* that are commonly isolated from soil. Photo A by D. M. Watson (Cornell University, Ithaca, New York), with permission; photo B by A. Tsuneda (Tottori Mycological Institute, Japan), from Ref. 58, with permission; photos C and D by R. G. Thorn, unpublished.

stage. In *Aspergillus* and *Penicillium*, for example, only 31 of 212 and 53 of 145–280 species, respectively, have known sexual stages (88–90). The majority of species in the Deuteromycotina are asexual forms of, or related to, the Ascomycota, and a few others are related to members of the Basidiomycota. Advances in molecular fungal phylogenies are gradually doing away with both the Deuteromycotina as a taxonomic category (widely accepted, e.g., [5]) and dual nomenclature of genera and species (less widely accepted, see [91–93]). However, sequence data required for molecular phylogenies are available for only 0.02%–0.5% of fungal species (based on the estimated and described numbers of species). Soil mycologists usually deal with cultures showing only asexual reproduction and so are most familiar with anamorph names such as *Aspergillus* and *Penicillium* (Fig. 1B). As long as there are hundreds of species that must be called *Aspergillus, Penicillium, Trichoderma, Fusarium*, or *Cephalosporium*, these anamorph names will remain, but it is not difficult to place most of these important genera close to their sexual relatives in a more natural classification (e.g., [94]).

The Ascomycota of soil (including their asexual forms) span the range from saprobes of plant or animal debris to parasites of plants, animals, fungi and microbes; many form lichens, and a few form ectomycorrhizae. Macroscopic forms include the cup fungi and highly prized edible fungi (truffles, morels) that are familiar to naturalist (95,96). Microscopic forms include pests of stored foods such as *Penicillium* and *Aspergillus*, sources of important pharmaceuticals such as penicillin (from *Penicillium chrysogenum* and other species of *Penicillium*), cyclosporin (from *Tolypocladium inflatum*), lovastatin (from *Aspergillus terreus*), and griseofulvin (from *Penicillium griseofulvum*), and human and animal pathogens such as *Histoplasma* and *Coccidioides*. Isolation and culture of all but specialized parasitic forms are simple. Identification is becoming increasingly feasible but can still be nightmarish and involves a vast and dispersed taxonomic literature, especially for fungi of tropical soils (for references, see the preceding list of guides for the identification of soil fungi.)

Yeasts are simply ascomycetes, basidiomycetes, or zygomycetes adapted by their unicellular reproduction to life in a liquid environment. A number of yeasts, including the well-known brewer's yeast *Saccharomyces cerevisiae*, are dimorphic: that is, produce yeast and mycelial phases under different conditions (97). Yeasts are particularly numerous in peat bogs (98) and in vertebrate dung (99). Black yeasts such as *Exophiala* are widespread in soils, and some species are the cause of human infections that are particularly prevalent in tropical areas (100). As in bacteriology, our knowledge of yeasts from soil or other natural environments is much less than that of medically important yeasts or yeasts involved in food production. Major references for the identification of yeasts include Barnett et al. (101), Kreger-van Rij (102), and von Arx et al. (103).

2.4 Basidiomycetes and Their Anamorphs

Lignin is the second most abundant naturally occurring polymer on earth and is the complex glue that holds the cellulose fibers of woody plant tissues together (14). The basidiomycetes include the primary agents of lignin degradation, the majority of ectomycorrhizal fungi, and some significant plant pathogens. Fruiting bodies of various soil-inhabiting basidiomycetes form umbrella-shaped mushrooms (Fig. 1 C) and puffballs, familiar from lawns, fields, and forests. Disorders of turf known as dry patch and fairy rings are caused by the growth of saprophytic basidiomycetes through soil below the turf, depleting soil nutrients and making the soil water-repellent (104). Although several thousand soil-inhabiting basidiomycetes are known (105–107), and their biomass can be considerable, e.g., as much as 60% of the living microbial biomass in all horizons (108), few are recorded in surveys of soil fungi based on isolation into culture (see Fig. 1 D). Chesters (109) has referred to basidiomycetes as the "missing link in soil mycology," and this comment is largely still true today. Some of the ectomycorrhizal fungi, such as species of *Inocybe* and *Russula*, are difficult or slow to grow in culture, and few basidiomycetes produce reproductive propagules in soil on the order of the asexual spores of ascomycetous or zygomycetous molds. This combination and the media generally used for isolation of soil fungi appear responsible for their tremendous underrepresentation in such surveys (110). Clamp connections (looping connections across septa, see Fig. 2 A–B) are considered diagnostic for basidiomycetes but are possessed by only a minority of species, and by these only when they are dikaryotic. A significant number of basidiomycetes isolated in previous studies may have been identified as sterile white mycelia or may not have been reported at all, because their lack of clamp connections and sporulation rendered them unidentifiable. Recently, identification of such isolates as ascomycetes or basidiomycetes has become possible using patterns of sensitivity or tolerance to sodium chloride and the fungicides benomyl and cycloheximide (actidione) (111), staining of mycelia with diazonium blue B following pretreatment with 1 *M* KOH (112,113), or visualization of the septal structures characteristic of ascomycetes or basidiomycetes using ammoniacal Congo red or trypan blue and light microscopy (114,115) or transmission electron microscopy (116–117; see Fig. 2).

Nonetheless, there have been some excellent recent studies of community ecology and autecology of decomposer and ectomycorrhizal basidiomycetes, particularly in Britain and Sweden (e.g., [1,118–124]). Some exciting developments include recent evidence that ectomycorrhizal basidiomycetes such as *Paxillus involutus* can tap organic and mineral sources of C, N, and other nutrients not accessible to their plant associates (125), short-circuiting the usual mineralization pathways involving decomposer fungi and bacteria (17,126). Demonstration that certain ligninolytic basidiomycetes (e.g., *Agaricus brunnescens* and

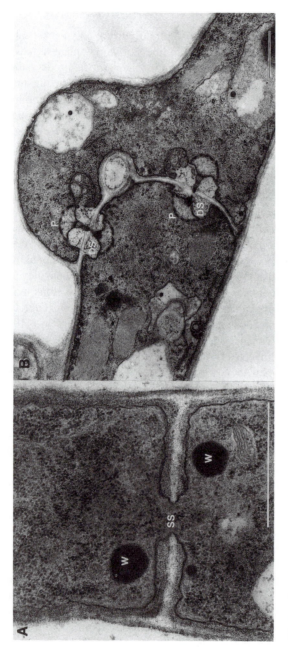

Figure 2. A. A simple septal pore (SS) and Woronin bodies (W) typical of ascomycete hyphae. B. A section through a clamp connection of the basidiomycete *Lentinula edodes* showing dolipore septa (DS) and parenthosomes (P) found in hyphae of many basidiomycetes. Scale bars 1 μm. Transmission electron microscopy and photos A and B by Y. Fukumasa-Nakai, from Ref. 58, with permission of A. Tsuneda (Tottori Mycological Institute, Japan).

Lepista nuda) can attack and degrade living bacteria as a source of nutrition (127) has led to investigations into microbial interactions (discussed later); however, many aspects of the role of basidiomycetes in nutrient cycling and the physiological ecology of soil are as yet unexplored.

3. COMMUNITIES OF SOIL FUNGI

A realization of the occurrence of communities or assemblages of soil fungi developed during the 1950s and 1960s (46–47,128). Communities vary with respect to major gradients in latitude and elevation, vegetation, and abiotic factors of the soil, including season (46,128–135). Soil fungal community analysis has been rare, largely because of the difficulties and scale involved in isolation and identification of the soil mycota (46,98,130–141). Nevertheless, surveys indicate that certain suites of fungi are characteristic of particular vegetation types or geographic areas: species of *Mortierella* and *Penicillium* are more common in temperate to high latitudes, especially in forest soils, whereas species of *Aspergillus* are more common in grasslands and deserts and at low latitudes; *Fusarium, Papulaspora,* and *Periconia* are characteristic of grassland soils, whereas *Mortierella, Mucor, Penicillium,* and *Oidiodendron* characterize forest soils (46). Despite generalizations such as these, Christensen and others have pointed out that greater resolution and more meaningful correlations result when lists of fungi identified to species are examined and when sampling and isolation procedures allow quantitative comparisons (38,46–47). However, these are communities of readily culturable fungi (142), and we have a long way to go yet to learn the true communities of soil fungi. The minute scale at which diversity occurs in the soil community is too seldom recognized; as Brock has reminded us, "Microbes are small, and their environments are also small," and "a single soil crumb . . . may contain numerous microenvironments differing in content, nutrient concentration, pH, or other factors" (142). The food base of a fungal individual may be as small as a single pollen grain, rotifer, or shred of decaying litter of similar size, or as large as the roots of trees covering several ha. The former case is likely common, while the latter is probably exceptional among soil fungi. Such investigations require detailed analyses of fungi colonizing minute soil particles (143,144), along with improved techniques for isolation and quantification of specialist fungi such as the parasites and predators of rotifers (145), and molecular analyses (see Chapters 13 and 14) or immunological techniques combined with direct observations (146,147).

Pugh (148) devised an ecological classification of fungal strategies, based on Grime (149), in which fungi ranged in their responses to stress (defined as limited resources) and disturbance (defined as events that reduce biomass). Fungi living under conditions of low stress (high food) and low disturbance were termed competitors, competing for occupation of the resource and space; the example given was fungi in logs and stumps. Pugh (148) has stated that the majority of fungi isolated from soil are ruderal organisms, suited by high rates of sporulation and growth to colonization of sites of low stress and high disturbance. He further suggested that large inputs of nutrients, such as the seasonal deposition of deciduous litter, could be considered to represent acts of disturbance. Fungi with simple nutritional requirements, efficiently able to use low concentrations of nutrients or recalcitrant substrates, were thought to be stress-tolerant, able to inhabit environments with high stress, low disturbance. Finally, survivors and escapes were fungi able to persist in the face of both high stress and high disturbance. The zymogenous (150) or sugar fungi (151) are roughly comparable to Pugh's ruderal fungi, while the stress-tolerant fungi are autochthonous. Pugh's choice of names for the categories may have been unfortunate, particularly the designation competitors for fungi of low stress and low disturbance, implying that fungi of other situations do not have to compete to acquire and defend their lesser resources. Again, a sense of scale is essential in considering the soil environment: good things come in small packages, and a tiny volume of substrate may represent an island of high nutrients (low stress) waiting for a ruderal colonist. In contrast, a recalcitrant food resource (such as cellulose) or one which a species is uniquely able to utilize (such as a living rotifer) would not represent a stress to the appropriate fungi—stress to such species would be a lack of cellulose or rotifers. Pugh's classification (148) is nonetheless useful as a point of departure. The extent of competition (and collaboration) in utilization of what are often called rich resources (carbon-rich, nitrogen-poor) such as woody litter will be discussed further in section 7, Interactions of Fungi with Other Soil Organisms.

4. FUNGAL NUMBERS AND BIOMASS

The dominance of most soils by fungi is impressive, whether measured by numbers of species, reproductive propagules (often 10^4–10^6 colony-forming units per gram dry weight of soil), hyphal lengths (commonly 100–1000 meters of hyphae per gram dry weight of soil), or biomass (37–184 g dry weight of mycelium per square meter) (10). However, the nature of soil as a medium for investigation makes realistic estimates of these values difficult. Not only is soil a complex of biotic and abiotic components and a myriad of predominantly microscopic organisms, but it is a dense, opaque medium as well, making direct microscopic observations difficult. For this reason, numbers, biomass, and activities of soil fungi have usually been calculated by a variety of indirect means. Brock (142) asked why microbiologists have been so slow to make use of direct methods, rather than relying on plate counts and indirect methods; the new methods are making direct examination easier.

Recent use of improved fluorescent stains (152) and confocal laser microscopy and digital image analysis (153) shows promise in overcoming some of the difficulties in direct microscopic visualization of hyphae in soil smears. A number of fluorescent stains used previously, including calcofluor white and fluorescein diacetate (FDA), either fail to stain a large (unknown) fraction of the hyphae in soil or detect dead hyphae better than living ones (152). Tsuji et al. (152) introduced a new fluorochrome, 5- (and 6-)sulfofluorescein diacetate, which is three times brighter than FDA, stains more organisms (but, as with FDA, not those with melanized walls), and better distinguishes living and dead hyphae. Empty and apparently dead hyphae are abundant in most soil preparations and may contribute significantly to previous estimates of hyphal lengths, biovolumes, and fungal biomass (154). Physical separation of hyphae from the soil matrix by elutriation (155) or aqueous two- (or more) phase separation (156) may simplify the microscopic observation in direct counts of fungi from soil but introduces an unknown factor in subsequent quantitation. Staining of hyphae separated from soil with diazonium blue B offers the ability to separate hyphae of ascomycetes (most unstained) from zygomycetes (stained, broad, mostly aseptate) or basidiomycetes (stained, narrow, septate, with or without clamp connections) (113). Fluorescently labeled antibodies specific to the fungus being studied have been used to assess the contribution to biomass and litter decay of mycelium of the basidiomycete *Mycena galopus* (146).

Among the indirect means of assessing fungal biomass and activity in soils, signature biomarkers, including both nucleic acids and certain fatty acids and phospholipid fatty acids, offer various levels of taxonomic discrimination within the fungi (157–161). In addition, certain enzymes may be produced predominantly or entirely in a particular soil or other substrate by a fungus under study, e.g., the lignin-modifying enzyme laccase by *Agaricus* (162,163). For a general view of fungal contributions, extraction of broadly fungus-specific chemical constituents (ergosterol or glucosamine), selective inhibition of fungal or bacterial components of the microbial biomass, or direct measures of fungal hyphae in soil can be combined with a chemical estimation of microbial biomass that includes bacterial, fungal, and micro- and mesofaunal components. The quantity of ergosterol ($C_{28}H_{44}O$, the most common fungal sterol and a precursor of vitamin D_2) is correlated with fungal growth in plant material (163–165), and, since ergosterol is rapidly mineralized upon death of the mycelium (166), has been related to the quantity of living mycelium in soil (166–169). Some doubts have been expressed about the suitability of ergosterol analysis for detecting the amount of fungal mycelia in soil, including reports of considerable amounts of ergosterol in soil-dwelling enchytraeid worms (which eat fungi) and indications that ergosterol may underestimate fungal biomass in highly organic soils (169).

A measure of the glucosamine ($C_6H_{13}NO_5$, an amino sugar) content of wood or fresh plant litter, derived by hydrolysis of the chitinous cell walls of fungi, may give an accurate estimate of fungal colonization (170,171). In materials colonized by fungi for more than a few days, however, the quantity of glucosamine, which is derived from walls of both living and dead hyphae, provides an overestimate of the quantity of viable hyphae. In more complex substrates such as soil or well-decayed litter, glucosamines may also be derived from bacterial cell envelopes and from arthropod exoskeletons. The latter may contribute a negligible proportion to the total, and the former can be corrected for by measuring the muramic acid ($C_9H_{17}NO_7$, an amino sugar) content and assuming a 1:1 molar ratio of glucosamine/muramic acid in bacterial cell envelopes (45).

Selective inhibition of bacteria or fungi with antibacterial (streptomycin sulfate) or antifungal (cycloheximide) antibiotics has been used to determine each group's contribution to the total microbial biomass (172–175). Using this technique, estimates of the fungal proportion of the microbial biomass (measured by response to glucose added to the antibiotic-amended soil) range from 11% to 90% (10). Other measures of microbial biomass in addition to the substrate-induced respiration measure include measures of phospholipid phosphate (176), chloroform fumigation–incubation or fumigation–extraction (177–179), adenosine triphosphate (ATP) (180,181), and ninhydrin-reactive N (182,183). Unfortunately, the relationships among several of these measures apparently vary, depending on soil type (184). Interested readers are referred to reviews by Parkinson and Coleman (43), Paul and Clark (44), Tunlid and White (45), and Newell (185).

5.   THE ROLES OF SOIL FUNGI

5.1   Litter Decomposition and Nutrient Cycling

The best known role of fungi in soil is nutrient cycling through the decomposition of plant litter. There are a number of good reviews of litter decomposition in terrestrial systems (14,15,52,56,186). What is perhaps not often appreciated is how long this process can be: humic remains (complex phenolic polymers) of plant litter in grasslands have been dated by using ^{14}C at over 2000 years old (187). The biochemical mechanism of plant litter degradation, whether trunks, needles, or leaves, is similar in that the cellulose microfibrils comprising 65%–80% of the dry weight of the litter are bound in an amorphous matrix of lignin (a complex polymer of aromatic alcohols) and hemicelluloses (complex polymers of sugars, particularly xylose). Physical and chemical access to cellulose is restricted by the lignin and hemicellulose matrix. The ability to digest cellulose to its glucose monomers is found in a wider variety of fungi and bacteria than is the ability to degrade native lignin (188); the latter is found primarily in the Basidiomycota and certain Ascomycota (Xylariales) and bacteria (Actino-mycetes). Several reviews of lignin degradation by fungi, of which certain mech-anistic details are still uncertain, are available (189–195). Key enzymes in the process of lignin degradation are lignin peroxidases (LiP, Enzyme Commission[2] number 1.11.1.14 [197]), manganese peroxidases (MnP, E.C. 1.11.1.13), and lac-cases (E.C. 1.10.3.2 [86]). The roles and differential production of these en-zymes by various wood-degrading basidiomycetes are reviewed by Hatakka (198), and good discussions of specific examples are provided by Galliano et al. (199) and Vares et al. (200). Particularly important are the findings of Vares et al. (200) and others that the types of lignin-modifying enzymes produced on natural solid substrates often differ from those known from synthetic liquid media. The best-known wood-degrading basidiomycete, *Phanerochaete chrysosporium*, usually considered to produce LiPs and MnPs but not laccases, was recently found to produce laccase when grown in a medium containing cellulose and am-monium tartrate (201). It is possible that the known enzyme profiles of other ligninolytic fungi similarly reflect the media on which we have grown them for study more than their true enzymatic potential. In the fungi, LiPs and MnPs are known primarily from basidiomycetes, whereas laccases and cellulases are widespread among ascomycetes.

The fungi causing decay of woody litter have been subject to numerous taxo-nomic and physiological studies (14,198,202,203), but remarkably little is known about the organisms responsible for decay of plant litter in soil. Soil-in-habiting, lignin-degrading basidiomycetes which have received most study are the commercial button mushroom *Agaricus brunnescens* (= *A. bisporus*) and the tree pathogens *Heterobasidion annosum* (= *Fomes annosus*), *Rigidoporus ligno-sus*, and the collective species *Armillaria mellea* (199,204–209). However, many saprophytic basidiomycetes inhabiting soil and fragmentary plant litter have sig-nificant ligninolytic abilities (210–212). A part of our failure to appreciate the importance of these fungi may be our general inability to isolate cultures of them from soil. In surveys for lignin-degrading fungi of soil, Arora and Sandhu (213) isolated two basidiomycetes (from 150 soil samples) and Falcón et al. (214) ob-tained only one. With more selective media and isolation techniques, it is possi-ble to increase the rate of isolation of saprophytic soil basidiomycetes. Thorn et al. (110) obtained 67 isolates of basidiomycetes from 64 soil samples using a combination of the particle filtration technique and a selective indicator medium.

---

[2]The Enzyme Commission number forms a classification system for enzymes, with the first number indicating the class of enzyme, the second a subclass, and so on. Lignin peroxidase (E.C. 1.11.1.14) and manganese peroxidase (E.C. 1.11.1.13) are both oxidoreductases acting on a peroxide as an electron acceptor, and laccase (E.C. 1.10.3.2) is an oxidoreductase acting on diphenols as electron donors and with oxygen as electron acceptor (196).

In addition to their likely importance in litter decomposition, such lignin-degrading soil fungi may provide valuable candidates for in situ bioremediation of soil contaminated with aromatic pollutants such as the polychlorinated biphenyls (PCBs) or pesticides such as dichlorodiphenyltrichloroethane (DDT) or 2,4-dichlorophenoxyacetic acid (2,4-D).

Litter decay fungi are involved in not only the cycling of nutrients but also their transport through the soil system. The mycelia of *Phanerochaete velutina* and *Hypholoma fasciculare* can behave as coordinated units in soil. Initial symmetrical radial growth can become directed toward food baits (pieces of wood), and the side opposite the bait gradually degenerates (120). Nutrients are frequently translocated from source to sink in basidiomycete mycelia, particularly through hyphal strands and rhizomorphs as in *Serpula lacrymans, Phanerochaete velutina*, and *Phallus impudicus* (215,216). Nitrogen may be transported via mycelium from soil to nitrogen-poor wood (217,218). Rhizomorphs and hyphal strands or cords are also important in the colonization of dispersed resources such as tree roots or dead wood in soil (14,151,219). In addition, rhizomorphs of fungi such as species of *Armillaria*, which have rinds of thick-walled, melanized hyphae, no doubt function in the persistence of the fungus through adverse conditions (dispersal through time as well as through space, [151]).

Nutrient regulation of decay rates has been the focus of both laboratory and field studies. Excess nitrogen limits LiP production in *Phanerochaete chrysosporium* grown in liquid media in vitro with glucose as a carbon source, but this suppression is not seen when grown in wood or in liquid media with cellulose as a carbon source (197,220). In field studies of litter degradation, C:N ratios and lignin:N ratios have been used to measure litter quality and to model decay rates (221–224). With litters of low lignin content or a broad range of lignin contents, lignin/N ratio may be a better predictor of decay rates than C:N (224); however, cellulose/lignin/N ratio may yield a still better prediction of decay rate (225). Litter quality alone may not be the key to decay rates: Hart et al. (226) found that although litter in an old-growth *Pinus ponderosa* stand had a lower decay rate than that in a young stand, reciprocal transplants of litter revealed no significant differences in decay rates of the two litters.

Some general patterns of nutrient changes during litter decay have been observed. As cellulose, lignin, and hemicelluloses are depolymerized, metabolized, and eventually respired as CO_2, the concentrations of most other elements such as N, P, Ca, Mg, and Mn increase with respect to C and as a percentage of dry weight. Some nutrients also increase in absolute quantity. Progressively decayed litter, including its microbial inhabitants, often shows a net accumulation of N and Ca from its surroundings (227–231). In contrast, K and Na often decrease in concentration early in decay or with mass lost (228,229). The accumulation of Ca as calcium oxalate on the surface of hyphae and rhizomorphs of litter decay fungi is both widespread and dramatic (232–234). Undecayed plant litter, especially wood, has much higher C:N ratios than the mycelium of litter decay fungi that grow in it (217,221,231,235,236).

Several mechanisms have been suggested by which wood decay fungi can acquire and conserve the nitrogen necessary for growth, production of ligninolytic enzymes, and production of often massive sporophores. Although it could be argued these mechanisms are not applicable to the soil system, where the C:N ratios are much lower, the competition for soluble nitrogen or other nutrients, their complexing with various phenolics, and the high C:N ratios in minute fragments of plant litter may make the same mechanisms of value to soil fungi. Thirty years ago, Cowling and Merrill (221) suggested three mechanisms by which fungi could conserve nitrogen during wood decay: (1) physiological adaptation of nitrogen metabolism, including reduction of nitrogenous structural compounds; (2) autolysis and reuse of nitrogen in the fungal mycelium; and (3) utilization of nitrogen sources outside the wood itself, including soil and the atmosphere. Their own studies substantiated the first two of these mechanisms—mycelia of *Trametes versicolor* and *Ganoderma applanatum* had reduced nitrogen content when grown on wood or synthetic media with higher C:N ratios (235), and various decay fungi were capable of growth using fractions of their own mycelia as the sole source of nitrogen (237). Subsequently, it was shown that *Schizophyllum commune* grown under conditions of nitrogen deprivation reused amino acids derived from proteolysis in older parts of the mycelium (238). As mentioned, translocation of nitrogen to wood from soil or synthetic media has been demonstrated in a number of fungi (217,218,239).

Mycoparasitism, or the use of nutrients in the mycelium of other wood-inhabiting fungi, was suggested as another means by which individual species could overcome the nitrogen limitations in wood (237,240). The discovery that wood-decay mushrooms in the genus *Pleurotus* attack and consume living nematodes (241) suggested a parallel with carnivorous plants. Carnivorous mushrooms might supplement their carbon-rich woody diet with nutrients from animals that migrated with their nutrients into wood from its surroundings. Subsequently, it was found that species of *Pleurotus* (242) and many other wood-decay fungi are capable of attacking and consuming microcolonies of living bacteria on agar (127,243,244; see Fig. 3) and in wood in vitro (245). *Pleurotus* attacks gram-positive and gram-negative bacteria, as well as nitrogen-fixing species of *Azotobacter* (127,245). It was recently shown that two species of *Pleurotus* and 78 of 149 other species of basidiomycetes tested attack and consume microcolonies and solitary cells of various yeasts, including species of *Candida, Cryptococcus, Pichia, Rhodotorula*, and *Sporidiobolus* (246). This ability was found among white-rot and brown-rot fungi, litter decomposers, and one ectomycorrhizal species (*Thelephora terrestris*) and 12 of 124 species of ascomycetes that were tested. Alternative avenues for nitrogen

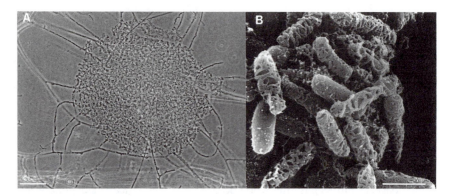

Figure 3. A. Attack of a colony of *Burkholderia cepacia* by the ligninolytic, nematophagous and bacterivorous mushroom *Hohenbuehelia pinacearum*, showing the directional hyphae that have grown toward the colony (scale bar 25 μm). B. Details of *Pseudomonas putida* attacked by *Pleurotus ostreatus*, showing perforations in the cell envelopes (scale bar 1 μm). Photos A and B by R. G. Thorn and A. Tsuneda (Tottori Mycological Institute, Japan), unpublished.

nutrition appear to be so widespread among wood- and litter-decay fungi that the interesting questions now center on those species that are not known to attack bacteria, nematodes, or other fungi, including yeasts. Where and how do they obtain their nitrogen?

Minor mechanisms by which the total nitrogen in decaying woody litter may be increased include the importation of nitrogen through fungal mycelia, in immigrating micro- or mesofauna, inputs of mineral nitrogen in soil leachates, and atmospheric deposition, but the major mechanism must surely be in situ nitrogen fixation. Diverse nitrogen-fixing bacteria have been found in decaying wood (247,248), and nitrogen-fixing activity demonstrated in woody litter (249,250). Associative, aerobic cellulolysis and nitrogen fixation occur in decaying plant litter when cellulolytic fungi provide species of *Azospirillum, Azotobacter*, or *Clostridium* protection from O_2 and the carbon required for nitrogen fixation (251–254). Attack and consumption of nitrogen-fixing bacteria, as seen in *Pleurotus* grown on agar in vitro (245), may not be the most profitable strategy, nor the actual outcome in vivo. Associations of various fungi with nitrogen-fixing cyanobacteria are numerous and widespread and range from loose symbioses in which fungus and cyanobacterium are morphologically unchanged, to highly specialized mutualisms that result in the production of a novel structure recognizable as a lichen thallus (255). The biological and biochemical nature of associations between lignocellulose-degrading fungi and cyanobacteria or green algae is an interesting, but largely unexplored topic. Nitrogen fixation by cyanobacteria, both lichenized and free-living, represents a major input of N to arctic and alpine tundra systems (7).

What emerges from the preceding discussion is a picture of the decay community as a consortium of members of the bacterial, fungal, protistan, and animal kingdoms. Consumption of fungi and microbes by soil invertebrates is discussed in Chapter 6; consumption of microbes and invertebrates by soil fungi will be discussed later in this chapter. Interfungal interactions will be discussed in Chapter 17, as the basis of biological control of soil-borne plant diseases, and briefly later in this chapter. The consortium concept of the decay community is not new, but its intricacies in nature have meant that controls, measures, and outcomes of interactions are difficult to determine experimentally.

A significant increase in dry weight loss and a marked stimulatory effect on mycelial growth of the decay fungi *Trametes versicolor, Trichaptum abietinum* and *Oligoporus placentus* occurred in wood chips coinoculated with bacteria (*Enterobacter*) and yeasts (*Zygosaccharomyces bailii* and *Pichia pinus*) (256). In contrast, *Pleurotus ostreatus* grown on oak wood caused greater weight loss in pure culture than with added bacteria and nematodes (257), and *Phlebia radiata* developed three-fold more CO_2 from wheat straw when in pure culture than in combination with the indigenous microflora or with the mycoparasite *Trichoderma harzianum* (258). In soil, litter, and well-decayed wood, the decay community may include ectomycorrhizal fungi, a topic that is expanded later.

The decay community is not static, but undergoes a succession in fungal species as the resource becomes depleted, colonized, and modified (56,151,259–262). In one such study, fungal isolates from living leaves of *Populus tremuloides* were dominated by dark ascomycetous fungi (possibly endophytes) and *Aureobasidium pullulans*; from net-caught fallen leaves by *Penicillium janthinellum*; from leaves in the fermentation layers (F_1 and F_2)[3] by *Trichoderma* species and *Penicillium syriacum*; and from the humus (H) layer by these latter two plus species of *Mortierella, Cylindrocarpon,* and *Phoma.* However, although basidiomycetes represented up to 30% of the total mycelia measured in agar films from the litter and fermentation (L_2, F_1, and F_2) layers, they were isolated from only 2% of particles of June leaf litter that were plated (265). In another study, cellulolytic fungi were dominant among isolates from litter of *Alnus* soon after leaf fall (November through January), but chitinolytic fungi represented as many as 30% of isolates recovered in February (266). The latter group, including species of *Mortierella*

---

[3]The following are terms used internationally (263) for the organic and upper mineral layers of soil, with those used in the United States (264) in parentheses: L (O) litter layer of freshly fallen leaves and twigs; F (O_c and O_i) fermentation layer of partly decomposed and comminuted litter, the plant material still recognizable; H (O_a) humus layer of well-decomposed litter; O (O_c, O_i, and O_a) peaty layer of plant remains accumulated under wet conditions; A (A) upper mineral soil horizon darkened by humified organic matter.

and *Verticillium*, were thought to be growing on the hyphae of previous colonists. No basidiomycetes were noted among the isolates recovered in any month. Robinson et al. (267) studied succession on substrates of varying resource quality buried in soil, in which species colonizing more lignified internodes of wheat straw (9.4% lignin, C:N = 109) were compared with those on leaves (4.9% lignin, C:N = 39). All fungi isolated were ascomycetous: *Penicillium hordei, Trichosporiella sporotrichoides*, and unidentified species of *Fusarium* and *Phoma* were more common on internodes than leaves, and of these, the *Fusarium* was more frequent early in succession and the others were more frequent later. *Epicoccum nigrum* and *Cladosporium* species were more frequent on leaves early in succession and were followed by the myoparasitic *Trichoderma harzianum*, which also occurred on internodes. Bowen and Harper (268) studied fungal populations on wheat straw that was first incorporated into agricultural soils, then retrieved, washed, and plated on ten different media, including five chosen to isolate lignin-decomposing species. They reported no *Trichosporiella, Epicoccum*, or *Phoma*, but found species of *Cladosporium* and *Fusarium* most common 2 months after incorporation of the straw into soil, and *Trichoderma harzianum* most common after 8 months. The basidiomycete *Typhula* was isolated from 4% to 6% of the straws from 1 to 12 months after incorporation. Slow-growing basidiomycetes were isolated from as many as 40% of the straws at 6–8 months after incorporation (in March and May), with high recoveries on malt agar (year 1) and p-hydroxybenzaldehyde agar (year 2). Their method, essentially a baiting technique (268) and use of selective media combined with particle filtration (110), appears promising in determining some of the natural basidiomycete members of the decay community in soils.

In soil and other complex substrates, organisms not traditionally considered to represent decay fungi may make a substantial contribution to the decay process. It seems probable that basidiomycetes, including the agarics, gasteromycetes, and even polypores, are important in the initial delignification of plant residues in agricultural soils as well as in soils of natural communities (110,269,270). Ascomycetes are usually regarded as secondary decomposers, and primarily cellulolytic. However, in addition to cellulases, many ascomycetes produce laccases and other phenol oxidases that are capable of metabolizing products of lignin degradation or even significantly degrading lignin itself (271), and ascomycetes of the Xylariales are often associated with a white-rot of wood (14). Lignin peroxidases have been found in the ascomycete *Neurospora sitophila* (anamorph *Chrysonilia s.*) and the actinomycete *Streptomyces viridosporus* (272). The ascomycetous mold *Aspergillus flavus* caused significant degradation of larch wood in vitro (273). Alone, or more likely in combination with the actions of basidiomycete decay fungi and the shredding of litter by soil macroinvertebrates, the ascomycetes and streptomycetes may have an important role in litter decay. Determination of these roles and observation of meaningful interactions during decay may require that coinoculation experiments be set up with a variety of permutations of organisms and inoculation times. In one study (274), wheat straws first colonized by *Trichoderma viride* (a mycoparasite) for 2–4 weeks prior to inoculation with basidiomycete strains that had been isolated from wheat straw showed weight losses equal to or less than straws incubated with the basidiomycete alone. However, simultaneous coinoculation of wheat straw with *Trichoderma viride* and either of two basidiomycete strains yielded greater weight loss than that induced by either the *Trichoderma* or basidiomycete alone, and subsequent inoculation with *Fusarium culmorum* or *T. viride* onto straws first decayed by basidiomycetes gave a further 20% to 30% reduction in dry weight (274). More studies of this sort are clearly warranted.

The decay community may also show succession based on season as well as substrate quality. Among the saprophytic fungi are some that tolerate or operate most efficiently at temperatures above 35°C (thermotolerant and thermophilic fungi), as well as those that tolerate or prefer temperatures below 5°C (psychrotolerant and psychrophilic fungi). Depending on the locale, these groups may assume greater or lesser roles at certain times of year. In a montane *Populus* forest in Alberta, Canada (51° N, altitude 1430 m), winter respiratory CO_2 evolution was estimated to be equivalent to 60% of the annual input of above- and below-ground litter (275). During this period (1 December to 31 March), soil temperatures fluctuated from −10°C to over 10°C, with 50 freeze–thaw cycles recorded in the litter (L) horizon. Decay community members active during this period were not identified, but isolation of soil fungi at low temperatures has identified a significant psychrophilic component in the fungal community elsewhere (276).

6. MYCORRHIZAE[4]

Of equal importance to the role that many fungi play in nutrient cycling through decomposition is the role that other soil fungi play in symbioses known as mycorrhizae with the roots of vascular and nonvascular plants. Mycorrhizal fungi provide plants with nutrients, of which N and P have been best studied, and protection from drought stress and root pathogens (16,277,278). Most mycorrhizal plants provide their associated fungi with photosynthetically derived carbon, and most mycorrhizal fungi are dependent upon this for survival. Approximately four-fifths of the world's vascular plants form what are known as arbuscular my-

---

[4]The word mycorrhiza is used for both the fungus–plant association and the fungus–root organ in which the connection is made. Note, however, that the mycorrhiza is neither the fungus nor the plant. The plural form may be found in the literature as mycorrhizae, mycorrhizas, or even mycorrhiza.

corrhizae or endomycorrhizae, many of the remainder form ectomycorrhizae, and only a minority of plants are nonmycorrhizal (33,279). There have been several recent reviews of the biology of mycorrhizae (16,17,126,280–286) and methods for their study (287–289). The literature for the identification of endomycorrhizal fungi is basically that of the Glomales (Zygomycota), whereas that for the ectomycorrhizal fungi is scattered in keys to both ascomycetes and basidiomycetes (consult [5]). Currah and Zelmer (290) provide a key to 15 genera of orchid mycorrhizal fungi. Identification of ectomycorrhizal roots, without associated fruiting bodies, can be approximated by using morphological features (291,292), cultural characters of fungi isolated from them (293), or analysis of ribosomal deoxyribonucleic acid (rDNA) isolated from the roots or fungal isolates and amplified by using the polymerase chain reaction (PCR) (159). The banding patterns of restriction fragments (RFLP analysis) of PCR-amplified rDNA may provide an exact match with those from identified ectomycorrhizal fruiting bodies in a restricted study with limited ectomycorrhizal diversity (159). Sequence analysis of the amplified rDNA will prove more useful in the long run and in larger studies by allowing the unknown ectomycorrhizal fungi to be placed in a phylogenetic framework and eventually identified as more sequences of known ectomycorrhizal fungi become available.

Table 2 indicates the five main types of mycorrhizae recognized, and their vascular and nonvascular plant symbionts. The types of mycorrhizae are defined on the basis of the location and morphological characteristics of the fungal hyphae in the plant roots, and also on the basis of the plant and fungal taxa involved. In ectomycorrhizae (including both arbutoid and monotropoid mycorrhizae), fungal hyphae form an external sheath around the infected root, then penetrate between root epidermal and cortical cells, forming what is called a Hartig net (Fig. 4 A–B). Intracellular penetration is rare and limited. In orchid and ericoid mycorrhizae, no hyphal sheath is formed, the infection is predominantly intracellular, and the internal hyphae form coils. Arbuscular mycorrhizae or endomycorrhizae are formed by members of the Glomales and are characterized by intracellular infections with the internal hyphae forming branched, shrublike structures known as arbuscules (Fig. 4 D), and sometimes globose structures called vesicles (Fig. 4 C).

Endomycorrhizae dominate in grasslands and species-rich forests, whereas ectomycorrhizae dominate in forests wherever a single or a few species of trees are dominant (33); this relationship correlates to the finding that endomycorrhizal associations are more abundant in low latitudes and ectomycorrhizal associations in high latitudes and at high elevations in the tropics (16). The ectomycorrhizal fungi include some of the most highly prized edible fungi, such as *Boletus, Cantharellus*, and *Tuber*. Many plants are able to form more than one type of mycorrhizal association: among others, *Ulmus, Tilia, Cupressus, Juniperus, Leptospermum, Acacia, Casuarina*, and *Populus* form both

Table 2. Types of Mycorrhizae and the Fungi Responsible[a]

Name	Fungal associates	Plant associates
Endomycorrhizae (arbuscular mycorrhizae [AM]) (formerly known as vesicular–arbuscular mycorrhizae [VAM])	Zygomycota: Glomales [*Glomus*, *Gigaspora*, *Acaulospora*, and others, (80, 81)]	Most vascular plants, including ferns, lycopods, horsetails, except those listed for other mycorrhiza types and exceptions in certain families, e.g., Cruciferae, Cyperaceae, Fumariaceae [but see (279)]; also liverworts (Metzgeriales and Marchantiales) and mosses
Ectomycorrhizae (including both arbutoid and monotropoid mycorrhizae)	Zygomycota: Endogonales (atypical) Ascomycota: Pezizales, Leotiales, and *Cenococcum* (anamorphic) Basidiomycota: see Table 3	Pinaceae, Betulaceae Salicaceae, Fagaceae Myrtaceae, Rosaceae, Ericaceae, Cystaceae, and others (16); in some ferns, but not known from bryophytes

Table 2. Continued

Name	Root morphology	Comments
Endomycorrhizae (Arbuscular mycorrhizae [AM] (formerly known as vesicular–arbuscular mycorrhizae [VAM])	Infection intracellular; forming arbuscules only (Gigasporineae: *Gigaspora* and *Scutellospora*) or both arbuscules and vesicles (Glomineae: *Glomus, Acaulospora*, etc.); no fungal sheath formed	Arbuscles may not always be formed (279); members of Salicaceae are frequently endomycorrhizal in youth, switching later to ectomycorrhizae (or both)
Ectomycorrhizae (including both arbutoid and arbutoid and monotropoid mycorrhizae)	Infection intercellular; fungal sheath formed around modified short roots, and Hartig net of fungal hyphae formed between root cortical cells	Infection in certain Ericaceae (termed arbutoid and monotropid mycorrhizae) by fungi causing typical ectomycorrhizae in other plants form both intercellular and limited intracellular infection

Name	Fungal associates	Plant associates
Ericoid Mycorrhizae	Ascomycota: Leotiales (*Hymenoscyphus ericae*); similar associations formed by species of *Oidiodendron* (anamorphic) Basidiomycota: Catharellales (*Clavaria* spp.), and others?	Ericaceae, particularly *Calluna, Vaccinium, Rhododendron*; and liverworts of the Jungermanniales (295)
Orchid Mycorrhizae	Basidiomycota: Ceratobasidiales (*Ceratobasidium, Thanatephorus*) Tulasnellales (*Tulasnella*), Poriales (*Armillaria, Loweporus, Microporus*), Ganodermatales (*Ganoderma*), and Hymenochaetales (*Phellinus, Erythromyces*)	Orchidaceae (only in the protocorm stage in some); and possibly in liverworts of the Jungermanniales and Metzgeriales
Ectendomycorrhizae (pseudomycorrhizae)	Ascomycota: anamorphic, *Phialocephala fortinii*; *P. dimorphospora, Phialophora finlandia, Chloridium paucisporum*, and possibly others, known collectively as MRA (*Mycelium radicis atrovirens*); and teleomorphic, *Wilcoxina* (Pezizales [297])	Known from Pinaceae: *Larix, Picea*, and *Pinus*

Table 2. Continued

Name	Root Morphology	Comments
Ericoid Mycorrhizae	Infection intracellular; internal hyphae form coils; no fungal sheath formed around roots	Cross-infectivity of *Hymenoscyphus ericae* to both ericoid and liverwort hosts shown (295)
Orchid Mycorrhizae	Infection intracellular; internal hyphae form coils; no fungal sheath formed around roots	Mycobionts form two groups: plant pathogenic *Rhizoctonia* (anamorphs of *Ceratobasidium* and *Thanatephorus*), *Tulasnella*, and *Armillaria*; and typical wood-decay fungi (polypores and Hymenochaetales)
Ectendomycorrhiza (pseudomycorrhizae)	Infection intercellular and intracellular; internal hyphae form coils; weak fungal sheath formed around roots	Infection may be beneficial or detrimental to the plant, depending on fungal strain and habitat; *Wilcoxina* forms typical ectomycorrhizae with hosts other than *Pinus* and *Larix* (286)

Data from Refs 16, 294–296.

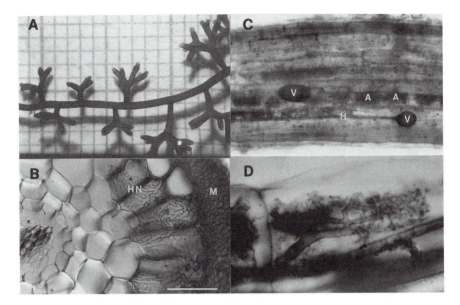

Figure 4. A. Dichotomously branched short roots of ectomycorrhizae formed between *Pinus strobus* and an unknown basidiomycete (grid = 1 mm). B. Cross section of an ectomycorrhizal short root of *Populus tremuloides*, showing the hyphal mantle (M) surrounding the root and Hartig net (HN) formed between cells of the root cortex (scale bar 50 μm). C. An endomycorrhizal root of *Artemisia tridentata*, showing the lack of a hyphal mantle surrounding the root, and the presence in the root of hyphae (H), vesicles (V), and arbuscules (A) of *Glomus macrocarpum*. D. Detail of an arbuscule in the root of *Artemisia tridentata*. Photos A and B from M. C. Brundrett, G. Murase, and W. B. Kendrick, Can. J. Bot. 68:551 (1990), with permission of M. C. Brundrett (CSIRO, Wembley, Western Australia) and the National Research Council of Canada; photo C by P. D. Stahl (University of Wyoming, Laramie, Wyoming), with permission; photo D by the late Mark Loree (University of Wyoming), provided by P. D. Stahl.

ecto- and endomycorrhizae; in the Ericaceae, members of *Arctostaphylos, Gaultheria, Kalmia, Ledum, Rhododendron,* and *Vaccinium* form both ecto- (or arbutoid) and ericoid mycorrhizae; and ectendomycorrhizae are found on normally ectomycorrhizal plants (16,294). In addition, roots of members of the Fabaceae are both endomycorrhizal and nodulated with nitrogen-fixing rhizobia, and roots of a variety of plant families are endo- or ectomycorrhizal and actinorhizal, nodulated with nitrogen-fixing species of *Frankia* (Actinomycetes); *Alnus, Myrica,* and *Shepherdia* may have both endo- and ectomycorrhizae in addition to actinorhizae (298). *Alnus*, in fact, has a fourth form of root association: *Penicillium nodositatum* forms nodules on the roots in a manner similar to *Frankia*, but the nature and significance of the association are not yet known (299–301).

Mycorrhizal symbioses are considered to be very old. Fungal endophytes thought to represent early endomycorrhizae have been found in the rootlike organs of *Asteroxylon* and *Rhynia* from the Devonian period (360–410 million years ago) and other possible fossil records date back to the Cambrian period (about 550 million years ago) (303). It has been speculated that endomycorrhizae were instrumental in colonization of land by primitive vascular plants in the late Silurian, about 420 million years ago (35,304). The date of origin of the endomycorrhizal fungi (Glomales) estimated from DNA sequence analysis ranges from 353 to 462 million years ago, which is consistent with this hypothesis (305). In contrast, other forms of mycorrhizae seem to have evolved independently many times in unrelated groups of fungi (33,34; see Tables 2 and 3). Under natural conditions both the plant and fungal partners are dependent upon the relationship for their survival, so the relationship is best described as one of mutualism. In situations of artificially high nutrient availability, as in agriculture and habitats made eutrophic through atmospheric inputs of NO_x and NH_4^+, ordinarily mycorrhizal plants are found to be nonmycorrhizal, and the mycobionts are unable to exist under these conditions. This lends support to the hypothesis that the origins of mycorrhizae may lie in the parasitism by plants of what were originally saprophytic fungi (34). Additional evidence is suggested by the fact that the mycobiont may be digested, consumed, and temporarily or permanently expelled by the photobiont in orchid mycorrhizae once the plant is photosynthetically self-sufficient (16), and in ericoid mycorrhizae during periods of drought stress (306). In addition, more than 400 species of vascular plants, gametophytes of several ferns and fern allies, and at least one liverwort sporophyte lack chlorophyll and rely on a mycorrhizal fungus for the supply of carbon and other nutrients. The mycobiont may in turn derive its carbon from a second mycorrhizal association with an autotrophic, chlorophyllous plant (307,308), or, in the case of orchid mycorrhizae, from saprotrophy or plant parasitism (16,309). Achlorophyllous mycorrhizal plants have been called mycotrophic (deriving nutrition from fungi) or mycoheterotrophic (dependent on nutrition derived from fungi) (307), but mycotrophy is much more widespread: mycorrhizal fungi provide essential carbon nutrition to plants ranging from germinating orchid seeds and protocorms to gametophytes of ferns and lycopods, and shaded seedlings of forest trees (310,312).

The mycorrhizal links between different plant species and generations are of great ecological interest (312–315). The parental investment of photosynthate, transferred via mycorrhizae, in seedling success provides support for K selection (in which relatively few progeny are produced, and effort is expended to ensure their survival, 316), even among tree species producing abundant annual seed crops. Seedlings in many forest types with deep litter layers often are most successful if growing on fallen, rotting logs, or on mineral soil exposed by light fires; rotting wood may provide both a stable, moist substrate and a source of mycorrhizal inoculum (315,317). It was recently found that some of the common leafy liverworts of heathlands share their mycorrhizal fungus with the dominant shrubs of the Ericaceae (295). This discovery led these authors to speculate that the liverworts may provide a source of mycorrhizal inoculum allowing ericaceous seeds to become quickly infected upon germination, improving their chances for survival and establishment. Other mosses and liverworts apparently harbor endomycorrhizae or basidiomycete mycelium (possibly mycobionts of orchid mycorrhizae) and may form a similar bridge for seedlings of compatible plants in the same habitat. Mycorrhizal mosses and liverworts may also provide an important bridge in time after a fire: mycorrhizal vascular plants recolonizing the area may find a ready inoculum of mycobionts among the more rapidly colonizing bryophytes. If shown, this would provide another example of a positive feedback mechanism called ecological bootstrapping, previously demonstrated in the shared ectomycorrhizal of Pacific coast conifers and ericoid shrubs (315,318). Shared mycorrhizal linking plants with symbiotic, nitrogen-fixing root associates (*Frankia* or rhizobia) may be essential in colonization of disturbed habitats such as mine tailings, talus slopes, and sites affected by glaciation or severe fires.

The nutrition of mycorrhizal fungi and their contribution to the nutrition of the mycorrhizal plant have been the subject of numerous studies and several reviews (16,126,319). Endomycorrhizae have been thought to benefit the plant primarily through provision of P, ectomycorrhizae through provision of N, and ericoid mycorrhizae through provision of both P and N (16,320,321), but the associations are undoubtedly more complex than that. In heathlands and coniferous forests, most of the N is present in organic form as amino acids, many of which become complexed with phenolic and humic residues of the litter (322). The apparent ability of arctic plants to take up amino-N has led Kielland to suggest that these plants short-circuit the ammonification step of the nitrogen cycle (323), but this hypothesis overlooks the fact that amino-N is the preferred form for the fungi of ericoid and ectomycorrhizae that dominate these habitats. Ectomycorrhizae and ericoid mycorrhizae are also thought to provide enzymatic as well as better spatial access to the large pool of complex nutrients in the litter (17,126,319). Reports referring to early studies on the saprophytic ability of mycorrhizal species of *Tricholoma* or *Lepista* (210,324) should be viewed with caution; some are saprophytic species incorrectly identified as mycorrhizal by incorrect placement in a mycorrhizal genus, and some are simply misidentified (325). Recent research using authenticated mycorrhizal isolates has provided evidence of mycorrhizal fungi gaining enzymatic access to complex organic nutrients. In one study, nonmycorrhizal plants metabolized 64% of the hide powder, 25% of the cotton, and 11% of the chitin provided as sources of C and N to their

Table 3. Groups of Basidiomycota and Their Biology[a]

Taxon	Predominant biology
Urediniomycetes (rusts) and Ustomycetes (smuts; some yeasts)	Plant parasites
Heterobasidiomycetes	
Septobasidiales	Insect parasites
Tremellales and Filobasidiales (some yeasts)	Mycoparasites and animal pathogens
Exobasidiales	Plant parasites
Homobasidiomycetes (diverse groups)	
Group 1 (no yeasts)[b]	
Cantharellales (including Gomphales)	Ectomycorrhizal (and saprophytes)
Thelephorales	Ectomycorrhizal
Cortinariales	Ectomycorrhizal, and saprophytes: white rot
Group 2 (no yeasts)	
Auriculariales	Saprophytes: white rot
Ganodermatales[c]	Saprophytes: white rot
Hymenochaetales[c]	Saprophytes: white rot
Polyporales (including Schizophyllales and Tricholomatales)[c]	Saprophytes: white (and brown) rots; ectomycorrhizal; plant parasites

Group 3 (no yeasts)	
Bondarzewiales	Saprophytes: white rot
Fistulinales	Saprophytes: brown rot
Hericiales (including Lachnocladiales)	Saprophytes: white rot
Russulales	Ectomycorrhizal
Stereales	Saprophytes: white (and brown) rots
Group 4 (Dacrymycetidae: Dacrymycetales; some yeasts)	Saprophytes: brown rot
Group 5 (no yeasts)	
Boletales	Ectomycorrhizal
Coniophorales	Saprophytes: brown rot
Sclerodermatales	Ectomycorrhizal (and saprophytes)
Group 6 (no yeasts)	
Agaricales (including Lycoperdales, Nidulariales, Phallales, Strophariales, Tulostomatales)	Saprophytes: white rot
Group 7 (Tulasnellomycetidae: Ceratobasidiales and Tulasnellales; no yeasts)[c]	Plant parasites

[a]Based in part on unpublished analyses of published and unpublished DNA sequences.
[b]No valid name for subclass; groups based in part on Jülich (302).
[c]Group includes some orchid mycorrhizae.

roots, and different replicates of *Suillus* mycorrhizae metabolized between 3% and 75% of the cotton, suggesting that contaminant microflora had greater effects than the experimental one (326). In another study, however, the ericoid mycorrhizal symbiont *Hymenoscyphus ericae* released $^{14}CO_2$ equivalent to 42% of ^{14}C-labeled lignin after 60 days, compared to 21%, 7%, and 5% for the ectomycorrhizal fungi *Paxillus involutus, Suillus bovinus*, and *Rhizopogon roseolus* (125). Four ectomycorrhizal fungi, grown for 240 days in association with *Pseudotsuga*, released $^{14}CO_2$ equivalent to 19%–28% of the ^{14}C-labeled hemicellulose, 5%–18% of the labeled cellulose, but only 1%–3% of labeled needles (327). Some of the fungi forming associations with germinating orchid seeds are typical white-rot basidiomycetes, such as the polypores *Ganoderma, Loweporus, Microporus*, and *Phellinus*. There is no question of the saprophytic ability of these mycorrhizal symbionts, but research is needed into the nature of the associations formed, and how they are controlled (296). Mycorrhizal fungi may also alter the pattern of litter decay and nutrient cycling without direct involvement. Ectomycorrhizal mats of *Hysterangium setchellii* form a soil microenvironment with increased microbial activity, resulting in faster lignin and cellulose decomposition (328).

Ecological studies on mycorrhizae, many of which unfortunately fall beyond the scope of this chapter, have focused on dispersal, succession, interactions with other soil organisms, and effects of disturbance. Spores of endomycorrhizal fungi may be dispersed by worms and worm-eating birds (329,330), and germinate after passage through a rodent digestive tract (331). There are numerous observations of mammals, and several of birds, feeding on fruiting bodies of ectomycorrhizal ascomycetes and basidiomycetes and effecting spore dispersal (332,333). *Glomus* spores isolated from mouse dung formed endomycorrhizae (334), and ectomycorrhizae were formed by *Pinus* seedlings inoculated with squirrel dung (335). Such observations lend support to claims of tripartite (plant–fungus–animal) mutualisms in nature.

Mushroom collectors have long recognized two forms of succession of species of ectomycorrhizal fungi: succession in relation to aging of a single plant associate and succession in parallel with succession in species composition of the plant community. The former has received formal study only in the last 15 years (336–340) and the latter only very recently, in connection with studies on old-growth forests. Mason et al. (336) found a succession of species associated with *Betula* planted in heathland in Scotland, ranging in age up to 28 years. Species characteristic of seedlings and young trees were identified and termed early-stage mycorrhizal fungi; other species that only became established with older trees were termed late-stage fungi. On the basis of such studies, the hypothesis was formed that species richness of ectomycorrhizal fungi rises with tree age until canopy closure and decreases thereafter (337,338). These studies, however, were conducted on young, planted stands, not representative in any way of an old-growth forest community. Visser (340) studied stands of *Pinus banksiana* up to 122 years in age that were established naturally after fire. No decline in species richness following canopy closure was observed, and it was stated that this may not be a general pattern in natural forests. More studies on ectomycorrhizal succession in association with a variety of tree species and including forests ranging from seedling to old growth status are clearly warranted. As with all ecological studies on mycorrhizae, this will require research on both above-ground patterns, using fruiting body surveys, and below-ground patterns, using identifications of mycorrhizal fungi based on cultures (293) or molecular techniques (159). Some studies have shown that succession in mycorrhizae may have a physiological basis. In two separate studies, early-stage mycorrhizal fungi with broad host ranges formed mycorrhizae in vitro with little or no glucose or P added to the nutrient solution, whereas late-stage, more host-specific mycorhizal fungi did not (341,342). Radial growth on agar by early successional species such as *Laccaria* and *Hebeloma* was nearly unaffected by low glucose levels, but late-stage fungi were severely repressed (341). Species of *Laccaria* and *Hebeloma* are among those found fruiting near sites of animal carcasses, animal wastes, or application of nitrogenous fertilizers (343,344). Some of the studies on interactions of mycorrhizal fungi with other soil organisms will be described later (and see Chapter 6). Likewise, a discussion of the possibility of local or widespread extinctions of ectomycorrhizal fungi is included in Section 8.1.

7. INTERACTIONS OF FUNGI WITH OTHER SOIL ORGANISMS

7.1 Interactions of Fungi with Soil Microfauna and Mesofauna

Fungal hyphae and spores form part of the diet of creatures ranging from amoebae and rotifers to nematodes, various kinds of mites (oribatid, mesostigmatid, prostigmatid, and astigmatid), collembolans (springtails), and earthworms (345–348). Lichens and fungal fruiting bodies provide food and shelter to a great variety of invertebrates such as mites, flies, and beetles (349,350), as well as vertebrates. The very large literature on soil micro- and mesofauna feeding on fungi and their effects on functions such as nutrient cycling and mycorrhizae will be left for Chapter 6. Soil fungi also act as parasites and predators of all manner of soil fauna and as agents of decay of their wastes and remains. This section will include a brief overview of soil fungi as parasites and predators of the micro- and mesofauna. There have been several reviews of fungi that attack nematodes, the nematophagous fungi (351–356), and fungi that attack other micro- or mesofauna such as amoeba and rotifers (71,145,243,357,358). Techniques for their discovery are generally based on one of two methods: extraction and maintenance of a natural population of soil fauna and observation of naturally infected members, or use of animals from a laboratory stock culture as bait for infective propagules from soil or other substrates (351,359).

Fungal parasites and predators of soil micro- and mesofauna span the taxonomic range from the motile fungi (Chytridiomycota and Oomycota), to Zygomycota (Entomophthorales and Zoopagales), ascomycetes and their anamorphs, and basidiomycetes and their anamorphs. Approximately 35 species of motile fungi, 100 species of zygomycetes, 250 species of ascomycetes, and 50 species of basidiomycetes are known to attack and consume soil micro- or mesofauna, but these numbers reflect how little, rather than how much, is known about the biology of this group of fungi. Some species and genera are specific to particular host groups, for example, *Ballocephala* (Entomophthorales) on tardigrades, but many others attack a variety of soil fauna. In *Dactylella*, some species attack amoebae while others attack rotifers, nematodes, or copepods (360), and the nematophagous *Dactylella cionopaga* is also capable of trapping and infecting two different astigmatid mites (361). Therefore, it is logical to consider the nematophagous fungi together with those attacking all other soil microfauna. No comprehensive key to these fungi exists, but the key of Cooke and Godfrey (362) to nematophagous fungi can be supplemented with recent treatments of nonentomogenous Entomophthorales (363), Zoopagales on nematodes and amoebae (364), *Arthrobotrys* (365,366), *Dactylella* (367), *Monacrosporium* (368), *Verticillium* (369), *Hohenbuehelia* and *Nematoctonus* (370), and a review of fungi associated with nematode cysts (371).

Fungi attacking micro- and mesofauna have been divided into two groups, the predators and the parasites (endoparasites). Predatory fungi develop extensive mycelia and are capable of consuming numerous animals, usually captured on specialized organs located throughout the mycelium. Endoparasitic fungi develop at most limited mycelia outside the infected host and produce propagules that act as agents of infection by penetrating, adhering to, or being ingested by host animals. The motile fungi attacking soil micro- and mesofauna are endoparasites, and the infective propagules are typically adhesive or modified zoospores. The zygomycete, ascomycete, and basidiomycete members of this group are divided among predators and endoparasites. It is generally supposed that endoparasitic species are the more nutritionally specialized and less capable of saprophytic growth in the absence of hosts. A number of predatory species of ascomycetes and basidiomycetes grow readily in culture and have been shown to degrade cellulose and lignin (243). Consumption of nematodes or other soil fauna by these species has been considered to provide a source of nitrogen, to supplement the high C:N ratio of lignocellulosic substrates (241,243). Several of the predatory nematophagous fungi can also attack other fungi or living bacteria, or both: species of *Arthrobotrys, Verticillium*, and *Pleurotus* can act as mycoparasites, and *Pleurotus* and *Hohenbuehelia* are bacteriolytic.

Until recently, all ascomycetous and basidiomycetous nematophagous fungi were known only as anamorphs, unconnected to any sexual state (teleomorphs). In part this is due to the fact that most discoveries of these fungi were made from observations of infected nematodes isolated from soil or other organic substrates. Drechsler illustrated a small ascomycete fruiting body suggestive of *Orbilia* that formed in one of his plates of *Arthrobotrys superba* (372) and also noted the similarity in shape of the adhesive knobs of *Nematoctonus* to illustrations of structures in cultures of *Pleurotus pinsitus* (a misidentified *Hohenbuehelia*) (373), but no definite connections were made. Discovery of the mushroom *Hohenbuehelia* as the sexual stage of *Nematoctonus* (374) was followed by connections between nine species of *Nematoctonus* and their teleomorphs in *Hohenbuehelia* (370). Connections have now been made between the ascomycete *Orbilia* (Leotiales) and nematophagous fungi identified as *Arthrobotrys, Dactylella,* and *Monacrosporium* (375). Maintenance of separate generic names for the anamorphs is both convenient and temporarily necessary, since there are nearly 150 species among these three genera, of which only three have been connected with a sexual stage. However, both they and another *Orbilia* anamorph, *Dicranidion*, are best thought of as one biological group. Species of *Dicranidion* are thought to be mycoparasites, like *Arthrobotrys*, but are not known to attack nematodes. Additional testing of cultures derived from teleomorphic genera for their ability to attack soil micro- and mesofauna may lead to discovery of new groups of fungi with this ability and to more anamorph–teleomorph connections among the known species.

The array of mechanisms by which the parasites and predators of microfauna attack their prey is a part of the appeal that this group of fungi possesses. Perhaps the most remarkable eukaryotic cell is the gun cell of the oomycete *Haptoglossa mirabilis*. When struck by a passing rotifer (or other microfauna), this cell first pierces the skin or membrane of the animal by means of a harpoonlike projectile and then injects infective protoplasm through a syringe formed by the everted lining of the cannon bore (376,377). All parts required for this process, from the harpoon to the swollen base that provides the turgor for its propulsion, are components of a single cell. At the other end of the taxonomic spectrum, 7 species of the wood- and litter-inhabiting basidiomycete genus *Hyphoderma* produce microscopic blisterlike cells called stephanocysts on their vegetative hyphae or germinating basidiospores, which adhere to and then infect passing nematodes (378–380). Another 11 species of *Hyphoderma* poison and then consume nematodes that have ingested their mycelium (379). Among *Arthrobotrys* and its relatives are the well-known adhesive nets and constricting and nonconstricting rings for trapping nematodes. Production of such traps on water agar is stimulated by the addition of nematodes or various peptides or amino acids, particularly valine (351). Recent studies have shown that the adhesive nets of

Arthrobotrys oligospora are also produced in media rich in nitrogen or carbon, if the other nutrient is lacking (381).

Various zygomycete, ascomycete, and basidiomycete nematode predators catch nematodes on adhesive pegs or knobs produced on their hyphae. In the basidiomycete *Hohenbuehelia* and the related anamorph *Nematoctonus*, a ball of adhesive mucus is produced from hourglass-shaped secretory cells. These are formed on conidia or basidiospores that germinate in the presence of nematodes and on the hyphae of predatory species. The writhing action of captured nematodes is slowed by the action of an as yet unidentified toxin (382). The nematode is penetrated at the site of the adhesive knob, but accessory branches also grow from hyphae in the vicinity and converge on the infection site or the victim's mouth. The basidiomycete *Pleurotus* produces smaller droplets of a nonadhesive substance from fine, tapered pegs on its mycelium. Nematodes that encounter these droplets are soon immobilized, and hyphae in the vicinity grow into its mouth or other orifices and eventually fill and lyse the body contents (242). Droplets of similar appearance are produced on the hyphae of a great variety of basidiomycetes, including species of *Psilocybe* (Strophariaceae, 383), *Schizophyllum* (Schizophyllaceae, 384), *Resupinatus* and *Stigmatolemma* (Pleurotaceae, 370), *Conocybe* (Bolbitiaceae), and *Panaeolina* (Strophariaceae, 385). The nature and function of these droplets are unknown in most cases, but in *Conocybe* and *Panaeolina* they contain nematotoxins that are described as antifeedants, since the nematodes are not colonized once immobilized (385). However, it seems likely that there would in many cases be an advantage for a fungus eventually to make use of a packet of nitrogen immobilized in its vicinity. Did fungi such as *Pleurotus*, growing in nitrogen-poor habitats such as wood, gain the nematophagous habit from what was originally a mechanism of defending its mycelium, or did *Conocybe*, growing in habitats such as nitrogen-rich pastures, lose the ability?

Nematodes held fast in traps of *Arthrobotrys* are eventually subdued by the effects of a secreted toxin, which has recently been identified as linoleic acid (386). A compound with similar toxic effects on nematodes to that shown by the droplets in *Pleurotus* was isolated from cultures of *Pleurotus ostreatus* grown on wheat straw and identified as trans-2-decenedioic acid (387). The nematicidal compounds produced by cultures of *Pleurotus pulmonarius* proved to be the fatty acids linoleic acid and *S*-coriolic acid (388). Various fatty acids have widespread roles in microbial interactions (389). Volatile fatty acids have been found in cultures of many wood decay basidiomycetes (390), and several have nematicidal effects (388). It would be interesting to determine the biological function of the droplets produced by *Psilocybe, Schizophyllum*, etc., listed previously, and to analyze them to determine whether they contain fatty acids similar to those in *Pleurotus*. Nematophagous and other fungi attacking soil microfauna may prove to be a rich source of secondary metabolites of pharmaceutical potential (391).

7.2 Fungi and Attine Ants

The leaf-cutting attine ants are the dominant herbivores and severe agricultural pests throughout the neotropics. In these areas, the trails along which they forage and return to their nests with bright green bits of cut foliage are a familiar sight. The main food of the ants is not the leaves themselves, but fungi that grow on them, cultivated by the ants in a subterranean fungus garden (392). The attines are a monophyletic group, and of those known, all but one genus (*Apterostigma*) apparently cultivate a monophyletic assemblage of fungi, related to the lawn mushrooms *Macrolepiota*, *Leucoagaricus*, and *Leucocoprinus* of the Agaricaceae (393). Members of *Apterostigma* have independently derived an association with fungi related to the litter fungi *Marasmius* and *Crinipellis* of the Tricholomataceae (393). Further research is required to verify previous reports of attines cultivating a wide diversity of basidiomycetes and ascomycetes (392,394). Somewhat similar associations occur in the Old World tropics between mound-building termites and the fungus *Termitomyces* of the Amanitaceae (395). *Termitomyces* are prized edible mushrooms in some areas, and fruiting bodies of some species may reach 1 m in diameter and weigh 10 kg or more.

7.3 Interactions of Soil Fungi with Bacteria and Other Fungi

Many soil- and litter-inhabiting fungi have been shown to produce bacteriolytic enzymes, including muramidases (396–399), and several have been shown capable of growth by using isolated bacterial envelopes as a sole source of carbon and nitrogen (398,400). Observations using scanning electron microscopy (SEM) of *Agaricus brunnescens* (the commercial button mushroom) growing in composted straw or casing soil have shown close associations of hyphae with numerous bacteria and bacterial colonies (401,402), and it has been suggested that bacterial biomass may make up a large part of the nitrogen nutrition for this mushroom (403,404). Barron discovered that *Pleurotus, Agaricus, Lepista*, and *Coprinus* could attack and consume living microcolonies of bacteria in vitro (127,242). Further studies have greatly expanded the list of fungi known to attack living bacteria in vitro (244,245,405). *Pseudomonas tolaasii* and *Burkholderia cepacia*, bacterial pathogens of cultivated mushrooms such as *Agaricus brunnescens* and *Lentinula edodes* (shiitake), are also attacked under some circumstances (245,406; see Fig. 3). The susceptibility to bacterial attack and ability to attack the bacteria depend on the species and strains of fungi and bacterial strains involved in the interactions (406). As yet, the ability to attack and consume colonies of living bacteria has been found only among ligninolytic basidiomycetes, but in all probability it is more widespread. Species of *Fusarium* and *Chaetomium* have been shown capable of bacterial cytolysis and lysis of isolated cell envelopes, respectively (407,408), and antibacterial antibiotics are produced by many ascomycetes.

In nature, why merely defend yourself from your competitors if you can eat them too? Further discussion of this topic is provided by Barron (243).

The large literature on mycoparasitism has recently been reviewed by Jeffries and Young (409); the subject of mycoparasitism as it relates to biological control of fungal plant pathogens is left for Chapter 17. A limited selection of the ecological aspects of interfungal interactions is presented here. A great many studies have investigated interactions of different species of fungi grown together in vitro (e.g., 121,410–420). Such studies conducted only on rich nutrient media should be discounted, and findings of in vitro studies on weak agar media should in general be corroborated with investigations conducted in natural substrates (421). In some instances, outcomes of interactions on agar and in natural substrates are parallel (422), but in many they are substantially different. Some generalities have emerged from the studies listed. Mycoparasitism is widespread among the fungi, found in all of the main taxonomic and substrate groups. Mycoparasitism can have the effect of capturing (or losing) both nutrients and territory (415) and affects the balance of vigor, maintenance, reproduction, or elimination of the individual. Fungal individuals can be minute or extensive, and the range of mycoparasitism can likewise be local or extensive, resulting in patchiness or complete replacement of the parasitized mycelium (121,414–416). Mycoparasitism may be broad or specific in host range, and mycoparasitic species may in turn be parasitized by other species that are or, more commonly, are not among their host range (e.g., 412). Mycoparasitism among at least some fungi seems to be most pronounced under nutrient-rich conditions, and less so under oligotrophic conditions (414,416); Stahl and Christensen (414) have speculated that mycelia become more territorial and combative when colonizing rich substrates or microhabitats.

Among the ligninolytic fungi, different species can be rated as to combativeness and outcomes of interactions between untested pairs predicted on the basis of their performance in other interactions (121,417,420,422). Combativeness appears to be positively correlated with production of hyphal cords (but not rhizomorphs) and with growth rates to a point, since the fastest-growing species tested, such as *Phanerochaete chrysosporium* and *Neurospora crassa*, are among the least combative. Many among the most combative are both strongly bacteriolytic in vitro (244) and also mycoparasitic, attacking yeasts in vitro (246). Species such as *Agrocybe gibberosa* with high combativeness in soil and good enzymatic capabilities (417) may be good candidates for efforts at in situ bioremediation of polluted soils.

Commercial cultivation of *Agaricus* represents a massive experiment in interfungal and fungal–bacterial interactions. During composting, the thermophilic fungus *Scytalidium thermophilum* completely colonizes the compost and modifies the bacterial flora to one that is selective for the growth of *Agaricus* by eliminating certain bacterial antagonists. When *Agaricus* is inoculated after the composting II phase, it rapidly replaces *Scytalidium*, although whether the mechanism of replacement involves mycoparasitism is unknown (423). Fruiting of the *Agaricus* is stimulated by the presence of certain pseudomonads, but *Pseudomonas tolaasii* and the mycoparasitic ascomycetes *Mycogone perniciosa* and the *Trichoderma harzianum* group cause significant disease losses (424,425). The stimulatory factor produced by favorable pseudomonads is unknown, but *P. tolaasii* produces a toxin, tolaasin, that disrupts fungal membranes (426) and subsequently degrades the fungal mycelium by enzymatic digestion (406). The thickwalled conidia of *Mycogone* are dormant until stimulated to germination by unknown compounds found in extracts of or intact fruiting bodies or mycelia of *Agaricus* and in a variety of other basidiomycetes (427). The degree of colonization by *Trichoderma harzianum* of *Agaricus* beds is related to the method and completeness of composting (425), which presumably affect both the microbial flora and the biochemical characteristics of the substrate.

Parallel observations have been made with other fungi, including other cultivated mushrooms. Colonization of oak bed logs of cultivated *Lentinula edodes* in Japan by *Trichoderma viride* and *T. harzianum* is affected by the degree to which the wood has been decayed by *L. edodes* (428). Coculture of *T. harzianum* with the wood decay *Trametes versicolor* has led to increased production by both fungi of laccase (410). In vitro interfungal interactions of many species, including *Agaricus brunnescens*, leads to elevated production of both laccases and tyrosinases (429,430), possibly related to the production of phenolic defense compounds (431–433). Fungus gnats (Mycetophilidae) were found to be attracted by an unidentified volatile attractant to zones of intraspecific fungal interactions on agar media (434). Dense, deeply pigmented zones called pseudosclerotial plates are formed when incompatible individuals or different species of various wood-inhabiting basidiomycetes meet (429,435). Hiorth (429) noted the correlation of laccase and tyrosinase production with these zones, and with infection by *Penicillium*. Melanization of pseudosclerotial plates is also correlated with the ability to sequester high concentrations of a variety of metal ions, thought to be protective against antagonistic microorganisms and grazing by microarthropods (433). Production of iron-binding compounds or siderophores that restrict access to iron in the substrate is a widespread, perhaps universal feature that is strongly related to interspecific interactions and has been found in both ligninolytic basidiomycetes (436) and bacteria (e.g., 437).

The ectomycorrhizal basidiomycetes have not generally been considered as mycoparasites and have not been found to attack bacteria or nematodes (127). However, it has been speculated that ready access to carbon supplied by the photobiont could make mycorrhizal fungi strong competitors with saprophytic fungi for the colonization of nutrient-rich resources (17,326), and this same argument could be used to suggest that they might have a combative advantage as mycoparasites of the same fungi. Zhao and Guo (413) showed that eight species of ectomycorrhizal fungi grown with the plant parasite *Rhizoctonia solani* grew appressed to or coiled around the *Rhizoctonia* hyphae. Penetration or lysis of the *Rhizoctonia* hyphae followed, with transfer of the nutrients P, S, K, and Ca to the mycelium of the mycorrhizal fungus. Hutchison and Barron (246) found that one of the eight species of ectomycorrhizal basidiomycetes that were tested, *Thelephora terrestris*, is capable of attacking and lysing cells of the yeast *Candida ernobii* in vitro. It is likely that more ectomycorrhizal fungi will be found to have such abilities when investigations are conducted with cultures of fungi grown in mycorrhizal association and in more natural substrates.

7.4   Tripartite (or More) Interactions and Food Webs

The pattern that begins to emerge is that soil fungi are interconnected in many ways with the diverse organisms sharing their environment. The true nature of fungal activities, whether apparently individualistic or symbiotic, can only be understood in light of these complex biotic interactions. A mycorrhizal fungus may derive the majority of carbon from a chlorophyllous photobiont, shunt some of that C to an achlorophyllous phytobiont or shaded seedling, and obtain C, N, P, and K used by all phytobionts from mycoparasitism of saprophytic or other mycorrhizal fungi in their environment. Saprotrophs may also be predators of soil fauna, fungi, or microbes and in turn be attacked by other soil fauna, fungi, and bacteria. Although ectomycorrhizal fungi protect the roots of associated plants from fungal root diseases (277), in vitro mycorrhization can be reduced by the presence of mycoparasitic species of *Trichoderma* (411). The nematode-trapping adhesive nets of *Arthrobotrys* and *Monacrosporium* can be blocked by cells of the yeast *Saccharomyces cerevisiae* that adhere when they are applied in vitro (438). However, the probability that the mycoparasitic *Arthrobotrys* would attack cells of *Saccharomyces* in its vicinity suggests that this effect may be unlikely to occur in vivo (416,418). The nematophagous *Dactylella cionopaga* is capable of capturing astigmatid mites, but in vitro these mites eliminated the *Dactylella* mycelium by grazing (361). Even complex illustrations of soil food webs (22,25) are simplified in order to be comprehensible and lack the double-headed arrows connecting fungi with soil fauna, roots, and bacteria that express some of these relationships. It is important to stress the continued need for fundamental field and laboratory studies of the biology, physiology, and autecology of individual species, combined with recognition of the complexities of interactions in nature.

In an interesting study of the effects of bacteria on fungal–fungal interactions (439), various bacteria (*Flavobacterium, Methanobacterium, Pseudomonas,* and *Staphylococcus*) inhibited growth on agar by the coprophilous ascomycetes *Chaetomium* and *Sordaria*, whereas growth by *Ascobolus* was enhanced by all bacteria. A similar pattern was seen when all three ascomycetes were grown together on rabbit pellets, with fruiting by *Chaetomium* and *Sordaria* reduced and *Ascobolus* enhanced by the addition of bacteria. However, when the basidiomycete *Coprinus* was added to the community with three ascomycetes (but no bacteria) on rabbit pellets, only the *Sordaria* fruited. Addition of *Flavobacterium* or *Methanobacterium* to the four-fungus community led to fruiting by all three ascomycetes, and *Staphylococcus* was most stimulatory to fruiting by *Chaetomium*. Several combinations of bacteria with single ascomycetes in rabbit pellets led to greater dry weight losses than those of controls without bacteria, but data were not presented for dry weight losses in mixed communities with bacteria (439).

Interactions may involve dispersal in addition to feeding (reviewed in [440]). Bacterivorous (saprophagous) nematodes have been found to be instrumental in the spread of bacteria through cultures of *Agaricus* grown in compost (441). Among the bacteria spread by nematodes are mushroom pathogens such as *Pseudomonas tolaasii* (441), bacteria such as species of *Bacillus* and *Enterobacter* that are inhibitory to mycelial growth of *Agaricus*, and species of *Pseudomonas* that may enhance mycelial growth of at least some strains of *Agaricus* (442). In a temperate forest, different feeding groups of insects may vector different groups of fungi: sap-feeding species of *Drosophila* mainly carry yeasts, fungal-feeding *Drosophila* mostly carry bacteria, and ants mainly carry molds (443).

8. MISCELLANEOUS TOPICS IN SOIL FUNGI

8.1 Biodiversity and Conservation

The numbers of species of fungi inhabiting even the smallest, best-known areas of the world have never been completely inventoried. After nearly 25 years of fungal surveys in Slapton Ley, a 211-hectare nature reserve in Devon, England, 2350 species of fungi were recorded, despite the fact that "almost no cultural work or soil isolations [were] carried out" (444). The estimated number of fungal species present, 2500, is 5.1 times that of the vascular plants known from the area (444). As mentioned, even the best inventories of local soil fungi primarily record members of the zygomycetes, the ascomycetes, and their anamorphs, with few or no members of the basidiomycetes nor other groups that together may account for 50%–80% of the fungal diversity in native soil communities. Depending upon what number of fungi are agreed to have been described, from 72,065 (5) to over 300,000 (445), estimates suggest that as much as 80%–95% of the world's fungal diversity is yet to be discovered and named (3). An ambitious project to locate and identify every species of all groups of organisms (an all taxa biodiversity inventory [ATBI]) in Guanacaste Conservation Area (GCA) of Costa Rica has been proposed, to take place between 1997 and 2004 (446,447). The GCA is an area of approximately 120,000 hectares with a diverse flora estimated at 10,000 species (compared to 211 hectares and 500 species of vascular plants in Slapton Ley). If sources of funding required for such a project are found, the concentration of mycological effort on such an ecologically rich area would considerably improve knowledge on the global diversity of fungi, including those in soil. There are many difficulties associated with assessment of the biodiversity of soil fungi, including problems with concepts of species and individuals and the fact that traditional isolation methods (or even the most innovative ones, e.g., [448]) yield only a small proportion of the organisms present and miss large taxonomic groups. Molecular methods for the assessment of biodiversity, such as the PCR amplification, cloning, and sequencing of 16S rDNA from bulk DNA extracted from soil, have proved to be powerful in prokaryotes (see Chapters 13 and 14), but similar techniques have not yet been developed for fungi (449). One of the great scientific benefits of the ATBI will be the opportunity to develop and test such techniques, progressing through synergies and cooperation with biologists in other fields.

There is a growing awareness of the value of biodiversity, even among speciose and uncharismatic groups such as the fungi (450,451). The tendency to arrange fungi into functional groups (decomposers, mycorrhizae, predaceous fungi) and to study only a small suite of species in the laboratory (*Neurospora crassa, Saccharomyces cerevisiae, Phanerochaete chrysosporium*) leads to the supposition that large numbers of species in nature may be functionally redundant, doing the same job as many others. However, the many links of the type described in sections 6 and 7, among a particular fungus and several generations or genera of plants, bacteria, soil fauna, and other fungi, suggest that the soil system is too complex to be run by only a few species of fungi. Perry et al. (315) described a positive feedback mechanism among ectomycorrhizal plants and fungi in forests of the Pacific Northwest, which they called "bootstrapping in ecosystems." Regeneration of forest trees following fire or clear cutting is dependent on mycorrhizal fungi surviving in symbiosis with ericaceous shrubs that survive these disturbances; presence in the undisturbed forest of a diversity of mycorrhizal species is essential for the survival of a few adapted to conditions following disturbance. Beare et al. (451) argue for the value of diversity among all groups of soil organisms for similar reasons: the suite of functions attributable to any one species is unlikely to be redundant, and the connections among species, functional groups, and kingdoms of organisms make the diverse overlapping of these functions essential in maintenance of stability. "Functionally similar organisms often have different environmental tolerances, physiological requirements and microhabitat preferences [and] as such they are likely to play quite different roles in the soil system" (451). Unfortunately, at present we know too little about even the species of fungi present in soil, let alone their biology, to know what these roles might be. Fundamental studies on the diversity and biology of soil fungi are urgently required.

There is growing evidence of a loss of biodiversity in certain areas among two groups of soil fungi, the ectomycorrhizal basidiomycetes and the soil-dwelling lichens. In the Netherlands, Arnolds (452–456) and others (457) have reported the decline in fruiting by many ectomycorrhizal species, particularly those growing on sandy soils. Atmospheric deposition of NH_4^+ from nearby agricultural sources has led to shifts in vegetation ecology, including a dramatic increase in forest ground-cover by the grass *Deschampsia*, and is thought to be the primary cause (455,456,458). As in some other parts of Europe, records from mushroom forays, commercial mushroom harvests, and systematic collections show a decline in fruiting quantity by some species and disappearance by others. Although some authors have spoken of this as a possible mass extinction of fungi (459,460), it is important to point out that the disappearance of fruiting bodies (mushrooms) does not necessarily mean extinction of the species. The reduction in fruiting has been dramatic and must be of ecological significance but is not equivalent to the loss of terricolous lichen diversity in response to pollution that has been noted in Belgium, Denmark, Germany, Russia, and England (461–465). Because disappearance of a lichen involves loss of the entire organism and not just its reproductive parts, these losses do represent local extinctions. The disappearance from Belgium of the ground-dwelling lichen *Peltigera* has been ascribed to air pollution, especially by nitrates, and to modern techniques of forest exploitation (461). Most losses in lichen flora due to atmospheric pollution, including those seen in the Netherlands (466), have been among epiphytic species. Research is urgently needed to determine whether the species of mycorrhizal fungi that are thought to have disappeared from the Netherlands are in fact surviving underground as mycorrhizae and mycelium. Isolation of mycorrhizal fungi or DNA from mycorrhizal roots and susequent identification using cultural characters (293) or molecular means (159) could answer this question. Preliminary evidence suggests that the number of mycorrhizal roots, mycorrhizal diversity, and mycorrhizal infectivity may be lower in forest plots with a thick *Deschampsia* sod layer than in those with the sod experimentally removed (458), and that litter removal may partially restore previous fungal diversity (467). However, reduction in emissions of NH_4^+ appears to be the only long-term solution.

8.2 Anaerobic Growth and Denitrification by Filamentous Fungi?

Conflicting views of the ability of filamentous fungi to grow anaerobically are expressed in the statements "All known anaerobic fungi are zoosporic" (468), and "Fungi continue to be regarded by many as exclusively aerobic organisms, despite considerable evidence to the contrary" (59). According to Griffin (469) "a few yeasts and filamentous fungi" show limited anaerobic fermentation in cells pregrown aerobically, but they cannot grow anaerobically; facultative anaerobic growth is limited to *Saccharomyces cerevisiae*, related yeasts, and some chytrids and oomycotans, and obligate anaerobic growth is limited to rumen chytrids. However, the filamentous ascomycete *Fusarium oxysporum* is capable of anaerobic growth if supplied with the electron acceptors MnO_2, nitrate, ferric or selenite ions, or with yeast extract (470). Other soil fungi, including *Geotrichum candidum, Fusarium solani, F. moniliforme, Mucor hiemalis*, and *Rhizopus* sp., have been reported to grow anaerobically (59). This capability may have the ecological significance of allowing the fungi to survive periods of soil flooding or situations in decomposition of rich, organic substrates leading to low oxygen tensions. The potential significance of this capability in nitrogen cycling has been suggested (471,472). Bollag and Tung (471) found that growing or resting cultures of *Fusarium oxysporum* or *F. solani* at low oxygen tensions reduced nitrite, with accompanying release of nitrous oxide. Shoun et al. (472) showed that several species of *Fusarium* and related teleomorphs, and also *Talaromyces* (= *Aspergillus*) *flavus*, were distinctly capable of denitrification from NO_2^- or NO_3^- when grown anaerobically for 7 days. Further studies on the distribution and significance of this phenomenon in nature are clearly warranted.

8.3   The Role of Soil Fungi in Bioremediation

There has been considerable interest in the use of fungi for bioremediation of anthropogenic pollutants, including persistent pesticides (473–477); polycyclic aromatic hydrocarbons (478); benzene, toluene, ethylbenzene, and xylenes (BTEX compounds, (479); dyes (480); and others (481–483). Most studies have concentrated on the bioremediation potential of white-rot, wood-inhabiting basidiomycetes, particularly *Phanerochaete chrysosporium*. Ligninolytic ability is broadly correlated with ability to degrade aromatic pollutants, although specific ligninolytic enzymes such as LiP may not be directly involved in all cases (475). Studies aimed at in situ bioremediation in soil or composted organic matter (477) would do well to investigate the potential of ligninolytic fungi indigenous to these habitats, since the wood-inhabiting species rarely grow or persist in nonsterile soils. As mentioned, combative species with good enzymatic potential are excellent candidates for such studies.

8.4   The Roles of Lichenized Fungi in Soil Formation, Stability, and Nutrition

The roles of lichens as pioneers in succession on rock and in soil formation are reviewed by Hale (7). Lichens are able to colonize on and beneath the surface of bare rocks (484) and break down the rock by mechanical and chemical means. Lichens begin the accumulation of organic matter through their own growth, can trap atmospheric nutrients deposited in precipitation and through symbiotic nitrogen fixation, and can also entrap fine wind-blown mineral and organic matter. Although these facts are widely known, the fact that lichens are fungi is too seldom recognized.

In polar and arid areas of the world, algae, cyanobacteria, lichens, and non-vascular plants form the dominant plant cover. In arid zones this cover is often in the form of a cryptogamic (or microphytic) soil crust, which has been the subject of a number of studies and several recent reviews (8,485,486). Filamentous cyanobacteria are apparently the dominant colonists of barren, arid sands or soils and act to form a surface mat binding particles together (486). Depending on particular features of the habitat, especially pH, salinity, and temperature, various lichens and mosses may become established; lichens also exude polysaccharides that bind soil particles, while mosses intertwine them with rhizoids (487). Nitrogen fixation during periods of moisture by cyanobacteria and lichens with cyanobacterial photobionts may contribute significantly to the nitrogen economies of soils in these habitats, although denitrifying bacteria associated with the cryptogamic crusts and high rates of ammonia volatilization in arid areas may reduce the impact of these contributions (8). Soil stabilization by cryptogamic crusts may survive up to 2 years after death of the crust organisms (487–489), but soil crusts are especially sensitive to physical disruption by grazing animals, tillage, or vehicular traffic, which leads to immediate and considerable erosion of the crust cover (8,490). Recovery by the lichen and moss components of the crust community following grazing disturbance is slow, with cover in previously grazed sites still significantly less than in ungrazed sites 7 years after cessation of grazing (490). Similar slow recoveries have been noted to follow wildfire (8).

9. CONCLUSIONS

Fungi are involved in virtually every soil process, and their importance needs to be clearly recognized. To paraphrase a former mentor, the fungi make the world go around, and a few lucky plants and animals get to live there. We need research into the diversity and biology of soil fungi, particularly in tropical areas, and into the extent and significance of biological interactions involving fungi in soil. Methods are now available for the isolation of saprophytic basidiomycetes from soil and for the identification of featureless, nonsporulating isolates as basidiomycetes or ascomycetes. Further work on cultural and molecular characterization of fungal cultures that are nonsporulating or produce only simple arthroconidia (fragments of hyphae) is also required; these make up a significant proportion of isolates from soil. The improvement in our ability to isolate and identify soil fungi will allow soil mycologists to go beyond the black box in studies of nutrient cycling, biological control, or other ecological questions. Apparent taxonomic confusion, including the renaming and reclassifying of fungi used in biochemical, agricultural, pharmaceutical, or other applied research fields, can lead to biological confusion and loss of information in these endeavors (86). Renaming and reclassifying of fungi will not decrease through the use of improved molecular techniques and analyses (93). Instead, the onus is on mycological taxonomists to communicate the rationale and ramifications of nomenclatural changes better. Search for fungal biodiversity will have among its many potential benefits the acquisition of new taxonomic knowledge; the ability to assess the correlation of soil fungal diversity with latitude, plant species diversity, and ecosystem function; and a great opportunity for biodiversity prospecting. Development of new techniques, or borrowing of techniques from the fields of microbiology and medicine (see Chapters 11, 12a, and 13), will be essential in future efforts to assess fungal biodiversity.

ACKNOWLEDGMENTS

Research that formed part of the background of this chapter was supported in part by an STA Fellowship to RGT at the Tottori Mycological Institute in Japan and by NSF grant BIR 9120006 to the Center for Microbial Ecology, Michigan State University. Support during its preparation was provided by the Department of Botany, University of Wyoming. Stimulating discussions were held with J. Baar, S. L. Miller, P. D. Stahl, and members of the Below-ground Processes seminar group at the University of Wyoming, and with G. F. Bills (Merck, Sharpe and Dohme, Rahway, NJ). M. Christensen (University of Wyoming) and D. Harris (Michigan State University) made helpful comments on the manuscript, and M. C. Brundrett (CSIRO, Wembley, Western Australia), P. D. Stahl (University of Wyoming) and A. Tsuneda (Tottori Mycological Institute, Japan) kindly provided photographs.

REFERENCES

1. Smith, M. L., J. N. Bruhn, and J. B. Anderson, *Nature* (London) *356*:428 (1992).
2. Domsch, K. H., W. Gams, and T. H. Anderson, *Compendium of Soil Fungi*, 2 vols., Academic Press, London (1980).
3. Hawksworth, D. L., *Mycol. Res. 95*:641 (1991).
4. Hawksworth, D. L., B. C. Sutton, and G. C. Ainsworth, *Ainsworth & Bisby's Dictionary of the Fungi*, 7th ed., Commonwealth Mycological Institute, Kew, England (1983).
5. Hawksworth, D. L., P. M. Kirk, B. C. Sutton, and D. N. Pegler, *Ainsworth & Bisby's Dictionary of the Fungi*, 8th ed., CAB International, Wallingford, England (1995).
6. Ahmadjian, V. *The Lichen Symbiosis*, John Wiley, New York (1993).
7. Hale, M., *Biology of Lichens*, 3rd ed., E. Arnold, London (1983).
8. West, N. E., *Adv. Ecol. Research, 20*:179 (1990).
9. Anderson, J. P. E. and K. H. Domsch, *Soil Biol. Biochem. 10*:207 (1978).

10. Kjøller, A., and S. Struwe, *Oikos, 39*:389 (1982).
11. Nannipieri, P., R. L. Johnson, and E. A. Paul, *Soil Biol. Biochem. 10*:223 (1978).
12. Yang, J. C., and H. Insam, *J. Trop. Ecol. 7*:383 (1991).
13. Bloem, J., G. Lebbink, K. B. Zwart, L. A. Bouwman, S. L. G. E. Burgers, J. A. de Vos, and P. C. de Ruiter, *Agric. Ecosyst. Environ. 51*:129 (1994).
14. Rayner, A. D. M., and L. Boddy, *Fungal Decomposition of Wood: Its Biology and Ecology*, Wiley, Chichester (1988).
15. Swift, M. J., *The Role of Terrestrial and Aquatic Organisms in Decomposition Processes* (J. M. Anderson and A. Macfadyen, eds.), Blackwell Scientific, Oxford, pp. 185–222 (1976).
16. Harley, J. L., and S. E. Smith, *Mycorrhizal Symbiosis*, Academic Press, London (1983).
17. Read, D. J., *Experientia, 47*:376 (1991).
18. Dickman, A., and S. Cook, *Can. J. Bot. 67*:2005 (1989).
19. Moore, J. C., and P. C. de Ruiter, *Agric. Ecosyst. Environ. 34*:371 (1991).
20. Moore, J. C., E. R. Ingham, and D. C. Coleman, *Annu. Rev. Entomol. 33*:419 (1988).
21. Elliott, E. T., H. W. Hunt, and D. E. Walter, *Agric. Ecosyst. Environ. 24*:41 (1988).
22. Hunt, H. W., D. C. Coleman, E. R. Ingham, R. E. Ingham, E. T. Elliott, J. C. Moore, S. L. Rose, C. P. P. Reid, and C. R. Morley, *Biol. Fertil. Soils 3*:57 (1987).
23. Ingham, R. E., *The Fungal Community, Its Organization and Role in the Ecosystem*, 2nd ed. (G. C. Carroll and D. T. Wicklow, eds.), Marcel Dekker, New York, pp. 669–690 (1992).
24. Bouwman, L. A., J. Bloem, P. H. J. F. van den Boogert, F. Bremer, G. J. H. Hoenderboom, and P. C. de Ruiter, *Biol. Fertil. Soils 17*:249 (1994).
25. de Ruiter, P. C., J. Bloem, L. A. Bouwman, W. A. M. Didden, G. H. J. Hoenderboom, G. Lebbink, J. C. Y. Marinissen, J. A. de Vos, M. J. Vreeken-Buijs, K. B. Zwart, and L. Brussaard, *Agric. Ecosyst. Environ. 51*:199 (1994).
26. Baker, R., *Can. J. Plant Pathol. 9*:370 (1987).
27. Becker, J. O., and F. J. Schwinn, *Pestic. Sci. 37*:355 (1993).
28. Deacon, J. W., and L. A. Berry, *Pestic. Sci. 37*:417 (1993).
29. Dreyfuss, M. M. and I. H. Chapela, *The Discovery of Natural Products with Therapeutic Potential* (V. P. Gullo, ed.), Butterworth-Heinemann, Boston, pp. 49–80 (1994).
30. Handelsman, J., and J. L. Parke, *Plant–Microbe Interactions: Molecular and Genetic Perspectives*, vol. 3 (T. Kosuge and E. W. Nester, eds.), McGraw-Hill, New York, pp. 27–61 (1989).

31. Rossman, A. Y., *Mycologia 88*:1 (1996).
32. Tkacz, J. S., *Encyclopedia of Microbiology*, vol. 1 (J. Lederberg, ed.), Academic Press, San Diego, pp. 331–337 (1992).
33. Malloch, D., K. A. Pirozynski, and P. H. Raven, *Proc. Natl. Acad. Sci. USA 77*:2113 (1980).
34. Malloch, D., *Can. J. Plant Pathol. 9*:398 (1987).
35. Pirozynski, K. A., and D. W. Malloch, *BioSystems 6*:153 (1975).
36. Barron, G. L., *Methods in Microbiology*, vol. 4 (C. Booth, ed.), Academic Press, London, pp. 405–427 (1971).
37. Bills, G. F., *Can. J. Bot. 73* (Suppl. 1):S33 (1995).
38. Bills, G. F., M. Christensen, M. J. Powell, and R. G. Thorn, *Assessing Fungal Biodiversity* (G. M. Mueller and A. Y. Rossman, eds.), Smithsonian Institution Press (in press).
39. Frankland, J. C., J. Dighton, and L. Boddy, *Methods in Microbiology*, vol. 22. *Techniques in Microbial Ecology* (R. Grigorova and J. R. Norris, eds.), Academic Press, London, pp. 343–404 (1990).
40. Gams, W., *Fungi in Vegetation Science* (W. Winterhoff, ed.), Kluwer Academic, Dordrecht, The Netherlands, pp. 183–223 (1992).
41. Kendrick, W. B., and D. Parkinson, *Soil Biology Guide* (D. L. Dindal, ed.), John Wiley & Sons, New York, pp. 49–68 (1990).
42. Parkinson, D., *Methods of Soil Analysis*, Part 2. *Microbiological and Biochemical Properties*, SSSA Book Series, No. 5. Soil Science Society of America, Madison, Wisc., pp. 329–350 (1994).
43. Parkinson, D., and D. C. Coleman, *Agric. Ecosyst. Environ. 34*:3 (1991).
44. Paul, E. A., and F. E. Clark, *Soil Microbiology and Biochemistry*, 2nd ed., Academic Press, San Diego (1996).
45. Tunlid, A., and D. C. White, *Soil Biochemistry*, vol. 7 (G. Stotsky and G. M. Bollag, eds.), Marcel Dekker, New York, pp. 229–262 (1992).
46. Christensen, M., *The Fungal Community* (D. T. Wicklow and G. C. Carroll, eds.), Marcel Dekker, New York, pp. 201–232 (1981).
47. Christensen, M., *Mycologia 81*:1 (1989).
48. Cooke, R. C., and J. M. Whipps, *Ecophysiology of Fungi*, Blackwell, Oxford (1993).
49. Dix, N. J., and J. Webster, *Fungal Ecology*, Chapman and Hall, London (1995).
50. Griffin, D. M., *Ecology of Soil Fungi*, Syracuse University Press, Syracuse (1972).
51. Hayes, A. S., *Sci. Prog. (London) 66*:25 (1979).
52. Heal, O. W. and J. Dighton, *Dev. Biogeochem. 3*:14 (1986).
53. Killham, K., *Soil Ecology*, Cambridge University Press, Cambridge (1994).

54. Parkinson, D., and J. S. Waide, eds., *The Ecology of Soil Fungi*, Liverpool University Press, Liverpool (1960).
55. Richards, B. N., *The Microbiology of Terrestrial Ecosystems*, Longman Scientific and Technical, Harlow, England (1987).
56. Swift, M. J., *Sci. Prog. (Oxford) 64*:175 (1977).
57. Swift, M. J., O. W. Heal, and J. M. Anderson, *Decomposition in Terrestrial Ecosystems*, Blackwell Scientific, Oxford (1979).
58. Tsuneda, A., *Fungal Morphology and Ecology: Mostly Scanning Electron Microscopy*, Tottori Mycological Institute, Tottori, Japan (1983).
59. Wainwright, M., *Trans. Br. Mycol. Soc. 90*:159 (1988).
60. Barron, G. L., *The Genera of Hyphomycetes from Soil*, Williams & Wilkins, Baltimore (1968).
61. Gams, W., *Supplement and Corrigendum to the Compendium of Soil Fungi*, IHW-Verlang. Eching, Germany (1993).
62. O'Donnell, K. L., *Zygomycetes in Culture*, Dept. of Botany, University of Georgia, Athens, Georgia (1979).
63. von Arx, J. A., *The Genera of Fungi Sporulating in Pure Culture*, J. Cramer, Vaduz, Germany (1981).
64. Watanabe, T., *Pictorial Atlas of Soil and Seed Fungi*, Lewis Publishers, Boca Raton, Fla. (1994).
65. Farr, D. F., G. F. Bills, G. P. Chamuris, and A. Y. Rossman, *Fungi on Plants and Plant Products in the United States*, American Phytopathological Society Press, St. Paul, Minn. (1989).
66. Rossman, A. Y., M. E. Palm, and L. J. Spielman, *A Literature Guide for the Identification of Plant Pathogenic Fungi*, American Phytopathological Society Press, St. Paul, Minn. (1987).
67. Martin, G. W., and C. J. Alexopoulos, *The Myxomycetes*, University of Iowa Press, Iowa City (1969).
68. Martin, G. W., C. J. Alexopoulos, and M. L. Farr, *The Genera of Myxomycetes*, University of Iowa Press, Iowa City (1983).
69. Raper, K. B., *The Dictyostelids*, Princeton University Press, Princeton, N.J. (1984).
70. Karling, J. S., *The Plasmodiophorales*, 2nd ed., Hafner Press, New York (1968).
71. Dick, M. W., *Straminipilous Fungi*, International Mycological Institute, Kew, England (in press).
72. Ali, N. M., and I. Kalyanasundaram, *Mycol. Res. 95*:885 (1991).
73. Amewowor, D. H. A. K., and M. F. Madelin, *FEMS Microbiol. Lett. 86*:69 (1991).
74. Margulis, L., J. O. Corliss, M. Melkonian, and D. J. Chapman, *Handbook of Protoctista*, Jones & Bartlett, Boston (1990).

75. Stamps, D. J., and F. J. Newhook, *Mycol. Pap. 162*:1 (1990).
76. van der Plaats-Niterink, A. J., *Stud. Mycol. 21*:1 (1981).
77. Sparrow, F. K., *Aquatic Phycomycetes*, University of Michigan Press, Ann Arbor (1960).
78. Karling, J. S., *Chytridiomycetarum Iconographia*, J. Cramer, Vaduz, Germany (1977).
79. Powell, M. J., *Mycologia 85*:1 (1993).
80. Morton, J. B., *Mycotaxon 32*:267 (1988).
81. Morton, J. B., and G. L. Benny, *Mycotaxon 37*:471 (1990).
82. Tzean, S. S., and G. L., Barron, *Can. J. Bot. 59*:1861 (1981).
83. Gams, W., *Persoonia 9*:381 (1977).
84. Benjamin R. K., *The Whole Fungus* (W. B. Kendrick, ed.), National Museums of Canada, Ottawa, pp. 573–621 (1979).
85. Schipper, M. A. A., and J. A. Stalpers, *Stud. Mycol. 25*:1 (1984).
86. Thurston, C. F., *Microbiology 140*:19 (1994).
87. Webster, J., *Introduction to Fungi*, 2nd ed., Cambridge University Press, Cambridge (1981).
88. Christensen, M., and D. E. Tuthill, *Advances in* Penicillium *and* Aspergillus *Systematics* (R. A. Samson and J. I. Pitt, eds.), Plenum Press, New York, pp. 195–209 (1985).
89. Pitt, J. I., *The Genus* Penicillium *and Its Teleomorphic States* Eupenicillium *and* Talaromyces, Academic Press, London (1979).
90. Ramirez, C., *Manual and Atlas of the* Penicillia, Elsevier Biomedical Press, Amsterdam (1982).
91. Bruns, T. D., T. J. White, and J. W. Taylor, *Annu. Rev. Ecol. Syst. 22*:525 (1991).
92. Berbee, M. L., and J. W. Taylor, *Exp. Mycol. 16*:87 (1992).
93. Samuels, G. J., and K. A. Seifert, *Annu. Rev. Phytopathol. 33*:37 (1995).
94. Blackwell, M., *Mycologia 86*:1 (1994).
95. Dennis, R. W. G. *British Ascomycetes*, rev. suppl. ed., J. Cramer, Vaduz, Germany (1981).
96. Breitenbach, J., and F. Kränzlin, *Fungi of Switzerland*, vol. 1. *Ascomycetes*, Verlag Mykologia, Lucerne (1981).
97. Szaniszlo, P. J., and J. L. Harris, *Fungal Dimorphism*, Plenum Press, New York (1985).
98. Christensen, M., and W. F. Whittingham, *Mycologia 57*:882 (1965).
99. Bills, G. F., and J. D. Polishook, *Nova Hedw. 57*:195 (1993).
100. de Hoog, G. S., and M. McGinnis, *Stud. Mycol. 30*:187 (1987).
101. Barnett, J. A., R. W. Payne, and D. Yarrow, *Yeasts: Characteristics and Identification*, 2nd ed., Cambridge University Press, Cambridge (1990).
102. Kreger-van Rij, N. J. W., ed., *The Yeasts: A Taxonomic Study*, 3rd ed., Elsevier, Amsterdam (1984).

103. von Arx, J. A., L. Rodriques de Miranda, M. T. Smith, and D. Yarrow, *Stud. Mycol. 14*:1 (1977).
104. Bond, R. D., *Modification of Soil Structure* (W. W. Emerson, R. D. Bond and A. R. Dexter, eds.), John Wiley, Chichester, pp. 285–288 (1978).
105. Singer, R., *The Agaricales in Modern Taxonomy*, 4th ed., Koeltz, Koenigstein, Germany (1986).
106. Moser, M., *Keys to Agarics and Boleti*, trans. S. Plant, Roger Phillips, London (1984).
107. Jülich, W., *Die Nichtblätterpilze, Gallertpilze und Bauchpilze: Aphyllophorales, Heterobasidiomycetes, Gastromycetes*, Kleine Kryptogamenflora, Band II b/1, Gustav Fischer Verlag, Stuttgart (1984).
108. Frankland, J. C., *Decomposer Basidiomycetes* (J. C. Frankland, J. N. Hedger and M. J. Swift, eds.), Cambridge University Press, London, pp. 241–261 (1982).
109. Chesters, C. G. C., *Trans. Br. Mycol. Soc. 32*:197 (1949).
110. Thorn, R. G., C. A. Reddy, D. Harris, and E. A. Paul, *Appl. Environ. Microbiol. 62*:4288 (1996).
111. Hutchison, L. J., *Can. J. Bot. 68*:2172 (1990).
112. Hagler, A. N., and D. G. Ahearn, *Int. J. Syst. Bact. 31*:204 (1981).
113. Summerbell, R. C., *Mycologia 77*:587 (1985).
114. Nawawi, A., J. Webster, and R. A. Davey, *Trans. Br. Mycol. Soc. 68*:59 (1977).
115. Sneh, B., L. Burpee, and A. Ogoshi, *Identification of* Rhizoctonia *Species*, American Phytopathological Society Press, St. Paul, Minn. (1991).
116. Khan, S. R., and J. W. Kimbrough, *Mycotaxon 15*:103 (1982).
117. Tsuneda, A., S. Murakami, L. Sigler, and Y. Hiratsuka, *Can. J. Bot. 71*:1032 (1993).
118. Coates, D., and A. D. M. Rayner, *New Phytol. 101*:173 (1985).
119. Dahlberg, A., and J. Stenlid, *New Phytol. 115*:487 (1990).
120. Dowson, C. G., A. D. M. Rayner, and L. Boddy, *J. Gen. Microbiol. 132*:203 (1986).
121. Dowson, C. G., A. D. M. Rayner, and L. Boddy, *New Phytol. 109*:423 (1988).
122. Dowson, C. G., P. Springham, A. D. M. Rayner, and L. Boddy, *New Phytol. 111*:501 (1989).
123. Frankland, J. C., A. D. Bailey, T. R. G. Gray, and A. A. Holland, *Soil. Biol. Biochem. 13*:87 (1981).
124. Thompson, W., and A. D. M. Rayner, *New Phytol. 92*:103 (1982).
125. Haselwandter, K., O. Bobleter, and D. J. Read, *Arch. Microbiol. 153*:352 (1990).
126. Read, D. J., *Frontiers in Mycology* (D. L. Hawksworth, ed.), CAB International, Wallingford, England, pp. 101–130 (1991).

127. Barron, G. L., *Can. J. Bot. 66*:2505 (1988).
128. Apinis, A. E., *Mycopathol. Mycol. Applic. 48*:93 (1972).
129. Orpurt, P. A., and J. T. Curtis, *Ecology 38*:628 (1957).
130. Christensen, M., *Ecology 50*:9 (1969).
131. Bissett, J., and D. Parkinson, *Can. J. Bot. 57*:1609 (1979).
132. Bissett, J., and D. Parkinson, *Can. J. Bot. 57*:1630 (1979).
133. Bissett, J., and D. Parkinson, *Can. J. Bot. 57*:1642 (1979).
134. Gochenaur, S. E., *Mycologia 70*:975 (1978).
135. Widden, P., *Can. J. Bot. 64*:1413 (1986).
136. Christensen, M., W. F. Whittingham, and R. O. Novak, *Mycologia 54*:374 (1962).
137. Gams, W., and K. H. Domsch, *Trans. Br. Mycol. Soc. 52*:301 (1969).
138. Gochenaur, S. E., *The Fungal Community* (D. T. Wicklow and G. C. Carroll, eds.), Marcel Dekker, New York, pp. 459–479 (1981).
139. Söderström, B. E., and E. Bååth, *Holarctic Ecol. 1*:62 (1978).
140. Wardle, D. A., and D. Parkinson, *Mycol. Res. 95*:504 (1991).
141. Widden, P., *Can. J. Bot. 64*:1402 (1986).
142. Brock, T. D., *Ecology of Microbial Communities* (M. Fletcher, T. R. G. Gray and J. G. Jones, eds.), Cambridge University Press, Cambridge, pp. 1–17 (1987).
143. Bååth, E., *Can. J. Bot. 66*:1566 (1988).
144. Kirby, J. J. H., J. Webster, and J. H. Baker, *Mycol. Res. 94*:621 (1990).
145. Barron, G. L., *Assessing Fungal Biodiversity* (G. M. Mueller and A. Y. Rossman, eds.), Smithsonian Institution Press, Washington D.C. (in press).
146. Frankland, J. C., *The Ecology and Physiology of the Fungal Mycelium* (D. H. Jennings and A. D. M. Rayner, eds.), Cambridge University Press, Cambridge, pp. 241–261 (1984).
147. Amman, R. I., W. Ludwig, and K. -H. Schleifer, *Microbiol. Rev. 59*:143 (1995).
148. Pugh, G. J. F., *Trans. Br. Mycol. Soc. 75*:1 (1980).
149. Grime, J. P., *Plant Strategies and Vegetation Processes*, John Wiley, Chichester (1979).
150. Winogradsky, S., *C.R. Acad. Sci. Paris 178*:1236 (1924).
151. Garrett, S. D., *New Phytol. 50*:149 (1951).
152. Tsuji, T., Y. Kawasaki, S. Takeshima, T. Sekiya, and S. Tanaka, *Appl. Environ. Microbiol. 61*:3415 (1995).
153. Bloem, J., M. Veninga, and J. Shepherd, *Appl. Environ. Microbiol. 61*:926 (1995).
154. Harris, D., Crop and Soil Sciences, Michigan State University, personal communication (1995).
155. Bingle, W. H., and E. A. Paul, *Can. J. Microbiol. 32*:62 (1986).

156. Smith, N. C., and D. P. Stribley, *Beyond the Biomass: Compositional and Functional Analysis of Soil Microbial Communities* (K. Ritz, J. Dighton and K. E. Giller, eds.), John Wiley, Chichester, pp. 49–55 (1994).

157. White, D. C., *Oikos 74*:177 (1995).

158. Sancholle, M., and Y. Dalpé, *Mycotaxon 49*:187 (1993).

159. Gardes, M., T. J. White, J. A. Fortin, T. D. Bruns, and J. W. Taylor, *Can. J. Bot. 69*:180 (1991).

160. Stahl, P. D., and M. J. Klug, *Appl. Environ. Microbiol. 62*:4136 (1996).

161. Olsson, P. A., E. Bååth, I. Jakobsen, and B. Söderström, *Mycol. Res. 99*:623 (1995).

162. Wood, D. A., *Biotechnol. Lett. 1*:255 (1979).

163. Matcham, S. E., B. R. Jordan, and D. A. Wood, *Appl. Microbiol. Biotech. 21*:108 (1985).

164. Seitz, L. M., D. B. Sauer, R. Burroughs, H. E. Mohr, and J. D. Hubbard, *Phytopathology 69*:1202 (1979).

165. Gessner, M. O., and A. L. Schmitt, *Appl. Environ. Microbiol. 62*:415 (1996).

166. West, A. W., W. D. Grant, and G. P. Sparling, *Soil Biol. Biochem. 19*:607 (1987).

167. Grant, W. D., and A. W. West, *J. Microbiol. Methods 6*:47 (1986).

168. Pietikäinen, J., and H. Fritze, *Soil Biol. Biochem. 27*:101 (1995).

169. Markkola, A. M., R. Ohtonen, O. Tarvainen, and U. Ahonen-Jonnarth, *New Phytol. 131*:139 (1995).

170. Swift, M. J., *Soil Biol. Biochem. 5*:321 (1973).

171. Eckblad, A., and T. Näsholm, *Plant Soil 178*:29 (1996).

172. Anderson, J. P. E., and K. H. Domsch, *Ark. Mikrobiol. 93*:113 (1973).

173. Anderson, J. P. E., and K. H. Domsch, *Can. J. Microbiol. 21*:314 (1975).

174. Stamatiadis, S., J. W. Doran, and E. R. Ingham, *Soil Biol. Biochem. 22*:81 (1990).

175. Landi, L., L. Badalucco, F. Pomare, and P. Nannipieri, *Soil Biol. Biochem. 25*:1771 (1993).

176. Hill, T. C. J., E. F. McPherson, J. A. Harris, and P. Birch, *Soil Biol. Biochem. 25*:1779 (1993).

177. Jenkinson, D. S., *J. Soil Sci. 17*:208 (1966).

178. Jenkinson, D. S., and D. S. Powlson, *Soil Biol. Biochem. 8*:209 (1976).

179. Vance, E. D., P. C. Brookes, and D. S. Jenkinson, *Soil Biol. Biochem. 19*:703 (1987).

180. Lee, C. C., R. F. Harris, J. D. H. Williams, D. E. Armstrong, and J. K. Syers, *Soil Sci. Soc. Am. Proc. 35*:82 (1971).

181. Paul, E. A., and R. L. Johnson, *Appl. Environ. Microbiol. 34*:263 (1977).

182. Amato, M., and J. N. Ladd, *Soil Biol. Biochem. 20*:107 (1988).

183. Mele, P. M., and M. R. Carter, *Can. J. Soil Sci. 76*:37 (1996).

184. Wardle, D. A., and A. Ghani, *Soil Biol. Biochem. 27*:821 (1995).
185. Newell, S. Y., *The Fungal Community: Its Organization and Role in the Ecosystem*, 2nd ed. (G. C. Carroll and D. T. Wicklow, eds.), Marcel Dekker, New York, pp. 521–561 (1992).
186. Eaton, R. A., and M. D. C. Hale, *Wood: Decay, Pests and Protection*, Chapman and Hall, London (1993).
187. Anderson, D. W., and E. A. Paul, *Soil Sci. Soc. Am. J. 48*:298 (1984).
188. Béguin, P., and J. -P. Aubert, *FEMS Microbiol. Rev. 13*:25 (1994).
189. Boominathan, K., and C. A. Reddy, *Handbook of Applied Mycology*, vol. 4, *Fungal Biotechnology* (D. K. Arora, R. P. Elander and K. G. Mukerji, eds.), Marcel Dekker, New York, pp. 763–822 (1992).
190. Buswell, J. A., *Handbook of Applied Mycology*, vol. 1, *Soil and Plants* (D. K. Arora, B. Rai, K. G. Mukerji, and G. Knudsen, eds.), Marcel Dekker, New York, pp. 425–480 (1991).
191. Crawford, R. L., *Lignin Biodegradation and Transformation*, John Wiley & Sons, New York (1981).
192. de Jong, E., J. A. Field, and J. A. M. de Bont, *FEMS Microbiol. Rev. 13*:153 (1994).
193. Eriksson, K -E., R. A. Blanchette, and P. Ander, *Microbial and Enzymatic Degradation of Wood and Wood Components*, Springer-Verlag, Berlin (1990).
194. Jeffries, T. W., *Biochemistry of Microbial Degradation* (C. Ratledge, ed.), Kluwer, Dordrecht, pp. 233–277 (1994).
195. Kirk, T. K., and R. L. Farrell, *Annu. Rev. Microbiol. 41*:465 (1987).
196. Webb, E. C., ed., *Enzyme Nomenclature 1992*, Academic Press, San Diego (1992).
197. Reddy, C. A., and T. D'Souza, *FEMS Microbiol. Rev. 13*:137 (1994).
198. Hatakka, A., *FEMS Microbiol. Rev. 13*:125 (1994).
199. Galliano, H., G. Gas, J. L. Seris, and A. M. Boudet, *Enzyme Microb. Technol. 13*:478 (1991).
200. Vares, T., M. Kalsi, and A. Hatakka, *Appl. Environ. Microbiol. 61*:3515 (1995).
201. Srinivasan, C., T. M. D'Souza, K. Boominathan, and C. A. Reddy, *Appl. Environ. Microbiol. 61*:4274 (1995).
202. Jiménez, M., A. E. González, M. J. Martinez, and B. E. Dale, *Mycol. Res. 95*:1299 (1991).
203. Otjen, L., R. Blanchette, M. Effland, and G. Leatham, *Holzforschung 41*:343 (1987).
204. Perry, C. R., M. Smith, C. H. Britnell, D. A. Wood, and C. F. Thurston, *J. Gen. Microbiol. 139*:1209 (1993).
205. Wood, D. A., *J. Gen. Microbiol. 117*:327 (1980).

206. Hsiang, T., R. L. Edmonds, and C. H. Driver, *Can. J. Bot. 67*:1262 (1988).
207. Rehman, A. U., and C. F. Thurston, *J. Gen. Microbiol. 138*:1251 (1992).
208. Stenlid, J., *Can. J. Bot. 63*:2268 (1985).
209. Stenlid, J., and A. D. M. Rayner, *New Phytol. 113*:245 (1989).
210. Lindeberg, G., *Ark. Botanik 33A(10)*:1 (1946).
211. Abbott, T. P., and D. T. Wicklow, *Appl. Environ. Microbiol. 47*:585 (1984).
212. Tanesaka, E., H. Masuda, and K. Kinugawa, *Mycologia 85*:347 (1993).
213. Arora, D., and D. K. Sandhu, *Proc. Plant Sci. Indian Acad. Sci. 94*:567 (1985).
214. Falcón, M. A., A. Rodriguez, A. Carnicero, V. Regaldo, F. Perestelo, O. Milstein, and G. de la Fuente, *Soil Biol. Biochem. 27*:121 (1995).
215. Jennings, D. H., *Nutrient Cycling in Terrestrial Ecosystems: Field Methods, Applications, and Interpretation* (A. F. Harrison, P. Ineson, and D. W. Heal, eds.), Elsevier Applied Science, New York, pp. 233–245 (1990).
216. Cairney, J. W. G., *Mycol. Res. 96*:135 (1992).
217. King, B., and J. Waite, *Int. Biodeterior. Bull. 15*:29 (1979).
218. Watkinson, S. C., E. M. Davison, and J. Bramah, *New Phytol. 89*:295 (1981).
219. Thompson, W., and L. Boddy, *New Phytol. 93*:277 (1983).
220. Buswell, J. A., and E. Odier, *CRC Crit. Rev. Biotechnol. 6*:1 (1987).
221. Cowling, E. B., and W. Merrill, *Can. J. Bot. 44*:1539 (1966).
222. Flanagan, P. W., and K. van Cleve, *Can. J. For. Res. 13*:795 (1983).
223. Flanagan, P. W., *Forest Ecosystems in the Alaskan Taiga* (K. van Cleve, F. S. Chapin III, P. W. Flanagan, L. A. Viereck, and C. T. Dryness, eds.), Springer-Verlag, Berlin, pp. 138–151 (1986).
224. Taylor, B. R., D. Parkinson, and W. F. J. Parsons, *Ecology 70*:97 (1989).
225. Entry, J. A., and C. B. Backman, *Can. J. For. Res. 25*:1231 (1995).
226. Hart, S. C., M. K. Firestone, and E. A. Paul, *Can. J. For. Res. 22*:306 (1992).
227. Dighton, J., and L. Boddy, *Nitrogen Phosphorus and Sulphur Utilization by Fungi* (L. Boddy, R. Marchant, and D. J. Read, eds.), Cambridge University Press, Cambridge, pp. 269–298 (1988).
228. Foster, J. R., and G. E. Lang, *Can. J. For. Res. 12*:617 (1982).
229. Yavitt, J. B., and T. J. Fahey, *Can. J. For. Res. 12*:745 (1982).
230. Edmonds, R. L., *Can. J. For. Res. 17*:499 (1987).
231. Lodge, D. J., *Aspects of Tropical Mycology* (S. Isaac, J. C. Frankland, R. Watling, and A. J. S. Walley, eds.), Cambridge University Press, Cambridge, pp. 37–57 (1993).
232. Graustein, W. C., K. Cromack, Jr., and P. Sollins, *Science 198*:1252 (1977).
233. Hintikka, V., K. Korhonen, and O. Näykki, *Karstenia 19*:58 (1979).

234. Horner, H. T., L. H. Tiffany, and G. Knaphus, *Mycologia 87*:34 (1995).
235. Merrill, W., and E. B. Cowling, *Phytopathology 56*:1083 (1966).
236. Edmonds, R. L., D. J. Vogt, D. H. Sandberg, and C. H. Driver, *Can. J. For. Res. 16*:822 (1986).
237. Levi, M. P., W. Merrill, and E. B. Cowling, *Phytopathology 58*:626 (1968).
238. Lilly, W. W., G. J. Wallweber, and S. M. Higgins, *Curr. Microbiol. 23*:27 (1991).
239. Worrall, J. J., and C. J. K. Wang, *Can. J. Microbiol. 37*:864 (1991).
240. Griffith, N. T., and H. L. Barnett, *Mycologia 59*:149 (1967).
241. Thorn, R. G., and G. L. Barron, *Science 224*:76 (1984).
242. Barron, G. L., and R. G. Thorn, *Can. J. Bot. 65*:774 (1987).
243. Barron, G. L., *The Fungal Community, Its Organization and Role in the Ecosystem*, 2nd ed. (G. C. Carroll and D. T. Wicklow, eds.), Marcel Dekker, New York, pp. 311–326 (1992).
244. Thorn, R. G., and A. Tsuneda, *Rept. Tottori Mycol. Inst. 30*:13 (1992).
245. Thorn, G. and A. Tsuneda, *Trans. Mycol. Soc. Jpn. 34*:449 (1993).
246. Hutchison, L. J., and G. L. Barron, *Can. J. Bot. 74*:735 (1996).
247. Aho, P. E., R. J. Seidler, H. J. Evans, and P. N. Raju, *Phytopathology 64*:1413 (1974).
248. Hendrickson, O. Q., *Can. J. For. Res. 21*:1299 (1991).
249. Jurgensen, M. F., M. J. Larsen, S. D. Spano, A. E. Harvey, and M. R. Gale, *For. Sci. 30*:1038 (1984).
250. Jurgensen, M. F., M. J. Larsen, R. T. Graham, and A. E. Harvey, *Can. J. For. Res. 17*:1283 (1987).
251. Lynch, J. M., and S. H. T. Harper, *J. Gen. Microbiol. 129*:251 (1983).
252. Veal, D. A., and J. M. Lynch, *Nature (London) 310*:695 (1984).
253. Hill, N. M., and D. G. Patriquin, *The Fungal Community, Its Organization and Role in the Ecosystem*, 2nd ed. (G. C. Carroll and D. T. Wicklow, eds.), Marcel Dekker, New York, pp. 783–796 (1992).
254. Abd-Alla, M. H., S. A. Omar, and A. M. Abdel-Wahab, *Folia Microbiol. 37*:215 (1992).
255. Hawksworth, D. L., *Bot. J. Linn. Soc. (London) 96*:3 (1988).
256. Blanchette, R. A., and C. G. Shaw, *Phytopathology 68*:631 (1978).
257. Thorn, R. G., and A. Tsuneda, Tottori Mycological Institute, Japan (unpublished).
258. Janzen, R. A., J. F. Dormaar, and W. B. McGill, *Soil Biol. Biochem. 27*:173 (1995).
259. Frankland, J. C., *J. Ecology 54*:41 (1966).
260. Frankland, J. C., *The Fungal Community* (D. T. Wicklow and G. C. Carroll, eds.), Marcel Dekker, New York, pp. 403–426 (1981).

261. Swift, M. J., *Organization of Communities Past and Present* (J. H. R. Gee and P. S. Giller, eds.), Blackwell Scientific, Oxford, pp. 229–253 (1987).
262. Swift, M. J., and O. W. Heal, *Microbial Communities in Soil* (V. Jensen, A. Kjöller, and L. H. Sorensen, eds.), Elsevier, London, pp. 115–131 (1986).
263. Bridges, E. M., *World Soils*, 2nd ed., Cambridge University Press, London (1978).
264. Fanning, D. S., and M. C. Balluff Fanning, *Soil: Morphology, Genesis, and Classification*, John Wiley & Sons, New York (1989).
265. Visser, S., and D. Parkinson, *Can. J. Bot. 53*:1640 (1975).
266. Struwe, S., and A. Kjöller, *Microbial Communities in Soil* (V. Jensen, A Kjöller, and L. H. Sorensen, eds.), Elsevier, London, pp. 149–162 (1986).
267. Robinson, C. H., J. Dighton, J. C. Frankland, and J. D. Roberts, *Soil Biol. Biochem. 26*:1053 (1994).
268. Bowen, R. M., and S. H. T. Harper, *Mycol. Res. 93*:47 (1989).
269. Domsch, K. H., and W. Gams, *Pilze aus Agrarböden* G. Fischer, Stuttgart (1970).
270. Bowen, R. M., and S. H. T. Harper, *Soil Biol. Biochem. 22*:393 (1990).
271. Rodriguez, A., M. A. Falcon, and J. Trojanowski, *Appl. Microbiol. Biotechnol. 45*:399 (1996).
272. Orth, A. B., D. J. Royse, and M. Tien, *Appl. Environ. Microbiol. 59*:4017 (1993).
273. Betts, W. B., R. K. Dart, and M. C. Ball, *Trans. Br. Mycol. Soc. 91*:227 (1988).
274. Bowen, R. M., *Soil Biol. Biochem. 22*:401 (1990).
275. Coxson, D. S., and D. Parkinson, *Soil Biol. Biochem. 19*:49 (1987).
276. Carreiro, M. M., and R. E. Koske, *Mycologia 84*:886 (1992).
277. Marx, D. H., *Annu. Rev. Phytopathol. 10*:429 (1972).
278. Dehne, H. W., *Phytopathology 72*:1115 (1982).
279. Tester, M., S. E. Smith, and F. A. Smith, *Can. J. Bot. 65*:419 (1987).
280. Harley, J. L., *Mycol. Res. 92*:129 (1989).
281. St. John, T. V., and D. C. Coleman, *Can. J. Bot. 61*:1005 (1983).
282. Saunders, F. E., B. Mosse, and P. B. Tinker (eds.), *Endomycorrhizas*, Academic Press, New York (1975).
283. Kottke, I., and F. Oberwinkler, *Trees 1*:1 (1986).
284. Bonfante-Fasolo, P., *Symbiosis 3*:249 (1987).
285. Allen, M. F., *The Ecology of Mycorrhizae*, Cambridge University Press, Cambridge (1991).
286. Peterson, R. L., and M. L. Farquhar, *Mycologia 86*:311 (1994).

287. Schenck, N. C., ed., *Methods and Principles of Mycorrhizal Research*, American Phytopathological Society, St. Paul, Minn. (1982).
288. Norris, J. R., D. J. Read, and A. K. Varma, eds. *Methods In Microbiology*, vol. 23, *Techniques for the Study of Mycorrhiza*, Academic Press, London (1991).
289. Brundrett, M. L. Melville and L. Peterson, *Practical Methods in Mycorrhizal Research*, Mycologue Publications, Waterloo, Canada (1994).
290. Currah, R. S., and C. Zelmer, *Rept. Tottori Mycol. Inst. 30*:43 (1992).
291. Agerer, R., ed., *Colour Atlas of Ectomycorrhizae*, Einhorn-Verlag, Schwäbische Gmünd, Germany (1987–1993).
292. Haug, I. and K. Pritsch, *Ectomycorrhizal Types of Spruce (Picea abies (L.) Karst.) in the Black Forest: a Microscopical Atlas*, Kernforschungazentrum Karlsruhe, Universität Tübingen, Germany (1992).
293. Hutchison, L. J., *Mycotaxon 42*:387 (1991).
294. Harley, J. L., and E. L. Harley, *New Phytol. 105* (Suppl.):1 (1987).
295. Duckett, J. G., and D. J. Read, *New Phytol. 129*:439 (1995).
296. Umata, H., *Mycoscience 36*:369 (1995).
297. Egger, K. N., *Can. J. Bot. 74*:773 (1996).
298. Rose, S. L., *Can. J. Bot. 58*:1449 (1980).
299. Capellano, A., B. Dequatre, G. Valla, and A. Moiroud, *Plant Soil 104*:45 (1987).
300. Valla, G., A. Capellano, R. Huguney, and A. Moiroud, *Plant Soil 114*:142 (1989).
301. Sequerra, J., A. Capellano, M. Faure-Raynard, and A. Moiroud, *Can. J. Bot. 72*:955 (1994).
302. Jülich, W., *Higher Taxa of Basidiomycetes*, J. Cramer, Vaduz, Germany (1981).
303. Pirozynski, K. A., and Y. Dalpé, *Symbiosis 7*:1 (1989).
304. Pirozynski, K. A., *Can. J. Bot. 59*:1824 (1981).
305. Simon, L., J. Bousquet, R. C. Levesque, and M. Lalonde, *Nature (London) 363*:67 (1993).
306. Read, D. J., *Microbial Ecology* (M. W. Loutit and J. A. R. Miles, eds.), Springer-Verlag, New York, pp. 324–328 (1978).
307. Leake, J. R., *New Phytol. 127*:171 (1994).
308. Cullings, K. W., T. M. Szaro, and T. D. Bruns, *Nature (London) 379*:63 (1996).
309. Hadley, G., *New Phytol. 69*:1015 (1970).
310. Finlay, R. D., and D. J. Read, *New Phytol. 103*:143 (1986).
311. Read, D. J., R. Francis, and R. D. Finlay, *Ecological Interactions in Soil* (A. H. Fitter, D. J. Read, and M. B. Lusher, eds.), Blackwell Scientific, Oxford, pp. 193–217 (1985).

312. Perry, D. A., T. Bell, and M. P. Amaranthus, *Ecology of Mixed-Species Stands of Trees* (M. G. R. Cannell, D. C. Malcolm and P. A. Robertson, eds.), Blackwell, London, pp. 152–178 (1992).

313. Read, D. J., and R. D. Finlay, "Mycorrhizas and inter-plant transfer of nutrients. I. Ectomycorrhizas," Proceedings of the 6th North American Conference on Mycorrhizae, Bend, Oreg., pp. 319–320 (1984).

314. Read, D. J., and R. Francis, "Mycorrhizas and inter-plant transfer of nutrients. II. Vesicular–arbuscular mycorrhizas," Proceedings of the 6th North American Conference on Mycorrhizae Bend, Oreg., pp. 321–323 (1984).

315. Perry, D. A., M. P. Amaranthus, J. G. Borchers, S. L. Borchers, and R. E. Brainerd, *Bioscience 39*:230 (1989).

316. MacArthur, R. H., and E. O. Wilson, *The Theory of Island Biogeography*, Princeton University Press, Princeton, N. J. (1967).

317. Kropp, B. R., *Can. J. For. Res 12*:428 (1982).

318. Molina, R., and J. M. Trappe, *New Phytol. 90*:495 (1982).

319. Dighton, J., *Experientia 47*:362 (1991).

320. Read, D. J., and D. P. Stribley, *Nature New Biol. 244*:81 (1973).

321. Read, D. J., *Can. J. Bot. 61*:985 (1983).

322. Kielland, K., *Biogeochemistry 31*:85 (1995).

323. Kielland, K., *Ecology 75*:2373 (1994).

324. Norkrans, B., *Symbolae Botanicae Upsalienses 2*:1 (1950).

325. Hutchison, L. J., *Mycologia 81*:587 (1989).

326. Dighton, J., E. D. Thomas, and P. M. Latter, *Biol. Fertil. Soils 4*:145 (1987).

327. Durall, D. M., A. W. Todd, and J. M. Trappe, *New Phytol. 127*:725 (1994).

328. Entry, J. A., P. K. Donnelly, and K. Cromack, Jr., *Biol. Fertil. Soils 11*:75 (1991).

329. McIlveen, W. D., and H. Cole, Jr., *Can. J. Bot 54*:1486 (1976).

330. Pirozynski, K. A., and D. W. Malloch, *Coevolution of Fungi with Plants and Animals* (K. A. Pirozynski and D. L. Hawksworth, eds.), Academic Press, London, pp. 227–246 (1988).

331. Trappe, J. M., and C. Maser, *Mycologia 68*:433 (1976).

332. Cázares, E., and J. M. Trappe, *Mycologia 86*:507 (1994).

333. Trappe, J. M., and D. L. Luoma, *The Fungal Community, Its Organization and Role in the Ecosystem*, 2nd ed. (G. C. Carroll and D. T. Wicklow, eds.), Marcel Dekker, New York, pp. 17–27 (1992).

334. Rothwell, F. M., and C. Holt, "Vesicular–arbuscular mycorrhizae established with *Glomus fasciculatus* spores isolated from the feces of cricetine mice," USDA Forest Serv. Res. Note NE-259, Broomall, Penn., (1978).

335. Kotter, M. M., and R. C. Farentinos, *Mycologia 76*:756 (1984).

336. Mason, P. A., F. T. Last, J. Pelham, and K. Ingleby, *Forest Ecol. Management 4*:19 (1982).
337. Dighton, J., and P. A. Mason, *Developmental Biology of Higher Fungi* (D. Moore, L. A. Casselton, D. A. Wood and J. C. Frankland, eds.), Cambridge University Press, Cambridge, pp. 117–139 (1985).
338. Last, F. T., J. Dighton, and P. A. Mason, *Trends Ecol. Evol. 2*:157 (1987).
339. Chu-Chou, M., and L. J. Grace, *Soil. Biol. Biochem. 20*:883 (1988).
340. Visser, S., *New Phytol. 129*:389 (1995).
341. Gibson, F., and J. W. Deacon, *Mycol. Res. 94*:166 (1990).
342. Hutchison, L. J., and Y. Piché, *Can. J. Bot. 73*:898 (1995).
343. Sagara, N., *Contrib. Biol. Lab. Kyoto Univ. 24*:205 (1975).
344. Sagara, N., *Can. J. Bot. 73* (Suppl. 1):S1423 (1995).
345. Old, K. M., and J. F. Darbyshire, *Soil Biol. Biochem. 10*:93 (1978).
346. Anderson, T. R., and Z. A. Patrick, *Soil Biol. Biochem. 12*:159 (1980).
347. Anderson, J. M., A. D. M. Rayner, and D. W. H. Walton, *Invertebrate–Microbial Interactions*, Cambridge University Press, Cambridge (1984).
348. Shaw, P. J. A., *The Fungal Community, Its Organization and Role in the Ecosystem*, 2nd ed. (G. C. Carroll and D. T. Wicklow, eds.), Marcel Dekker, New York, pp. 295–310 (1992).
349. Seyd, E. L., and M. R. D. Seaward, *Zool. J. Linn. Soc. (London) 80*:369 (1980).
350. Fogel, R., *Insect Mycophagy: A Preliminary Bibliography*, U.S. Dept. Agric. For. Serv. Gen. Tech. Rept. RNW-36 (1975).
351. Barron, G. L., *The Nematode-Destroying Fungi*, Canadian Biological Publications, Guelph, Ontario (1977).
352. Gray, N. F., *Biol. Rev. Camb. Philos. Soc. 62*:245 (1987).
353. Gray, N. F., *Diseases of Nematodes*, vol. 2 (G. O. Poinar and H. -B. Jansson, eds.), CRC Press, Boca Raton, Fla., pp. 3–38 (1988).
354. Nordbring-Hertz, B., *Adv. Microb. Ecol. 10*:81 (1988).
355. Dackman, C., H. -B Jansson, and B. Nordbring-Hertz, *Soil Biochemistry*, vol. 7 (G. Stotsky and J. M. Bollag, eds.), Marcel Dekker, New York, pp. 95–130 (1992).
356. Dijksterhuis, J., M. Veenhuis, W. Harder, and B. Nordbring-Hertz, *Adv. Microb. Physiol. 36*:111 (1994).
357. Drechsler, C., *Biol. Rev. 16*:265 (1941).
358. Barron, G. L., *Biology of Conidial Fungi*, vol. 2 (G. T. Cole and B. Kendrick, eds.), Academic Press, New York, pp. 167–200 (1981).
359. Bailey, F., and N. F. Gray, *Ann. Appl. Biol. 114*:125 (1989).
360. Barron, G. L., *Can. J. Bot. 68*:691 (1990).
361. Walter, D. E., and D. T. Kaplan, *Pedobiologia 34*:281 (1990).
362. Cooke, R. C., and B. E. S. Godfrey, *Trans. Br. Mycol. Soc. 47*:61 (1964).
363. Tucker, B. E., *Mycotaxon 13*:481 (1981).

364. Dayal, R., *Sydowia 27*:293 (1973).
365. Schenck, S., W. B. Kendrick, and D. Pramer, *Can. J. Bot. 55*:977 (1977).
366. van Oorschot, C. A. N., *Stud. Mycol. 26*:61 (1985).
367. Zhang, K. -Q., X. -Z. Liu, and L. Cao, *Mycosystema 7*:111 (1995).
368. Liu, X. -Z., and L. -Q. Zhang, *Mycol. Res. 98*:862 (1994).
369. Gams, W., *Neth. J. Plant Pathol. 94*:123 (1988).
370. Thorn, R. G., and G. L. Barron, *Mycotaxon 25*:321 (1986).
371. Carris, L. M., and D. A. Glawe, *Fungi Colonizing Cysts of* Heterodera glycines, Bull. Univ. Illinois Urbana-Champaign Agric. Expt. Stn. #786 (1989).
372. Drechsler, C., *Mycologia 29*:447 (1937).
373. Drechsler, C., *Phytopathology 31*:773 (1941).
374. Barron, G. L., and Y. Dierkes, *Can. J. Bot. 55*:3054 (1977).
375. Pfister, D. H., and M. E. Liftik, *Mycologia 87*:684 (1995).
376. Robb, E. J., and G. L. Barron, *Science 218*:1221 (1982).
377. Barron, G. L., *Mycologia 79*:877 (1987).
378. Liou, J. Y., and S. S. Tzean, *Mycologia 84*:786 (1992).
379. Tzean, S. S., and J. Y. Liou, *Phytopathology 83*:1015 (1993).
380. Hallenberg, N., *Mycol. Res. 94*:1090 (1990).
381. Scholler, M., and A. Rubner, *Microbiol. Res. 149*:145 (1994).
382. Giuma, A. Y., and R. C. Cooke, *Trans. Br. Mycol. Soc. 56*:89 (1971).
383. Heim, R., and R. G. Wasson, *Les Champignons Hallucigènes du Mexique*, Muséum National d'Histoire Naturelle, Paris (1959).
384. Parag, Y., *Israel J. Bot. 14*:192 (1965).
385. Hutchison, L. J., S. E. Madzia, and G. L. Barron, *Can. J. Bot. 74*:431 (1996).
386. Stadler, M., O. Sterner, and H. Anke, *Arch. Microbiol. 160*:401 (1993).
387. Kwok, O. C. H., R. Plattner, D. Weisleder, and D. T. Wicklow, *J. Chem. Ecol. 18*:127 (1992).
388. Stadler, M., A. Mayer, H. Anke, and O. Sterner, *Planta Medica 60*:128 (1994).
389. Alexander, M., *Annu. Rev. Microbiol. 25*:361 (1971).
390. Hanssen, H. -P., "Volatile metabolites from some Aphyllophorales," Aphyllophorales Symposium 1982 in Eisenstadt, Austria, pp. 29–35 (1988).
391. Anke, H., M. Stadler, A. Mayer, and O. Sterner, *Can. J. Bot. 73* (Suppl. 1):S932 (1995).
392. Weber, N. A., *Gardening Ants: the Attines*, American Philosophical Society, Philadelphia (1972).
393. Chapela, I. H., S. A. Rehner, T. R. Schulz, and U. G. Mueller, *Science 266*:1691 (1994).
394. Kermarrec, A., M. Decharme, and G. Febvay, *Fire Ants and Leaf-Cutting Ants* (C. S. Lofgren and R. K. VanderMeer, eds.), Westview, Boulder, Colo., pp. 231–246 (1986).

395. Heim, R., *Termites et Champignons*, Societé nouvelle des éditions Boubée, Paris (1977).
396. Fermor, T. R., *Trans. Br. Mycol. Soc. 80*:357 (1983).
397. Grant, W. D., T. R. Fermor, and D. A. Wood, *J. Gen. Microbiol. 130*:761 (1984).
398. Grant, W. D., L. L. Rhodes, B. A. Prosser, and R. A. Asher, *J. Gen. Microbiol. 132*:2353 (1986).
399. Grant, W. D., B. A. Prosser, and R. A. Asher, *J. Gen. Microbiol. 136*:2267 (1990).
400. Fermor, T. R., and D. A. Wood, *J. Gen Microbiol. 126*:377 (1981).
401. Atkey, P. T., and D. A. Wood, *J. Appl. Bacteriol. 55*:293 (1983).
402. Masaphy, S., D. Levanon, R. Tchelet, and Y. Henis, *Appl. Env. Microbiol. 53*:1132 (1987).
403. Sparling, G. P., T. R. Fermor, and D. A. Wood, *Soil Biol. Biochem. 14*:609 (1982).
404. Fermor, T. R., D. A. Wood, S. P. Lincoln, and J. S. Fenlon, *J. Gen. Microbiol. 137*:15 (1991).
405. Madzia, S. E., *Bacteria as a Nutrient Source for Fungi*, MSc. Thesis, University of Guelph, Guelph, Ontario (1992).
406. Tsuneda, A., and R. G. Thorn, *Can. J. Microbiol. 40*:937 (1994).
407. Grant, W. D., B. A. Prosser, and S. J. Wakefield, *J. Gen. Microbiol. 137*:287 (1991).
408. Imada, A., K. Nakahama, S. Igarashi, and M. Isono, *Archiv. Mikrobiol. 91*:41 (1973).
409. Jeffries, P., and T. W. K. Young, *Interfungal Parasitic Relationships*, CAB International, Cambridge, (1994).
410. Freitag, M., and J. J. Morrell, *Can. J. Microbiol. 38*:317 (1992).
411. Summerbell, R. C., *New Phytol. 105*:437 (1987).
412. Lifshitz, R., M. Dupler, Y. Elad, and R. Baker, *Can. J. Microbiol. 30*:1482 (1984).
413. Zhao, Z., and X. Guo, *Acta Microbiologica Sinica 29*:170 (1989).
414. Stahl, P. D., and M. Christensen, *Soil Biol. Biochem. 24*:309 (1992).
415. Rayner, A. D. M., L. Boddy, and C. G. Dowson, *FEMS Microbiol. Ecol. 45*:53 (1987).
416. Persson, Y., and E. Bååth, *FEMS Microbiol. Lett. 101*:11 (1992).
417. Robinson, C. H., J. Dighton, and J. C. Frankland, *Mycol. Res. 97*:547 (1993).
418. Tzean, S. S., and R. H. Estey, *Can. J. Microbiol. 24*:780 (1978).
419. Tzean, S. S., and R. H. Estey, *Phytopathology 68*:1266 (1978).
420. Owens, E. M., C. A. Reddy, and H. E. Grethlein, *FEMS Microbiol. Ecol. 14*:19 (1994).
421. Tsuneda, A., N. Maekawa, I. Ohira, and I. Furukawa, *Can. J. Bot. 69*:2792 (1991).

422. Rayner, A. D. M., *Ann. Appl. Biol. 89*:131 (1978).
423. Straatsma, G., J. P. G. Gerrits, M. P. A. M. Augustijn, H. J. M. Op Den Camp, G. D. Vogels, and J. L. D. Van Griensven, *J. Gen. Microbiol. 135*:751 (1989).
424. Fletcher, J. T., P. F. White, and R. H. Gaze, *Mushrooms: Disease and Pest Control*, Intercept, Newcastle upon Tyne (1989).
425. Stamets, P., and J. S. Chilton, *The Mushroom Cultivator*, Agarikon Press, Olympia, Wash., (1983).
426. Rainey, P. B., C. L. Brodey, and K. Johnstone, *Phys. Mol. Plant Pathol. 39*:57 (1991).
427. Holland, D. M., and R. C. Cooke, *Mycol. Res. 94*:789 (1990).
428. Tsuneda, A., and R. G. Thorn, *Can. J. Bot. 73*(Suppl. 1):S1325 (1995).
429. Hiorth, J., *Meddeleser Norske Skogsfors. 20*:251 (1965).
430. Thorn, R. G., and E. A. Paul, Center for Microbial Ecology, Michigan State University (unpublished).
431. Lynn, D. G., and M. Chang, *Annu. Rev. Plant Physiol. Plant Mol. Biol. 41*:497 (1990).
432. Appel, H., *J. Chem. Ecol. 19*:1521 (1993).
433. McDougall, D. N., and R. A. Blanchette, *Mycologia 88*:98 (1996).
434. Boddy, L., D. Coates, and A. D. M. Rayner, *Trans. Br. Mycol. Soc. 81*:149 (1983).
435. Rayner, A. D. M., and N. K. Todd, *Adv. Bot. Res. 7*:333 (1979).
436. Fekete, F. A., V. Chandhoke, and J. Jellison, *Appl. Environ. Microbiol. 55*:2720 (1989).
437. Bossier, P. M. Hofte, and W. Verstraete, *Adv. Microb. Ecol. 10*:385 (1988).
438. Rosenzweig, W. D., and D. Ackroyd, *Can. J. Microbiol. 30*:1437 (1984).
439. Safar, H. M., and R. C. Cooke, *Trans. Br. Mycol. Soc. 91*:73 (1988).
440. Malloch, D. W., and M. Blackwell, *The Fungal Community, Its Organization and Role in the Ecosystem*, 2nd ed. (G. C. Carroll and D. T. Wicklow, eds.), Marcel Dekker, New York, pp. 147–171 (1992).
441. Grewal, P. S., *Ann. Appl. Biol. 118*:47 (1991).
442. Grewal, P. S. and P. Hand, *J. Appl. Bacteriol. 72*:173 (1992).
443. Gilbert, D. G., *Oecologia 46*:135 (1980).
444. Cannon, P. F., and D. L. Hawksworth, *Adv. Plant Pathol. 11*:277 (1995).
445. Rossman, A. Y., *Biodiversity and Terrestrial Ecosystems* (C. -I. Peng and C. H. Chou, eds.), Institute of Botany, Academica Sinica Monogr. Ser. 14, pp. 169–194 (1994).
446. Janzen, D. H., and W. Hallwachs "All taxa biodiversity inventory (ATBI) of terrestrial systems," Report to NSF, Biodiversity and Biological Collections Gopher, Harvard University, huh.harvard.edu (1994).
447. Cannon, P. F., *Inoculum 46*(4):1 (1995).
448. Bills, G. F., and J. D. Polishook, *Mycologia 86*:187 (1994).

449. O'Donnell, A. G., M. Goodfellow, and D. L. Hawksworth, *Phil. Trans. Roy. Soc. London, Biol. 345*:65 (1994).
450. Freckman, D. W., ed., "Life in the soil: Soil biodiversity: its importance to ecosystem processes," Report of a workshop held at the Natural History Museum, London, England, Aug. 30–Sept. 1, 1994, Natural Resource Ecology Laboratory, Colorado State University, Fort Collins, Colo. (1994).
451. Beare, M. H., D. C. Coleman, D. A. Crossley, Jr., P. F. Hendrix, and E. P. Odum, *Plant Soil 170*:5 (1995).
452. Arnolds, E., *Trans. Br. Mycol. Soc. 90*:391 (1988).
453. Arnolds, E., *Nova Hedw. 48*:107 (1989a).
454. Arnolds, E., *Persoonia 14*:77 (1989b).
455. Arnolds, E., *Agric. Ecosyst. Environ. 35*:209 (1991).
456. Arnolds, E., *Nova Hedw. 55*:325 (1992).
457. Nauta, M. M., and E. C. Velinga, *Atlas van Nederlandse Paddestoelen*, A. E. Balkema, Rotterdam (1995).
458. Baar, J., *Ectomycorrhizal Fungi of Scots Pine as Affected by Litter and Humus*, Thesis, Landouwuniversiteit Wageningen, CIP-Data Koninklijke Bibliotheek, Den Haag, The Netherlands (1995).
459. Jaenike, J., *Trends Ecol. Evol. 6*:174 (1991).
460. Cherfas, J., *Science 254*:1458 (1991).
461. Goffinet, B., E. Sérusiaux, and P. Diederich, *Belg. J. Bot. 127*:184 (1994).
462. Alstrup, U., *Int. J. Mycol. Lichenol. 5*:1 (1992).
463. Ernst, G., *Ber. Bot. Vereins Hamburg 13*:100 (1993).
464. Gorshkov, V. V., *Aerial Pollution in Kola Peninsula* (M. V. Kozlov, E. Haukioja and V. T. Yarmishko, eds.), Apatity, St. Petersburg, pp. 290–298 (1993).
465. Newsham, K. K., A. R. Watkinson, H. M. West, and A. H. Fitter, *Functional Ecol. 9*:442 (1995).
466. van Dobben, H. F., *Nova Hedw. 37*:691 (1983).
467. de Vries, B. W. L., E. Jansen, H. F. van Dobben, and T. W. Kuyper, *Biodiversity Conservation 4*:156 (1995).
468. Orpin, C. G., *Anaerobic Fungi: Biology, Ecology, and Function* (D. O. Mountfort and C. G. Orpin, eds.), Marcel Dekker, New York, pp. 1–45 (1994).
469. Griffin, D. H., *Fungal Physiology*, 2nd ed., Wiley-Liss, New York (1994).
470. Gunner, H. B., and M. Alexander, *J. Bacteriol. 87*:1309 (1964).
471. Bollag, J. -M, and G. Tung, *Soil Biol. Biochem. 4*:271 (1972).
472. Shoun, H., D. -H. Kim, H. Uchiyama, and J. Sugiyama, *FEMS Microbiol. Lett. 94*:277 (1992).
473. Bumpus, J. A., and S. D. Aust, *Appl. Environ, Microbiol. 53*:2001 (1987).
474. Bumpus, J. A., M. Tien, D. Wright, and S. D. Aust, *Science 228*:1434 (1985).

475. Köhler, A., A. Jäger, H. Willershausen, and H. Graf, *Appl. Microbiol. Biotechnol. 29*:618 (1988).
476. Ryan, T. P., and J. A. Bumpus, *Appl. Microbiol. Biotechnol. 31*:302 (1989).
477. Michel, F. C., C. A. Reddy, and L. J. Forney, *Appl. Environ. Microbiol. 61*:2566 (1995).
478. Bumpus, J. A., *Appl. Environ. Microbiol. 55*:154 (1989).
479. Yadav, J. S., and C. A. Reddy, *Appl. Environ. Microbiol. 59*:756 (1993).
480. Bumpus, J. A., and B. J. Brock, *Appl. Environ. Microbiol. 54*:1143 (1988).
481. Aust, S. D., *Microb. Ecol. 20*:197 (1990).
482. Haider, K. M., and J. P. Martin, *Soil Biol. Biochem. 20*:425 (1988).
483. Lamar, R. T., M. W. Davis, D. M. Dietrich, and J. A. Glaser, *Soil Biol. Biochem. 26*:1603 (1995).
484. Gerrath, J. F., J. A. Gerrath, and D. W. Larson, *Can. J. Bot. 73*:788 (1995).
485. Harper, K. T., and J. R. Marble. *Application of Plant Sciences to Rangeland Management and Inventory* (P. T. Tueller, ed.), Kluwer Academic, Boston, pp. 135–169 (1988).
486. Johansen, J. R., *J. Phycol. 28*:139 (1993).
487. Schulten, J. A., *Am. J. Bot. 72*:1657 (1985).
488. Williams, J. D., J. P. Dobrowolski, N. E. West, and D. E. Gillette, *Trans. A.S.A.E. 38*:131 (1995).
489. Williams, J. D., J. P. Dobrowolski, and N. E. West, *Trans, A.S.A.E. 38*:139 (1995).
490. Johansen, J. R., and L. L. St. Clair, *Great Basin Naturalist 46*:632 (1986).

5

Ecology and Biology of Soil Protozoa, Nematodes, and Microarthropods

RICHARD D. BARDGETT University of Manchester, Manchester, United Kingdom

BRYAN S. GRIFFITHS Scottish Crop Research Institute, Dundee, United Kingdom

1. INTRODUCTION

The soil biota is composed of an extremely diverse and interactive assemblage of organisms which regulate processes of organic matter (OM) decomposition and nutrient turnover and alter physical properties of soil. In addition to soil microorganisms, including bacteria, fungi, actinomycetes, and algae, the soil biota contains populations of invertebrates whose feeding activities govern the functioning of decomposer microflora and regulate the flow of energy and nutrients within detritus food webs. Such regulation occurs in both natural and managed terrestrial ecosystems. This chapter considers the biological processes and ecological significance of the soil-dwelling invertebrate fauna, in particular the protozoa, nematodes, and microarthropods.

2. PROTOZOA

Soil protozoa are single-celled eukaryotic organisms (1), which are commonly subdivided into four groups: flagellates, naked amoebae, shelled or testate amoe-

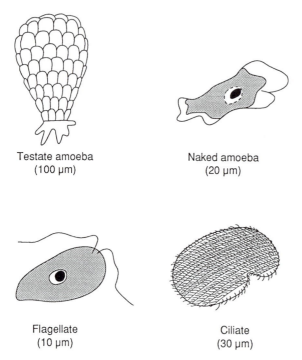

Testate amoeba
(100 μm)

Naked amoeba
(20 μm)

Flagellate
(10 μm)

Ciliate
(30 μm)

Figure 1.   Diagram of the four basic groups of soil protozoa.

bae (Testacea), and ciliates (2,3). The general morphological characteristics of these groups are shown in Fig. 1. As protozoa require a water film for locomotion and feeding, activity is limited to the water-filled pore space in soil. They can withstand drying of the soil and other adverse conditions by forming resistant cysts. Reproduction is usually asexual by binary fission, but sexual reproduction also occurs in all groups.

2.1   Ecological Importance

The biomass within a soil is dominated by the primary decomposers, bacteria and fungi, but of the invertebrate fauna protozoan biomass sometimes exceeds that of earthworms (Table 1). Biomass, however, is not a good indicator of activity or importance. Thus, while protozoa comprised a small percentage of the total biomass (Table 1) and their respiration losses were similarly a small percentage of the total soil CO_2 output (4,5), they consumed 22% of the total C input to a beech-forest soil and 63% of the standing microbial biomass (6). Respiratory energy losses by protozoa were 73% (7) and annual production

Table 1. The Biomass (kg C ha^{-1}) of Soil Invertebrates from Contrasting Sites

Site/reference	Protozoa	Nematodes	Earthworms	Others[a]
Beech forest[b] [6]	6.1	4.3	39	NA
Meadow[b] [7]	2.1	3.3	15	22
Wheat field[b] [7]	1.8	2.2	4.0	0.7
Barley field [9]	31	0.9	14	4.3
Mixed arable [29]	13	1.2	15	1.2
Mixed arable [33]	50	2.0	20	1.0

[a]Includes micro- and macro arthropods, enchytraeid worms, insect larvae.
[b]Calculated from data presented as milligrams per square meter (mg m^{-2}), assuming a C content of 0.4 g g^1.

39% to 98% of that of the total soil fauna (8), while their contribution to total net nitrogen mineralization was 12% to 30% (9–12). These figures show that protozoa play a substantial part in the energy and nutrient flows of terrestrial ecosystems.

The role of protozoa in soil systems has been summarized as regulation and modification of the size and composition of the microbial community; acceleration of the turnover of microbial biomass/soil organic matter/nutrients; direct excretion of nutrients; and inoculation of new substrates by the phoretic transport (i.e., microorganisms adhering to the surface of larger organisms) or excretion of viable microorganisms (13). Soil protozoa have been reviewed extensively (1,2,13–22,27–32) and general conclusions can be drawn from this information.

The importance of protozoa in terrestrial ecosystems results mainly from their feeding activities, because they are too small to influence the structure of the soil directly. They may, however, have an indirect effect on soil structure through their interactions with bacteria and fungi (33). The feeding habits of protozoa in soil are uncertain and the majority are considered to feed on bacteria, although mycophagous (i.e, fungal feeders such as *Thecamoeba granifera minor, Grossglockneria acuta*, and an unidentified flagellate) (19,34,35), predatory (i.e., those that eat other protozoa, such as *Didinium nasutum*–eating *Paramecium*) and saprophytic protozoa which absorb soluble compounds are also found (3). Protozoa are a key link in the terrestrial food chain as they are the major predators of bacteria in many soils and form a prey for larger organisms (see Chapter 9).

Protozoa have a direct and an indirect effect on energy and nutrient flows in soil. Their direct effect stems from the fate of ingested food; their indirect effects are covered in Chapters 6 and 9. Approximately 40% of the ingested nutrients

Table 2. Effect of C:N Ratio of Prey and Protozoa on the Fate of Ingested N, Following the Consumption of Prey Containing 100 mg C by Protozoa with an Ecological Growth Efficiency of 40%

C:N		N (mg) utilized for			Consumed N
Prey	Protozoa	Consumption	Production	Excretion	(%) excreted
3	3	33.3	13.3	20	60
5	3	20	13.3	6.7	34
10	3	10	10	—	—
3	5	33.3	8	25.3	76
5	5	20	8	12	60
10	5	10	8	2	20
3	10	33.3	4	29.3	88
5	10	20	4	16	80
10	10	10	4	6	60

From Ref. 13.

are used for the production of protozoan biomass, and 60% is excreted (31). The excreted nutrients form a direct flow of nutrients into the soil system, which are available for plant uptake and microbial growth. The extent of excretion is largely dependent on the quality of the substrate, which is usually considered in terms of its carbon (C)/nitrogen (N) ratio, as shown in Table 2. Experimental results have shown that nutrient regeneration was reduced when the marine flagellate *Paraphsomonas imperforata* was fed nutrient-limited prey (36), and nitrogen (N) excretion was correlated with the amount of bacterial N ingested by a freshwater flagellate (37). There was a time lag of 8 days before net N mineralization occurred when the soil ciliate *Colpoda steinii* was fed N-limited bacteria (38). The population of protozoa, and therefore the amount of nutrients they excrete, also increases as substrate quality (%N) increases (39).

2.2   Protozoan Ecological Characteristics

Variations of protozoan numbers and biomass are largely explicable in terms of resource quantity and quality. In woodland ecosystems, for example, there was a tendency for protozoan production to increase with increasing litter quality (4,8,40). It was concluded that production was dependent upon food supply and habitat structure (8). This conclusion supports laboratory observations that food supply is the major factor limiting protozoan populations in soil (41). A threshold amount of 400 μg C g^{-1} was required for a detectable increase in protozoan biomass in a clay loam soil (42). Soil structure is less important than food supply. Reduced pore size constrained the development of *Tetrahymena pyriformis* populations only when food was abundant (43), and soil structure was shown to influence protozoa indirectly through effects on soil moisture and aeration status (42). There was also no correlation between protozoa and pore size distribution in a range of grassland soils (44).

The impact of various environmental and land management factors on protozoa can be understood in terms of their effects on substrate supply (which is essentially equivalent to OM input) and, secondarily, on soil structure, moisture, and climatic factors. Thus, a sandy clay loam and a sandy loam soil had populations of protozoa (10^8 and 211×10^9 m^{-2}, respectively) which were significantly different per gram of soil carbon (19 and 88×10^6 g^{-1} C, respectively) (45). The differences were due to the quality of the OM (substrate supply), soil type, and environmental conditions.

The characteristic decrease in the number of protozoa with soil depth, from approximately 10^6 g^{-1} at 10 cm to 101 g^{-1} below 1 m, is likely to result from decreasing levels of OM down the soil profile (46). Protozoa can survive at considerable depths and viable cells have been recovered from 7.4 m below the surface (47), where occurrence may be linked to particular parts of the soil structure as no protozoa were recovered from clay-rich soil fractions (48). A transition from grassland (permanent plant cover) to arable land (annual cultivation) reduced the energy input to the soil and, consequently, protozoan biomass (7). Different methods of cultivation, however, tend not to affect numbers of protozoa, as there were no differences between shallow and deep ploughing (49), conventional tillage (i.e., with soil inversion by rotovation or plowing), or no-tillage (i.e., soil undisturbed apart from direct drilling of seed) (33,50), conventional (i.e., reliance upon inorganic fertilizers and routine use of pesticides), or integrated farming (i.e., reliance on organic manures and reduced pesticide use) (51), and stubble mulch or no-tillage (52), for example. Similar protozoan populations in four different arable cropping systems in Sweden were attributed to similar inputs of OM to the different systems (46).

The application of organic manures to soil is an obvious source of added substrate and consequently increases protozoan populations (53–55). Plant residues decomposing on the soil surface were colonized by fungi and had significantly lower populations of protozoa than buried residues, which were colonized by bacteria (50). The application of inorganic fertilizers generally increased protozoan populations, because of the increased plant growth, which led to a greater OM input to the soil (53), and because of the greater bacterial populations in the rhizosphere (56). The application of elemental sulfur to a Canadian pasture, however, increased acidification of the soil from pH 5.7 to 4.7 and so reduced numbers of soil microorganisms, particularly fungi and mycophagous protozoa (57).

2.3 Distribution of Protozoan Populations

Protozoa in soil are characterized by rapid changes in population density. Amoebal numbers increased 5- to 10-fold within 1 week (5), and there were equally large increases after rain caused by a stimulation of bacterial prey (58,59). Protozoan and bacterial populations were correlated on the majority of occasions when measured every day for a whole year (60), revealing characteristic predator–prey relationships between protozoa and bacteria (61,62). These marked short-term changes make it questionable to present data on seasonal trends by connecting recorded values based on samplings separated by long intervals (46). The seasonal maxima sometimes observed can also be related to food supply, such as the spring and autumn peaks of *Dictyostelium mucoroides* (62) and the peak of testacea following autumn leaf fall (4). Population minima are more likely to be related to unfavorable environmental conditions, such as the low numbers of protozoa observed during dry summers (46) and periods of low temperature.

There may be rapid changes in protozoan activity as well as total numbers. Under changing environmental conditions, particularly drying/wetting of the soil, protozoa survive in resistant structures termed cysts; they can rapidly develop cysts (encyst) and equally rapidly emerge from them (excyst) (63,64). In a desert soil, for example, 100% of the protozoa were active 6 days after watering, compared to 0% before (65). On decomposing plant residues in soil the total number of protozoa rapidly reached a peak, but numbers declined only slowly (66–68), probably because of the survival of protozoan cysts.

Protozoa have a clumped horizontal distribution (21) and maxima are again linked to substrate availability. Population sizes in the rhizosphere are on the average six times higher than those in the surrounding soil (range 1–35 times higher) (32) and also vary with plant age and species, and nutrient status of the soil (31,32,56,69). Protozoa tend to be more numerous at sites of decomposing OM than in the rhizosphere (31) because of increased concentrations and absolute amounts of substrate. A 90-fold increase on decomposing grass residues in soil, for example, has been demonstrated (68). Patches of OM are potentially capable of supplying a large proportion of mineralized N in soil, and both protozoa and nematodes are directly involved in that process (13).

Changes in numbers or biomass are only a gross indication of changes in protozoan populations. Alterations in the species composition of the population can be used to monitor changes in the environment or differences between environments (1). Ciliates (i.e., *Colpoda steinii*), for example, require a much higher soil moisture to be active than flagellates (i.e., *Cercobodo crassicauda*) or amoebae (i.e., *Vahlkampfia aberdonica*) (29,64,70). Naked amoebae are generally the most important group of protozoa in soil (42,58,59), although larger biomasses of flagellates (55) and ciliates (42) have been recorded. There can be marked differences in the species of protozoa present in different habitats, although some are universally distributed (71–73). Irrigation of a spruce forest soil had no effect on numbers of protozoa but increased the number of species present, indicating that certain species required the higher soil moisture content normally only found after rain (74). An interesting development has been the observation that, among the ciliates, Colpodea (C) such as *Colpoda aspera* are more r-selected (r-selection = selection of early colonizers that respond to substrate availability with high growth rate) than Polyhymenophora (P) such as *Blepharisma undulans* and the C:P ratio indicates the level of disturbance of a biotope (75). This finding was used to demonstrate that the initial recolonization of sterilized soil was by r-strategists (initial colonizers with a high growth rate) (76). An increase in numbers and species of K-selected (species with low growth rates and long-term survival characteristics) ciliates led to the conclusion that organic fertilizers improved soil conditions more than mineral fertilizers (77). The concept of protozoa as environmental bioindicators has been adapted for their use in laboratory bioassays. These include a toxicity test using soil ciliates (*Colpoda cucullus* and *Blepharisma undulans*) and a flagellate (*Oikomonas termo*) for hebicides (Chlorex, MCPA, dichlorprop, Matrigon), fungicides (Benlate), and insecticides (Sumicidin) (78) and a bioassay for heavy metals using the soil ciliate *Colpoda steinii* (79). This latter test showed that the test population of 9.8×10^4 *C. steinii* ml^{-1} surviving in soil extract from uncontaminated soil was significantly reduced to 6.1×10^4 ml^{-1} in an extract from nickel-contaminated soil.

2.4 Methods for Studying Protozoa

The study of protozoa in soil is hampered by their relatively small size, low numbers, and variable morphological characteristics. Bacteria, although smaller than protozoa, occur at 10^9 g^{-1} soil and can be counted directly with a microscope, with a ×40 or ×100 magnification objective lens; often, soil dilutions in which most of the soil particles have been diluted out (80) can be used. Nematodes, although only occurring at 10^1 or 10^2 g^{-1} soil, are large enough (25 × 1500 mm) to be extracted and counted directly from soil. Protozoa, however, at 10^4 g^{-1} soil, can only be observed directly with difficulty because the dilutions required to have protozoa present still contain a great many soil particles.

Procedures for extraction of protozoa from soil have not been fully exploited and culturing techniques are commonly used. Direct enumeration has been most successfully used for testate amoebae, because the test (structure in which the protozoan lives) provides a definite means of identification. Differential staining can distinguish living (cytoplasm-filled) from dead/empty test results. Variations on a filtration technique provide practical means to measure directly the abundance, activity, and species of testate amoebae (81,82). Soil ciliates have also been observed directly, either as active organisms in a soil suspension (83) or with a filtration technique (84). The direct observation of other protozoa, in either agar films (85) or soil smears (86,87), is hindered by the masking of animals by soil particles. A technique to remove soil particles by density-gradient centrifugation (88) allowed the direct observation of metabolically active protozoa. The application of more discriminating centrifugation procedures (89), possibly coupled with concentration protocols such as electromigration (in which protozoa are attracted to an electrode) (90), may further improve the direct observation of soil protozoa. An important objective for soil protozoologists would be the measurement of in situ bacterivory (consumption of bacteria), as has been done for aquatic protozoa (91,92).

Culturing techniques are based on the ability of protozoa to grow in a culture of food microorganisms, such as bacteria or fungal spores. Generally, a suspension of soil is serially diluted by using the dilution or most probable number (MPN) technique (93,94). The dilutions are monitored for the appearance of protozoa and the dilution at which there is no growth of protozoa used to calculate the original population. An MPN technique for protozoa using agar plates was originally developed (95) and later modified by the use of glass rings embedded in the agar surface (96). An adaptation was to observe zones of amoebal growth in a semisolid agar (97). The MPN technique has been reduced to a microscale method by utilizing commercially available microtiter plates, and this method has become widely used with many variations (98). The dilutant can be a nonspecific general growth medium, such as diluted nutrient broth or tryptone soya broth (56,58,98), and the soil microorganisms which develop serve as food for the protozoa. Alternatively, the dilutions can be made in a saline solution and specific food organisms, *Pseudomonas fluorescens*, for example, added to the wells for the protozoa to feed on (51,64). Mycophagous protozoa (i.e., protozoa which eat fungi) can also be enumerated and isolated by using a suspension of fungal conidia (34,99,100). It should be noted that both the food microorganism (101) and the growth medium (102) used can affect the MPN determination. Mathematical simulation of the appearance of protozoa in the dilutions has been used to study the structure of the protozoan community (103).

Pretreatment of soil samples prior to dilution has been used to distinguish active protozoa (trophozooites) from resting structures (cysts). The protozoa which grow in a dilution originate from both trophozooites and cysts and yield an estimate of the total protozoan population. In a dilution after pretreatment (designed to kill trophozooites) protozoa only develop from cysts, so the difference between the two dilution series estimates the population of trophozooites. Hydrogen chloride (HCl) (2%) has been used to destroy trophozooites (95) but overestimated the degree of activity by destroying some of the cysts; therefore, procedures based on freezing, desiccation, and heating have been used (64,104,105). Culturing techniques are also used to estimate the numbers of protozoan species (1,2,106) rather than simply total abundance.

Culturing methods, despite their flexibility and widespread use, underestimate the true population of protozoa. Studies in aquatic systems have shown that, compared with direct observation, the MPN technique only accounted for 1% of the flagellate population (107). This is similar to the differences between cultural and direct counts of soil bacteria (108). The estimation of protozoan biomass from total numbers is important in the calculation of nutrient fluxes, rates of transfer of carbon, nitrogen, and phosphorous, in particular, through the soil fauna (see Chapter 9). The size of the biomass of species within a class of protozoa can vary by up to three orders of magnitude (109), yet the most practical approach is to assume an average size of flagellates, ciliates, and amoebae for calculations. Typically used figures are 50 mm³, 3000 mm³, and 400 mm³, respectively (14), although lower (14 mm³ per flagellate and 180 mm³ per amoeba [51]) and higher estimates (600 mm³ per flagellate and 2250 mm³ per amoeba, calculated from [50]) are used.

Specific biochemical markers are used for the estimation of bacteria and fungi (110), and particular fatty acids, indicative of certain microbial groups, have been identified in soil (111,112). Phosphatidyl choline (PC) is a marker for protozoa in the rumen of sheep (*Ovis aries*) (113); its applicability for soil is uncertain, because it is common in other eukaryotic organisms, such as fungi, yeasts, and algae, as well as some gram-negative bacteria (114). Protozoan glycolipids might eventually yield useful taxonomic markers (114). Another method involves the determination of a β-N-acetyl-glucosaminidase–like enzyme (which catalyzes the glycolysis of N-acetylglucosamine, a component of bacterial cell walls), which has been identified as a putative marker for protozoan bacterivory (115). Molecular techniques, using specific nucleic acid probes, have been used for the study of parasitic protozoa (116,117) and it certainly is only a matter of time before they are used for soil protozoa.

3. NEMATODES

Nematodes are a phylum of aquatic roundworms (24). Like protozoa they require a water film around soil particles in which to move, feed, and reproduce. They have therefore similarly evolved mechanisms to survive unfavorable conditions (cryptobiosis, a complete cessation of metabolism, and dormant juvenile stages) (118). Nematodes in soil may be grouped according to the type of food that they consume, which is based on the morphological characteristics of their mouthparts (Figs. 2 and 3). As many as eight categories are recognized, feeding on plants, hyphae, bacteria, organic matter, fauna, eukaryotic cells, as well as infective stages of animal parasites (many of which have a soil-dwelling phase to allow transmission between hosts) and omnivores (which ingest a diverse range of materials) (119). Reproduction is either sexual or asexual and there are four

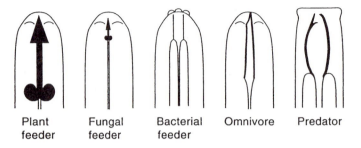

| Plant | Fungal | Bacterial | Omnivore | Predator |
| feeder | feeder | feeder | | |

Figure 2.   Schematic mouthparts of the nematode trophic groups most commonly found in soil.

juvenile stages between egg and adult in the nematode life cycle. These show considerable differences in metabolic rate and trophic status (23,120). The juvenile stages of some predatory nematodes, for example, are bacterial feeders (120). Feeding by nonplant-parasitic (non-stylet-bearing) nematodes is thought to be indiscriminate, with ingestion only limited by particle size (23,121). Any food specificity, such as the better growth of the bacterial-feeding *Caenorhabditis elegans* on *Enterobacter cloacae* than on *Bacillus* sp. (122), is thought to oc-

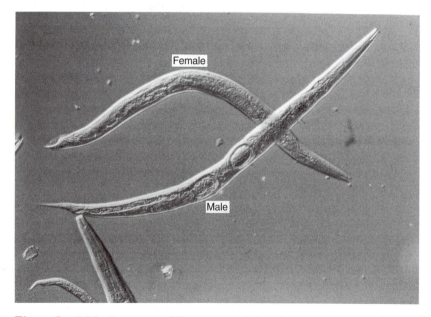

Figure 3.   Male (bottom) and female (top, 1.5 × 12 mm) bacteria-feeding nematodes (*Caenorhabditis* sp) from decomposing grass residues in soil.

cur during the digestive process (121). Plant-parasitic nematodes have received considerable attention because of their economic importance. They also differ from the other trophic groups of nematodes in that they are part of the grazing food chain (i.e., plant–herbivore), rather than the detritus food chain (dead organic matter–microorganism–microbial feeding fauna), in the sense of Odum (123). However, through excretory products, cadavers, and the induction of premature root decomposition, strictly plant-parasitic nematodes do contribute to below-ground nutrient cycling (120). This chapter will concentrate on the microbial-feeding nematodes, while nematodes at higher trophic levels will be dealt with in Chapters 6 and 9.

The role of nematodes in soil systems has been summarized as (1) the regulation and modification of the size and composition of the microbial community, (2) the acceleration of the turnover of microbial biomass/soil organic matter/nutrients, (3) the direct excretion of nutrients, and (4) the inoculation of new substrates by the phoretic transport (i.e, microorganisms adhering to the surface of larger organisms) or excretion of viable microorganisms (13). Soil nematodes have been reviewed extensively (17,18,20,22–26,28–32) and general conclusions can be drawn from this information.

3.1 Ecological Importance

In terms of biomass, nematodes are generally less abundant than protozoa (Table 1), contributing 1% or less to total soil respiration (26,124). Consumption by nematodes is generally lower than that by protozoa, with microbial feeders consuming 5% to 8% of OM input and 5% to 25% of the bacterial standing crop (51,125,126), while their contribution to total net N mineralization was 4% to 22% (9,10,12), which exceeded that of protozoa in a North American grassland (11).

Direct comparison of nematode and protozoan biomass is misleading because of differences in their ecological growth efficiency (production/consumption). This is approximately 40% for protozoa (31), but only about 10% for nematodes (124,127). This means that nematodes excrete more nutrients per unit biomass than protozoa (128), and they can be as important as protozoa in nutrient mineralization (13). The effect of resource quality on nematodes has been studied in greater detail than that on protozoa. Thus, nematode numbers increase more on residues with a low C:N ratio than on material with high C:N ratios (37,129,130). There was also a positive correlation between the amount of added nutrient (particularly N) and nematode numbers (131).

Primary productivity has generally correlated well with total nematode numbers (26), although this correlation does not always hold in forest ecosystems (132), an effect which may be related to litter quality. Substrate availability is of prime importance for nematodes, as it is for protozoa. Nematode abundance in a beech (*Fagus silvatica*) forest, for example, was influenced more by litter fall than climate (133). In arable cropping systems, nematode numbers are markedly reduced by a prolonged fallow period, because of minimum OM input (134,135), and are correlated with root density (136).

3.2 Nematode Ecological Characteristics

Nematodes appear to be more sensitive to edaphic factors than protozoa. In a survey of Dutch grasslands, for example, nematode biomass was correlated with the abundance of soil pores (spaces between soil particles) 30–90 μm in diameter, and the contribution of nematodes to the total soil biomass was greater in sandy soils (0.6% of biomass C) than in loamy soils (0.3%) or clay soils (0.1%) (44). This sensitivity is related to the movement of nematodes through soil pores (137) and the limited width range of soil nematodes (30–100 μm, calculated from [138]). Soil pores bare the channels between solid components of soil, such as sand grains, clay platelets, and organic matter. Protozoa, as a group, are much more variable in size, and their ability to change dimension (particularly amoebae) means that they are less affected by soil structure than nematodes. Thus, while there were no differences in numbers of protozoa (see previous discussion), there were more nematodes in a sandy clay loam (30×10^5 m^{-2}) than in a sandy loam (7×10^5) (45). The addition of ionic polyacrylamide soil conditioners, which alter the pore size distribution and aggregation properties of soil, had variable effects, depending on the parent soil. Thus, conditioners increased nematode numbers in sandy soils and decreased numbers in clay soils (139).

Site (edaphic and climatic factors) was most important in determining the nematode fauna of grazed pastures (140). Nematodes tend to decrease with depth, either down the soil profile (141,142), or down the different layers in forest litter (143). The nature of the nematodes, such as the proportions of gravid (egg-bearing) females and juveniles (sexually immature stages) and their size, indicated that there was an increasing food shortage with depth (144). Nematode numbers decreased from forest > meadow > prairie, although there were more nematodes in grassland than in adjacent woodlands (142). Woodlands consistently favored bacteria-feeders while grasslands had relatively more plant-feeders (142,144). There was a general finding that perennial grasslands supported a larger nematode population than did annual crops (7,125,136,144,146,147).

Resource and environment interactions determine the effects of different management practices. For example, increased grazing intensity by sheep reduced numbers of nematodes; that reduction related to a 64% decrease in litter (OM) input and changes in living space and microclimate (148). However, there were more nematodes on a grazed than on an ungrazed annual grassland, where the controlling factors were temperature and moisture (141). There was a greater population of bacteria-feeding nematodes under North American prairie-dog colonies than in the surrounding prairie (149). Grazing of the plant cover increased soil temperature and moisture content, while the input of extra N from excreta increased the supply of microorganisms. Mowing of a tallgrass prairie in the United States increased populations of microbe-feeders and reduced that of plant-feeders, in line with the increased input of detrital root material (150). Nematode numbers do not appear to differ between monocultures and varied rotations of crop plants (146,151). Different methods of cultivation had varying effects on nematodes. Tillage increased fungus-feeders, compared to a no-till treatment, but had no effect on bacteria-feeders (50) and microbe-feeders were favored by an integrated farming system compared to conventional practices (152). Plowing of a 5-year grass ley tended to increase nematodes and corresponded with an increase in microbial production, but there was only a small effect on the types of nematodes present; this was probably due to their being already adapted to disturbance (153). The application of organic manures generally increased nematode numbers, through increased substrate, plant growth, and altered soil conditions (55,134,154,155,157). The nature of the OM was important, as farm-yard manure had no effect on nematode populations in its first year of application while poultry manure significantly increased total numbers of nematodes and the proportion of bacteria-feeders (158). Inorganic fertilizers had a more variable effect, with reports of both reduced (143,159) and enhanced nematode populations (129,136,160–162). Inorganic fertilizers alter the pH balance of the soil, potentially affecting nematodes.

Experimental acidification of Norway spruce (*Picea abies*) and Scots pine (*Pinus silvestris*) forests generally reduced nematode numbers after 5 months, but numbers had recovered 7 years after the treatment (163). Liming of Scots pine forests reduced total numbers and altered the nematode community structure (164), although numbers of bacteria-feeders were increased in another study (162).

Seasonal changes in nematode numbers are related to both substrate and climatic factors. There was a strong association between nematode numbers and herbage production in New Zealand pastures, suggesting that N might be more important than temperature or moisture (165). In more arid climates temperature and moisture seem to be the controlling factors (141), although seasonal fluctuations in moisture also affect plant production (166). Nematode numbers were increased when pasture was irrigated; in dry pastures, numbers were correlated with temperature and moisture (26). Numbers were also related to moisture in a Scots pine forest (161). Interestingly, nematode populations were more strongly related to previous environmental conditions than to ambient conditions at the time of sampling (167).

3.3 Distribution of Nematode Populations

Nematode numbers in soil associated with decomposing OM can change as rapidly as protozoan numbers, in part because of their ability to move relatively long distances through soil, whereas protozoa tend not to migrate (41,64,68).

Thus, nematode populations can increase faster than their natural generation time would allow during the initial stages of decomposition (68,168). The rapid decline in nematode numbers after the population maximum has been reached (68,169–171) results from both emigration away from the active site (157) and death of nematodes, which probably occurs as resources become limiting. Emigration encourages the colonization of new resources and the spread of microorganisms associated with the nematodes (172), as discussed in the next chapter.

Some nematodes enter a resistant state to withstand adverse environmental conditions. Most work has been done on the anhydrobiotic state resistant to drought, when nematodes have no detectable metabolism and are characteristically coiled but can rapidly, within 24 hours, resume normal activity (173). Abiotic controls on nematodes and protozoa have been further reviewed (174,175).

The rhizosphere contains 3 to 70 times as many nematodes as the surrounding bulk soil, depending on plant and soil factors (13). The rhizosphere tends to favor plant-feeders rather than microbe-feeders (176), in much the same way grassland soils contain a majority of plant-feeders, as discussed. The biomass of bacteria-feeding nematodes in the rhizosphere exceeded that of protozoa (56), although rhizosphere nematodes may be deleterious to plant growth overall, as judged by increased plant growth following the application of nematicides (177,178). More bacteria-feeders were found on dead roots than on live roots (166,179), presumably reflecting the greater input of resources from decaying roots than from root exudation. This is also reflected in the generally higher populations of nematodes associated with decomposing OM than with the rhizosphere (13). The factors determining the relative sizes of the protozoan and nematode populations are unclear. In one experiment, the relative importance of nematodes was greatest at the moisture tension at which their movement was also greatest (Griffiths et al., unpublished data), suggesting that edaphic factors may be important.

3.4   Nematodes as Environmental Indicators

The use of nematodes as environmental bioindicators is more advanced than that of protozoa, probably because of their relative ease of extraction and identification. Nematodes can be used as indicators at several levels. The biomass of bacteria- and fungus-feeders, for example, indicates the preceding amount of microbial production (54,180) and is a better indicator of bacterial production than is bacterial biomass (181). The relative contribution of bacteria-feeding nematodes belonging to the families Cephalobidae and Rhabditidae is variable (13), with cephalobid nematodes (for example, *Acrobeloides* sp.) more tolerant of drier soil conditions than rhabditid nematodes like *Caenorhabditis* sp. There is a potential to use the ratio of these different groups to infer soil conditions, but more needs to be discovered about their biological characteristics. There are seasonal (drought-related) changes in the abundance of many nematode species which, in a conceptual model of grassland nematodes, can be related to food resources (182). Nematode community structure can be used to assess changes due to pollution, for instance, as a result of tributyl tin contamination (183).

An ecological measure of environmental disturbance can be given by the so-called maturity index, based on the species composition of non-plant-feeding nematodes (184). This has been used to characterize disturbed soils (185) and further modified to include plant parasites (186). The limitations of such indices, including biogeographic, pollution, climatic, and edaphic factors (186), are becoming evident. In a study of different meadows in Poland, nematode trophic structure and generic diversity were more dependent on habitat conditions than successional stage (187). The usefulness of diversity indices may be limited to within rather than between particular soil types/geographical areas (188). Work on nematode species diversity has also been reviewed in greater detail (189). The extinction of nematode species from particular sites may contribute to the population structure of natural communities and has been related to low population size and deficiency of resources (190). Future areas for emphasis in nematological research should aim to understand the biological mechanisms of ecological systems (191).

3.5 Extraction and Identification of Nematodes

Effective sampling of soil is crucial in obtaining an accurate estimate of nematode populations, particularly as they often have a clumped and localized distribution (25,192,193). Some direct observational techniques have been developed (83,194,195), but these are not in widespread use and most commonly used methods rely on extracting nematodes from soil. The extraction of active nematodes relies on their ability to wriggle through a mesh. A soil slurry is placed on a mesh immersed in water; the active nematodes move through the mesh and can be counted while the vast majority of the soil is retained on the mesh. Good examples are given in several papers (25,193,196–198); the factors which affect extraction efficiency, such as soil type, species of nematode, and apparatus, have also been discussed (199). Modifications also exist for extracting nematodes from forest litters and other OM-rich substrates (200). It is vitally important with all methods to estimate the efficiency of nematode extraction, usually by adding a known number of nematodes to the soil prior to extraction (201). There is probably not a single method which is equally suited to all groups of nematodes and to all soils (124).

Methods based on the flotation of nematodes potentially extract all (both active and inactive) nematodes and their eggs. The specific density of nematodes is <1.1 g cm^{-3}, while that of soil particles is >2.0 g cm^{-3}, and when mixed with a solution of density 1.1–2.0 g cm^{-3} nematodes and OM float and soil sinks. The nematodes can then be collected on a filter and counted. The optimal procedure depends on the nature of the sample and the objectives of the study (201–203). Elutriation is a form of flotation in which an up-current of water separates nematodes from soil particles and holds them in suspension while heavier particles sink. Various protocols for elutriation which are quantitatively superior to active extraction methods have been described (198). The use of molecular techniques is more advanced for soil nematodes than for soil protozoa, and data from proteins, deoxyribonucleic acid (DNA), and ribonucleic acid (RNA) can be used in conjunction with morphological data for identification, classification, and phylogenetic analysis (204–206).

The identification of nematodes to trophic group or to species requires that individual specimens be examined by light microscopy. The nematodes are usually preserved (with chemicals such as formaldehyde) and permanent mounts made on microscope slides (25,207). To determine nematode biomass, a representative number (ca. 50) of nematodes from each sample are measured, with either an eyepiece graticule or a digitizer linked to a microcomputer (208). Wet weight is derived from the equation; $mg = 0.096345 + 0.00023 \times area$ (mm^2) (209), and biomass is determined by using a wet weight–dry weight conversion factor of 0.24 (26). In the equation, the numerical constants represent the intercept and slope of a linear relationship. The estimate of nematode biomass is more precise than that of protozoa because the actual nematodes extracted from the sample are measured, rather than assuming an arbitrary size.

4. MICROARTHROPODS

Soil microarthropods, belonging to the phylum Arthropoda, are organisms which have a segmented exoskeleton similar to that of insects. The subclass Apterygota contains the Collembola, which are small, wingless insects, rarely exceeding 5 mm (length), with six abdominal segments and biting mouthparts which can be withdrawn into the head (210) (Fig. 4). Reproduction in collembolans is considered to be predominantly sexual, but parthenogenesis is common, especially among eudaphic species (211). The Acari (mites), of the class Arachnida, have saclike bodies, which may be divided by a furrow into anterior propodosoma and posterior hysterosoma (210) (Fig. 4). As with Collembola, reproduction is predominantly sexual, with the developmental stages of egg, hexapod larva, one to three nymphal stages, and adult (211). The classification of mites is complicated, with seven orders currently described; however, most soil forms belong to the four orders Crytostigmata, Mesostigmata, Prostigmata, and Astigmata. Other, less numerous soil microarthropods, within the subclass Apterygota, include the Protura. Further information about the classification, life cycles, and phenological features of soil microarthropods is available (211–213).

(a)

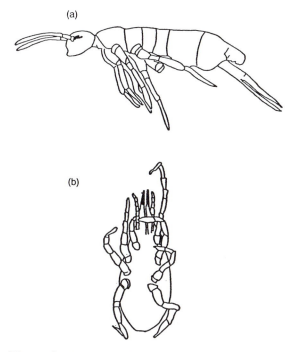

(b)

Figure 4.   Diagram of the external morphological characteristics of (a) Collembola and (b) Acarina.

4.1 Ecological Importance

Numerically, microarthropods (Collembola, mites, protura) are the most abundant nonaquatic faunal group in soils of most ecosystems (214). Recorded densities of soil microarthropods are as high as 300,000 m⁻² in temperate deciduous forests, old grasslands, and other well-drained aerobic soils with dense root systems and high organic matter content. In contrast, densities less than 50,000 m⁻² have been recorded in acidic peats and soils of low organic matter content, such as those of tropical, semiarid, and arid environments. Collembola (springtails) and Acari (mites) usually account for up to 95% of total numbers of microarthropods (215), and their relative abundance varies, depending on ecosystem type, soil conditions, and land management. In temperate grasslands and tundra ecosystems, the biomasses of Collembola and Acari are often reported to be similar (approximately 100–150 mg dry wt m⁻² (214,216). However, in tropical grasslands and both deciduous and coniferous forest soils, the biomass of Acari can be as much as three to five times as great as that of Collembola (214). In intensively managed agroecosystems, the biomass of Acari has been estimated to

Table 3.   Biomass Estimates (mg dry wt m^{-2}) of Collembola and Acari in Different Ecosystems

Ecosystem	Collembola	Acari
Tundra	50	90
Temperate grassland	90	120
Tropical grassland	10	80
Temperate coniferous forest	80	500
Temperate deciduous forest (mor)	130	900
Temperate deciduous forest (mull)	110	300
Tropical forest	20	100
Agroecosystem	17	120

Adapted from Ref. 214.

be around 120 mg and of collembolans only 17 mg dry wt m^{-2} (217). Typical biomass estimates of both Collembola and mites in different ecosystems are shown in Table 3.

The direct contribution of soil microarthropods to decomposition, as measured by their respiratory activity, is viewed to be quite low (211,214). The relative contributions of mite and Collembola respiration to total invertebrate respiration vary in different ecosystems. In moorland soils, mites and collembolans contributed 1% to 1.6% and 0.4%, respectively, to total soil invertebrate respiration (218,219). In an Australian pasture soil, however, mites and collembolans contributed 1.9% and 26%, respectively, to total soil invertebrate respiration (220).

4.2   Distribution of Microarthropod Populations

The abundance of soil microarthropod populations shows considerable spatial and temporal variation. Direct observation of microarthropods in rhizotrons (exposed in situ soil sections) in a mixed deciduous forest has shown that densities were at least twice as high around roots as in the bulk soil (221). Similarly, core samples collected between Norway spruce trees showed the highest densities of Collembola to be in areas with large masses of fine, mycorrhizal roots (222). With respect to seasonal variation, microarthropod numbers in most ecosystems are greatest in summer (223–227). However, winter maxima of populations of Collembola and other soil microarthropods are also widely reported (228–231). Seasonal variations are related to the suitability of soil microclimatic conditions for microarthropod reproduction, and to variations in optimal soil climatic requirements of different species of microarthropod that may be present in different soils (230). In general, seasonal variations are related to soil moisture status, temperature, and availability of plant residues; rapid increases in microarthropod numbers are not uncommon in surface soil after episodic rainfall events (232,233).

Numbers of microarthropods are generally greater in surface soil than in the lower horizons. In a range of upland grassland soils, it was found that 92% to 98% of Acari and Collembola were extracted from the upper 0–2 cm of soil, which included the litter layer (227). Similarly, in a study of north Queensland (Australia) rain forests, it was found that 53% to 75% of the Acari and Collembola were present in the 0–4 cm soil layer, and 3% to 20% in the surface litter (234). Other studies of grassland and forest soils have also shown that microarthropods tend to concentrate in the upper soil horizons and litter layers (230,235–237), and only certain microarthropod groups, such as the Prostigmata (Acari), are found at greater depths (238).

Vertical migrations of both Collembola and Acari are known to occur in response to extreme surface temperatures and rainfall. A study of an aspen poplar woodland showed that the collembolan *Onychiurus subtenuis* moves to the lower layers of the litter profile during the summer to prevent desiccation (239). Similar migrations have been reported for other species of Collembola (230,240,241). In desert and rain forest ecosystems (242,234), however, Collembola and other microarthropods can return rapidly to the litter layer after rainstorms (239). Such migrations have been attributed to rainfall-induced bursts in microbial activity in the litter layer, and hence increased food resource availability for microbe-feeding microarthropods (239). Microarthropods are also known to have a nonrandom horizontal distribution in the soil and to aggregate in response to patchiness of food resources and soil water (230,243,244), clumping of eggs by certain species (245), and production of aggregation pheromones (246).

The species composition of the microarthropod community is also strongly influenced by soil conditions, in particular soil depth. Two morphological types of microarthropods which dwell at different depths in the soil can be differentiated (247). In general, larger, more active genera, adapted to an open and often dry habitat, predominate in the surface soil and/or litter layer. These genera (e.g., Collembola of the genus *Entomobrya* and *Tomocerus*, and Acari of the genus *Hermannia*) have an ability to withstand desiccation and have strong pigmentation, well-developed eyes, and long appendages. At greater soil depth, genera are generally smaller, are weakly pigmented, often lack eyes, and have soft bodies with short appendages. Such soil-dwelling organisms include Collembola of the genera *Onychiurus* and *Tullbergia* and the Cryptostigmatid or oribatid mites.

4.3   Microarthropod Feeding Habits

Like most soil faunal groups, microarthropods occupy all trophic categories of below-ground detritus food webs (248). However, most microarthropods can be regarded as either microphytophages, feeding on soil microflora (fungi, bacteria, actinomycetes, and algae), and/or detrivores, scavenging on dead organic matter and plant litter. Microarthropods have been estimated to obtain 51% of their energy from the bacterial pathway and 26% and 24% from the fungal and root pathways in shortgrass steppe, North America (248). Some microarthropods are also known to be predators, e.g., certain prostigmatic mites and carnivorous Collembola (*Isotoma* spp.). The latter feed on rotifers and eggs of nematodes (247). An extensive review of the feeding characteristics of microarthropods has been prepared (214).

It is well established that field populations of many soil microarthropods feed preferentially on soil microorganisms, in particular fungal hyphae (249–255). In addition, soil microarthropods, in particular Collembola, are known to select between species of fungi both in the laboratory (255–258) and in the field (259,260). Indeed, most microarthropods are well adapted to graze on fungi and bacteria in the soil, having small bodies and well-developed mouthparts for either plucking and scraping microbes off surfaces (engulfers) or piercing fungal cell walls to drain fluid contents (fluid-feeders) (248,261). Microarthropods also exhibit a range of life histories, such as explosive growth or population persistence, reflecting their strategy for utilizing soil microorganisms or detritus (261).

The grazing activities of both mycophytophagous (fungus-feeding) and detrivorous (detritus-feeding) microarthropods result in the fragmentation, dissemination, and comminution (division into small fragments) of organic matter and plant litter. Such physical destruction of organic material has important consequences for decomposition and nutrient cycling, and the regulation of microbial populations in soil. The direct (physical disruption) and indirect (microbial interactions) roles of soil microarthropods in decomposition and nutrient cycling processes have been reviewed (215,248,262). Microarthropod–microbe interactions and their ecological significance will be considered in more detail in other sections of this text.

4.4   Impacts of Land Management

Land management affects soil microarthropod populations by altering the quantity and quality of detritus input and soil microhabitat stability and complexity in terms of soil/litter physical and chemical conditions. The diversity of feeding habits and habitat preferences of soil microarthropods implies that their responses to management will be complex and may be limited to various specific groups of organisms (263).

Perhaps the most destructive agricultural practice is tillage or cultivation of land. In general, cultivation of soil is known to reduce the abundance and genetic diversity of soil microarthropods (214,264,265). Reduced- and nontillage management, however, is known to result in an increase in populations of both Collembola (217,266) and Acari (267). This effect has been attributed to improved moisture conditions, increased fungal growth, and favorable habitat near surface residues. Tillage with different plows resulted in a sharp decline in microarthropod populations; however, it was followed by rapid increases in numbers toward the end of the growing season (268). Similarly, tillage with a moldboard plow caused acarine and collembolan populations to be significantly reduced initially (269). However, population recovery to control levels differed among the various orders and suborders of microarthropods. Mesostigmatic mite populations recovered after 3 months, whereas Collembola and cryptostigmatic mite populations showed signs of recovery after 4 and 5 months, respectively. Prostigmata and total Acarina failed to show any signs of recovery by the end of the growing season. Rapid declines in populations of microarthropods following cultivation are attributed to changes in soil temperature and moisture, and to mechanical damage (268).

The application of inorganic fertilizers to grasslands has been shown to affect microarthropod abundance. Addition of nitrogenous fertilizer to moorland peat soils increased total Collembola numbers by 67% (270). Likewise, it was found that Collembola and Acari numbers were approximately twice as high in the surface soil of a fertilized (NPK) upland grassland as in adjacent unfertilized grasslands (227). Increases in microarthropod numbers upon application of phosphate (271,272) and nitrogenous (273,274) fertilizers to lowland grassland have been shown. Negative effects of inorganic fertilizers have also been reported. The abundance and species diversity of microarthropods declined after the application of nitrogenous fertilizer to grassland (275). Similarly, in a study of a coniferous forest soil, the application of urea halved the population of microarthropods; however, numbers began to increase 4 months after application (276).

Studies of lowland grasslands (273) and forest soils (277) suggest that liming has little effect on microarthropod populations; the effects of liming are often restricted to different groups or species. For example, liming an acidic spruce forest soil resulted in an increase in total Collembola, and the effect was restricted to one species (*Folsomia quadrioculata*) (278). In the same study, numbers of microphytophagous Oribatei mites and acidophilous Collembola species declined after liming. Changes in livestock grazing pressure substantially modify the environment for soil- and litter-inhabiting microarthropods, affecting their abundance, diversity, and spatial distribution. Increased sheep stocking density of an Australian pasture (10, 20, 30 sheep ha^{-1}) severely reduced numbers of Collembola in the surface soil (223,279). Likewise, reductions in Collembola numbers were associated with increased sheep stocking density of a lowland perennial ryegrass (*Lolium perenne*) grassland (280). These authors attributed the responses to changes in abiotic factors, in particular soil pore space and surface litter, both of which were reduced with increased sheep grazing intensity.

In a study of sheep-grazed upland grasslands, however, the converse was found be true (227). Numbers of Collembola and Acari in the surface soil were found to decline along a gradient toward less grazing. In addition, the short- and long-term cessation of sheep grazing on a wide range of upland grassland types resulted in significant reductions in the abundance of total Collembola and the dominant collembolan species, *Onychiurus procampatus* (216,227) (Fig. 5). In contrast to the finding of those studies of lowland grasslands, grazing pressures on the upland sites were considerably lower (1–8 sheep ha^{-1}), having relatively little effect on soil pore space. Reductions in microarthropod numbers were attributed largely to increases in soil pH and moisture content, induced by less intensive management and cessation of sheep grazing (227), and to associated changes in the abundance of microbial food sources, in particular fungal hyphae (227,281,282). Changes in microarthropod and microbial abundance were also attributed to management-induced alterations in the quantity and quality of litter input into the soil (227).

Pesticide effects on soil microarthropod populations are often variable and differ for different groups of organisms (247,263). For example, the application of the insecticide azinphosphomethyl (Guthion) resulted in a decline in numbers of soil Acaridae and Prostigmata mites, whereas numbers of oribatid and mesostigmata mites and Collembola increased (283). Various pesticides, herbicides, and fungicides applied to field crops lead to reductions in populations of predaceous mites, but increases in numbers of saprophagous Collembola (284). Other studies, however, report reductions in numbers of both Collembola and Acari following the application of pesticides and herbicides (285–287). It has been suggested that certain sensitive species of Collembola (e.g., *Folsomia quadrioculata*) would be well suited for testing the environmental compatibility of pesticides (287).

4.5   Methods for the Assessment of Microarthropods

The assessment of soil-inhabiting microarthropod populations is dependent on their quantitative separation from samples of soil and organic debris. The wide range of methods available for the assessment of soil invertebrate populations have been reviewed (288,289). For the quantitative assessment of microarthropods, two types of methodology are used: active (behavioral) techniques, which manipulate the activity and behavioral characteristics of animals, and passive techniques, which rely primarily on mechanical separation of animals from soil, and differences in specific gravity or sedimentation of soil and animals (289).

The most commonly used active techniques for extraction of soil microarthro-

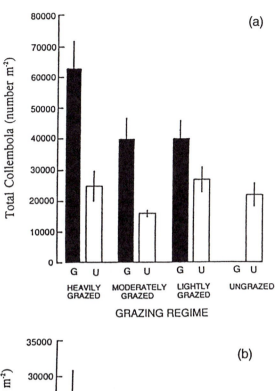

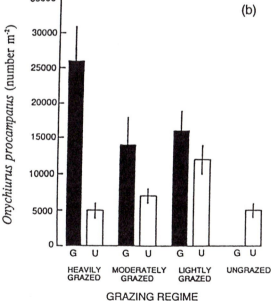

Figure 5. The effects of removing sheep grazing for 2 years on (a) the abundance (mean number $m^{-2} \pm$ SE) of total Collembola and (b) the abundance (mean number $m^{-2} \pm$ SE) of the fungi-feeding collembolan *Onychiurus procampatus*, in the surface soil of four upland grassland sites, representing a gradient of sheep management intensity, in Cumbria. (From Ref. 227.)

pods are modifications of the Berlese or Tullgren funnel. At their simplest, these dry funnel methods use desiccation to drive animals from soil, relying on heat produced from a light bulb suspended over a soil sample, funnel, and collecting jar. More sophisticated "dry funnel" techniques have been developed (229,290,291), in which intact soil cores are exposed to controlled gradients of temperature, light, and humidity. When using dry funnel techniques, extraction efficiency can vary, depending on soil moisture content, clay content, and use of intact or broken-up samples (289,290,292). In addition, biases in extraction of animals, based primarily on body size, may occur (293).

Passive techniques for extracting microarthropods depend on gentle mechanical disruption of soil to free animals (e.g., wet sieving), followed by flotation in a solution with sufficient specific gravity to suspend animals, but allow soil particles to settle out. Soil washing and flotation techniques used to extract arthropods from soil have been reviewed (288,294). Passive techniques generally require more labor and are less cost-effective than active methods. However, they are particularly suitable for heavy clay soils, which may harden under dry funnels and trap animals within soil. Passive techniques are also less dependent on temperature and therefore can recover inactive taxa and quiescent life stages, including eggs (289).

ACKNOWLEDGMENTS

The authors would like to thank Gregor Yeates (Landcare Research Ltd, New Zealand) for comments on the manuscript and Myron Hobbelen (IGER) for clerical assistance. This work was partly funded by the Scottish Office Agriculture, Environment and Fisheries Department, and the Biotechnology, and Biological Sciences Research Council.

REFERENCES

1. W. Foissner, *Prog. Protistol.* 2:69–212 (1987).
2. J. D. Lousier and S. S. Bamforth, *Soil Biology Guide* (D. L. Dindal, ed.), Wiley, New York, p. 97 (1988).
3. M. Sleigh, *Protozoa and Other Protists*, Edward Arnold, London (1989).
4. J. D. Lousier and D. Parkinson, *Soil Biol. Biochem.* 16:103–114 (1984).
5. R. Meisterfeld, *Ver. Gres. Ökol.* 17:221–227 (1989).
6. R. Meisterfeld, *Symp. Biol. Hung.* 33:291–299 (1986).
7. L. Ryszkowski, *Modern Ecology, Basic and Applied Aspects* (G. Esser and D. Overdieck, eds.), Elsevier, Amsterdam, p. 443 (1991).
8. W. Schönborn, *Acta Protozool.* 31:11–18 (1992).
9. T. Rosswall and K. Paustian, *Plant Soil 76*:3–21 (1984).

10. H. W. Hunt, D. C. Coleman, E. R. Ingham, R. E. Ingham, E. T. Elliott, J. C. Moore, S. L. Rose, C. P. P. Reid, and C. R. Morley, *Biol. Fertil. Soils 3*:57–68 (1987).

11. E. T. Elliott, H. W. Hunt, and D. E. Walter, *Agric. Ecosyst. Environ. 24*:41–56 (1988).

12. P. C. de Ruiter, J. C. Moore, K. B. Zwart, L. A. Bouwman, J. Hassink, J. Bloem. J. A. de Vos, J. C. Y. Marinissen, W. A. M. Didden, G. Lebbink, and L. Brussaard, *J. Appl. Ecol. 30*:95–106 (1993).

13. B. S. Griffiths, *Soil Protozoa* (J. F. Darbyshire, ed.), CAB International, Wallingford, England, p. 65 (1994).

14. J. D. Stout and O. W. Heal, *Soil Biology* (A. Burgess and F. Raw, eds.), Academic Press, London, p. 149 (1967).

15. J. D. Stout, *Adv. Microb. Ecol. 4*:1–50 (1980).

16. S. S. Bamforth, *J. Protozool. 32*:404–409 (1985).

17. R. E. Ingham, J. A. Trofymow, E. R. Ingham, and D. C. Coleman, *Ecol. Monogr. 55*:119–140 (1985).

18. S. Visser, *Ecological Interactions in Soil* (A. H. Fitter, D. Atkinson, D. J. Read, and M. B. Usher, eds.), Blackwell Scientific, Oxford, p. 297 (1985).

19. K. M. Old, *Prog. Protistol. 1*:163–194 (1986).

20. J. M. Anderson, *Ecology of Microbial Communities* (M. Fletcher, T. R. G. Gray, and J. G. Jones, eds.), Cambridge University Press, Cambridge, p. 1257 (1987).

21. A. Feest, *Prog. Protistol. 2*:331–361 (1987).

22. D. C. Coleman, C. P. P. Reid, and C. V. Cole, *Adv. Ecol. Res. 13*:1–55 (1983).

23. D. W. Freckman and E. P. Caswell, *Ann. Rev. Phytopathol. 23*:275–296 (1985).

24. D. W. Freckman. *Agric. Ecosyst. Environ. 24*:195–217 (1988).

25. D. W. Freckman and J. G. Baldwin, *Soil Biology Guide* (D. L. Dindal, ed.), Wiley, New York, p. 155 (1988).

26. G. W. Yeates, *J. Nematol. 11*:213–229 (1979).

27. R. Vargas, *Agronomia Costarricense 14*:121–134 (1990).

28. H. A. Verhoef and L. Brussaard, *Biogeochem. 11*:175–211 (1990).

29. K. B. Zwart and L. Brussaard, *The Ecology of Temperate Cereal Fields* (L. G. Firbank, N. Carter, J. F. Darbyshire, and G. R. Potts, eds.), Blackwell Scientific, Oxford, p. 139 (1991).

30. V. V. S. R. Gupta, *Soil Biota: Management in Sustainable Farming* (C. E. Parkhurst, B. M. Doube, V. V. S. R. Gupta, and P. R. Grace, eds.), CSIRO, Melbourne, p. 107 (1994).

31. B. S. Griffiths, *Plant, Soil 164*:25–33 (1994).

32. K. B. Zwart, P. J. Kuikman, and J. A. van Veen, *Soil Protozoa* (J. F. Darbyshire, ed.), CAB International, Wallingford, England, p. 93 (1994).

33. P. F. Hendrix, D. A. Crossley Jr., D. C. Coleman, R. W. Parmelee, and M. H. Beare, *INTECOL Bull. 15*:59–63 (1987).
34. W. E. Hekman, P. J. H. F. Boogert, and K. B. Zwart, *FEMS Microb. Ecol. 86*:255–265 (1992).
35. W. Petz, W. Foissner, and H. Adam, *Soil Biol. Biochem. 17*:871–875 (1985).
36. D. A. Caron and J. C. Goldman, *J. Protozool. 35*:247–249 (1988).
37. S. Nakano, *Arch. Hydrobiol. 129*:257–271 (1994).
38. J. F. Darbyshire, M. S. Davidson, S. J. Chapman, and S. Ritchie, *Soil Biol. Biochem. 26*:1193–1199 (1994).
39. M. M. Côuteaux, M. Mousseau, M.-L. Célérier, and P. Bottner, *Oikos 61*:54–64 (1991).
40. W. Schönborn, *Pedobiologia 15*:415–424 (1975).
41. R. Vargas and T. Hattori, *FEMS Microb. Ecol. 38*:233–242 (1986).
42. B. S. Griffiths and I. M. Young, *Eur. J. Soil Sci.* (in press).
43. I. M. Young, A. Roberts, B. S. Griffiths, and S. Caul, *Soil Biol. Biochem. 26*:1173–1178 (1994).
44. J. Hassink, L. A. Bouwman, Z. B. Zwart, and L. Brussaard, *Soil Biol. Biochem. 25*:47–55 (1993).
45. P. M. Rutherford and N. G. Juma, *Biol. Fertil. Soils 8*:144–153 (1989).
46. J. Schnurer, M. Clarholm, and T. Rosswall, *Biol. Fertil. Soils 2*:119–126 (1986).
47. J. Sinclair and W. C. Ghiorse, *Appl. Environ. Microbiol. 53*:1157–1163 (1987).
48. J. L. Sinclair, D. H. Kampbell, M. L. Cook, and J. T. Wilson, *Appl. Environ. Microbiol. 59*:467–472.
49. D. Miteva, *God. Sof. Univ. 69*:97–102 (1975).
50. M. H. Beare, R. W. Parmelee, P. F. Hendrix, W. Cheng, D. C. Coleman, and D. A. Crossley Jr., *Ecol. Monogr. 62*:569–591 (1992).
51. L. Brussaard, L. A. Bouwman, M. Geurs, J. Hassink, and K. B. Zwart, *Neth. J. Agric. Sci. 38*:283–302 (1990).
52. E. T. Elliott, K. Horton, J. C. Moore, D. C. Coleman, and C. V. Cole, *Plant Soil 76*:149–155 (1984).
53. B. N. Singh, *J. Gen. Microbiol. 3*:204–210 (1949).
54. B. Sohlenius, *Biol. Fertil. Soils 9*:168–173 (1990).
55. M. H. Opperman, M. Wood, and P. J. Harris, *Soil Biol. Biochem. 21*:263–268 (1989).
56. B. S. Griffiths, Biol. Fertil. Soils, 9:83–88 (1990).
57. V. V. S. R. Gupta and J. J. Germida, *Soil Biol. Biochem. 20*:787–792 (1988).
58. M. Clarholm, *Microb. Ecol. 7*:343–350 (1981).
59. M. Clarholm, *Biol. Fertil. Soils 8*:373–378 (1989).

60. D. W. Cutler, L. M. Crump, and H. Sandon, *Phil. Trans. R. Soc. 211*:317–350 (1922).
61. J. F. Darbyshire, K. B. Zwart, and D. A. Elston, *Soil Biol. Biochem. 25*:1583–1589 (1993).
62. F. T. Kuserk, *Ecology 61*:1474–1485 (1980).
63. R. J. Bryant, L. E. Woods, D. C. Coleman, B. C. Fairbanks, J. F. McLellan, and C. V. Cole, *Appl. Environ. Microbiol. 43*:747–752 (1982).
64. P. J. Kuikman, A. G. Jansen, and J. A. Van Veen, *Soil Biol. Biochem. 23*:193–200 (1991).
65. L. W. Parker, D. W. Freckman, V. Steinberger, L. Driggers, and W. G. Whitford, *Pedobiologia 27*:185–195 (1984).
66. S. Christensen, B. S. Griffiths, F. Ekelund, and R. Rønn, *FEMS Microb. Ecol. 86*:303–310 (1992).
67. B. S. Griffiths, F. Ekelund, R. Rønn, and S. Christensen, *Soil Biol. Biochem. 25*:1293–1295 (1993).
68. B. S. Griffiths and S. Caul, *Biol. Fertil. Soils 15*:201–207 (1993).
69. J. F. Darbyshire and M. P. Greaves, *Can. J. Microbiol. 13*:1057–1068 (1967).
70. J. F. Darbyshire, *J. Soil Sci. 27*:369–376.
71. J. D. Stout, *J. Anim. Ecol. 32*:281–287 (1963).
72. J. D. Stout, *Soil Biol. Biochem. 16*:121–125 (1984).
73. S. S. Bamforth, *Soil Biol. Biochem. 16*:133–137 (1984).
74. W. Petz and W. Foissner, *Ver. Ges. Ökol. 17*:397–399 (1989).
75. G. Lüftenegger, W. Foissner, and H. Adam, *Oecologia 66*:574–579 (1985).
76. G. W. Yeates, S. S. Bamforth, D. J. Ross, K. R. Tate, and G. P. Sparling, *Biol. Fertil. Soils 11*:181–189 (1991).
77. E. Aescht and W. Foissner, *Pedobiologia 37*:321–335 (1993).
78. B. Schreiber and N. Brink, *Biol. Fertil. Soils 7*:289–296 (1989).
79. T. A. Forge, M. L. Berrow, J. F. Darbyshire, and A. Warren, *Biol. Fertil. Soils 16*:282–286 (1993).
80. A. J. Ramsay and A. D. Bawden, *Soil Biol. Biochem. 15*:263–268 (1983).
81. M. M. Coûteaux, *Revue Öcol. Biol. Sol. 4*:593–596 (1967).
82. J. D. Lousier and D. Parkinson, *Soil Biol. Biochem. 13*:209–213 (1981).
83. G. Lüftenegger, W. Petz, W. Foissner, and H. Adams, *Pedobiologia 31*:95–102 (1988).
84. M. M. Coûteaux and L. Palka, *Soil Biol. Biochem. 20*:7–10 (1988).
85. P. C. T. Jones and J. E. Mollison, *J. Gen. Microbiol, 2*:54–69. (1948).
86. J. S. Bunt and Y. T. Tchan, *Proc. Linn. Soc. N. S. W. 80*:148–153 (1955).
87. G. A. Korganova and Ju G. Geltser, *Pedobiologic 17*:222–225 (1977).
88. B. S. Griffiths and K. Ritz, *Soil Biol. Biochem. 20*:163–173 (1988).
89. M. Starink, M.-J. Bar-Gilissen, R. P. M. Bak, and T. E. Cappenberg, *Appl. Environ. Microbiol. 60*:167–173 (1994).

90. C. A. M. Broers, H. O. F. Molhuizen, C. K. Stumm, and G. D. Vogels, *J. Microb. Methods 14*:217–220 (1992).

91. E. B. Sherr and B. F. Sherr, *Appl. Environ. Microbiol. 46*:1388–1393 (1983).

92. B. F. Sherr, E. B. Sherr, and R. D. Fallon, *Appl. Environ. Microbiol. 53*:958–965 (1987).

93. M. H. McCrady, *J. Infect. Dis. 17*:183–212 (1915).

94. C. Lamanna and M. F. Mallette, *Basic Bacteriology, Its Biological and Chemical Background*, Williams & Wilkins, Baltimore, p. 243 (1953).

95. D. W. Cutler, *J. Agric. Sci. Camb. 10*:135–143 (1920).

96. B. N. Singh, *Ann. Appl. Biol. 33*:112–119 (1946).

97. D. Menapace, D. A. Klein, J. F. McClellan, and J. V. Mayeux, *J. Protozool. 22*:405–410 (1975).

98. J. F. Darbyshire, R. E. Wheatley, M. P. Greaves, and R.H.E. Inkson, *Revue Écol. Biol. Sol. 11*:465–474 (1974).

99. S. Chakraborty, *Soil Biol. Biochem. 15*:661–664 (1985).

100. T. R. Anderson, *Soil Biol. Biochem. 25*:223–226 (1993).

101. L. E. Casida Jr., *Appl. Environ. Microbiol. 55*:1857–1859 (1989).

102. R. Rønn, F. Ekelund, and S. Christensen, *Pedobiologia 39*:10–19 (1994).

103. R. Vargas and T. Hattori, *Bull. Jpn. Soc. Microb. Ecol. 2*:53–55 (1988).

104. M. Pussard and F. Delay, *Protistologica 21*:5–16 (1985).

105. P. M. Rutherford and N. G. Juma, *Can. J. Soil Sci. 72*:183–200 (1992).

106. F. C. Page, An Illustrated Key to Freshwater and Soil Amoebae, Freshwater Biological Association Scientific Publication No. 34 (1976).

107. D. A. Caron, P. G. Davis, and J. M. Sieburth, *Microb. Ecol. 18*:89–104 (1989).

108. D. W. Hopkins, S. J. MacNaughton, and A. G. O'Donnell, *Soil Biol. Biochem. 23*:227–323 (1991).

109. J. S. Frey, J. F. McClellan, E. R. Ingham, and D. C. Coleman, *Biol. Fertil. Soils 1*:73–79.

110. A. W. West, W. D. Grant, and G. P. Sparling, *Soil Biol. Biochem. 19*:607–612 (1987).

111. E. Bååth, Å. Frostegård, and H. Fritze, *Appl. Environ. Microbiol. 58*:4026–4031 (1992).

112. L. Zelles, Q. Y. Bai, and F. Beese, *Soil Biol. Biochem. 24*:317–323 (1992).

113. A. John and M. J. Ulyatt, *J. Agric. Sci. Camb. 102*:33–44 (1984).

114. H. Lechavalier and M. P. Lechavalier, *Microbial Lipids* vol. 1 (C. Ratledge and S. G. Wilkinson, eds.), Academic Press, London, pp. 869–902 (1988).

115. J. Vrba, K. Simek, J. Nedoma, and P. Hartman, *Appl. Environ. Microbiol. 59*:3091–3101 (1993).

116. P. D. Butcher and M. J. G. Farthing, *Biochem. Soc. Trans. 17*:363–369 (1989).

117. D. Fong, M. M. Y. Chan, R. Rodriguez, C. C. Chen, Y. Liang, and D. T. J. Littlewood, *Mol. Biochem. Parasitol.* *62*:139–142 (1993).
118. W. L. Nicholas, *The Biology of Free-Living Nematodes*, 2nd ed., Methven, London (1975).
119. G. W. Yeates, T. Bongers, R. G. M. De Goede, D. W. Freckman, and S. S. Georgieva, *J. Nematol.* *25*:315–331 (1993).
120. G. W. Yeates, *Biol. Fertil. Soils* *3*:143–146 (1987).
121. F. H. Wood, *Soil Biol. Biochem.* *5*:593–602 (1973).
122. P. S. Grewal, *Nematologica* *37*:72–82 (1991).
123. E. P. Odum, *Fundamentals of Ecology*, 3rd ed., W. B. Saunders, Philadelphia, p. 63 (1971).
124. B. Sohlenius, *Oikos* *34*:186–194 (1980).
125. B. Sohlenius, S. Boström, and A. Sandor, *Biol. Fertil. Soils* *6*:1–8 (1988).
126. B. Sohlenius, *Holarct. Ecol.* *2*:30–40 (1979).
127. D. C. Coleman, R. V. Anderson, C. V. Cole, E. T. Elliott, L. Woods, and M. K. Campion, *Microb. Ecol.* *4*:373–380 (1978).
128. B. S. Griffiths, *Ann. Appl. Biol.* *22*:141–145 (1989).
129. M. H. Beare, J. M. Blair, and R. W. Parmelee, *Soil Biol. Biochem.* *21*:1027–1036 (1989).
130. R. W. Parmelee, M. H. Beare, and J. M. Blair, *Soil Biol. Biochem.* *21*:97–103 (1989).
131. G. B. Pradhan, B. K. Senapati, and M. C. Dash, *Revue, Écol. Biol. Sol.* *25*:59–76 (1988).
132. G. W. Yeates and D. C. Coleman, *Nematodes in Soil Ecosystems* (D. W. Freckman, ed.), University of Texas Press, Austin, p. 55 (1982).
133. H. Zell, *Verh. Ges. Ökol.* *17*:125–130 (1989).
134. D. C. M. Corbett, R. D. Winslow, and R. M. Webb, *Report Rothamsted Experimental Station for 1968. Part 2*, p. 157–174 (1969).
135. G. W. Yeates, J. D. Stout, D. J. Ross, M. E. Dutch, and R. F. Thomas, *NZ. J. Agric. Res.* *19*:51–61 (1976).
136. B. Sohlenius, S. Böstrom, and A. Sandor, *J. Appl. Ecol.* *24*:131–144 (1987).
137. H. R. Wallace, *Ann. Appl. Biol.* *46*:74–85 (1958).
138. J. B. Goodey, *Soil and Freshwater Nematodes*, 2nd ed., Methven, London (1963).
139. A. Nadler and Y. Steinberger, *Soil Sci.* *155*:114–122 (1993).
140. G. W. Yeates, *Soil Biol. Biochem.* *16*:95–102 (1984).
141. D. W. Freckman, D. A. Duncan, and J. R. Larson, *J. Range Mgmt.* *32*:418–422 (1979).
142. Y. Pick-Hoong, *Nematologica* *12*:195–214 (1966).
143. B. Sohlenius and L. Wasilewska, *J. Appl. Ecol.* *21*:327–342 (1984).
144. B. Sohlenius and A. Sandor, *Biol. Fertil. Soils* *3*:19–25 (1987).

145. E. R. Ingham, D. C. Coleman, and J. C. Moore, *Biol. Fertil. Soils 8*:29–37 (1989).

146. O. Andrén and J. Lagerlöf, *Acta Agric. Scand. 33*:33–52 (1983).

147. G. W. Yeates and K. A. Hughes, *Pedobiologia 34*:379–387 (1990).

148. K. L. King and K. J. Hutchinson, *Aust. J. Ecol. 8*:245–255 (1983).

149. R. E. Ingham and K. J. Detling, *Oecologia 63*:307–313 (1984).

150. T. C. Todd, S. W. James and T. R. Seastedt, *Plant Soil 144*:117–124 (1992).

151. S. M. Baird and E. Bernard, *J. Nematol. 16*:379–386 (1984).

152. A. El Titi and U. Ipach, *Agric. Ecosyst. Environ. 27*:561–572 (1989).

153. B. Sohlenius and A. Sandor, *Pedobiologia 33*:199–210 (1989).

154. E. Aescht and W. Foissner, *Biol. Fertil. Soils 13*:17–24 (1992).

155. W. Dmowska and J. Kozlowska, *Pedobiologia 32*:323–330 (1988).

156. G. W. Yeates, *NZ. J. Agric. Res. 21*:321–330 (1978).

157. M. H. Opperman, M. Wood, P. J. Harris, and C. P. Cherrett, *Soil Biol. Biochem. 25*:19–24 (1993).

158. B. S. Griffiths, K. Ritz, and R. E. Wheatley, *Soil Use Manage. 10*:20–24 (1994).

159. H. Berger, W. Foissner and H. Adam, *Pedobiologia 29*:261–272 (1986).

160. E. Bååth, U. Lohm, B. Lundgren, T. Rosswall, B. Söderström, B. Sohlenius, and A. Wirén, *Oikos 31*:153–163 (1978).

161. R. Ohtonen, A. Ohtonen, H. Luotonen, and A. M. Markkola, *Biol. Fertil. Soils 13*:50–54 (1992).

162. R. Hyvönen and V. Huhta, *Pedobiologia 33*:129–143 (1989).

163. R. Hyvönen and T. Persson, *Biol. Fertil. Soils 9*:205–210 (1990).

164. R. G. M. De Goede and H. H. Dekker, *Pedobiologia 37*:193–209 (1993).

165. G. W. Yeates, *Pedobiologia 22*:312–338 (1981).

166. T. R. Seastedt, T. C. Todd, and S. W. James, *Pedobiologia 30*:9–17 (1987).

167. G. W. Yeates, D. A. Wardle, and R. N. Watson, *Soil Biol. Biochem. 25*:869–876 (1993).

168. C. de Guiran, L. Bonnel, and M. Abirached, *Effluents from Livestock* (J. K. R. Gasser, ed.), Applied Science, London, p. 109 (1980).

169. L. Wasilewksa, E. Papiska and J. Zieliski, *Pedobiologia 21*:182–191 (1981).

170. B. Sohlenius and S. Boström, *Pedobiologia 26*:67–78 (1984).

171. L. A. Bouwman, J. Bloem, P. H. J. F., Van den Boogert, F. Bremer, G. H. J. Hoenderboom, and P. C. de Ruiter, *Biol. Fertil. Soils 17*:249–256 (1994).

172. J.-C. Cayrol, J.-P. Frankowski, and C. Quiles, *Bull. Fed. Synd. Agric. Cultiv. Champ. NS 11*:301–312 (1981).

173. Y. Demeure, D. W. Freckman, and S. D. Van Gundy, *J. Nematol. 11*:189–195 (1979).

174. H. R. Wallace, *Plant Parasitic Nematodes*, vol. 1 (B. M. Auckerman, W. F. Mai and R. A. Rhode, eds.), Academic Press, New York, p. 257 (1971).

175. W. G. Whitford, *Biol. Fertil. Soils 8*:1–6 (1989).
176. B. S. Griffiths, I. M. Young, and B. Boag, *Pedobiologia 35*:265–272 (1991).
177. D. H. Wright and D. C. Coleman, *Biol. Fertil. Soils 7*:46–52 (1988).
178. R. E. Ingham and J. K. Detling, *Plant Soil 121*:279–281 (1990).
179. T. W. Hofman and J. J. Jacob, *Ann. Appl. Biol. 115*:291–298 (1989).
180. H. Christensen, B. S. Griffiths, and S. Christensen, *Soil Biol. Biochem. 24*:703–709 (1992).
181. O. Andren, K. Paustian, and T. Rosswall, *Agric. Ecosyst. Environ. 24*:57–67 (1988).
182. M. Hodda and F. R. Wanless, *Nematologica 40*:116–132 (1994).
183. R. M. Warwick, H. M. Platt, K. R. Clarke, J. Agard, and J. Gobin, *J. Exp. Mar. Biol. Ecol. 138*:119–142 (1990).
184. T. Bongers, *Oecologia 83*:14–19 (1990).
185. C. H. Ettema and T. Bongers, *Biol. Fertil. Soils 16*:79–85 (1993).
186. G. W. Yeates, *Pedobiologica 39*:97–101 (1994).
187. L. Wasilewska, *Pedobiologia 38*:1–11 (1994).
188. G. W. Yeates and A. F. Bird, *Fundam. Appl. Nematol. 17*:133–145 (1994).
189. E. C. Bernard, *Biol. Fertil. Soils 14*:99–103 (1992).
190. D. H. Wright and D. C. Coleman, *Oikos 67*:563–572 (1993).
191. H. Ferris, *J. Nematol. 25*:374–382 (1993).
192. P. B. Goodell, *Nematodes in Soil Ecosystems* (D. W. Freckman, ed.), University of Texas Press, Austin, p. 178 (1982).
193. B. Boag, *Plant Virus Epidemics, Monitoring, Modelling and Predicting Outbreaks*, Academic Press, p. 119 (1986).
194. G. Minderman, *Nematologica 1*:216–226 (1956).
195. R. V. Anderson, W. D. Gould, R. E. Ingham, and D. C. Coleman, *Trans. Am. Microsc. Soc. 98*:213–218 (1979).
196. M. Oostenbrink, *Methods of Study in Quantitative Soil Ecology: Population, Production and Energy Flow* (J. Phillipson, ed.), Blackwell Scientific Publications, Oxford, p. 72 (1971).
197. D. J. F. Brown and B. Boag, *Nematol. Medit. 16*:93–99 (1988).
198. D. J. Hooper, *Laboratory Methods for Work with Plant and Soil Nematodes* (J. F. Southey, ed.), HMSO, London, MAFF reference book 402, p. 5 (1986).
199. D. R. Viglierchio and R. V. Schmitt, *J. Nematol. 15*:438–444 (1983).
200. A. J. Schouten and K. K. M. Arp, *Pedobiologia 35*:393–400 (1991).
201. B. S. Griffiths, B. Boag, R. Neilson, and L. Palmer, *Nematologica 36*:465–473 (1990).
202. D. R. Viglierchio and T. T. Yamashita, *J. Nematol. 15*:444–449 (1983).
203. D. W. Freckman, D. T. Kaplan, and S. D. van Gundy, *J. Nematol. 9*:176–181 (1977).

204. J. Curran, *Nematology, from Molecule to Ecosystem* (F. J. Gomers and P. W. T. Maas, eds.), Dekker and Huisman, Wildervank, The Netherlands, p. 83 (1992).
205. V. R. Ferris and J. M. Ferris, *Nematology, from Molecule to Ecosystem* (F. J. Gomers and P. W. T. Maas, eds.), Dekker and Huisman, Wildervank, The Netherlands, p. 92 (1992).
206. E. Van der Knaap, R. J. Rodriguez, and D. W. Freckman, *Soil Biol. Biochem. 25*:1141–1151 (1993).
207. D. J. Hooper, *Laboratory Methods for Work with Plant and Soil* (J. F. Southey, ed.), HMSO, London. MAFF reference book 402, p. 599 (1986).
208. B. S. Griffiths, *Soil Biol. Biochem. 18*:637–641 (1986).
209. I. Andrassy, *Acta Zool. Hung. 2*:1–15 (1956).
210. F. G. W. Jones and M. Jones, *Pests of Field Crops*. Edward Arnold, (1964).
211. J. P. Curry, *Grassland Invertebrates*, Chapman and Hall; (1994).
212. R. D. Barnes, *Invertebrate Zoology*, 4th ed., Holt Saunders, (1980).
213. A. D. Imms. *A General Textbook of Entomology*, 9th ed., Methuen, (1957).
214. H. Petersen and M. Luxton, *Oikos 39*:287–388 (1982).
215. T. R. Seastedt, *Annu. Rev. Entomol. 29*:25–46 (1984).
216. R. D. Bardgett, *The Effects of Changes in Sheep Management Intensity on Faunal/Fungal Interactions Related to Nutrient Cycling in Upland Soils*. Unpublished PhD thesis, Lancaster University (1991).
217. P. F. Hendrix, R. W. Parmelee, D. A. Crossley, D. C. Coleman, E. P. Odum, and P. M. Groffman, *Bioscience 36*:374–380 (1986).
218. J. C. Coulson and J. B. Whittaker, Ecology of moorland animals. *The Ecology of British Moors and Montane Grasslands* (O. W. Heal and D. F. Perkins, eds.), Springer Verlag, Berlin, pp. 52–93 (1978).
219. O. W. Heal and D. F. Perkins, *Philo. Trans. R. Soc. London 274B*:295–314 (1976).
220. K. J. Hutchinson and K. L. King, *J. Appl. Ecol. 17*:369–387 (1980).
221. J. Lussenhop, R. Fogel, and K. Pregitzer, *Agric. Ecosystems Environ. 34*:235–250 (1991).
222. T. B. Poole, *Pedobiologia 4*:35–42 (1964).
223. K. L. King and K. J. Hutchinson, *J. Appl. Ecol. 13*:41–55 (1976).
224. E. C. Adams, *Pedobiologia 11*:321–337 (1971).
225. W. G. Hale, *Pedobiologia 6*:65–99 (1966a).
226. W. G. Hale, *Rev. Ecol. Biol. Sol. 3*:97–122 (1966b).
227. R. D. Bardgett, J. C. Frankland, and J. B. Whittaker, *Agric. Ecosytems Environ. 45*:25–45 (1993a).
228. J. P. Glasgow, *J. Anim. Ecol. 8*:323–369 (1939).
229. A. Macfadyen, *J. Anim. Ecol. 21*:87–117 (1952).
230. S. Milne, *Pedobiologia 2*:41–52 (1962).
231. N. Hijii, *Ecol. Res. 2*:159–173 (1987).

232. P. Greenslade, *J. Arid Environ. 4*:219–228 (1981).
233. J. P. Winter, R. P. Voroney, and D. A. Ainsworth, *Can. J. Soil Sci. 70*:641–653 (1990).
234. J. A. Holt, *Aust. J. Ecol. 10*:57–65 (1985).
235. M. J. Mitchell, *Ecology 57*:302–312 (1978).
236. M. Luxton, *Pedobiologia 21*:365–386 (1981).
237. B. R. Hutson and L. G. Veitch, *Aust. J. Ecol. 8*:113–126 (1983).
238. J. C. Perdue and D. A. Crossley, *Biol. Fertil. Soils 9*:135–138 (1990).
239. M. Hassall, S. Visser and D. Parkinson, *Pedobiologia 29*:175–182 (1986).
240. V. G. Marshall, *Can. J. Soil Sci. 54*:491–500 (1974).
241. H. Takeda, *Pedobiologia 18*:22–30 (1978).
242. W. G. Whitford, D. W. Freckman, N. L. Elkins, L. W. Parker, R. Parmalee, J. Phillips, and S. Tucker, *Soil Biol. Biochem. 13*:417–425 (1981).
243. M. B. Usher, *The Role of Terrestrial and Aquatic Organisms in Decomposition Processes* (J. M. Anderson and A. Macfadyen, eds.), Blackwell, Oxford, pp. 61–94 (1976).
244. M. B. Usher, R. G. Booth, and K. E. Sparkes, *Pedobiologia 23*:126–144 (1982).
245. P. W. Murphy, *J. Soil Sci. 4*:155–193 (1953).
246. H. A. Verhoef, C. J. Nagelkerke, and E. N. G. Joosse, *J. Insect Physiol. 23*:1009–1013 (1977).
247. J. A. Wallwork, *The Distribution and Diversity of Soil Fauna*, Academic Press, London (1976).
248. J. C. Moore, D. E. Walter, and H. W. Hunt, *Annu. Rev. Entomol. 33*:419–439 (1988).
249. J. M. Anderson and I. N. Healey. *J. Anim. Ecol. 41*:359–368 (1972).
250. M. Luxton, *Pedobiologia 12*:434–463 (1972).
251. J. H. McMillan, *Rev. Ecol. Biol. Sol. 12*:449–457 (1975).
252. M. J. Mitchell and D. Parkinson, *Ecology 57*:302–312 (1976).
253. J. B. Whittaker, *Oikos 36*:203–206 (1981).
254. P. Reutimann, *Pedobiologia 30*:425–433 (1987).
255. R. D. Bardgett, J. B. Whittaker, and J. C. Frankland, *Biol. Fertil. Soils 16*:296–298 (1993).
256. S. Visser and J. B. Whittaker, *Oikos 29*:320–325 (1977).
257. J. C. Moore, E. R. Ingham, and D. C. Coleman, *Biol. Fertil. Soils 5*:6–12 (1987).
258. P. J. A. Shaw, *Pedobiologia 31*:179–187 (1988).
259. K. Newell, *Soil Biol. Biochem. 16*:227–234 (1984a).
260. K. Newell, *Soil Biol. Biochem. 16*:235–240 (1984b).
261. J. Lussenhop, *Adv. Ecol. Res. 23*:1–34 (1992).
262. M. J. Swift, O. W. Heal, and J. M. Anderson, *Decomposition in Terrestrial Ecosystems*, Blackwell, London (1979).

263. V. V. S. R. Gupta, *Soil Biota: Management in Sustainable Farming Systems* (C. E. Pankhurst, B. M. Doube, V. V. S. R. Gupta, and P. R. Grace, eds), CSIRO Australia, pp.107–124 (1994).
264. C. L. Rockett, *Int. J. Acarol. 12*:175–180 (1986).
265. G. J. House, M. Del Rosario Alugaray, and M. Del Rosario Alzugaray, *Environ. Entomol. 18*:302–307 (1989).
266. M. H. Beare, R. W. Parmelee, P. F. Hendrix, and W. Cheng, *Ecol. Monog. 62*:569–591 (1992).
267. G. J. House and R. W. Parmelee, *Soil Tillage Res. 5*:351–360 (1985).
268. S. J. Loring, R. J. Snider, and L. S. Robertson, *Pedobiologia 22*:172–184 (1981).
269. D. Mallow, R. J. Snider, and L. S. Robertson, *Pedobiologia 28*:115–131 (1985).
270. J. C. Coulson and J. E. L. Butterfield, *J. Ecol. 66*:631–650 (1978).
271. S. B. Wakerley, *Ann. Appl. Biol. 51*:399–401 (1963).
272. K. L. King and K. J. Hutchinson, *J. Appl. Ecol. 17*:581–591 (1980).
273. C. A. Edwards and J. R. Lofty, *The Soil Ecosystem* (J. R. Sheal, ed.), Systematics Association, London, pp. 237–247 (1969).
274. C. A. Edwards and J. R. Lofty, *Rothamsted Report 1974*, part 2 pp. 133–154 (1974).
275. H. Siepel and C. F. Bund van de, *Pedobiologia 31*:339–354 (1988).
276. V. M. Behan, S. B. Hill, and D. K. M. Kevan, *Pedobiologia 18*:249–263 (1978).
277. T. Persson, *National Swedish Environment Protection Board, Report 3418*, (1988).
278. P. Hartmann, R. Fischer, and M. Scheidler, *Verhandlungen Gessssellschaft Okologie 17*:585–589 (1989).
279. K. L. King, K. J. Hutchinson, and P. Greenslade, *J. Appl. Ecol. 13*:731–739 (1976).
280. J. H. Waisingham, *Society Proceedings of the Royal Dublin Society A6 11*:297–304 (1976).
281. R. D. Bardgett, J. B. Whittaker, and J. C. Frankland, *Biol. Fertil. Soils 16*:255–262 (1993c).
282. R. D. Bardgett and D. K. Leemans, *Biol Fertil Soils*
283. G. Lara, E. Parada, N. Butendieck, and R. Covarrubias, R. *Ciencia Investigación Agrari 13*:81–89 (1986).
284. W. Wegorek and H. Trojanowski, *Colloques I'INRA 36*:27–37 (1986).
285. D. Conrady, *Pedobiologia 29*:273–284 (1986).
286. C. A. Edwards, *CRC Crit. Rev. Plant Sci. 8*:221–253 (1989).
287. J. Vogel, W. Funke, and N. Wilhelm, *Verhandlungen Gesellschaft Okologie 18*:469–472 (1989).
288. C. A. Edwards. *Agric. Ecosystem Environ. 34*:145–176 (1991).

289. R. McSorley and D. E. Walter, *Agric. Ecosystem Environ. 34*:201–207 (1991).

290. T. R. Seastedt and D. A. Crossley Jr., *J. Georgia Entomol. Soc. 13*:338–344 (1978).

291. D. A. Crossley Jr. and J. M. Blair, *Agric. Ecosystems Environ. 34*:187–192 (1991).

292. M. Bieri, V. Delucchi, and M. Stadler, *Pedobiologia 30*:127–135 (1986).

293. H. Tamura, *Rev. Ecol. Biol. Sol. 13*:21–34 (1976).

294. T. R. E. Southwood, *Ecological Methods with Particular Reference to the Study of Insect Populations*. Chapman and Hall, London (1978).

6

Interactions Between Microbe-Feeding Invertebrates and Soil Microorganisms

BRYAN S. GRIFFITHS   Scottish Crop Research Institute, Dundee, United Kingdom

RICHARD D. BARDGETT   University of Manchester, Manchester, United Kingdom

1. INTRODUCTION

In contrast to that of soil microflora, the contribution of invertebrates to total soil respiration is viewed as low (1–4). In recent years, however, it has been increasingly recognized that the invertebrate soil fauna have an important regulatory role in the processes of organic matter decomposition and nutrient cycling through its influence on the composition and activity of soil microbial communities. The soil fauna affect microbial populations directly, by feeding (grazing) on fungi and bacteria, and indirectly by comminution of organic matter, dissemination of microbial propagules, and alteration of nutrient availability. The total of all the below-ground interactions leads to the characteristic flow of energy and nutrients through the food web. There are interactions between invertebrates and microorganisms which are too detailed to be included in food web studies but which are important in the functioning of soil ecosystems. Many of these are involved in the turnover and cycling of nutrients.

The action of invertebrates generally leads to increased nutrient turnover and availability, as a result of both direct and indirect effects. Their direct effect is excretion of excess ingested nutrients that are not required for production. Sixty percent of N ingested by protozoa, and 90% by bacteria-feeding nematodes is not used in production and is excreted into the soil environment. This N is directly biologically available, usually in the form of ammonium-N, although nematodes can excrete up to 50% as amino acid-N (5). Indirect effects of microbe-feeding invertebrates are due to (1) modification of the microbial community, (2) accelerated turnover of microbial cells, and (3) inoculation of new substrates. It is difficult to separate the direct and indirect effects experimentally, but the former can be estimated by calculating amounts of excretion from observed changes in invertebrate populations. This approach is usually taken in food web analyses, which have shown that the direct effects of invertebrates account for approximately 30% of total net nitrogen mineralization (6). Such calculations have also been included in mathematical simulations of nutrient release from microbial grazing (7) and have demonstrated that grazing can release 100 to 700 μg N g^{-1} soil at certain active sites (6). The indirect stimulation of microbial activity results from a greater abundance of young microbial cells, defined as cells in the log growth phase rather than senescent cells (8,9), and maintenance of bacterial abundance at a low enough level to minimize substrate limitation (10). Experiments on bacterial processes (i.e., nitrification and nitrogen fixation) and productivity have shown that grazed bacteria are more active than ungrazed bacteria (8,11–17).

2. MICROFAUNA (NEMATODES AND PROTOZOA)–MICROBIAL INTERACTIONS

Experimental systems of increasing complexity have clearly demonstrated the enhancing effect of microbe-feeding microfauna (mainly protozoa and nematodes) on nutrient flows in soil. The rate of net mineralization (i.e., the accumulation of nutrients resulting from the difference between the release of inorganic nutrients [mineralization] and their uptake by microorganisms) in sterile soil inoculated with single species of microbe and grazer is faster in the presence of nematodes or protozoa (18–24). This is also the case if single species of bacteria, fungi, and their respective nematode grazers are added together (23), and if bacterial and protozoan populations extracted from soil are used to inoculate sterile soil (25,26). In N-limited soil, the extra N mineralized by grazing activity may be rapidly immobilized by microbial growth (21,27). This result is probably more likely in natural soil, with nutrients mineralized by grazing activity used for microbial or plant growth and not accumulating to any great extent in the soil. Several studies have shown that N released by microfaunal grazing is subsequently taken up by plants, although this mechanism has not yet been determined for phosphorus in soil (28). Plants grown in the presence of microbe-feeding protozoa and nematodes contain more N than plants grown in their absence (23,29,30) and can also have greater dry weight (31). The source of the extra N available to plants has been shown to be both N directly released from bacterial cells and N mineralized from soil organic matter as a result of the higher N turnover (32). Grazing can release 30% to 50% of the N immobilized in the microbial biomass for plant uptake (31–33). The enhancement of N transfer to plants by grazing was greater when soil moisture content was allowed to fluctuate than when relatively constant soil moisture was maintained (32,34).

Feeding by protozoa and nematodes is selective, in terms of the prey species consumed, and complex, in view of the interactions that occur. Feeding results in activity in the observed microbial community. It is well documented that bacteria can persist even in the presence of actively feeding protozoa and are not completely eliminated by the grazer. This occurs both in liquid culture (35,36) and in soil (37,38) and has been reviewed (39). The effect is partly density-dependent, as there is a threshold prey density above which growth of the grazer occurs (40) and below which protozoa encyst (41). The persistence of bacteria is also due to their ability to reproduce and replace those cells consumed by predation (35,36). In a mixture of prey bacteria, more akin to real communities, survival of a bacterial species is related to its ability to grow, the hydrophobicity (water-repellent nature) of its cell surface, and its initial population density (42). Species will be eliminated if there is an alternative prey species present at a concentration greater than that necessary for active feeding and if its rate of growth is less than the rate of predation (43). These studies assume nonselective ingestion of different bacteria, but aquatic protozoa exhibit a general preference for large bacteria (44–47). This preference can lead to the selective removal of actively dividing cells, because they are larger than undivided cells, and therefore to the direct cropping of bacterial production (46). In an experimental aquatic system, protozoa removed a significant fraction of daily bacterial production (60% to 100%) (48). There is no evidence that nematodes select particular size classes of prey, other than the limits imposed by the physical dimensions of their mouthparts. While prey geometric characteristics are important (i.e., size and shape in relation to their mouthparts as for nematodes) for protozoa, there is clear evidence that chemical clues are used to discriminate between bacterial and algal prey species by a range of flagellate and ciliate protozoa (49). Invertebrates grow better on some prey species than others, while some prey, such as the bacterium *Chromobacterium violaceum*, are toxic or support no growth at all. This has been shown for protozoa (50–59), nematodes (60,61), and nematodes as prey for collembola (62). There was, for example, a 100-fold difference in amoebal yield on 10 bacterial prey species and a 100-fold difference in the yield of different amoebae growing on the same bacterial species (63).

The chemical clues to prey selection are also evident in the attraction of predator to prey. The differential attractiveness of two bacterial species to protozoa was due to an as yet unidentified, high-molecular-weight compound extracted from the cell wall (64). There were significant differences in the numbers of the flagellate *Bodo saltans* which migrated toward cultures of different bacteria (54). Bacterial colonies could be either attractive, repulsive, or neutral to the nematode *Caenorhabditis elegans*, although only live bacteria elicited a response (65). The chemoattraction of this nematode varied, depending on whether it had been reared in axenic or xenic culture (66), and volatile chemicals have been shown to attract or repel *C. elegans* at different concentrations (67). Chemoattraction is probably important for soil nematodes because of their relatively low numbers and long generation times and has been reviewed (68,69). The release of extracellular enzymes by the ciliate *Tetrahymena thermophila* may play a role in its utilization of complex nutrients (70) and would lead to more complex interactions if it were a widespread phenomenon in soil. Rhizosphere- (i.e., the soil immediately adjacent to, and influenced by, a plant root) dwelling nematodes were thought to graze bacteria in suspension preferentially rather than those attached to particles (15), while some protozoa are adapted for the preferential removal of bacteria from surfaces (71).

This selective grazing determines the observed microbial community structure. In the sheep (*Ovis aries*) rumen, for example, the presence of protozoa increased the proportion of xylanolytic bacteria, which degrade xylan, a plant cell wall component, and fungal zoospores (72). Grazing by protozoa and nematodes in the rhizosphere reduced mycorrhizal colonization of tree roots and may affect the composition of the microbial community (73,74). Protozoa in sewage sludge eliminated slow-growing bacteria (43), and after inoculation with three fast- and three slow-growing bacterial species, only two of the former survived in detectable numbers (75). In an aquatic ecosystem, however, protozoan preference for larger bacterial cells allowed the coexistence of both fast- and slow-growing bacteria (46). Grazing by flagellates in marine enclosures greatly modified the composition of the bacterial community by allowing growth of uncommon bacteria (76) and affected bacterial size distribution because the largest and smallest cells escaped grazing (77). A bacterial species responded to grazing by developing cells 20 μm long that were resistant to grazing but had a lower growth rate than the parental 1.5 μm cells (78). Small rods and cocci predominated in aquatic enclosures when protozoa were reduced, but when protozoa increased, long rods and filaments predominated (79). The bacterium *Leptospirillum ferrooxidans* became the dominant iron oxidizer in a coal biotreatment plant at a much earlier stage when flagellates were present (80). In activated sludge, protozoan grazing caused *Cytophaga* to be replaced by a grazing-resistant *Microcyclus* strain (81). It was also noted that flocculation and filamentous growth offered protection from grazing and that grazed bacterial communities were morphologically more diverse than ungrazed communities (82). The effects of protozoan and nematode grazing on soil microbial communities have yet to be demonstrated, and some of the novel techniques outlined in the following chapters may allow these events to be studied.

An equally important problem is the relation between microbial community structure and function. Do the observed morphological changes induced by grazing affect the functioning of the community? An indication of altered potential within a community in response to grazing was obtained from a preliminary laboratory experiment. Flasks containing sterile barley (*Hordeum vulgare*) leaves as substrate were inoculated with a suspension of bacteria isolated from soil and half were also inoculated with the ciliate *Tetrahymena pyriformis*. The bacterial community was sampled after incubation for 18 days at 15°C, and its metabolic potential for growth on 95 different carbon sources assessed using commercially available plates (BIOLOG) (83). Principal component analysis clearly separated the grazed from the ungrazed communities (Fig. 1). There is a possibility of controlling microbial activity by using protozoa to alter microbial community structure, although changes in activity patterns with and without protozoa have not been demonstrated (59). The rumen contains a complex microbial community, and here the presence of protozoa can increase enzyme activity, although it is not certain whether this is a direct or an indirect effect (72,84,85). In cattle (*Bos taurus*) fed a high-grain diet, however, rumen ciliates acted to moderate the rate of rumen fermentation (86).

The majority of studies into the effects of grazing on microbial activity deal with single-species interactions. The production of stimulatory factors by protozoa has been demonstrated by using filtrates from protozoan cultures to enhance bacterial growth (87) and activity (84,88). Protozoa tended to increase the acclimation period for the mineralization of xenobiotics and to affect the extent of mineralization (89). Grazing also tends to increase microbial production (12,13,15) and the turnover rate of microbial protein (90). There were more bacteria in sewage sludge in the presence of nematodes; their presence was related to increased oxygen levels caused by nematode bioturbation (17). Thus, while it can be demonstrated that microbial community structure is altered by grazing, more research into the subsequent effects on the activity of the resulting community in soil is required.

Dispersal of inocula, of either prey or grazer, is important in the colonization of new substrates in soil (91,92). Microorganisms are able to survive ingestion by protozoa and nematodes and to become active after defecation in a new location. This capacity has been proposed as a mechanism for the survival of fastidious microorganisms in hostile environments (93), and some, such as the *Legionella* bacteria, are adapted for replication within autochthonous amoebae (94). In another example, *Escherichia coli* decreased more rapidly than *Enterococcus faecalis* when added to aquatic systems containing protozoa, even though both bacteria were ingested at the same rate, because *E. faecalis* was digested more slowly and a higher proportion were still viable after defecation

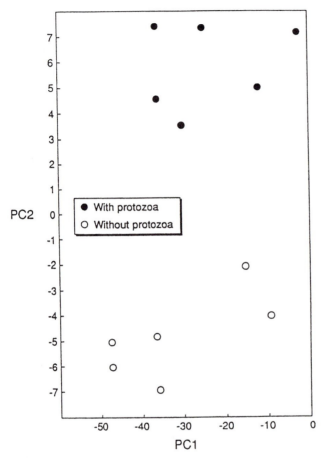

Figure 1. Results of principal component analysis (PCA) of the metabolic profile of bacterial communities developing on decomposing barley leaves, in six replicate flasks with and without added protozoa. PCA is a statistical technique for analyzing multivariate data (each metabolic profile consists of 95 variables, i.e. C sources) that transforms the original variables into a smaller number of new variables called principal components. In this case, the treatments could be separated by using the first two principal components (PCs), PC1 and PC2.

(95). The nematode *C. elegans* was responsible for spread of the bacterium *Pseudomonas fluorescens* through a mushroom crop (61). Indeed, nematodes have been used as carriers of bacterial inoculants under field conditions (96). The grazers themselves can also be transported to new locations; either by ingestion and defecation, such as amoebal spores by woodlice (*Armadillidium nasatum* and *A. vulgare*) and earthworms (*Aporrectodea caliginosa* and *Octolasion tyr-*

taeum) (97,98), or phoretically, such as the transport of nematodes by oribatid, endeostigmatid, mesostigmatid, and astigmatid mites and a collembolan (*Hypogastura scotti*) (99).

Other aspects of interactions concern coexistence, competition, and symbiosis. Bacteria-feeding flagellates and ciliates were able to coexist when apparently competing for the same bacterial resource, because the flagellate was able to occupy two trophic levels at once (100). It was able to absorb organic nutrients from solution directly as well as to consume bacteria. Competition is affected by the existence of both r-selected (i.e., initial colonizers with a high growth rate) and K-selected (i.e., slower-growing species characteristic of stable ecosystems) species of protozoa (101) and nematodes (60). Nematodes are rapidly out-competed by protozoa for a bacterial food source (22,102), yet bacteria-feeding nematodes and protozoa coexist at active sites in soil (103,104). This effect may be due to nematode predation on protozoa (105) or physical separation of the grazers by their ability to access soil pores of different dimensions. There were more bacteria-feeding nematodes in the rhizosphere when plant-feeding nematodes were also present, because of a greater allocation of resources below-ground (106). There seems to be a complex series of interactions between earthworms and nematodes. There are reports of more microbe-feeding and fewer plant-feeding nematodes in soil containing earthworms (such as *Agrotoreutus nyongi* and *Lampito mauritii*) and in worm casts than in the surrounding soil (107,108). There has been, however, a tendency for earthworms (such as *Dendrobaena octaedra*) to reduce nematode numbers in limed coniferous forest plots (109) and agricultural soil (110). Many symbiotic interactions occur between nematodes and bacteria (92); in a recently described example the ectosymbiotic bacteria associated with nematodes inhabited the oxygen sulfide chemocline in marine sands (111). Protozoa exhibit a similar range of symbioses with, for example, ectosymbiotic bacteria on the ciliate *Kentrophoros* oxidizing hydrogen sulfide to sulfur, anaerobic protozoa harboring endosymbiotic methanogens, and aquatic ciliates containing photosynthetically active algae (112). Amoebae have also been shown to acquire mercury resistance from their harbored bacteria (113). In this symbiosis, bacteria (*Aeromonas* sp.) containing Hg resistance genes were provided with a stable environment for growth and survival within the cytoplasm of an amoeba (*Acanthamoeba* sp.), while the bacterial metabolism enabled *Acanthamoeba* to survive at otherwise toxic concentrations of mercury. Symbiotic relationships in soils, however, have not, to the best of our knowledge, been identified.

3. MICROARTHROPOD (COLLEMBOLA AND MITE)– MICROBIAL INTERACTIONS

It has been established that field populations of soil microarthropods (Collembola and mites) feed preferentially on fungal hyphae, and that Collembola in particular will select between fungal species in the laboratory (114–117) and in the field (118,119). The selection of food by microarthropods is largely based on prey species; however, age may also influence prey choice. The collembolan *Folsomia candida*, for example, fed on metabolically active hyphae in preference to inactive and dead hyphae of common soil fungi (120,121). It has been suggested that Collembola preferentially graze regions of fungal thallus with high nitrogen content (120). Soil inhabiting species of Collembola have also been shown to locate and select fungal food sources by volatile compounds released from the mycelium (122).

Selective microarthropod grazing may have a strong influence on the abundance, activity, distribution, and morphological characteristics of soil microorganisms. Laboratory studies showed that fungus-feeding by Collembola (*F. candida*) decreased fungal populations, with a subsequent increase in bacterial numbers, on decomposing oak (*Quercus robur*) litter (123). Similarly, fungus grazing by the collembolan *F. candida*, on oak leaf litter, reduced fungal standing crop (124). In a more recent study, laboratory microcosms containing litter from upland grasslands were used to assess the impact of grazing by a collembolan, *Onychiurus procampatus*, on the activity of a common saprotrophic fungus, *Phoma exigua* (125). Fungal biomass and respiration rate were reduced only when Collembola were present in excess of mean field densities (Figs. 2 and 3), but perhaps more typical of spatial aggregations in the soil.

The effects of collembolan grazing on microbial biomass and activity are dependent in part on the age of the animal. In the presence of juvenile Collembola (*F. candida*), microbial respiration was enhanced but microbial biomass was unchanged. However, when adult Collembola were added to microcosms, microbial biomass was increased and respiration decreased (126). An explanation for these results was an age-specific food-selection strategy by the collembolan. It is known that the respiration rate of bacteria is considerably lower than that of fungi when biomasses are equal (127); therefore, it was suggested that juvenile animals fed mainly on bacteria, while adults fed mainly on fungi (126). This was consistent with the observed increase in bacterial biomass and reduction in fungal biomass as a result of grazing by older collembola (123).

As well as controlling biomass, microarthropod grazing is known to stimulate fungal growth and activity. Respiration rates in intact soil cores from a high arctic site containing both microorganisms and Collembola were significantly higher than in those containing microorganisms alone, and this enhancement was greater than could be accounted for by the respiration of the Collembola themselves (128). Under laboratory conditions, grazed fungus had a lower respiration rate than ungrazed fungus when Collembola were present on mycelium (129). However, fungal respiration was increased in Collembola when they were periodically removed from the mycelium. Other studies, however, have demonstrated variable effects of grazing by soil microarthropods on fungal activity. Grazing by the collembola *Onychiurus subtenuis* had no effect on the respiration

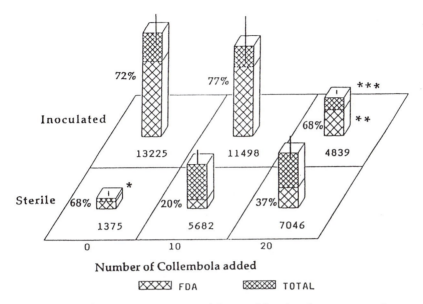

Figure 2. Total hyphal length (m g⁻¹ litter) of fungi and percentage of fluorescein diacetate active hyphae, on 3 g sterile grass litter (from a mixture of species) inoculated with (inoculated) or without (sterile) fungus (*Phoma exigua*) prior to the addition of 0, 10, or 20 Collembola at the end of a 12 week experimental period. Collembola additions significantly different within the same inoculation treatments: *, **, ***, $P < 0.05$, $P < 0.01$, $P < 0.001$, respectively. Fluorescein diacetate is a vital stain used to distinguish live from dead hyphae. (From Ref. 125.)

of a sterile dark fungus in microcosms containing aspen poplar (*Populus tremuloides*) leaf litter (130). Similarly, grazing by the collembolan *Folsomia fimetaria* had no effect on the respiration of soil microorganisms decomposing barley straw (131).

Stimulation of fungal activity has been referred to as compensatory growth, driven by mechanisms of fungal growth after senescent hyphae are grazed, and regrowth after periodic grazing of actively growing mycelia (132). Such compensatory growth was illustrated in grazing of the fungus *Mortierella isabellina* by the collembolan *Onychiurus armatus*, which induced switching from a "normal" hyphal mode, with appressed growth and sporulating hyphae, to fan-shaped sectors of fast growing and nonsporulating mycelium with extensive areas of aerial mycelium (133). In addition, specific amylase (starch-degrading enzymes) activities were several times higher in grazed cultures where switching occurred than in plates without switching.

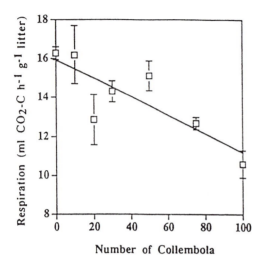

Figure 3. Relationship between the total number of Collembola added to microcosms, containing 3 g sterile grass litter (from a mixture of species) inoculated with the fungus *Phoma exigua*, and respiration rate (CO_2 evolution) ($R^2 = 0.703$; $P < 0.001$). (From Ref. 125.)

Selective grazing by Collembola has been shown to alter the competitive ability of fungi. In a study of two unidentified fungi growing in aspen poplar leaves at snow melt, there were a competitively inferior, sterile dark fungus that was grazed by the collembolan *O. subtenius*, and a competitively superior basidiomycete fungus, which was unpalatable and toxic to the collembolan (134). It was shown that collembolan grazing reinforced the competitive ability of the basidiomycete. Subsequent studies confirmed these interactions in the field (135). Selective grazing by *Onychiurus latus* (collembolan) altered the competitive ability of two basidiomycete fungi involved in the decomposition of Sitka spruce (*Picea sitchensis*) litter (118,119). A combination of laboratory and field exclusion experiments suggested that higher densities of the collembolan resulted in a reduction in the activity of the palatable fungus *Marasmius androsaceus* and an increase in that of the unpalatable *Mycena galopus*, and vice versa. Selective grazing may be an important factor determining the field distribution of these fungi, since *M. androsaceus* was restricted to the uppermost layer of freshly fallen litter (L litter horizon), where the density of *O. latus* was low, and *M. galopus* predominated in the lower layer of partially decomposed litter (F horizon) (Fig. 4). Additional laboratory studies showed that grazing by the collembolan stimulated decomposition of Sitka spruce litter by *M. androsaceus*, whereas that of *M. galopus* declined.

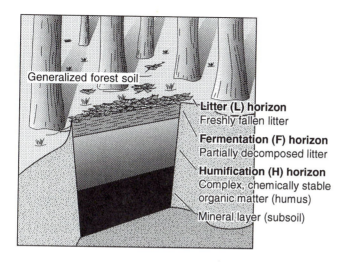

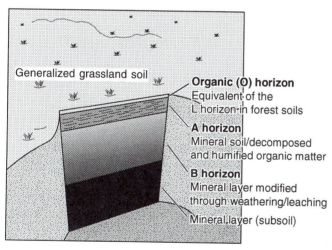

Figure 4.   Schematic diagrams to illustrate the horizons (layers) within soil profiles from under forest or grassland. For more detail refer to Ref. 156.

It is well established that the feeding activities of soil microarthropods enhance nutrient mineralization in soils. As microorganisms are digested and metabolized in the animal gut, nutrients which were immobilized by the microbial biomass are mineralized and excreted into the soil environment (136). Through this mechanism, collembolan grazing of fungi growing on oak litter increased leaching of ammonium, nitrate, and calcium (124). It is likely, however, that grazing also enhances nutrient mineralization indirectly through the stimulation of microbial biomass and activity in soil. For example, microcosm studies showed that the stimulation of microbial activity, induced by the feeding of the collembolan *Tomocerus minor* on pine (*Pinus nigra*) litter, increased exchangeable phosphate and nitrate concentrations, and cellulase (cellulose-degrading enzymes) and dehydrogenase (enzymes catalyzing redox reactions, indicative of microbial activity) enzyme activities (137,138). Similarly, the collembolan *T. minor* was associated with enhanced nitrogen mineralization in the F horizon of a pine forest, attributed to the animal excretion products and the stimulation of fungal growth within the horizon (139). Collembolan grazing activities, however, had no effect on nutrient release from fungi growing on aspen poplar leaves and upland grassland litter (125,130). In both studies the lack of significant effects was attributed to experimental limitations.

The guts and feces of grazing animals provide an additional environment in the soil where alterations in microbial community structure and activities occur. Such changes are likely to have a profound influence on the distribution of microorganisms in the soil and on spatial processes of nutrient mineralization. Microbial dynamics in microarthropod guts and feces and their implications have been recently reviewed (132,140). The gut and feces of microarthropods appear to provide a favorable environment for enhancement of microbial activity. There were, for example, five times as many chitinolytic (degrade chitin, key component of fungal cell walls) bacteria in the gut of *F. candida* as were present in soil (141). It has also been suggested that microarthropods facilitate the activity of microbial enzymes, such as cellulases, in their guts by maintaining a basic pH (132).

As well as feeding on saprophytic microorganisms, microarthropods interact with both pathogenic organisms and beneficial mycorrhizal fungi in the rhizosphere. Microarthropod grazing activities appear particularly effective at suppressing pathogens, since many pathogenic fungi lack specific defenses against fungus-feeding animals and are a preferred food source for many Collembola (142,143). However, microarthropods are also recognized as transporters of pathogenic inoculum to roots, thereby increasing disease (144). The role of microarthropod activities in plant pathogen suppression has been reviewed extensively (132,144,145).

It is well established that microarthropods, particularly collembolans, selectively graze vesicular–arbuscular mycorrhizal (VAM) spores, germ tubes, and hyphae (121,146), and that these activities reduce the benefits of mycorrhizal infection (nutrient inflow and dry-matter production) to plants (147,148). However, mycorrhizal fungi which fruit only below-ground appear to be in part dependent on soil fauna to disperse VAM fungal propagules through the soil to new sites (149). The significance of mycorrhizal–microarthropod interactions has also been reviewed extensively (132,149,150).

Microarthropods are mobile and therefore passively transport bacteria, fungi, and protozoa in their gut or on their cuticle through the soil to new microsites and substrates (so-called phoretic transport). Both mites and Collembola have been reported to carry propagules from several species of saprotrophic fungi (151–153) and VAM mycorrhizal fungi (154). Recent laboratory experiments, for example, showed that the collembolan *Heteromurus nitidus* was able to transport viable propagules of three fungi (*Mucor hiemalis, Chaetomium globosum, Sphaerobolus stellatus*) beyond the mycelial front to uncolonized soil (S. Moody, University of Lancaster, United Kingdom, personal communication). However, viable propagules of the fungus *Agrocybe gibberosa* were not readily transported by the collembolan. It has also been speculated that some fungi are in fact adapted for dispersal by microarthropods (155).

ACKNOWLEDGMENTS

This work was funded by the Scottish Office Agriculture Environment and Fisheries Department, and the Biotechnology and Biological Sciences Research Council.

REFERENCES

1. A Macfadyen, *Soil Organisms* (J. Doeksen and J. van der Drift, eds.) North Holland, Amsterdam, p. 3 (1963).
2. C. A. Edwards, D. E. Reichle, and D. A. Crossley, *Analysis of Temperate Forest Ecosystems* (D. E. Reichle, ed.), Springer-Verlag, Berlin, p. 147 (1970).
3. D. E. Reichle, *Soil Organisms as Components of Ecosystems* (U. Lohm and T. Persson, eds.), Ecological Bulletin No. 25, Swedish Natural Science Research Council, Stockholm, p. 145 (1977).
4. H. Petersen and M. Luxton, *Oikos 39*:287 (1982).
5. R. V. Anderson, W. D. Gould, L. E. Woods, R. E. Cambardella, R. E. Ingham, and D. C. Coleman, *Oikos 40*:75 (1983).
6. B. S. Griffiths, *Plant Soil 164*:25 (1994).
7. P. M. Rutherford and N. G. Juma, *Can. J. Soil Sci. 72*:201 (1992).
8. D. W. Cutler and D. V. Bal, *Ann. Appl. Biol. 13*:516 (1926).
9. R. E. Johannes, *Advances in Microbiology of the Sea*, vol. 1 (M. R. Droop and E. J. Ferguson Wood, eds.), Academic Press, London, p. 203 (1965).
10. J. M. Sieburth and P. M. Davis, *Ann. Inst. Oceanogr. 58*:285 (1982).
11. J. F. Darbyshire, *Soil Biol. Biochem. 4*:359 (1972).
12. M. Pussard and J. Rouelle, *Protistologica 22*:105 (1986).
13. B. Riemann, N. O. G. Jorgensen, W. Lampert, and J. A. Fuhrman, *Microb. Ecol. 12*:247 (1986).

14. B. S. Griffiths, *Soil Biol. Biochem. 21*:1045 (1989).
15. P. Sundin, A. Valeur, S. Olsson, and G. Odham, *FEMS Microbiol. Ecol. 73*:13 (1990).
16. F. J. M. Verhagen and H. J. Laanbroek, *Appl. Environ. Microbiol. 58*:1962 (1992).
17. B. J. Abrams and M. J. Mitchell, *Oikos 35*:404 (1980).
18. J. Meiklejohn, *Ann. Appl. Biol. 17*:614 (1930).
19. D. C. Coleman, C. V. Cole, R. V. Anderson, M. Blaha, M. K. Campion, M. Clarholm, E. T. Elliott, H. W. Hunt, B. Schaefer, and J. Sinclair, *Ecol. Bull. (Stockholm) 25*:299 (1977).
20. C. V. Cole, E. T. Elliott, H. W. Hunt, and D. C. Coleman, *Microb. Ecol. 4*:381 (1978).
21. L. E. Woods, C. V. Cole, E. T. Elliott, R. V. Anderson, and D. C. Coleman, *Soil Biol. Biochem. 14*:93 (1982).
22. B. S. Griffiths, *Soil Biol. Biochem. 18*:637 (1986).
23. R. E. Ingham, J. A. Trofymow, E. R. Ingham, and D. C. Coleman, *Ecol. Monogr. 55*:119 (1985).
24. P. M. Rutherford and N. G. Juma, *Can. J. Soil Sci. 72*:183 (1992).
25. J. S. Frey, J. F. McClellan, E. R. Ingham, and D. C. Coleman, *Biol. Fertil. Soils 1*:73 (1985).
26. P. J. Kuikman and J. A. van Veen, *Biol. Fertil. Soils 8*:13 (1989).
27. J. L. Sinclair, J. F. McClellan, and D. C. Coleman, *Appl. Environ. Microbiol. 42*:667 (1981).
28. B. S. Griffiths, *Soil Protozoa* (J. F. Darbyshire, ed.), CAB International, Wallingford, England, p. 65 (1994).
29. E. T. Elliot, D. C. Coleman and C. V. Cole, *The Soil-Root Interface* (J. L. Harley and R. Scott-Russell, eds.), Academic Press, London, p. 221 (1979).
30. B. S. Griffiths, *Roots and the Soil Environment, Aspects of Applied Biology 22*, The Association of Applied Biologists, Wellesbourne, p. 141. (1989).
31. M. Clarholm, *Soil Biol. Biochem. 17*:181 (1985).
32. P. J. Kuikman, M. M. I. van Vuuren, and J. A. van Veen, *Agric. Ecosyst. Environ. 27*:271 (1989).
33. K. Ritz and B. S. Griffiths, *Plant Soil 102*:220 (1987).
34. P. J. Kuikman, A. G. Jansen, and J. A. van Veen, *Soil Biol. Biochem. 23*:193 (1991).
35. M. Habte and M. Alexander, *Soil Biol. Biochem. 10*:1 (1978).
36. A. Sambanis and A. G. Fredrikson, *Microb. Ecol. 16*:197 (1988).
37. M. J. Acea and M. Alexander, *Soil Biol. Biochem. 20*:703 (1988).
38. C. E. Heynen, J. D. van Elsas, P. J. Kuikman, and J. A. van Veen, *Soil Biol. Biochem. 20*:483 (1988).
39. M. Alexander, *Annu. Rev. Microbiol. 35*:113 (1981).
40. S. K. A. Danso and M. Alexander, *Appl. Microbiol. 29*:515 (1975).

41. B. F. Sherr, E. B. Sherr, and T. Berman, *Appl. Environ. Microbiol. 45*:1198 (1983).
42. K. R. Gurijala and M. Alexander, *Appl. Environ. Microbiol. 56*:1631 (1990).
43. L. M. Mallory, C. S. Yuk, L. N. Liang, and M. Alexander, *Appl. Environ. Microbiol. 46*:1073 (1983).
44. J. M. González, E. B. Sherr, and B. F. Sherr, *Appl. Environ. Microbiol. 56*:583 (1990).
45. S. S. Epstein and M. P. Shiaris, *Microb. Ecol. 23*:211 (1992).
46. B. F. Sherr, E. B. Sherr and J. McDaniel, *Appl. Environ. Microbiol. 58*:2381 (1992).
47. K. Šimek and T. H. Chrzanowski, *Appl. Environ. Microbiol. 58*:3715 (1992).
48. S. Findlay, L. Carlough, M. T. Crocker, H. K. Gill, J. L. Meyer, and P. J. Smith, *Limnol. Oceanogr. 31*:1335 (1986).
49. P. G. Verity, *J. Protozool. 38*:69 (1991).
50. B. N. Singh, *Ann. Appl. Biol. 29*:18 (1942).
51. O. W. Heal and M. J. Felton, *Animal Populations in Relation to Their Food Resources* (A. Watson, ed.), Brit. Ecol. Soc. Symp. 10:145 (1970).
52. S. G. Berk, R. R. Colwell, and E. B. Small, *Trans. Am. Microsc. Soc. 95*:514 (1976).
53. W. D. Taylor and J. Berger, *Can. J. Zool. 54*:1111 (1976).
54. G. C. Mitchell, J. H. Baker and M. A. Sleigh, *J. Protozool. 35*:219 (1988).
55. L. E. Casida, Jr., *Appl. Environ. Microbiol. 55*:1857 (1989).
56. C. G. Ogden and P. Pitta, *Biol. Fertil. Soils 9*:101 (1990).
57. S. Schulz, S. Wagener, and N. Pfennig, *Eur. J. Protistol. 26*:122 (1990).
58. J. F. Darbyshire, D. A. Elston, A. E. F. Simpson, M. D. Robertson, and A. Seaton, *Soil Biol. Biochem. 24*:827 (1992).
59. S. McGinness and D. B. Johnson, *Microb. Ecol. 23*:75 (1992).
60. R. V. Anderson and D. C. Coleman, *Nematologica 27*:6 (1981).
61. P. S. Grewel, *Nematologica 37*:72 (1991).
62. S. K. Gilmore and D. A. Potter, *Pedobiologia 37*:30 (1993).
63. P. H. H. Weekers, P. L. E. Bodelier, J. P. H. Wijen, and G. D. Vogels, *Appl. Environ. Microbiol. 59*:2317 (1993).
64. R. A. Snyder, *Hydrobiologia 215*:205 (1991).
65. P. A. Andrew and W. L. Nicholas, *Nematologica 22*:451 (1976).
66. H-B. Jansson, A. Jeyaprakash, N. Marban-Mendoza, and B. M. Zuckerman, *Exptl. Parasitol. 61*:369 (1986).
67. C. I. Bargmann, E. Hartwieg, and H. R. Horvitz, *Cell 74*:515 (1993).
68. D. B. Dusenbery, *J. Nematol. 15*:168 (1983).
69. D. R. Viglierchio, *Rev. Nematol. 13*:425 (1990).

70. J. Florin-Christensen, M. Florin-Christensen, A. Tiedtke, and L. Rasmussen, *Microb. Ecol. 19*:311 (1990).
71. D. A. Caron, *Microb. Ecol. 13*:203 (1987).
72. A. G. Williams and S. E. Withers, *J. Appl. Bact. 70*:144 (1990).
73. S. Chakraborty, G. Theodorou and G. D. Bowen, *Can. J. Microbiol. 31*:295 (1985).
74. R. E. Ingham, *Agric. Ecosyt. Environ. 24*:169 (1988).
75. J. L. Sinclair and M. Alexander, *Can. J. Microbiol. 35*:578 (1988).
76. M. Bianchi, *Microb. Ecol. 17*:137 (1989).
77. S. S. Epstein and M. P. Shiaris, *Microb. Ecol. 23*:211 (1992).
78. S. Shikano, L. S. Luckinbill, and Y. Kurihara, *Microb. Ecol. 20*:75 (1990).
79. K. Jurgens, H. Arndt, and K. O. Rothhaupt, *Microb. Ecol. 27*:27 (1994).
80. D. B. Johnson and L. Rang, *J. Gen. Microbiol. 139*:1417 (1993).
81. H. Güde, *Microb. Ecol. 5*:225 (1979).
82. H. Güde, *Microb. Ecol. 11*:193 (1985).
83. J. L. Garland and A. L. Mills, *Appl. Environ. Microbiol. 57*:2351 (1991).
84. R. D. Yoder, A. Trenkle, and W. Burroughs, *J. Anim. Sci. 25*:609 (1966).
85. K. Ushida, T. Kaneko and Y. Kojima, *Jpn. J. Zootech. Sci. 58*:893 (1987).
86. T. G. Nagaraja, G. Towne, and A. A. Beharka, *Appl. Environ. Microbiol. 58*:2410 (1992).
87. K. Gustafsson, *Can. J. Microbiol. 35*:1100 (1989).
88. P. Levrat, M. Pussard and C. Alabouvette, *Eur. J. Protistol. 28*:79 (1992).
89. B. A. Wiggins and M. Alexander, *Can. J. Microbiol. 34*:661 (1988).
90. C. J. Newbould and K. Hillman, *Lett. Appl. Microbiol. 11*:100 (1990).
91. S. Visser, *Ecological Interactions in Soil* (D. Atkinson, D. J. Read and M. B. Usher, eds.), Blackwell Scientific, Oxford, p. 297 (1985).
92. G. O. Poinar and E. L. Hansen, *Helminthol. Abstr. (Ser. B.). 55*:61 (1986).
93. C. H. King, E. B. Shotts, R. E. Wooley Jr., and K. G. Porter, *Appl. Environ. Microbiol. 54*:3023 (1988).
94. G. N. Sanden, W. E. Morrill, B. S. Fields, R. F. Breiman, and J. M. Barbaree, *Appl. Environ. Microbiol. 58*:2001 (1992).
95. J. M. González, J. Iriberri, L. Egea, and I. Barcina, *Appl. Environ. Microbiol. 56*:1851 (1990).
96. J. C. Cayrol, J. P. Frankowski, and C. Quiles, *Rev. Nematol. 10*:57 (1987).
97. M. J. Huss, *Mycologia 81*:677 (1989).
98. S. S. Bamforth, *Agric. Ecosyst. Environ. 24*:229 (1988).
99. N. D. Epsky, D. E. Walter, and J. L. Capinera, *J. Econ. Entomol. 81*:821 (1988).
100. P. P. Umorin, *Oikos 63*:175 (1992).
101. E. G. Lüfteneggar, W. Foissner, and H. Adam, *Oecologia 66*:574 (1985).
102. B. Sohlenius, *Pedobiologia 8*:340 (1968).
103. B. S. Griffiths, *Biol. Fertil. Soils 9*:83 (1990).

104. B. S. Griffiths, F. Ekelund, R. Rønn, and S. Christensen, *Soil Biol. Biochem. 25*:1293 (1993).
105. E. T. Elliot, R. V. Anderson, D. C. Coleman, and C. V. Cole, *Oikos 35*:327 (1980).
106. R. E. Ingham and D. C. Coleman, *Oikos 41*:227 (1983).
107. B. K. Senapati, *Soil Biol. Biochem. 24*:1441 (1992).
108. Z. Russom, R. A. Odihirin, and M. M. Matute, *Ann. Appl. Biol. 123*:331 (1993).
109. R. Hyvönen, S. Andersson, M. Clarholm, and T. Persson, *Biol. Fertil. Soils 17*:201 (1994).
110. G. W. Yeates, *Pedobiologia 22*:191 (1981).
111. M. F. Polz, H. Felbeck, R. Novak, M. Nebelsick, and J. A. Ott, *Microb. Ecol. 24*:313 (1992).
112. B. Finlay and T. Fenchel, *New Scientist 123*(1671):66 (1989).
113. C. Hagnere and C. Harf, *Eur. J. Protistol. 29*:155 (1993).
114. S. Visser and J. B. Whittaker, *Oikos 29*:320 (1977).
115. J. C. Moore, E. R. Ingham, and D. C. Coleman, *Biol. Fertil. Soils 5*:6 (1987).
116. P. J. A. Shaw, *Pedobiologia 31*:179 (1988).
117. R. D. Bardgett, J. B. Whittaker, and J. C. Frankland *Biol. Fertil. Soils 16*:296 (1993).
118. K. Newell, *Soil Biol. Biochem. 16*:227 (1984).
119. K. Newell, *Soil Biol. Biochem. 16*:235 (1984).
120. M. A. Leonard, *Pedobiologia 26*:361 (1984).
121. J. C. Moore, T. V. St. John, and D. C. Coleman, *Ecology 66*:1979 (1987).
122. G. Bengtsson, A. Erlandsson and S. Rundgren, *Soil Biol. Biochem. 20*:25 (1988).
123. R. D. Hanlon and J. M. Anderson, *Oecologia 38*:93 (1979).
124. P. Ineson, M. A. Leonard, and J. M. Anderson, *Soil Biol. Biochem. 14*:601 (1982).
125. R. D. Bardgett, J. B. Whittaker, and J. C. Frankland, *Biol. Fertil. Soils 16*:255 (1993).
126. G. Bakonyi, *Biol. Fertil. Soils 7*:138 (1989).
127. J. P. E. Anderson and K. H. Domsch, *Can. J. Microbiol. 21*:314 (1975).
128. J. A. Addison and D. Parkinson, *Oikos 30*:529 (1978).
129. G. Bengtsson and S. Rundgren, *Soil Biol. Biochem. 15*:469 (1983).
130. S. Visser, J. B. Whittaker and D. Parkinson, *Soil Biol. Biochem. 13*:2 (1981).
131. O. Andrén and J. Schnurer, *Oecolgia 68*:57 (1985).
132. J. Lussenhop, *Adv. Ecol. Res. 23*:1 (1992).
133. K. Hedlund, L. Boddy, and C. M. Preston, *Soil Biol. Biochem. 23*:361 (1991).

134. D. Parkinson, S. Visser, and J. B. Whittaker, *Soil Biol. Biochem. 11*:529 (1979).

135. J. B. Whittaker, *Oikos 36*:203 (1981).

136. R. V. Anderson, D. C. Coleman, and C. V. Cole, *Terrestrial Nitrogen Cycles* (F. E. Clark and T. Rosswall, eds.), Ecological Bulletin 33, Swedish Natural Science Research Council, Stockholm. p. 201 (1981).

137. A. Teuben and T. A. P. J. Roelofsma, *Biol. Fertil. Soils 9*:145 (1990).

138. J. H. Faber, A. Teuben, M. P. Berg, and P. Doelman, *Biol. Fertil. Soils 12*:233 (1992).

139. H. A. Verhoef, F. G. Dorel and H. R. Zoomer, *Biol. Fertil. Soils 8*:255 (1989).

140. M. M. Coûteaux and P. Bottner, *Beyond the Biomass* (K. Ritz, J. Dighton, and K. E. Giller, eds.), John Wiley and Sons, Chichester, England, p. 159 (1994).

141. H. Borkott and H. Insam, *Biol. Fertil. Soils 9*:126 (1990).

142. E. A. Curl, R. T. Gudauskas, J. D. Harper and C. M. Peterson, *Ecology and Management of Soil Borne Plant Pathogens* (C. A. Parker, A. D. Rovira, K. J. Moore, and P. T. W. Wong, eds.), American Phytopathological Society, St. Paul, Minn., p. 20 (1983).

143. R. T. Lartey, E. A. Curl, C. M. Peterson, and J. D. Harper, *Environ. Entomol. 18*:334 (1989).

144. M. K. Beute and D. M. Benson, *Annu. Rev. Phytopathol. 17*:485 (1979).

145. E. A. Curl and J. D. Harper, *The Rhizosphere* (J. M. Lynch, ed.), John Wiley and Sons, Chichester, England, p. 369 (1990).

146. P. A. Kaiser and J. Lussenhop, *Soil Biol. Biochem. 23*:307 (1991).

147. A. J. Wamock, A. H. Fitter, and M. B. Usher, *New Phytol. 90*:285 (1982).

148. R. D. Finlay, (1985) *Ecological Interactions in Soil* (A. H. Fitter, ed.), Blackwell, Oxford, p. 319 (1985).

149. S. C. Rabatin and B. R. Stinner, *Microbial Mediation of Plant-Herbivore Interactions* (P. Barbosa, V. A. Krischik and C. J. Jones, eds.), John Wiley & Sons, New York, p. 141 (1991).

150. J. C. Moore, D. E. Walter, and H. W. Hunt, *Annu. Rev. Entomol. 33*:419 (1988).

151. V. M. Behan and S. B. Hill, *Rev. Ecol. Biol. Sol. 15*:497 (1978).

152. M. Hassall, S. Visser, and D. Parkinson, *Pedobiologia 29*:175 (1986).

153. M. Hassall, D. Parkinson, and S. Visser, *Pedobiologia 29*:209 (1986).

154. S. C. Rabatin and B. R. Stinner, *Agric. Ecosyt. Environ. 24*:135 (1988).

155. D. A. Pherson and A. J. Beattie, *Rev. Ecol. Biol. Sol. 16*:325 (1979).

156. E. A. Fitzpatrick, *Soils: Their Formation, Classification and Distribution*, Longman, London, p. 1 (1983).

7

Microbial Processes Within the Soil

JAMES I. PROSSER   University of Aberdeen, Aberdeen, United Kingdom

1. INTRODUCTION

Soil contains a high number and diversity of microorganisms with a wide range of metabolic activities and physiological properties. Most agricultural soils contain on the order of 10^8 culturable bacterial cells g^{-1} soil dry weight. This probably represents less than 10% of the total population, as many organisms cannot be cultured in the laboratory. If we assume a bacterial cell mass of 10^{-11} g, it can be estimated that 1% of the soil dry mass is bacterial biomass. Bacteria are most abundant in the surface horizons, and, on an area basis, bacterial biomass has been estimated at 1–2 tons ha^{-1}. The total microbial biomass is even greater if we also consider contributions from fungi (whose biomass can exceed that of bacteria), protozoa, and algae. Coupled with these high biomass concentrations is the enormous metabolic diversity within the microbial kingdom. Microorganisms, as a group, are capable of metabolizing all naturally occurring compounds, in addition to the vast majority of anthropogenic compounds. The potential for processing of plant, animal, and microbial material within the soil is therefore considerable and microorganisms are of major importance in the cycling of nutrients and flow of energy on a global scale. Indeed, many key steps in nutrient cycling, e.g., nitrogen fixation, denitrification, and nitrification, can only be carried out by microorganisms. The range of environmental conditions over which microorganisms can maintain growth and activity increases their importance. Extreme conditions may reduce the activity of some microorganisms but will stimulate that of others, and in many extreme environments microorganisms are the only contributors to nutrient cycling.

Despite their key and essential role in the cycling of nutrients within the soil, the importance of microorganisms is frequently undervalued because of difficulties in detecting their presence and activity. With the exception of macroscopic structures, such as rhizomorphs and fruiting structures of fungi, special techniques are required for detection of cells. Special techniques are also required to determine whether particular components of the microbial biomass are metabolically active.

Difficulties in detection of microbes involved in specific processes have led many soil chemists, and even soil biologists, to consider soil as a black box. Microbially mediated processes within the soil are measured in terms of rates of input, output, and change in form of material, with no understanding of the organisms carrying out conversions or of the factors which control their growth and activity. Soil can be considered as a collection of enzymes or catalysts mediating chemical, rather than biological reactions. While this approach may be acceptable for some studies in soil, it is obviously severely limited in its ability to predict the impact of environmental change on soil processes and does not increase our understanding of the mechanisms controlling the global cycling of nutrients. The importance of such an understanding has been highlighted by the recent concern regarding the production of methane and nitrogen oxides and more general aspects of global warming. Microorganisms are essential for removal of pollutants from soil, e.g., chemical pesticides and oil, and for the correct functioning of unpolluted soil environments and those incorporating agricultural practices. Microbial activity is also responsible for the formation and maintenance of soil structure, the chemical composition of soil, and soil fertility.

It is clear, therefore, that identification of the nature and measurement of the rate of microbial processes in the soil are necessary. This chapter discusses the problems in achieving these aims and the traditional and modern techniques used to assess microbial soil processes. The complexity and heterogeneity of the soil environment provide a challenge to such studies, but recent technical advances should generate information which is currently lacking and will allow the dissection and understanding of the individual components of the soil biomass. A major challenge is the coordination/linking of studies carried out at different levels, as soil microbiology now involves molecular techniques, biochemical and physiological studies, and population and whole system approaches. Emphasis here will be placed on methodology and techniques, as nutrient cycling is considered in Chapter 9.

2. MICROCOSM OR FIELD MEASUREMENTS

The ultimate aim of all studies on microbially mediated processes in soil is to increase the understanding of the activities of microorganisms in the environment. Unfortunately, practical problems prevent field measurements of the majority of processes and most of the information on soil microbial processes arises from either laboratory analysis of samples or from studies carried out in microcosms or laboratory experimental model systems (1). This applies both to general microbial activity and that of specific groups of organisms. Frequently, there is no direct evidence for the role of a particular organism in the soil, and its activity can only be extrapolated from pure cultures observed in the laboratory. However, the disadvantage of this approach is that interactions between microorganisms are not taken into account.

The distinction between microcosms and experimental model systems is somewhat arbitrary and it is not always clarified, but they can be defined on the basis of the experimental approach being adopted. A soil microcosm is an experimental system that aims to mimic, as closely as possible, the natural soil environment while providing monitoring and control of experimental conditions. In addition, microbial activities and other properties of the system may be assessed to a much higher degree than in field experiments. Such systems are invaluable in assessing, for example, the effect of environmental factors on microbial processes, the impact of pesticides or pollutants on microbial communities, or the effects of changes in population structure. Their worth lies in their ability to represent accurately the soil environment, or at least those features of the environment which are considered to be important for the processes under investigation. Laboratory experimental systems fulfill a different role. They make no effort to mimic the soil environment in its entirety but are designed to test experimentally hypotheses based on particular simplifying assumptions or to determine the properties of microorganisms in isolation from the complexity of the soil environment.

In practice, the structure of microcosms and experimental models overlap and the distinction is one of function rather than design. However, an extreme example of an experimental model is a chemostat inoculated with a pure culture of a single organism. The chemostat allows examination of microbial physiological characteristics at constant but variable specific growth rates, below the maximum rate, and under limitation by a nutrient, determined by the composition of the inflowing medium. Although a chemostat may have some features in common with the soil environment, it does not mimic the soil and would not usually be classed as a microcosm. Nevertheless, chemostats provide an enormous amount of information of value for studies of soil processes. Pure culture studies provide information on the effects of limitation by different nutrients which may be limiting growth in the soil. Growth parameters of ecological importance may be determined and quantified, e.g., activity per cell, maximum specific growth rate, saturation constants for different substrates present in the soil, and maintenance energy requirements. The effects of environmental factors on microbial physiological characteristics, activity, and growth parameters can also be assessed and quantified. Extrapolation to the natural environment is possible but must be done with care. There is no reason to suggest that the effect of glucose on the specific growth rate of an organism in a chemostat will differ from that of the same organism in the soil, if all other environmental conditions are identical and if there are no significant interactions with other organisms. Specific growth rate will, however, be affected by other features of the soil environment. For example, the presence of clay minerals may affect activity (2), and spatial heterogeneity will lead to gradients in substrate and product concentrations and environmental conditions such that the effect of glucose on specific growth rate may differ from that determined in a chemostat. Any difference would not invalidate the chemostat experiments, or the kinetic constants obtained from them, but, importantly, would provide information that such soil factors are of significance and merit further investigation.

The application of chemostats to ecology in this way purposely ignores, or rather eliminates, the natural heterogeneity and complexity of natural systems to permit detailed and mechanistic studies which would otherwise be impossible. Subsequent incorporation of these factors is then possible, giving more complex experimental systems. A simple example would be the inclusion of a second organism with which a first is thought to interact to study the mechanistic basis of their interaction. More complex systems allow inclusion of a greater number of complicating soil factors and these systems overlap with microcosms. One example is the use of continuous flow sand or soil columns, termed packed column reactors (3). These have been used, in particular, to study nitrogen transformations; results of such studies will be discussed later. They consist of a tube packed with sand or soil. A nutrient solution is supplied continuously to the top of the column, is transformed during passage through the sand or soil, and is analyzed in effluent collected from the base of the column. Aeration may be achieved by pumping air or oxygen through the column, and more sophisticated systems also allow accurate control and modification of temperature, moisture content, and pH. This degree of control allows use of such systems as experimental models. For example, sterilization of soil prior to inoculation with pure cultures allows precise measurement of growth constants. These systems also mimic many of the features of the natural soil environment and, in that sense, may also be considered microcosms. The activity of natural communities can be studied under realistic environmental conditions, and the rates of processes and effects of environmental conditions can be readily monitored.

3. SOIL HETEROGENEITY

We are accustomed to studying growth and activity of pure cultures of microorganisms in liquid culture or on agar plates, with relatively homogeneous and defined conditions. In nature, growth is affected and limited by a wide range of environmental (abiotic) factors, in addition to direct and indirect biological interactions. The heterogeneity of the soil environment requires consideration when assessing microbial activity, from both conceptual and methodological viewpoints.

3.1 Physical Heterogeneity

Structural heterogeneity in soil arises from the presence of particulate material with different shape, surface property, and size distributions. Thus, there is a basic heterogeneity, due to the existence of gas, liquid, and solid phases and interfaces between these phases. While it is difficult to demonstrate conclusively, most microorganisms are thought to be attached to particulate material, which may significantly alter their physiological properties. The nature of the surface may be important in attachment and colonization. For example, the majority of clay minerals are negatively charged, favoring initial attachment of positively charged nutrients (e.g., NH_4^+, Ca^{2+}) and cells with positive cell surface charge, while the existence of a double charge layer can lead to attraction of negatively charged particles, which is more common for bacteria. Thus, both nutrients and cells may be concentrated at the surface of inert material, such as clay minerals, while other particulate organic matter is itself a substrate. Attachment and biofilm formation are known to affect cell activity in a number of ways.

One example is nitrification. The oxidation of ammonia by autotrophic nitrifying bacteria in liquid media occurs optimally within the pH range 7.0–8.5 and rarely below pH 7, but its occurrence has been reported in soils with pH values as low as 3.7 (4,5). Nitrification in such acid soils may be due to urease production by autotrophic nitrifiers, to the possible existence of acidophilic strains, or to heterotrophic nitrification (6). Nevertheless, there is evidence for high concentrations of autotrophic ammonia oxidizers, and of their activity, in acid soils. An additional explanation is protection due to surface growth and biofilm formation. To test this, Allison and Prosser (7) examined surface growth of a strain of *Nitrosomonas europaea* with a pH optimum for growth in liquid culture of 7. *Nitrosomonas europaea* was allowed to colonize sand and vermiculite in continuous flow packed columns supplied with medium, containing ammonium, at low pH. Steady states were established in nitrite concentrations in effluent from the columns at a pH value of 6, and high rates of activity were measured at pH 5. In addition, a multispecies ammonia-oxidizing biofilm was formed as wall growth in an ammonia-limited chemostat inoculated with acid soil known to exhibit nitrification. This biofilm was capable of oxidizing ammonia at pH 5 and growth was possible at pH 5.5. The study demonstrated that conditions that exist in the soil, in particular surface growth and mixed communities, may allow growth and activity of ammonia oxidizers at pH values significantly lower than those possible by suspended cells.

A mechanistic explanation for this phenomenon is indicated by the work of Armstrong and Prosser (8), who observed a decrease in the pH minimum for growth of ammonia oxidizing bacteria attached to clay minerals. This involved a local buffering effect in which NH_4^+ adsorbed to the clay surface exchanged with H^+ produced during ammonia oxidation. There is also strong evidence for protection of biofilm populations from inhibition and stress. For example, populations of ammonia oxidizers in soil are protected from the action of specific inhibitors of ammonia oxidation when attached to soil particles or glass surfaces (9,10). Other biofilm populations are protected from the activity of antibiotics. This has significance for the techniques for measuring microbial processes which involve specific inhibitors (discussed later).

A major consequence of physical heterogeneity is restriction of movement of microorganisms, for which soil acts as a filter, and of substrates and metabolic products. Transport of both cells and soluble nutrients can be achieved in water films and through bulk flow, with additional movement by higher organisms, e.g., soil animals and plant roots. For microbial cells, movement depends on the pore structure of soil and in particular the distribution of pore size and pore neck diameter. Movement into and out of soil pores is strongly dependent on whether a pore is filled with soil water, which in turn depends on pore size. For example, at a matric potential of -147 kPa, pores with neck diameters of 1 μm and less will be water-filled, while larger pores will be filled with air. Bacterial cells of size >1 μm will be excluded from such pores, on the basis of size alone, while those in the larger pores will be unable to move, through lack of water. Equally, substrates and products of metabolism will not diffuse between pores in such a soil, reducing activity by limiting supply of substrates. At -14.7 kPa, pores of neck diameter <10 μm will be water-filled. This will allow greater mobility of nutrients and entry of bacterial cells, but not of fungi, yeasts, or some soil protozoa. While this may reduce competition and other interactions with these organisms, it will also restrict turnover of nutrients which might result from predation of bacteria by protozoa.

Unicellular flagellated microorganisms can effect their own movement through soil to some extent, if the water content is sufficient for motility, but nonmotile cells are dependent on outside forces, e.g., bulk flow, and movement on the surfaces of higher organisms. For multicellular microorganisms such as actinomycetes and filamentous fungi, the mycelial growth form allows growth between nutrient sources across barren areas, with transport of nutrients along hyphae. This is seen in its most highly developed form in the rhizomorphs of mycorrhizal fungi, in which transport of nutrients can occur in both directions along specialized hyphal structures (11).

3.2 Chemical Heterogeneity

The chemical composition of soil is also heterogeneous, ranging from inert clay minerals to organic material which may consist of soluble, diffusible, low-molecular-weight organic compounds or complex, high-molecular-weight polymeric material at different stages of decomposition. The concentrations of inorganic nutrients will vary and large-scale variations will result from physical heterogeneity. Transport of soluble nutrients depends mainly on diffusion through water films or water-filled pores or on bulk flow of soil water. On a smaller scale, ion exchange processes will lead to gradients in inorganic nutrients around charged particulate material. Within organic particulate material, gradients in nutrients and oxygen will result from microbial activity. These will lead to activities which would not be predicted from measurement of bulk soil properties. For example, even in well-aerated soils, anaerobic processes will take place within particulate material as a result of microbial oxygen uptake at particle surfaces and limitations to diffusion of oxygen into particles.

Gradients in nutrient concentrations will lead to establishment of interactions which would not be possible for freely suspended organisms in well-mixed liquid culture. For example, close interaction between nitrifiers and denitrifiers in soil is possible at aerobic/anaerobic interfaces. Of equal importance are the changes in metabolic potential resulting from nonstandard conditions. The majority of ammonia-oxidizing bacteria are considered to be aerobic organisms, but their activity changes at low oxygen concentrations, with a greater proportion of ammonia converted into nitrous oxides. In addition, nitrite-oxidizing bacteria can assimilate organic material and can reduce nitrate to ammonia at low oxygen concentrations. This presents the possibility of internal cycling of nitrogen, in the presence of organic material at low oxygen concentrations, with aerobic ammonia and nitrite oxidation and anaerobic reduction of nitrate to ammonia. Such activities would not be detectable by measurement of changes in ammonia, nitrite, and nitrate but could be significant in terms of the growth and activity of these organisms.

3.3 Biological Heterogeneity

Biological heterogeneity occurs both within the microbial soil community and through the presence of higher organisms. The latter contribute to microbial processes in a number of ways. One obvious effect is that of root exudates in the rhizosphere (see Chapter 2). Approximately 5% of fixed carbon is released from plant roots and much of the organic material is readily utilizable by microorganisms. As a consequence, the concentrations of microorganisms within the rhizosphere are considerably greater than those in the bulk soil. Quantifying the activity of this population is technically difficult, in part because of difficulties in distinguishing plant and microbial activity. It is difficult to remove soil for measurement of activity without also removing root cells, and it is difficult to obtain the quantities of soil required for some activity measurement techniques. The effects of material released by the root will vary with increasing distance, making definition of the rhizosphere difficult. The rhizosphere also affects soil activity qualitatively by selection of microorganisms with properties different from those of bulk soil populations. For example, microorganisms isolated from the rhizosphere tend to have greater phosphatase activity. Root growth may also be a major means of cell dispersal through the soil.

Soil animals have three major effects on microbial activity. The first is turnover of material immobilized by bacteria, for example, through predation by protozoa. Although there are several reports of increased bacterial metabolic activity resulting from predation, it is difficult to distinguish bacterial and protozoal activity by using traditional techniques. However, luminescence-based techniques (described in Section 5.19) have been used to determine bacterial activity during predation of *lux*-marked *Pseudomonas fluorescens* by *Colpoda steinii* (12). Viable cell concentrations of *P. fluorescens* decreased at a greater rate in the presence of *C. steinii*, as a result of predation, but this was associated with an increase in luminescence activity per cell as the surviving bacteria benefited from the nutrients released after predation (Fig. 1).

A second effect of soil animals is in providing different and unique environments for microbial activity. For example, earthworm guts contain relatively high concentrations of microbial cells and of nutrients, in comparison with those in the soil. It is not evident how selective this environment is, but rates of activity are higher that those in the bulk soil. The third role of soil animals is indirect, in providing a means of dispersal. For example, addition of earthworms to soil microcosms has a dramatic effect on movement of *P. fluorescens*, increasing, by several orders of magnitude, cell concentrations in effluent from intact soil cores supplied with simulated rainwater (13). Soil animals also increase microbial activity by releasing nutrients during turnover of soil material.

There will also be heterogeneity within the microbial population. Different groups of organisms naturally vary in their ability to utilize different compounds

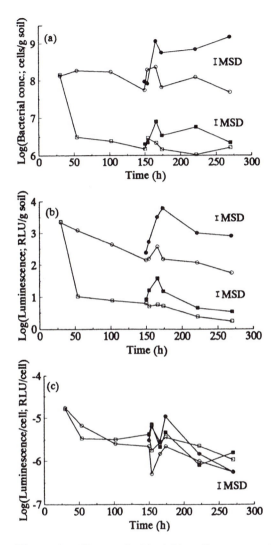

Figure 1. Changes in (a) viable cell concentration, (b) luminescence, and (c) luminescence viable cell^{-1} after soil inoculation of a luminescence-marked strain of *Pseudomonas fluorescens* into small pores and introduction of sterile water (○,●) or *Colpoda steinii* (□,■) into large pores (neck diameter 30–60 μm). Microcosms were amended, at 150 h, with either sterile water (○,□) or double-strength complex medium (●,■). (From Ref. 12.)

and in the range of environmental conditions over which they are active. There will also be considerable variation in the physiological state of individual cells within a population. Some may be present as spores, rather than vegetative cells, but there will also be significant heterogeneity in the latter. Cells supplied with nutrients and experiencing favorable environmental conditions will be active, but in the bulk soil, where nutrients are scarce or unavailable, the majority of cells are likely to be dormant or to have a low metabolic rate. The growth rates and cellular activities of microorganisms are such that readily available substrates are utilized rapidly. Thus, even when nutrients become available, their rapid consumption by microorganisms will lead to starvation and a return to dormancy. Assessment of microbial activity rarely takes this heterogeneity into account.

Another important factor is the need for rapid recovery when conditions become favorable for growth after starvation. Laboratory studies indicate a lag period prior to growth which increases with increasing starvation. The same situation frequently occurs in the soil, although there is evidence that surface growth and biofilm formation may increase the speed with which resuscitation can take place. This is particularly important in the soil environment, where the supply of many nutrients is transient and sporadic and competition is intense. Hence, a rapid response to environmental change is important in soil because of the temporal heterogeneity with respect to both biological and physicochemical properties.

These factors must all be considered when attempting to assess the activity of soil microorganisms, their ability to respond to environmental change, and the effects of such changes on activity. Microorganisms are unique in their metabolic versatility, the range of environmental conditions over which they remain active, their survival capabilities, and their dispersal capabilities. If nutrients become available, therefore, it is only a matter of time before microorganisms colonize and utilize those nutrients, and before microbial cells themselves die and are decomposed to released immobilized material. The challenge is to determine the speed with which nutrients are accessed and colonized by microorganisms and to define and quantify the speed with which utilization and turnover of nutrients occur.

4. GENERAL ACTIVITY

In many studies, information is required on overall microbial activity within a soil, rather than the activity of a particular microorganism or group of microorganisms or the rate of a particular process. This information is most usually obtained by measurement of population activity, although an alternative is to examine the activities of individual cells using activity stains.

A distinction must be made between potential and actual, or in situ, activity, although this distinction is often not entirely clear. In situ activity may be defined as the activity of the population in the field, under the prevailing environmental conditions. As this is usually impossible to determine in the field, activity is usually measured in soil samples in the laboratory. The important feature is that activity is measured without exogenous addition of substrates and under environmental conditions (e.g., temperature, matric potential) pertaining in the natural environment. Potential activity is activity measured under optimal conditions and is usually determined after addition of substrate and under environmental conditions which are considered ideal. However, precise identification of the latter is not possible for all members of a population. Also changes in environmental conditions may increase activity without addition of substrate. Here, however, potential activity will be considered as that measured during incubation of soil in the laboratory after addition of a substrate.

4.1 Population Activity

The most common measure of nonspecific microbial activity is respiration. This may be determined by following O_2 uptake or more commonly by CO_2 evolution, with or without amendment of soil with nutrients. CO_2 may be measured by titration of CO_2 trapped in alkali or by infrared gas analysis, and measurements can be made in soil cores, in sieved soil, or in the field. As with most methods, problems can arise through alteration of soil conditions, with subsequent effects on activity. For example, passage of gas over or through soil will increase mixing of the gas phase, will change concentrations of O_2 and CO_2, and will also increase evaporation of soil water. Significant increases in sensitivity can be achieved by the use of radiolabeled substrates, e.g., ^{14}C-labeled glucose, and quantification of [14]CO_2.

An important example of measurement of potential activity is the substrate-induced respiration (SIR) technique of Anderson and Domsch (14), which is used to estimate the biomass concentration. This involves measurement of soil respiration during incubation with glucose. A major limitation of this general activity measurement and others is heterogeneity in the microbial population. Not all microorganisms may be able to use glucose, and, in fact, some may be killed by high concentrations of nutrients. The incubation period is chosen to maximize the number of organisms generating activity and to obtain sufficient CO_2 for measurement, while minimizing the risk of significant growth. This, however, limits the technique to measurement of organisms which become activated within the incubation period chosen. In addition, respiration will depend on soil type, pH, temperature, and other environmental conditions, making comparisons difficult. As discussed later, SIR can be made selective by the addition of inhibitors.

The overall activity of the total soil microbial population may also be assessed by measuring adenosine triphosphate (ATP) levels. Like SIR, this technique is used as a measure of microbial biomass. The ATP is assayed by reaction of a soil sample with reduced luciferin and luciferase. The reaction is ATP-dependent and the amount of light emitted is measured by luminometry or scintillation counting. Measurement can be made with or without amendment with nutrients to provide actual and potential activities. One advantage of this method over SIR is that nutrient amendments are not necessary, eliminating limitations associated with the activation of populations by added substrate. After cell death, ATP is degraded quickly in soil, and measurements will therefore be indicative of recent metabolic activity, rather than historical activity. A practical limitation of the technique is that ATP can be difficult to extract from soil. In addition, some of the ATP, e.g., in rhizosphere soil, may not be of microbial origin.

The amount of ATP will depend not only on the biomass but also on the activity and physiological state of cells. A more valid measure of activity is therefore adenylate charge (AC), defined as the ratio of ATP to the total adenylate pool:

$$\frac{ADP + ATP}{AMP + ADP + ATP}$$

where AMP = adenosine monophosphate and ADP = adenosine diphosphate.

Metabolically active cells in liquid culture have AC values of 0.8–0.9, which fall to 0.5–0.8 in inactive cells and <0.5 in dying cells. For soil microbial populations values of 0.3–0.4 have been obtained, as expected for a nutrient-limited environment, but values of 0.85 have also been reported, reducing to 0.46 when soil is dried (15). These data are difficult to explain, although one suggestion is that cells with low AC are poor survivors and die first, leaving those with higher values.

General microbial activity may also be assessed by measurement of soil enzyme activity. Some enzymes are indicative of specific activity and will be discussed later. Others measure activities carried out by all microorganisms, the most obvious examples are dehydrogenase and phosphatase. Enzyme assays are equivalent to those used for in vitro assays of purified enzymes but with procedures to remove soil particles when determining color reactions. They may involve incubation with substrate, equivalent to potential activity, or may depend on substrate present in the soil or on endogenous metabolic activity. The latter is preferable for assessing in situ activity but the technique requires large and active cell populations for detection and quantification of activity. For example, Meikle et al. (16) required 10^8 cells g^{-1} soil to detect dehydrogenase activity of *P. fluorescens* in the absence of added substrate.

Activity measurements should be carried out under environmental conditions similar to those prevailing in the natural environment; this restriction will also limit detection as these are rarely optimal. Amendment with substrate increases sensitivity but also introduces the possibility for growth of the population. Incubation periods must therefore be chosen to reduce the possibility of growth. Enzyme activity may derive solely from microbial activity but is maintained in soil for long periods following cell lysis, through protection of enzymes following adsorption to clay minerals and other soil components. Enzyme activity may therefore represent previous rather than current or recent microbial activity.

4.2   Cellular Activity

The measurement of population activities in soil may be useful for black box types of studies, where detailed information is not required, but it gives no information on the sites of activity. This type of information can be obtained by microscopic examination of samples treated with activity stains.

Fluorescein diacetate (FDA) has been used to determine the activity of, in particular, filamentous fungi. It is converted by esterases to fluorescein in active cells or active regions of cells and can be detected by fluorescence microscopy. In pure culture, for example, this technique shows the tips of fungal hyphae to be significantly more active than distal hyphal regions. Although the stain can be used on soil samples, care must be taken to reduce damage to fungal hyphae when homogenizing the soil, as this affects their activity.

Fluorescein diacetate is less useful for unicellular bacteria but other methods are available. The most commonly used methods are based on tetrazolium salts for detection of dehydrogenase activity in single cells (17). The water-soluble stain is converted in active cells to an insoluble colored or fluorescent formazan compound, and the method can be used to determine in situ or potential cell activity. 2-(4-Iodophenyl)-3-(4-nitrophenyl)-5 phenyltetrazolium chloride (INT) has been used most frequently as an activity stain, but 5-cyano-2,3-ditolyl tetrazolium chloride (CTC) appears to correlate better with cell culturability (18).

An alternative technique is the direct viable count (DVC) method of Kogure et al. (19). This method involves incubating cells in the presence of yeast extract and nalidixic acid and examining changes in cell length microscopically. Nalidixic acid inhibits deoxyribonucleic acid (DNA) gyrase and prevents cell division. Active cells are seen as those which have elongated during the incubation period. This technique measures potential activity and suffers from the limitations discussed above; e.g., some cells may not utilize yeast extract, some may not be activated during the incubation period, and high nutrient concentrations may lead to substrate-accelerated death (20). Nevertheless, the technique has been used to estimate the metabolic activity of *Flavobacterium* in bulk soil and in the rhizosphere by Heijnen et al. (21). Soil and rhizosphere samples were incubated in the presence and absence of nutrient broth and nalidixic acid, and, after cell extraction, the proportion of cells of length >2 μm was determined microscopically. This measure of activity was compared with CTC staining and, for both, was made specific to *Flavobacterium* by use of fluorescent antibodies. The DVC technique measures potential activity, because of the addition of nutrient broth, but both techniques indicated a decrease in activity with time in the bulk soil (Fig. 2).

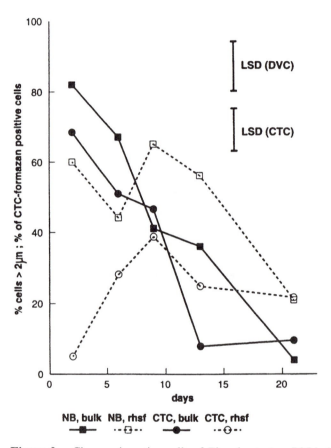

Figure 2.   Changes in active cells of *Flavobacterium* P25 in bulk soil and rhizosphere soil measured using the potential direct viable count (DVC) technique and by staining with CTC. Solid symbols represent the percentage of elongated cells (>2 μm) after extraction and incubation with nutrient broth and nalidixic acid for 24 h. Open symbols represent the percentage of CTC-positive cells. Squares and circles represent bulk and rhizosphere soils, respectively. (From Ref. 21.)

In the rhizosphere, activity increased until day 13 and then decreased to a level approaching that of the bulk soil. A related technique is the microcolony assay, in which microbial cells from soil are incubated in agar slide culture or on filters placed on agar (18). Viable cells go through several rounds of cell division to form microscopic colonies which may be enumerated microscopically. This technique is now capable of detecting 10^4 cells ml^{-1} and can be used in conjunction with fluorescent antibodies to allow detection of specific organisms (18).

Although these techniques are satisfactory for pure cultures, analysis of soil microbial populations presents a number of problems. A disadvantage of fluorescent stains is background fluorescence and autofluorescence from soil particles. Some bacterial cells are not sensitive to nalidixic acid, limiting its application to mixed communities, and the mechanism of action of activity stains should be known. The ability of cells to take up stains may be affected by cell physiological features, and the decrease in cell size associated with starvation in some organisms makes detection of stain difficult.

These techniques give an indication of the proportion of active cells within a microbial population. The development of image analysis techniques may provide a means of quantifying single-cell activity, but these have not yet been used to measure activity in the total soil population. They do, however, provide information on the sites at which cells are active.

Individual cell activity can also be measured by microautoradiography (22,23). This involves incubation with a radiolabeled substrate and subsequent detection by exposure of cells to photographic film, or coating on a slide with a photographic emulsion. Microscopic examination of the film reveals active cells surrounded by dark silver grains. Nonspecific detection of active cells can be achieved by incubation with "universal" substrates, e.g., glucose; the limitations of this approach have been discussed. Alternatively, radiolabeled thymidine or glutamic acid may be used to detect cells in which DNA and protein synthesis (respectively) are occurring. The technique can detect active regions of fungal hyphae but measures potential activity, and extensive incubation periods will lead to growth. This, however, can be used to advantage if measurement of specific growth rate is required. Exponential growth will be accompanied by exponential increases in incorporation of labeled thymidine or glutamic acid; use of this approach for measurement of microbial growth rates in soil is described in Section 6.

5. SPECIFIC ACTIVITY

While activity of the microbial population as a whole may provide useful information for many studies, a full understanding of microbial ecological characteristics requires knowledge of the extent to which particular groups of microorganisms or particular processes are contributing to overall activity. Activity may therefore be considered as specific to particular microorganisms or to specific processes. In practice such a distinction may be difficult to draw and is essentially artificial. In this section, however, the ability to measure the activity of particular populations and cells, and of particular processes, will be considered.

5.1 Population Activity of Specific Organisms

Molecular techniques for detection of microbial activity are discussed in Chapter 14, but one technique of particular relevance to this chapter is the use of luminescence-based marker systems to determine the activity of specific organisms. These involve the introduction of genes for light emission cloned from naturally luminescent organisms, such as the prokaryotic *lux* genes, from *Vibrio fischeri* or *Vibrio harveyi*, or eukaryotic *luc* genes of the firefly (see 24,25). The microorganism under study must be cultured, to allow construction of marked strains, and may be marked with the structural genes for luciferase production or with the complete *lux* system, including regulatory genes and genes responsible for production of the luciferase substrate. If cells are marked with *luc* genes or with the *lux* structural genes only, substrate must be added exogenously. Luciferase production is constitutive and the light emitted can be detected visually or can be quantified by luminometry, while the activity of individual cells can be determined by charge-coupled device (CCD) microscopy.

Measurement of luminescence by luminometry is rapid, does not involve extraction of cells from the soil, and has been used to detect the activity in liquid culture of as few as 100 cells ml^{-1} of luminescence-marked *Escherichia coli* (26). Detection of luminescence following introduction of marked cells into soil is less sensitive because of quenching and masking of light by soil particles. Nevertheless, lower detection levels are commonly in the order of 10^3–10^4 cells g soil^{-1}, and, immediately after inoculation, luminescence is proportional to cell number or biomass concentration over several orders of magnitude.

The direct proportionality between luminescence and cell concentration is lost as cells become inactive, but luminescence provides a measure of the in situ population metabolic activity which correlates with other measures of activity. For example, Meikle et al. (16) compared increases in luminescence and dehydrogenase activity of a luminescence-marked strain of *P. fluorescens* inoculated into sterile soil. Luminescence and dehydrogenase activity were detectable after incubation for 24 h and increased in parallel after addition of complex medium (Fig. 3). Assessment of activity by luminometry is considerably more rapid, convenient, and sensitive than traditional techniques and, importantly, is selective for the marked microorganism only. Luminometry can therefore be used to determine the effects of environmental factors on the activity of marked inoc-

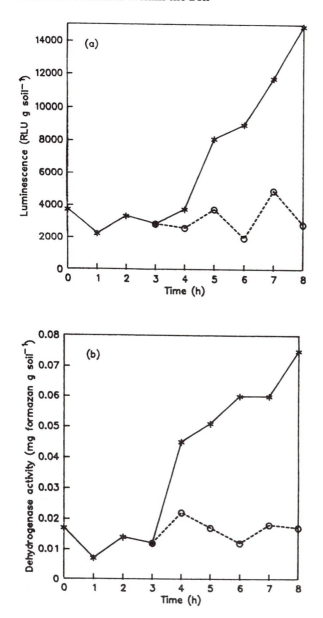

Figure 3. A luminescence-marked strain, *Pseudomonas fluorescens* 10586s/FAC510, was inoculated into autoclaved soil and incubated for 24 h. Luminescence (a) and dehydrogenase activity (b) were then measured for a period of 8 h. At 4 h soil was amended with either phosphate buffer (o) or double-strength 523 medium (*). (From Ref. 16.)

ula. For example, Rattray et al. (27) followed reductions in the activity of *lux*-marked *E. coli* in soil at a range of matric potentials and Meikle et al. (28) used luminescence to follow the activity of *P. fluorescens* in soil during starvation for several months. In both cases, activity decreased rapidly after inoculation, particularly in nonsterile soils and at low matric potentials (Fig. 4). Meikle et al. (29) also found luminescence to be a more sensitive and reliable measure of activity than respiration rate and dehydrogenase activity for cells introduced into sterile soil.

Luminometry may also be used to determine potential luminescence, in which luminescence is measured during incubation of samples with complex medium (29) for, typically, 2 h. This technique is equivalent to the SIR method of Anderson and Domsch (14). The kinetics of light output provides information on the time taken to recover activity, while the final luminescence value is a measure of the marked population capable of activation within the incubation period. Meikle et al. (28) found that an increased period of starvation of *P. fluorescens* in soil led to a decrease in final luminescence values and an increase in the lag period prior to maximum luminescence. These effects were greater in nonsterile soil and at lower matric potentials. Importantly, the technique indicated the inability of this organism to respond rapidly to increased substrate supply after extended starvation, even though dilution plate counts gave high viable cell counts. Luminometry is also able to quantify the activity of cells which cannot be cultured using traditional plating techniques but retain activity, as indicated by cell activity stains and DVC (30). Detection of activity of these populations in nonsterile samples is not possible as their activity cannot be distinguished from that of the background population, but luminometry allows selective quantification of activity of *lux*-marked cells.

The involvement of particular organisms is often based on circumstantial evidence. The activity of a microorganism capable of carrying out a process is suspected if that organism is found in an environment where that process can be measured. In adopting this approach, the microorganisms considered are usually those which can be cultivated in the laboratory and may not be the dominant or significant population in the environment. Similarly, rates of activity under laboratory conditions may be significantly different from those in the soil. Much of the work on experimental laboratory systems has been targeted to determining the effects of such factors on process rates in an attempt to determine the potential role of particular organisms in soil.

The contribution made to soil processes by specific groups of organisms can sometimes be assessed by addition of inhibitors specific to those groups. For example, the SIR method of Anderson and Domsch (14) can be modified to determine the roles of fungi and bacteria by addition of actidione and strep-

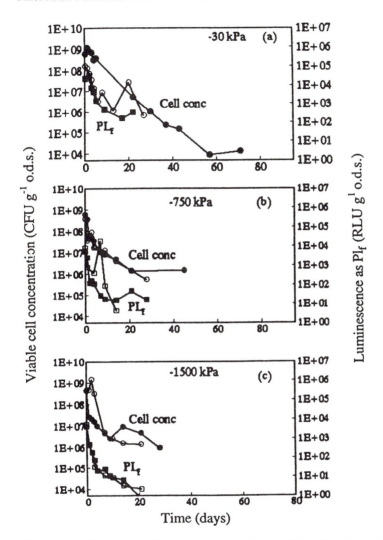

Figure 4. Changes in viable cell concentration (o,●) and final potential luminescence value (PL$_f$) (□,■) in nonautoclaved soil microcosms inoculated with a luminescence-marked strain of *Pseudomonas fluorescens* 10586s/FAC510. Microcosms were adjusted to a matric potential of –30 kPa (a), –750 kPa (b), or –1500 kPa (c). (From Ref. 28.)

tomycin, respectively. While this approach may be applicable in laboratory studies, the use of antibiotics in the field, and in soil in general, is questionable. Their worth depends to some extent on the particular antibiotic, but in general soil populations are significantly less sensitive to effects of antibiotics than the same organisms in liquid culture. This difference is due to degradation of the antibiotic within the soil, adsorption and inactivation on soil particulate material, and protective effects of surface attachment and biofilm formation.

5.2 Measurement of Specific Processes

Carbon transformations

The most common approach to measurement of a microbially mediated process is measurement of changes in concentrations of substrates or products which are unique to the process. This is most straightforward when the substrate is a starting point for metabolism, rather than an intermediate. Thus, lignin or cellulose degradation can be assessed by amending soil with these substrates and measuring the rate of disappearance or the formation of product. If the measured product is CO_2, corrections must be made for the proportion of substrate carbon assimilated into biomass, and this proportion differs significantly between different microbial groups. For many processes, particularly for compounds that are transformed slowly, decreases in substrate concentrations must be determined by using radiolabeled substrates which are mixed into the soil. Soil is then sampled and residual label measured. This increases sensitivity and also reduces problems associated with amendment with unrealistically high substrate concentrations. Alternatively, radiolabeled products can be measured, typically $^{14}CO_2$ from ^{14}C-labeled organic substrates. This is satisfactory unless CO_2 is turned over, through CO_2 fixation by plants or autotrophic bacteria.

These techniques have been invaluable in determining rate constants for the decomposition of organic matter, whose chemical assay in complex environments such as soil is difficult and whose breakdown is slow. They can be applied in laboratory microcosms or the soil may be sampled as soil core, mixed with substrate, and replaced to provide more realistic assessment of decomposition rates in the field.

Decomposition of radiolabeled (^{14}C- or ^{15}N-) plant material may be studied as described, but a more direct way of investigating degradation of plant material is the use of litter bags. This technique, which applies to any particulate substrate, was developed to determine the rate of decomposition of leaves. Leaf litter is placed in bags with a mesh size chosen to eliminate some or all of the soil biota. For example, a mesh size of 0.5 mm will allow entry of microorganisms and small animals while a mesh size of 0.003 mm will permit entry of bacteria only (see 31). A suitable number of replicate weighed litter bags are then placed on, or buried in, the soil. After sampling, the degradation rate is determined by weight loss. If required, other properties, such as surface area, tensile strength, and chemical composition, can be determined. Microscopic examination may also indicate the microorganisms responsible for degradation.

The majority of primary production in terrestrial environments is carried out by photosynthetic higher organisms. CO_2 fixation by autotrophic bacteria can be determined by incubation of samples with $^{14}CO_2$ or ^{14}C-labeled bicarbonate in the dark, to eliminate photosynthesis, with subsequent quantification of radiolabel in organic carbon. The incubation period must be long enough to provide measurable quantities of ^{14}C-labeled organic material but must prevent turnover and assimilation of this material by heterotrophs.

Nitrogen transformations

Assay of the activity of enzymes specific to particular processes may be used to determine the activity of organisms involved in those processes. The best known and most widely used is the acetylene (C_2H_2) reduction test, which assays nitrogenase to determine rates of nitrogen fixation. Nitrogenase can utilize either nitrogen or acetylene as a substrate, converting the latter to ethylene (C_2H_4), which can be assayed by flame ionization gas chromatography.

Rates of nitrification are most conveniently measured by determining reduction in ammonia concentrations or increases in concentration of nitrate. This is often done in soil slurries, although care must be taken to prevent oxygen limitation, which will limit nitrification and increase rates of denitrification. This illustrates a problem with this approach. Ammonia may be assimilated by heterotrophs and is produced by mineralization of organic nitrogen, while nitrate may be assimilated or reduced by denitrifiers. Similarly, although denitrification can be assessed by measuring reductions in nitrate concentration, analysis of data is complicated by potential for production of nitrate by nitrifiers and nitrate assimilation by heterotrophs. The situation is complicated further in the presence of plants, which also assimilate ammonia and nitrate. Thus, rates of nitrification and denitrification may be measured by changes in ammonia and nitrate concentrations, but both compounds are utilized in a number of other processes.

One solution to this problem is the use of inhibitor block experiments. For example, autotrophic nitrification is inhibited by a number of compounds, including nitrapyrin (2-chloro-6-trichloromethyl pyridine), allylthiourea, and acetylene. The target for these inhibitors is ammonia monooxygenase, which catalyzes the first step in ammonia oxidation. Inhibition is believed to involve chelation with copper components of this key enzyme. Measurement of changes in ammonia and nitrate concentrations in the presence and absence of such inhibitors provides rates of autotrophic nitrification by difference. Similarly, rates of denitrification may be measured in the presence of acetylene, which blocks the terminal nitrous oxide reductase enzyme, leading to accumulation of nitrous oxide (N_2O). While this is satisfactory for measurement of potential rates, after amendment with nitrate, in situ rates may be underestimated because acetylene inhibition of nitrification leads to a reduced supply of nitrate.

Inhibitor block experiments suffer from the problems associated with reduced sensitivity of soil microorganisms to inhibitors. For example, 0.5 µg ml^{-1} nitrapyrin will completely inhibit the growth of *N. europaea* in liquid culture. This concentration does not perceptibly affect growth of the same organism inoculated into soil, where concentrations one order of magnitude greater are required for inhibition. Protection appears to be afforded by surface attachment and growth (32) and varies with clay type and hence, presumably, with soil type.

The most important development in determining rates of nitrogen transformations has been the application of ^{15}N-based techniques (see 33). This involves supplying ^{15}N in a form appropriate for the process being studied and measuring the appearance of the isotope in an appropriate product, or determining the reduction in its concentration, in an approach termed pool dilution. The most straightforward examples of its use are addition to samples of $^{15}N_2$ or $^{15}NO_3^-$. The rate of appearance of ^{15}N in plant material supplied with $^{15}N_2$ is then related to the rate of nitrogen fixation. Changes in $^{15}N_2O$ or $^{15}N_2$, after amendment with $^{15}NO_3^-$, will measure rates of denitrification. An alternative, cheaper approach for measurement of nitrogen fixation is to compare plant uptake of $^{15}NH_4^+$ in the presence and absence of nitrogen fixers; the difference is equivalent to the rate of nitrogen fixation. A third approach is to determine the natural abundance of ^{15}N in plant material, as nitrogenase activity leads to a depletion of ^{15}N.

Changes in $^{15}N_2O$ or $^{15}N_2$, following amendment with $^{15}NO_3^-$, will measure rates of denitrification. Nitrification rates can be determined by incubation with $^{15}NO_3^-$ and measurement of depletion of ^{15}N in the nitrate pool. Information on the relative rates of heterotrophic and autotrophic nitrification can be obtained by incubating with $^{15}NH_4^+$. The appearance of ^{15}N in nitrate indicates autotrophic nitrification, while dilution of $^{15}NO_3^-$ would indicate heterotrophic nitrification in which nitrate is produced from organic N. Addition of ^{15}N-labeled fertilizer can also be used to measure rates of ammonia volatilization by determination of $^{15}NH_4^+$ in ammonia trapped in acid.

An alternative approach is the application of ^{15}N pulse-dilution methods to determine rates of mineralization and immobilization of nitrogen. $^{15}NH_4^+$ is added and its dilution by $^{14}NH_4^+$ is related to the rate of mineralization, i.e., production of ammonia, from unlabeled organic N. The approach can be extended by adding

^{15}N-labeled organic matter and measuring the appearance of ^{15}NH$_4^+$ in an initially unlabeled ammonia pool. This has the added advantage that different fractions of organic matter can be labeled, allowing assessment of the contribution to mineralization from, e.g., crop residues and microbial biomass. Rates of immobilization of nitrogen can also be determined by analysis of the ^{15}N:^{14}N ratio in microbial biomass. This requires a reliable method for extraction of N from microbial biomass. The approach is also complicated by the potential for turnover of nitrogen between different pools as incubation times increase, changes in natural pool abundances following amendment, and differences in rates of conversion of ^{15}N and ^{14}N.

Phosphorus and sulfur transformations

Phosphorus and sulfur are found in soil in organic and inorganic forms and as components of microbial biomass. Rates of microbially mediated transformations between organic and inorganic forms are measured by the techniques described. For example, addition of ^{32}P-labeled inorganic phosphorus to soil has been used to study its uptake, by plants and microorganisms, and its mineralization, by measuring dilution of the isotope. Similar approaches have been adopted for measurement of sulfur transformations using ^{35}S. Process rates are also determined by measurement of concentrations of substrates and products. For example, rates of sulfate reduction are measured by incubation of soil samples in the presence of sulfate and measurement of sulfide production. The sensitivity of these measurements can be increased significantly by use of ^{35}SO$_4^-$. For both approaches, however, difficulties arise in distinguishing between different pools of organic phosphorus or sulfur due to technical problems associated with chemical separation and analysis of different groups of organic compounds.

5.3 Activity of Individual Cells

The techniques described for measurement of cell activity can often be modified to determine the proportion, or number, of active cells of a particular type. For example, DVC can be used in combination with immunofluorescence to detect specific bacteria in mixed communities (34), although this has been used mainly for detection in aquatic environments. Similarly, microautoradiography has been linked to immunofluorescence to determine the number of active nitrifiers fixing ^{14}CO$_2$ (35).

Luminescence-based techniques also provide the potential for in situ detection of active and potentially active cells within soil and on plant material. Visual detection of emitted light, or use of photographic film, was used by Shaw and Kado (36) to follow infection of potato tubers by *lux*-marked *Erwinia carotovora* subsp. *carotovora.* Detection by luminescence was possible before other visible signs of tissue maceration were apparent, and light output of tuber samples, measured by luminometry, was directly related to viable cell concentration. The advantage of this technique is the ability to localize plant pathogens and to follow colonization and infection without disrupting the infection process. A combination of photography, for localization of colonies, and luminometry, to measure activity, has been used by McLennan et al. (37) to follow infection of tuber tissue by *E. carotovora* and by Fravel et al. (38) to identify regions of rhizosphere colonization by plasmid-marked *Enterobacter cloacae.*

The sensitivity of detection of light can be improved significantly by the use of charge-coupled device (CCD) image enhancement microscopy, which allows detection and quantification of single photons. For marked strains with high levels of luminescence per cell, individual cells can be detected by CCD microscopy. Examples are detection of single cells of *lux*-marked *Pseudomonas syringae* in soil suspensions (39) and of *lux*-marked *E. cloacae* colonizing wheat roots (40). Image analysis software also allows quantification of light output by individual cells. For *P. syringae,* cell activity following nutrient amendment was found to be less in nonsterile soil than in sterile soil, because of competition for nutrients by indigenous microorganisms. While the previous approach can be used to determine actual and potential activity, detection of individual cells without nutrient amendment is only possible for the brightest strains. The CCD detection of microcolonies and larger accumulations of cells is, however, feasible with the majority of luminescence-marked organisms studied. Shaw et al. (41) detected *lux*-marked *Xanthomonas campestris* infecting cabbage plants, and Beauchamp et al. (42) investigated the colonization of plant roots after germination of seeds inoculated with luminescence-marked rhizobacterial pseudomonads using CCD imaging, luminometry, and dilution plate enumeration. Luminescence was detected in the crown regions but generally not in the tip or midroot region. Although CCD imaging was the least sensitive technique, it did allow detection of the marked population without disturbing the plant material.

6. GROWTH RATES

Soil microorganisms do not participate in nutrient cycling processes to provide substrates for other organisms, but to survive and support their own growth. Activity is therefore linked to maintenance and growth. Growth in natural environments, when it occurs, is frequently limited by low concentrations of nutrients, in particular energy substrates. Although organic carbon levels are high in the soil, available carbon is generally low. In addition, there will be strong competi-

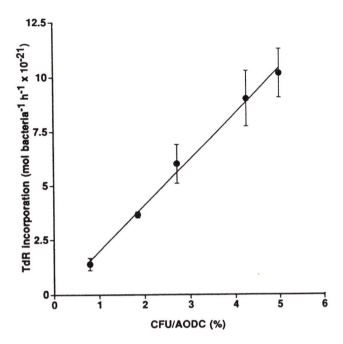

Figure 5. The relationship between specific thymidine incorporation and the number of culturable bacteria, expressed as the percentage of acridine orange staining cells forming colonies, after size fractionation of cells. (From Ref. 45.)

turnover times increasing from 4 days for unfiltered samples to 32 days for the smallest size class.

Flow cytometry has also been used to demonstrate differences in growth in cells of different size (46). Cells were extracted from soil and examined by flow cytometry for cell number, cell size, and DNA content. They were then incubated in soil extract medium, and increases in numbers and respiration rate were determined by using traditional techniques, demonstrating population growth. Flow cytometry and direct microscopy were also used to determine changes in numbers of cells in different size classes, and both techniques showed a significant increase in the number of large cells, while the number of small cells remained approximately constant. There were, however, differences between the techniques, with flow cytometry giving lower counts for small cells and higher counts for large cells (Fig. 6).

Although this study takes into account heterogeneity within the bacterial population, the majority of these techniques measure average growth rates and do not consider environmental heterogeneity. For example, it is possible to envisage

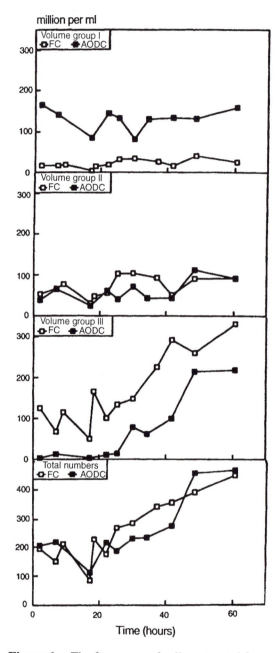

Figure 6. The frequency of cells, extracted from soil, falling into three volume classes, and the total number of cells, determined by flow cytometry (FC) and acridine orange direct counts (AODC). Volume classes were (I) <0.065 µm³, (II) 0.065–0.18 µm³, (III) >0.18 µm³. (From Ref. 46.)

regions where growth rates are approaching the maximum, while, in others, microorganisms will be essentially inactive for long periods.

7. CONCLUSIONS

A wide variety of techniques and approaches has been adopted for the measurement of microbial processes in the soil. The choice of technique depends heavily on the purpose of the experimental study for which measurements are required. Despite the limitations of the techniques described in this chapter, many microbially mediated soil transformations can be quantified with a high degree of confidence and accuracy. Measurement of population activity is more straightforward than determination of activity of individual cells. The latter is, however, increasingly important as more information is required about the detailed structure of microbial populations in natural environments, the way the physical and chemical heterogeneity of the soil environment leads to biological heterogeneity, and the way it, in turn, affects transformations both qualitatively and quantitatively. The challenge is, therefore, to develop techniques which take into account soil heterogeneity and its effects on microbial activity within the soil.

REFERENCES

1. J. W. T. Wimpenny (ed.), *A Handbook of Laboratory Systems for Microbial Ecosystem Research* CRC Press, Boca Raton, Fla. (1988).
2. G. Stotzky and R. G. Burns, In: *Experimental Microbial Ecology* (R. G. Burns and J. H. Slater, ed.), Blackwell Scientific, Oxford, p. 105 (1982).
3. J. I. Prosser and M. J. Bazin, In: *A Handbook of Laboratory Systems for Microbial Ecosystem Research* (J. W. T. Wimpenny, ed.), CRC Press, Boca Raton, Fla., p. 31 (1988).
4. J. G. Boswell, *New Phytol, 54*:311 (1955).
5. D. F. Weber and P. L. Gainey, *Soil Sci. 94*:138 (1962)
6. J. I. Prosser, *Adv. Microbial Physiol. 30*:125 (1989).
7. S. M. Allison and J. I. Prosser, *Soil Biol. Biochem. 25*:935 (1993).
8. E. F. Armstrong and J. I. Prosser, *Soil Biol. Biochem. 20*:409 (1988).
9. S. J. Powell and J. I. Prosser, *Appl. Environ. Microbiol. 52*:782 (1986).
10. S. J. Powell and J. I. Prosser, *Microbial Ecol. 24*:43 (1992).
11. R. C. Cooke and A. D. M. Rayner, *Ecology of Saprophytic Fungi* Longman, London and New York, (1984).
12. D. Wright, K. Killham, L. A. Glover, and J. I. Prosser, *Appl. Environ. Microbiol. 61*:3537 (1995).
13. L. L. Daane, J. A. E. Molina, E. C. Berry, and M. J. Sadowsky, *Appl. Environ. Microbiol. 62*:515 (1996).

14. J. P. E. Anderson and K. H. Domsch, *Soil Biol. Biochem. 10*:215 (1978).
15. M. Tateno, *Soil Biol. Biochem. 17*:387.
16. A. Meikle, K. Killham, J. I. Prosser, and L. A. Glover, *FEMS Microbiol. Lett. 99*:217 (1992).
17. R. Zimmermann, R. Iturriaga, and J. Becker-Birck, *Appl. Environ. Microbiol. 36*:926 (1978).
18. O. Nybroe, *FEMS Microbiol. Ecol. 17*:77 (1995).
19. K. Kogure, U. Simidu, and N. Taga, *Can. J. Microbiol. 25*:415 (1979).
20. P. H. Calcott and J. R. Postgate, *J. Gen. Microbiol. 70*:115 (1972).
21. C. E. Heijnen, S. Page, and J. D. van Elsas, *FEMS Microbiol. Ecol. 18*:129 (1995).
22. M. L. Brock and T. D. Brock, *Mitteilungen Iunternationale Vereiningung Theoretische angewandte Limnologie 15*:1 (1968).
23. J. S. Waid, K. J. Preston, and P. J. Harris, *Bull. Ecol. Res. Comm. (Stockholm) 17*:317 (1973).
24. J. I. Prosser, *Microbiol. 140*:5 (1994).
25. J. K. Jansson, *Curr. Opinion Biotechnol. 6*:275 (1995).
26. E. A. S. Rattray, J. I. Prosser, K. Killham, and L. A. Glover, *Appl. Environ. Microbiol. 56*:3368 (1990).
27. E. A. S. Rattray, J. I. Prosser, L. A. Glover, and K. Killham, *Soil Biol. Biochem. 24*:421 (1992).
28. A. Meikle, S. Amin-Hanjani, L. A. Glover, K. Killham, and J. I. Prosser, *Soil Biol. Biochem. 27*:881 (1995).
29. A. Meikle, L. A. Glover, K. Killham, and J. I. Prosser, *Soil Biol. Biochem. 26*:747 (1994).
30. S. Duncan, L. A. Glover, K. Killham, and J. I. Prosser, *Appl. Environ. Microbiol. 60*:1308 (1994).
31. Swift, In: *Experimental Microbial Ecology*. (R. G. Burns, and J. H. Slater, ed.), Blackwell Scientific, Oxford, p. 164 (1982).
32. Powell and J. I. Prosser, *J. Gen. Microbiol. 137*:1923 (1991).
33. Killham, *Soil Ecology*. Cambridge University Press, Cambridge (1994).
34. S. Xu, N. Roberts, F. L. Singleton, R. W. Atwell, D. J. Grimes, and R. R. Colwell, *Microbial Ecol. 8*:313 (1982).
35. C. B. Fliermans and E. L. Schmidt, *Appl. Microbiol. 30*:674 (1975).
36. J. J. Shaw and C. I. Kado, *Bio/Technol. 4*:560 (1986).
37. K. McLennan, L. A. Glover, K. Killham, and J. I. Prosser, *Lett. Appl. Microbiol. 15*:121 (1992).
38. D. R. Fravel, R. D. Lumsden, and D. P. Roberts, *Plant Soil 125*:233 (1990).
39. D. Silcock, R. N. Waterhouse, L. A. Glover, J. I. Prosser, and K. Killham, *Appl. Environ. Microbiol. 58*:2444 (1992).
40. E. A. S. Rattray, L. A. Glover, J. I. Prosser, and K. Killham, *Appl. Environ. Microbiol. 61*:2950 (1995).

41. J. J. Shaw, F. Dane, D. Geiger, and J. W. Kloepper, *Appl. Environ. Microbiol.* *58*:267 (1992).
42. J. W. Beauchamp, J. W. Kloepper, and P. A. Lemke, *Can. J. Microbiol.* *39*:434 (1993).
43. D. B. Nedwell, and T. R. G. Gray, In: *Ecology of Microbial Communities* (M. Fletcher, T. R. G. Gray, and J. G. Jones, ed.), Cambridge University Press, Cambridge, p. 21 (1987).
44. D. S. Jenkinson and L. C. Parry, *Soil Biol. Biochem.* *21*:535 (1989).
45. E. Baath, *Microbial Ecol.* *27*:267 (1994).
46. R. A. Christensen, R. A. Olsen, and L. R. Bakken, *Microbial Ecol.* *29*:49 (1995).

8

Microbial Interactions in Soil

JACK T. TREVORS   University of Guelph, Guelph, Ontario, Canada

JAN DIRK VAN ELSAS   Research Institute for Plant Protection (IPO-DLO), Wageningen, The Netherlands

1.   INTRODUCTION

Microbial interactions are central to microbial ecology, the study of interrelationships among microorganisms as well as with eukaryotic organisms like protozoa, plants, and animals and with abiotic components of the environment. Microbial ecology focuses on understanding the multitude of microbial species present, their numbers or densities, and interrelationships with other organisms and nonliving components of the ecosystem. This area has long been studied by using methods and techniques based on cultivation of microorganisms from the environment. However, using these techniques, it has not been possible to ascertain the roles of the substantial nonculturable fraction of the viable microbial populations in the environment. Therefore, in recent years direct molecular biology techniques such as extraction of deoxyribonucleic acid (DNA) from soil, DNA:DNA probing, and polymerase chain reaction (PCR) amplification of target DNA have been used to detect and amplify, respectively, specific DNA sequences in DNA extracts from soil. To understand the roles of microorganisms in nutrient and mineral cycles, plant growth and disease, and the degradation of chemicals, numerous studies are performed in aquatic environments, agricultural and forest soils, and nonagricultural (e.g., chemically contaminated) soils. Several texts on microbial ecology (1–8) are excellent sources of information upon which to build a good understanding of this area. Texts on soil microbiology and biochemistry (9,10) and the extraction, purification, and amplification via PCR of nucleic acids from soils (11) are also recommended references.

Microorganisms in soil can participate in genetic as well as other interactions, such as microbial symbioses, mutualisms, and antagonisms. Moreover, they can provide a food source for protozoa or other microbivorous grazers (see also Chapters 6 and 9).

Microbial cells in soil often are adsorbed to negatively charged clays and soil organic matter and sometimes transform or accumulate metal forms in soil. These microbial actions can be essential components of biogeochemical cycles in the biosphere.

Current knowledge of microbial interactions in soil has been obtained from laboratory, microcosm, greenhouse, and field studies. These studies are very different from those on single species performed in microbial physiology/biochemistry, genetics, and molecular biology work, where the complexity and heterogeneity of the soil environment, as well as spatial and temporal variabilities, are basically absent.

Microorganisms survive in nature by growing at a rate sufficient to balance the death of cells caused by starvation and environmental changes like substrate limitations, extremes of temperature and pH, desiccation, limiting O_2 concentrations, and ingestion by protozoans (12). Interactions between viable microbial cells in soil therefore depend on their ability to survive by obtaining adequate nutrients, usually in the aqueous phase. For instance, in plasmid transfer via conjugation in soil, viable donor and recipient cells are required for formation of the conjugation bridge and subsequent plasmid transfer. Whereas bulk soil is often severely limited in the amount of substrate available to microorganisms, the rhizosphere of plants provides root exudates and surfaces for microbial colonization and growth. As stated by Alexander (12), a vast community of metabolically active microorganisms is commonly associated with the root system of higher plants. Hence, the rhizosphere behaves differently from bulk (nonrhizosphere) soil in terms of both the numbers and types of microorganisms present and their metabolic activities. In recent years, the rhizosphere of a plant species like wheat (*Triticum aestivum*) has been shown to be a location where, for instance, genetically modified microorganisms (GMMs) such as *Pseudomonas fluorescens* can survive for an enhanced period following introduction. The rhizosphere has also been shown to stimulate conjugal plasmid transfer between *Pseudomonas* strains (see the section on Conjugation).

In any discussion of microbial interaction in soils, it is necessary to define the soil studied (e.g., rhizosphere or nonrhizosphere soil), as well as its textural class (based on percentages of sand, silt, and clay), pH, organic matter (OM) content, moisture content, and past history, such as treatment with pesticides, fertilizers, animal wastes, sewage sludges, and compost. All these factors can have effects on the soil microflora and therefore on microbial interactions.

In this chapter, microbial interactions in soil will be discussed with an emphasis on the role of the microorganisms involved. Interactions between microbes and protozoans, *Bdellovibrio*, fungi, and metals are also examined. In environmental applications such as biological control of plant diseases in agriculture and forestry or the biodegradation of toxic chemical(s) in contaminated soils, GMMs will be subjected to interactions with neighboring organisms. Hence, a thorough understanding of microbial interactions occurring in situ is important when releasing GMMs into soil (13–16).

The rhizosphere and plant–microbe interactions have been discussed in Chapter 2 and therefore will only be briefly treated here, mainly with respect to functions (e.g., nitrogen fixation) relevant to the ecological characteristics of the microorganisms involved.

2. INTERACTIONS BETWEEN MICROORGANISMS IN SOIL

Interactions between microorganisms can take place at different levels in soil communities, i.e., bacteria, fungi, protozoa, and nematodes can all interact with members of their own grouping as well as with those of other groups. Furthermore, interactions can be genetic, resulting in horizontal gene transfers, or mutualistic, symbiotic, or antagonistic. In this section, we will examine interactions mainly between bacteria in soil. First, the three possible modes of genetic interaction will be discussed, after which other interactions will be briefly addressed.

2.1 Genetic Interactions

An important mechanism by which microbes can interact in soil is gene transfer. The ability of soil microorganisms (in particular bacteria) to exchange DNA or release DNA or (transducing) bacteriophages into soil that transform or transduce other bacterial cells enables populations to adapt and evolve. In recent years, research on gene transfer in soil (17,18) has taken on more importance from environmental and human health perspectives in light of the planned releases of GMMs into soils.

Table 1 summarizes factors that affect gene transfer in soil. Some of these factors have not been investigated thoroughly; however, their study is important, given the need to fill gaps in our knowledge about the ecological impact of gene transfers and GMM releases in soils. These factors may differentially affect the

Table 1.   Some Factors That Affect Gene Transfer in the Soil Environment[a]

1. Nutrient status and fluxes in soil, including organic soil amendments such as compost and animal wastes
2. Soil water content and drying–rewetting cycles
3. Soil temperature and freezing–thawing cycles
4. Soil texture
5. Soil pH
6. Oxygen status of soil
7. Rhizosphere versus bulk (nonrhizosphere) soil
8. Microbial species present
9. Density and distribution of microbial species present
10. Density of donor and recipient cells
11. Host range of plasmids and their ability to be mobilized by indigenous plasmids in soil microorganisms
12. Plasmid incompatibility in microbial cells
13. Length of mating period during conjugation
14. Retrotransfer (back-transfer) of a *tra* plasmid from recipient to donor cells
15. Physiological status of microorganisms (e.g., competence for transformation)
16. Concentration of extracellular transforming DNA
17. Lysis of bacteria in soil by specific bacteriophage(s)
18. Presence of toxic pollutants or other selection pressures (e.g., antibiotics, pesticides)
19. Predation of bacteria by protozoa
20. Matrix in which microbial cells are added to soils (e.g., in solid carriers such as peat, alginate and K-carrageenan, or as a liquid inoculum)

[a]These factors affect the ecological and physiological characteristics of microorganisms, and hence gene transfer processes among these microorganisms.

gene transfer processes in different soils. For this reason, the outcome of gene transfer events is not always predictable and straightforward when soil characteristics are known. On the other hand, some dominating factors that significantly affect gene transfer frequencies (as well as bacterial survival) in soil can often be identified. These factors relate to the creation of favorable niches for bacteria in soil (18). Mathematical models predictive of gene transfer rates in soil have been based on these factors (18). Data that predict gene transfer and the fate of introduced GMMs in soil are essential to develop good regulatory tools and guidelines for the use of GMMs in environmental applications.

Figure 1 is a diagrammatic representation of the horizontal gene flow in soil

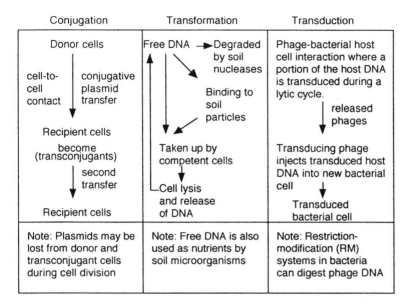

Figure 1.   Gene flow in soil by conjugation, transformation, and transduction.

by conjugation, transformation, and transduction. These processes are central to the molecular evolution of soil microorganisms (17,18,42). To date, the majority of reports on gene transfer in soil have addressed only conjugation and transformation as relevant processes, which is partly due to the fact that these processes are easiest to measure in soil. The following sections will discuss the current knowledge on conjugation, transformation, and transduction in soil. Additional information on gene transfer and bacterial genetics in natural environments can be found in texts by Fry and Day (37), Levin et al. (38), Trevors and van Elsas (39), and relevant reviews (40,41).

Conjugation

Conjugation is the transfer of plasmid and/or chromosomal DNA from donor to recipient cells to produce transconjugant cells. The primary transconjugants can again act as donor cells and the process can be repeated. Conjugation in gram-negative bacteria requires cell-to-cell contact by a conjugation bridge or pilus (bacterial surface appendage originating in the cell membrane and extending through the cell wall, composed of pilin proteins). Following cell-to-cell contact, the DNA is transported across the cell membranes, from the donor to the recipient cell. Transfer-proficient plasmids (Tra⁺) can be responsible for plasmid and even chromosomal transfers. Chromosomal transfer occurs when a conjugative plasmid integrates into and carries part of the donor bacterial chromosome during conjugation. Nonconjugative plasmids can be mobilized by integration with a conjugative plasmid, or via a process called mobilization, in which they are transferred as separate entities by the conjugation apparatus. Conjugative plasmids may have a broad or narrow range of hosts in which they can reside. Much of the current knowledge on bacterial conjugation has been obtained from experiments conducted in vitro, in broth media, or on the surface of agar plates or membrane filters. The physical support provided by agar or filter surfaces often enhances plasmid transfer rates and prevents the pilus from becoming disturbed during conjugation. Therefore, it is not surprising that soil particles, organic matter, and root surfaces provide excellent sites for plasmid transfers in soil.

In laboratory and soil studies, selective markers being transferred are often resistances to antibiotics or heavy metals; upon acquiring the plasmid, the transconjugant cells become resistant to the antibiotic(s) or metals. The frequency of plasmid transfer (number of transconjugants per donor or recipient cell), an important parameter characterizing the process, can range from a low of 10^{-8} (barely detectable) to a theoretical high of even 1. Transconjugants can be analyzed for the presence of the plasmid transferred by standard plasmid extraction and analysis methodology, as reviewed (19), or via molecular methods (PCR, hybridization).

Conjugation in gram-positive species like *Streptococcus* probably does not involve a pilus (20). Some *Streptococcus* plasmids are transferred in nutrient broth because recipient cells excrete sex pheromones (sometimes referred to as clumping inducing agents [CIAs]). Donor cells produce the pheromone (a specific protein), that can cover the entire cell surface. The protein causes the donor and recipient cells to form mating cell aggregates. Although mating cell aggregation has been described in in vitro studies, it has not been adequately described in the soil environment.

Belliveau and Trevors (21) reported that a strain of the gram-positive soil bacterium *Bacillus cereus* containing a large mercury resistance plasmid (pGB130) was able to transfer this plasmid to mercury-sensitive strains of *B. cereus* and the related *B. thuringiensis*. Upon receiving the plasmid, both species became resistant to mercury. The mechanism of mercury resistance was the enzymatic volatilization of Hg^{2+} to Hg° by mercuric reductase encoded on the plasmid. The frequency of plasmid transfer ranged from 10^{-4} to 10^{-5}. *Bacillus* species are spore-forming gram-positive bacteria common in soil, and it is possible that such transfers do play a role in *Bacillus* populations in mercury-stressed soils. For further reading on conjugal interactions in gram-positive bacteria, the article by Dunny (22) is recommended.

Numerous reports have shown that conjugation in nonsterile soils can occur at frequencies roughly between 10^{-6} and 10^{-8} transconjugants per recipient, but often only when nutrients are present or added. The rhizosphere (e.g., that of wheat) has been shown to represent a site in soil where conjugal transfer is greatly stimulated (16). This stimulus has been credited to the enhanced availability of nutrients to bacteria, as well as to the increased possibilities for bacterial movement and sites for their colonization. Physicochemical factors in soil also influence conjugation. Favorable temperature and pH values have been shown to stimulate conjugal transfers (16). In addition, a clay like montmorillonite added to soil also enhanced the rates of plasmid transfer between introduced bacteria (16).

As mentioned, mobilization of a nonconjugative plasmid can occur when a conjugative plasmid enters its host. As this process requires two steps and three hosts (in fact, it has been called triparental mating), transfer frequencies can be lower than those in simple one-step matings. However, triparental mating may be a significant mechanism for transfer of DNA between microorganisms in soil. In the wheat rhizosphere, a mobilizable IncQ plasmid was shown to be mobilized by an IncP mobilizer plasmid present in a different strain, showing the feasibility of triparental mobilization in the rhizosphere (41). Conversely, Henschke et al. (23) could not detect any transfer of a nonconjugative, recombinant *Escherichia coli* plasmid (pUC19) to indigenous soil bacteria, although laboratory strains of *E. coli* could act as recipients of this plasmid. As *E. coli* is not considered a normal inhabitant of most soils, there may have been too few compatible conjugative plasmids in soil to facilitate the plasmid transfer.

Top et al. (24), studying the conjugal transfer of heavy metal resistance genes from *Escherichia coli* to *Alcaligenes eutrophus* in sterile soil samples, found transfer occurred at frequencies of 10^{-5} transconjugants per recipient cell. This transfer was the result of the mobilization of a nonconjugative plasmid (pDN705) by a conjugative plasmid (RP4 or pULB113) present in another bacterial strain (triparental mating) or the recipient strain (retromobilization).

Knudsen et al. (25) developed a computer simulation model to predict the survival and conjugation of *Pseudomonas cepacia* (containing recombinant plasmid R388:Tn*1721*) in rhizosphere and phyllosphere microcosms. This model described plasmid transfer as a function of the size of the effective donor and recipient cell populations. It was capable of adequately simulating bacterial donor, recipient, and transconjugant populations at hourly intervals. However, conjugal gene transfer rates in soil are heavily dependent on local soil conditions. Factors based on soil conditions important for a mathematical model have been suggested (18). Factors that elicit a linear or log-linear transfer rate response are bacterial cell density, cellular activity, and nutrient fluxes. Factors such as soil clay content, temperature, pH, and moisture content were suggested to relate to plasmid transfer via a hyperbolic response curve, whereas biotic interactions, dependent on the type of interaction, were found to elicit variable responses.

Transformation

Transformation is the uptake and expression by bacteria of genes located on extracellular (naked) DNA. It allows integration of foreign DNA into the chromosome of a recipient cell or its carriage on a replicating plasmid. Hence, recombinant DNA, whether on a plasmid or on the chromosome, can be transferred to other bacteria via transformation. For natural transformation to occur, recipient cells have to develop a state of competence, which indicates their ability to bind and take up DNA in a form resistant to degradation by deoxyribonucleases (DNases). Most bacteria regulate competence. Gram-positive *Bacillus subtilis* cells can release so-called competence factors (proteins that induce the state of competence) into their growth media. When cell concentrations are high, there are sufficiently high levels of competence factors present to induce competence in the population. Hence, as far as current knowledge goes, competence in gram-positive organisms is a cell-density-dependent phenomenon. Gram-negative species such as *Haemophilus influenzae* and *Pseudomonas stutzeri* do not appear to release competence factors. In these organisms, competence is induced by slowing the growth rate, switching from a rich medium to a nutrient-depleted medium, and increasing the cyclic adenosine monophosphate (cAMP) level or blocking DNA synthesis while allowing protein and ribonucleic acid (RNA) synthesis to continue. It has been suggested that bacteria become competent during the stationary phase to obtain a greater catabolic range (ability to degrade or metabolize substrates) or greater tolerance to waste metabolites, as well as to obtain precursors for DNA synthesis (26).

There are several reports of natural transformation of bacteria in the environment. Table 2 lists some soil microorganisms that can become naturally competent and are thus capable of being transformed by extracellular DNA. One of the earliest evidences for the transformation of *B. subtilis* in soil was reported by

Table 2. Some Genera of Soil Microorganisms (Bacteria) Capable of Natural Genetic Transformation

Gram-negative	Gram-positive
Achromobacter	Bacillus
Acinetobacter	Micrococcus
Azotobacter	Mycobacterium
Moraxella	Streptomycetes
Pseudomonas	Streptococcus
Actinomycetes	

Graham and Istock (27). Frequencies of transformation of three antibiotic-resistance markers ranged from 10^{-1} to 10^{-8} per recipient cell, depending on whether
mixed or single cultures were used and how the sampling was timed. Bacteria
were thought to be transformed by extracellular DNA released from *B. subtilis*
cells. Recent work in our laboratory has shown that the naturally transformable
species *Acinetobacter calcoaceticus* can be transformed by naked DNA in soil
microcosms (27a). Transformation, however, was dependent on nutrients added
to soil, which could be mimicked by added maize root exudate. Hence, the rhizosphere of maize can be suggested to represent a site in soil where conditions
conducive to transformation events occur.

It has been shown that naked DNA is stabilized when attached to sand or clay
surfaces because of an enhanced resistance to DNases (28,29). Moreover, the potential for sand-adsorbed DNA to transform both gram-positive and gram-negative bacteria has been analyzed (28,30). Whereas *B. subtilis* was transformed at
25- to 50-fold higher frequencies by sand-adsorbed DNA than by dissolved
DNA (28), *P. stutzeri* was transformed at a less efficient rate by sand-adsorbed
DNA. Since *B. subtilis* cells attach to DNA-coated sand better (10% of added
cells) than *P. stutzeri* (0.8% of added cells), it was postulated that the former
species had a greater chance of contacting the DNA, which would facilitate its
transformation. Transformation of *B. subtilis* at the sand surface was not inhibited by the presence of DNase I at concentrations up to 1 μg/ml, whereas that of
P. stutzeri cells was inhibited by 50 ng/ml DNase I. The attached *B. subtilis* cells
were therefore likely to contact the sand-adsorbed DNA without interference
from enzymatic digestion.

Transduction

The third mechanism of bacterial gene transfer is transduction, in which a bacteriophage (short form: phage) transports genes from one bacterial cell to another.
Phages, when replicating, can insert up to a certain amount of DNA into their
protein capsid head. This DNA may be of phage or bacterial origin. When a
newly assembled phage that contains bacterial DNA with or without additional
phage DNA (transducing phage particle) infects a new bacterial host, it introduces the foreign DNA (along with the eventual phage DNA) into the new cell.
If the foreign DNA is integrated into the host chromosome, or if the transducing
phage becomes temperate (integrates as a whole into the new host genome), the
new host may express the foreign gene. Phages may also reside in the new host
as a plasmid and replicate in the circular form. Transduction may be one mechanism by which bacteria such as *E. coli* and *Salmonella typhimurium* have acquired resistances to antibiotics, as it has been shown that R-plasmids can be
transferred by phages (31).

Because transduction is a gene transfer process mediated by phages and most phages have narrow host ranges, it is generally believed transduction only occurs between related microorganisms. For instance, Ruhfel et al. (32) showed that bacteriophage CP-51 could transduce the related species *B. anthracis, B. cereus*, and *B. thuringiensis*. Additionally, phage PBL1c can transduce protoplasts of *B. larvae, B. subtilis*, and *B. popilliae* (33). These and other data have suggested that it is unlikely transduction can extend much beyond the species level or that of very closely related genera such as *E. coli* and *S. typhimurium*.

Transduction of *E. coli* in soil has first been reported by Germida and Khachatourians (34). Zeph et al. (35) also reported the transduction of *E. coli* by bacteriophage P1 in both sterile and nonsterile soils. In sterile soil, the rate of transduction was enhanced by montmorillonite or kaolinite clays. Furthermore, there were three to four orders of magnitude more transductants in sterile soil than in nonsterile soil, showing the inhibiting effect of the soil biota to transduction. To date, there has been no evidence of transduction to the indigenous microflora in soil. Studies on transduction in soil have been hampered by the often low frequencies of transduction and potential interference with successful transduction by adsorption of phage to soil particles as well as the lack of energy sources for bacterial hosts, both donors and potential recipients. Under such nutrient-limited conditions, many host bacteria will not lyse upon phage infection and produce phage progeny, but rather maintain the phage in a carrierlike state without permitting phage propagation. Hence, even though transduction might represent a gene transfer mechanism relevant to soil bacteria under certain conditions (high cell densities and nutrient-rich conditions), its real ecological role remains to be elucidated.

2.2 Other Interactions

Microorganisms in soil, in particular bacteria, are frequently involved in a multitude of nongenetic interactions with other microorganisms (12). Many microbial cells in soil, in particular in the rhizosphere, live in close proximity to other cells, and such cells obviously can influence their neighbor cells in numerous unique ways. Such interactions can, for instance, be nutritional, since microbes may depend on other microbes for specific substrates or degradation products, or they can compete for the same substrates. On the other hand, microbes can exert detrimental influences on other microbes, for instance, via the production of antibiotic or toxic compounds. Alexander (12) has categorized the possible microbe–microbe interactions in soil as follows: (a) neutralism, in which two species live together but have no obvious beneficial or detrimental interactions; (b) symbiosis, in which the coexisting partners rely on each other, and both profit from the interaction (e.g., nutritionally); (c) protocooperation, in which a symbiosis exists but is not obligatory; (d) commensalism, in which only one partner benefits from the interaction, whereas the other remains unaffected; (e) competition (mostly for substrate), in which one or both species may suffer from the presence of the other one; (f) amensalism (due to antibiotic or toxin production), in which one species is suppressed and the other is not affected; and (g) parasitism and predation, the direct attack by one species of the other one.

Whereas interactions (b) through (d) are beneficial to either both partners or just one of them, those under (e) to (g) can be characterized as detrimental to at least one partner. There are many possible examples that demonstrate the occurrence of all of these interactions in a natural soil. For instance, a typical commensalistic interaction is the degradation of cellulose by cellulolytic fungi that produce organic acids which serve as substrates for noncellulolytic bacterial and fungal species (12). Another example is the liberation of growth factors by certain microbes in soil, which allows for the growth of other, fastidious microbes. Symbiotic interactions can be seen in the algal–fungal associations called lichens (see Section 3.1 and Chapter 4).

A typical example of an economically relevant competitive interaction occurs between rhizobia in soil and those used as seed inoculants. As the most competitive type may form nodules more frequently on the plant, it is important that the strains effective in nitrogen fixation used as inoculants are also the fittest in nodulation (see Section 3.1).

The inhibition of pathogenic fungi by fluorescent pseudomonads due to the in situ production of antibiotics of the phenazine type (36) is an example of an amensalistic microbe–microbe interaction, in which the bacterial partner acts against the fungal one by producing an inhibiting compound. This principle is of importance in the development of novel strategies of biological control of pathogenic fungi. For instance, a novel biotechnological approach in our laboratory attempts to improve phenazine-producing *Pseudomonas* strains for use as inoculants antagonistic to plant pathogenic *Pythium* sp. (see Chapter 14). Finally, typical predatory and parasitic microbe–microbe interactions are the attacks on soil bacteria by protozoa on the one hand and bdellovibrios on the other hand. These interactions will be treated in Sections 4 and 5.

3. INTERACTIONS BETWEEN MICROORGANISMS AND PLANTS

This section addresses selected beneficial as well as detrimental interactions between microorganisms (bacteria and fungi) and plants, with an emphasis on the ecological characteristics of the interaction. First, the association of bacteria with plants in a mutually beneficial symbiosis leading to nitrogen fixation is discussed, then the significance of the invasion of plants by microorganisms leading to damage (pathogenicity) is treated, and finally a fungus–plant interaction (mycorrhiza) leading to mutual benefits is addressed.

3.1 Nitrogen-Fixing Bacteria

In this section, the significance of the nitrogen-fixing *Rhizobium*–legume and actinorhizal (*Frankia*)–plant associations is treated. Other interactions with plants are further discussed in Chapters 2 and 17. The *Rhizobium*–legume as well as the *Frankia*–plant symbioses are of considerable economic and environmental importance in agriculture. The actinomycete *Frankia* forms nitrogen-fixing nodules in association with 170 plant species (perennial woody shrubs and trees) belonging to 17 genera, 8 families, and 7 orders (43). Nodules on legumes are formed by symbioses between the host plants and rhizobia which consist of the genera *Rhizobium, Bradyrhizobium*, and *Azorhizobium* (44). Although not all rhizobia depend on an interaction with the plant for their survival, some are poor survivors in soil unless a carbon and energy source is added (6).

Rhizobia contain the nitrogenase enzyme system (*nif*) responsible for nitrogen fixation. Legumes consist of a group of plants of about 700 different genera, with 14,000 species (43). The formation of the legume–*Rhizobium* symbiosis (Table 3) starts by rhizobial cells being attracted to the host plant roots as a result of plant-released signal compounds that trigger a bacterial response leading to chemotaxis toward the root and subsequent root colonization. Further, the presence of lectins (carbohydrate–protein compounds) on the host plant's roots is important, since rhizobial cells specifically produce an acidic polysaccharide that can interact with the plant lectin. This chemical interaction is remarkable considering that it occurs in the soil matrix in the presence of millions of other cells. The nodulation process starts with the curling of root hairs; in particular the root hair tip forms a tight curl called the shepherd's crook (43). An infection thread is then formed by enzymatic degradation of the root hair cell wall. The plant root encases the rhizobia with the plant cell nucleus controlling the growth of the infection thread. This process may last up to 2 days, after which the bacterial cells are released into the cortical cells of the plant root. The bacterial cells then divide rapidly and lose their rod-shaped morphological features, becoming pleomorphic (taking on a club shape, referred to as bacteroids).

The functional nitrogenase enzyme complex is synthesized and the nodule is surrounded by plant membrane envelopes. The reversible oxygen carrier leghemoglobin controls the supply of oxygen, at the same time protecting the oxygen-sensitive nitrogenase from excess oxygen. This is an excellent example of biochemical cooperation in the soil/plant environment. Sources of carbon and energy (amino acids) are supplied by the plant to the bacteroids and fixed nitrogen is provided by the nitrogen-fixing nodule to the host plant. It is important to stress the complexity of these interactions and their significance in providing fixed N for legume crop productivity.

When fixed N, such as nitrate (NO_3^-), is present in soil, both nodulation and nitrogen fixation are inhibited. Nitrate assimilation, which requires carbon sub-

Table 3.   Symbiotic Nitrogen-Fixing Bacterial Species and Their Host Plants

1. *R. leguminosarum* bv. *viciae*	Pea (*Pisum*)
2. *R. leguminosarum* bv. *trifolii*	Clover (*Trifolium*)
3. *R. leguminosarum* bv. *phaseoli*	Bean (*Phaseolus*)
4. *R. meliloti*	Alfalfa (*Medicago*)
5. *R. loti*	Lotus (*Lotus*)
6. *R. fredii*	Soybean (*Glycine*)
7. *R. galegae*	Goat's rue (*Galega*)
8. *Rhizobium* strain NGR234	Sirato (*Macroptilium*)
9. *Azorhizobium caulinodans*	Sesbania (*Sesbania*)
10. *Bradyrhizobium japonicum*	Soybean (*Glycine*)
11. *Bradyrhizobium* sp.	Parasponia (*Parasponia*), Cowpea (*Vigna*), Peanut (*Arachis*), Lupine (*Lupinus*)

strates, may deprive the nitrogen-fixing system of carbon and energy (43). Alternatively, when nitrate is transformed to nitrite (NO_2^-), nitrite competes with leghemoglobin and destroys it. Low soil pH and waterlogging of soils (anaerobic soil condition) are also detrimental to symbiotic nitrogen fixation, which requires a low level of oxygen.

An article by Roth and Stacey (45) on *Rhizobium*–legume symbiosis is recommended for further detailed information, in particular on the regulation and function of *nod* (nodulation) gene expression. There is increasing evidence that the nodulation genes are involved in the production of a phytohormone, the chemical substance that induces root hair curling and division of plant cortical cells (45).

The most studied *Frankia* (sporulating gram-positive, filamentous actinomycete capable of N_2 fixation and formation of root nodules on dicots)–plant association is that with alder (*Alnus* sp.) shrubs and trees, which are often found along streams and in high organic matter soils. The biochemical and genetic characteristics of this interaction are not as well understood as those of the *Rhizobium*–legume symbiosis. However, it is clear that, much as in the *Rhizobium*–legume interaction, there is a regulated series of events leading to a close association between *Frankia* and its host plant. From an ecological perspective, the *Frankia*–alder interaction is important because of the vast habitats where this interaction is found, ranging from the arctic to the tropics and semidesert to rain forests. The contribution of this interaction to nitrogen fixation has been estimated to amount up to 300 kg/ha/yr (43), depending on habitat and climatic conditions.

Several groups of free-living nitrogen-fixing bacteria are further able to establish a loose interaction with host plants, resulting in their nitrogen-fixing activity associated with plant roots. Conceivably, root-associated bacteria in their ecological processes profit from the presence of the root surface and root exudates. An example of associative nitrogen fixers are *Azospirillum* spp. (e.g., *A. brasilense* and *A. lipoferum*), which are commonly found in the rhizospheres of several tropical crop plants. It is still unclear how loose this plant–bacterium interaction is and whether fixation of atmospheric nitrogen is the only bacterial process of importance to the plant.

The association of specific algae or cyanobacteria with fungi results in so-called lichens. The physiological and systematic characteristics of lichens are discussed in detail in Chapter 4 and will not be further treated here. Several lichens have nitrogen-fixing ability. Such lichens have cyanobacteria as the phycobiont, which receives physical protection and nutrients while the fungal mycobiont obtains carbon and fixed nitrogen from the phycobiont. Examples of nitrogen-fixing lichens include *Collema, Nephroma, Peltigera, Solorina, Stereocaulon, Sticta, Placynthium, Lobaria, Lichina, Lepogium, Epheba*, and *Dendriscolaulon* (43). Hence, lichens are obviously of ecological importance because of their nitrogen-fixing ability and their capability to establish in harsh environments such as poor soils and even surfaces of rocks, where they cause weathering under low nutrient C and N conditions.

3.2 Plant-Pathogenic Microorganisms

Plant pathogens and their control are treated extensively in Chapter 17 and hence will not be discussed here. Rather, the ecological features of the interactions and their significance for the organism's life cycle will be shortly addressed.

Numerous soil bacteria, fungi, viruses, *Mycoplasma*, and *Rickettsia* can cause diseases in plants (7). These microorganisms, in particular the bacteria and fungi, have in their evolution developed specialized mechanisms by which they are attracted to plant roots, interact and invade the plant, and, once inside, cause a pathogenic reaction; the whole process often allows their own proliferation. Hence, pathogenicity is part of their life cycle, since it allows self-perpetuation. We will not further discuss the often complex plant–pathogen interactions and the mechanisms involved, since they are beyond the scope of this chapter. The symptoms of the diseases caused by phytopathogenic microorganisms appear as necroses, soft or wet rots, vascular disorders, tumors, and/or lesions (7). Common bacterial plant pathogens are diverse, and strains belonging to the genera *Pseudomonas, Xanthomonas, Erwinia, Clavibacter, Agrobacterium*, and *Streptomyces* can be involved (7). Soil is a natural reservoir for these bacteria. Therefore, the plant seeds, roots, stems, and leaves are often in contact with plant pathogens by contact with soil or splash water during rainfall. However, plants do not always develop symptoms of disease even in the presence of pathogenic organisms, since factors such as pathogen population density, presence of pathogen-suppressive organisms, and abiotic factors can impede the pathogenic process. It is known that planting the same crop in a soil site in consecutive years can lead to the buildup of a pathogen complex (enhancing the pressure by pathogens on plants), but also of suppressive (antagonistic) organisms which counteract pathogen pressure. Hence, a delicate balance may be struck between the two complexes, and conditions of the soil environment dictate the balance.

From an economic perspective, the interaction of plant-pathogenic microorganisms with agricultural crops and forests is a significant problem, which is complicated by a lack of adequate control methods for bacteria once they have infested a crop or forest site. For example, antibiotics added to soils are usually ineffective against bacteria as a result of adsorption effects in soil. Furthermore, the widespread use of antibiotics in soil would not be acceptable as this may contribute to selection pressure that increases the antibiotic resistance gene pool in nature. Bacterial populations under antibiotic stress are known to respond by becoming antibiotic-resistant through mutations or acquisition of resistance plasmids that are often self-transmissible.

Fungi contain plant pathogens that are devastating from an economic perspective (see Chapter 4). Plant-pathogenic fungi belong to the ascomycetes, basidiomycetes, Peronosporales, and Chytridiomycetes (7). Fungicidal chemicals are often employed in agriculture to control fungal diseases of crops. A natural host response to fungal disease is the production of fungistatic compounds known as phytoalexins, which may assist the plant in preventing a localized fungal infection (7). Phytoalexin production may be associated with a hypersensitive reaction, which is defined as rapid death of plant cells at the infection site. This prevents further spread of the disease, since as a result of localized cell death, small necrotic lesions are formed (7).

3.3 Mycorrhiza

Mycorrhiza–plant relationships are among the most significant plant-beneficial interactions in soil (Chapter 4). The interaction can involve two or even three (and possibly more) living components. The most commonly described interaction is that between two partners, the mycorrhizal fungus and the plant. In symbiotic nitrogen fixation, the partners are the mycorrhizal fungus, a specific (nitrogen-fixing) *Rhizobium* sp., and the host plant. In some nonlegume shrubs, a nitrogen-fixing association is found among the shrub, the mycorrrhizal fungus, and the actinomycete *Frankia* (43). The Mycorrhiza–plant interaction is mutualistically beneficial and can be characterized as symbiotic, since the plant is often aided in its uptake of essential compounds from soil (such as phosphate), whereas the mycorrhizal fungus receives carbonaceous substrate from the plant which it would otherwise not obtain from soil.

Mycorrhizal fungi which live in symbiosis with the root surface of plants are ectotrophic or sheathing (ectomycorrhizas [ECM]), while those fungi that enter the plant host are designated as endotrophic or vesicular–arbuscular [VAM]) (5). Chapter 4 provides more details about mycorrhiza. In addition, another reference (46) is recommended for further reading. The ECMs are septate fungi that infect the roots of shrubs and temperate trees of the north and south temperate and sub-arctic regions (43). These fungi form a sheath over the surface of the roots and penetrate between cells of the root cortex to form what is designated the Hartig net. For example, ECMs are associated with roots of pine trees, larch, and hemlock, as well as of oak trees (43). Fungi that act as ECMs include basidiomycetes, ascomycetes, phycomycetes, agaricaceae, boletaceae, russulaceae, and cortinariaceae. The ECM fungi increase the uptake of nutrients by the host plants and produce antibiotics that may eliminate microbial plant pathogens or form a barrier between the root and plant-pathogenic fungi (such as *Pythium, Rhizoctonia*, and *Phytophthora*). The ECM fungi also produce auxins (plant growth hormones) which may enhance plant growth.

The VAM mycorrhizal fungi are found in soil as chlamydospores from which hyphae or germ tubes emerge and invade the roots. This process forms an appressorium, permitting the fungi to penetrate between root cells. The intracellular fungi, known as arbuscules, then branch and form a mass of hyphae. Vesicles that may contain lipid globules form (5). The arbuscules remain for about 4 to 10 days, at which time they are digested by the plant cells, while new ones are formed in other cells (43). Transfer of nutrients may occur between the fungal mycelium and the plant cell membranes. The VAMs are the most common mycorrhizal types and include fungi classified as Zygomycetes. The five major genera of fungi that VAM comprises are *Glomus, Gigaspora, Acaulospora, Scerocystis*, and *Scutellospora* (43). Mycorrhizal fungi are associated with most plants and depend on the host plant for their supply of carbon as they are usually unable to compete for carbon substrates with other soil microorganisms. The VAMs are associated with cultivated plants, forest and shade trees, and shrubs.

4. PROTOZOA AS PREDATORS OF BACTERIA

Protozoa are unicellular eukaryotic forms, are generally aerobic, are mostly microscopic in size, and belong to the kingdom Protista. Depending on their cell form and behavior, they can be separated in flagellates (containing flagellae for movement), ciliates (containing ciliae), and amoebae (Chapter 5). The numbers of protozoa in soil have been estimated to range between a few thousand and 5×10^5 per g of soil (47). Protozoa in soil are largely responsible for an enhancement of the soil's N, P, and S mineralization by their ingestion and digestion of bacteria and fungi. Moreover, they can enhance soil nitrification, immobilizing C, N, and P in their own biomass. They further participate in soil respiration and act as food sources for arthropods and nematodes (see Chapters 5 and 6). The free-living phagotrophic protozoa exhibit three different feeding habits: filter feeding, raptorial feeding, and diffusion feeding (48). Filter feeding (by some flagellates and ciliates) involves the filtering of suspended food particles, such as bacteria. A filter-feeding ciliate such as *Paramecium caudatum* can ingest 400 bacterial cells per second (49). Raptorial feeding (by small flagellates and amoebae) is the individual ingestion of suspended bacteria and of bacteria surface-bound to clays and organic matter. Diffusion feeding (by some amoeboid protozoa such as foraminifera and small heliozoans) involves extension of the pseudopodium to engulf prey, and then the formation of a food vacuole (50).

Protozoa must move through soil to seek bacteria and to avoid unfavorable soil conditions. Bacterial cells are often found in small colonies (usually less than 10 cells) (3) that cover about 0.1% of the available soil surface area (8). Cultivation of soils will assist in bringing bacterial and protozoan populations into contact with each other (8).

Danso et al. (51) studied the interactions between *Rhizobium* cells and protozoa in natural soil. Rhizobial cell numbers declined in the presence of protozoa, but they were not totally eliminated. A concomitant increase in protozoan density was observed. The inability of protozoa to remove all bacterial prey cells was not attributed to another predator's attacking the protozoa, since in sterile soil inoculated with bacteria and protozoa, protozoa also did not eliminate all the prey cells (51). The persistence of some prey cells was probably the result of protozoan numbers reaching a population density where the energy expended in finding prey equalled that gained from feeding.

Protozoa do not respond to all types of bacteria added to soil. Casida (52) did not observe an increase in protozoan numbers upon the addition of *Arthrobacter globiformis* and *Bacillus thuringiensis* cells to soil. However, *B. mycoides* spores added to soil caused an increase in the numbers of selected protozoa; although it is tempting to speculate, it is not known whether this response was associated with germination of the *Bacillus* spores. Even though *Escherichia coli* is generally not considered to represent a common soil microorganism, protozoa increased their numbers in response to its addition, in a manner similar to that when *B. mycoides* was added. It is likely that protozoa exhibit specificity in their predatory (grazing) activity (Chapter 5), and this may explain the differences observed.

Free-living phagotrophic protozoa may not distinguish between clay particles and bacteria and ingest both. Fenchel (49) reported that filter-feeding ciliates ingested latex beads (diameters, 0.09 to 5.7 μm) as readily as bacteria. The optimal particle size ingested by 14 different ciliates ranged from 0.3 to 1 μm (diameter). This size range is similar to the size of some bacterial cells. For example, *Rhizobium* spp. are rod-shaped cells ranging in size from 0.5 to 0.9 μm diameter by 1.2 to 3.0 μm length. Members of the genus *Pseudomonas* are straight or curved

rods of 0.5 to 1.0 μm by 1.5 to 4.0μm. The ranges of particle sizes of montmorillonite, illite, and kaolinite clays are 0.01 to 1.0, 0.1 to 2.0, and 0.1 to 5.0 μm, respectively (53). Hence, the 1 to 5 μm clay sizes are in the same magnitude range
as some common soil bacteria. Fenchel (49) suggested size selection may be a
function of the size of the protozoan mouth. Furthermore, Heijnen et al. (54) observed that *Bodo saltans* fed almost solely on *R. leguminosarum* biovar *trifolii* in
the presence of other bacterial species in demineralized water. This finding supports the concept of preferential grazing. However, prior to inoculation into water, *B. saltans* was maintained on *R. leguminosarum* biovar *trifolii* as a prey. It is
not known whether this maintenance regime influenced subsequent feeding behavior of protozoans during the experiments.

   If selective feeding by indigenous protozoans is common, it may be possible to
select bacterial strains that are not ingested, or are ingested at a low frequency, for
use in soil. Moreover, the presence in soil of clays is another important parameter
which controls the outcome of protozoan predation. Bacterial cells adsorbed to
clay particles exhibit an increase in total volume. This can affect the feeding behavior of predatory protozoans because the combined clay–cell complex may be
too large to ingest. Clays may also protect some bacterial cells from predation by
protozoa by increasing the number of protective soil microhabitats available to
bacteria (55). Also, protozoa may ingest clay particles indiscriminately, thereby
sparing bacterial cells. This reduces their food intake and decreases their reproductive potential. In recent years, information on clay–bacteria–protozoan interactions have been obtained by using sterile and nonsterile soils amended with this
mixture. Clays, in particular montmorillonite, were found to protect introduced
bacteria from protozoan predation (55). Although the approach provided useful
information, the laboratory conditions used may not reflect those found in clayey
field soils. The possibility that, in addition to protozoa, slime molds, bacteriophages, *Bdellovibrio*, and myxobacteria might be partially responsible for the observed declines in bacterial densities in soil has been addressed, but evidence for
such mechanisms is lacking (56).

   The presence of protozoa in soil certainly increases the turnover and mineralization of organic carbon and nitrogen (57,58, Chapter 5), by the consumption of
bacteria and release of compounds, in particular nitrogen. The activity of bacterial
populations is often increased, thereby increasing the amount of carbon metabolized by the bacteria. Therefore, bacterial–protozoan interactions in soil are of central importance to soil nutrient cycles, since they act as catalysts of these cycles.

5. PREDATORY *BDELLOVIBRIO*

The genus *Bdellovibrio* is a group of predatory bacteria found in soil and aquatic
environments (59). Their prey often consists of other (saprophytic) bacteria that
occur in the same habitat. *Bdellovibrio* spp. have a predatory life cycle that com-

prises two developmental phases. During the free-living attack phase, the cells are small (0.2 to 0.5 μm in width by 0.5 to 1.4 μm in length) and motile as a result of the presence of a single polar flagellum. During the second phase, designated the attack phase, *Bdellovibrio* encounters a suitable prey cell and attaches at its nonflagellated pole. It then penetrates the outer membrane and cell wall of the (gram-negative) prey cell, losing its own flagellum during the process. Once the predator cell enters the periplasm of the prey cell, it alters the prey cell biochemically in such a way that it cannot be attacked by other attack-phase cells. The prey cell becomes a spherical structure referred to as a bdelloplast (59). The prey cell cytoplasm is a rich source of nutrients. The *Bdellovibrio* elongates into a coiled, aseptate filament. When the nutrients are depleted, the filament divides into single cells which then develop a new flagellum required for the next attack phase. The bdelloplast is lysed and the cycle can be repeated. The entire cycle can take place in 4 hours (30°C) with *E. coli* as the prey organism. However, the cycle duration in soil is not known. The ability to attack and utilize other cells offers a considerable advantage to *Bdellovibrio* cells in the nutrient-limiting soil environment. Hence, the *Bdellovibrio*–bacterium interaction is a typical example of bacterial parasitism in soil, with one partner, the *Bdellovibrio*, taking full advantage of the nutritional value of its prey bacterium.

6. BACTERIAL–FUNGAL INTERACTIONS

Interactions between bacteria and fungi can vary in accordance with the division given in Section 2.2. In the first place, the interaction can be amensalistic in that bacteria can inhibit fungi by excreting compounds such as phenazines. Second, there can be direct competition between bacteria and fungi for nutrients. Such competition probably plays an important role in the interaction of both organisms with plant roots, and a possible strategy to control plant-pathogenic fungi might be based on the use of highly competitive bacteria that outcompete the fungi. Third, bacteria can be directly parasitized by fungi. Barron (60–62) has described several soil fungi, *Pleurotus ostreatus*, *Lepista nuda*, *Agaricus brunnescens*, and *Coprinus quadrifidus*, that attack soil bacteria (e.g., *Pseudomonas* and *Agrobacterium* spp.) and use the cells as sources of nutrients (Fig. 2). The fungi sense the bacterial colonies (by some chemoattractive mechanism) in their vicinity and initiate specialized directional hyphae that can grow toward them. The fungi then secrete compounds (unknown at this stage) in advance of the directional hyphae that lyse the bacterial cells. After penetrating the bacterial colony, the directional hyphae produce a coralloid mass of assimilative hyphae. After colonization, there is no further proliferation of hyphae in the vicinity of the bacterial colony. Nutrients in excess of immediate fungal requirements are translocated through the directional hyphae for utilization elsewhere. The bacterial colonies disappear within a period as short as 24 hours. This interaction be-

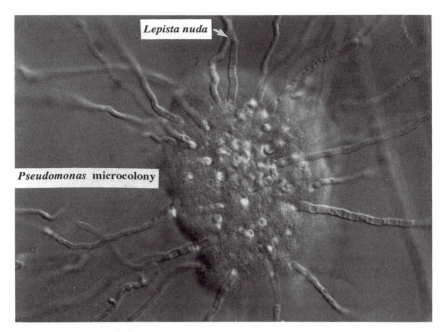

Figure 2. Microcolony (in center) of *Pseudomonas* sp. cells being attacked by the soil fungus *Lepista nuda*. (Photograph supplied by Professor G. Barron, University of Guelph, Ontario, Canada.)

tween fungi and bacterial colonies gives the fungi a considerable advantage in obtaining nutrients, especially under nutrient-limiting conditions that are prevalent in soil.

Agaricus bisporus, the cultivated mushroom, is able to mineralize dead *Bacillus subtilis* cells, using the cellular components as sole sources of N and C for growth. This capacity illustrates the significant role that microbial biomass has in providing nutrients for this basidiomycete and other fungi. Moreover, over 150 species of fungi are known to attack nematodes or their eggs (60). Nematode-destroying fungi such as *Stylopage hadra*, *Dactylella gephyropaga*, and *Arthrobotrys oligospora*, which are predatory, produce an extensive hyphal system and at intervals along the hyphae, trapping devices are produced that catch and hold the live nematodes. The nematode is then penetrated by the hyphae and the body contents are metabolized. Endoparasitic fungi such as *Verticillium*, *Harposporium anguillulae*, and *Meria* do not have extensive hyphal development outside the body of the host. Instead, conidiophores and conida attach to the nematodes, germinate, penetrate the cuticle of nematodes, and form an infection hypha in the body cavity. In several days, the body of the nematode is filled with fungal hyphae, some of which can break out. A commonly studied soil nematode, *Panagrellus redivivus*, can be totally destroyed by fungi.

Hence, many fungi are specialized in obtaining nutrients through the attack on organisms in their vicinity, such as bacteria, nematodes, or their eggs, or even other fungi. In using such strategies, fungi are probably among the most proficient microbial colonizers of the soil (see Chapter 4).

7. BACTERIAL–ANIMAL INTERACTIONS

Interactions between bacteria and soil animals (invertebrates) are particularly important when soil bacteria are able to colonize and profit from the surfaces of soil invertebrates. As soil is the habitat of many invertebrate species (Chapters 5 and 6), there is a great diversity of possible bacterium–invertebrate interactions. Most of these interactions seem fortuitous; i.e., bacteria coincidentally in a condition where invertebrate surfaces can be colonized will do so. Some bacteria, however, have developed highly specialized systems which enable and often oblige them to interact with soil animals; these bacteria sometimes have even lost their ability to live in the open soil environment; i.e., they became obligatory parasites. An example of bacteria considered symbiotic are the entomopathogenic *Photorhabdus* and *Xenorhabdus* spp., which are preferentially carried by nematodes. Following contact by the nematode of insect larvae, the bacteria are injected and the larvae are subsequently killed with concomitant proliferation of the bacteria.

External and internal surfaces of invertebrate species in soil can thus be inhabited by numerous microorganisms. Many herbivorous animals, particularly insects, have a hindgut that contains microorganisms that degrade cellulose, which is not digestible by the host. They thus provide essential growth factors like amino acids and vitamins to the insect (7). An example of this type of nutritional interaction is the ecological system of the flagellate protozoa that live as ectosymbionts within sacciform dilations of termites' guts. The protozoa are dependent on the production of cellulose-degrading enzymes by endosymbiotic bacteria in the gut (7). A second example is leaf-cutting ants that make piles of leaves on which the fungus *Hypomyces hippomoeae* grows. The ants cultivate the fungus by adding leaf material. The developing fungal mycelium is used as a food source by the ants. A similar type of fungal cultivation is carried out by termites.

Bacteria can also be associated with the guts of earthworms as well as with their casts. These interactions are mainly nutritional; i.e., the associated bacteria profit from the substrates that become more easily available in their new niches. Bacterial cells added to soils can be later detected in earthworms and casts in the same soil site (63). Earthworms are known to enhance the rate of nitrogen and phosphorus cycles, and it is likely they do so by allowing microorganisms to mineralize these components in soil better, via mixing. The ingested and transformed soil C and N compounds are mostly rejected in their casts. Earthworms further increase soil stability and provide channels for movement of water through soil. They are also capable of burying litter in soil, where it can decompose by microbial enzymatic action.

8. MICROORGANISMS–METAL INTERACTIONS

In many soils, metals are important components of the habitat, occurring either in a free (unbound) form or bound to soil particles, often clay or organic matter complexes. The ecological characteristics of the interaction of microorganisms with these metals are briefly discussed here, as a paradigm of the interaction of soil microbes with abiotic stressful components of their environment. A more substantial discussion of microbe–metal interactions can be found in Chapter 20.

Most studies on microbe–metal interactions have been conducted under defined laboratory conditions, and in the last 15 years the physiological, genetic, and molecular biological characteristics of metal resistance mechanisms have been elucidated, particularly for bacteria, fungi, and algae. The metal-resistant microorganisms studied were commonly isolated from metal-contaminated soils, sediments, water, sewage, and the hospital environment. In such bacteria, metal-resistance genes were often encoded on plasmids (64–66).

Mechanisms by which bacteria can be resistant to metals are as follows:

(a) Inactivation or complexation of the metal: for example, some bacteria produce hydrogen sulfide (H_2S), which can bind tightly to metal(s) and form an insoluble metal-sulfide complex.
(b) Alteration of the cellular site of inhibition by the metal.
(c) Impermeability of bacterial cells to metals: As a result, metals cannot enter the bacterial cells. The impermeability may be due to the presence of an extracellular capsular layer which acts as a metal trap.
(d) Efflux pumps: as the metal usually enters the cells by a transport system for an existing essential element, it is actively effluxed or pumped from the cells. Example of this resistance mechanism are efflux pumps for cadmium and arsenic ions.
(e) Volatilization of the toxic metal by an enzymatic reaction, such as the volatilization of Hg^{2+} to $Hg°$ by a mercuric reductase (*mer*): this enzyme is found in gram-negative (*Pseudomonas*) as well as gram-positive (*Bacillus cereus*, *Streptomyces* spp.) soil bacteria.

Bacterial strains resistant to silver, e.g., *Pseudomonas stutzeri*, have been isolated from a silver mine (67). The silver resistance in this species is plasmid-encoded. Silver is highly toxic to bacteria, inhibiting respiration and phosphate uptake (68,69), and hence the resistance mechanism is capable of warding off these effects. *Thiobacillus ferroxidans* and *T. thiooxidans* can accumulate silver during the oxidation of sulfide-containing ores. This may be a potential mechanism for the bacterial leaching of sulfide ores that also contain bound silver (70).

Two other metal–microbe interactions are the reduction of hexavalent chromium (chromate) to the less-toxic trivalent chromium by *P. putida* PRS2000 (71) and the isolation of *P. syringae* pv. *tomato* strains resistant to copper as a result of the presence of copper-resistance plasmids (72). Chromium and copper may be factors of relevance in some soils, with copper entering agricultural soils via manure of copper-treated cattle. Both *Pseudomonas* species are commonly found in soil.

Jordan and Lechevalier (73) reported that microbial populations in soils close to a zinc smelter contained high proportions of bacteria capable of growing on media amended with zinc. The possibility exists that part of the indigenous soil bacterial population was inherently resistant to this and other metals while other species acquired resistance through plasmid-mediated gene exchange. Research by M. Mergeay and coworkers has convincingly shown the relevance of plasmids in the spread among soil bacteria of, for instance, genes encoding resistance to zinc, cadmium, and cobalt (74) (Chapter 20).

An anaerobic mechanism not leading to detoxification is the methylation of mercury to dimethylmercury by bacterial species (64–66). The production of dimethylmercury is a serious environmental problem that has had devastating effects on human and environmental health. Problems associated with mercury pollution became known after the discovery of high levels of methylmercury in fish and shellfish in Minamata Bay, Japan. This toxicity problem, which became known as Minamata disease, resulted in a large number of confirmed cases of methylmercury poisonings and human deaths between 1953 and 1970. Mercury is found in the terrestrial environment at levels of about 50 µg/kg soil (65). Therefore, when soil is anaerobic as a result of high moisture levels (impeding oxygen diffusion), the potential exists for mercury methylation to occur.

Another mechanism of microbe–metal interactions is the anaerobic corrosion of iron mediated by the anaerobic, sulfate-reducing bacterium *Desulfovibrio*. This organism produces S^{2-}, which reacts with Fe^{2+}, giving FeS. Two environmental conditions necessary for this reaction are a lack of oxygen and the presence of sulfate. As stated by Paul and Clark (43), under these conditions a 3 mm thick iron pipe can be corroded in 5 to 7 years.

Toxic metals can further be remobilized from bacterial wall–clay composites (75). Clays and bacterial cell walls usually have a net negative charge over a wide pH range. Metals like silver, copper, and chromium can bind to cell walls of *Bacillus* and *Escherichia* species, and organic matter–clay aggregates. The

Table 4.   Some Soil Microbe–Metal Interactions

1. Intracellular accumulation: active or passive transport of metals into cells
2. Cell wall–associated metals: metals bind to microbial cell walls and with
   some cells, divalent metals exchange for protons
3. Metal–siderophore interactions: siderophores are chelating compounds
   excreted by some microorganisms that assist in uptake of iron
4. Extracellular immobilization of metals by a bacterial metabolite like
   hydrogen sulfide (H_2S)
5. Extracellular polymer–metal interactions: extracellular bacterial polymers
   are usually polysaccharides that strongly complex metals
6. Enzymatic transformations such as methylation, demethylation, and
   volatilization of mercury and reduction of chromate

sorbed metals can be removed or remobilized with, for instance, an acid such as
fulvic acid, which is a normal component of soil. The remobilization of metals
has important implications for the movement of metals in soil, and there is a potential movement from metal-polluted soils to groundwater.

There are several other examples of microbe–metal interactions that are important in soil microbiology. *Azotobacter vinelandii*, a gram-negative, free-living, nitrogen-fixing aerobic bacterial species commonly found in soils, produces three
primary siderophores (iron-chelating compounds), azotobactin, azotochelin, and
aminochelin, as well as 2,3-dihydroxybenzoic acid, which may also function in
iron uptake by this microorganism (76). These compounds are produced in response to low iron concentrations. The absence of iron derepresses siderophore
production, which is followed by excretion of the compounds into the extracellular
medium to chelate and then solubilize iron required for growth and other enzymatic functions such as nitrogen fixation. The competition between microorganisms for available iron in soil is a mechanism which can determine the fitness of a
soil inhabitant; it is therefore a mechanism widely studied with respect to its possible use in biological control of pathogenic fungi or other pathogens in soil (77).

A summary of metal–microbe interactions is presented in Table 4. Additional
information on metals and geomicrobiology can be found in an excellent review
by Ehrlich (78) and articles on metal resistance (24,64–68,74,75).

9.   CONCLUSIONS

The interactions discussed in this chapter represent a few examples of the high
diversity of known and potential interactions between organisms in soil as well
as with components of this habitat. The interactions are quite commonly essen-

tial components of the life cycles of the soil microorganisms, either in their evolution or adaptation to changing environmental conditions (via horizontal gene transfer, i.e. by genetic interactions) or in their ecological characteristics (via any of the other beneficial or detrimental interactions). Both types of interactions are relevant to components of nutrient cycles that sustain the biosphere and give soil its high microbial diversity. It is essential that these soil processes not be affected in such a way that permanent adverse ecological effects are produced. Considerable effort should still be expended on the study of microbial gene transfer in soil (41,79,80), as well as on the adverse effects of agrochemicals (81), acid precipitation, and industrial pollution. Further, forest soil and agricultural management practices affect soil microbial diversity and interactions and processes in soil (e.g., the colonization of plant roots by bacteria (82)), and hence warrant study.

The use of molecular biology techniques in soil microbiology (9,11,41,83) is currently a common practice in many laboratories. These molecular techniques are discussed in several other chapters (chapters 12a, 12b, 13, 14). As classical techniques either have been cultivation-based and thus only took into account the culturable microorganisms or studied soil processes with an overall approach, the direct molecularly based approach will bring forth new information on microbial diversity as well as on shifts in microbial communities caused by microbial interactions and stress factors. For instance, selected essential rhizosphere microorganisms or microbial strains able to degrade toxic pollutants can be studied in their natural habitat. Molecular techniques further permit the study of the fate and effects of genetically modified organisms (GMOs) introduced into the soil environment. It is expected that the novel protocols also will allow the detection of plant pathogens at previously undetectable levels. The application of molecular techniques to soil microorganisms assists in understanding the molecular evolution of all these organisms, their genome sizes and rearrangements of gene orders, as a result of their presence in a natural soil setting (84,85). These selected examples all are part of the network of microbial interactions that will be unraveled in greater detail in forthcoming years.

ACKNOWLEDGMENT

Appreciation is expressed to NSERC (Canada) for support of research by J.T.T.

REFERENCES

1. R. M. Atlas and R. M. Bartha, *Microbial Ecology: Fundamentals and Applications*, Addison-Wesley, Reading, Mass. (1981).
2. R. G. Burns and J. H. Slater (eds.), *Experimental Microbial Ecology*, Blackwell Scientific, Oxford, (1982).

3. R. Campbell, *Microbial Ecology*, Blackwell Scientific Oxford (1993).
4. J. M. Lynch, *Soil Biotechnology*, Blackwell Scientific Oxford (1983).
5. J. M. Lynch and J. E. Hobbie (eds.), *Micro-Organisms in Action: Concepts and Applications in Microbial Ecology*, Blackwell Scientific Oxford (1988).
6. J. M. Lynch and N. J. Poole (eds.), *Microbial Ecology: A Conceptual Approach*, John Wiley & Sons, New York (1979).
7. H. Stolp, Cambridge University Press, Cambridge (1988).
8. M. Wood, *Soil Biology*, Chapman & Hall, New York (1989).
9. K. Alef and P. Nannipieri (eds.), *Methods in Applied Soil Microbiology and Biochemistry*, Academic Press, San Diego, pp. 1–576 (1995).
10. *Methods of Soil Analysis*, Part 2, *Microbiological and Biochemical Properties*, Soil Science Society of America, Inc., Madison, Wisc. pp. 1–1121 (1994).
11. J. T. Trevors and J. D. van Elsas (eds.), *Nucleic Acids in the Environment*, Springer-Verlag, Berlin, pp. 1–256 (1995).
12. M. Alexander, *Soil Microbiology*, John Wiley & Sons, New York (1977).
13. S. C. Jackman, H. Lee, and J. T. Trevors, *Microbial Rel. 1*:125 (1992).
14. M. Levin and H. Strauss (eds.), *Risk Assessment in Genetic Engineering*, McGraw-Hill, New York (1991).
15. National Research Council, *Field Testing Genetically Modified Organisms*, National Academy Press, Washington, D. C. (1989).
16. J. D. van Elsas and J. T. Trevors, *Bacterial Genetics in Natural Environments* (M. J Day and J. C. Fry, eds.). Marius Press, Carnforth, England, p. 188 (1990).
17. M. J. Gauthier (ed.), *Gene Transfers and Environment*, Springer-Verlag, New York (1992).
18. M. J. Bazin and J. M. Lynch (eds.), *Environmental Gene Release*, Chapman & Hall, New York, pp. 1–166 (1994).
19. J. T. Trevors, *J. Microbiol. Methods 3*:259 (1985).
20. P. M. Bennett and J. Grinsted (eds.), *Methods in Microbiology*. Vol. 17, *Plasmid Technology*, Academic Press, New York, (1984).
21. B. H. Belliveau and J. T. Trevors, *Biol. Metals 3*:188 (1990).
22. G. M. Dunny, *Microbial Cell-Cell Interactions* (M. Dworkin, ed.), American Society Microbiology, Washington, D.C. p. 9 (1991).
23. R. B. Henschke, E. J. Henschke, and F. R. J. Schmidt, *Appl. Microbiol. Biotechnol. 35*:247 (1991).
24. E. Top, M. Mergeay, D. Springael, and W. Verstraete, *Appl. Environ. Microbiol. 56*:2471 (1990).
25. G. R. Knudsen, M. V. Walter, L. A. Porteous, V. J. Prince, J. L. Armstrong, and R. J. Seidler, *Appl. Environ. Microbiol. 54*:343 (1988).
26. G. J. Stewart and C. A. Carlson, *Ann. Rev. Microbiol. 40*:211 (1986).
27. J. B. Graham and A. A. Istock, *Mol. Gen. Genet. 166*:287 (1978).

27a. K. M. Nielsen, M. D. M. van Weerelt, T. N. Berg, A. M. Bones, A. N. Hagler, and J. D. van Elsas. *Appl. Environ. Microbiol. 63* (1997), in press.

28. M. G. Lorenz, B. W. Aardema, and W. Wackernagel, *J. Gen. Microbiol. 134*:107 (1988).

29. G. Romanowski, M. G. Lorenz, and W. Wackernagel, *Appl. Environ. Microbiol. 57*:1057 (1991).

30. M. G. Lorenz and W. Wackernagel, *Appl. Environ. Microbiol. 57*:1246 (1990).

31. K. Mise and R. Nakaya, *Mol. Gen. Genet. 157*:131 (1977).

32. R. E. Ruhfel, N. J. Robillard, and C. B. Thorne, *J. Bacteriol. 157*:708 (1984).

33. N. Bakhiet and D. P. Stahly, *Appl. Environ. Microbiol. 49*:577 (1985).

34. J. J. Germida and G. G. Khachatourians, *Can. J. Microbiol.* 34:190 (1988).

35. L. R. Zeph, M. A. Onaga, and G. Stotzky, *Appl. Environ. Microbiol. 54*:1731 (1988).

36. L. S. Thomashow, D. M. Weller, R. F. Bonsall, and L. S. Pierson III, *Appl. Environ. Microbiol. 56*:908 (1990).

37. J. C. Fry and M. J. Day (eds.), *Bacterial Genetics in Natural Environments.* Chapman & Hall, London (1989).

38. M. A. Levin, R. J. Seidler, and M. Rogul (eds.), *Microbial Ecology: Principles, Methods and Applications*, McGraw-Hill, New York (1992).

39. J. T. Trevors and J. D. van Elsas, *Can. J. Microbol. 35*:895 (1989).

40. J. T. Trevors, T. Barkay, and A. Bourquin, *Can. J. Microbiol. 32*:191 (1987).

41. J. T. Trevors and J. D. van Elsas, *Manual of Environmental Microbiology*, (C. J. Hurst, G. R. Knudsen, M. J. McInerney, L. D. Stetzenbach, and M. V. Walter, eds.) (1997) Amer Soil Microbiol., p. 505–508.

42. J. T. Trevors, *Antonie van Leeuwenhoek 67*:315 (1995).

43. E. A. Paul and F. E. Clark, *Soil Microbiology and Biochemistry*, Academic Press, New York (1989).

44. J. I. Sprent and P. Sprent, *Nitrogen Fixing Organisms*, Chapman & Hall, London (1990).

45. L. E. Roth and G. Stacey, *Microbial Cell–Cell Interaction* (M. Dworkin, ed.) American Society of Microbiology, Washington, D.C., p. 255 (1991).

46. J. L. Harley and S. E. Smith, *Mycorrhizal Symbiosis*, Academic Press, London (1983).

47. M. A. Sleigh, *The Biology of Protozoa*, Edward Arnold, London (1973).

48. L. S. England, H. Lee, and J. T. Trevors, *Soil Biol. Biochem.* (1994).

49. T. Fenchel, *Microb. Ecol. 6*:1–11 (1980).

50. T. Fenchel, *Ecology of Protozoa: The Biology of Free-Living Phagotrophic Protists*, Springer-Verlag, New York (1987).

51. S. K. A. Danso, S. O. Keya, and M. Alexander, *Can. J. Microbiol. 21*:884 (1975).

52. L. E. Casida, Appl. *Environ. Microbiol. 55*:1857 (1989).
53. H. O. Buckman, and N. C. Brady, *The Nature and Properties of Soils*, MacMillan, New York (1964).
54. C. E. Heijnen, C. H. Hok-A-Hin, and J. A. van Veen, *FEMS Microbiol. Ecol. 85*:65 (1991).
55. C. E. Heijnen and J. A. van Veen, *FEMS Microbiol. Ecol. 85*:73 (1991).
56. M. J. Acea, C. R. Moore and M. Alexander, *Soil Biol. Biochem. 20*:509 (1988).
57. P. J. Kuikman, A. G. Jansen, J. A. van Veen, and A. J. B. Zehnder, *Biol. Fertil Soils 10*:22 (1990a).
58. P. J. Kuikman and J. A. van Veen, *Biol. Fertil Soils 8*:13 (1989).
59. K. M. Gray and E. G. Ruby, *Microbial Cell–Cell Interactions*. (M. Dworkin, ed.) American Society of Microbiology, Washington, D.C., p. 333 (1991).
60. G. L. Barron, *Experimental Microbial Ecology* (R. G. Burns and J. H. Slater, eds.), Blackwell Scientific Oxford, pp. 533–552 (1982).
61. G. L. Barron, *Can. J. Bot. 66*:2504 (1988).
62. G. L. Barron, *The Fungal Community: Its Organization and Role in the Ecosystem*, 2nd ed. (C. C. Carroll and D. T. Wicklow, eds.), Marcel Dekker, New York, p. 311 (1992).
63. C. D. Clegg, J. M. Anderson, H. M. Lappin-Scott, J. D. van Elsas, and J. M. Jolly, *Soil Biol. Biochem. 27*:1423 (1995).
64. B. H. Belliveau, M. E. Starodub, C. Cotter, and J. T. Trevors, *Biotechnol. Adv. 5*:101 (1987).
65. B. H. Belliveau and J. T. Trevors, *Appl. Organometallic Chem. 3*:283 (1989).
66. J. T. Trevors, K. M. Oddie, and B. H. Belliveau, *FEMS Microbiol Letts. 32*:39 (1985).
67. C. Haefeli, C. Franklin, and Kimber Hardy, *J. Bacteriol. 158*:389 (1984).
68. R. W. Slawson, H. Lee, and J. T. Trevors, *Biol. of Metals 3*:151 (1990).
69. R. W. Slawson, M. I. Van Dyke, H. Lee, and J. T. Trevors, *Plasmid 27*:72 (1992).
70. F. D. Pooley, *Science 296*:642 (1982).
71. Y. Ishibashi, C. Cervantes, and S. Silver, *Appl. Environ. Microbiol. 56*:2268 (1990).
72. C. L. Bender and D. A. Cooksey, *J. Bacteriol. 165*:534 (1986).
73. M. J. Jordan and M. P. Lechevalier, *Can. J. Microbiol. 21*:1855 (1975).
74. M. Mergeay, D. Springael, and E. Top, *Bacterial Genetics in Natural Environments* (J. C. Fry and M. J. Day, eds.), Chapman & Hall, London, p. 152 (1990).
75. C. A. Flemming, F. G. Ferris, T. J. Beveridge, and G. W. Bailey, *Appl. Environ. Microbiol. 56*:3191 (1990).
76. M. Huyer and W. J. Page, *Appl. Environ. Microbiol. 54*:2625 (1988).

77. B. Schippers, A. W. Bakker, P. A. H. M. Bakker, P. J. Weisbeek, and B. Lugtenberg. *Microbial Communities in Soil* (V. Jensen and L. H. Sörensen, eds.), Elsevier, Amsterdam, p. 55, 1986.

78. H. L. Ehrlich, *Geomicrobiology*, Marcel Dekker, New York (1990).

79. G. Stotzky, *Gene Transfer in the Environment* (S. B. Levy and R. V. Miller, eds.) McGraw-Hill, New York, p. 165 (1989).

80. J. D. van Elsas and J. T. Trevors, *J. Environ. Sci. Health A26*:981 (1991).

81. L. Somerville and M. P. Greaves (eds.), *Pesticide Effects on Soil Microflora*, Taylor and Francis, London (1987).

82. J. D. van Elsas, J. T. Trevors, D. Jain, A. C. Wolters, C. E. Heijnen, and L. S. van Overbeek, *Biol. Fertil Soils 14*:14 (1992).

83. E. M. H. Wellington and J. D. van Elsas (eds.) *Genetic Interactions Among Microorganisms in the Natural Environment*, Pergamon Press, New York, pp. 1–303 (1992).

84. J. T. Trevors, *Antonie van Leeuwenhoek 69*:293 (1996).

9

Soil Food Webs and Nutrient Cycling in Agroecosystems

JAAP BLOEM, PETER DE RUITER, and LUCAS BOUWMAN   DLO Research Institute for Agrobiology and Soil Fertility (AB-DLO), Haren, The Netherlands

1. INTRODUCTION

A soil food web can be defined as a network of consumer–resource interactions among different functional groups of soil organisms. Bacteria and fungi are the primary consumers of dead organic matter (OM) in soil. Dead organic matter originates from plant residues, decaying roots, decaying microorganisms, root exudates, and animal manure. The breakdown of complex biopolymers into CO_2, H_2O, mineral nitrogen (N), phosphorus (P), and other mineral elements is called mineralization. Mineralization is performed not only by microbes but also by microbivores (grazers) and predators which decompose microbes and other organisms. Mineral nutrients released by decomposers are available for uptake by plants and microbes for production of new biomass. Thus, nutrients are cycled through ecosystems.

In natural ecosystems, mineralization of organic matter is the main source of nutrients for plant growth, and there is a reasonable balance between mineralization and nutrient uptake. In the present conventional agricultural systems, a large part of the nutrients is supplied by chemical fertilizers. In addition, animal manure—which is produced in large amounts from fodder often imported from other areas—is also added to soil. Increased inputs of external nutrients into agriculture have greatly increased crop yields, but they have also increased nutrient losses to the environment. N losses may increase strongly when N fertilization becomes higher than a certain threshold value, often near the optimum for crop production (1). The results are increased nitrate and phosphate concentrations in ground- and surface water, abundant growth of algae, poor water quality, and high costs for production of clean drinking water. High mineral nitrogen concentrations in soil can lead to denitrification and loss of nitrogen in the form of nitrous oxide (N_2O) to the atmosphere, especially during wet periods with low oxygen concentrations in the soil. N_2O can contribute to global warming in the lower atmosphere and to the decline of the ozone layer through photochemical reactions of its decomposition products in the stratosphere. Land spreading of animal manure leads to volatilization of ammonia (NH_3), which is partly deposited on nearby sites and contributes to soil acidification upon transformation into HNO_3 through nitrification.

The environmental problems caused by conventional agriculture have stimulated biological research into nutrient cycling in different agricultural systems, with the final aim of evaluating farming systems with reduced inputs of nutrients and chemicals. From 1985 to 1992, the Dutch Programme on Soil Ecology of Arable Farming Systems was carried out at the Lovinkhoeve experimental farm (Marknesse, Noordoostpolder, The Netherlands) (2). A conventional farming system was compared with a so-called integrated farming system (3,4). Integrated farming involves a shift from directly feeding the plant with mineral nutrients to feeding the soil organisms with organic matter, thereby indirectly feeding the plant through nutrient mineralization by soil organisms (5). This means the use of organic manure, compost, and mineral N fertilizer has to be coherent with crop (residue) management. An extra input of stabilized organic matter in integrated farming is meant to increase or conserve a large pool of young humus in the soil, which might act as a biological slow-release fertilizer. Integrated farming contributes to waste recycling by the use of compost and processed manure solids. To prevent nitrate leaching during the winter period, green manures should be grown after the early harvest of cereal crops.

Integrated farming at the Lovinkhoeve was characterized by approximately 40% lower N inputs, partial replacement of chemical fertilizers by processed organic manure and compost, reduced soil tillage, reduced use of chemicals for crop protection, and no soil fumigation against phytophagous (plant-eating) nematodes (6,7). It was hypothesized that the reduced input of fertilizers can be compensated for by a higher mineralization of N and other nutrients from organic matter by the soil organisms. This was expected to result in a crop yield amounting to at least 80% of the yield in the conventional system. One of the objectives was to quantify the role of different functional groups of soil organisms in N mineralization under integrated and conventional management.

Here, we first discuss effects of microbes and their grazers on mineralization in soil as determined in experiments with liquid cultures and with soil microcosms. Next, field observations of the dynamics of microorganisms, microbivores, and nitrogen mineralization during a full year at the Lovinkhoeve site in the Netherlands are discussed. It is shown how mineralization can be calculated by using a food web model, and the simulated mineralization is compared with the mineralization observed by in situ incubations of soil cores. With the model, the contributions of different groups of soil organisms to the total nitrogen mineralization are estimated. Finally, the food web at the Lovinkhoeve experimental farm is compared with food webs of other agricultural systems in the United States and Sweden.

2. MICROBES, MICROBIVORES, AND MINERALIZATION IN MICROCOSMS

2.1 Mineralization, Immobilization, and Grazing

Heterotrophic bacteria and fungi derive energy and nutrients from the decomposition of organic matter. Part of the decomposed material is used for production of microbial biomass (growth), and another part is mineralized to CO_2, H_2O, mineral N, P, and other nutrients. The fraction of consumed C that is converted to bacterial biomass C (the gross growth efficiency or yield) can range from 0% to 70% (8,9). The wide range may be explained by differences in organic substrate quality (C:N ratio and molecular complexity) and availability of inorganic nutrients (especially N). Under extreme nutrient limitation, organic carbon may be respired away without any net production of biomass (10). However, a high yield may be observed under N limitation if bacteria accumulate reserve carbohydrates (11,12). High yields of 50% to 70% have been found in glucose-amended soil samples (12,13), and with dissolved organic matter leached from macrophytes (water plants) (14). Bacteria growing on macrophyte-derived particulate organic matter showed a low yield of 10%. For fungi, the same wide range of yields has been found as for bacteria.

Depending on the carbon/nutrient ratios of their substrate, bacterial C:N ratios can vary from 3.7 to 17 (15–17) and C:P ratios from 30 to 500 (17). Relatively high bacterial carbon/nutrient ratios may occur when bacteria store glycogen-like reserve material, for example, under N limitation (11). Most studies report relatively constant bacterial C:N ratios between 3.7 and 6.7. For fungi a much wider range (between 7.1 and 44) has been found (18,19). If the decomposed organic matter contains considerably less N and P than bacteria, all N and P released during decomposition is used for bacterial growth and no net mineralization occurs. For example, if bacteria with a biomass C:N ratio of 5 are growing with a yield of about 30% and use an organic substrate with a C:N ratio of 15, the substrate will match their N requirement and there will be no net mineralization. If the C:N ratio of the substrate is higher, the bacteria may take up mineral N from their environment and immobilize it in their biomass. It has generally been observed that no ammonium is released by bacteria when the C:N ratio of the decomposed organic matter is higher than about 15 (17,20,21). Similarly, P mineralization by bacteria appears to cease above a substrate C:P ratio of 60 (17).

Mineral nutrients immobilized in microbial biomass are released when microbes are grazed by microbivores such as protozoa and nematodes. Their C:N ratios are similar to or even higher than those of their prey and their growth efficiencies are generally 40% or less (22). Thus, predation will result in excretion of surplus N and P in the form of ammonium and phosphate. It has been known for a long time that the presence of bacterivorous grazers can accelerate the decomposition of organic matter and the remineralization of inorganic nutrients (23,24). Besides their direct contribution by excretion of mineral nutrients, microbivores may enhance mineralization indirectly by stimulating the microbial activity. Generally, it has been proposed that grazers may keep bacterial numbers well below the carrying capacity of the environment and thereby relieve bacteria from density-dependent limiting factors, such as limitation by inorganic nutrients, organic substrates, low oxygen concentrations, toxic metabolites, and space. Thus, bacteria would be kept in a prolonged state of physiological youth and their rate of assimilation of organic matter would be greatly increased (23). The most favored hypothesis is that recycling of mineral nutrients from microbial biomass by grazers may promote the growth of nutrient-limited bacteria. Also, grazers may produce (organic) metabolites that stimulate the growth and activity of bacteria (25). Further, grazers may disseminate microbes through the soil, thereby promoting the colonization of substrates and, consequently, mineralization (26).

2.2 Effects of Microbivores in Water Cultures

The effect of grazers on decomposition and mineralization depends on the growth conditions and differs for C, N, and P. Bloem et al. (27) investigated effects of protozoa on bacterial activity and mineralization of C, N, and P under controlled steady-state conditions in two-stage continuous cultures. In a continuous culture the growth rate of the bacteria is kept constant and is determined by the inflow rate of fresh medium. In a two-stage system, bacteria are grown in the first stage in the absence of grazers and are fed to the second stage, where grazers are growing on the bacteria. Such an artificial system facilitates accurate determination of mineralization rates and of bacterial and protozoan numbers and activity. Two parallel two-stage systems were used. Bacteria from culture vessel 1 (first stage) were fed to vessel 2 (second stage), which contained no protozoa and served as a grazer-free control. In the parallel system, bacteria from vessel 3 (first stage) were fed to vessel 4 (second stage), which contained protozoa (heterotrophic nanoflagellates of about 6 μm diameter). In the first stages (vessels 1 and 3) bacteria were grown on a yeast extract medium (10 mg l^{-1}) with a molar C:N:P ratio of 100:15:1.2. This medium was assumed to approach dissolved complex natural substrates. After 4 weeks at a dilution rate of 0.65 day^{-1}, corresponding with a generation time of 1 day, the bacteria reached a steady state at 83×10^9 bacteria l^{-1} (Fig. 1). In the absence of grazers during days 31 to 35 (phase I), all vessels showed the same bacterial numbers. This indicates that no net bacterial growth or mortality occurred in the second stages in the absence of protozoa. To prevent bacterial growth on the walls of the culture vessels, after day 35 the culture systems were broken down, cleaned, sterilized, and restarted at the same dilution rate. After 3 weeks the bacteria showed the same stable numbers as in phase I. Then, on day 26 protozoa were inoculated in vessel 4. After day 33 (phase II) the protozoa reached a steady state at 240×10^6 flagellates l^{-1} and strongly reduced the bacterial numbers from 83×10^9 to 2.3×10^9 cells l^{-1}. When the substrate concentration was increased by a factor of 5, the bacteria also increased by a factor of 5 in the first stage, but not in the second stage. In the second stage the bacterial numbers remained constant and the protozoa increased by a factor of 5 (27, results not shown here). Thus, the protozoa clearly controlled bacterial numbers and an increased production of bacteria was directly converted to protozoan biomass. The protozoa consumed 10 bacteria flagellate^{-1} h^{-1} and were growing with a yield of 50% (C:C, μg carbon produced/μg carbon consumed). The yield of the bacteria was 18%(C:C).

C was mineralized mainly by the bacteria in the first stages (vessels 1 and 3, Fig. 2A) at a rate of 310 μg C l^{-1} h^{-1}. Some C mineralization (30 μg C l^{-1} h^{-1}) occurred in the second stages (vessels 2 and 4), but this was not significantly increased by protozoa (vessel 4, phase II). In contrast with C, P was hardly mineralized by bacteria (Fig. 2B). The protozoa strongly increased the P mineralization rates from 0.5 to 3 μg P l^{-1} h^{-1}. Also N mineralization in the second stages was strongly increased in the presence of protozoa, from 2 to 20 μg N l^{-1} h^{-1} (Fig. 2C). However, most N was mineralized by the bacteria in the first stages at rates of 60 μg N l^{-1} h^{-1}. Thus, the effects of protozoa on mineralization were different for C, P, and N.

The relative importance of bacteria and protozoa as nutrient remineralizers in food webs depends largely on the C:N:P ratio of the substrate (28). If the substrate contains N and P in excess of the amount incorporated in bacterial biomass, the bacteria will function as remineralizers. This was obviously the case with N in the continuous cultures (Fig. 2C). From experiments with N-limited batch cultures, Caron et al. (28) concluded that the role of protozoa as remineral-

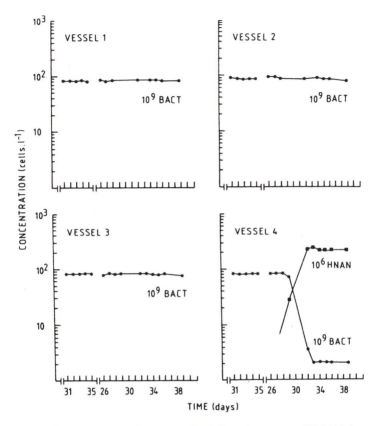

Figure 1. Numbers of bacteria (BACT) and protozoa (HNAN, i.e., heterotrophic nanoflagellates) in two parallel two-stage continuous cultures, during steady-state periods of two successive runs (days 31–35 and days 26–40). Bacteria grown in vessel 1 were fed to vessel 2, where bacterial numbers remained constant. Bacteria grown in vessel 3 were fed to vessel 4, where bacterial numbers also remained constant when no protozoa were present. When protozoa were added to vessel 4, bacterial numbers were reduced from 83×10^9 to 2.3×10^9 cells 1^{-1}. (From Ref. 27.)

izers of a growth-limiting nutrient is greatest when the carbon/nutrient ratio is high, i.e., under severe nutrient limitation. In our continuous cultures P was not limiting because the C:P ratio of the substrate was 83, which is not high. Nevertheless, the protozoa were the main P mineralizers. This was due to the fact that the bacteria had a very low C:P ratio of 21 and appeared to immobilize rather than mineralize P. Similar results were reported by Güde (29), Thingstad (30), and Bloem et al. (31). Tezuka (17) found P mineralization by bacteria only at

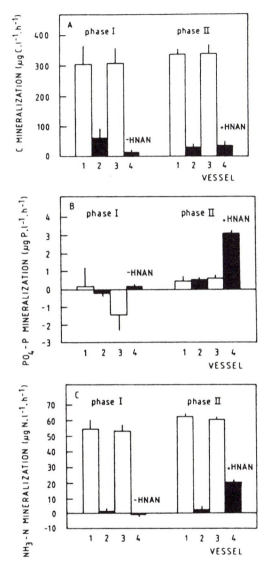

Figure 2. Carbon (A), phosphorus (B), and nitrogen (C) mineralization rates
in the continuous-culture vessels (1 through 4) in the absence (phase I) and in
the presence (phase II) of protozoa (HNAN, i.e., heterotrophic nanoflagellates)
in vessel 4. Error bars indicate 1 SD; $n = 3$. (From Ref. 27.)

substrate C:P ratios below 60. This was attributed to the capability of bacteria to store excess P within the cell. Since most natural substrates have a C:P ratio higher than 60, bacteria often immobilize P, and then grazing by bacterivores is the important mechanism of P remineralization. In our cultures, P mineralization could be explained completely by consumption of bacteria and excretion of surplus P by protozoa, which had the same C:P ratio as the bacteria. Grazers may also have stimulated bacterial activity. In the second stages, incorporation of tritium-labeled thymidine into bacterial macromolecules, per cell, was two times higher in the presence of grazers, indicating an increased specific growth rate. However, the grazers reduced the bacterial numbers by a factor of 36 (Fig. 1), and overall there was no significant increase in C mineralization. An accelerated degradation of organic carbon by protozoa, as observed by Pengerud et al. (32), may be restricted to mineral-poor substrates and may be explained mainly by protozoan nutrient regeneration.

2.3 Effects of Microbivores in Pots with Soil

The results discussed were obtained with bacteria and protozoa in water cultures. In soil, nematodes are important bacterivores in addition to protozoa. Moreover, interactions between microbes and grazers are more complicated in soil because bacteria may be protected from grazing by small soil pores, especially in loam and clay soils (33,34). Bouwman et al. (35) found significant effects of nematodes on C and N mineralization in microcosms with silt loam soil amended with 2 mg lucerne (*Medicago sativa*) meal and 0.22 mg wheat (*Triticum aestivum*) straw meal g^{-1} soil. This amendment with a molar C:N ratio of 13 was used to simulate the addition of crop residues in a common quantity of 3200 kg C ha^{-1}. After amendment, the presterilized soil was inoculated with microorganisms alone or with microorganisms and bacterivorous nematodes (*Rhabditis* sp. and *Acrobeloides bütschlii*) and incubated at 20°C for up to 6 months. The average number of bacteria in the treatment with nematodes ($1.18 \pm 0.29 \times 10^9$ bacteria g^{-1} soil, mean $\pm$ SD, $n = 15$) was 18% lower than in the treatment without nematodes ($1.43 \pm 0.31 \times 10^9$ bacteria g^{-1} soil). However, the difference was not statistically significant. In the first month, both C and N mineralization were increased in the presence of nematodes (Fig. 3). The increase in C mineralization was explained mainly (73%) by direct effects of the nematodes mineralizing bacterial biomass, and partly (27%) by indirect effects such as an advanced dissemination of bacteria through the soil and an increased colonization of substrates. Dissemination of bacteria by nematodes was also indicated by a strong increase in nitrification in the presence of nematodes, whereas in the absence of nematodes the nitrification remained very low (35). The increase in N mineralization was explained completely by direct mineralization of bacterial biomass N by the nematodes. After the initial stimula-

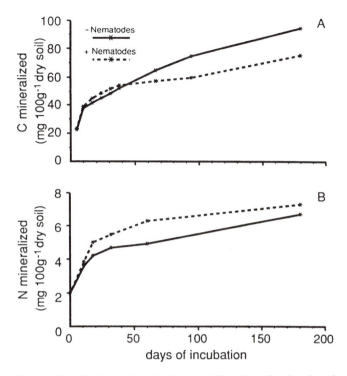

Figure 3. Carbon (A) and nitrogen (B) mineralization in microcosms with soil amended with crop residues in the absence and the presence of nematodes during 180 days of incubation. (From Ref. 35.)

tion in the first month, in the second month the nematodes considerably reduced bacterial activity, and consequently C mineralization, possibly as a result of overgrazing. At the same time, N mineralization was still increased because of the relatively high N content of the bacteria consumed. After 2 months, both C and N mineralization were drastically reduced by about 40% in the presence of nematodes, compared to the treatment without nematodes. Probably because of a reduction in the active part of the bacterial population, in this period mineralization by the nematodes fell far below their basic maintenance requirements. At the end of the experiment, after 6 months, in the treatments with nematodes considerably less C (20%) and more N (13%) had been mineralized than in the treatments without nematodes. Thus, effects of grazers may vary strongly in time as a result of temporal variance in domination of the different mechanisms: dissemination, digestion, and overgrazing of bacteria.

In contrast with the protozoa in the water cultures (Fig. 1), the nematodes did not greatly reduce the total number of bacteria in the soil of the microcosms. In water, all bacteria are surrounded by substrates, and grazer and food are not separated by particles. In soil, microbial activity is limited to spots where substrate is present and bacteria may be separated both from substrate and from predators. From microscopic observations, Clarholm and Rosswall (36) concluded that even under favorable conditions only 15% to 30% of the bacteria in soil were active, and these bacteria were also relatively large, with lengths greater than 4 μm and widths greater than 1.4 μm. If active bacteria are selectively grazed because they are bigger or because they are not protected by small pores (33,34), grazing may affect total bacterial activity much more than total bacterial numbers. It has been shown that protozoa in water selectively feed on bigger cells and on dividing cells (37–39). Thus, protozoa can crop the bacterial production without strongly reducing numbers.

Whereas Bouwman et al. (35) found no significant reduction of total bacterial numbers by nematodes in microcosms with silt loam, Kuikman et al. (40–42) reported up to 50% reduction of bacteria (colony-forming units) by protozoa in microcosms with loamy sand. The latter soil contains fewer small pores and may provide less protection of bacteria. The activity of protozoa increased the turnover of ^{14}C- and ^{15}N-labeled substrates (bacterial biomass) compared to that of microcosms without protozoa. After 33 days of incubation the accumulated ^{14}C-CO_2 evolved from soils with protozoa was 50% higher than that from soils without protozoa. With protozoa, nitrogen uptake by wheat plants was increased by 20%, and total dry matter production by 50%. It was concluded that protozoa increased N mineralization by stimulating the turnover of bacterial biomass.

3. MICROBES, MICROBIVORES, AND N MINERALIZATION IN ARABLE FIELDS UNDER CONVENTIONAL AND INTEGRATED MANAGEMENT

Microcosms are simple ecosystems which facilitate investigation of the role of specific functional groups of organisms in mineralization. The situation in the field is more complex because many functional groups of organisms are present which can interact in different ways. Figure 4 shows the soil food web of the Lovinkhoeve experimental farm (43). The most important predator–prey relationships (trophic interactions) are represented by arrows between different functional groups of organisms. The functional groups were defined mainly by food choice. The biomasses of all these functional groups and N mineralization were monitored during a full year in an integrated and a conventional field under

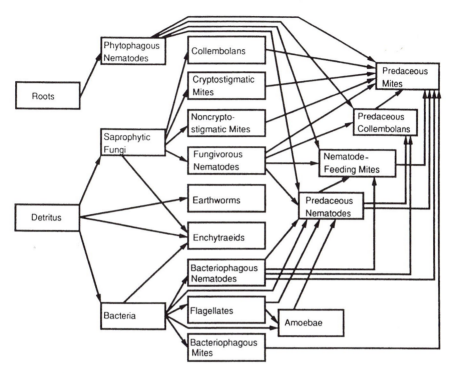

Figure 4.   Diagram of the soil food web at the Lovinkhoeve farm. (From Ref. 43.)

winter wheat (*Triticum aestivum*) (44,45). The results were used as input for a food web model which calculates N mineralization from biomasses and trophic interactions (43). Finally, the simulated N mineralization was compared with the observed N mineralization.

At the Lovinkhoeve, a 4-year crop rotation of winter wheat, sugar beet (*Beta vulgaris*), spring barley (*Hordeum vulgare*), and potato (*Solanum tuberosum*) is practiced on a calcareous silt loam soil (pH KCl 7.5). Between 1966 and 1985, the integrated field received on average 5700 kg organic matter ha^{-1} yr^{-1} as farmyard manure, crop residues, and green manure, whereas the conventional field received on average 3200 kg ha^{-1} yr^{-1} as crop residues only. This resulted in a significantly higher content of organic matter (2.8% vs. 2.2%) and total soil N (0.14% vs. 0.10%) in the integrated field compared with the conventional field (1,7). In 1990, the integrated field received 40% less fertilizer N than the conventional field (144 kg N ha^{-1} vs. 237 kg N ha^{-1}). The integrated field was plowed to 10 cm depth and the conventional field to 20 cm depth. The grain yield from the integrated field was 77% of the yield from the conventional field (5.7×10^3 kg ha^{-1} vs. 7.4×10^3 kg ha^{-1} dry weight).

Samples were taken at 1 to 6 week intervals from 20 November 1989 (day -42) to 30 October 1990 (day 302; day 0 is 1 January 1990). The 0–10 cm and 10–25 cm depth layers were analyzed separately. To facilitate comparison of the layers the results are expressed per hectare per centimeter depth. Bacteria (Fig. 5a) constituted 94% and 75% of the total biomass of all functional groups of organisms determined by direct microscopic counts in the conventional and the integrated field, respectively. The average biomass of the fungi (Fig. 5b) was estimated to be 0.11 kg C ha^{-1} cm^{-1} depth. This was only 1% of the average bacterial biomass of 9.5 kg C ha^{-1} cm^{-1} depth. Therefore, here we focus on bacteria and their main grazers, the protozoa (Fig. 5c and d) and the nematodes (Fig. 5e). Because the microbial biomass was strongly dominated by bacteria it was assumed that the protozoa were mainly bacterivorous. This assumption was supported by the observation that also the nematodes were mainly bacterivorous. The biomass of bacterivorous nematodes was 10 times greater than that of fungivorous nematodes as determined by their morphological characteristics. Of the protozoa, amoebae were quantitatively the most important bacterivores with an average biomass of 0.6 kg C ha^{-1} cm^{-1} depth, followed distantly by flagellates with an average biomass of 0.023 kg C ha^{-1} cm^{-1} depth. The average biomass of the bacterivorous nematodes was 0.013 kg C ha^{-1} cm^{-1} depth. In the integrated field, the N mineralization rate, as determined by in situ incubations for 6 week periods, was on average 30% higher than in the conventional field (101 kg N ha^{-1} yr^{-1} vs. 78 kg N ha^{-1} yr^{-1}). The bacterial biomass was not significantly higher in the integrated field, but the biomasses of amoebae and nematodes were 64% and 22% higher, respectively. A higher biomass of bacterivores should imply a higher turnover of the bacterial biomass to balance higher grazing losses. This hypothesis was supported by a 24% higher frequency of dividing cells, which is an index of the bacterial growth rate; a 20% higher active bacterial biomass, as determined by the ratio between red cells (with a high nucleic acid content, presumably active) and blue cells (with a low nucleic acid content, presumably inactive) after staining with europium chelate (stains nucleic acids red) and fluorescent brightener (stains polysaccharides in cell walls blue) (44); and a 19% higher potential O_2 consumption rate. However, only the latter difference was statistically significant. Thus, the higher input of organic matter in the integrated field appeared to result in a higher activity rather than a higher biomass of bacteria (44). Similar results were reported by Schnürer et al. (46) and Paustian et al. (47).

The bacterial biomass was relatively low in December (days -31 to -1) and

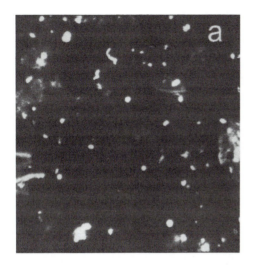

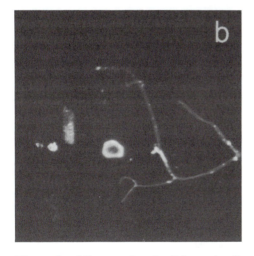

Figure 5. Micrographs of soil bacteria, diameter about 1 μm (a); fungal hyphae, diameter about 3 μm (b); protozoa, heterotrophic flagellates, diameter about 5 μm (c); protozoa, amoebae, diameter about 10 μm (d); and nematodes, length about 700 μm (e).

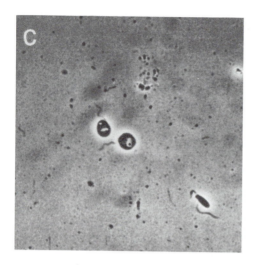

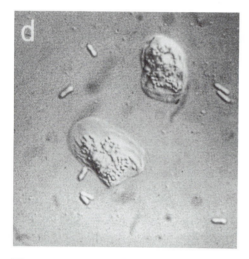

Figure 5. Continued.

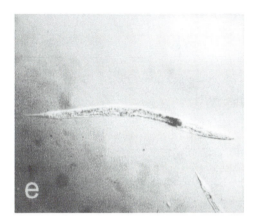

Figure 5.   Continued.

January (days 0–30) (Fig. 6). This may have been caused by soil temperatures just above freezing point. At relatively low soil temperatures of about 5°C in February (days 31–58) and March (days 59–89) the bacterial biomass increased strongly. The bacterial biomass stabilized in April (days 90–119) and May (days 120–150), probably as a result of limited availability of easily decomposable organic matter. Application of fertilizer N on day 81 on both fields, and on day 122 on the conventional field, did not seem to affect the bacteria; that finding may suggest that the bacteria were not N-limited. In summer, the bacterial biomass showed no net growth. However, harvest on 15 August (day 226) and skim plowing on 23 August (day 234) were followed by steep peaks in the bacterial frequency of dividing cells (44), bacterial biomass (Fig. 6), and protozoan biomass (Fig. 7). These peaks in microbial growth were attributed to release of easily decomposable organic matter from plant roots after harvest, and to increased availability of organic matter that resulted from soil mixing by skim plowing (stubble mulching). On the conventional field on 24 August (day 235) soil fumigation with 1,3-dichloropropene was applied against phytophagous nematodes, in particular potato cyst nematodes (eelworm), which cause potato sickness. Consequently, the bacterivorous protozoa (Fig. 7) and nematodes (Fig. 8) in the conventional field were drastically reduced for several weeks. The strongest reduction occurred in the 10–25 cm layer, where the fumigant was injected at about 18 cm depth. The fumigation did not affect bacterial biomass. In the integrated field, where bacterivores

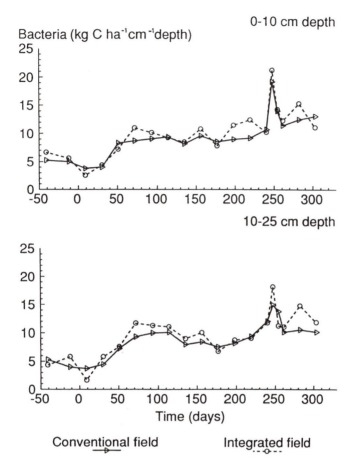

Figure 6. Biomass of bacteria in the 0–10 and 10–25 cm depth layers of the conventional and the integrated field at the Lovinkhoeve from 20 November 1989 (day –42) to 30 October 1990 (day 302; day 0 = 1 January 1990). The least significant difference of the interaction depth × field at $P = 0.05$ ($LSD_{depth \times field}$) is 2.45 kg C ha^{-1} cm^{-1} depth. (From Ref. 44.)

were not reduced, the peaks in bacterial biomass around day 246 were as high as or even higher than in the conventional field with reduced bacterivores. This suggests that the bacterivores did not control the bacterial numbers, maybe because of protection of bacteria by the silt loam soil. On the other hand, in the integrated field the availability of bacteria must have been an important factor for the growth of the protozoa because the biomass of the amoebae closely followed the bacterial peak (Figs. 6 and 7).

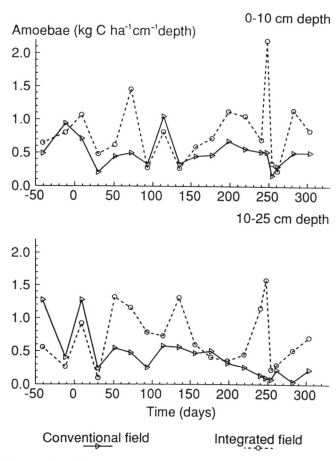

Figure 7. Biomass of protozoa (amoebae) in the 0–10 and 10–25 cm depth soil layers of the conventional and the integrated field at the Lovinkhoeve (day 0 = 1 January 1990). $LSD_{depth \times field}$ = 0.250 kg C ha^{-1} cm^{-1} depth. (From Ref. 44.)

Nitrogen mineralization was low in winter (Fig. 9). During the increased bacterial growth in February (days 31–58) and March (days 59–89) (Fig. 6), mineralization remained very low. During the bacterial growth peak in early September (around day 246), a strong reduction in N mineralization was found; in the conventional field even immobilization was observed. Increased bacterial growth apparently resulted in immobilization rather than mineralization of N. Immobilization of mineral N due to bacterial growth in soil was also reported by Elliott et al. (12), Ingham et al. (48), and Rutherford and

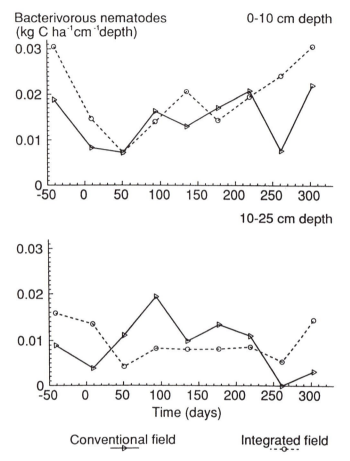

Figure 8. Biomass of bacterivorous nematodes in the 0–10 and 10–25 cm depth layers of the conventional and the integrated field at the Lovinkhoeve (day 0 = 1 January 1990). LSD$_{depth \times field}$ = 0.00207 kg C ha^{-1} cm^{-1} depth. (From Ref. 44.)

Juma (49). The N immobilization during the bacterial peak was not followed by a flush of N remineralization during the steep bacterial decline after 4 September (day 246) (Figs. 6 and 9). This may be explained by N immobilization into nonmicrobial N such as microbial metabolites and recently stabilized materials which are not easily remineralized (49). From April (days 90–119) to the end of August (days 212–242), bacterial growth was relatively low and N mineralization was relatively high. This may be explained by three possible mechanisms.

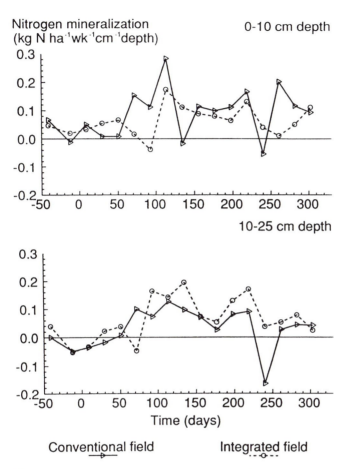

Figure 9. Nitrogen mineralization rates in the 0–10 and 10–25 cm depth layers of the conventional and the integrated field at the Lovinkhoeve (day 0 = 1 January 1990). The points at each sampling date represent averages obtained by in situ exposure of 10 soil cores per field during 6 weeks after the sampling date. $LSD_{depth \times field}$ = 0.0224 kg N ha^{-1} wk^{-1} cm^{-1} depth. (From Ref. 44.)

1. Decomposition of organic matter by slowly growing bacteria with a low growth efficiency. A low yield can result from the depletion of organic or inorganic nutrients essential for the synthesis of cell material. In this case, a relatively small part of the organic substrate is used for synthesis of biomass and a larger part is used for maintenance, being mineralized to CO_2 and NH_4^+ (50).

2. Endogenous respiration of soil bacterial biomass during starvation: If the availability of organic substrates is not sufficient to meet the maintenance requirements of the bacterial biomass, bacteria can respire part of their cell material, such as ribonucleic acid (RNA) and proteins, resulting in excretion of NH_4^+ (51,52).
3. Consumption of microbial biomass and excretion of NH_4^+ by bacterivores (53,48).

The observation that bacterivores apparently did not control bacterial numbers during the period after harvest (day 226) does not imply they played an unimportant role in N mineralization. Whether or not protozoa and nematodes prevent increases in bacterial numbers or whether they only follow increases in bacterial numbers, they will always mineralize N. In periods with relatively high bacterial growth rates and net immobilization of N into bacterial biomass, grazing may be the major mechanism of N mineralization.

4. SIMULATION OF N MINERALIZATION IN THE FOOD WEBS OF ARABLE FIELDS UNDER CONVENTIONAL AND INTEGRATED MANAGEMENT

N mineralization can be calculated from feeding rates by using a food web model which was originally developed by Hunt et al. (54). For description of the food web model, the terms "predator" and "prey" are used also for microbivorous grazers and microbes. The model calculates N mineralization by following a scheme (Fig. 10) in which the feeding rate, i.e., the rate at which material is taken from an energy source, is split into a rate at which organic material is returned to the environment in the form of feces or prey residues, a rate at which material is incorporated into the biomass of the consumer, and a rate at which material is released in inorganic form. Average annual feeding rates of the functional groups in the food web (Fig. 4) were calculated using the steady-state assumption; i.e., the production rate of a group balances the rate at which material is lost through natural death and predation

$$F = \frac{d_{nat}B + P}{e_{ass}e_{prod}} \tag{1}$$

where F is the feeding rate (kg C ha^{-1} yr^{-1}), d_{nat} is the specific natural death rate (yr^{-1}), B is biomass (kg C ha^{-1}), P is the death rate due to predation (kg C ha^{-1} yr^{-1}), e_{ass} is assimilated carbon per unit consumed carbon, and e_{prod} is biomass production per unit assimilated carbon.

Simulation of the dynamics of C and N flows within 1 year required that

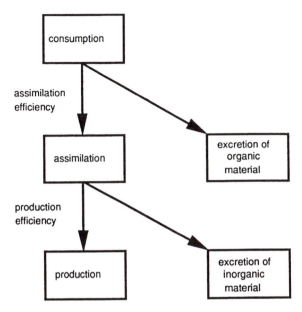

Figure 10. Scheme relating consumption, biomass production, excretion of organic material, and excretion of inorganic material. (From Ref. 43.)

the biomass dynamics be incorporated in the model. This was done by adding the rate of change in biomass of a functional group to the rate of material loss (55)

$$F = \frac{d_{nat}B + P + \Delta B/\Delta t}{e_{ass}\, e_{prod}} \tag{2}$$

where $\Delta B/\Delta t$ is the rate of change in biomass (kg C ha^{-1} wk^{-1}), and other terms are as in equation (1).

If a predator was considered to feed on more than one prey type, then both the preference of the predator for a given prey and the relative abundances of the prey types were taken into account

$$F_i = \frac{w_i B_i}{\displaystyle\sum_{i=1}^{n} w_i B_i}\, F \tag{3}$$

where F_i is the feeding rate on prey i (kg C ha^{-1} wk^{-1}), w_i is the preference for prey i relative to other prey types, and n is the number of functional groups in the web.

N mineralization was calculated per trophic interaction, depending on feeding rate, assimilation efficiency, production efficiency, and C:N ratios of food and consumer

$$N_{min} = e_{ass}\left(\frac{1}{r_{prey}} - \frac{e_{prod}}{r_{pred}}\right)F \tag{4}$$

where N_{min} is the N mineralization rate resulting from a trophic interaction (kg N ha^{-1} wk^{-1}), r_{prey} is the C:N ratio of prey, and r_{pred} is the C:N ratio of predator. To account for the effect of temperature on the feeding and mineralization rates, mean soil temperatures per week were related to the specific death rates by means of a Q_{10} of 3, which was similar to the value used by Johnson et al. (56). The Q_{10} (temperature quotient) is the rate of a chemical or biological process at temperature $T + 10$ divided by the rate at temperature T. Many biological processes are two to three times faster at 20°C than at 10°C.

The calculations started with the feeding rates of the top predators which experienced only natural death, i.e., $P = 0$ (see equation [2]). The predatory losses in the groups of one trophic level lower were calculated from the feeding rates of the top predators, using equation [3]. These losses were added to the nonpredatory losses in order to calculate the feeding rates of the groups at this level. All feeding rates were subsequently calculated throughout the food web, working back to the primary consumers, i.e., microorganisms and saprotrophic animals feeding on dead organic matter, such as earthworms and enchytraeids (small 10 to 20 mm long white or transparent worms). Thus, N mineralization was calculated by using biomass dynamics, specific death rates, preference weighing factors, assimilation efficiencies, production efficiencies, C:N ratios, and soil temperature (43).

The N mineralization rates calculated by the model and the observed rates are presented in Fig. 11 for the conventional and the integrated field, and for the 0–10 and 10–25 cm depth layers. For the simulation it was assumed that bacteria immobilized N and that N was mineralized by predators only. This was implemented by choosing a C:N ratio of substrate for bacteria of 12, a bacterial C:N ratio of 4, and a bacterial production efficiency of 0.45. Consequently, the model simulated net N immobilization in both depth layers of the conventional field and net N mineralization in both depth layers of the integrated field, in the period following harvest (day 226) and soil tillage (day 234). Thus, the model produced a reasonable simulation of the observed pattern. This indicates that an important part of the observed N mineralization can be explained by the grazing activity of microbivores. The observed pattern could not be simulated when it was assumed that not only grazers but also bacteria mineralized N. The latter assumption was made in earlier simulations when a substrate C:N ratio of 10, a bacterial C:N ratio of 5, and a bacterial production efficiency of 0.30 were chosen (57). This illustrates that the model is especially sensitive to values of the C:N ratio of the substrate and the C:N ratios, specific death rates, and production efficiencies of the bacteria and protozoa (43). The observed N mineralization pattern (Fig. 9) may be ex-

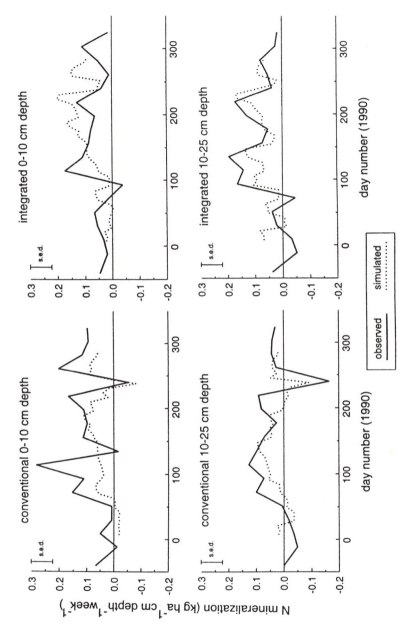

Figure 11. Simulated and observed dynamics in N mineralization in the 0–10 and 10–25 cm depth layers of the conventional and the integrated field at the Lovinkhoeve (day 0 = 1 January 1990). Vertical bar denotes the standard error of differences of means (SED). (From Ref. 43.)

plained as follows: After harvest (day 226) and soil tillage (day 234), an increased availability of organic matter (crop residues) caused relatively high bacterial growth rates, as indicated by a peak in bacterial biomass (Fig. 6) and in frequencies of dividing cells (44). The high bacterial growth rates caused net N immobilization in the conventional field (Fig. 9), where the microbivores were strongly reduced by soil fumigation (Figs. 7 and 8). In the integrated field, where no fumigation was applied and faunal densities were not reduced, the N immobilization by the bacteria was obscured by the relatively high N mineralization by the microbivores, resulting in net mineralization. The relatively high net mineralization rates in April (days 90–119) were not simulated by the model if it was assumed that all N mineralization was due to microbivores only. In this period the bacteria may have been growth-limited by organic substrate or inorganic nutrients (probably not N). Consequently, the bacteria may have grown with a low efficiency and may have caused an important part of the observed N mineralization.

The fact that the patterns in N mineralization are only partly mimicked by the simulated rates indicates that some of the model parameters cannot be treated as constants when temporal variability within seasons is taken into account. In particular, the chemical composition of the substrate used by the microbes might vary during the season, for example, by means of root exudates or by a decrease in the easily decomposable fractions, which in turn may affect the microbial carbon/nutrient ratios and growth efficiencies. Further improvement of the model therefore depends on whether or not empirical data on the temporal variability of this kind of model parameter become available (58). Also, the model treated bacteria as one homogeneous group of organisms, while the bacterial populations may vary strongly in time with respect to species composition and substrate utilization (59). This may also affect the efficiency with which the bacteria utilize N from their substrate and may lead to alternating periods of net N mineralization and immobilization by the bacteria (44). More knowledge about the precise factors which limit bacterial growth in the field and which determine the balance between mineralization and immobilization may facilitate reduction of N losses to the environment. The need for N conservation was confirmed by the considerable mineralization rates which were still found after harvest in the period without crop uptake (after day 250). N can be conserved by sowing of catch crops which start growing after harvest or by application of crop residues with a high C:N ratio to promote microbial N immobilization (1).

5. CONTRIBUTIONS OF VARIOUS ORGANISMS TO N MINERALIZATION IN DIFFERENT FARMING SYSTEMS

The food web model was applied to a set of seven below-ground food webs from one native and six farming systems (60), using one set of parameter values which have been proposed by Hunt et al. (54), with a few exceptions. The following webs were used: Central Plains Experimental Range, Colorado, United States, a shortgrass prairie (*Bouteloua gracilis*) (54); Lovinkhoeve Experimental Farm, Marknesse, the Netherlands, winter wheat under conventional and integrated farming (45); Horseshoe Bend Research Site, Georgia, United States, sorghum (*Sorghum bicolor*)/soybean (*Glycine max*)/rye (*Secale cereale*), conventional tillage using mold-board plowing and no-tillage where plant residues remain on the surface without soil disturbance (61,62); Kjettslinge Experimental Field, Uppsala, Sweden, barley with no nitrogen fertilizer and barley with 120 kg N ha^{-1} yr^{-1} for 6 years (63).

The model was used to compare the webs with respect to overall N flux rates and to the contribution of the various groups of organisms to N mineralization. The latter estimates have to be treated with caution because the results of the model cannot be validated at the level of the contributions of the individual groups of organisms, and because the estimates of the contributions are much more sensitive to the physiological parameter values than the overall N mineralization rates. The simulated contribution of the various groups of organisms varied strongly among the food webs (Table 1). Bacteria were the most important group in the Central Plains and in the conventional field at the Lovinkhoeve, in both webs accounting for 52% of the overall N mineralization rate. In most webs, bacteria were found to contribute much more to N mineralization than fungi. Fungi only contributed considerably in the Kjettslinge webs, as a result of the high fungal biomass in these webs. The contribution of fungi ranged from 1% at the Lovinkhoeve to 34% in the fertilized field at Kjettslinge. The model calculations suggest that in some of the webs protozoa may contribute more to N mineralization than bacteria and fungi. In the integrated field at the Lovinkhoeve, in Horseshoe Bend, and in the unfertilized field at Kjettslinge, protozoa were the most important group, accounting for 46%, 48%, and 45% of the N mineralization rate, respectively. This large contribution of protozoa was caused by their relatively high specific death rate (6 yr^{-1}) and the low C:N ratio of their food (4). If, for example, the C:N ratio of bacteria was increased from 4 to 5, the contribution of protozoa in the integrated field at the Lovinkhoeve decreased from 55 to 41 kg N ha^{-1} yr^{-1} and the contribution of bacteria increased from 51 to 82 kg N ha^{-1} yr^{-1}. Thus, the outcome of the model is relatively sensitive to the value of the bacterial C:N ratio. Although earthworms may contribute considerably to the overall N flux rates, their contribution to N mineralization was small, ranging from 1% in the conventional tillage field at Horseshoe Bend to 6% in the integrated field at the Lovinkhoeve. However, this small contribution depended strongly on the assumed diet of the earthworms, i.e., detritus with a C:N ratio of 10, and on the value used for their production efficiency (0.40) (64). The total contribution of the remaining fauna was small (approximately 1%) in all webs, except in the Central Plains, where it amounted to 18%, of which 13% was due to the activity of the bacterivorous nematodes. The contribution of the nema-

Table 1. Nitrogen Mineralization Rates (kg ha^{-1} yr^{-1}) of the Functional Groups in the Different Food Webs, as Calculated by the Food Web Model.[a]

	CPER	LH-IF	LH-CF	HSB-NT	HSB-CT	KS-BO	KS-B120
Microbes							
Bacteria	45.44	51.59	40.15	97.40	136.95	37.31	20.50
Fungi	7.61	0.91	0.82	14.66	12.67	19.03	23.61
VAM	4.27						
Protozoa							
Amoebae	11.05	54.69	33.37	115.71[c]	144.64[c]	54.51[c]	16.80[c]
Flagellates	0.49	2.20	1.73				
Nematodes							
Herbivores	1.12	0.16	0.12	0.11	0.10	0.004	0.06
Bacteriovores	12.93	0.88	0.88	0.87	2.27	2.03	2.38
Fungivores	0.22	0.08	0.06	0.05	0.03	0.26	0.16
Predators[b]	0.35	0.06	0.07			0.25	0.25
Arthropods							
Herbivorous herbage arthropods						0.23	0.28
Predatory herbage arthropods						0.08	0.10
Herbivorous macroarthropods						0.02	0.02
Microbivorous macroarthropods						0.42	0.41
Predatory macroarthropods						0.34	0.33

Predatory mites	0.06	0.03	0.02	0.08^d	0.01^d	0.12^e	0.18^e
Nematophagous mites	0.05	0.003	0.002	0.23	0.05		
Cryptostigmatic mites	0.37	0.001	0.002	0.35	0.12		
Noncryptostigmatic mites	0.44	0.015	0.01				
Bacterivorous mites		0.0004	0.001				
Fungivorous collembola	0.15	0.16	0.19	0.12	0.03	0.44^f	0.42^f
Predatory collembola		0.003	0.01				
Annelids							
Enchytraeids		0.31	0.42	0.12	0.38	3.32	2.56
Earthworms		7.62	^g	12.00	2.40	1.56	1.56

[a] Values refer to the 0–25 cm depth layer, except for the Horseshoe Bend webs (0–5 cm). CPER, Central Plains Experimental Range (Colorado, U.S.A.); LH, Lovinkhoeve (NL); IF, integrated farming; CF, conventional farming; HSB, Horseshoe Bend (Georgia, USA); CT, conventional tillage; NT, no-tillage, KS, Kjettslinge (Sweden); BO = B zero, without fertilizer; B120, with fertilizer.
[b] Including predators and omnivores.
[c] Including amoebae and flagellates.
[d] Including all predatory arthropods.
[e] Including all predatory microarthropods.
[f] Including all microbivorous microarthropods.
[g] Not found.

Source: Reprinted from Ref. 60 by permission of Kluwer Academic Publishers.

todes may have been underestimated because of the relatively low values used for their specific death rates (1–3 yr^{-1}), which may have been higher in the Lovinkhoeve (60) and Kjettslinge (63) webs.

The group contributions as presented in Table 1, however, only refer to N mineralization resulting directly from the consumption by the group, while the group also affects the functioning of other groups in the web and hence their contribution to nutrient cycling. Many studies have indicated indirect contributions of the fauna to C and N mineralization rates by affecting limiting factors for microbial growth (65–67), which were not explicitly taken into account in the present food web model. However, if the fauna stimulates microbial growth, such indirect effects are implicitly taken into account in the model because this will result in a higher microbial biomass or shorter microbial turnover times and a higher biomass of predators.

To understand the role of the various groups of organisms, and their interactions, in nutrient cycling, detailed experimental studies have proven to be useful (66) and are still required. However, the large amount of data on the soil organisms required is a serious drawback of the food web approach. Given the quantitative importance of the microorganisms in agricultural ecosystems (Table 1) it may be possible to restrict the analyses in such systems to microbes and protozoa. In natural ecosystems, the role of the soil fauna may be more important than in agricultural systems (68).

The results of the food web model are especially sensitive to values of the C:N ratio of the substrate, and to the size of the biomass, C:N ratios, specific death rates, and production efficiencies of the bacteria and protozoa. Reliable measurement of these important parameters is still a problem, especially in soil. Therefore, methods have to be improved or developed for accurate measurement of the amount and quality of available substrate (69,70), and of the amount, composition, and turnover rates of bacterial (71–74) and protozoan biomass in soil. A better fit between simulated and observed dynamics in N mineralization may be obtained by including the dynamics of soil organic matter in the model, resulting in simulation of alternating periods of mineralizing and immobilizing microbial populations. Moreover, it is desirable to define functional groups within the quantitatively dominating microbial biomass, which until now has been treated as one pool. New methods are now emerging from molecular biology which may be applied in microbial ecology to identify and quantify different groups of microorganisms in relation to their functional role in nutrient cycling. The use of denaturing gradient gel electrophoresis to analyze polymerase chain reaction–amplified genes coding for 16S ribosomal RNA (rRNA) (Chapter 12b) (75) and the use of fluorescent 16S rRNA-targeted oligonucleotide probes for identification of individual cells (Chapter 10) (76,77) may be combined with automatic image analysis (74). These techniques may offer the opportunity to measure shifts in the field between slowly growing bacteria with low growth efficiency and high mineralization, and faster growing bacteria with high growth efficiency and low mineralization.

Six years of comparison between conventional and integrated arable farming at the Lovinkhoeve farm showed that a reduction of mineral fertilizer N by, on average, 35% could be compensated for by addition of stabilized organic N in the form of compost and processed animal manure. This resulted in increased amounts of N in the pool of young humus in the soil (78) and in increased N mineralization, which was mainly attributed to increased bacterial activity and increased protozoan biomass (44). The contribution of bacteria to N mineralization in 1990 was estimated to be 40 kg N ha^{-1} yr^{-1} in the conventional field and 52 kg N ha^{-1} yr^{-1} in the integrated field (Table 1). For amoebae the difference was even greater. Their contribution was estimated to be 33 kg N ha^{-1} yr^{-1} in the conventional field and 55 kg N ha^{-1} yr^{-1} in the integrated field. Similar differences were found for the bacteria and protozoa in the fertilized and unfertilized fields in Kjettslinge. The higher mineralization in the integrated field at the Lovinkhoeve supported crop yields amounting to, on average, 90% of those from the conventional field (78). For an entire rotation cycle from 1988 to 1991, calculated N losses to the environment were less than 289 kg N ha^{-1} in the conventional field (field 12B) and less than 181 kg ha^{-1} in the integrated field (field 16A) (78). Higher inputs of mineral N on the conventional field in spring, when N uptake by the crop was still low, increased the risk of high N losses. The higher mineralization of N by the food web in the integrated field supplied the crop with mineral N throughout the growing season, without much accumulation of mineral N, thus with a lower risk of losses to the environment.

6. CONCLUSIONS

Bacteria and fungi are the primary consumers of dead organic matter, such as manure and crop residues, in soil. Part of the decomposed material is used for production of microbial biomass, and another part is mineralized to CO_2, H_2O, mineral nitrogen (N), phosphorus (P), and other nutrients. Mineral nutrients are required for plant growth, but in the present conventional farming systems high inputs of fertilizers and manure cause serious losses of nitrate and phosphate to ground- and surface water, and of nitrous oxide and ammonia to the atmosphere. The resulting environmental problems have stimulated research into nutrient cycling in integrated (reduced-input) farming systems. Integrated farming involves a shift from directly feeding the plant with mineral nutrients to feeding the soil organisms with organic matter, thereby indirectly feeding the plant through nutrient mineralization by soil organisms.

If the decomposed organic matter contains considerably less N and P than the microbial biomass, all the N and P released during decomposition is used for microbial growth and no net mineralization occurs. Generally, no ammonium is released by microbes when the C:N ratio of the decomposed organic matter is higher than about 15. Similarly, P mineralization appears to cease above a substrate C:P ratio of 60. Mineral nutrients immobilized in microbial biomass are released when microbes are grazed by microbivores such as protozoa and nematodes. Microbivores may also enhance decomposition and mineralization indirectly by stimulating the microbial activity.

The effect of grazers on decomposition and mineralization depends on growth conditions and differs for C, N, and P. In liquid continuous cultures with bacteria and protozoa grown on yeast extract with a molar C:N:P ratio of 100:15:1.2, bacterial numbers were greatly reduced by protozoan grazing. Although P was not growth-limiting, bacteria immobilized P, and the major mechanism of P mineralization appeared to be consumption of bacteria by protozoa. N mineralization was performed mainly (70%) by bacteria but was increased 30% by protozoa. Protozoa did not enhance C mineralization. An accelerated decomposition of organic carbon may be restricted to mineral-poor substrates. In soil, nematodes are important microbivores in addition to protozoa. Interactions between microbes and grazers are more complicated in soil than in water because bacteria may be protected from grazing by small soil pores, especially in loam and clay soils. In microcosms with loam soil amended with crop residues with a C:N ratio of 13, bacterial numbers were not significantly reduced by bacterivorous nematodes. In the presence of nematodes, C mineralization was increased during the first month and subsequently reduced. N mineralization was increased during the first 2 months and then reduced. After 6 months, in the treatments with nematodes considerably less C (20%) and more N (13%) had been mineralized than in the treatments without nematodes. Effects of grazers may vary strongly in time and are correlated with the type of organic matter decomposed.

In a field study, microbes, microbivores, and N mineralization were monitored in a winter wheat field under conventional and integrated management. The microbial biomass was strongly dominated by bacteria and was not significantly higher in the integrated field than in the conventional field, whereas protozoa and nematodes were 64% and 22% higher, respectively. Average N mineralization was 30% higher in the integrated field. The differences were attributed to the approximately 30% higher soil organic matter content of the integrated field, which appeared to increase activity of bacteria and biomasses of protozoa and nematodes. The contributions of different functional groups to the total N mineralization were calculated by using a food web model. The model produced a reasonable simulation of the observed mineralization pattern and indicated that an important part of the observed N mineralization can be explained by the grazing activity of protozoa and nematodes. Simulation of seven food webs from different farming systems in the United States, Sweden, and the Netherlands indicated that in most webs bacteria contributed much more to N mineralization than fungi. In some of the webs, protozoa may contribute more to N mineralization (up to 50%) than bacteria and fungi.

Comparison of conventional and integrated farming showed that a 35% reduction of mineral fertilizer N could be compensated for by addition of stabilized organic N in the form of compost and processed animal manure. The higher mineralization of N by the food web in the integrated field supplied the crop with mineral N throughout the growing season, without much accumulation of mineral N, thus with a lower risk of losses to the environment.

REFERENCES

1. H. G. van Faassen, and G. Lebbink, *Neth. J. Agric. Sci. 38*:265 (1990).
2. L. Brussaard, J. A. van Veen, M. J. Kooistra, and G. Lebbink, *Ecol. Bull. 39*:35 (1988).
3. P. Vereijken, *Neth. J. Agric. Sci. 34*:387 (1986).
4. C. A. Edwards, *Sustainable Agricultural Systems* (C. A. Edwards, L. Lal, P. Madden, R. H. Miller, and G. House, eds.), Soil and Water Conservation Society, Ankeny, IA, p. 249 (1990).
5. J. M. Lopez-Real, *The Role of Microorganisms in a Sustainable Agriculture* (J. M. Lopez-Real, and R. D. Hodges, eds.), AB Academic Publishers, Berkhamstead, England, p. 1 (1986).
6. M. J. Kooistra, G. Lebbink, and L. Brussaard, *Agric. Ecosyst. Environ. 27*:361 (1989).
7. G. Lebbink, H. G. van Faassen, C. van Ouwerkerk, and L. Brussaard, *Agric. Ecosyst. Environ. 51*:7 (1994).
8. E. A. Holland and D. C. Coleman, *Ecology 68*:425 (1987).
9. S. Schwaerter, M. Søndergaard, B. Riemann, and L. M. Jensen, *J. Plankton Res. 10*:515 (1988).
10. I. Martinussen and T. F. Thingstad, *Mar. Ecol. Prog. Ser. 37*:285 (1987).
11. S. J. Chapman and T. R. G. Gray, *Soil Biol. Biochem. 13*:11 (1981).
12. E. T. Elliott, C. V. Cole, B. C. Fairbanks, L. E. Woods, R. J. Bryant, and D. C. Coleman, *Soil Biol. Biochem. 15*:85 (1983).
13. B. Behera and G. H. Wagner, *Soil Sci. Soc. Am. Proc. 38*:591 (1974).
14. S. Findlay, L. Carlough, M. T. Crocker, H. K. Gill, J. L. Meyer, and P. J. Smith, *Limnol, Oceanogr. 31*:1335 (1986).
15. L. R. Bakken, *Appl. Environ. Microbiol. 49*:1482 (1985).
16. S. Lee and J. A. Fuhrman, *Appl. Environ. Microbiol. 53*:1298 (1987).
17. Y. Tezuka, *Microb. Ecol. 19*:227 (1990).
18. J. A. van Veen and E. A. Paul, *Appl. Environ. Microbiol. 37*:686 (1979).
19. J. P. E. Anderson and K. H. Domsch, *Soil Sci. 130*:211 (1980).
20. J. Z. Castellanos and P. F. Pratt, *Soil Sci. Soc. Am. J. 45*:354 (1981).
21. E. G. Beauchamp and J. W. Paul, *Nitrogen in Organic Wastes Applied to*

Soils (J. A. Hansen and K. Henriksen, eds.), Academic Press, London, p. 140 (1989).

22. B. S. Griffiths, *Soil Protozoa* (J. F. Darbyshire, ed.), CAB International, Oxon, p. 65 (1994).

23. R. E. Johannes, *Limnol. Oceanogr. 10*:434 (1965).

24. T. Fenchel and P. Harrison, *The Role of Terrestrial and Aquatic Organisms in Decomposition Processes* (J. M. Anderson and A. Macfadyen, eds.), Blackwell Scientific, Oxford, p. 285 (1976).

25. M. Pussard, C. Alabouvette, and P. Levrat, *Soil Protozoa* (J. F. Darbyshire, ed.), CAB International, Oxon, p. 123 (1994).

26. A. F. Bird, *J. Nematol. 19*:514 (1987).

27. J. Bloem, M. Starink, M. J. B. Bär-Gilissen, and T. E. Cappenberg, *Appl. Environ. Microbiol. 54*:3113 (1988).

28. D. A. Caron, J. C. Goldman, and M. R. Dennett, *Hydrobiologia 159*:27 (1988).

29. H. Güde, *Microb. Ecol. 11*:193 (1985).

30. T. F. Thingstad, *Mar. Ecol. Prog. Ser. 35*:99 (1987).

31. J. Bloem, C. Albert, M. J. B. Bär-Gilissen, T. Berman, and T. E. Cappenberg, *J. Plankton Res. 11*:119 (1989).

32. B. Pengerud, E. F. Skjoldal, and T. F. Thingstad, *Mar. Ecol. Prog. Ser. 35*:111 (1987).

33. J. Postma and J. A. van Veen, *Microb. Ecol. 19*:149 (1990).

34. J. Hassink, L. A. Bouwman, J. Bloem, and L. Brussaard, *Geoderma 57*:105 (1993).

35. L. A. Bouwman, J. Bloem, P. H. J. F. van den Boogert, F. Bremer, G. H. J. Hoenderboom, and P. C. de Ruiter, *Biol. Fertil. Soils 17*:249 (1994).

36. M. Clarholm and T. Rosswall, *Soil Biol. Biochem. 12*:49 (1980).

37. J. M. González, E. B. Sherr, and B. F. Sherr, *Appl. Environ. Microbiol. 56*:583 (1990).

38. B. F. Sherr, E. B. Sherr, and J. McDaniel, *Appl. Environ. Microbiol. 58*:2381 (1992).

39. K. Šimeki, J. Vrba, and P. Hartman, *FEMS Microbiol. Ecol. 14*:157 (1994).

40. P. J. Kuikman, M. M. I. Van Vuuren, and J. A. van Veen, *Agric. Ecosyst. Environ. 27*:271 (1989).

41. P. J. Kuikman, A. G. Jansen, J. A. van Veen, and A. J. B. Zehnder, *Biol. Fertil. Soils 10*:22 (1990).

42. P. J. Kuikman, A. G. Jansen, and J. A. van Veen, *Soil Biol. Biochem. 23*:193 (1991).

43. P. C. de Ruiter, J. Bloem, L. A. Bouwman, W. A. M. Didden, G. H. J. Hoenderboom, G. Lebbink, J. C. Y. Marinissen, J. A. de Vos, M. J. Vreeken-Buijs, K. B. Zwart, and L. Brussaard, *Agric. Ecosyst. Environ. 51*:199 (1994).

44. J. Bloem, G. Lebbink, K. B. Zwart, L. A. Bouwman, S. L. G. E. Burgers, J. A. de Vos, and P. C. de Ruiter, *Agric. Ecosyst. Environ. 51*:129 (1994).
45. K. B. Zwart, J. Bloem, L. A. Bouwman, L. Brussaard, W. A. M. Didden, G. Lebbink, J. C. Y. Marinissen, M. J. Vreeken-Buijs, S. L. G. E. Burgers, and P. C. de Ruiter, *Agric. Ecosyst. Environ. 51*:187 (1994).
46. J. Schnürer, M. Clarholm, and T. Rosswall, *Biol. Fertil. Soils 2*:119 (1986).
47. K. Paustian, L. Bergström, P.-E. Jansson, and H. Johnsson, *Ecol. Bull. 40*:153 (1990).
48. E. R. Ingham, J. A. Trofymow, R. N. Ames, H. W. Hunt, C. R. Morley, J. C. Moore, and D. C. Coleman, *J. Appl. Ecol. 23*:597 (1986).
49. P. M. Rutherford and N. G. Juma, *Biol. Fertil. Soils 12*:228 (1992).
50. H. Veldkamp, *Unifying Concepts in Ecology* (W. H. van Dobben, and R. H. Lowe-McConnel, eds.), Dr. W. Junk, The Hague, p. 44 (1975).
51. C. W. Boylen and M. N. Mulks, *J. Gen. Microbiol. 105*:323 (1978).
52. U. Wanner and T. Egli, *FEMS Microbiol. Rev. 75*:19 (1990).
53. E. T. Elliott, K. Horton, J. C. Moore, D. C. Coleman, and C. V. Cole, *Plant Soil 76*:149 (1984).
54. H. W. Hunt, D. C. Coleman, E. R. Ingham, R. E. Ingham, E. T. Elliott, J. C. Moore, S. L. Rose, C. C. F. Reid, and C. R. Morley, *Biol. Fertil. Soils 3*:57 (1987).
55. R. V. O'Neill, *J. Theor. Biol. 22*:284 (1969).
56. H. Johnsson, B. Bergström, P.-E. Jansson, and K. Paustian, *Agric. Ecosyst. Environ. 18*:333 (1987).
57. P. C. de Ruiter, J. C. Moore, K. B. Zwart, L. A. Bouwman, J. Hassink, J. Bloem, J. A. de Vos, J. C. Y. Marinissen, W. A. M. Didden, G. Lebbink, and L. Brussaard, *J. Appl. Ecol. 30*:95 (1993).
58. P. C. de Ruiter, A. M. Neutel, and J. C. Moore, *TREE 9*:378 (1994).
59. J. Hassink, J. H. Oude Voshaar, E. H. Nijhuis, and J. A. van Veen, *Soil Biol. Biochem. 23*:515 (1991).
60. P. C. de Ruiter, J. A. van Veen, J. C. Moore, L. Brussaard, and H. W. Hunt, *Plant Soil 157*:263 (1993).
61. P. F. Hendrix, R. W. Parmelee, D. A. Crossley Jr., D. C. Coleman, E. P. Odum, and P. M. Groffman, *BioScience 36*:374 (1986).
62. G. J. House and R. W. Parmelee, *Soil Till. Res. 5*:351 (1985).
63. O. Andrén, T. Lindberg, U. Boström, M. Clarholm, A.-C. Hansson, G. Johansson, J. Lagerlöf, K. Paustian, J. Persson, R. Petterson, J. Schnürer, B. Sohlenius, and M. Wivstad, *Ecol. Bull. 40*:85 (1990).
64. J. C. Y. Marinissen and P. C. de Ruiter, *Agric. Ecosyst. Environ. 47*:59 (1993).
65. D. C. Coleman, C. P. P. Reid, and C. V. Cole, *Adv. Ecol. Res. 13*:1 (1983).
66. H. A. Verhoef, and L. Brussaard, *Biogeochemistry 11*:175 (1990).

67. M. H. Beare, R. W. Parmelee, P. F. Hendrix, W. Cheng, D. C. Coleman, and D. A. Crossley Jr., *Ecol. Monogr. 62*:569 (1992).
68. D. A. Crossley Jr., D. C. Coleman, and P. F. Hendrix, *Agric. Ecosyst. Environ. 27*:47 (1989).
69. C. A. Cambardella and E. T. Elliott, *Soil Sci. Soc. Am. J. 58*:123 (1994).
70. F. W. Meijboom, J. Hassink, and M. J. van Noordwijk, *Soil Biol. Biochem. 27*:1109 (1995).
71. E. Bååth, *Soil Biol. Biochem. 22*:803 (1990).
72. P. H. Michel and J. Bloem, *Soil Biol. Biochem. 25*:943 (1993).
73. J. Bloem, P. R. Bolhuis, M. R. Veninga, and J. Wieringa, *Methods in Applied Soil Microbiology and Biochemistry* (K. Alef and P. Nannipieri, eds.), Academic Press, London, (1995) p. 162.
74. J. Bloem, M. R. Veninga, and J. Shepherd, *Appl. Environ. Microbiol. 61*: (1995) 926.
75. G. Muyzer, E. C. de Waal, and A. G. Uitterlinden, *Appl. Environ. Microbiol. 59*:695 (1993).
76. E. F. DeLong, G. S. Wickham, and N. R. Pace, *Science 243*:1360 (1989).
77. R. I. Amman, L. Krumholz, and D. A. Stahl, *J. Bacteriol. 172*:762 (1990).
78. H. G. van Faassen, and G. Lebbink, *Agric. Ecosyst. Environ. 51*:209 (1994).

10

Direct Approaches for Studying Soil Microbes

ANTON HARTMANN, BERNHARD AßMUS, GUDRUN KIRCHHOF, and MICHAEL SCHLOTER   GSF-National Research Center for Environment and Health, Neuherberg, Germany

1. INTRODUCTION

Understanding microorganisms in their habitats and their interactions with other biota and their chemical and physical environment is a major goal of ecological studies. For some time, soil microbiology generally was unable to work at this level. Methods for selective identification of certain microbes in their complex environment were not available or available only to a very limited extent. Most studies used cultivation techniques, such as plate counts or most probable number (MPN) protocols to enumerate microbial populations in soils. However, it became apparent that many microbes from soil environments do not respond readily to cultivation techniques (1) and therefore population data cannot be considered as accurate numbers. Depending on the soil under study, the culturable portion of microorganisms underestimates the resident population of active and quiescent microbes by at least one to two orders of magnitude.

Indirect quantification of microbial populations through their physiological activity is a widely used practice in soil microbiology. Microbial biomass is estimated by several methods, such as respiration; metabolic heat production;

adenylate energy content; adenosine triphosphate (ATP) content, enzymatic activities such as dehydrogenase and phosphatase activity, or soil fumigation techniques (2). At best, these methods provide a measure for the active or activatable population, but they cannot provide information on the exact population levels. In particular, spatial discrimination down to the micrometer level is not possible. Genetic methods using specific oligonucleotide primers for polymerase chain reaction (PCR) also are not able to indicate spatial resolution because they use extracts of deoxyribonucleic acid (DNA) or ribonucleic acid (RNA) of at least 0.5 g of soil or greater (3).

In this chapter we will discuss methods of direct in situ analysis of microbes in soils and related environments. In addition, we report on studies of microbial populations extracted soil suspensions to which cultivation or amplification techniques have not been applied. We discuss current methods of staining microbes, either nonspecifically with fluorescent dyes or specifically with marked antibodies, oligonucleotide probes, or introduced marker genes. The application of microscopic and cytometric methods such as scanning confocal laser microscopy and flow cytometry greatly aids studies of microbial populations in complex environments and is also mentioned briefly.

2. CONVENTIONAL STAINING METHODS FOR MICROORGANISMS

For direct enumeration of microorganisms, DNA-binding fluorochromes such as acridine orange (3,6-bis[dimethylamino]acridinium chloride [AO]) (4) and 4′, 6-diamidino-2-phenylindole (DAPI) (5) have been used for many years. Cell numbers derived from these DNA-binding fluorochromes usually show a reasonable correlation with viable counts when applied to growing laboratory cultures but can exceed plate counts by several orders of magnitude when used to examine bacterial populations in natural environments such as soil, sediment, and aquatic habitats (6). However, the whole microbial community, including the part that cannot be cultured, is of significant interest for the understanding of processes in soils and therefore is examined by direct observation.

The direct microscopic viable count (DVC) method (7) has been employed successfully to determine viable bacteria in environmental samples. In the DVC or Kogure method, samples are incubated with yeast extract and nalidixic acid for a period that is sufficient to induce growth. Yeast extract provides nutrients, while nalidixic acid, a DNA gyrase inhibitor, blocks DNA synthesis. Thus, cell replication is inhibited, while cell growth and enlargement are promoted. The result is the formation of cell filaments or enlarged cell forms in the case of substrate-responsive cells. Subsequent acridine orange staining of these preparations, followed by epifluorescence microscopy, allows enumeration of responsive cells to provide an estimate of the viable population. Using radiolabeled substrates such as methyl-[^{3}H]thymidine or [^{14}C]glutamic acid, a good correlation of cell-associated metabolic activity, measured by microautoradiography, and substrate responsiveness (DVC method) was obtained in survival experiments (6). In microautoradiography, the uptake and incorporation of radiolabeled compounds into microbial cells are determined by exposure to a highly sensitive film. However, when it is used in soil samples, the interpretation of the DVC method may be complicated by the occurrence of pleomorphic (variably shaped) and filamentous cell forms of indigenous bacteria. Furthermore, cell elongation or swelling may be too slow or too limited to obtain a reliable determination of substrate-responsive cells in environmental samples.

Alternative approaches for determining active microbes in soils employ the fluorogenic ester fluorescein diacetate (FDA) (8) or fluorescent redox probes such as 5-cyano-2,3-ditolyl tetrazolium chloride (CTC) (9). In the FDA method, metabolically active microbes hydrolyze FDA, yielding the bright green fluorescing molecule fluorescein. However, some bacteria are evidently unable to take up FDA efficiently and concomitantly show weak fluorescence development. In addition, hydrolysis of FDA is primarily a function of cellular esterase activity and does not necessarily provide information concerning the growth potential or respiratory status of a microbe in its soil habitat.

Like the redox dye 2-(p-iodophenyl)-3(p-nitrophenyl)-5-phenyltetrazolium chloride (INT), CTC allows direct microscopic visualization of actively respiring bacteria in native and nutrient-amended environmental samples. However, CTC produces a bright red fluorescent formazan (emission maximum 602 nm) when it is chemically or biologically reduced (9). Actively respiring microbes were readily distinguishable from abiotic particles and other background substances, which typically show fluorescence at a shorter wavelength. In addition to CTC, Rh 123, a cationic fluorescent dye which is accumulated in an uncoupler-sensitive fashion via transmembrane potential in bacteria, has been successfully used to assess physiological activity of bacteria rapidly in environmental samples (10).

The fluorescent brightener calcofluor white M2R (CFW), also called cellufluor, binds to polysaccharide structures of microbial cell surfaces and can thus be used for direct staining (4). Bacterial and fungal hyphae are stained bright blue using ultraviolet excitation and can readily be distinguished from other particles. However, the staining intensity was variable in a natural population of bacteria, and smaller coccoid, starving cells, and bacterial spores were unstained. It was demonstrated that CFW can be used in thin sections (15 μm thickness) to observe microbes in undisturbed soil (11). This method of preparation preserves the natural structure of soil and allows observation of the stained microorganisms and plant roots in soil samples. However, attempts to stain bacteria specifically in these thin soil sections with a fluorescein-coupled antiserum were unsuccessful, probably because of the properties of the resin used. The application of specific fluorescent stains in native or fixed but not embedded preparations of environmental samples is described in detail in later sections.

To facilitate efficient counting of cells in complex microbial communities, a program for use on an MS-DOS personal computer was developed (12). When it is applied to bacteria in soil smears, up to 10 subpopulations can be registered and results can be statistically processed in real time. During counting, three files are created and the numbers of individual cells are stored in a data file. The second file contains the coefficient of variation (CV) of the mean number per field, which is used to determine the optimum number of fields to be counted in order to obtain a given variance. The third file contains the final results and statistics.

3. MOLECULAR DETECTION METHODS FOR MICROORGANISMS IN SOIL

3.1 Immunological Techniques

Immunological detection methods are based on the ability of antibodies to recognize specific three-dimensional structures (e.g., parts of proteins or polysaccharides) of biological macromolecules. These techniques play an important role as diagnostic tools in medicine and food technology (13). In soil microbiology, they are becoming increasingly important for tracking of specific microorganisms and for microbial community analysis (14).

Monoclonal antibodies are produced in vitro by cell culture techniques (15): B-lymphocytes of an immunized animal, e.g., a mouse, are isolated and fused with myeloma cells, which enable them to grow continuously. One fused hybridoma clone produces one type of antibody. Polyclonal antisera (raised in animals, e.g. rabbits), in contrast, are produced by immunized animals and consist of different antibodies, as a result of the mixture of antigenic determinants mainly used for immunization (16).

Characterization of the quality of antibodies/antisera for in situ detection

Four features must be taken into account when considering whether an antibody or antiserum is suitable for in situ detection of microorganisms: cross-reactivity, affinity characteristics, and stability and localization of the antigenic determinants.

Cross-reactivity

Polyclonal antisera contain a variety of antibodies against typical and less typical structures of the target. The latter can lead to unspecific binding (cross-reaction) with strains of other species. The degree of cross-reactivity depends on the number of antibodies in the serum, which recognize nonspecific structures. If polyclonal antisera are produced, subfractions of microorganisms, e.g., cell wall extracts, are used for immunization (17). Schröder and Kunz (18) described a method to remove nonspecific compounds from polyclonal antisera. Use of monoclonal antibodies ensures less cross-reactivity with nontarget microorganisms, because these unique antibodies are directed against one antigenic determinant only. Depending on the distribution of the particular epitope (surface-exposed immunogenic group), a monoclonal antibody can show strain- (19), species- (20), or family-specificity (21). An antigenic epitope is a three-dimensional structure formed by a small number of amino acids or sugars, which is recognized by the antibody. However, even cross-reaction of strain-specific monoclonal antibodies must be taken into account because the organic fraction of soils, especially humic acids, contains a large variety of antigenic determinants, which can be identical to the antigenic epitopes (22).

Affinity characteristics

In the interaction of antibodies with the antigen after a definite period of time, an equilibrium is reached between the molecules present at the beginning of the reaction and the complex. With respect to the concentrations, the equilibrium state is characterized by the binding constant k_b (23), defined as the quotient of free antibody and antigen molecules in a solution compared to the molecules which formed an antigen–antibody complex. k_b, therefore, is an important quality characteristic of antibodies. As polyclonal antisera are mixtures of different antibodies, it is difficult to determine a k_b value for the whole serum. This is in contrast to monoclonal antibodies; the k_b of a monoclonal antibody with high affinity to the antigen can be about 10^{-10} M^{-1}, when optimal temperature and medium conditions for the immunological reaction are used.

Stability of antigenic determinants

As detection with antibodies is based on the bacterial phenotype, only a stable expression of antigenic determinants makes the detection sensitive and reliable. If an epitope is expressed only during certain stages of growth or under certain environmental conditions, the target microorganism can be detected only during these stages (22). Therefore, it is important that the recognized antigenic determinant be expressed constitutively, especially if monoclonal antibodies are used.

Localization of the antigenic determinants

Depending on the location of the antigenic determinant, microorganisms must be pretreated before immunological detection is possible; e.g., for the detection of the enterobacterial common antigen (ECA), Obst et al. (21) described a pretreatment of the bacteria by boiling water. As the ECA is not localized directly on the cell surface, this pretreatment is necessary to allow detection of the antigenic epitope by the antibody. If the antigenic determinant is localized directly on the cell surface, immunological detection of the microorganisms is possible without further treatment (24).

Detection methods

Agglutination techniques and agar precipitation tests

Agglutination techniques have visible effects during the formation of an antibody–antigen complex. These agglutination reactions are based on agglutination of the antigens with the antibody. In most cases antibodies are further coupled with latex particles, resulting in a better signal (25). The detection limit of these techniques is about 10^8 bacteria/ml (26). Because of the high detection limit, these techniques cannot be used without selective enrichment of the microorganism to be detected. Some examples of the application of this technique to detect bacteria in soil and rhizosphere samples are given in Table 1.

With the introduction of the agar double-diffusion precipitin test by Ouchterlony (27), use of agar precipitation tests (APTs) became widespread in microbiology. For example, APTs have been used to identify strains of *Pseudomonas syringae* (28) and *Pseudomonas pisi* (29). As the detection limit is no better than that of classic agglutination techniques, selective enrichment of the microorganisms is necessary when APT techniques are applied.

Immunofluorescence and immunogold staining for in situ detection of soil microorganisms

Immunofluorescence (IF) staining was suggested for identification of bacteria by Kikumoto and Sakamoto (30) to identify *Enterobacter aroideae* in host tissues of different vegetables and soil using epifluorescence microscopy. The antibodies can be coupled directly with a fluorescence marker (31) or marked with a secondary antibody, which binds to the F_c-part of the primary antibody and carries the fluorescence marker (32). Fluorescein and rhodamine derivatives are mainly used as fluorescence markers, which can be visualized by using an epifluorescence microscope and the corresponding wavelength filters (33). Immunofluorescence staining has equal sensitivity to PCR for the detection of target bacteria at a level of 10^3 to 10^4 cells (34). It can be used in soil extracts or for in situ studies of bacteria, e.g., on roots or substrate particles. To reduce non-specific binding of the antibody to soil or roots, Bohlool and Schmidt (35) preincubated the soil samples with 2% gelatin. The gelatin adsorped non-specifically to root or soil particles and could only be displaced in regions where the antibodies had an affinity to the specific epitope. Use of other protein solutions as blocking reagents has also been described (19,24). Immunofluorescence staining can be used for quantitative in situ studies but does not distinguish between culturable and nonculturable cells. Direct confirmation of positive cell results is not possible. Immunofluorescence colony (IFC) staining has been developed for the detection of culturable bacteria in the range of 10 to 100 bacteria/ml extract (36); it allows direct confirmation of positive culture results and can be efficiently applied for in situ detection of culturable bacteria (37). Fluorescence-labeled antibodies can also be used for flow cytometric analysis (see later discussion).

Immunogold detection is mainly used for in situ detection of bacteria with electron miroscopy techniques in root tissues (38). The fixed root material is embedded in resin and cut with a microtome. The sections are treated with the primary specific antibodies and a secondary antibody, which is labeled with gold particles. The detection limit is about 10^5 bacteria/g dried root. If immunogold detection is combined with silver enhancement techniques, light microscopy can be used (39). Some examples of the application of immunogold technique to detect specific bacteria in root tissue are given in Table 1.

Detection of bacteria with ELISA techniques

The most commonly used methods for the detection of bacteria with immunological techniques are based on enzyme-linked antibodies. There is a distinction between enzyme-linked immunosorbent assay (ELISA) and immunoblotting techniques. Bacteria from soil extracts are adsorbed to microtiter plates (ELISA) or nitrocellulose membranes (immunoblot) (24). After blocking nonspecific binding sites (discussed earlier) the primary specific antibody is added, followed by a secondary antibody, which carries one enzyme molecule per antibody. A direct coupling of the primary antibodies is not recommended, as the ideal 1:1 ratio of antibody to enzyme is difficult to achieve without loss of specific activity of either partner (40). Because of the intrinsic amplification of the signal by the enzyme action, even relatively low numbers of bound antibody molecules can be detected. The ideal enzyme should have a high catalytic activity and a range of substrates that yield both soluble products (for ELISA) and insoluble products (for immunoblotting). Enzymes linked to the antibodies are mainly peroxidase (24) or alkaline phosphatase (41). For all systems using enzyme-linked antibodies, detection is based on a substrate which is converted by the enzyme to a colored product. For detailed information about different substrates see Harlow and Lane (42).

The detection limit for ELISA depends on the antibody and the substrate for the enzyme. For polyclonal antisera and colorimetric detection of the bound enzyme, a detection limit of 10^6 bacteria/ml has been reported (43). When the streptavidin–biotin system was used as enhancement factor, an increase in sensitivity was reported (44). The streptavidin–biotin system combines the specificity of highly purified antibody probes with the high-affinity binding of the small water-soluble vitamin biotin to the bacterial protein streptavidin. The secondary antibody is coupled covalently with biotin. Streptavidin is labeled with a high number of enzymes, yielding a higher signal in ELISA and an increase in sensitivity that is due to the presence of more than one enzyme molecule per secondary antibody (see previous discussion). To track bacteria from soil extracts with high-affinity monoclonal antibodies and sensitive detection of the bound enzyme based on luminol and chemoluminescence (45), we observed a detection limit of 10^2 cells/ml (24). Hydrogen peroxide (H_2O_2) is converted by the antibody-bound peroxidase to water and oxygen radicals. These radicals oxidize luminol to aminophthalic acid. During this reaction photons are emitted. As the number of photon counts is proportional to the number of bacteria, this system can be used for a quantitative determination of bacteria in environmental samples such as soil and water. The sensitivity of the immunoblot is comparable to that of the classic ELISA, depending on the enzyme/substrate system in use (46).

Application to soil and related habitats

Table 1 gives an overview of the current status of antibodies developed for microorganisms which are important in soil, sediment, and the rhizosphere. In most cases, immunological techniques have been applied to soil to investigate the interaction between roots and plant-pathogenic or plant-growth-promoting (rhizo) bacteria (PGPR) or fungi in the rhizosphere. Mainly monoclonal antibodies have been used successfully. When the cross-reactivity of polyclonal antisera is below the detection limit or nonspecific antibodies have been removed successfully, good results are also obtained with polyclonal antisera. There are few data available for the application of immunological techniques in undisturbed soil or sediment to detect microorganisms. Some groups use the antibody approach to control the release of genetically engineered bacteria in soil (see Table 1).

Immunoseparation techniques

A serological method for immunoisolation of *Bacillus polymyxa* from root-free soil and from the rhizosphere of wheat was developed by Mavingui et al. (47). Microtiter plates were coated with anti-*Bacillus polymyxa* CF43 antibody. After the addition of soil samples and several incubation and washing steps, the bound bacteria were desorbed with 0.1 M KCl. Using classic methods of enumeration, the population size of *Bacillus polymyxa* was less than 0.1% of the total microflora. When the immunotrapping method was used for enrichment, this percentage was increased to 20%.

   Alun et al. (48) used an immunocapture technique based on antibodies coupled with (para)magnetic particles. This technique was originally developed for the isolation of lymphocyte subsets directly from blood samples. Alun et al. described the use of this method for the isolation of *Pseudomonas putida* directly from lake water by using bacterial flagella as a target antigen for capture. The paramagnetic particles had a size of 4.5 μm. The bacteria marked with the coupled antibody were isolated from the lake sample with a magnetic particle concentrator. Important considerations for this technique were not only the density

Table 1. Examples of Antibodies for Sediment and Soil Relevant Microorganisms and Their Applications

Antibody specificity	Application[c]	Detection limit	Reference
Acetobacter diazotrophicus[a]	IG, TEM	—	(51)
Agrobacterium tumefaciens[a]	IF	10^5 cells/ml	(52)
Azoarcus spp.[a]	ELISA, IF	10^2 cells/ml	(19)
Azospirillum brasilense[b]	IG	—	(53)
Bacillus anthracis[a]	IF	—	(54)
Bradyrhizobium japonicum[b]	TEM	—	(39)
Bradyrhizobium sp.[a]	ELISA	—	(55)
Burkholderia cepacia[b]	ELISA	10^2 cells/ml	(58)
Enterobacter aroidae[a]	IF	—	(30)
Enterobacter cloacae[a]	ELISA		(56)
Erwinia carotovora[a]	IFC, ELISA		(36;37)
Erwinia chrysanthemi[b]	IF	10^3 cells/ml	(37)
Flavobacterium sp.[b]	IF	20 cells/g dry weight	(32)
Gaeumannomyces graminis[a]	ELISA	10^6 cells/ml	(17)
Gliocladium roseum[b]	ELISA	—	(57)
Pantoea agglomerans[a]	ELISA,	10^6 cells/ml	(41)
Pseudomonas fluorescens[b]	ELISA	10^4 cells/ml	(59)
Pseudomonas pisi[a]	APT	10^8 cells/ml	(29)
Pseudomonas putida,[b]	ELISA, IB	10^2 cells/dot	(46)
Pseudomonas syringae[a]	APT	10^8 cells/ml	(28)
Pseudomonas syringae[a]	AT	—	(60)
Rhizobium japonicum[a]	ELISA, IB	10^6 cells/ml	(43)
Rhizobium spp.[a]	ELISA	10^5 cells/ml	(61)
Rhizoctonia solania[a]	IF	10^3 cells/ml	(62)
Thiobacillus ferrooxidans[b]	IB	10^3 cells/dot	(63)
Vibrio cholerae[a]	IF	10 cells/ml	(64)
Xanthomonas campestris[b]	IF	10^5 cells/ml	(65)

[a]Polyclonal antiserum
[b]Monoclonal antibody
[c]APT, agar precipitin test; AT, agglutination technique; ELISA, enzyme-linked immunsorbent assay; IB, immunoblot; IF, immunofluorescence; IFC, immunofluorescence colony staining; IG, immunogold; TEM, transmission electron microscope.

of the antigen on the cell surface but also the firmness of binding of the antigen to the bacterial surface needed to tolerate the shear forces during the washing steps. The recovery of target cells released into lake water at concentrations of 10^2 to 10^5 CFUs ml^{-1} was about 20%. The nonspecific binding of cells to uncoupled paramagnetic particles varied between 0.1% and 1%. Christensen et al. (49) used this technique to recover thermophilic sulfate-reducing bacteria from oil field waters. Antibodies directed against lipopolysaccharides of *Thermodesulfobacterium mobile* were coupled to paramagnetic particles with a size of 2.8 μm. The prevailing cells immunocaptured with this antibody were morphologically and serologically similar to *Thermodesulfobacterium mobile* type strain cells. *Thermodesulfobacterium mobile* was not detected in these oil field waters by classic isolation procedures. Extraction with antibody-coated paramagnetic particles allowed pure strains to be isolated directly from primary enrichment cultures without prior time-consuming subculturing and consecutive transfers to selective media. Dye (50) reported the use of immunomagnetic separation for the enrichment of *Rhizobium* strains from soil samples. By use of antibodies coupled to 2.8 μm paramagnetic particles and some optimization of Christensen's procedure, a 200-fold enrichment of *Rhizobium* was achieved.

3.2 Nucleic Acid Probes

Introduction

rRNA sequences, a source for specific probes

Comparative sequencing of ribosomal RNA (rRNA) molecules and their genes increasingly provides a molecular approach for prokaryotic phylogeny (66). The rRNA molecules, structural components of the ribosomes, are valuable indicators of phylogeny for the following reasons. They are essential elements of protein synthesis and therefore present in all living organisms. They seem to lack lateral gene transfer events that would blur evolutionary relationships. Higher-order structures similar throughout, secondary folding of the rRNA molecule in helical segments as a result of intramolecular interactions, and the tertiary structure responsible for protein binding are regarded as phylogenetically proven (67). The primary structures of 16S rRNA (approximately 1500 base length), 23S rRNA (between 2740 and 3130 base length), and, to a lesser extent, 5S rRNA (about 115 to 120 base length) exhibit segments conserved on genus up to domain levels (68,69). They also provide sequences unique to individual species or subspecies (67,70). These regions are excellent targets for synthetic oligonucleotide probes (single-stranded DNA stretches of 15 to 20 base length with base composition complementary to the target site) because of their natural amplification in the cell (a fast-growing bacterial cell harbors 10^3 to 10^5 ribosomes) and their variable specificity at different taxonomic levels (71). Probes directed to the ribosomal RNA have been successfully used in different fields of microbial ecology (72). It is possible to inspect whole community RNA for the occurrence of specific rRNA sequences, as demonstrated for phototrophic bacteria in a hot spring environment (73) and sulfate-reducing bacteria in coastal sediments (74). Specific rDNA (rRNA gene)–directed oligonucleotide probes are useful as primers in polymerase chain reactions (PCRs), providing evidence for the presence of certain taxa in total environmental DNA as proposed for streptomyces in soil DNA (7). In situ investigations of (microbial) biofilms (76), endosymbiontic bacteria in insects (77), and protozoa (78) or plant-colonizing bacteria (79) have been made possible by the development of single-cell detection with rRNA-targeted fluorescently labeled probes (80).

With a daily increasing data source and improved sequence retrieval, rRNA-based approaches bear immense potential for further analyses of community structures in the near future.

Construction of probes

Using the PCR (81) and sequencing techniques, it is now routine to obtain rRNA sequence data of both culturable and unculturable microorganisms within a few weeks (78,80,82,83). Specific probes can be designed through comparative analysis of these sequences. Reference data for rRNA sequence comparison are continually growing and are now easier to exploit through centralized databases like the GenBank, National Center for Biotechnology Information (Bethesda, Maryland, United States: http://www.ncbi.nlm.nih.gov) (84) and European Bioinformatics Institute (EBI) (Cambridge, England: http://www.ebi.ac.uk) (85) databases or more specialized sources, the Ribosomal RNA Database Projects at the University of Antwerp (Belgium: http://www.uiam3.uia.ac.be) (86) and the University of Illinois (Urbana, Illinois, United States: http://www.rdp.life.uiuc.edu) (87).

The now-widespread automated synthesis of oligonucleotides (88) facilitates production of optimally matching single-strand probes with the base composition of choice. Furthermore, oligonucleotide strands coupled with linkage groups (e.g., biotin), fluorescent labels (e.g., fluorescein, rhodamine), enzymes (e.g., peroxidase, alkaline phosphatase), or antigenic determinants (e.g., digoxigenin) are commercially available.

Probe design must take into consideration the avoidance of self-complementary stretches within one probe molecule and the selection of probe length (usually 15 to 20 bases) and base composition (optimally between 40% and 60% guanine plus cytosine [G + C] content) for common hybridization protocols.

Ecological restrictions in the application of phylogenetic probes

Phylogenetic probes are useful to elucidate the microbial structure of ecosystems. With a few exceptions, such as clusters of methylotrophic bacteria (89) and gram-negative sulfate-reducing bacteria (90), similar ecologically relevant functions are usually exhibited by microbial groups of polyphyletic origin. There-fore, the combined application of phylogenetic probes and function-specific gene probes as well as specific antibodies may be useful for the simultaneous as-sessment of bacterial community structure and physiological activities. In situ monitoring of messenger RNA (mRNA) levels of ecologically important func-tional genes in parallel with phylogenetic affiliation of microorganisms in their natural habitats will be useful in addressing specific physiological activities of autochthonous populations (80,91)

Whole cell hybridization with rRNA-directed oligonucleotide probes for detection of microorganisms in situ

Advantages and limitations of rRNA-directed fluorescently labeled probes

Since the first report on staining single microbial cells with fluorescence-labeled rRNA-directed oligonucleotides by DeLong et al. (82), this technique has rapidly spread throughout the field of microbial ecology (80). Its widespread ac-ceptance is due to a number of remarkable advantages provided by this strategy: These rRNA-derived probes, usually with a length of 15 to 20 nucleotides, allow detection of single microbial cells as well as their affiliation to a certain taxo-nomic unit (82,92) e.g., order, even if the microorganisms have never been cul-tured (78,93). Fluorescent dyes make target microorganisms visible by using an epifluorescence microscope, i.e., a light microscope equipped with a number of filter sets to select appropriate excitation and emission wavelengths for the re-spective dyes. Thus, the stained bacteria can be viewed directly, a less time-con-suming, more convenient method than detecting the binding of radioactively labeled probes after exposure using x-ray sensitive films. Finally, the amount of specifically hybridizing probe is directly correlated to the ribosome content of the target cells (82,94) and thus a measure of protein synthesis capacity (95). Al-though the data are still limited, the signal intensity of rRNA-targeted probes has been used in several investigations to estimate in situ metabolic activity (96,97).

However, there are also major restrictions on this technique when applied to microorganisms in their natural habitats: First, thick bacterial cell walls or the protein or polysaccharide capsules surrounding many bacteria may severely hamper penetration of probes into the cytoplasm. Reliable pretreatment of vari-ous gram-positive species which resist standard fixation procedures still has to be investigated; observation of endophytic bacteria in roots or nodules requires additional sample processing (79,98). Second, probes may fail to bind to their target site in situ, if access to this site is hindered by other ribosome components. Furthermore, if the physiological activity of the cells examined is too low, the fluorescence resulting from a specifically bound probe may be too weak for de-tection, especially in natural soil (96) or water (99) samples, whereas in sludge flocs from aerated basins of sewage treatment plants up to 90% of the bacteria stained by 4',6-diamidino-2-phenylindole (DAPI) can also be detected with a fluorescent bacterial rRNA probe (100). Plant material and organic soil particles tend to exhibit fluorescence without any staining. In many natural samples, this autofluorescence severely affects microscopic examination of hybridized samples, because it leads to blurring and a signal barely above the level of the background fluorescence (96,98). These problems may be overcome at least in part by new microscopic techniques such as scanning confocal laser microscopy (discussed later).

Strategies to enhance fluorescence signal and the use of other labels for nucleotide probes

Several attempts have been made to enhance the fluorescence intensity of labeled cells. Labeling of one oligonucleotide probe with more than one dye molecule did not improve signal intensity but increased background fluorescence (94). The simultaneous use of several fluorescently labeled oligonucleotides, each hybridizing to another target site, was more promising (94,101). However, depending on the target microorganism, it may be difficult to find several rRNA sequences that exhibit the desired level of specificity. Ribonucleic acid polynucleotide probes against rRNA sequences were synthisized by in vitro transcription of rDNA using a mixture of fluorescence or digoxigenin-labeled and unlabeled nucleotides for incorporation into the growing nucleotide chain (102). The probes seemed to work specifically and gave an enhanced signal (103), but no use in complex habitats was described.

Zarda et al. (104) have linked oligonucleotide probes to reporter molecules such as the steroids biotin and digoxigenin (DIG). Such probes can be detected after hybridization by a fluorescence-labeled specific antibody against the reporter steroid. As one antibody can be labeled with several dye molecules, the signal intensity is increased several-fold (104). Detection of probes can also be performed by a color-producing reporter enzyme, e.g., horseradish peroxidase or alkaline phosphatase. For these enzymes, substrates which result in a colored product after cleavage by the enzyme are commercially available. This can be done both directly, by enzyme-conjugated oligonucleotides (105), and indirectly, with a DIG-labeled probe, and the enzyme linked to the anti-DIG antibody (104). The use of such detection systems has been suggested for samples with elevated background fluorescence. The DIG-labeled probes have been shown to be superior to fluorescent probes in identifying *Frankia* strains in pure cultures and root nodule homogenates (98). However, for the high-molecular-weight molecules necessary for the alternative detection methods, e.g., enzymes or antibodies, penetration of the cell wall of target cells is much more difficult than it is for a fluorescence-labeled oligonucleotide and requires additional sample pretreatment, e.g., enzymatic cell wall digestion (98). Therefore, fluorescently labeled probes still are applied in most studies.

The latest development which includes both target amplification and phylogenetic labeling in one procedure at the single-cell level is the introduction of in situ PCR using fluorescence- or DIG-labeled nucleotides (106). This method has only recently been applied to bacteria but seems promising for circumventing the detection problems due to low ribosome content in many resting soil bacteria.

Current status of whole cell hybridization for identification and localization of soil and rhizosphere bacteria

Ribosomal RNA–directed probes have been used to detect bacteria in a wide range of complex habitats, e.g., aquatic microbial communities attached to surfaces (76,107), sediments (93,108), and sewage sludge flocs (100,109). However, little information has been reported from investigations of soil. Bacteria from laboratory cultures were fixed and mixed with soil extracts. Hybridizing such a mixture with oligonucleotides, Schleifer et al. (110) demonstrated that the detection of stained bacteria was not hampered by soil components when signal strength was sufficient. Hahn et al. (96) found that within natural soil microflora, only a small part of DAPI-stained cells could be detected after staining with a bacterial probe, and even for an introduced *Pseudomonas* sp. recovery by in situ hybridization after 1 hour was less than recovery by culturing. These authors, however, could significantly increase the number of labeled soil bacteria by activating growth with a small amount of liquid medium added to the soil sample and incubation for 16 hours (96). In our laboratory, we have investigated the colonization of the rhizosphere of wheat seedlings by *Azospirillum brasilense* (typical rhizosphere bacterium which can fix atmospheric nitrogen associated with the root) in tubes filled with unsterile soil. The bacteria could be detected and localized on root samples after several weeks by using fluorescence-labeled probes and scanning confocal laser microscopy (discussed later) (79). This suggests that at least in microenvironments with elevated metabolic activity, the direct investigation of soil bacteria with rRNA-targeted probes should be possible.

Table 2 provides an overview of the current status of the development of probes for bacteria which may be important in soil, sediment, and rhizosphere investigations and the reported use of these probes for whole cell hybridization in laboratory cultures or complex habitats. However, in the past the specificity of the suggested probes has not always been examined exhaustively. In more recent publications, the possibility of probe hybridization to rRNA of microorganisms other than the desired ones is taken into account by testing with a wide range of reference bacterial strains of different taxa and discussion of stringent hybridization conditions (e.g., 68). In some of these cases, the staining of undesired bacteria is reported. Probe ALF1b, for example, which had been developed to detect members of the alpha subclass of *Proteobacteria*, turned out also to cover some members of the delta subclass, as well as most representatives of the phylum *Spirochaeta* and relatives (68). So far, there has been no application of rRNA-targeted probes to monitor the occurrence of soil fungi. However, the first rRNA or rDNA sequence analyses of soil relevant fungi have been made (111,112), and specific rRNA properties have already been used for identification of endomycorrhizal fungi (113). Therefore, the extension of the in situ hybridization technique with rRNA-directed oligonucleotides to the field of fungal ecology can be expected soon.

The newly developed potential for mRNA detection in single bacterial cells with polynucleotide probes and in situ hybridization (114) is an advanced strategy relying on the same methodological principles as the phylogenetic probes discussed previously. It offers the possibility of monitoring gene expression in individual cells, thus increasing the information that a microscopic sample can yield. Hönerlage et al. (115) followed the expression of an extracellular protease in *Bacillus megaterium* in pure culture and after inoculation in sterilized soil.

Introduced reporter systems for in situ studies

Advantages and drawbacks

Besides the serologically or phylogenetically based in situ techniques already discussed, introduced marker genes (genes conferring distinctive physiological properties on a host microorganism for which a sensitive and well-defined detection assay is available) offer the possibility of monitoring microorganisms in natural ecosystems. Provided that the activity of the introduced gene is not exhibited by the indigenous population, several systems can be applied for direct in situ identification (130,131). One essential assumption for assaying microorganisms in situ is the use of methodologies which leave the bacteria and the environment undisturbed. Nucleic acid–based techniques with introduced marker genes, such as the *lux* system (bacterial bioluminescence genes of *Vibrio harveyi, Vibrio fischeri, Xenorhabdus luminescens*, or *Photobacterium leiognathi*) (132,133) and the *gusA* assay (*Escherichia coli* β-glucuronidase activity which oxidizes X-5-chloro-4-bromo-3-indolyl glucuronide [GlcA] to an indigo dye [134]), meet these requirements (135–137). The completely undestructive nature of light measurement with the *lux* system allows not only the determination of spatial distribution of microorganisms (138,139) but the recording of temporal changes in vivo (on-line monitoring, real time analysis) (140,141).

The outer membrane protein FhuA of *Escherichia coli* has been used as a reporter protein, which is detectable on the bacterial cell surface by the monoclonal antibody Fhu3.1. This allows quantification of expression with ELISA techniques as well as in situ detection with fluorescently labeled antibodies. After fusion of the *fhuA* gene with the *umuD* promoter responding to activation by the SOS operon, this reporter protein was successfully used to indicate genotoxic stress (142). Recombinant reporter plasmids harboring the *umu–fhuA* operon fusion were successfully transformed into different soil bacteria, which can also be

Table 2. Examples of rRNA-Targeted Oligonucleotides for Sediment and Soil Relevant Microorganisms and Their Applications in Whole Cell Hybridization

Probe specificity	Reference for sequence	Applications in whole cell hybridization
Universal	(116)	Cell cultures (82)
Domain *Archaea*	(82)	Cell cultures (82)
Domain *Bacteria*	(101)	Biofilms (76), sediments (93), soil (96)[a], rhizosphere (79)
Domain *Eucarya*	(101)	Sea water (117)
Different orders and families of methanogenic *Archaea*	(118)	Partially in anaerobic sewage sludge (119)
Alpha subclass of *Proteobacteria*	(68)	Activated sludge (100), rhizosphere (79)
Beta subclass of *Proteobacteria*	(68)[b]	Biofilms (120), activated sludge (100)
Gamma subclass of *Proteobacteria*	(68)[b]	Biofilms (120), activated sludge (100)
Gram-positive bacteria with high GC content	(121)[b]	Activated sludge (109)
Different groups of *Streptomyces* and *Streptoverticillium* species	(75)	Not reported
Genus *Frankia*	(122)	Nodule homogenates (98)
Different ecotypes of *Frankia*	(123)	Not reported
Bacillus macerans	(124)	Cell cultures (124)

Bacillus polymyxa	(124)	Cell cultures (124)
Different groups and species of genuine pseudomonads and related organisms	(110)	Cell cultures (110), *P. diminuta*–specific probe in soil extracts (110)[a]
Burkholderia cepacia	(104)[b]	Soil (96)[a]
Methylotrophic bacteria (serine pathway)	(89)	Cell cultures (89)
Methylotrophic bacteria (RuMP pathway)	(89)	Cell cultures (89)
Different subtypes of methylotrophic bacteria	(125)	Not reported
Gram-negative sulfate-reducing bacteria	(101)	Biofilms (76)
Genus *Desulfobacter*	(101)	Biofilms (107)
Desulfovibrio vulgaris	(76)	Biofilms (104)
Shewanella putrefaciens	(108)	Sediments (108)
Bradyrhizobium japonicum	(127)[b]	Not reported
Genus and five species of *Azospirillum*	(128)[b]	*A. basilense*–specific probe in rhizosphere (79)
Genus *Azoarcus*	(129)	Not reported
Acinetobacter calcoaceticus	(109)	Activated sludge (109)
Cytophaga–Flavobacterium cluster	(109)	Activated sludge (109)

[a]See text for details.
[b]Target sequence located on 23S rRNA.

used as in situ biomonitoring bacteria. Further applications may well be developed by inserting other ecologically relevant promoters upstream of the *fhuA* gene.

The green fluorescent protein (GFP) seems to be another promising reporter for in situ studies. The gene encoding this protein (*gfp*) is isolated from the bioluminescent jellyfish *Aequorea victoria* and emits green light (maximum at 509 nm) during excitation with blue light (maximum at 395 nm) (143). Green fluorescent protein requires only irradiation of blue light and is not limited by availability of substrates, as the *lux* and *gusA* assays are. Its fluorescence persists after treatment with formaldehyde; therefore, fixed preparations can also be examined. This stable fluorescence (no photobleaching is observed) of the *gfp* gene product is capable of being expressed in heterologous organisms and can be used in expression level studies as well as in subcellular localization of proteins. For example, fusion of *gfp* and *cotE*, a gene of a coat protein of *Bacillus subtilis*, results in the production of the CotE–GFP fusion protein in a recombinant *B. subtilis* strain. Fluorescence can be observed at different cell sites with varying intensities during sporulation (144). So far, however, no studies in natural habitats have been published, but this reporter system seems to offer new potential for in situ investigations.

The application of introduced marker gene systems is restricted to culturable organisms with well-established recombination techniques (transformation with plasmids, transposons) since the reporter gene cassette has to be introduced artificially into the genome of the microorganism of interest.

Applications and detection limits

Examples of the application of the marker systems discussed in soil and rhizosphere and their reported detection sensitivities are listed in Table 3. While working with bioluminescence in soil, it was observed several times that the sensitivity decreased one order of magnitude as a result of quenching of light by soil particles (131,145). One promising advance to overcome this drawback may be the use of charge-coupled device (CCD) enhanced microscopy (146), which is further discussed later.

4. NEW MICROSCOPIC TECHNIQUES AND IMAGE ANALYSIS

4.1 Image Analysis and the Use of CCD Cameras

The introduction of image processing and analysis broadened the field of microscopic examinations in microbial ecology a few years ago. The benefit provided by these computer-assisted methods ranges from simple contrast enhancement to automated cell counting and determination of biomass (158). Any computing

Table 3. Application of Introduced Marker Genes in Soil and Rhizosphere in situ Investigations

Reporter system	Object of study	Detection limit	Reference
gusA Gene	Plant–microbe interactions		(147)
	Root colonization and nitrogenase activity		(135)
	Temporal and spatial regulation of symbiotic genes of Rhizobium		(148)
	Rhizobial ecology		(149)
	Fungal growth in roots		(130)
lux Genes	Plant and root colonization	10^3 CFU/g seed	(150)
	Monitoring of biocontrol bacteria	10^4 Cells/g soil	(141)
	Detection of GMM[a]s in soil	Single cells	(146)
	Detection of E. coli in soil	10^3 Cells/g soil	(131)
	Effects of matric potential on survival and activity of GMM[a]s in soil		(151)
	Phytopathogenic processes	Single cells	(140)
	Persistence and movement of plant surrounding soil	50 Cells/g soil	(152)
	Detection of Bacillus subtilis in soil		(153)
	Enumeration of bacteria in soil	10^4 CFU/g soil	(154)
	Bacterial association in rotting processes	10^3 Cells/g soil	(155)
	Root colonization by Pseudomonas spp	10^3–10^4 CFU/g soil	(139)
	Activity measurement of Pseudomonas fluorescens in soil	10^2–10^3 Cells/g soil	(156)
	Protozoal grazing on a GMM[a] inoculum		(157)
fhuA Gene	Monitoring of ecotoxicological responses		(142)
gfp Gene	In development stages		(144)

[a]GMM, genetically modified microorganism

analysis of photomicrographs needs a digitized image, which can be obtained by converting images recorded with a video camera. Sieracki et al. (159) used this method to enumerate planktonic bacteria and compared different algorithms for accurate cell sizing (160). A similar approach has been used to estimate cell sizes of bacteria in different soils (161). For the biomass determination of fungal hyphae in soil and on synthetic polymers, adequate software routines are required (162). Jurtshuk et al. (124) digitized epifluorescence images by using an image cytometer to quantify fluorescence intensity of *Bacillus* cells conferred by rRNA-targeted probes.

Comparison of different analog-to-digital converters revealed that CCD cameras were superior to video-based systems in detecting low signals of small fluorescence-stained bacterial cells (163). These cooled and slowly scanning cameras (164) provide both enhanced sensitivity and geometric stability (i.e., reproducibility of the exact localization of a certain object in repeated scans) and therefore significantly improve the signal-to-noise ratio. These cameras have been used to detect genetically modified bacteria in plant material and have reached a sensitivity limit of 1.5×10^4 colony-forming units (CFUs) per leaf (152). In another experiment, even single cells carrying the *lux* marker gene could be detected after inoculation in soil samples (146).

4.2 Scanning Confocal Laser Microscopy and Three-Dimensional Study of Bacterial Communities and Habitats

When viewing thick samples in conventional microscopy, results are frequently affected by the limited focus area. Exact localization of investigated microorganisms is impossible in many cases. The detection of fluorescence signals is often hampered by high levels of background fluorescence conferred by, e.g., plant material, other organic matter (OM), or clay particles. Optical sectioning methods have been discussed as an alternative to the thin-sectioning of specimens, which may introduce various artifacts. Elimination of out-of-focus information can be accomplished by deconvolution (165,166): A series of images from a sample is recorded conventionally in different focal planes. Then the object is reconstructed by applying mathematical operations that consider the point-spread functions of light from an illuminated source. This strategy, however, although proposed to be equally powerful at a lower cost level (167), has been outcompeted by the use of scanning confocal laser microscopy (SCLM). The physical properties (168) as well as biological applications of this technique (169) have been described recently. Briefly, single spot illumination by a focused laser beam as well as a pinhole allowing only in-focus signals to be detected lead to an accurate optical section of a specimen and virtually eliminate blur originating from unfocused parts. The instruments available at the present time use software that provides image processing routines as well as the ability to reconstruct an object that has been analyzed by a series of optical sections three-dimensionally. Scanning laser microscopy has been used to study complex microbial biofilms (170,171) or to detect bacteria in the rhizoplane (172). Bloem et al. (173) have determined cell numbers and volumes of bacteria as well as frequencies of dividing cells in soil smears with SCLM and an adapted software package and thus have provided the most direct approach for estimating bacterial biomass and activity to date. The combination of SCLM with fluorescence-labeled molecular probes allows in situ monitoring of specifically identified bacteria. By this means, microbial communities in activated sludge flocs have been analyzed (174). In our group, colonization patterns of *Azospirillum brasilense* strains in the rhizosphere have been compared by using SCLM and monoclonal antibodies (33) or rRNA-directed probes (79). For the first time, a nondestructive technique for identification and exact localization of microorganisms in complex habitats is now available (Fig. 1). Recently SCLM has also been recommended for fungus research (175).

5. FLOW CYTOMETRIC ANALYSIS OF MICROBIAL POPULATIONS

5.1 Principles of Flow Cytometry

Flow cytometry (FC) is an appropriate tool to study microbial populations provided in the form of a suspension of single cells. In a flow cytometer, a stream of cells moving in single file crosses a focused laser beam. A number of arranged photomultiplier tubes record a couple of parameters simultaneously, e.g., scattered light or the signals of fluorescent stains and probes. Although recommended several times (176,177), FC is rarely applied in microbial ecology, because of its relatively high cost and the fact that bacterial size is close to the detection limits of most instruments used in cell biology (177).

5.2 Applications in Microbial Ecology

Flow cytometry requires no prior cultivation of microorganisms and offers great advantages in characterizing mixed microbial communities because several thousand single cells can be analyzed in a few minutes, yielding much better statistical information than microscopic counting. Most applications reported are in aquatic microbial ecology, where FC allows both the monitoring of an introduced microorganism (178) and the investigation of natural microflora (179,180). Recently the use of FC in aquatic microbiology has been reviewed (181), as well as its use in other microbiological applications (182,183).

The contribution of a specific microorganism to a whole community can be

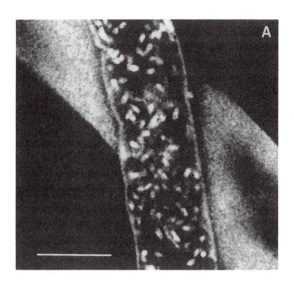

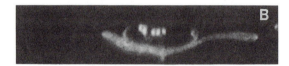

Figure 1. Scanning confocal laser microscopy (SCLM) images of a wheat root hair, internally colonized by bacteria. The sample was hybridized with a ribosomal ribonucleic acid–(rRNA) targeted probe covering bacteria which was linked to tetramethylrhodamine-isothiocyanate (TRITC). Excitation wavelength 543 nm. A: Single optical section. B: Optical cross section (z-scan) of the same root hair, close to the center of the image in panel A, demonstrating the location of the bacteria inside the lumen of the root hair cell. Bar applies to both panels and equals 10 µm.

determined by FC, if this microorganism is specifically stained, for example, with fluorescence-labeled oligonucleotide probes (94,101,184) or antibodies (178,185). Specific labeling is also a prerequisite for specific cell sorting and enrichment, which can also be performed by a flow cytometer (185).

Little information is available on flow cytometric analysis of soil microorganisms. Page and Burns (186) have described the recovery of an inoculated *Flavobacterium* from soil by antibody labeling and FC. However, the cell numbers achieved were lower by one order of magnitude than in direct counts, and thus the authors concluded that further development of the method was required. In our own experience, the recovery of a significant part of soil microbes in suspension and separation from soil particles are the crucial steps in sample preparation for flow cytometry.

6. CONCLUSIONS

By using techniques for in situ detection of single microbial cells, studies of microbial populations in their microhabitats including interactions with other organisms and their environment became possible. Because soil habitats are complex and diverse, few applications have yet been reported. Inherent problems are still the high percentage of quiescent cells, especially in nutrient-poor soil habitats, and the high numbers of microorganisms not yet described. Using molecular approaches, a steadily growing variety of probes will be available to stain certain uncultivated microbes specifically so that successful cultivation and physiological characterization may be achieved. Care has to be taken that the rRNA-directed probes are optimally designed for efficient in situ binding. Interpretation of in situ labeling experiments has to take into account that very different activity levels of bacteria exist and that in complex communities unexpected hybridization may occur. Since the population of soil microbes is very diverse, the number of cells belonging to specific groups is usually expected to be small. The single-cell-based microscale identification is limited by severe sensitivity, statistical, and representativity limitations, because colonizable soil surfaces are large and microhabitat characteristics are very diverse. This is also important for studies of the fate of specific microbes, e.g., genetically engineered strains, which cannot cross a relatively low sensitivity margin. Therefore, direct approaches to study microbes in soil have to be supported by other techniques and may be limited to active microhabitats. Nevertheless, the combination of general and activity stains, including reporter constructs, fluorescently labeled oligonucleotide probes, and antibodies, with digital imaging microcopy such as scanning confocal laser microscopy opens a new and fascinating area for ecological studies of microbial populations in soil habitats.

REFERENCES

1. Torsvik, J. Goksoyr, and F. L. Daae, *Appl. Environ. Microbiol.* 56:782 (1990).
2. A. Paul and F. E. Clark, *Soil Microbiology and Biochemistry*, Academic Press, San Diego (1988).
3. W. Pickup, *J. Gen. Microbiol. 137*:1009 (1991).

4. W. West, *J. Microbiol. Meth.* 5:125 (1986).
5. H. Huber, G. Huber, and K. O. Stetter, *System. Appl. Microbiol.* 6:105 (1985).
6. B. Roszak and R. R. Colwell, *Appl. Environ. Microbiol.* 53:2889 (1987).
7. K. Kogure, U. Simidu, and N. Taga, *Arch. Hydrobiol.* 102:117 (1984).
8. B. Lundgren, *Oikos* 36:17 (1981).
9. G. Rodriguez, D. Phipps, K. Ishiguro, and H. F. Ridgway, *Appl. Environ. Microbiol.* 58:1801 (1992).
10. P. Yu and G. A. McFeters, *J. Microbiol. Meth.* 20:1 (1994).
11. J. Postma and H.-J. Altemüller, *Soil Biol. Biochem.* 22:89 (1990).
12. J. Bloem, D. K. Van Mullem, and P. R. Bolhuis, *J. Microbiol. Meth.* 16:203 (1992).
13. S. Kerr, H. I. Ball, D. P. Mackie, D. A. Pollock, and D. A. Finlay, *J. Appl. Bacteriol* 72:302 (1992).
14. N. W. Schaad, *Annu. Rev. Phytopathol.* 17:123 (1979).
15. G. Köhler and C. Milstein, *Nature* 256:495 (1975).
16. M. Chase, *Meth. Immunol. Immunochem.* 1:197 (1967).
17. H. El-Nassar, L. L. Moore, and R. George, *Phytopathol.* 76:1319 (1986).
18. J. Schröder and W. Kunz, *Bio Forum* 7:235 (1991).
19. M. Schloter, S. Moens, C. Croes, G. Reidel, M. Esquenet, R. DeMot, A. Hartmann, and K. Michiels, *Microbiol.* 140:823 (1994).
20. Y. Bagger *APMIS* 99:972 (1991).
21. U. Obst, I. Huebner, M. Wecker, and D. Bitter-Suermann, *Aqua* 38:136 (1989).
22. S. C. Jackman, H. Lee, and J. T. Trevors, *Microb. Releases* 2:120 (1993).
23. N. Ermakov, A. Ishkov, E. Miroshnichenko, O. Plyusheh, E. Reminzova, P. Svershnikov, and E. Chenchikova, *Biokhimiya* 55:2163 (1990).
24. M. Schloter, W. Bode, A. Hartmann, and F. Beese, *Soil Biol. Biochem.* 24:399 (1992).
25. R. Hampton, E. Ball, and D. DeBoer, *Serological Methods for Detection and Identification of Viral and Bacterial Plant Pathogens* (S. DeBoer, ed.), APS Press, St. Paul, Minnesota, p. 389 (1990).
26. Z. Klement, *Methods in Phytobacteriology* (D. Sands, ed.), Akademia Kiado, Budapest, p. 568 (1990).
27. Ö. Ouchterlony, *Kemi. Mineral Geol.* 26:1 (1949).
28. T. Tominaga, *Bull. Natl. Inst. Agric. Sci. Jpn. Ser. C.* 205 (1971).
29. J. Taylor, *NZ J. Agric. Res.* 15:421 (1972).
30. T. Kikumoto and M. Sakamoto, *Ann. Phytopathol. Soc. Jpn.* 33:181 (1967).
31. J. Goding, *J. Immunol. Meth.* 20:215 (1976).
32. J. Mason and R. G. Burns, *FEMS Microbiol. Ecol.* 73:299 (1990).

33. M. Schloter, R. Borlinghaus, W. Bode, and A. Hartmann, *J. Microsc.* *171*:173 (1993).
34. J. Van Vuurde and N. Roozen, *J. Plant Pathol. 96*:75 (1990).
35. B. Bohlool and E. Schmidt, *Advances in Microbial Ecology* (M. Alexander, ed.), Academic Press, London, p. 203 (1980).
36. M. Leeman, J. Raaijmakers, P. Bakker and B. Schippers, *Biotic Interactions and Soil Borne Diseases* (A. Beemster, B. Bollen, M. Gerlagh, M Ruissen, B. Schippers, and A. Tempel, eds.), Elsevier, Amsterdam, p. 374 (1991).
37. H. Underberg and J. Van Vuurde, *Proc. 7th Int. Conf. Plant Pathol. Bact.* (Z. Klement, ed.), Akademia Kiado, Budapest, p. 937 (1990).
38. H. Levanony and Y. Bashan, *Curr. Microbiol. 18*:145 (1989).
39. E. K. James, J. I. Sprent, F. R. Michin, and N. J. Brewin, *Plant Cell Environ. 14*:467 (1991).
40. A. G. Farr and P. K. Nakane, *J. Immunol. Meth. 47*:129 (1981).
41. S. Ruppel, C. Hecht-Buchholz, R. Remus, U. Ortmann, and R. Schmelzer, *Plant Soil 145*:261 (1992).
42. E. Harlow and D. Lane, *Antibodies*, Cold Spring Harbour Laboratory, New York (1988).
43. S. Asanumu, G. Thottappilly, A. Ayanaba, and R. Ranga, *Can. J. Microbiol. 31*:524 (1985).
44. H. Levanony and Y. Bashan, *Curr. Microbiol. 20*:91 (1990).
45. G. H. G. Thorpe and L. Krika, *Bioluminescence and Chemoluminescence*, Springer, New York (1987).
46. M. I. Ramos-Gonzalez, F. Ruiz-Cabello, I. Brettar, F. Garrido, and J. L. Ramos, *J. Bacteriol. 174*:2978 (1992).
47. P. Mavingui, O. Berge, and T. Heulin, *Symbiosis 9*:215–221 (1990).
48. A. Alun, L. Hellmann, and F. Vartdal, *J. Clin. Microbiol. 26*:2572–2575 (1988).
49. B. Christensen, T. Torsvik, and T. Lien, *Appl. Environ. Microbiol. 58*:1244 (1992).
50. M. Dye, *J. Microbiol. Meth. 19*:235 (1994).
51. E. K. James, V. M. Reis, F. L. Olivares, J. I. Baldani, and J. Döbereiner, *J. Exp. Bot. 45*:757 (1994).
52. C. Depierreux, M. T. LeBris, M. F. Michel, B. Valeur, M. Monsigny, and F. Delmotte, *FEMS Microbiol. Lett. 67*:237 (1990).
53. T. Hurek, B. Reinhold-Hurek, M. vanMontagu, and E. Kellenberger, *J. Bacteriol. 176*:1913 (1994).
54. A. C. Phillips and K. M. Martin, *J. Immunol. Meth. 106*:109 (1988).
55. M. Sinclair and A. Eaglesham, *Soil Biol. Biochem. 22*:69 (1990).
56. J. C. Pedersen and C. S. Jacobsen, *Appl. Environ. Microbiol. 59*:1560 (1993).

57. C. Breuil, B. T. Luck, L. Rossignol, J. Little, C. J. Echeverri, and D. L. Brown, *J. Gen. Microbiol. 138*:3211 (1992).

58. K. Tsuchiya, Y. Takahashi, K. Shohara, Y. Homma, and T. Suzui, *Ann. Phytopathol. Soc. Jpn. 42*:247 (1995).

59. O. Nybroe, A. Johansen, and M. Laake, *Lett. Appl. Microbiol. 11*:293 (1990).

60. L. Lovrekovich and Z. Klement, *Phytopathol. 47*:19 (1963).

61. C. Scholz, R. Remus, and R. Zielke, *Zentralbl. Mikrobiol. 146*:197 (1991).

62. C. R. Thornton, F. M. Dewey, and C. A. Gilligan, *Plant Pathol. 42*:763 (1993).

63. R. Arredondo and C. A. Jerez, *Appl. Environ. Microbiol. 55*:2025 (1989).

64. R. P. Bryton and R. R. Colwell, *J. Microbiol. Meth. 6*:309 (1987).

65. A. Alvarez, A. A. Benedict, and C. Y. Mizumoto, *Phytopathol. 75*:722 (1985).

66. J. Olsen, C. R. Woese, and R. Overbeeck, *J. Bacteriol. 176*:1 (1994).

67. W. Ludwig and K.-H. Schleifer, *FEMS Microbiol. Rev. 15*:155 (1994).

68. W. Manz, R. I. Amann, W. Ludwig, M. Wagner, and K. -H. Schleifer, *System. Appl. Microbiol. 59*:1647 (1992).

69. S. Winker and C. R. Woese, *System. Appl. Microbiol. 14*:305 (1991).

70. D. A. Stahl, B. Flesher, H. R. Mansfield, and L. Montgomery, *Appl. Environ. Microbiol. 54*:1079 (1988).

71. D. A. Stahl and R. I. Amann, *Nucleic Acid Techniques in Bacterial Systematics* (E. Stackebrandt and M. Goodfellow, eds.), John Wiley & Sons, New York, pp. 205–248 (1991).

72. D. M. Ward, M. M. Bateson, R. Weller, and A. L. Ruff-Roberts, *Advances in Microbial Ecology*, Vol. 12 (K. C. Marshall, ed.), Plenum Press, New York, pp. 219–286 (1992).

73. A. L. Ruff-Roberts, J. G. Kuenen, and D. M. Stahl, *Appl. Environ. Microbiol. 60*:697 (1994).

74. R. Devereux and G. W. Mundform, *Appl. Environ. Microbiol. 60*:3437 (1994).

75. D. Witt, W. Liesack, and E. Stackebrandt, *Recent Advances in Microbial Ecology* (T. Hattori, Y. Ishida, Y. Maruyama, R. Y. Morita, and A. Uchida, eds.), Japan Scientific Societies Press, Tokyo, pp. 679–684 (1989).

76. R. I. Amann, J. Stromley, R. Devereux, R. Key, and D. A. Stahl, *Appl. Environ. Microbiol. 58*:614 (1992).

77. P. Baumann, M. A. Munson, C.-Y. Lai, M. A. Clark, L. Baumann, N. A. Moran, and B. C. Campbell, *ASM News 59*:21 (1993).

78. R. I. Amann, N. Springer, W. Ludwig, H.-D. Görtz, and K.-H. Schleifer, *Nature 351*:161 (1991).

79. B. Aβmus, P. Hutzler, G. Kirchhof, R. Amann, J. R. Lawrence, and A. Hartmann, *Appl. Environ. Microbiol. 61*:1013 (1995).

80. R. I. Amann, W. Ludwig, K. -H. Schleifer, *Microb. Rev. 59*:143 (1995).
81. R. K. Saiki, D. H. Gelfand, S. Stoffel, S. J. Scharf, R. Higuchi, G. T. Horn, K. B. Mullis, and H. A. Ehrlich, *Science 293*:487 (1988).
82. E. F. DeLong, G. S. Wickham, and N. R. Pace, *Science 243*:1360 (1989).
83. R. Weller, J. W. Weller, and D. M. Ward, *Appl. Environ. Microbiol. 57*:1146 (1991).
84. D. Benson, M. Boguski, D. J. Lipman, and J. Ostell, *Nucl. Acids Res. 22*:3441 (1994).
85. D. B. Emmert, P. J. Stoehr, G. Stoessner, and G. N. Cameron, *Nucl. Acids Res. 22*:3445 (1994).
86. J.-M. Neefs, Y. Van De Peer, P. De Rijk, A. Goris, and R. De Wachter, *Nucl. Acid. Res. 19*:1987 (1991).
87. G. J. Olsen, N. Larsen, and C. R. Woese, *Nucl. Acid Res. 19*:2017 (1991).
88. M. D. Mateucci and M. H. Caruthers, *J. Am. Chem. Soc. 103*:3185 (1981).
89. C. Tsien, B. J. Bratina, K. Tsuji, and R. S. Hanson, *Appl. Environ. Microbiol. 56*:2858 (1990).
90. R. I. Amann, B. J. Binder, R. J. Olson, S. W. Chisholm, R. Devereux, and D. A. Stahl, *Appl. Environ. Microbiol. 56*:1919 (1990).
91. D. A. Stahl and M. D. Kane, *Curr. Opin. Biotechnol. 3*:244 (1992).
92. R. I. Amann, L. Krumholz, and D. A. Stahl, *J. Bacteriol. 172*:762 (1990).
93. S. Spring, R. Amann, W. Ludwig, K.-H. Schleifer, and N. Petersen, *System. Appl. Microbiol. 15*:116 (1992).
94. G. Wallner, R. Amann, and W. Beisker, *Cytometry 14*:136 (1993).
95. P. F. Kemp, *Microb. Ecol. 28*:159 (1994).
96. D. Hahn, R. I. Amann, W. Ludwig, A. D. L. Akkermans, and K.-H. Schleifer, *J. Gen. Microbiol. 138*:879 (1992).
97. L. K. Poulsen, G. Ballard, and D. A. Stahl, *Appl. Environ. Microbiol. 59*:1354 (1993).
98. D. Hahn, R. I. Amann, and J. Zeyer, *Appl. Environ. Microbiol. 59*:1709 (1993).
99. R. E. Hicks, R. I. Amann, and D. A. Stahl, *Appl. Environ. Microbiol. 58*:2158 (1992).
100. M. Wagner, R. Amann, H. Lemmer, and K. -H. Schleifer, *Appl. Environ. Microbiol. 59*:1520 (1993).
101. R. I. Amann, B. J. Binder, R. J. Olson, S. W. Chisholm, R. Devereux, and D. A. Stahl, *Appl. Environ. Microbiol. 56*:1919 (1990).
102. W. Ludwig, S. Dorn, N. Springer, G. Kirchhof, and K.-H. Schleifer, *Appl. Environ. Microbiol. 60*:3236 (1994).
103. K. Trebesius, R. Amann, W. Ludwig, K. Mühlegger, and K.-H. Schleifer, *Appl. Environ. Microbiol. 60*:3228 (1994).
104. B. Zarda, R. Amann, G. Wallner, and K.-H. Schleifer, *J. Gen. Microbiol. 137*:2823 (1991).

105. R. I. Amann, B. Zarda, D. A. Stahl, and K.-H. Schleifer, *Appl. Environ. Microbiol. 58*:3007 (1992).
106. R. E. Hodson, W. A. Dustman, R. P. Garg, and M. A. Moran, *Appl. Environ. Microbiol. 61*:4074 (1995).
107. N. B. Ramsing, M. Kühl, and B. B. Jörgensen, *Appl. Environ. Microbiol. 59*:3840 (1993).
108. T. J. DiChristina and E. F. DeLong, *Appl. Environ. Microbiol. 59*:4152 (1993).
109. M. Wagner, R. Erhart, W. Manz, R. Amann, H. Lemmer, D. Wedi, and K.-H. Schleifer, *Appl. Environ. Microbiol. 60*:792 (1994).
110. K.-H. Schleifer, R. Amann, W. Ludwig, C. Rothemund, N. Springer, and S. Dorn, *Pseudomonas: Molecular Biology and Biotechnology* (S. Galli, S. Silver, and B. Wisholt, eds.), American Society for Microbiology, Washington, D.C., pp. 127–134 (1992).
111. R. N. Nazar, X. Hu, J. Schmidt, D. Culham, and J. Robb, *Physiol. Mol. Plant Pathol. 39*:1 (1991).
112. L. Simon, M. Lalonde, and T. D. Bruns, *Appl. Environ. Microbiol. 58*:291 (1992).
113. L. Simon, R. C. Levesque, and M. Lalonde, *Appl. Environ. Microbiol. 59*:4211 (1993).
114. D. Hahn, R. I. Amann, and J. Zeyer, *Appl. Environ. Microbiol. 59*:2753 (1993).
115. W. Hönerlage, D. Hahn, and J. Zeyer, *Arch. Microbiol. 163*:235 (1995).
116. S. J. Giovannoni, E. F. DeLong, G. J., Olsen, and N. R. Pace, *J. Bacteriol. 170*:720 (1988).
117. E. L. Lim, L. A. Amaral, D. A. Caron, and E. F. DeLong, *Appl. Environ. Microbiol. 59*:1647 (1993).
118. L. Raskin, J. M. Stromley, B. E. Rittmann, and D. A. Stahl, *Appl. Environ. Microbiol. 60*:1232 (1994).
119. L. Raskin, L. K. Poulsen, D. R. Noguera, B. E. Rittmann, and D. A. Stahl, *Appl. Environ. Microbiol. 60*:1241 (1994).
120. W. Manz, U. Szewzyk, P. Ericsson, R. Amann, K.-H. Schleifer, and T.-A. Stenström, *Appl. Environ. Microbiol. 59*:2293 (1993).
121. C. Roller, M. Wagner, R. Amann, W. Ludwig, and K. -H. Schleifer, *Microbiol. 140*:2849 (1994).
122. D. Hahn, M. J. C. Starrenburg, and A. D. L. Akkermans, *Appl. Environ. Microbiol. 56*:1342 (1990).
123. D. Hahn, M. Dorsch, E. Stackebrandt, and A. D. L. Akkermans. *Plant Soil 118*:211 (1989).
124. R. J. Jurtshuk, M. Blick, J. Bresser, G. E. Fox, and P. Jurtshuk Jr., *Appl. Environ. Microbiol. 58*:2571 (1992).
125. G. A. Brusseau, E. S. Bulygina, and R. S. Hanson, *Appl. Environ. Microbiol. 60*:626 (1994).

126. M. D. Kane, L. K. Poulsen, and D. A. Stahl, *Appl. Environ. Microbiol. 59*:682 (1993).

127. N. Springer, W. Ludwig, and G. Hardarson, *System. Appl. Microbiol. 16*:468 (1993).

128. G. Kirchhof and A. Hartmann, *Symbiosis 13*:27 (1992).

129. T. Hurek, S. Burggraf, C. R. Woese, and B. Reinhold-Hurek, *Appl. Environ. Microbiol. 59*:3816 (1993).

130. Y. Couteaudier, M.-J. Daboussi, A. Eparvier, T. Langin, and J. Orcival, *Appl. Environ. Microbiol. 59*:1767 (1993).

131. E. A. S. Rattray, J. I. Prosser, K. Killham, and A. Glover, *Appl. Environ. Microbiol. 56*:3368 (1990).

132. E. A. Meighen, *Microbiol. Rev. 55*:123 (1991).

133. S. A. B. Stewart and P. Williams, *J. Gen. Microbiol. 138*:1289 (1992).

134. R. A. Jefferson, *Nature 342*:837 (1989).

135. A. Vande Broek, J. Michielis, A. Van Gool, and J. Vanderleyden, *Mol. Plant Microbe Inter. 6*:592 (1993).

136. K. J. Wilson, A. Sessitsch, and A. D. L. Akkermans, *Beyond the Biomass* (K. Ritz, J. Deighton, and K. E. Giller, eds.), J. Wiley & Sons, Chichester, England, pp. 1447–1456 (1994).

137. K. J. Wilson, A. Sessitsch, J. C. Corbo, K. E. Giller, A. D. L. Akkermans, and R. A. Jefferson, *Microbiol. 141*:1691 (1995).

138. D. J. O'Kane, W. L. Lingle, J. E. Wampler, M. Legozki, and A. A. Szalay, *Plant Mol. Biol. 10*:387 (1988).

139. L. A. De Weger, P. Dunbar, W. F. Mahafee, B. J. J. Lugtenberg, and G. S. Sayler, *Appl. Environ. Microbiol. 57*:3641 (1991).

140. J. J. Shaw and C. I. Kado, *Bio/Technology 4*:560 (1986).

141. D. R. Fravel, R. D. Lumsden, and D. P. Roberts, *Plant Soil 125*:233 (1990).

142. S. Stubner, M. Schloter, G. Moeck, J. W. Coulton, F. Ahne, and A. Hartmann, *Environ. Tox. Water Quality 9*:285 (1994).

143. M. Chalfie, Y. Tu, G. Euskirchen, W. W. Ward, and D. C. Prasher, *Science 263*:802 (1994).

144. C. D. Webb, A. Decatur, A. Teleman, and R. Losick, *J. Bacteriol. 177*:5906 (1995).

145. J. I. Prosser, *Microbiol. 140*:5 (1994).

146. D. J. Silcock, R. N. Waterhouse, L. A. Glover, J. I. Prosser, and K. Killham, *Appl. Environ. Microbiol. 58*:2444 (1992).

147. K. J. Wilson, K. E. Giller, and R. A. Jefferson, *Advances in Molecular Genetics of Plant-Microbe Interactions* (H. Hennecke and D. P. S. Verma, eds.), Kluwer, Dordrecht, pp. 226–229 (1991).

148. S. B. Sharma and E. R. Signer, *Gene Dev. 4*:344 (1990).

149. K. J. Wilson, *Soil Biol. Biochem. 27*:501 (1995).

150. W. F. Mahaffee, P. A. Backman, and J. J. Shaw, *Plant Growth-Promoting*

Rhizobacteria—Progress and Prospects, Proceedings of 2nd International Workshop on Plant Growth-Promoting Rhizobacteria, Interlaken, Switzerland, pp. 248–251 (1991).

151. A. S. Rattray, J. I. Prosser, L. A. Glover, and K. Killham, *Soil Biol. Biochem.* *24*:421 (1992).

152. J. J. Shaw, F. Dane, D. Geiger, and J. W. Kloepper, *Appl. Environ. Microbiol.* *58*:267 (1992).

153. N. Cook, D. J. Silcock, R. N. Waterhouse, J. I. Prosser, L. A. Glover, and K. Killham, *J. Appl. Bacteriol.* *75*:350 (1993).

154. F. A. Grant, J. I. Prosser, K. Killham, and L. A. Glover, *Soil Biol. Biochem.* *24*:961 (1992).

155. K. McLennan, L. A. Glover, K. Killham, and J. I. Prosser, *Lett. Appl. Microbiol.* *15*:121 (1992).

156. A. Meikle, K. Killham, J. I. Prosser, and L. A. Glover, *FEMS Microbiol. Lett.* *99*:217 (1992).

157. D. A. Wright, K. Killham, L. A. Glover, and J. I. Prosser, *Geoderma 56*:633 (1993).

158. D. E. Caldwell, D. R. Korber, and J. R. Lawrence, *Adv. Microb. Ecol. 12*:1 (1992).

159. M. E. Sieracki, P. W. Johnson, and J. M. Sieburth, *Appl. Environ. Microbiol.* *49*:799 (1985).

160. M. E. Sieracki, S. E. Reichenbach, and K. L. Webb. *Appl. Environ. Microbiol.* *55*:2762 (1989).

161. J. Bloem, P. C. de Ruiter, G. Koopman, G. Lebbink, and L. Brussaard, *Soil Biol. Biochem. 24*:655 (1992).

162. P. Morgan, C. J. Cooper, N. S. Battersby, S. A. Lee, S. T. Lewis, T. M. Machin, S. C. Graham, and R. J. Watkinson, *Soil Biol. Biochem. 23*:609 (1991).

163. C. L. Viels and M. E. Sieracki, *Appl. Environ. Microbiol. 58*:584 (1992).

164. Y. Hiraoka, J. W. Sedat, and D. A. Agard, *Science 238*:36 (1987).

165. A. Agard, *Annu. Rev. Biophys. Bioeng. 13*:191 (1984).

166. A. Agard, Y. Hiraoka, P. Shaw, and J. W. Sedat, *Meth. Cell. Biol. 30*:353 (1989).

167. G. L. Gorby, *J. Histochem. Cytochem. 42*:297 (1994).

168. T. Wilson and C. Sheppard, *Theory and Practice of Scanning Optical Microscopy*, Academic Press, London (1984).

169. D. Shotton and N. White, *Trends Biochem. Sci. 14*:435 (1989).

170. J. R. Lawrence, D. R. Korber, B. D. Hoyle, J. W. Costerton, and D. E. Caldwell, *J. Bacteriol. 173*:6558 (1991).

171. G. M. Wolfaardt, J. R. Lawrence, R. D. Robarts, S. J. Caldwell, and D. E. Caldwell, *Appl. Environ. Microbiol. 60*:434 (1994).

172. F. Dazzo, P. Mateos, G. Orgambide, S. Philip-Hollingsworth, A. Squartini,

N. Subba-Rao, H. S. Pankratz, D. Baker, R. Hollingsworth, and J. Whallon, *Trends in Microbial Ecology* (R. Guerro and C. Pedros-Alio eds.), Spanish Society for Microbiology, Barcelona, pp. 259–262 (1993).

173. J. Bloem, M. Veninga, and J. Shepherd. *Appl. Environ. Microbiol. 61*:926 (1995).
174. M. Wagner, B. Aβmus, R. Amann, P. Hutzler, and A. Hartmann. *J. Microsc. 172*:181 (1994).
175. K. J. Czymmek, J. H. Whallon, and K. L. Klomparens, *Exp. Mycol. 18*:275 (1994).
176. C. M. Yentsch and C. S. Yentsch, *Recent Advances in Microbial Ecology* (T. Hattori, Y. Ishida, Y. Maruyama, R. Y. Morita, and A. Uchida, eds.), Japan Scientific Society Press, Tokyo, pp. 707–711 (1989).
177. H. M. Shapiro, *ASM News 56*:584 (1990).
178. J. P. Diaper and C. Edwards, *Microbiol. 140*:35 (1994).
179. R. Robertson and D. K. Button, *Cytometry 10*:70 (1989).
180. P. Monfort and B. Baleux, *Cytometry 13*:188 (1992).
181. M. Troussellier, C. Courties, and A. Vaquer, *Cell Biol. 78*:111 (1993).
182. P. Fouchet, C. Jayat, Y. Hechard, M.-H. Ratinaud, and G. Frelat, *Biol. Cell 78*:95 (1993).
183. C. Edwards, J. Diaper, J. Porter, D. Deere, and R. Pickup, *Beyond the Biomass* (K. Ritz, J. Deighton, and K. E. Giller, eds.), J. Wiley & Sons, Chichester, England, pp. 57–65 (1994).
184. G. Wallner, R. Erhart, and R. Amann, *Appl. Environ. Microbiol. 61*:1859 (1995).
185. J. Porter, C. Edwards, J. A. W. Morgan, and R. W. Pickup, *Appl. Environ. Microbiol. 59*:3327 (1993).
186. S. Page and R. G. Burns, *Soil Biol. Biochem. 23*:1025 (1991).

11

Indirect Approaches for Studying Soil Microorganisms Based on Cell Extraction and Culturing

ELIZABETH M. H. WELLINGTON, PETER MARSH, JOY ELIZABETH MARGARET WATTS, and JOHN BURDEN University of Warwick, Coventry, United Kingdom

1. INTRODUCTION

The intimate association between microbial cells and the soil matrix requires the use of specific extraction procedures to allow efficient recovery of cells from sample material. In addition, some types of molecular analysis of the soil microbial community start with a concentrated cell fraction for extraction of high-purity deoxyribonucleic acid (DNA) and allow recovery of large DNA fragments, following chemical lysis of cells. These more recent molecular methods for analyzing microbial community structure in soil have focused attention on biomass extraction techniques. However, the approach to this is well established and was originally developed to facilitate physiological studies of indigenous soil bacteria (1,2).

The traditional approach to monitoring microbial populations in natural environments relied on the dilution plate procedure, as outlined later in this chapter. A combination of gentle dispersion and serial dilution of the sample in a suitable diluent allowed relatively sensitive detection of soil microorganisms when combined with the use of selective culturing techniques (3). Although this approach still provides useful data for detection and monitoring of microbial propagules in soil, it provides no information about the physiological characteristics of the cells that give rise to colonies on plates other than that they are capable of growth. Therefore, the method does not allow enumeration of nonculturable cells or of propagules which are unable to grow on highly selective media or fail to compete with other groups when plated on nonselective media.

The provision of microbial cells extracted directly from the field without cultivation creates a unique opportunity to examine the metabolic state of those cells if the extraction procedures are not damaging or selective (4). Alternatively, specific stages of the life cycle of a particular group of microorganisms may be required for study. For example, specific extraction methods have allowed the differentiation of bacterial spores from vegetative cells (5). The combined use of cell sorting devices with differential markers such as oligofluors (fluorescent oligonucleotides) applied to extracted cell fractions has the potential to provide data on the numbers of certain taxonomic groups and allow their physical separation from other cells (6). In addition, autecological studies have been facilitated by the use of immunological detection and capture methods (7), which have an improved resolution if the particulate soil matter is removed from the extract as this interferes with antibody binding (8,9).

This chapter will address the questions that arise when microbial cells have to be dislodged from soil particles for further analyses. After a short discussion of a selected suite of approaches to cell extraction from soil, some cultivation-based analysis methods for soil bacteria will be considered.

2. APPROACHES TO MICROBIAL BIOMASS EXTRACTION

The initial stage for the efficient extraction of cells from soil requires application of physical and/or chemical methods for dispersion, resulting in the breaking of bonds between cells and soil components. These bonds are highly diverse and encompass electrostatic and van der Waals forces, hydrogen bonding, physical entrapment, and many others, depending on the nature of the soil and its clay and due to organic matter content. Entrapment due to the activity of other bacteria with the formation of aggregates and slime layers also occurs. The forces required to break these bonds and disperse the cells will, therefore, be dependent on the type of soil and the nature of the microbial population to be extracted. Much of the effort in the development of biomass extraction techniques has focused on separation and concentration of bacterial cells (or at least nonfilamentous microorganisms) from soil. Even microbial spores are often excluded when techniques such as density are employed to concentrate the cells.

There are two main approaches used to extract biomass from soil. In both, some form of dispersion is required first to dislodge the attached cells, and it may be brought about by physical or chemical means or by a combination of both. In addition, a concentration step is often required to purify and separate the cellular fraction from soil debris, colloids, and organic matter. The purification step can be achieved by several methods, and again the sample handling depends on the final objective. For example, if mycelial biomass from either fungi or actinomycetes is required then a quite different approach is needed as the majority of methods described so far have focused on unicellular biomass. Hyphae become readily trapped within soil particles unless they are fragmented. The mycelial fraction has been recovered usually in association with the soil debris after dispersion and centrifugation (1,9). This mixture has proved difficult to separate and few attempts have been made to provide a pure mycelial fraction from soil.

2.1 Dispersion of Soil

Methods for soil dispersion have been applied to bulk soil, rhizosphere, and plant material as treatments to free the bacterial cells that occur in close association with the substrate. Methods are evaluated by comparing measurements of microbial biomass in the original sample and after extraction into the various fractions. Biomass determination in soil is fraught with difficulties, and in addition the efficacy of cell recovery is often difficult to measure. However, a range of techniques, including direct microscopic counts, microbial respiration, adenosine triphosphate (ATP) levels, viable counts, and biomarkers such as lipids, has been used (10). In addition to measuring recovery, such methods can be useful for detection of cell damage (11). Both indigenous populations and pure culture inoculants have been used to evaluate the cell extraction methods. However, bacteria introduced into soil can behave differently from indigenous cells, as demonstrated by soil fractionation studies where inoculant cells were recovered in different soil aggregate fractions (12).

As a first step, the soil sample must be suspended in a suitable diluent such as water, phosphate buffer, or physiological saline solution. Some diluents, such as pyrophosphate, are also effective for disruption of soil aggregates; however, they may alter the metabolic properties of the recovered cells (13). The original water content and matric potential of the sample should be noted, as dried soil may respond very differently to dispersion when compared to the same sample extracted prior to drying.

Physical dispersion

The physical treatments to disperse soil and soil organisms fall into three categories, i.e., shaking, blending (homogenization), and ultrasonication. Traditionally, blending is most often used for pretreatment of soil samples, by use of either a blender (e.g., Waring) or a stomacher. The latter may be more useful if certain groups of gram-negative bacteria predominate in the sample as these are more susceptible to lysis by blending and the more abrasive action of the blender. *Salmonella* species have been recovered from soil and plant material efficiently using a stomacher for dispersion combined with chemical dispersion and centrifugation (14). Shaking the sample material is probably the least effective method for dispersion, but is suitable for highly sensitive cells and bacteriophages. It is most often used with dilution plate isolations, where the sample is diluted to allow enumeration as a viable count (see Section 5). Ultrasonic treatment is the most disruptive physical force for soil samples. It may be a useful pretreatment for heavy clay soils (15), but sensitive gram-negative cells may be prone to lysis. Various studies have shown that ultrasonication is likely to cause more extensive lysis than the other dispersion methods, and this effect has been prevented by the use of mild ultrasonic treatment (16).

Chemical dispersion

One of the perceived disadvantages of the use of physical dispersion techniques is the likelihood of damage caused to fragile cells. Larger microbial cells, and protozoans in particular, are less likely to survive physically abrasive extraction techniques (17). If microbial cells can be desorbed by chemical means from soil particles, then a gentle shaking regimen will suffice for efficient extraction. Chemical dispersion of soil can be achieved by the use of chelating agents which exchange polyvalent cations surrounding clay particles for monovalent cations. This method reduces the electrostatic attraction between soil particles and cells. Iminodiacetic acid ion exchange resins have been widely used (e.g., Dowex A1 [17], Chelex-100 [5]). The resin exchanges Na^+ ions for Ca^{2+} ions of clay particles in the soil matrix, thus bringing about the disruption of the electrostatic forces that bind soil aggregates. Maximal dispersion of mineral particles without the destruction or irreversible alteration of soil particles was obtained by gentle shaking with Dowex A1 for 2 hours (4°C) (18). The ion-exchange resin Chelex-100 has been widely used for the efficient extraction of microbial biomass; other dispersants used include Tris buffer and sodium hexametaphosphate (19).

Soil treated with an ion exchange resin may still contain microbial cells attached to soil particles by means of polymeric gums, thus making inorganic extractants redundant. These extracellular polymers which bind cells to soil surfaces and to one another are a complex mixture of polysaccharides of plant and microbial origin. The action of detergents dissolves extracellular hydrophobic material such as gums. MacDonald (17) reported that the use of several different detergents in conjunction with Dowex A1 increased the yield of microbial cells from soil. It was found that 0.1% sodium deoxycholate gave the highest yield (84%). This method was further developed by Herron and Wellington (5), and Dowex was substituted for Chelex-100 and combined with the same detergent and polyethylene glycol (PEG) 6000, used for dissolving of hydrophobic material and phase separation. This method allowed extraction of streptomycete spores from soil when used in conjunction with centrifugation and filtration for the purification step. It also proved effective for biomass extraction prior to DNA extraction (20). Some chemical dispersants may be toxic to cells. Hence, for the extraction of *Salmonella* cells from soil, sodium deoxycholate was excluded and the optimal method for recovery included the use of Chelex-100 and polyethylene glycol (14). An outline of the method and its use for different purposes is given in Fig. 1.

The use of resins probably results in the most efficient methods for soil dispersion and subsequent cell extraction, as Hopkins et al. (10) found that shaking with a chelating ion exchange resin gave consistently greater dispersion of soil aggregates than shaking with Tris buffer or distilled water or use of ultrasonic treatment.

A further development of dispersion techniques using ion exchange resins was to disperse soil during the initial stages of an extraction, followed by progressively more rigorous treatments (e.g., homogenization, ultrasonication), in order to maximize the recovery of the more fragile members of the soil microbial community. Ion exchangers may not provide the most effective method for dispersion in all soils, as their action depends on the amount of divalent cations in the sample. Lindahl and Bakken (11) compared chemical and physical dispersion techniques for biomass extraction from a clay loam and concluded that use of a blender was the most efficient approach for dispersion of their soil.

2.2 Concentration and Purification of Biomass Fraction

Differential centrifugation

One of the advantages of biomass extraction techniques is that they reduce the final volume of the extract so that the microbial population is concentrated. This means that the detection limit of the method for enumeration of specific cells in soil is improved (lowered), especially when larger volumes of soil can be handled (14). The main aim of biomass extractions is often to study the characteristics of cells as they existed in the environment. Therefore, a purification step is required, in addition to concentration, to remove soil particles and colloids which remain suspended after the initial low-speed centrifugation to remove soil debris. The approach most frequently used is to combine low- and high-speed centrifugations for removal of soil particles and concentration of cells following physical or chemical dispersion. The initial concentration step is a low-speed spin (around 1000 g) for varying lengths of time, depending on the subsequent steps and target biomass extracted. These low-speed spins produce pellets con-

Concentration of cells from soil for monitoring specific populations

10 g soil + 10 g Chelex 100 (BioRad) + 10 ml 2.5% polyethylene glycol 6000
shake gently for 2 hrs

⇓

centrifuge at 177 x g 30 sec

⇓

recentrifuge liquid phase at 3500 x g 15 min
three alternative analyses

⇓
⇓

DNA extraction	**Viable count**	**Immunomagnetic capture**
⇓	⇓	⇓
resuspend pellet in 1 ml 20 mg ml^{-1} lysozyme 37°C for 30 min	plate dilutions of extract on agar	resuspend pellet in phosphate buffered saline pH 7.2
SDS lysis or physically disrupt cells		rotate 30min add magnetic beads coated with antibody (Dynal)
precipitate salt with potassium acetate		capture bead-cell complex by holding magnet against tube
ethanol precipitation of DNA		
PCR		wash and process as required (PCR, selective plating *etc.*)

Figure 1.   The use of soil dispersion and concentration combined with DNA extraction and immunomagnetic capture for detection and monitoring of *Salmonella* species in soil.

taining soil debris, fungal mycelium, and the more dense spores (excluding actinomycete spores), whereas bacterial cells tend to stay in suspension. Herron and Wellington (5) developed a method for spore extraction and these dense forms were recovered by two low-speed spins after chemical dispersion. The initial pellet was reextracted and supernatants filtered to remove the ion exchange resin (Chelex-100) used for dispersion. A final higher-speed spin (2830 g) produced a pellet of soil colloids and spores. This crude extract could be used for viable and direct counts and DNA extractions. A faster final spin (>15,000 g) may be necessary for sedimentation of less dense cells; a spin at 17,700 g for 15 min proved optimal for the recovery of *Salmonella* cells from the supernatant (14).

Further purification of the cell fraction following the initial low-speed centrifugation step can be achieved by density gradient centrifugation (21). The use of buoyant density to separate soil particles and cells relies on the assumption that most cells are not aggregated and are detached from soil particles. The soil type is therefore highly influential in the success of this approach; highly organic soils, for example, will have less dense particles and these may be impossible to separate from the cells. A number of gradient media have been used; for a long time, Percoll has set the standard as it allows the recovery of a clean bacterial fraction for biochemical studies (21). A single-step method for purification may also be possible with the density material Nycodenz (22), where the direct application of a soil homogenate to the gradient can provide a bacterial layer after spinning at 10,000 g for 20–60 min. This technique omitted the low-speed spin step for separation of the larger soil particles. The use of density gradient centrifugation can provide the purest biomass fraction. However, it has been limited to the recovery of bacteria from soil, and inevitably many cells are not recovered because of entrapment within the soil and aggregation. In addition, most spores would be too dense to be recovered in one fraction with vegetative cells.

Elutriation

Soil particles and unicells (single cells) may also be separated by a sedimentation process in which suspended particles sediment at a velocity governed by their size and density as defined by Stoke's law. This technique was applied to soil biomass extraction by Macdonald (17), who used chemical dispersion of the samples prior to introducing them to specially designed elutriators for soil washing experiments using saline solution. One of the main advantages of the approach was the avoidance of centrifugation and the accompanying loss of cells which were still firmly associated with soil particles. However, aggregates of cells and particles will also be sedimented at a rate different from that of detached cells, and so this method also results in the loss of (clumped or) attached cells. The method of dispersion using an anionic detergent and chelating ion-exchange resin was only 40% to 48% efficient at dispersing selected soils (23). Serious practical problems occur when handling the large volumes required for elutriation and as a result of the need for concentration of the elutriate. Some form of filtration would be required to harvest cells; the use of ultrafiltration has been recommended, but this method may result in cell damage (23).

Other methods

The complete avoidance of any form of sedimentation for biomass purification has been proposed by the use of flow cytometry (6), which separates mixtures of particles by size and light scattering as the specimen is passed through an aperture and presented to a focused light beam. This method requires very highly dispersed extracts and has not been applied in sufficient detail to determine its efficiency. Other approaches for avoiding density-based techniques include the use of aqueous two-phase partitioning (24). This method of separation depends on differences in hydrophobicity of particles and possibly charge. A chemical dispersion method was required and this may prove to be a limitation if such methods are only 40%–50% efficient in certain soils. The partitioning method needs to be applied to a wide range of soil types to evaluate the applicability of the approach for biomass extraction.

3. IMMUNOMAGNETIC CAPTURE

Immunomagnetic capture (IMC) has greatly assisted in the efficient recovery of cells from environmental samples in recent years (25,26). The method is based on the use of microscopic magnetic beads (such as Dynabeads [Dynal]), which are coated with monoclonal or polyclonal antibodies against bacteria of interest. Their major use so far has been in the detection of medically important species in food and water, such as *Salmonella, Vibrio*, and *Yersinia* species. Hence, their use in ecology is primarily in autecology, where defined population(s) within a sample are of interest, rather than in general community analysis. The IMC approach provides a useful alternative to the more traditional isolation procedures. In addition, it does not require cultivation of the target cells. Using IMC, a highly purified cell extract is produced, as the captured magnetic beads can be washed and soil particles removed.

The majority of capture techniques have used a direct IMC approach in which either monoclonal antibodies or specific polyclonal antibodies were used to coat magnetic beads. Both direct (27) and indirect capture methods (9) have been applied to soil extracts. The direct method may be preferable if monoclonal antibodies are used, since polyclonal purified immunoglobulin G (IgG) fraction may contain <30% specific antibodies. The indirect approach was used to recover an actinomycete group, *Streptosporangium*, from soil. Antibodies to spores and mycelium were raised in rabbits and the immunoglobulin G (IgG) fraction purified, then added directly to the soil extract following addition of partially hydrolyzed gelatin. For indirect capture, sheep antirabbit Dynabeads were added to the soil sample after chemical dispersion using the method of Herron and Wellington (5). In contrast to recovery from food and water samples, the capture of cells from soil samples requires dispersion of cells and addition of blocking agents to overcome problems of nonspecific binding due to effects of clays and organic matter. The addition of partially hydrolyzed gelatin to the dispersed soil extract prior to the addition of the antispore antibody was very effective against nonspecific binding. The extract was then mixed with sheep antirabbit beads for recovery of the antibody–spore complexes (9). The direct method was employed for the recovery of streptomycete spores from soil using magnetic beads coated with a specific monoclonal antispore antibody (27). A nonionic detergent (Nonidet P40) reduced nonspecific binding, as did inclusion of blocking agents such as skimmed milk.

Immunomagnetic capture techniques have recovered very pure fractions of specific cells from dispersed soil extracts, but the recovery efficiency can vary widely, e.g., from 4% to 90%, depending on the type of soil and the technique used. The method is ideal for recovery of a population where specific antibodies are available or can be prepared. An approach to the detection of a specific bacterial group, *Salmonella* species, in soil is given in Fig. 1. The combined use of soil dispersion and concentration can provide a soil extract for highly sensitive and selective detection using DNA extraction and analysis, selective viable counts, and immunomagnetic capture used for enzyme-linked immunoassay (ELISA), DNA, and viable plate analysis.

4. SOIL FRACTIONATION

Soil fractionation is a technique that allows the separation of soil into different-sized particles. This allows a description of soil structure mainly at the aggregate/microaggregate scale. Aggregates are heterogeneous mixtures of soil particles that are classed into two main groups, macroaggregates (which are larger than 250 μm) and microaggregates (less than 250 μm). Macroaggregates are formed as a result of temporary associations between microaggregates. Microaggregates also have greater stability than macroaggregates because of strong binding between the organic and mineral colloids present (28). Aggregates are essential in soils as they increase soil fertility by improving aeration and allowing greater drainage. They also improve the soil's resistance to erosion. Aggregates provide various locations for microbial habitation, and two main areas have been identified:

1. Inner area, the inside of the aggregate structure that may be entered via small pores
2. Outer area, consisting of the larger pores and the surface of the aggregate

Hattori (29) found by the washing–sonic oscillation method that the inner and outer areas have distinct and different microbial communities (Fig. 2). The inner aggregate fractions contained higher levels of gram-negative bacteria, while the outer areas contained higher levels of gram-positive organisms. This may be due to a number of causes such as polymer formation, motility, surface charge, and life cycle of the bacteria involved. The aggregate provides protection in a number of ways: First, it gives bacteria protection from predation by protozoans. Second, aggregates can serve as a 'buffer' against changing environmental conditions such as water levels and pH. If bacteria have different physical locations in soil aggregates, then the physical separation and concentration of these aggregates can serve as a further attempt for biomass extraction. This gives a crude soil extract which can be subjected to further purification as discussed in the previous sections.

Many approaches have been taken to fractionate soil into its constituent components. Most have involved wet sieving followed by sedimentation steps. A modified method (Fig. 3), derived from that of Jocteur-Monrozier et al. (28), is

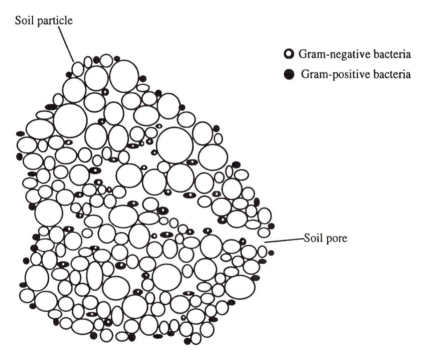

Soil particle

O Gram-negative bacteria
● Gram-positive bacteria

Soil pore

Figure 2.   Location of gram-positive and gram-negative bacteria in a soil aggregate. (Adopted from Ref. 29.).

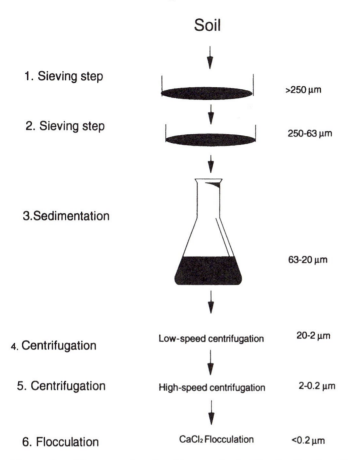

Figure 3. The stages involved in wet sieving for the fractionation of water-stable soil aggregates. The sizes of each aggregate fraction are indicated on the right.

used for illustration of the approach (P. Baker and J. Watts, unpublished observations). First, a known weight of soil is placed on a sieve and wetted slowly overnight to prevent breakage of the stable aggregates. The soil and sieve are then placed on a shaker for 10 min, with enough distilled water to cover the bottom of the sieve (Fig. 2). The sieve is then removed from the water and allowed to drain for 5 min, after which the soil is collected and weighed; this is the >250 μm fraction. The remaining soil suspension can be poured onto the next sieve and the process repeated. The soil suspension is then poured into bottles and allowed to sediment for a given time according to Stokes's law. After this step, the soil suspension is removed by vacuum, leaving the third fraction. The soil suspension (supernatant) is then treated with a low-speed centrifugation step and the pellet collected to yield the fourth fraction. The supernatant fraction is then treated with a high-speed centrifugation step, which gives the fifth fraction. Finally, calcium chloride is added at very low levels to the soil suspension, which is left to flocculate soil particles overnight, and the suspension is then treated with a high-speed centrifugation step to isolate the smallest fraction (<0.2 μm).

Amounts of each fraction will vary with the type of soil being analyzed. For example, a high-clay soil will contain a higher level of the smaller fractions than a sandy soil. The method provides fractions of differently sized soil particles, which can then be analyzed in a number of ways to characterize the microbial community present. Finally, it should be noted that there are certain problems associated with wet sieving as the process itself may break the water-stable aggregates into smaller components. The sedimentation steps are based on Stokes's law, which assumes all particles have a certain density. As the particle sizes and the size distribution of soil are largely heterogeneous, sedimentation as described by Stokes's law can only be an approximation.

5. ENUMERATION AND ISOLATION OF SOIL BACTERIA

Since soil provides a habitat for a large variety of bacteria, each with different nutritional requirements, it is impossible to devise a single method by which all soil bacteria can be isolated. Evidence has been presented that traditional methods for monitoring microorganisms in their habitat (such as the total viable count [TVC] approach described later) only allow the detection of a small fraction of the total microbial community (30). Compared to methods such as direct microscopic counts and molecular counting methods, plate counts of soil microorganisms may yield data that can be as much as 1000 times lower. This underestimation occurs for several reasons. First, the bacteria in soil exist as clumps of cells, most of the time attached to soil particles often within the pores of soil aggregates. The shaking component of the method is usually insufficient to disrupt such clumps and aggregates, and the subsequent colony formed on plating out will have arisen from a number of organisms, rather than the single cell which the technique assumes. This was confirmed by a recent study (10) in which a large proportion of bacterial cells remained associated with soil particles even after rigorous multistep dispersion and differential centrifugation procedures.

5.1 The Total Viable Count Method

The simplest method for isolating and enumerating bacteria from soil, and the basis of most other isolation methods, is the total viable count (TVC). The total viable count method is based on unselective plating. Among the soil biota, bacteria are particularly well suited to isolation and quantification by plating. However, they are never uniformly distributed in soil, and their levels may fluctuate wildly over short periods (31).

The TVC method involves disrupting soil aggregates and releasing microorganisms into a suitable diluent, which is then diluted decimally and plated out onto a suitable growth medium. The number of colonies which grow on the plates are counted after incubation, each one being assumed to have arisen from an individual bacterial cell, and the TVC is calculated with reference to the dilution (31).

The counts obtained by this technique are expressed in terms of colony-forming units (CFUs) per gram of soil, rather than cells per gram. As outlined previously, to overcome the problem of counting several cells in one colony dispersing agents such as surfactants (Tween 80) can be incorporated into the diluent used. These agents help to break up the soil particles and cell clusters and release individual cells. In addition, they can also break up spore chains commonly produced by organisms such as *Streptomyces* spp. into individual spores. Sedimentation of particle-bound cells during dilution and plating out can also contribute to errors in the enumeration of bacterial cells in soil by TVC. Constant agitation of samples during these stages is recommended to help prevent this outcome.

Another reason for underestimation using this method may be that bacterial cells die, either during the dilution or as a result of desiccation on the plate surface, or that they are isolated in a condition known as viable but nonculturable (VBNC). Such cells are alive and viable but do not grow on nutrient media for a variety of reasons, such as desiccation, pH, temperature shifts, light damage, and starvation responses (32). Furthermore, propagules may fail to germinate on the medium used for isolation, or slow-growing organisms, such as actinomycetes, may be outgrown by faster-growing organisms, such as *Pseudomonas* spp., whose large mucoid colonies will obscure other smaller colonies. The use of the TVC method for enumerating filamentous organisms such as actinomycetes gives an inaccurate picture of the numbers of these organisms in soil, since, being filamentous, a single CFU of these organisms on a plate may represent the biomass equivalent of many thousands of unicellular organisms. Obligately anaerobic organisms will not show up on aerobically grown TVC plates, and these organisms form another component of the underestimation of soil populations using this method. As these organisms (e.g. *Clostridium* spp.) are common in soils as both spores and vegetative cells, the underestimation of biomass can be serious.

Despite the limitations of the TVC method, it remains the most common method for enumerating soil bacteria. It offers many advantages, including convenience, simplicity, and ability to process a large number of samples quickly and with minimum equipment. However, as mentioned, the main disadvantage of the technique is the underestimation of microbial numbers. Also, the TVC method provides little information about biomass in soil, because it is unable to differentiate between actively growing vegetative cells and spores.

5.2   The Most Probable Number (MPN) Technique

To overcome some of the limitations of the TVC method, other enumeration techniques for soil bacteria have been developed. The most probable number (MPN) technique can be used to estimate microbial numbers in soil when quantitative measurement of individual cell numbers is not possible. The MPN technique relies on detection of specific qualitative attributes of the organism of interest, in particular on growth of the target organisms in liquid medium, using serial decimal dilutions of the soil in a diluent as an inoculum. These dilutions are then cultured in replicates of either 3, 5, or 10 samples. After incubation, the growth in each culture is scored as positive or negative; a positive reading indicates that at least one organism capable of growth is present in that sample. The results for the highest dilution containing at least one positive result can then be used to calculate the most probable number of organisms in the original sample, using an MPN table based on the mathematical work of Halvorson and Ziegler (33) and more recent work by Cochran (34).

A full review of the mathematics of MPN techniques is given by Woolmer (35). An important aspect of the MPN technique is the ability to estimate microbial population size on the basis of a functional characteristic of that population, such as the detection of plant-nodulating bacteria in soil by linking MPN methodology to plant infection methods (36). This technique can be used in conjunction with a selective medium to enumerate a specific organism or group of organisms within the soil.

The MPN technique continues to be an important method for enumerating microbial populations in soil. It has a distinct advantage over direct methods of enumeration, such as microscopy and staining, as it only estimates live cell numbers and prevents the overestimation of population numbers common to direct counting techniques. It is also less sensitive to differences in soil mineralogical characteristics than methods which use filtration of soil organisms (37), and provides much more realistic estimates of the number of viable propagules of vesicular–arbuscular (VA) mycorrhizae than spore count methods do (38). The MPN method suffers from the disadvantage of requiring considerably more time and materials than direct counts, as well as being less precise than a well-replicated direct count. However, the precision of MPN estimates can be improved by increasing the number of replicates at each dilution.

5.3 Selective Isolation of Soil Bacteria

The ability to isolate specific group or species of bacteria in soil samples is of great use to the soil microbiologist. Many commercially important bacteria, such as antibiotic-producing actinomycetes, are common in soil, and the ability to isolate such organisms selectively is of great benefit.

Furthermore, selective isolation allows the detection and tracking of pathogenic organisms in soil, such as shown in studies on the fate of *Salmonella* in sewage sludge applied to land.

The isolation of specific bacteria from soil can also be carried out by using the dilution and plating techniques used in TVC, but it requires a selective treatment which can be used to select for the specific required organism. In many cases, selection merely requires the use of a selective medium specific for the target organism, some of which are shown in Table 1. Plating using specific agar is useful in the isolation of a given organism but may contribute to the underestimation of the numbers of the organism in the soil. This is because the selection pressure used may have a lethal effect on the cells it is supposed to select for, since cells growing in soil are usually stressed and have sublethal injuries due to factors such as starvation, light damage, and desiccation. All of these factors may cause cell damage and render the cells susceptible to selection treatments. This can be seen with *Salmonella* spp. isolated from soil, which may not be able to grow on the *Salmonella*-selective XLD medium when plated directly from soil, as a result

Table 1. Selective Media for the Enumeration and Isolation of Soil Bacteria[a]

Genus	Selective medium
Pseudomonas	*Pseudomonas* CFC agar
	Kings A agar
Salmonella	XLD agar
	Salmonella Shigella agar
	Brilliant green agar
Micromonospora	Colloidal chitin agar
	Starch-casein agar
Streptomyces	RASS agar (5)
Escherichia coli	McConkey agar
	EMB agar
	Tryptone bile agar

[a]The composition of these media can be found in any microbiology textbook; their use is described in Ref. 45.

of damage to the outer membrane that makes the cells susceptible to inhibition by the bile salts in this medium. This may be overcome by preenriching the soil with a nonselective medium before plating onto a selective medium. However, such a method can only be used if only isolation, not enumeration, is required. Other selective treatments can also be used, in conjunction with or separately from selective media, such as antibiotic selection and heat treatments.

Antibiotic selection relies on the use of resistance profiles for particular groups of organisms, for example, the inclusion of rifampicin and ampicillin in RASS medium (see table 1) for the selective isolation of streptomycetes. Heat treatments generally select for spore-forming bacteria such as *Bacillus, Streptomyces*, and *Micromonospora* spp.

In such selective treatments, the soil is generally dried before being heated, and this has the effect of killing any vegetative cells in the soil, leaving only the heat-resistant spores viable; they can then be revived and isolated by plating out. Any physiological characteristic of an organism, provided it is unique to that organism, can be used to select for it, and such features may include tolerance to extremes of pH, salt tolerance, or resistance to heavy metals.

One of the most commercially important groups of soil bacteria is the actinomycetes. These organisms are well-known producers of antibiotic compounds, and consequently much work has been done on selective isolation of them. The isolation of actinomycetes from soil has traditionally been done by using conventional plate dilution techniques. Such methods, however, often lead to a serious underestimation of the actinomycete population present, because of either of inhibition or outcompetition by other more abundant or more proliferate microorganisms. Many attempts have been made to overcome these problems by the use of pretreatment techniques such as antibiotic selections and heat treatments (39). Prior to the mid-1960s, published information concerning the isolation of actinomycetes dealt almost exclusively with methods for *Streptomyces* spp. (40), reflecting their growing importance to the pharmaceutical industry as producers of antibiotics. However, as the search for novel bioactive compounds increased (41), and with it the discovery of commercially successful antibiotics from other rarer genera such as *Micromonospora* (41) and *Actinoplanes* (42), more emphasis was placed on the development of selective methods for the isolation of nonstreptomycete actinomycetes (40,43,44). Because many actinomycetes produce spores, selective isolation techniques for these organisms can often be developed on the basis of the properties of these spores. Spores of *Micromonospora* are very resistant to heat, so many selection methods for this genus rely on pretreatment of the soil by heating. Spores of *Streptomyces* can be selectively extracted from soil by using anionic exchange resin (Chelex-100) (5), as described. Actinomycetes can also be selected for by pretreatment of the soil, enriching it with those substrates, such as chitin, which these organisms are able to break down, thus increasing their numbers in the soil before the selection is made. These enrichment techniques also form the basis (for baiting as a method) of isolating specific groups of organisms. In baiting, a substrate, such as a cellulose film or hair, is introduced into the soil and left for a period, after which it is removed and placed onto an appropriate agar. In this way, organisms from the soil which are able to grow on the substrate are captured and transferred to the agar.

We will not further address all possible strategies for the selection of bacteria from soil. A full review of the range of specific isolation methods for particular bacteria is given in Weaver et al (45).

6. CONCLUSIONS

Because of the heterogeneity of soil, no single biomass extraction approach can be recommended as suitable for all soil types. Much of the conflict that exists in the published appraisal of methods is probably due to varying efficiencies of different methods when they are applied to differing soils. Whichever methods are used, biomass extraction is an important tool for the study of microbial activity and survival in soil. It provides a unique opportunity to study cells without further physiological alterations due to cultivation. In addition, it provides an opportunity to study the majority of cells in the soil which have so far not been cultured (uncultured or unculturable microbial fraction). Further work is still needed to develop more efficient dispersion and concentration techniques which can provide pure cell extracts but avoid time-consuming protocols.

After the cell extractions, approaches to microbiological analyses are often based on cultivation. The total viable count, in which dilution plating to nonselective media is used, is still of relevance, since it provides an overall picture of the culturable fraction of microbes in a specific soil. The efficiency of cell enumeration of the TVC method is obviously dependent on the efficiency of soil dispersion and cell extraction methods applied prior to plating. The TVC approach should ideally be combined with DNA-and (microcopic) cell count-based methods in order to obtain a full understanding of the microorganisms of interest in the soil. However, the latter approaches are also critically dependent on soil dispersion method, and it is important to calibrate this when performing the analyses.

REFERENCES

1. A. Fægri, V. L. Torsvik, and J. Goksoyr, *Soil Biol. Biochem. 9*:105 (1977).
2. L. R. Bakken and R. A. Olsen, *Soil Biol. Biochem. 21*:789 (1989).
3. R. Herbert, *Methods in Microbiology*, vol. 22 (R. Grigorova and J. R. Norris, eds.) Academic Press, London, p. 1.

4. H. Christiansen, L. R. Bakken, and R. A. Olsen, *FEMS Microbiol. Ecol.* *102*:129 (1993).

5. P. R. Herron and E. M. H. Wellington, *Appl. Environ. Microbiol.* *56*:1406 (1990).

6. C. Edwards, J. Diaper, J. Porter, D. Deere, and R. Pickup, *Beyond the Biomass* (K. Ritz, J. Dighton, and K. E. Giller, eds.), John Wiley & Sons, Chichester, England, p. 57, (1994).

7. M. Schloter, R. Borlinghaus, W. Bode, and A. Hartmann, *J. Microscopy* *171*:173 (1993).

8. E. L. Schmidt, R. O. Bankole, and B. B. Bohlool, *J. Bacteriol.* *95*:1987 (1968).

9. P. H. Mullins, H. Gürtler, and E. M. H. Wellington, *Microbiology 141*:2149 (1995).

10. D. W. Hopkins, S. J. MacNaughton, and A. G. O'Donnell, *Soil Biol. Biochem.* *23*:217 (1991).

11. V. Lindahl and L. R. Bakken, *FEMS Microbiol. Ecol.* *16*:135 (1995).

12. G. A. Recorbet, A. Richaume, and L. Jocteur-Monrozier, *Lett. Appl. Microbiol 21*:38 (1995).

13. V. Lindahl, *J. Microbiol. Meth.* *25*:279 (1996).

14. P. E. Turpin, K. A. Maycroft, C. L. Rowlands, and E. M. H. Wellington, *J. Appl. Bacteriol.* *74*:181 (1993).

15. T. Ozawa and M. Yamaguchi, *Appl. Environ. Microbiol.* *52*:911 (1986).

16. D. W. Hopkins and A. G. O'Donnell, *Genetic Interactions Between Microorganisms in the Environment.* (E. M. H. Wellington and J. D. Van Elsas, eds.) Manchester University Press, Manchester, England, p. 104 (1992).

17. R. M. MacDonald, *Soil Biol. Biochem.* *18*:407 (1986).

18. A. P. Edwards and J. M. Bremner, *Nature 205*:208 (1965).

19. F. Niepold, R. Conrad, and H. G. Schlegel, *Antonie van Leeuwenhoek 45*:485 (1979).

20. C. S. Jacobsen and O. F. Rasmussen, *Appl. Environ. Microbiol.* *58*:2458 (1992).

21. L. R. Bakken, *Appl. Environ. Microbiol.* *49*:1482 (1985).

22. L. R. Bakken and V. Lindahl, *Nucleic Acids in the Environment*, (J. T. Trevors and J. D van Elsas, eds.) Springer, Berlin, p. 9 (1995).

23. D. W. Hopkins, A. G. O'Donnell, and S. J. Macnaughton, *Soil. Biol. Biochem.* *23*:227 (1991).

24. N. C. Smith and D. P. Stribley, *Beyond the Biomass* (K. Ritz, J. Dighton and K. E. Giller, eds.), John Wiley & Sons, Chichester, England p. 49 (1994).

25. J. A. W. Morgan, C. Winstanley, R. W. Pickup, and J. R. Saunders, *Appl. Environ. Microbiol.* *57*:503–509 (1991).

26. A. C. Fluit, M. N. Widojojoatmodjo, A. T. A. Box, R. Torensma, and J. Verhoef, *Appl. Environ. Microbiol.* *59*:1342 (1993).

27. A. Wipat, E. M. H. Wellington, and V. A. Saunders, *Microbiology 140*:2067 (1994).
28. L. Jocteur-Monrozier, J. N. Ladd, R. W. Fitzpatrick, R. C. Foster, and M. Raupauch, *Geoderma 49*:37 (1991).
29. T. Hattori, *Rep. Inst. Agric. Res. Tohoky Univ. 37*:23 (1988).
30. W. Liesack and E. Stackebrandt, *Biodiversity and Conservation 1*:250 (1992).
31. F. Schinner, *J. Industrial Microbiol. 14*:213. (1995).
32. S. P. Oliver, K. D. Payne, and P. M. Davidson, *J. Food Protection 57*:62 (1994).
33. H. O. Halvorsson and N. R. Ziegler, *J. Bacteriol. 25*:101 (1933).
34. W. G. Cochran, *Biometrics 6*:305 (1950).
35. P. L. Woomer, *Methods of Soil Analysis. Part 2. Microbiological and Biochemical Properties.* (R. W. Weaver, S. Angle, P. Bottomley, D. Bezdicek, S. Smith, A. Tabatabai, and A. Wollum, eds.) SSSA, Madison, Wisconsin, p. 60 (1994).
36. J. Brockwell, A. Diatloff, A. L. Garcia, and A. C. Robinson, *Soil Biol. Biochem 7*:305 (1975).
37. P. L. Woomer, *Appl. Environ. Microbiol. 54*:1494 (1988).
38. W. M. Porter, *Aust. J. Soil, Res. 17*:515 (1979).
39. S. T. Williams and F. L. Davies, *J. Gen. Microbiol. 38*:251 (1965).
40. M. Shearer, *Devel. Industrial Microbiol. 28*:91 (1987).
41. G. M. Luedemann and B. C. Brodsky, *Antimicrob. Agents Chemoth. 1963*:116 (1964).
42. N. J. Palleroni, *Arch. Microbiol. 110*:13 (1979).
43. H. Nonomura and Y. Ohara, *Hakku Kogaku Zasshi 47*:463 (1969).
44. H. Nonomura and Y. Ohara, *J. Ferment Technol. 49*:904 (1971).
45. R. W. Weaver, S. Angle, P. Bottomley, D. Berdicek, S. Smith, A. Tabataba, and A. Wollum, *Methods of Soil Analysis Part 2–Microbiological and Biochemical Properties*, SSSA, Madison, Wisconsin (1994).

12a

Microbial Biomarkers

J. ALUN W. MORGAN Horticulture Research International, Wellesbourne, United Kingdom

CRAIG WINSTANLEY Coventry University, Coventry, United Kingdom

1. INTRODUCTION

Microbial ecology aims to provide an understanding of the role of microorganisms in nature and in relationship to human society (1). A comprehension of the interactions within microbial communities and between communities and their environment is central. The main obstacles to increasing our knowledge in this field remain methodological (2). Cell enumeration techniques and identification procedures are often difficult or tedious, and the collection of relevant samples or the simulation of natural conditions in the laboratory can be problematical. In addition, the exclusive use of culture methods restricts the numbers of cells included in any study since a considerable proportion of the microbial biomass of soil, water, or sediment is unculturable on standard media (3). Once it is isolated, it is relatively simple to obtain further information on a microorganism, but the limitations on numbers that can be dealt with by this approach are often restrictive. Direct microscopic analysis of natural populations can produce very useful information on biomass, cell size, and numbers, but this approach provides little or limited information on the types of organisms present. To overcome these difficulties, methods involving direct detection and separation of biochemical and genetic components of mixed populations to use as biomarkers have been developed with a view to identification and quantification of the biomass in the environment. If specific biochemical constituents are restricted to particular subsets of the microbial community, then the analysis of these constituents can give valuable information about the composition of a microbial community (4). We have considered the term 'biomarker' as any biological component that can be used to indicate a useful feature of a particular microbial community. Biomarkers could therefore be biochemical components of cells such as deoxyribonucleic acid (DNA), ribonucleic acid (RNA), fatty acids, and sterols, or even the cells themselves, where, for example, *Escherichia coli* could be used as a biomarker for the presence of fecal contamination. In this way, specific taxonomic features of any microorganism could be used to provide information on its presence, or on the composition of a community. In this chapter we describe the use of selected biomarkers in microbial ecology, and illustrate the methodology with examples of both whole population analysis and strain-specific detection. Our choice of examples is itself necessarily selective because of the increasing amount of research now being undertaken with the type of experimental techniques described. Although we have sought to highlight those approaches targeting populations in natural environments, the rapidly developing use of biomarkers to identify isolated microorganisms is worthy of some attention. Subcellular biomarkers have been classified loosely into a number of categories depending on their location. For the purposes of this chapter the categories of cell envelope, general biochemical, and nucleic acid biomarkers will be addressed.

2. CELL ENVELOPE BIOMARKERS

2.1 Fatty Acids

Fatty acids, a group of chemicals especially useful as biomarkers, have been of great value in determining bacterial history or development (phylogeny) and also provide a useful set of features for characterizing strains (taxonomy) (5,6). Specific fatty acids, especially phospholipids, which are the major constituents of the membranes of all living cells (with the exception of archaebacteria), have the useful property that they are degraded rapidly after cell death, are not found in storage lipids or in anthropogenic contaminants, and usually have a comparatively rapid turnover. Bacteria also contain phospholipids as a relatively constant proportion of their biomass. This makes the analysis of the phospholipid fraction of mixed communities a useful measure of the viable cellular biomass which can complement other traditional methods such as enzyme activities, muramic acid levels, and total adenosine triphosphate (ATP) (7). Lipid biomarkers can be recovered from isolates and environmental samples by a single-phase chloroform/methanol extraction, fractionation of the lipids on columns containing silicic acid, and derivatization prior to analysis by capillary gas–liquid chromatography/mass spectroscopy (GC–MS). Analysis of phospholipid ester-linked fatty acids is presently one of the more sensitive chemical methods to determine microbial biomass and community structure (8,9).

Quantitative estimates of microbial biomass and community structure by means of analysis of the phospholipid fraction have been performed on sediments (10–15), soil columns enriched with methane and propane (16), and various soil types (9,17). The method is applicable to the study of mixed populations of varying degrees of complexity and is relatively straightforward to perform. Usually the straight chain fatty acids have the least taxonomic potential for bacteria. Since they are ubiquitous, fatty acids such as palmitic acid (16:0) can be regarded as good indicators of bacterial biomass (17–19). Most actinomycetes also contain *iso* and *anteiso* fatty acids, which can also be used for biomass estimations (20).

Fatty acids in the range of C12 to C19 are reported to be characteristic of bacteria (15,21) since they are found in higher concentrations in bacteria than other organisms. In sediments the presence of saturated fatty acids, branched fatty acids, monounsaturated fatty acids, and polyunsaturated fatty acids in this range has been used to indicate the contribution of bacterial populations to the total population (15,22,23). Findlay et al. (24) used biomarker fatty acids extracted from sediments to classify distinct microbial groups: microeukaryotes (polyunsaturated fatty acids); aerobic prokaryotes (monounsaturated fatty acids); gram-positive and other anaerobic bacteria (saturated and branched fatty acids in the range of C14 to C16); and sulfate-reducing bacteria including other anaerobic bacteria (saturated and branched fatty acids in the range of C16 to C19). Branched fatty acids have been used as biomarkers for bacteria including anaerobic bacteria and the sulfate-reducing bacterium *Desulfovibrio*. Detection can be highly specific for a single strain. For example, the fatty acid 10Me16:0 can be used as a signature for the presence of *Desulfovibrio* since it is not detected in other sulfate-reducing bacteria. The presence of monounsaturated fatty acids has been used to determine the dominance of prokaryotic organisms in both freshwater and marine sediments. In general, unsaturated fatty acids in the range C12 to C19 are indicative of gram-negative bacteria. However, the monoeonic fatty acids 16:1d9 and 18:1d9 that are found in bacteria are also present in microalgae. To determine the origin of such fatty acids, analysis of the levels of polyunsaturated fatty acids (18:2, 20:4, and 20:5) which are characteristic of microeukaryotes can be used. High levels of the unsaturated fatty acids with low levels of the polyunsaturated fatty acids support the conclusion of bacterial dominance (15,25). In addition, the differences in the relative proportions of branched and monounsaturated fatty acids have been used as a marker for the proportion of gram-positive and gram-negative bacteria in marine sediments (15). Baath et al.

(26) used fatty acids 18:1ω7 and *cy*17:0 to monitor gram-negative bacteria, and fatty acids *i*15:0, 10Me16:0, and 18.1ω9, to monitor gram-positive bacteria in soil polluted with alkaline dust.

Cyclopropyl fatty acids (C17 and C19) have also been proposed as bacterial signatures in sediments (27–29), and have been reported as major constituents in large numbers of gram-negative bacteria and *Desulfobacter* (spp.,) but are rare in eukaryotes. Parks and Taylor (14) suggest that cyclopropyl fatty acids are characteristic of aerobic sediment bacteria, being present in oxidized sediment surface layers, and decrease with increasing depth as anaerobic conditions prevail. Sediment samples where cyclopropyl fatty acids were not detected at the surface have been used to infer that a shift in the community from domination by aerobic to anaerobic bacteria has occurred through pollution of the site (15).

The presence of long-chain fatty acids, such as fatty acid 24:0, when detected in water and sediments has been used to indicate the level of terrestrial organic matter. This information has been used to determine the amount of terrestrial input of organic matter into a study area (15,30).

In soil, Zelles et al. (17) separated microbial populations on the concentrations of monoenoic versus normal fatty acids. Two categories of populations were identified: in the first, equal amounts of both groups were present; in the second, monoenoic fatty acids were present in increased amounts compared to normal fatty acids. Gillan and Hogg (25) also divided soil bacteria into 'chemotypes' defined by their fatty acid compositions. Six groups were identified: (a) straight chain, (b) straight chain *cis* monoenoic, (c) branched-chain saturated, (d) straight-chain *trans* monoenoic, (e) cyclopropyl, and (f) branched-chain monoenoic acids. Each of the monoenoic (b, d) and the cyclopropyl (e) fatty acids are produced by two chemotypes, 16:1, 18:1 producers, and 17:0, 19:0 producers, respectively. The branched saturated fatty acids (c) were found to be produced by one or more chemotypes and Zelles et al. (17) subdivided these further into several groups according to the position of the methyl branching on the fatty acid chain. This classification was used to give a relatively simple and fast overview of the diversity of fatty acid composition in differently managed cropped soils. This approach is limited since little information is presented on the relationship between the fatty acid profiles and the types of microorganisms present.

Changes in the environment that result in an alteration in a microorganism's physiological characteristics can sometimes be detected in differences in the lipid profile. This factor is often mentioned when using fatty acid profiles for identification purposes where growth conditions must be kept constant. However, these changes may be exploited to determine the metabolic status of an organism and the environmental conditions it encounters (9). The presence of *trans* fatty acids has been associated with the physiological status of the microorganism (31–33) and the ratio of *trans* fatty acids to *cis* isomers of monounsaturated fatty acids has been used as a measure of physiological stress (10). Changes in fatty acid composition of phospholipids have been observed as bacteria enter a nonculturable but viable condition (34) where cells are unable to grow on laboratory media but are able to take up vital stains (Fig. 1). These changes in the fatty acid composition may be linked with a survival strategy whereby alterations are occurring to a cell for the efficient uptake of nutrients or for strengthening of the membrane. Changes in the fatty acid profiles of mixed microbial communities may be used in a similar way to detect stress responses or periods of activity.

2.2 Sterols and Quinones

Ergosterol is a predominant sterol which is limited to the true fungi (35). The content of ergosterol in fungal membranes depends on the species and can vary with the physiological state of a fungus. Factors such as age, developmental stage, and general growth conditions can cause variation in ergosterol levels. It is, however, possible to use measurements of ergosterol content as an indicator of fungal biomass (36). Suberkropp et al. (37) compared ATP and ergosterol as indicators of fungal biomass associated with decomposing leaves (*Liriodendron tulipifera L.*). They concluded that ergosterol concentrations provided a more accurate measure of fungal biomass when other organisms were present and likely to contribute to the ATP pool. This approach is limited to coarse determinations of fungal biomass and cannot be used to distinguish fungal species or define differences in fungal communities.

Quinones, lipid-soluble substances involved in electron transport, can also be used as biomarkers. Lipski et al. (38) used quinone analyses, physiological tests, and fatty acid profiles to differentiate gram-negative nonfermentative bacteria isolated from biofilters. Quinone type was found to be an efficient method to prearrange isolates prior to the analysis of results from the physiological tests. The detection of quinones appears to be restricted to the discrimination of isolated colonies and has limited potential to the analysis of mixed populations.

2.3 Lipopolysaccharides and Lipoproteins

Lipopolysaccharides (LPS) are major constituents of the outer membrane of gram-negative cell walls. They consist of an outer O antigen, a middle core of L,D-heptose, and occasionally D,D-heptose, and an inner lipid A region which consists of phosphorylated glucosamine disaccharide with covalently linked fatty acids. Both L,D-heptose and 3-hydroxyacid levels associated with LPS are potential biomarkers. Indeed, Fox et al. (39) used GC–MS analysis of dust to study bacterial levels by targeting the hydroxy fatty acids (3-OH 12:0 and 3-OH 14:0), L-glycero-D-mannoheptose and muramic acid (a chemical marker for peptidoglycan). It may be that variability in LPS, lipoproteins, or outer membrane proteins which has been used to distinguish between bacterial isolates (40–42), will be exploited to study natural populations in the future.

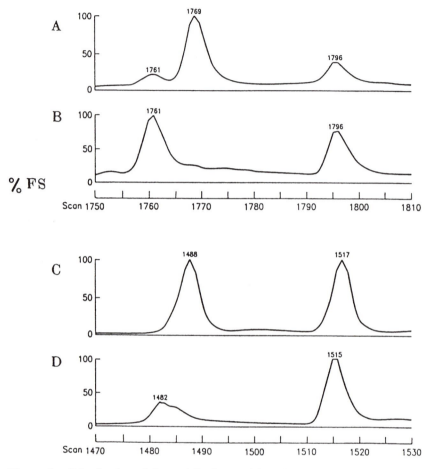

Figure 1. Distribution of C_{18} and C_{16} fatty acid methyl esters (FAMEs) isolated from fresh cells (A and C), and nonculturable cells (B and D) of *Aeromonas salmonicida*. The positions at which specific FAMEs elute are as follows: $C_{18:1}$ 1769 (type 1) and 1761 (type 2), $C_{18:0}$ 1796 (A and B); $C_{16:1}$, 1488 (type 1) and 1482 (type 2); and $C_{16:0}$, 1517 (C and D). A change in the retention time of $C_{16:1}$ and $C_{18:1}$ FAMEs was found to be due to shift in the position of the double bond; %FS, percentage of the full scale of the total ionization current.

3. GENERAL BIOCHEMICAL MARKERS

3.1 Volatile Metabolites

There are a number of biochemical markers not associated with the cell envelope that allow the specific detection of individual microorganisms in environmental samples. These include secondary alcohols. For example, *Mycobacterium xenopi* can be detected through the hydrolysis of wax ester mycolates, which liberates 2-docosanol, a characteristic and dominant secondary alcohol which can be detected at low levels by GC–MS. This biomarker was found to be very useful for the rapid detection of *M. xenopi* in drinking water (43). Results from the GC–MS detection of 2-docosanol were obtained within 2 days compared to the 12 weeks required for culturable detection of *M. xenopi*. The detection limit for this type of approach was also found to be 10^3 CFU ml^{-1} drinking water.

Volatile fungal metabolites can also be used as indicators of fungal growth in samples such as stored cereals and wheat. Metabolites 3-octanone, 1-octen-3-ol, and 3-methyl-1-butanol, 3-methylfuran, or specific groups of compounds, like the carbonyl compounds, provide a number of potential targets (44–47). In general, the production of a volatile metabolite which is to be used as a biomarker must not change with substrate type or level, an essential feature that needs to be addressed in such studies. The product 1-octen-3-ol is produced during the breakdown of lipids and varies with different lipid content in the substrate (48). This limits its use as a biomarker of biomass; however, this type of approach will provide useful information on the type of substrate being attacked in undefined substrates. Since the production of volatile metabolites can be influenced by the duration of fungal growth (44), it could be essential to consider the stage of fungal growth in any study when these products are used as biomarkers. Borjesson et al. (44) indicated that 3-methylfuran could be used as a good indicator of fungal growth with little variation related to fungal species or growth substrate. This study also illustrated that terpenes (molecules composed of C$_5$ isoprene units) were the most suitable compounds for differentiating between fungal species and could be used to differentiate between fungi growing on different substrates. The value of these approaches is, however, questionable, since studies on mixed bacterial and fungal populations were not performed.

Pyrolysis mass spectroscopy has been used for the rapid classification of microorganisms. The complex molecules present in a microorganism are degraded through cleavage at their weakest point to produce smaller, volatile fragments such as methane, ammonia, water, methanol, and H$_2$S, which are separated in a mass spectrometer to produce a pyrolysis mass spectrum that can be used as a chemical profile or fingerprint of the complex material analyzed (49). This method has been used in microbial identification, detection of chemical constituents of microbial samples (production of indole), and monitoring of production of recombinant proteins or secondary metabolites (50). Further advances in the application of this method to the analysis of complex microbial mixtures could provide methods that could make a significant contribution to the study of natural communities in the future.

3.2 Biochemical and Immunological Tests

There are abundant biochemical methods which can be used to differentiate between cultured organisms, including the substrate-utilization Biolog system (51) or the API system (52,53), which can be varied to assist identification of particular target organisms. To overcome the technical problems of isolating individual colonies, Garland and Mills (54) investigated the overall pattern of sole-carbon-source utilization of mixed populations from environmental samples by placing cell suspensions directly onto the test media. This type of approach is limited since nonculturable cells are not included in the study, and the conditions within the test strip may favor certain types of organisms and so bias the results. Morgan and Pickup (55) in a similar study used the overall pattern of microbial peptidases (important in protein degradation), oxidases (enzymes that use molecular oxygen as an electron acceptor), and esterases (a large group of enzymes that hydrolyze ester bonds) to characterize mixed microbial populations on the basis of their enzyme activities. This approach is not applicable to soil samples, unless a suitable cell extraction method is developed. Specific enzymes can also be used as biomarkers for the detection of certain microbes; for example, β-galactosidase can be used to determine the presence of *Escherichia coli* in water (56).

There are also numerous immunologically based approaches to strain identification with monoclonal antibodies obtained from a cell line producing a single antibody molecule. Among many examples of bacteria targeted by this approach are *Vibrio* spp. (57) and *Salmonella* spp. (58). By using fluorescent antibody labeling techniques to study cells immobilized in gel beads, Hunik et al. (59) were able to study the spatial distribution in a biofilm of *Nitrosomonas europaea* and *Nitrobacter agilis* cells. There are other examples of fluorescently labeled antibodies used in conjunction with microscopic techniques to study bacterial populations in the environment (60,61). Many biochemical and antibody-based tests are of limited value when it comes to the study of natural populations in situ. The main problems with this technique are associated with background fluorescence, nonspecific binding to other cells, length of time required to visualize samples using a microscope, and the provision of a range of suitable antibodies. The flow cytometer may overcome some of these problems since it measures the physical or chemical characteristics of individual cells as they move in a fluid stream past optical or electronic sensors (62). The flow cytometer can analyze samples at a rate of over 2000 cells s^{-1} and uses a laser set at a specific wavelength which can reduce background fluorescence. Therefore, the technique offers a rapid and precise means for detecting, counting, and characterizing cells in mixed populations. Through the measurement of the fluorescence characteristics of cells, the flow cytometer can be used in a way comparable to the standard immuno-fluorescence method with improvements in speed and reliability, in addition to the ability to process numerous samples extremely quickly. Porter et al. (63) used flow cytometry to detect fluorescently labeled *Escherichia coli* and *Staphylococcus aureus* cells in lake water. The method was taken further by using the flow cytometer to sort cells from the water sample to produce a target population of greater than 70% purity. Therefore, this method could produce cell samples to which analyses designed for monocultures could be applied.

4. NUCLEIC ACID BIOMARKERS
4.1 Whole Genome Studies

Phenotypic assessment of microbial diversity only measures a restricted part of the total genetic information (genome) present. By using methods involving the detection of nucleic acids this problem can be overcome. One DNA-based method used traditionally to compare genome sequence similarities involves re-association of denatured DNA performed with mixtures of DNA of two organisms. Torsvik et al. (64,65) explored this technique to test whether reassociation of denatured DNA from mixtures of soil microorganisms could provide a measure of genotypic diversity (the range of different microorganisms present). It was found that the major part of DNA isolated from the bacterial fraction is very heterogeneous, with a $Cot_{1/2}$ (50% reassociation) of 4600, equivalent to 4000 completely different genomes of soil bacteria. The method also indicated that there may be populations of plasmids and/or bacteriophages which showed very rapid reassociation (5% of total DNA). Their results also suggested that most of the diversity was located in the part of the community that cannot be isolated and cultured by standard techniques. One of the problems associated with this method is that a small number of samples have to be maintained in a spectrophotometer for a long period under standard conditions, a situation that is not convenient in most laboratories.

Holben et al. (66) reported a method for obtaining a profile of a bacterial community by studying the guanine and cytosine percentage (% G + C) of DNA. The dye bisbenzimide binds preferentially to A + T base pairs of DNA and alters its buoyant density. A sample of DNA obtained from a bacterial community is mixed with the dye and subjected to equilibrium density centrifugation. The DNA bands according to its % G + C content. The DNA profile of the gradient is recorded and reflects the relative proportions of microorganisms with differing % G + C content in the sample. The gradient can also be extracted and each fraction analyzed further by polymerase chain reaction (PCR) or reassociation studies. Ultimately, a detailed profile of the community based on the % G + C content can be obtained.

4.2 Target Genes

Although Cot or % G + C content determinations provide information on the whole population, information concerning the types or identity of the cells present is not obtained. The analysis of genes encoding ribosomal RNA (rRNA) can overcome this. Sequence divergence among molecules of rRNA has served to define the primary lines of evolutionary descent and provided a framework for a natural classification of microorganisms. The largest data sets of complete sequences are for the 5S and 16S rRNAs. The first ecological studies were performed on 5S rRNA ($\equiv$120 nucleotides) where total rRNA was isolated directly from the environment and the various types of 5S rRNA were separated by high-resolution gel electrophoresis. After rRNA types were extracted from the gel and sequenced, the phylogenetic affinities of the contributing organisms could be determined from sequence comparisons with the databases. To date, 5S rRNA studies have been restricted to bacteria in environments of limited species diversity such as hot springs in Yellowstone National Park, United States (67), the microbial components of marine invertebrates in a hydrothermal vent (68), and bacteria inhabiting a copper recovery pond (69). The 16S rRNA gene ($\equiv$1600 nucleotides) allows a broader phylogenetic analysis. The approaches commonly adopted with 16S rRNA for the study of diversity within natural microbial communities are as follows:

rRNA gene isolation and sequence analysis

Total DNA is extracted from the sample, the 16S rRNA genes are amplified by PCR, and the products are cloned into a suitable plasmid vector. One colony developing on a selective plate containing a single 16S rRNA gene insert (clone) therefore represents an individual rRNA gene present within the total population. To distinguish them from each other in a rapid fashion, the clones are restriction-mapped. For more detailed studies, the clones are sequenced. In this way, comparisons of DNA profiles or sequence information are used to determine either the diversity or the identity of clones (70). Each individual clone is acting as a biomarker for an individual cell. Organisms that are dominant in the total population will have a better chance of being represented among the clones and a picture of the population is achieved through the analysis of numerous clones (>100). One complication associated with this approach is the generation of 16S rRNA molecules that did not exist in the original sample but were created by the PCR process. These products can be created by an error in the copying process so that the incorrect nucleotide is inserted into the sequence. This occurs at a predictable frequency depending on the type of enzyme used and the number of cycles. In general, this is considered a minor problem in the context of the scale of these studies. However, during PCR amplification two different 16S rRNA molecules can combine to produce a hybrid (chimeric) molecule with areas representing part of each of the original molecule. The frequency of production of chimeric molecules within a standard reaction is approximately 10% for templates that are 82% similar but rises to 30% of the final product with coamplification of two nearly identical sequences (71). In DNA samples isolated from mixed populations, diverse genomes are present, and the formation of chimeric molecules leading to the description of nonexistent species is a problem that still needs to be addressed.

RFLP analysis of mixed rRNA gene pool

The PCR amplification of 16S rRNA genes leads to a mixed PCR product containing fragments of similar size. The product is digested with restriction enzymes that cleave DNA at specific sequences and the banding pattern of the DNA is analyzed by agarose gel electrophoresis that separates the DNA fragments on the basis of their size. The complexity of the banding pattern is used to determine the complexity of the original population (72). In this simple approach, many more samples can be processed since only limited gel analysis is required for each population. However, sequence information to aid identification or assess species similarities will not be obtained and the resolution of this method has not been fully determined.

In situ hybridization

In situ hybridization is often used to confirm the presence of microorganisms in samples where 16S rRNA sequence analysis has indicated their existence. This method is especially important when sequences that do not resemble any microorganism that has been cultured have been obtained. The method is also used to detect specific species or strains present in samples. The 16S rRNA gene sequence is used to design fluorescently labeled synthetic oligonucleotide probes (15 to 22 bases long) which are hybridized to the rRNA present within fixed cells. Those cells containing 16S rRNA regions homologous to the labeled probe retain the fluorescent tag and are detected by microscopy or flow cytometry. The probes can be designed at different levels of identity: kingdom, genus, or species (73–75). Embley and Findlay (76) used such an approach to analyze the relationship between methanogenic archael endosymbionts within their anaerobic ciliate hosts. Such applications illustrate the exciting potential for this technique, which can provide information on both the number and activity of cells. Confocal laser microscopy is a new version of fluorescence microscopy that uses a series of lasers to provide a light source at a specific wavelength and focal distance. This method allows the quantification of the fluorescence of individual cells, three-dimensional imaging, and reduced background noise. With these advantages, this method could be an important development to the use of fluorescent tags in microbial studies.

Denaturing gradient gel electrophoresis

Muyzer et al. (77) have developed an interesting alternative to the analysis of PCR products obtained from samples by using denaturing gradient gel electrophoresis (DGGE). In this method, PCR-amplified fragments are separated from one another on the basis of their DNA composition. As the PCR product migrates through a gel containing a chemical gradient of formamide/urea, the two DNA strands begin to melt at the denaturant concentration equal to its melting temperature; PCR products with slightly different base compositions melt at different locations within the gel. The position and number of each band on a gel reflect the sequence diversity within the PCR product. The method is fully described in Chapter 12b. Initially, DGGE was developed to identify mutations in specific genes. Muyzer et al. (77) used it to differentiate amplified products from 16S rRNA genes where the products from the strains were identical in length but different in sequence composition. It can also provide information on the sequence diversity of a wide variety of other genes that produce PCR products of a similar size. Therefore, it provides an interesting alternative to the use of restriction enzyme site analysis or DNA sequencing, which are commonly employed to differentiate PCR products and screen 16S rRNA clone banks (78–80). The method has also been applied to the direct analysis of microbial communities in situ, where the initial mixed PCR product was analyzed directly (77). Using DGGE in this way it was possible to identify members down to 1% of the total population.

Approaches to targeting of DNA

There are many approaches for the targeting of DNA as a biomarker. If cells in a microbial population are capable of growth on conventional media, then colony hybridization, using specific labeled probes, can be used to identify target organisms in a mixed population. Often, areas of the 16S rRNA gene provide the necessary specific probe target sequence. Such studies may, however, require detection without the necessity to culture. Depending on the sensitivity required, it may be possible to apply nucleic acid hybridization by treating an environmental sample directly. If the target gene is not present in a significant quantity, the nucleic acids can be purified and concentrated from the sample prior to hybridization. With modifications both approaches can be made quantitative. However, this can be extremely problematical under certain circumstances. Approximate detection limits of 10^3 cells per milliliter (water) or gram (soil or sediment) of sample have been achieved (81). The PCR amplification of DNA sequences improves the detection of genes considerably, with sensitivity to one cell per 100 ml or 10 g of sample. Indeed, the in situ detection of pathogens such as *Legionella* spp. (82,83) and *Salmonella* (84) in environmental samples is becoming routine. Fluit et al. (85) were able to detect *Listeria monocytogenes* in cheese by combining PCR with the initial separation of the target population using antibody-coated iron beads which are attracted to a strong magnet during separations.

Although the 16S rRNA gene is often the target for such studies, because of its strictly essential nature there are limitations to its use. Often there is little 16S rRNA gene divergence between closely related species. Other areas of the microbial chromosome show greater variation and could be used to differentiate closely related species. However, for an approach to be applicable to numerous strains there is the requirement for a conserved region to be present at either end of the amplified sequence to allow the PCR primers to bind. The spacer regions between rRNA genes offer a suitable target. Conserved areas within the rRNA genes can be used for primer binding and allow amplification of the variable spacer region between the genes. The sequence between the 16S rRNA and 23S rRNA genes has been used for this type of study and gives better resolution between strains than the 16S rRNA gene alone (86). Alternatively, variable coding sequences have been used as targets for PCR. The numerous examples often include genes encoding interesting metabolic activities or virulence factor, but can involve other variable genes, such as the flagellin gene, which encodes the structural protein present in bacterial flagella (87). Figure 2 shows an agarose gel of PCR-amplified *Pseudomonas* flagellin genes. Diversity can be assessed by variations in product size, restriction sites, and DNA sequence. Other examples include the detection of polychlorinated biphenyl (PCB) degradation genes in polluted sediments (88), the hypersensitivity–pathogenicity *(hrp)* genes in phytopathogenic *Xanthomonas* (89), the pectate lyase *(pel)* gene in *Erwinia carotovora* (90), the invasive associated protein *(iap)* gene encoding an extracellular protein P60 in *Listeria* spp, (91) and a virulence gene in *Aeromonas salmonicida* (92). For use in PCR-based detection systems, the only prerequisites are that target genes should contain conserved regions from which oligonucleotide primers can be designed, as well as variable regions to give a measure of diversity, and that the target gene is universally present in all individuals of a species or strain. This method can be used for cultured isolates or used directly, employing DNA obtained from the environment; the attraction is that only DNA from the target population is amplified.

4.3 Miscellaneous Methods

Other DNA-based methods for the resolution of cultured isolates include comparisons of genome size using pulsed-field gel electrophoresis (PFGE) (93) and the use of repetitive DNA sequences for fingerprinting (94) including the genera-

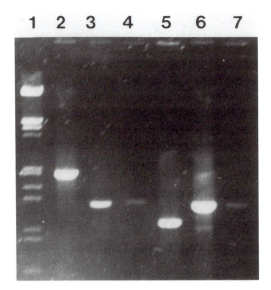

Figure 2. Polymerase chain reaction (PCR) amplification of flagellin gene sequences. The agarose gel shows flagellin gene PCR products obtained from *Pseudomonas putida* Paw8 (lane 2), *P. putida* PRS2000 (lane 3), *P. aeruginosa* PAO1 (lane 4), *P. aeruginosa* NCIB82295 (lane 5), *P. fluorescens* FH1 (lane 6), and *P. putida* FBA11 (lane 7). Lambda DNA digested with *Hind*III and *Eco*RI was used as a size marker (lane 1), 21.2, 5.1 (4.9), 4.2, 3.5, 2.0, 1.9, 1.5, 1.3, 0.9, 0.83, 0.56 kb. The primers used were CW45 and CW46 (87), which bind to conserved regions within the flagellin gene.

tion of random amplified polymorphic DNA (RAPD) patterns (95,96). Jensen et al. (97) used the amplification of the 16S rRNA-23S rRNA spacer region to develop a rapid identification system for bacteria based on spacer polymorphisms. A single primer set was used to amplify 16S–23S spacer regions for 300 strains of bacteria. They found that the size of the amplified products generally clustered isolates at the species level. Therefore, this method offers the potential to group and identify rapidly isolates obtained from mixed samples.

4.4 Reporter DNA

It is clear that DNA provides a very useful and versatile target as a biomarker. Genetic engineering now makes it possible to insert specific genes into prokaryotic, eukaryotic, or viral hosts. Many versatile marker systems have been developed and introduced into different microorganisms to allow assessment of the survival and spread of strains, gene transfer, and cell activity in model environments and field studies. A candidate marker gene must fulfill certain criteria. It has to be absent from the strain used in the study and either absent or sufficiently low in abundance in the microbial population under study to allow detection of marked cells at a level of sensitivity appropriate for the intended study. Potential markers include antibiotic resistance genes, which in specific combinations can be used to overcome a high background of naturally occurring antibiotic-resistant bacteria. However, selection of a strain carrying antibiotic resistance markers on highly selective media can reduce host pathogenicity or persistence, and the introduction of antibiotic resistance into natural systems may not be considered desirable. For these reasons it has been necessary to develop marker systems that do not rely upon antibiotic resistance. Marker systems can vary considerably. The *xylE* gene from *Pseudomonas putida* (98) encodes the enzyme catechol 2,3-dioxygenase, which can be detected by a simple colorimetric test. Bacterial colonies marked with this gene can be identified after application of catechol, which is colorless. The appearance of a yellow colored product (2-hydroxymuconic semialdehyde) indicates expression of *xylE*. In a similar way, the *lacZY* genes from *Escherichia coli* can be inserted in other lactose-minus species and used as a colorimetric marker (colorless to blue transitions; see Chapter 14) for their detection (e.g., *Pseudomonas* spp.) (99). More recently, the β-glucuronidase gene (*gusA*) from *E coli* has become a very useful marker (100), partly because of the range of fluorescent substrates available for the enzyme β-glucuronidase and its rare occurrence in environmental samples. This has allowed the detection of single cells in situ by fluorescence microscopy (101). The *gusA* gene has also been introduced into fungal strains, such as *Fusarium oxysporum* (102). The luciferase-encoding genes (*luxAB*) isolated from a species of *Vibrio* (103), when expressed under the correct conditions and in the presence of an aldehyde, allow bacteria to produce light. This marker has proved to be very useful for in situ detection when combined with luminometry or specialized light-sensitive cameras fitted to a microscope. The *lux*AB marker is widely used since it is uncommon to find large numbers of indigenous microorganisms that emit light. Simple cell detection methods can be used, and very sensitive detection methods are available, if needed.

Reporter genes can be manipulated to give an indication of the activity of a desirable phenotype. King et al. (104) constructed a bioluminescent reporter bacterium for the monitoring of naphthalene and salicylate bioavailability and catabolism. The system involved the use of a reporter plasmid encoding naphthalene degradation and the bioluminescence reporter function. A transcriptional fusion between the *luxCDABE* gene cassette of *Vibrio fischeri* and the *nahG* gene from the naphthalene degradation lower operon ensured that in the presence of salicylate or naphthalene the lower operon is induced, leading to bioluminescence. On-line monitoring of catabolic activity in waste streams can be achieved by using an optical whole-cell biosensor to monitor light output continually (105). Such methods offer the potential to monitor a variety of environmental samples for a range of chemicals for which an understanding of gene regulation is available.

There are limitations to the use of reporter genes. In the case of the *lacZY* marker gene, cultures of the microorganism are essential for detection, whereas the use of *lux* marker systems may be restricted by the low metabolic activity of cells in situ. In some environments, the presence of *xylE*-containing pseudomonads capable of the catabolism of aromatic compounds precludes the use of any PCR-based method for the detection of released strains marked with this gene. However, the inherent limitation of all introduced genetic markers is that they are limited to studies on culturable microorganisms that are amenable to genetic manipulation.

5. MICROBIAL POPULATIONS AS BIOMARKERS

A number of studies have been conducted to use changes in the fauna of soils to serve as indicators of disturbance. These have included examining soils in regions of radioactive contaminations such as Chernobyl (106) and in areas contaminated with toxic chemicals (107). Microbial populations can be used as biomarkers in a similar way. This type of approach has formed a significant part of risk assessment studies that investigate the release of genetically modified microorganisms (GMMs) in the environment. It involves the use of colony-forming units as biomarkers of community structures. Such markers include estimates of species diversity, detection of nutritional groups, and antibiotic-resistance profile of populations (108). In a typical study, selective media or discriminatory tests are used to monitor groups of organisms such as pseudomonads (P1 medium, Pseudomonas isolation agar), members of the enterobacteriaceae (e.g., *E. coli*, MacConkey agar), spore formers (*e.g., Bacillus* spp. using heat treatments), chitin-utilizing bacteria (clearing zones on chitin-containing agar), cellulose-utilizing bacteria (clearing zones on cellulose-containing agar), denitrifying bacteria and nitrate-reducing bacteria (most probable number [MPN] systems), yeasts and filamentous fungi (potato dextrose agar), and protozoa (microscopic counts). This approach offers considerably more information than total microbial counts alone but is still restrictive, compared to the techniques described earlier in this chapter. De Leij et al. (109) have further developed the dilution plate count technique by measuring the rate at which colonies develop on each medium. The population appearing as colonies was split into *r* strategists (fast-growing opportunists) and *K* strategists (slower-growing specialists). This simple alteration of a standard method provides information on the microbial community, without becoming so labor-intensive that it severely restricts the number of samples that can be processed. A combination of culture and analysis of other cellular biomarkers may be useful in future studies on natural microbial populations. This approach is inherently limited, since it cannot be used to detect nonculturable microorganisms which normally predominate in these types of samples.

6. CONCLUSIONS

It has taken time to develop methods that meet the desire of microbial ecologists to learn more about microbial communities in natural environments like soil. In recent years, new techniques have offered the potential to answer some of the questions commonly asked. Although questions about the interpretation and relevance of some modern methods regarding the real makeup and interactions in mixed populations remain, there is no doubt that the use of biochemical or molecular signatures provides a great opportunity to advance our knowledge in this area.

ACKNOWLEDGMENT

This work was supported by the Biotechnology and Biological Sciences Research Council BBSRC, United Kingdom.

REFERENCES

1. T. J. Brock, *Principles of Microbial Ecology*, Prentice-Hall, Englewood Cliffs, New Jersey (1966).
2. J. A. W. Morgan, *Sci. Progress 75*:265 (1991).
3. D. B. Roszak and R. R. Colwell, *Microb. Rev. 51*:365 (1987).
4. D. C. White, D. B. Ringelberg, D. B. Hedrick, and D. E. Nivens, *ACS Symposium Series 541*:8 (1994)
5. M. P. Lechevalier, *Crit. Rev. Microbiol. 5*:109 (1977).
6. L. M. Mallory and G. S. Sayler, *Microb. Ecol. 10*:283 (1984).
7. D. C. White, *Soc. Gen. Microb. Symp. 34*:37 (1983).
8. D. C. White, *Arch. Hydrobiol. Beih. Ergebn. Limnol. 31*:1 (1988).
9. A. Frostegard, A. Tunlid, and E. Baath, *Appl. Environ. Microbiol. 59*:3605 (1993).
10. D. C. White, *Microbes in Their Natural Environments* (L. H. Slater, R. Whittenbury, and J. W. T. Wimpenny, eds.) SGM, London, Vol. 34, p. 37 (1983).
11. R. J. Bobbie and D. C. White, *Appl. Environ. Microbiol. 39*:1212 (1980).
12. J. B. Guckert, C. B. Antworth, P. D. Nichols, and D. C. White, *FEMS Microb. Ecol. 31*:147 (1985).

13. J. Boon, J. W. Leeuw, and P. A. Schenk, *Geochem. Cosmochem. Acta 39*:1559 (1975).
14. R. J. Parkes and J. Taylor, *Estuarine Coastal Shelf Sci. 16*:173 (1983).
15. N. Rajendran, O. Matsudu, N. Imamura, and Y. Urushigawa, *Appl. Environ. Microbiol. 58*:562 (1992).
16. P. D. Nichols, J. M. Henson, C. P. Antworth, J. Parson, J. T. Wilson, and D. C. White, *Environ. Toxicol. Chem. 6*:89 (1987).
17. L. Zelles, Q. Y. Bai, T. Beck, and F. Beese, *Soil Biol. Biochem. 24*:317 (1992).
18. J. L. Harwood and N. J. Russel, *Lipids in Plants and Microbes*, Allen & Unwin, London (1984).
19. B. H. Baird and D. C. White, *Benthic Ecology and Sedimentary Processes of Venezuela Basin: Past and Present* (D. K. Young and M. D. Richardson, eds.), Elsevier, Amsterdam, p. 217 (1985).
20. M. P. Lechevalier, *Crit. Review. Microbiol. 5*:109 (1976).
21. M. Kates, *Adv. Lipid Res. 3*:17 (1977).
22. R. J. Parkes and J. Taylor, *Estur. Coastal Shelf Sci. 16*:173 (1983).
23. R. J. Bobbie and D. C. White, *Appl. Environ. Microbiol. 39*:1212 (1980).
24. R. H. Findlay, J. W. Fell, N. K. Coleman, and J. R. Vestal, *Biology of Marine Fungi* (S. T. Moss, ed.), Cambridge University Press, Cambridge (1986).
25. F. T. Gillan and R. W. Hogg, *J. Microb. Methods 2*:275 (1984).
26. E. Baath, A. Frostegard, and H. Fritze, *Appl. Environ. Microbiol. 58*:4026 (1992).
27. B. H. Baird, and D. C. White, *Mar. Geol. 68*:217 (1985).
28. J. P. Bowman, J. H. Skerratt, P. D. Nichols, and L. I. Sly, *FEMS Microb. Ecol. 85*:15 (1991).
29. G. J. Perry, J. K. Volkman, and R. B. Johns, *Geochim. Cosmochim. Acta 43*:1715 (1979).
30. T. Seiki, E. Date, and H. Izawa, *Mar. Pollut. Bull. 23*:95 (1991).
31. J. B. Guckert, C. P. Antworth, P. D. Nichols, and D. C. White, *FEMS Microb. Ecol. 31*:147 (1985).
32. J. B. Guckert, M. A. Hood, and D. C. White, *Appl. Environ. Microbiol. 52*:794 (1987).
33. P. D. Nichols, A. C. Palmisano, G. A. Smith, and D. C. White, *Phytochemistry 25*:1649 (1986).
34. J. A. W. Morgan, P. A. Cranwell, and R. W. Pickup, *Appl. Environm. Microbiol. 57*:1777 (1991).
35. J. E. Nylund and H. Wallander, *Meth. Microbiol. 24*:77 (1992).
36. M. O. Gessner and E. Chauvet, *Appl. Environ. Microbiol. 59*:502 (1993).
37. K. Suberkropp, M. O. Gessner, and E. Chauvet, *Appl. Environ. Microbiol. 59*:3367 (1993).

38. A. Lipski, S. Klatte, B. Bendinger, and K. Altenhorf, *Appl. Environ. Microbiol. 58*:2053 (1992).
39. A. Fox, R. M. T. Rosario, and L. Larsson, *Appl. Environ. Microbiol. 59*:4354 (1993).
40. J. T. Belisle, M. E. Brandt, J. D. Radolf, and M. V. Norgard, *J. Bacteriol. 176*:2151 (1994).
41. T. Ojanen, I. M. Helander, K. Haahtela, T. K. Korhonen, and T. Laakso, *Appl. Environ. Microbiol. 59*:4143 (1993).
42. F. Siverio, M. Cambra, M. T. Gorris, J. Corzo, and M. M. Lopez, *Appl. Environ. Microbiol. 59*:1805 (1993).
43. S. Alugupalli, L. Larsson, M. Slosarek, and M. Jaresova, *Appl. Environ. Microbiol. 58*:3538 (1992).
44. T. Borjesson, U. Stollman, and J. Schnurer, *Appl. Environ. Microbiol. 58*:2599 (1992).
45. D. Abramson, R. N. Sinha, and J. T. Mills, *Cereal Chem. 57*:346 (1980).
46. E. Kaminski, S. Stawicki, and E. Wasowicz, *Acta Aliment. Pol. 1*:153 (1975).
47. D. Tuma, R. N. Sinha, W. E. Muir, and D. Abramson, *Int. J. Food Microbiol. 8*:11 (1989).
48. M. Wurzenberger and W. Grosch, *Z. Lebensm.-Unters.-Forsch. 175*:186 (1982).
49. R. C. W. Berkeley, R. Goodacre, R. J. Helyer, and T. Kelley, *Lab Pract. 39*:81 (1990).
50. R. Goodacre, *Microbiol. Eur. 2*:19 (1994).
51. Klingler, J. M., R. P. Stowe, D. C. Obenhuber, T. O. Groves, S. K. Mishra, and D. L. Pierson, *Appl. Environ. Microbiol. 58*:2089 (1992).
52. Breschel, T. S. and F. L. Singleton, *Appl. Environ. Microbiol. 58*:21 (1992).
53. Bille, J., B. Catimel, E. Bannerman, C. Jacquet, M.-N. Yersin, I. Caniaux, D. Monget, and J. Rocourt, *Appl. Environ. Microbiol. 58*:1857 (1992).
54. J. L. Garland and A. L. Mills, *Appl. Environ. Microbiol. 57*:2351 (1991).
55. J. A. W. Morgan and R. W. Pickup, *Can. J. Microbiol. 39*:795 (1993).
56. R. Nir, Y. Yisraeli, R. Lamed, and E. Sahar, *Appl. Environ. Microbiol. 56*:3861 (1990).
57. D. Chen, P. J. Hanna, K. Altmann, A. Smith, P. Moon, and L. S. Hammond, *Appl. Environ. Microbiol. 58*:3694 (1992).
58. R. Torensma, M. J. C. Visser, C. J. M. Aarsman, M. J. J. G. Poppelier, R. van Beurden, and A. C. Fluit, *Appl. Environ. Microbiol. 58*:3868 (1992).
59. J. H. Hunik, M. P. van den Hoogen, W. de Boer, M. Smit, and J. Tramper, *Appl. Environ. Microbiol. 59*:1951 (1993).
60. G. Muyzer, A. C. de Bruyn, D. J. M. Schmedding, P. Bos, P. Westbrok, and G. J. Kuenen, *Appl. Environ. Microbiol. 53*:660 (1987).
61. P. R. Brayton, M. L. Tamplin, A. Huq, and R. R. Colwell, *Appl. Environ. Microbiol. 54*:2862 (1987).

62. M. R. Melamed, T. Lindmo, and M. L. Mendelsohn, *Flow Cytometry and Cell Sorting*, Wiley-Liss, New York (1990).

63. J. Porter, C. Edwards, J. A. W. Morgan, and R. W. Pickup, *Appl. Environ. Microbiol. 59*:3327 (1993).

64. V. Torsvik, J. Goksoyr, and F. L. Daae, *Appl. Environ. Microbiol. 56*:782 (1990).

65. V. Torsvik, K. Salte, R. Sorheim, and J. Goksoyr, *Appl. Environ. Microbiol. 56*:776 (1990).

66. W. E. Holben, V. G. M. Calabrese, D. Harris, J. O. Ka, and J. M. Tiedje, *Trends in Microbial Ecology* (R. Guerrero and C. Pedros-Alio, eds). p. 367 (1993).

67. D. A. Stahl, D. J. Lane, G. J. Olsen, and N. R. Pace, *Appl. Environ. Microbiol. 49*:1379 (1985).

68. D. J. Lane, D. A. Stahl, G. J. Olsen, and N. R. Pace, *Biol. Soc. Wash. Bull. 6*:389 (1985).

69. G. J. Olsen, D. J. Lane, S. J. Giovannoni, and N. R. Pace, *Annu. Rev. Microbiol. 40*:337 (1986).

70. D. M. Ward, R. Weller, and M. M. Bateson, *Nature 345*:63 (1990).

71. G. C. -Y. Wang and Y. Wang, *Microbiology 142*:1107 (1996).

72. E. Hicks, R. I. Amann, and D. A. Stahl, *Appl. Environ. Microbiol. 58*:2158 (1992).

73. R. I. Amann, L. Krumholz, and D. A. Stahl, *J. Bacteriol. 172*:762 (1990).

74. E. F. DeLong, G. S. Wickham, and N. R. Pace, *Science 243*:1360 (1989).

75. H. C. Tsien, B. J. Bratina, K. Tsuji, and R. S. Hanson, *Appl. Environm. Microbiol. 56*:2858 (1990).

76. M. Embley and B. Findlay, *Microbiol. 140*:225 (1994).

77. G. Muyzer, E. C. de Waal, and A. G. Uitterlinden, *Appl. Environ. Microbiol. 59*:695 (1993).

78. Y. Berthier, V. Verdier, J-L. Guesdon, D. Chevrier, J-B. Denis, G. Decoux, and M. Lemattre, *Appl. Environ. Microbiol. 59*:851 (1993).

79. G. Laguerre, M-R. Allard, F. Revoy, and M. Amarger, *Appl. Environ. Microbiol. 60*:56 (1994).

80. L. Bernier, R. C. Hamelin, and G. B. Ouellette, *Appl. Environ. Microbiol. 60*:1279 (1994).

81. J. A. W. Morgan, C. Winstanley, R. W. Pickup, J. G. Jones, and J. R. Saunders, *Appl. Environ. Microbiol. 55*:2537 (1989).

82. M. Koide, A. Saito, N. Kusano, and F. Higa, *Appl. Environ. Microbiol. 59*:1943 (1993).

83. C. J. Palmer, Y-L. Tsai, C. Paszko-Kolva, C. Mayer, and L. R. Sangermano, *Appl. Environ. Microbiol. 59*:3618 (1993).

84. J. S. Way, K. L. Josephson, S. D. Pillai, M. Abbaszadegan, C. P. Gerba, and I. L. Pepper, *Appl. Environ. Microbiol. 59*:1473 (1993).

85. A. C. Fluit, R. Torensma, M. J. C. Visser, C. J. M. Aarsman, M. J. J. P. Poppelier, B. H. I. Keller, P. Klapwijk, and J. Verhoef, *Appl. Environ. Microbiol. 59*:1289 (1993).

86. R. Frothingham and K. H. Wilson, *J. Bacteriol. 175*:2818 (1993).

87. C. Winstanley, J. A. W. Morgan, R. W. Pickup, and J. R. Saunders, *Microbiol. 140*:2019 (1994).

88. R. W. Ebb, and I. Wagner-Dobler, *Appl. Environ. Microbiol. 59*:4065 (1993).

89. R. P. Leite, G. V. Minsavage, U. Bonas, and R. E. Stahl, *Appl. Environ. Microbiol. 60*:1068 (1994).

90. A. Darrasse, S. Priou, A. Kotoujansky, and Y. Bertheau, *Appl. Environ. Microbiol. 60*:1437 (1994).

91. A. Bubert, S. Kohler, and W. Goebel, *Appl. Environ. Microbiol. 58*:2625 (1992).

92. C. E. Gustafson, C. J. Thomas, and T. J. Trust, *Appl. Environ. Microbiol. 58*:3816 (1992).

93. K. D. Harsono, C. W. Kaspar, and J. B. Luchansky, *Appl. Environ. Microbiol. 59*:3141 (1993).

94. R. N. Waterhouse and L. A. Glover, *Appl. Environ. Microbiol. 59*:1391 (1993).

95. L. M. Lawrence, J. Harvey, and A. Gilmour, *Appl. Environ. Microbiol. 59*:3117 (1993).

96. R. C. Hamelin, G. B. Ouellette, and L. Bernier, *Appl. Environ. Microbiol. 59*:1752 (1993).

97. M. A. Jensen, J. A. Webster, and N. Straus, *Appl. Environ. Microbiol. 59*:945 (1993).

98. C. Winstanley, J. A. W. Morgan, R. W. Pickup, J. G. Jones, and J. R. Saunders, *Appl. Environ. Microbiol. 55*:771 (1989).

99. D. J. Drahos, B. C. Hemming, and S. McPherson, *BioTechnol. 4*:439 (1986).

100. R. A. Jefferson, *Nature 342*:837 (1989).

101. P. J. Robinson, J. T. Walker, C. W. Keevil, and J. Cole, *FEMS Microbiol. Lett. 129*:183 (1995).

102. Y. Couteaudier, M-J. Daboussi, A. Eparvier, T. Langin, and J. Orcival, *Appl. Environ. Microbiol. 59*:1767 (1993).

103. J. J. Shaw, F. Dane, D. Geiger, and J. W. Kloepper, *Appl. Environ. Microbiol. 58*:267 (1992).

104. J. M. H. King, P. M. DiGrazia, B. Applegate, R. Burlage, J. Sanseverino, P. Dunbar, F. Larimer, and G. S. Sayler, *Science 249*:778 (1990).

105. A. Heitzer, K. Malachowsky, J. E. Thonnard, P. R. Bienkowski, D. C. White, and G. S. Sayler, *Appl. Environ. Microbiol.* *60*:1487 (1994).

106. D. A. Krivolutzkii, and A. D. Pokarzhevskii, *Bioindicators and Environmental Management* (D. W. Jeffrey, and B. Modden, eds.), Academic Press, London (1991).

107. H. Koehler, *Agric. Ecosystem. Environ., 40*:193 (1992).

108. J. D. Doyle and G. Stotzky, *Microb. Releases.* 2:63 (1993).

109. F. A. A. M. de Leij, E. J. Sutton, J. M. Whipps, J. M. Lynch, *FEMS Microb. Lett. 13*:249 (1994).

12b

Application of Denaturing Gradient Gel Electrophoresis and Temperature Gradient Gel Electrophoresis for Studying Soil Microbial Communities

HOLGER HEUER and KORNELIA SMALLA Federal Biological Research Center for Agriculture and Forestry, Braunschweig, Germany

1. INTRODUCTION

In the past, studies on the diversity of natural microbial populations have been restricted to culturable microorganisms. Therefore, the understanding and knowledge of the dynamics of natural microbial communities have remained limited because only a minor fraction of all cells in natural ecosystems is accessible to cultivation techniques. Only recently have methods for direct extraction of nucleic acids from different environmental samples become available, allowing a cultivation-independent analysis of microbial communities. Direct deoxyribonucleic acid (DNA) extraction is based on direct lysis in the presence of the soil matrix (1–9) or, after extraction of the bacterial fraction by repeated differential centrifugation steps (10–13). Protocols for extraction of ribosomal ribonucleic acid (RNA) from soils using direct cell lysis or ribosome extraction have been developed as well (14–16). Each method has certain advantages and the choice depends on the question being addressed.

The DNA reassociation studies performed by Torsvik et al. (17–19) using total DNA extracted directly from soil indicated an extremely high diversity in soil microbial communities. In fact, DNA analysis showed that the genetic diversity of the total microbial population in soil was 200 times higher than the diversity in the population of bacteria isolated from the same soil (19). The new molecular techniques based on total nucleic acid extraction followed by direct shotgun cloning, cloning of cDNA fragments after transcription of 16S ribosomal RNA (rRNA) or polymerase chain reaction (PCR) amplification of the 16S rRNA gene, cloning, and sequencing, have provided exciting cultivation-independent data on bacterial constituents of aquatic and terrestrial habitats. Pace et al. (20) were the first to suggest the use of the 16S rRNA gene as a molecular marker for cultivation-independent studies of microbial populations in environmental samples. The 16S and 23S rRNA molecules are ideal phylogenetic marker molecules because of their universal distribution, structural and functional conservation, and size, which allows for sufficient sequence divergence (21). Several studies proved that the dominant microorganisms isolated by cultivation from environmental samples do not match the most frequent species identified after sequencing of the PCR-amplified 16S rRNA gene from total DNA extracted directly from environmental samples (22–26). The conclusion drawn from these studies was that microorganisms isolated by cultivation might not represent the most dominant and significant microorganisms in the respective environmental samples. However, the cloning and sequencing strategies are rather time- and labor-consuming and thus not suitable for monitoring a large number of samples, e.g., in studies on the succession of microbial communities during the growing season, or following shifts of microbial communities after perturbations.

A new approach to the study of the structural diversity of microbial communities is the analysis of PCR-amplified DNA fragments by denaturing gradient gel electrophoresis (DGGE) or temperature gradient gel electrophoresis (TGGE). Denaturing gradient gel electrophoresis has been recently introduced in microbial ecology by Muyzer et al. (27). This technique has the potential needed for monitoring techniques since it offers the chance to analyze the major constituents of microbial communities by generating fingerprints. The intention of this chapter is to describe the potential of DGGE of rDNA or other functional genes amplified by PCR from mixed communities for soil and rhizosphere microbial ecology and to evaluate the limitations of this approach.

2. ANALYSIS OF PCR-AMPLIFIED DNA

2.1 How Do DGGE and TGGE Function?

Gradient gel electrophoresis was originally developed and used in medical research to detect point mutations (28–32). In the two types of gradient gel electrophoresis, TGGE and DGGE, DNA fragments of the same length but of different sequences can be separated according to their melting properties. When

DNA is electrophoresed through a linearly increasing gradient of denaturants, the fragments remain double-stranded until they reach the conditions that cause melting of the lower temperature melting domains. Branching of the molecule caused by partial melting of the regions with low melting temperature (T_m) sharply decreases the mobility of the DNA fragments in the gel. A GC-clamp (40–45 bases GC-rich sequence) attached to the 5′end of the forward primer prevents the complete melting of the PCR product (30,32). With DGGE, double-stranded DNA is separated in a linearly increasing denaturing gradient of urea and formamide at elevated temperature (60°C). In contrast, a linearly increasing temperature gradient in the presence of a constant high concentration of urea and formamide is used for separation of PCR products in TGGE. The gradient which allows an optimal separation of bands is generally determined by running so-called perpendicular gels and/or running a time course in a parallel gel using different gradients (see Fig. 1). The perpendicular gel has an increasing gradient of the denaturants which is perpendicular to the electrophoresis direction. The DNA is loaded in a slot across the entire width of the gel. A sigmoid-shape curve is visible after DNA staining. At the left side of the gel (low denaturant concentrations) the PCR products have high electrophoretic mobility since the DNA remains double-stranded, while migration at the right side of the gel (high denaturant concentration) is virtually stopped by high concentration of the denat-

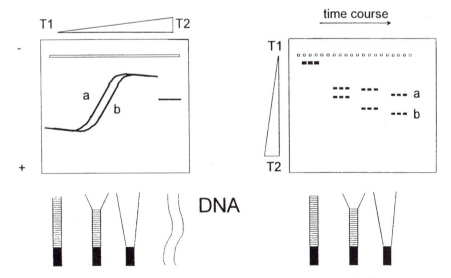

Figure 1. Schematic description of TGGE analysis of a 16S rDNA fragment amplified from a mixture of two strains (e.g., a, *Pseudomonas fluorescens*; b, *Pantoea agglomerans*) in a perpendicular gel (left side) and in a parallel gel (right side).

urant (urea/formamide or temperature), causing a very quick branching of the DNA molecule when entering the gel. The intermediate range of denaturant concentrations where different electrophoretic mobilities between the PCR products are seen can be used for separation of PCR products based on their different melting behavior in a 'parallel' gel. In parallel gels, the denaturant gradient (increasing from the top to the bottom) is parallel to the electrophoresis direction. The banding pattern visible after staining reflects the different melting behavior of the DNA sequences amplified from mixed communities. Running of time travels as described by Muyzer et al. (27) at different gradient concentrations is an alternative approach to finding the optimal conditions for separation of the different products.

Both TGGE (Diagen) and DGGE (Biorad) systems are commercially available. The separation capacity and the handling of both systems are similar. The advantage of the DGGE (Biorad) system is that two gels can be run in parallel and that the patterns are quite reproducible in different runs. We have compared both systems using PCR products amplified from potato rhizosphere DNA. Optimal separation of 16S rDNA fragments (products between nt 968 and 1401) was achieved in a DGGE gel with a 40% to 60% (100% denaturant is 7 M urea and 40% v/v deionized formamide) gradient and 5 hours running time at 60°C. A comparable separation was achieved for the same DNA sample in TGGE run with a temperature gradient from 38°C to 52°C, 8 M urea, and 20% formamide. Gels can be stained by ethidium bromide, Sybrgreen, or silver nitrate. The highest sensitivity can be reached with silver staining, which, however, can impede subsequent blotting or sequencing of bands (unpublished results). A further disadvantage of silver staining is that single-stranded DNA is stained as well. Since single-stranded DNA often represents the strongest band, automated evaluation of the banding patterns can be complicated.

In principle, DGGE or TGGE can be used for analysis of PCR-amplified ribosomal or any other functional genes from mixed communities, pure cultures, or clones. For studies on the structural diversity of microbial communities the 16S rRNA gene is especially useful because of the existence of alternating sequences of invariant, more or less conserved and highly variable regions. Polymerase chain reaction products obtained with various primer sets annealing to the conserved regions of the 16S rRNA gene and spanning one or up to three variable regions have been separated by DGGE or TGGE. Thus, primer sets 341 F/ 534 R (27), 341F/ 927 R (33), 968 F/ 1401 R (34), and 1055 F/ 1406 R (35) have been successfully employed. Separation of, e.g., 16S rDNA fragments amplified from community DNAs using such primers annealing to the conserved regions of the 16S rRNA gene is achieved by different melting behavior of the DNA fragments that is due to sequence differences in the variable regions.

When the diversity of other functional genes in mixed populations is intended to be studied, the bottleneck often is the availability of sequence information.

Ideally, several genes should be sequenced and aligned to obtain consensus sequences for the design of oligonucleotide primers which then anneal to regions conserved within a family of similar but not identical sequences (36).

2.2 Does the DGGE or TGGE Pattern Reflect the Composition of the Microbial Community?

Muyzer et al. (27) suggested that DGGE and TGGE analyses of PCR-amplified 16S rDNA fragments generate fingerprints of the most dominant constituents of mixed microbial populations when primers annealing to conserved regions of the 16S rRNA gene are used. This can be easily shown by 10-fold dilution series of the template (rhizosphere DNA) before PCR. The main bands of 16S rDNA amplified from potato rhizosphere DNA were still visible after diluting the templates 1:1000 (see Fig. 2). However, few bands appeared at higher dilutions which were not visible before (lanes 1, 2, 6, 7, and 11–13), indicating that these sequences were less efficiently amplified at higher concentrations of competing DNA.

For microbial communities which have a large number of equally abundant

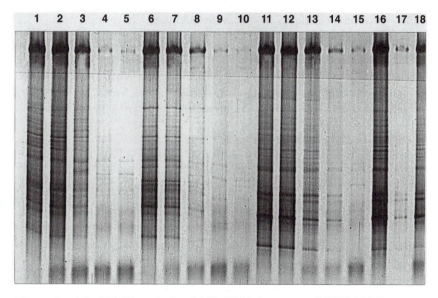

Figure 2. The TGGE analysis of 16S rDNA fragments (968F/1401R) amplified from potato rhizosphere DNAs serially diluted: Lanes 1–5 sample a: 1: 10^{-1}; 2: 10^{-2}; 3: 10^{-3}; 4: 10^{-4}; 5: 10^{-5}; lanes 6–10 sample b: 6: 10^{-1}; 7: 10^{-2}; 8: 10^{-3}; 9: 10^{-4}; 10: 10^{-5}; lanes 11–15 sample c: 11: nondiluted; 12: 10^{-1}; 13: 10^{-2}; 14: 10^{-3}; 15: 10^{-4}.

species, complex banding patterns will be observed. However, when only a few species dominate in numbers, the banding patterns will become less complex. When monitoring the fate of *Pseudomonas stutzeri* KC inoculated into a groundwater aquifer by DGGE and plate counts in parallel, the KC-specific band could still be detected when the strain represented approximately 1% of the total culturable bacterial population, and this can be suggested to represent an empirical threshold for efficient detection via PCR followed by DGGE or TGGE (Smalla et al., in preparation).

2.3 How Can the Resolution and Information of DGGE or TGGE Fingerprints Be Increased?

When DGGE or TGGE analyses of PCR products amplified with primers homologous to conserved regions are performed to detect shifts of the structural diversity of a soil or rhizosphere community, changes will only be detected when predominant populations are affected. If, for instance, the number or the composition of a rhizobial population in a bulk soil is changed, this may not be detected since the proportion of rhizobia is often less than 0.01% of the total bulk soil population. On the other hand, very complex banding patterns may be difficult to evaluate and thus techniques to reduce the complexity have to be applied. To overcome these limitations, several strategies might be successful:

   Fractionation of the DNA according to its % G + C prior to PCR and DGGE or TGGE
   Development of group-specific primers
   Hybridization of DGGE or TGGE patterns with probes

Fractionation of the DNA according to its % G + C

The separation of the soil DNA according to its % G + C using bisbenzimide (37,38) prior to PCR and DGGE analysis can be used to increase the sensitivity by reducing the complexity of the template DNA. The dye bisbenzimide (Hoechst 33258) binds to double-stranded DNA, preferentially to A–T base-pair-rich regions. The buoyant density of the DNA–bisbenzimide complex is reduced in proportion to the dye bound (38), and hence these complexes can be separated according to the % G + C by isopycnic centrifugation in CsCl gradients. Most aerobic soils studied by Harris have shown similar patterns with a dominant peak at about 68% G + C. These DNA fractions with a different % G + C can be used as templates for PCR with DGGE/TGGE primers (39).

   Recently, Wawer et al. (40) suggested the use of agarose gels containing bisbenzimide to which polyethylene glycol 6000 is bound to separate PCR products according to their % G + C, which could be also exploited for the purpose of decreasing the complexity of templates used for DGGE/TGGE.

Application of group-specific primers for PCR

The design and application of primers which allow the amplification of the 16S rDNA fragment of certain bacterial groups from mixed communities seem to provide the most promising approach to dissect complex communities and specifically track the dynamics of certain groups, e.g., agrobacteria/rhizobia (Smalla et al., in preparation), Actinomycetales (Heuer et al., in preparation), or ammonia oxidizers (41). The application of group-specific primers permits one to follow specific groups even if their abundance is below the threshold of template sequences needed to generate a band mentioned.

Two strategies are possible:

The PCR product obtained with group-specific primers is analyzed directly via DGGE or TGGE. This requires that one of the primers have a GC-clamp.

The DNA from mixed communities is first amplified with group-specific primers which also span the region of the 16S rDNA used for amplification with the universal DGGE/TGGE primers. No GC-clamp is needed for the group-specific primers since the PCR product obtained will be used as a template for a second PCR with DGGE/TGGE primers (e.g., 968F/1401R). To allow only the amplification of DNA amplified with the group-specific primers, the PCR product is purified via preparative gel electrophoresis and excised from an agarose gel before the second PCR with the DGGE/TGGE primers. In parallel, the community DNA can be amplified directly with the DGGE/TGGE primers, thus allowing comparison of the fingerprints of the dominant constituents of a certain group with the pattern obtained for the dominant constituents of the community.

Hybridization with Taxon- or Strain-Specific Probes

More detailed information can be obtained by hybridizing electroblotted DGGE/TGGE gels with taxon-specific oligonucleotide probes (27,42). Muyzer et al. (27) used an oligonucleotide probe specific for sulfate-reducing bacteria (corresponding to positions 385–402 in the 16S rDNA of *Escherichia coli*) to hybridize electroblotted DGGE gels of PCR products (341–534) from mixed microbial communities. However, prerequisites for designing appropriate oligonucleotide probes are the availability of sequence data and the presence of sufficiently specific DNA within the DNA fragment amplified for DGGE/TGGE analysis.

Recently, an alternative, straightforward approach to develop probes even without 16S rDNA sequence data for the respective strain has been applied in our laboratory. We have used universal primers spanning the hypervariable V6 region (971 F; 1057 R) of the 16S rRNA molecule to amplify this region from respective strains. The PCR products were digoxigenin-labeled during the PCR reaction. The specificity of the probes was tested using dot-blotted PCR products obtained with the DGGE/TGGE primers (968–1401) from bacteria belonging to α, β, or γ proteobacteria, cytophaga, high GC, or low GC gram-positive bacteria. Hybridizations were performed at high stringency (90% to 100%) according to the method of Fulthorpe et al. (43). Dot blot hybridizations are shown for several probes in Fig. 3. To study the specificity of the approach for more closely related species, strains of the *Agrobacterium/Rhizobium* group were also used as templates for PCR amplification with DGGE/TGGE primers (968–1401). In relation to these hybridization studies, it is assumed that the V6 probes (region 971–1057) are specific for the strain used as the template for the PCR amplification of the V6 region and closely related strains. The V6 probes generated either from isolates or from clones can be used for hybridization of electroblotted DGGE or TGGE gels. This approach may be of interest for following an inocu-

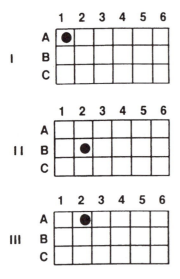

Figure 3. Specificity of different V6-derived probes I: *Agrobacterium tumefaciens* probe; II: *Burkholderia solanacearum* probe; III: *Rhizobium meliloti* probe. Dot blot hybridizations of PCR products generated from different strains using the DGGE/TGGE primers (968F/1401R) with Dig-labeled probes (I–III) at high stringency (90%–100%): Dot blots: A1 *Agrobacterium tumefaciens*; A2 *Rhizobium meliloti*; A3 *Sphingomonas paucimobilis*; B1 *Alcaligenes eutrophus*; B2 *Burkholderia solanacearum*; B3 *Comamonas acidovorans*; C1 *Pantoea agglomerans*; C3 *Pseudomonas fluorescens*; C4 *Pseudomonas putida*; C5 *Xanthomonas campestris*; C6 *Acinetobacter calcoaceticus*.

lated strain in complex DGGE/TGGE patterns or confirming the assignment of bands from PCR products of strains or clones to a band with the same DGGE/TGGE electrophoretic mobility in a pattern amplified from environmental DNA.

Furthermore, individual bands excised from the gel can be reamplified and sequenced, thus allowing their phylogenetic affiliation (35,44–46).

3. APPLICATION OF DGGE OR TGGE PROFILING

The DGGE or TGGE analysis of 16S rDNA fragments amplified from total community DNA extracted from different environments such as hydrothermal vents, hot spring microbial mats, stratified marine water, biodegraded wall paintings, bulk and rhizosphere soils, and the phyllosphere of potatoes revealed the presence of distinguishable and reproducible band patterns (27,34,35,42,46).

In the following, some applications of DGGE/TGGE are described to illustrate the potential of applying the technique in advanced soil microbiological studies.

3.1 Structural Analysis of Soil Microbial Communities by DGGE or TGGE Analysis of 16S rDNA Amplified from Soil or Rhizosphere DNA

The TGGE analysis of PCR amplified 16S rDNA fragments has been applied to compare the rhizosphere and phyllosphere bacterial populations of transgenic potatoes expressing T4-lysozyme with those of nontransgenic potatoes. Five different composite rhizosphere samples of transgenic potatoes and the respective nontransgenic potatoes (roots with adhering soil particles) were used for extraction of the bacterial fraction with a protocol based on the use of a cation exchange resin (Chelex-100) (47,48). Community DNA was extracted from the bacterial pellet according to the technique of Smalla et al. (7). Purified rhizosphere DNA was used as a template for PCR amplification with the DGGE/TGGE primers (968 F; 1401 R). The TGGE analysis showed highly complex and reproducible patterns. The TGGE patterns were almost identical within the five samples of transgenic and nontransgenic potatoes but also between the two groups compared. Only a few bands showed some fluctuations of major constituents of the potato rhizosphere. In contrast, the fingerprints generated from phyllosphere DNA were less complex and much more variable (see Fig. 4).

Recently TGGE analysis has been applied to study the microbial community recovered from Zn-contaminated soils (Brim et al. in preparation). The TGGE patterns derived from DNA extracted directly from soil samples were compared

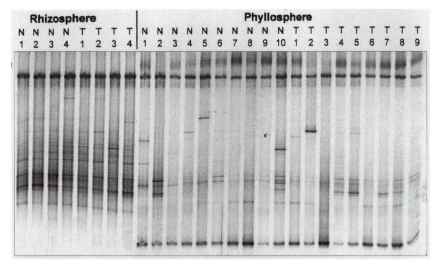

Figure 4. The TGGE analysis of 16S rDNA fragments (968F/1401R) amplified from different potato rhizosphere and phyllosphere DNA samples: N, nontransgenic Désirée; T, transgenic Désirée.

with bands generated with the most dominant culturable bacteria and 16S rDNA clones. The band of the most frequently isolated *Arthrobacter*-like strains was one of the major bands of the TGGE pattern derived from the direct DNA extract. One of the bands derived from clones also showed the same electrophoretic mobility as the product of the *Arthrobacter* strains. However, it became evident that the 16S rDNA clones sequenced do not include all bands of TGGE patterns derived from soil DNA.

3.2 Application of DGGE/TGGE to Follow the Extraction of Bacteria from Soil/Rhizosphere Samples and the Lysis Efficiency of Different DNA Extraction Protocols

The efficient and representative extraction of indigenous microorganisms from soil matrices or the rhizosphere has been a major problem in soil microbiology. Bacteria are not only heterogeneously distributed in soil (49) but bound to the soil matrix by varying interactions (48). Thus, most methods used for the recovery of the microbial fraction from soil samples involve repeated homogenizations with different buffer solutions, ion exchange resins, or detergents, and separation of the released cells from soil particles by differential centrifugation. The attempts to dislodge bacterial cells from soil are rather empirical. Uncertainties exist concerning the reproducibility of the extraction procedures as well as the representativeness of the bacteria extracted. We have found that TGGE analysis is extremely useful for monitoring the efficiency of the different extraction steps applied to dislodge bacteria from soil matrices as well as determining the reproducibility of the protocol (see Fig. 5). We have also applied TGGE analysis of 16S rDNA fragments amplified from DNA extracted and purified with different protocols from the same soil samples. Figure 6 shows that major bands in the fingerprints obtained from DNA extracted after a harsh lysis (lysozyme, bead beating, [SDS]; Gebhard and Smalla, in preparation) are not present or are weaker in the pattern of DNA recovered after a soft lysis (lysozyme, alkaline SDS).

One of the missing or weaker bands has the same electrophoretic mobility as the *Arthrobacter* strains isolated by cultivation from the same soil. Obviously,

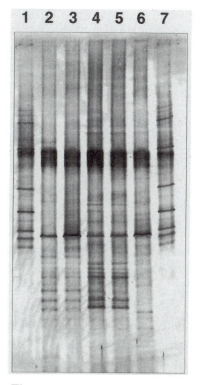

Figure 5. Reproducibility of DNA extraction from Zn-contaminated soil samples using a modified Chelex protocol for extraction of bacterial cells: TGGE analysis of 16S rDNA fragments amplified from soil community DNAs extracted from the same soil samples using the identical protocol on different days (A/B): Lanes 1 and 7: standard; lane 2: soil 4A; lane 3: soil 4B; lane 4: soil 5A; lane 5: soil 5B; lane 6: soil 6A.

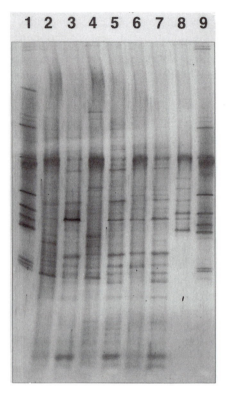

Figure 6. The TGGE analysis of 16S rDNA fragments amplified from Zn-contaminated soil samples using two different DNA extraction and purification protocols: A: Dijkmans et al. (1993), B: Gebhard and Smalla, in preparation: Lanes 1/9: standard; lane 2: soil 2A; lane 3: soil 2B; lane 4: soil 3A; lane 5: soil 3B; lane 6: soil 5A; lane 7: soil 5B; lane 8: culturable fraction obtained after NaCl extraction.

the lysozyme alkaline/SDS lysis (50) was not harsh enough to release DNA from *Arthrobacter* strains.

3.3 Application of DGGE/TGGE for Comparison of the Most Dominant Culturable Bacteria with the Dominant Constituents Amplified from Directly Extracted DNA

Uncertainties on whether the bacteria recovered by cultivation from an environmental sample resemble the major constituents of the sample can be easily addressed by the application of DGGE or TGGE. The approach presently used in our laboratory is plating of the bacterial fraction recovered from the environmental sample on different nutrient media such as R2A, plate count agar, King's B, or potato extract agar. After incubation for 5–14 days, the colonies grown on the plates are resuspended en masse using a Drigalski spatula. The DNA is released from the bacterial cell pellet obtained by centrifugation using a DNA extraction method which is also applied to extract the DNA from the environmental sample directly. The purified DNAs are used as templates for PCR amplification with the DGGE/TGGE primers (968F; 1401R). To prevent the DGGE/TGGE fingerprints generated from the culturable fraction from being biased as a result of the different size of colonies, an alternative strategy is to use the dominant colonies selected by their different colony morphological characteristics as a template for PCR amplification with the DGGE/TGGE primers.

Preliminary results indicate that the patterns generated from DNA extracted directly from rhizosphere soils have only a few bands in common with the respective patterns of the culturable fraction.

Recently, Rölleke et al. (46) sequenced major bands excised from DGGE gels of 16S rDNA fragments (341F; 534R) amplified from DNA extracted from wall paintings. According to the sequence data, major constituents of the microbial community isolated from biodegraded wall paintings were identified as members or close relatives of the genera *Halomonas, Clostridium*, and *Frankia*. Bacteria of these groups have not yet been isolated from wall paintings since the cultivation techniques used have not been appropriate. The approach used by Rölleke et al. (46) demonstrates the potential of DGGE/TGGE to select the appropriate cultivation conditions for recovery of the major constituents of the microbial community.

Furthermore, the use of ribosomal RNA after reverse transcription as a template for PCR amplification with DGGE or TGGE primers offers the chance to generate patterns of the dominant metabolically active bacteria. Since metabolically active cells generally have a much higher ribosomal RNA content and rRNA/DNA ratio than resting cells, the fingerprints derived from rRNA templates should resemble the metabolically active cells. Teske et al. (45) compared DGGE patterns derived from reverse transcribed 16S rRNA obtained from fjord water samples with those from the corresponding genes. The 16S rDNA DGGE patterns amplified from rDNA markedly differed from those of reversely transcribed PCR-amplified 16S rRNA. The results indicate a difference between the presence of 16S rDNA genes (population structure) and their expression (metabolically active fraction). Similar observations were made by Felske et al. (51) for soil microbial communities. The TGGE fingerprints derived from rRNA from ribosomes extracted directly from soil and used after reverse transcription for PCR amplification with the TGGE primers (968F/1401R) were shown to be different from the 16S rDNA fingerprints amplified from the respective community DNA.

3.4 Application of DGGE and TGGE to Analyze PCR or Cloning Biases

Besides the application of DGGE and TGGE in the study of the microbial diversity of soil or rhizosphere, denaturing gradient techniques offer a straightforward approach to analyze PCR and cloning biases.

Preferential PCR amplification was identified as one of the most severe problems for microbial diversity studies based on sequencing of PCR-amplified ribosomal DNA from total community DNA. Several authors demonstrated that preferential amplification occurs (52–55). This may lead to an incorrect picture of the composition of the microbial community.

The PCR-amplified ribosomal DNA which is planned to be used for cloning and sequencing or for hybridization studies with 16S or 23S rDNA directed oligonucleotide probes can be analyzed by DGGE or TGGE for preferential amplification as recently performed in our laboratory (Smalla et al., in preparation). Community DNAs which were extracted directly from different environmental samples such as soil, rhizosphere, manure, epilithon, or river sediments have been used for amplification of the 1.5 kb fragment (16S rRNA gene) or a 4 kb region spanning the 16S and the 23S rRNA genes. To check the resulting PCR products for PCR biases the 1.5 kb or 4 kb PCR products as well as the corresponding community DNA were amplified with the DGGE/TGGE primers (968F, 1401R). When the resulting PCR products are analyzed side by side on DGGE or TGGE gels, bands present in both lanes can be compared. Bands which occur only in the lanes in which PCR products were amplified from the 1.5 or 4 kb PCR products indicated preferential amplification (see Fig. 7). With this approach, PCR products and genomic DNA from various environments (soil, aquifer, river sediment, and epilithon) did not provide evidence for major shifts of the bacterial composition due to the first PCR amplification. However, in some samples, bands came up when the 4 kb PCR product was used as a template. Hence, these templates may have been preferentially amplified in the first PCR. Although this approach is also based on PCR amplification, DGGE/TGGE analyses offer an elegant tool to analyze preferential amplification.

In addition, 16S rDNA clones obtained from total community DNA directly or after PCR amplification of the 16S rRNA gene can be used as templates for PCR amplification with DGGE/TGGE primers. When the PCR products from clones and directly extracted community DNA are analyzed side by side on a DGGE/TGGE gel it is possible to check whether the clones analyzed belong to these major constituents of the microbial community analyzed. This is extremely important since the cloning step is not free of bias. Rainey et al. (56) reported that clone libraries differed, depending on the cloning vectors used.

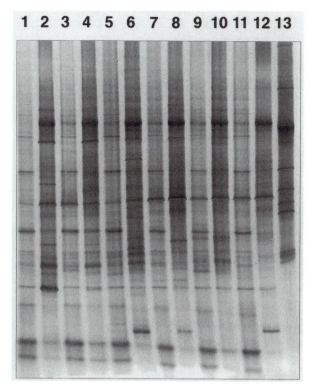

Figure 7. Analysis of preferential amplification using soil community DNA (A) or the corresponding 4 kb fragment spanning the 16S and 23S rDNA (B) as template for PCR amplification with TGGE/DGGE primers (968F/1401R): Lane 1: 1A; lane 2: 1B; lane 3: 2A; lane 4: 2B; lane 5: 3A; lane 6: 3B; lane 7: 4A; lane8: 4B; lane 9: 5A; lane 10: 5B; lane 11: 6A; lane 12: 6B; lane 13: standard.

4. POTENTIAL PITFALLS OF THE DGGE OR TGGE ANALYSIS

As with any other technique applied to study microbial communities in their natural habitat, there are limitations to the DGGE/TGGE approach which need to be recognized and understood. Only when these limitations are taken into consideration is meaningful exploitation of the great potential of the DGGE/TGGE approach possible.

4.1 Limitations Not Specific for DGGE/TGGE

The representativeness of the soil or rhizosphere community DNA used as a template for PCR amplification is critical to DGGE/TGGE analysis of PCR products from mixed communities. Crucial points for the recovery of community DNA representative for the habitat sampled are the following:

Representative soil or rhizosphere sampling
Efficient dislodging of cells from soil particles or roots
Complete lysis of bacterial cells, including those which are known to be more difficult to lyse such as gram-positives in general, spores, dwarf, or resting cells (57,58)

As in all PCR-based approaches, selective amplification of genes from mixed communities by PCR may bias the analysis. Selective amplification can be due to differences in:

$G + C$ content of the template DNA
Accessibility of the templates to primer hybridization after denaturation (secondary structures)
Efficiency of primer–template hybrid formation (primer preference)
Concentration of templates: at a high product concentration the reannealing of PCR products becomes more efficient than the priming reaction (52–55).

4.2   Limitations Specific for DGGE/TGGE

Ideally one species yields one band in a DGGE/TGGE gel. However, for some strains two or even more bands have been detected. This may be due to sequence heterogeneities of 16S rRNA operons (59). Thus, DGGE/TGGE can be used for studies on 16S rRNA operon heterogeneities within strains. On the other hand, examples exist where similar electrophoretic mobilities were observed for closely related and even for phylogenetically unrelated strains (Smalla, unpublished results). This must be taken into account when numbers of bands are used as a measure for diversity. For ecosystems containing large numbers of equally abundant microorganisms, highly complex banding patterns will be obtained, impeding the qualitative and quantitative analysis by the DGGE/TGGE gels.

Furthermore, only limited sequence information can be obtained with the DGGE/TGGE approach because the separation of PCR products larger than 500 bp is reduced (60).

5.   CONCLUSION AND COMPARISON OF DGGE/TGGE COMMUNITY ANALYSES WITH OTHER FINGERPRINTING TECHNIQUES

Community fingerprinting techniques have become rather popular during the last few years. Fingerprinting techniques are used to describe the structural or metabolic diversity of microbial communities. Metabolic fingerprints using microtiter plates (BIOLOG)—an approach which was first applied by Garland and Mills (61) to characterize microbial communities by their substrate utilization patterns—have been shown to be cultivation-dependent. In contrast, the community fingerprinting techniques based on the analyses of DNA, RNA, or fatty acids extracted directly from environmental samples allow a cultivation-independent analysis of the structural diversity of microbial communities. Whole community fatty acid profiles derived from phospholipid components of cellular membranes of microorganisms directly extracted from environmental samples have been used to study microbial communities from different environments independent from cultivation (62,63). However, community fatty acid profiles have to be interpreted cautiously since the knowledge of qualitative and quantitative distribution of fatty acids in environmental microorganisms and the effects of growth conditions on their distribution is rather limited (63,64). Alternative cultivation-independent community profiling approaches are PCR-based fingerprinting of community DNA such as amplified DGGE or TGGE analysis of PCR-amplified genes from microbial community DNA, restriction analysis (ARDRA) of the 16S rRNA gene and the intergenic regions of the 16S and the 23S amplified mixed community rRNA (65), and random amplified polymorphic DNA (RAPD) fingerprinting of community DNA (66). All these PCR fingerprinting techniques generate habitat-specific patterns. However, except for DGGE/TGGE fingerprints of 16S rDNA fragments, information on the major species is not known or is difficult to obtain. Compared with the other fingerprinting techniques DGGE/TGGE analysis offers the advantage that rapid fingerprinting in order to follow temporal or spatial changes of microbial communities can be combined with deeper insights based on sequencing of selected bands or hybridization with group- or strain-specific probes. The advantage of looking at the total DNA is that information can be gained on the major constituents, including nonculturables. The advantage of analyzing rRNA is that fingerprints of the most dominant metabolically active species can be obtained. Presently, DGGE/TGGE fingerprinting of 16S rDNA is one of the most promising approaches to study the structural diversity of microbial communities. After solving the technical problems to establish the system, it quickly becomes a valuable tool in microbial ecology with versatile application potentials.

ACKNOWLEDGMENTS

The TGGE work in our laboratory was funded by BMBF grant 0310582A. The work on high GC specific primers was performed in collaboration with Martin Krsek and Paul Baker from the laboratory of E. Wellington, University of Warwick, Coventry, United Kingdom. Characterization of Zn-contaminated soils was done in collaboration with Hassan Brim from Max Mergeay's group, Mol, Belgium. The DGGE work was performed during a stay in Larry Forney's laboratory at the

Center for Microbial Ecology in East Lansing, Michigan. Studies on preferential amplification studied by TGGE were done in collaboration with I. Wagner-Döbler's group at GBF, Braunschweig. Furthermore, the authors are grateful for the EU grant which enabled them to organize, together with G. Muyzer, MPI, Bremen, the EU workshop, Applications of DGGE and TGGE in Microbial Ecology.

REFERENCES

1. A. Ogram, G. S. Sayler, and T. J. Barkay, *J. Microbiol. Meth.* 7:57 (1987).
2. L. A. Porteous, and J. L. Armstrong, *Curr. Microbiol.* 22:345 (1991).
3. L. A. Porteous and J. L. Armstrong, *Curr. Microbiol.* 27:115 (1993).
4. Y. L. Tsai and B. H. Olsen, *Appl. Environ. Microbiol.* 57:1070 (1991).
5. Y. L. Tsai and B. H. Olsen, *Appl. Environ. Microbiol.* 58:2292 (1992).
6. C. Picard, C. Ponsonnet, E. Paget, X. Nesme, and P. Simonet, *Appl. Environ. Microbiol.* 58:2717 (1992).
7. K. Smalla, N. Cresswell, L. C. Mendonca-Hagler, A. Wolters, and J. D. van Elsas, *J. Appl. Bacteriol.* 74:78 (1993).
8. C. C. Tebbe and W. Vahjen, *Appl. Environ. Microbiol.* 59:2657 (1993).
9. J. D. van Elsas and K. Smalla, *Molecular Microbial Ecology Manual* (A.D.L. Akkermans, J. D. van Elsas, and F. J. de Bruijn, eds.), 1.3.3., Kluwer, Dordrecht, The Netherlands (1995).
10. V. L. Torsvik, *Soil Biol. Biochem.* 12:15 (1980).
11. W. E. Holben, J. K. Jansson, B. K. Chelm, and J. M. Tiedje, *Appl. Environ. Microbiol.* 54:703 (1988).
12. C. S. Jacobsen and O. F. Rasmussen, *Appl. Environ. Microbiol.* 58:2458 (1992).
13. V. Torsvik, *Molecular Microbial Ecology Manual* (A. D. L. Akkermans, J. D. van Elsas and F. J. de Bruijn, eds.), 1.3.1., Kluwer, Dordrecht, The Netherlands (1995).
14. M. A. Moran, V. L. Torsvik, T. Torsvik, and R. E. Hodson, *Appl. Environ. Microbiol.* 59:915 (1993).
15. S. Selenska-Pobell, *Molecular Microbial Ecology Manual* (A. D. L. Akkermans, J. D. van Elsas, and F. J. de Bruijn, eds.), 1.5.1., Kluwer, Dordrecht, The Netherlands (1995).
16. D. M. Ward, A. L. Ruff-Roberts, and R. Weller, *Molecular Microbial Ecology Manual* (A. D. L. Akkermans, J. D. van Elsas, and F. J. de Bruijn, eds.), 1.2.3., Kluwer, Dordrecht, The Netherlands (1995).
17. V. Torsvik, K. Salte, R. Sørheim, and J. Goksøyr, *Appl. Environ. Microbiol.* 56:776 (1990a).
18. V. Torsvik, J. Goksøyr, and F. L. Daae, *Appl. Environ. Microbiol.* 56:782 (1990b).

19. V. Torsvik, J. Goksøyr, F. L. Daae, R. Sørheim, J. Michalsen, and K. Salte, *Beyond the Biomass* (K. Ritz, J. Dighton, and K. E. Giller, eds.), p. 39, John Wiley & Sons, New York (1994).

20. N. R. Pace, D. A. Stahl, D. J. Lane, and G. Olson, *Adv. Microbiol. Ecol. 9*:1 (1985).

21. W. Ludwig and K. H. Schleifer, *FEMS Microbiology Reviews 15*:155 (1994).

22. S. J. Giovannoni, T. B. Britschgi, C. L. Moyer, and K. G. Field, *Nature 345*:60 (1990).

23. D. M. Ward, R. Weller, and M. M. Bateson, *Nature 345*:63 (1990).

24. R. Weller, J. W. Weller, and D. M. Ward, *Appl. Environ. Microbiol. 57*:1146 (1991).

25. J. A. Fuhrman, K. McCallum, and A. A. Davis, *Nature (London) 356*:148 (1992).

26. E. Stackebrandt, W. Liesack, N. Ward, and B. M. Goebel, *Proceedings of ISME-6* (R. Guerrero and C. Pedrós-Alió, eds.), p. 567, Barcelona, Spain (1993).

27. G. Muyzer, E. C. de Waal, and A. Uitterlinden, *Appl. Environ. Microbiol. 59*:695 (1993).

28. S. G. Fischer and L. S. Lerman, *Cell 16*:191 (1979).

29. S. G. Fischer and L. S. Lerman, *Proc. Natl. Acad. Sci. USA 80*:1579 (1983).

30. R. M. Myers, S. G. Fischer, L. S. Lerman, and T. Maniatis, *Nucleic Acids Res. 13*:3131 (1985).

31. R. M. Myers, T. Maniatis, and L. S. Lerman, *Methods Enzymol. 155*:501 (1987).

32. V. C. Sheffield, D. R. Cox, L. S. Lerman, and R. M. Myers, *Proc. Natl. Acad. Sci. USA 86*:232 (1989).

33. G. Muyzer, S. Hottenträger, A. Teske, and C. Wawer, *Molecular Microbial Ecology Manual* (A. D. L. Akkermans, J. D. van Elsas and F. J. de Bruijn, eds.), 3.4.4.: 1, Kluwer, Dordrecht, The Netherlands (1996).

34. H. Heuer, K. Hartung, B. Engelen, and K. Smalla, *Proceedings of the Ninth Forum for Applied Biotechnology*, Med. Fac. Landbouww., Univ. Gent, *60/4b*, 2639 (1995).

35. M. J. Ferris, G. Muyzer, and D. M. Ward, *Appl. Environ. Microbiol. 62*:340 (1996).

36. C. Wawer and G. Muyzer, *Appl. Environ. Microbiol. 61*:2203 (1995).

37. W. E. Holben, V. G. M. Calabrese, D. Harris, J. O. Ka, and J. M. Tiedje, *Trends in Microbial Ecology* (R. Guerrero and C. Pedrós-Alió, eds.), p. 367, Spanish Society for Microbiology (1993).

38. D. Harris, *Beyond the Biomass* (K. Ritz, J. Dighton, and K. E. Giller, eds.), p. 111, John Wiley & Sons, New York (1994).

39. L. Øvreås, T. Castberg, and V. Torsvik, *Proceedings of the Workshop on Ap-*

plication of DGGE and TGGE in Microbial Ecology (K. Smalla and G. Muyzer, eds.) Braunschweig (1995).

40. C. Wawer, H. Rüggeberg, G. Meyer, and G. Muyzer, *Nucl. Acids Res.* 23:4928 (1995).

41. K. P. Witzel, *Proceedings of the Workshop on Application of DGGE and TGGE in Microbial Ecology* (K. Smalla and G. Muyzer, eds.) (1995).

42. G. Muyzer, A. Teske, C. O. Wirsen, and H. W. Jannasch, *Arch. Microbiol.* *164*:165 (1995).

43. R. R. Fulthorpe, C. McGowan, O. V. Maltseva, W. H. Holben, and J. M. Tiedje, *Appl. Environ. Microbiol. 61*:3274 (1995).

44. G. Muyzer and E. C. de Waal, *Structure Development and Environmental Significance of Microbial Mats* (L. J. Stal and P. Caumette, eds.), NATO ASI Series *G35*, Springer Verlag, Berlin (1994).

45. A. Teske, C. Wawer, G. Muyzer, and N. B. Ramsing, *Appl. Environ. Microbiol. 62*:1405 (1996).

46. S. Rölleke, G. Muyzer, C. Wawer, G. Wanner, and W. Lubitz, *Appl. Environ. Microbiol. 62*:2059 (1996).

47. P. R. Herron and E. M. H. Wellington, *Appl. Environ. Microbiol. 56*:1406 (1990).

48. D. W. Hopkins and A. G. O'Donnell, *Genetic Interactions Among Microorganisms in the Natural Environment* (E. M. H. Wellington and J. D. van Elsas, eds.), p. 104, Pergamon Press, Elms Ford, New York (1992).

49. T. Hattori, *Rep. Inst. Agric. Res.* Tohoku University *37*:23 (1988).

50. R. Dijkmans, A. Jagers, S. Kreps, J. -M. Collard, and M. Mergeay, *Microb. Releases* 2:29 (1993)

51. A. Felske, B. Engelen, U. Nübel, and H. Backhaus, accepted for publication by *Appl. Environ. Microbiol.*

52. A. -L. Reysenbach, L. J. Giver, G. S. Wickham, and N. R. Pace, *Appl. Environ. Microbiol. 58*:3417 (1992).

53. C. M. Dutton, C. Paynton, and S. Sommer, *Nucl. Acids Res. 21*:2953 (1993).

54. C. Morrison and F. Gannon, *Biochim. Biophys. Acta 1219*:493 (1994).

55. M. T. Suzuki and S. J. Giovannoni, *Appl. Environ. Microbiol. 62*:625 (1996).

56. F. A. Rainey, N. Ward, L. I. Sly, and E. Stackebrandt, *Experientia 50*:796 (1994).

57. M. Moré, J. B. Herrick, M. C. Silva, W. C. Ghiorse, and E. L. Madsen, *Appl. Environ. Microbiol. 60*:1572 (1994).

58. J. D. van Elsas and K. Smalla, *Transgenic Organisms: Biological and Social Implications* (J. Tomiuk, K. Wöhrmann, and A. Sentker, eds.), p. 127, Birkhäuser Verlag, Basel, Switzerland (1996).

59. U. Nübel, B. Engelen, A. Felske, J. Snaidr, A. Wieshuber, R. I. Amann, W. Ludwig, and H. Backhaus, *J. Bacteriol. 178*:5636 (1996).

60. G. Muyzer and N. B. Ramsing, *Wat. Sci. Tech. 32*:1 (1995).

61. J. L. Garland and A. L. Mills, *Appl. Environ. Microbiol. 57*:2351 (1991).

62. S. O. Petersen and M. J. Klug, *Appl Environ. Microbiol. 60*:2421 (1994).

63. S. K. Haack, H. Garchow, D. A. Odelson, L. J. Forney, and M. J. Klug, *Appl. Environ. Microbiol. 60*:2483 (1994).

64. H. J. Heipieper, G. Meulenbeld, Q. van Oirschot, and J. A. M. de Bont, *Appl. Environ. Microbiol. 62*:2773 (1996).

65. A. A. Massol-Deya, D. A. Odelson, R. F. Hickey, and J. M. Tiedje, *Molecular Microbial Ecology Manual* (A. D. L. Akkermans, J. D. van Elsas, and F. J. de Bruijn, eds.), 3.3.2., Kluwer, Dordrecht, The Netherlands (1995).

66. M. Malik, J. Kain, C. Pettigrew, and A. Ogram, *J. Microbiol. Meth. 20*:183 (1994).

13

Microbial Diversity in Soil: The Need for a Combined Approach Using Molecular and Cultivation Techniques

WERNER LIESACK   Max-Planck-Institut für terrestrische Mikrobiologie, Marburg, Germany

PETER H. JANSSEN   University of Melbourne, Parkville, Victoria, Australia

FREDERICK A. RAINEY, NAOMI L. WARD-RAINEY, and ERKO STACKEBRANDT   Deutsche Sammlung von Mikroorganismen und Zellkulturen GmbH, Braunschweig, Germany

1. SOIL MICROBIAL COMMUNITIES

Soil is composed of organic and inorganic matrices formed by the combined action of biotic and abiotic processes. Dependent on the geographical, geological, hydrological, climate, vegetation, fauna, and anthropogenic influences, the soil exhibits properties which determine its microbial community (1,2). The mineral composition, salinity, pH, nutrient availability, organic input, temperature, and water content, among other factors, determine which ecological niches are available. In addition, local variations caused by the low rate of mixing, very high surface area, and the influence of plants result in a high spatial heterogeneity. The presence of plants, resulting in an additional input of organic material into the soil, creates new habitats, the rhizoplane and rhizosphere. The increased metabolic activity and complexity (3) of the system at the root surface and in the proximal soil result in changes in the types of microorganisms found in comparison with the bulk soil (4), the so-called rhizosphere effect. Seasonal changes and vegetation cycles also superimpose a temporal heterogeneity, so that no two soil samples have the same microbial community. Just as plant communities show local variations in species composition within one vegetation type (5), so does the microbial community within one soil type (6). These factors are among those faced by microbiologists before they even take a sample to begin cultivation or molecular biological studies.

Many studies carried out to investigate the dynamics of a microbial population on the rhizoplane or in the rhizosphere inoculate the root system of aseptically grown plants with a pure culture or use a genetically modified microorganism with a marker that can be followed. Studies on the dynamics of the indigenous community are much more difficult because of their complexity and the lack of knowledge of what to expect. If the ecological characteristics of soil microorganisms are to be investigated, then it is necessary to be able to carry out a number of analyses to investigate the influence of various physical parameters, biological activities, and interactions with other microorganisms and with plants and to ascertain the degree of spatial and temporal heterogeneity. Thus methods which allow fast, accurate, and comprehensive analyses of a microbial community within a soil sample need to be developed.

Two goals, requiring somewhat different strategies, can be envisaged in the study of the microbial community of a soil: the study of, 1, the diversity and 2, the community structure. Although the term 'diversity' has been defined in a number of ways (5,7), in microbiological terms it is often used to describe the number of different species or, in molecular terms, the number of different sequence types present in a habitat, whereas 'community structure' includes quantitative information on the number of individuals of different taxa or physiological groups. A complete set of qualitative and quantitative data would give us knowledge of both, but this is not at present possible in the microbiological study of soils. In one investigation, the use of deoxyribonucleic acid (DNA) reassociation kinetics led to an estimate of approximately 4000 different-size microbial genomes in 1 gram of soil (8), representing perhaps as many as 13,000 different species (9). Obviously, these are not all present in equal numbers but vary from 1 cell to perhaps 10^8 cells per gram of soil. In a similar study employing DNA hybridization techniques, it was shown that the organismic diversity varied with soil type (10). A study of the diversity will give an idea as to what sorts of microorganisms are present, but this will not reveal what the importance of these microorganisms is within the soil, and certainly not what their roles are. The study of the community structure will, at least in the near future, for methodological reasons, tend to concentrate on the (relatively) few abundant species but will not tell us about the rich variety of other microorganisms found in lesser numbers but making up a far greater range of species. These microorganisms may be important in acting as a reservoir of slightly to greatly different physiological types (11), which may respond by a population size increase due to changes in the habitat. Knowledge of the composition of this reservoir may help to make sense of what will otherwise appear to be complete shifts in the community structure with time or within short distances, when only the abundant microorganisms are considered.

The microbial population of a soil consists of members of all three major branches of life: 1, bacteria, including the often separately listed actinomycetes (domain *Bacteria*); 2, methanogens and other archaea (domain *Archaea*); and 3, fungi, microalgae, and protozoa (domain *Eucarya*). These groups together make up 90% or more of the total biomass of soil microbiota and microfauna (nematodes, arthropods, molluscs, oligochetes) (12). Members of the *Bacteria* and *Archaea* dominate in waterlogged soils, while fungi tend to be more important in aerobic soils (1,13,14). Prokaryotes tend to dominate in the rhizosphere (15). On the basis of traditional cultivation studies, typical soil prokaryotes are considered to belong to the lineages of gram-positive bacteria, especially *Clostridium* spp., *Bacillus* spp., and actinomycetes (*Arthrobacter* spp., *Brevibacterium* spp., *Corynebacterium* spp., and *Micrococcus* spp.) and to the heterogeneous grouping often combined in cultivation studies as *Pseudomonas* spp. (1,16). Other genera commonly isolated from soil include *Acinetobacter, Agrobacterium, Alcaligenes, Caulobacter, Flavobacterium, Hyphomicrobium, Metallogenium, Sarcina, Staphylococcus, Streptococcus,* and *Xanthomonas* (1,17,18), although in the face of rapidly changing taxonomies and better, molecularly based identification methods, such listings of diversity are being rendered increasingly worthless. Soils also contain a great diversity of protozoa, and their biological characteristics have been reviewed (19) (see Chapter 5). Fungi (treated in Chapter 4) tend to be found predominantly as mycelia (3), and thus identification and enumeration are made very difficult. Apart from groups which can be microscopically identified, as is often possible for protozoa, enumeration and identification, until the beginning of the 1990s, have had to rely mainly on the cultivation and isolation of soil microorganisms. The problem inherent in this approach is the ability to investigate only those microorganisms that can be cultivated. Other, perhaps completely unsuspected groups which may be abundant or particularly active will not even be considered, and a perhaps totally false picture of the soil microbial community will emerge. Since the mid-1980s, the use of 16S ribosomal ribonucleic acid– (rRNA)–based techniques has facilitated the molecular identification of a wide variety of as yet uncultivated microorganisms and novel microbial groups in various environments. In addition, the development of methods allowing the detection of functional genes is beginning to reveal the heterogeneity of microbial subpopulations within one physiological group in a habitat. The following chapter focuses on molecularly based investigations into the diversity of the prokaryotic domains *Archaea* and *Bacteria* in soils, and the need for strategies combining both molecular approaches and traditional cultivation studies.

2.   THE 16S rRNA GENE AS A BIOLOGICAL MARKER

2.1   General Aspects

The use of genetic markers to investigate community structure and diversity of microorganisms has provided a new approach to studying aspects of the microbial ecological characteristics of complex habitats (11,20–24). Genes providing information on the presence and diversity of a single physiological group (via a functional gene) or indicating the phylogenetic relationship of detected microorganisms with all other forms of life (via ubiquitously distributed genes) have been used. Among the most widely used are those of the RNA of the small subunit of the ribosome (SSU rRNA), the 16S rRNA gene (16S rDNA), and 18S rRNA gene (18S rDNA) of prokaryotes and eukaryotes, respectively. Since this review deals mainly with prokaryotic microorganisms, the term '16S rRNA' is preferred to the more general term 'SSU rRNA.' The original method was the extraction of the rRNA from pure cultures, followed by comparative analysis of 16S rRNA oligonucleotide catalogues (25–27). This approach was superseded by direct analysis of the 16S rRNA via dideoxy sequencing using reverse transcriptase (28). With the availability of the polymerase chain reaction (PCR), direct sequencing of 16S rRNA genes rapidly became the method of choice. Comparative analysis of the primary structure of these rRNA genes not only resulted in the realization that all forms of life are separated into three major evolutionary lines, the three so-called domains *Bacteria*, *Archaea*, and *Eucarya* (29), but transformed microbial taxonomy from a pure identification system to an evolutionarily-based systematic framework (30–32). The 16S rRNA gene (and the homologous 18S rRNA gene) has regions which are highly conserved, even among members of the three known domains of life, while other regions display considerable sequence variation even within closely related taxa (30,33). This characteristic allows the inference of phylogenies based on the comparative sequence analysis of the 16S rRNA gene, from the deepest separation of the different branches of life to the genus or sometimes even species or strain level, and facilitates the identification and classification of a microorganism with relatively little effort (21,31–33). The power and comparative ease of this method, compared to those of traditional methods, have been demonstrated (34–36). Strains which have 70% or more homology in DNA–DNA hybridization studies are defined, on a genotypic basis, as belonging to the same species (37). The rationale for this definition was the high degree of correlation found between DNA homology and chemotaxonomic, genomic, serological, and phenotypic characteristics. Fox et al. (38) and Stackebrandt and Goebel (39) have shown that strains with 99% to 100% identity of 16S rRNA sequence do not necessarily belong to the same species, since in some cases these shared less than 70% homology in terms of DNA–DNA hybridization data. However, strains with 16S rDNA sequence similarities below 97% never had more than 60% DNA–DNA hybridization homology (39). This means that differentiation of species based on comparative 16S rDNA analysis is often not possible. In addition, a survey of the 16S rDNA entries in the international sequence data bases revealed an intraspecific variation in the 16S rRNA sequences which may represent 1, different genomic loci within one microorganism, 2, homologous regions within closely related strains, or 3, methodological errors such as sequencing errors or human error in data handling (40). These results indicate that the formal description of new species is only possible on the basis of a polyphasic approach (41) using both genetic analyses (16S rDNA sequencing, DNA–DNA hybridization) and characterization of the phenotypic traits. However, the power of the 16S rDNA-based phylogenetic framework to suggest possible phenotypic similarities between apparently physiologically different microorganisms was clearly demonstrated when it was shown that members of the genus *Pelobacter* originally described to have an obligately fermentative metabolism, were capable of growth with Fe(III) and S^0, as predicted by the close phylogenetic relationship to species of the Fe(III)- and S^0-reducing genera *Geobacter* and *Desulfuromonas* (42).

A further advantage of 16S rRNA-based techniques is that microorganisms do not need to be cultivated and can be detected in environmental samples (20,43,44). By estimating the phylogenetic relatedness to known microorganisms in terms of the homology of the gene sequence, the identity of a newly isolated or molecularly detected microorganism can be ascertained. However, sometimes these microorganisms represent new lines of descent without close known relatives. To date, 16S rRNA sequences from more than 6000 microorganisms are available for comparison (45,46). With the current rapid expansion of the data base, a steady improvement in the degree of matching to known microorganisms can be expected. Such data can be easily stored and stringently compared in later studies, given the power of data transfer in global computer networks and worldwide access to data bases (47).

The analysis of the total pool of 16S rRNA gene sequences in a sample from a habitat under study provides a means with which to investigate communities at the genus and perhaps species level, rather than simply counting numbers belonging to various physiological groups. Since a high homology between sequences is indicative of closely related species, the methodology should also be applicable to studying changes within a community. The major hurdles here include eliminating methodological problems involved in obtaining the genetic information (see Section 4), dealing with the high heterogeneity of the soil habitat, and obtaining a suitable data base of sequences from cultivated microorganisms with known physiological characteristics for comparison. This may entail parallel isolation and physiological studies from the same habitat (see Section 7).

Investigations of a hot spring environment (48,49), using the 16S rRNA as a genetic marker, revealed new groups of microorganisms for which no pure cultures were known and also detected representatives of cultivated groups that were not identical to available pure cultures. The study of marine environments (50–53) produced similar results and expanded this observation, and it was found that different oceans contained unknown but related microorganisms, and also new lineages of evolution unrelated to known microorganisms (54–56).

The lack of correlation between data obtained from 16S rDNA-based molecular studies and known microorganisms has been reviewed (11,57). This was illustrated in a medically related molecular study to unravel the genetic diversity of cultured and as yet uncultured oral spirochetes from a patient with severe periodontitis, which revealed an unexpectedly high diversity of as yet unknown and uncultured spirochete species, even in the one gingival crevice investigated (58). The general conclusion is that the knowledge of microbial diversity obtained after more than 100 years of pure culture study is incomplete, and that very few of the total number of microbial species are in culture (8,20,59).

2.2   The Molecular Toolbox for Ecological Studies in Soil

Introduction

Not only for cultivation studies, but also for molecular investigations, the possible effects of the sample handling procedure on the detectable microbial diversity should not be neglected. Deep marine sediment obtained from below the sea floor of the Japan Sea was treated in different ways prior to the extraction of nucleic acids, for example, 1, incubated aerobically for up to 24 h at 16°C before freezing, 2, incubated anaerobically at 16°C before freezing, or 3, frozen within 2 h of sampling. The different handling procedures resulted in the dominance of completely different 16S rDNA sequence types within the clone libraries generated (60). From these investigations, it was concluded that immediate freezing should always be employed when sediment samples are to be used to investigate microbial diversity by molecular methods. A prerequisite for approaches directed towards the analysis of the genetic diversity or, more ambitiously, the structure of the microbial community, is the extraction of the total pool of RNA or DNA ('environmental RNA or DNA') from the sample under investigation. For the subsequent specific retrieval of 16S rDNA sequences from the environmental nucleic acid pool extracted, various different strategies have to date been applied.

The biodiversity of planktonic marine microorganisms was investigated by shotgun cloning of partially digested environmental DNA into lambda phage (43,44,53). The advantage of this strategy is that the rRNA genes are flanked by additional sequences which allow the correlation of phylogenetic data with further genetic characterization of the molecularly detected microorganisms (see also Section 5). The disadvantage is the laborious and time-consuming need to screen the genomic libraries to identify clones containing rRNA sequences. The latter made up only about 0.125% to 0.3% of the total clones (43,53).

Environmental RNA was isolated for the analysis of a hot spring mat community (48,49,61,62). Ribosomal copy DNA (rcDNA) of the 16S rRNA fraction was synthesized by a reverse transcriptase reaction. The priming site for the rcDNA synthesis was the universally conserved region at positions 1392–1406 (*Escherichia coli* numbering [63]). The rcDNA was used for the generation of a clone library containing almost exclusively rRNA genes (48,49,61). One major disadvantage of this strategy was that the reverse transcriptase tends not to make full-length copies because of breaks caused by the secondary structure or post-transcriptional base modifications often present in the 16S rRNA, for example, that postulated to occur at position 966/967 for members of the indigenous hot spring microbial community (62) or at position 965 of most streptomycetes (64) (*E. coli* numbering), which limited the amount of retrievable sequence information. An alternative approach used a mixture of hexanucleotides of random sequence composition as primers to overcome this problem. This permitted the recovery of over 900 nucleotides of sequence information from 16S rRNA types beyond the modification sites at positions 966/967 (62). One major advantage of the rRNA approach in comparison to environmental rDNA-based strategies discussed later may be the exclusive detection of populations viable at the time of sampling, i.e., those with sizable rRNA pools. The relative proportions of the different sequence types in the rRNA fraction extracted should reflect a function of cell numbers and metabolic activity of the microbial populations present. The latter conclusion can be deduced from the close correlation of the cellular ribosome content and the growth rate of individual cells of some organisms (65–68). However, dormant populations of *Vibrio* spp. (69,70) and *Desulfobacter latus* (71) maintained 30% of the maximum ribosome content of the respective exponentially growing cells, even after a prolonged period of starvation. Similar results were found for ammonia-oxidizing bacteria after specific inhibition of their metabolic activity (72). The maintenance of a substantial number of active ribosomes may be an important survival strategy for dormant stages of indigenous populations often present in environments with poor nutrient supply like soil but may give an impression of activity based on their presence in the environmental rRNA pool.

Nowadays, the most widely used approach is methodologically rather straightforward. It applies the technology of the PCR (73) in combination with a primer system targeting highly conserved regions of the 16S rRNA genes (74,75). This allows the amplification of 16S rDNA from a phylogenetically wide range of microorganisms in one single reaction. For marine environments, the use of this approach resulted in the detection of a variety of as yet uncultured *Bacteria* and *Archaea* (51,55,76). Quantitative dot blot hybridization with environmental rRNA extracted from the same sample under study was used to determine the relative abundance and/or metabolic activity of these molecularly detected microorganisms (51,76). The analysis of archaeal populations present in a thermal environment located in Yellowstone Park revealed an unexpectedly high number of previously unknown lineages branching deeply between *Euryarchaeota* and *Crenarchaeota* (77). Moreover, uncultured bacterial and archaeal symbionts of protozoa (78–80), bacterial symbionts of a gutless marine worm (81), and uncultured spirochetes in the hindgut of termites (82,83), as well as in clinical samples (58), could be identified by a so-called full-cycle rRNA approach (20). This requires first the retrieval of 16S rRNA sequence information, followed by in situ analysis via whole-cell hybridization using sequence-specific probes to confirm and visualize the presence of the molecularly detected microorganisms.

Compared to that for aquatic systems, the methodological toolbox available for the analysis of terrestrial environments is still under development. Protocols for the quantitative isolation of rRNA and messenger RNA (mRNA) from soils have become available (84–87). The in situ detection of microbial populations in bulk soil via whole-cell hybridization is still faced with some methodological problems, but its use in model systems has already been demonstrated (88,89). The use of new, more sensitive fluorescent dyes seems to facilitate the application of this important approach to terrestrial environments. For example, a specific oligonucleotide probe labeled with Cy3 reactive dye allowed the in situ detection of spores and vegetative cells of *Bacillus megaterium* in soil (90).

Molecular strategies to date applied to analyze indigenous populations in terrestrial environments have focused mainly on the extraction of the environmental DNA fraction, followed by PCR-mediated amplification of the rRNA gene pool, generation of clone libraries, and comparative sequence analysis of individual 16S rDNA fragments.

DNA extraction techniques

The extraction and subsequent purification of environmental DNA from soil are crucial for all molecular studies focusing on the indigenous microbial community. It has to reflect as much as possible the genetic diversity and the community structure that exist at the time of sampling. In principle, two different procedures can be applied for the isolation of environmental DNA.

Cell extraction

The first approach uses differential centrifugation steps to separate the cell fraction from the soil matrix. The cell fraction is then later lysed, followed by the isolation and purification of the genomic DNA (cell extraction procedure) (91–95). Steffan et al. (95) reported a maximal retrieval of up to 35% of the microbial population from the soil sample under study. It has been suggested that the cell extraction procedure selects for those groups of microorganisms which are less strongly attached to the soil particles (96–98). Several physical and chemical dispersion techniques which attempt to disrupt the close association of cells to soil or sediment particles have been described, for example, the use of a cation-exchange resin (99). This technique allows a faster processing of samples in comparison to the dispersion of the soil by homogenization in a blender (Waring) and results in environmental DNA of high molecular weight. It was successfully used to study the survival of a 2,4-dichlorophenoxyacetic acid– (2,4-D)-degrading strain of *Pseudomonas cepacia* inoculated into soil (99). No empirical data are available as to how completely these methods will extract the DNA of the indigenous microbial community from soil samples.

Direct cell lysis

The second approach uses direct lysis of the cells in the soil matrix, followed by extraction and purification of the environmental DNA (direct lysis procedure) (95,97,100–104). This approach may detect the overall genetic potential of an environmental sample, including theoretically dormant forms, but may also encompass moribund populations or even naked DNA from naturally lysed cells (see also Section 3.3).

One critical point in the direct lysis procedure is the choice of the lysis protocol. Mechanical disruption of the cells should facilitate a quantitative extraction of environmental DNA (95,97). Zhou et al. (104) reported DNA recovery from soils seeded with gram-negative bacteria of 92% to 99% by using only a high-salt SDS-based treatment combined with extended heating (2 to 3 h). For gram-positive bacteria, significantly higher DNA yields were obtained when the high-salt SDS method was combined with mortar-and-pestle grinding and freeze-thawing. Similarly, Moré et al. (105) reported improved extraction efficiencies of environmental DNA from sediments by using a combination of different lysis procedures. In this study, the sequential application of SDS treatment and bead-mill homogenization proved to be more efficient than SDS treatment followed by three cycles of freezing and thawing. The densitometrically determined environmental DNA yield was doubled after use of bead-mill homogenization (11.8 µg DNA per g [dry weight] of sediment versus 5.2 µg of DNA obtained after use of the freeze-thawing method). The improved extraction efficiency was attributed to increased cell lysis. Viable counts of sediment microorganisms revealed that 2% and 8%, respectively, survived the bead-mill homogenization and freeze-thaw procedures. *Bacillus* endospores were not lysed using freeze-thawing treatment (94% survival rate), but lysis was almost quantitatively obtained by using mechanical disruption (2% survival rate). However, the intensive use of bead-mill homogenization increases the risk of extracting severely sheared environmental DNA (96,97), but unfortunately no data were presented for the correlation between bead-mill homogenization and fragmentation of the DNA extracted in the studies published by Moré et al. (105). The use of DNA of low molecular weight as the target for the amplification of the 16S rRNA gene pool will result in an increased formation of chimeric 16S rDNA fragments during the PCR reaction (see Section 4.2). The exclusive use of soft lysis techniques, for example, lysozyme, proteinase K, SDS treatment, or three cycles of freeze-thawing, permits the retrieval of environmental DNA of high molecular weight. These methods were successfully used to track defined microbial populations inoculated into soil (100,102) but these procedures probably select for the easily lysed microbial populations rather than extracting a representative fraction of the total gene pool. An alternative method for the lysis of cells is based on their disruption by grinding in liquid nitrogen with the natural abrasives of the soil fraction (103).

Comparison of lysis and extraction methods

Leff et al. (96) performed a comparative study of different DNA extraction protocols on stream sediments. The three approaches compared were 1, the direct lysis procedure via bead-mill homogenization published by Ogram et al. (97), 2, 'soft' lysis using freeze-thawing (102), and 3, the cell extraction procedure published by Jacobsen and Rasmussen (99). The three DNA fractions were hybridized with a radioactively labeled bacterial oligonucleotide probe to test for the relative amount of environmental DNA originating from members of the domain *Bacteria*. The relative contribution of bacterial DNA to the total was lowest using bead-mill homogenization and highest with the soft lysis (freeze-thawing) method. This observation does not necessarily point to lower lysis efficiency of the bead-mill homogenization protocol for bacterial microorganisms but may instead reflect a relatively higher input of eukaryotic and archaeal DNA into the gene pool extracted.

The similarity and complexity of DNA extracted via a microbial cell extraction method (95) versus a direct lysis protocol were compared by using a community DNA hybridization technique (10). The two methods yielded DNAs which were 75% similar to each other and equally complex.

A completely new approach for a fast comparative analysis of the efficiency of different extraction protocols applied to the same sample is denaturing gradient gel electrophoresis (DGGE) (106). This methodology, which is described in Chapter 12b, allows the separation of a mixture of short double-stranded DNA fragments of the same length but different primary structure, for example, a PCR product of partial 16S rDNA fragments amplified from environmental DNA, in a polyacrylamide gel containing a linearly increasing gradient of denaturants (107–109). The separation in DGGE is based on the different electrophoretic mobilities of a partially melted DNA molecule and the completely helical state of the molecule. Environmental DNA extracted from the same soil sample using different lysis protocols (bead-mill homogenization alone versus a combination of methods) produced different 16S rDNA profiles when analyzed by DGGE (Fig. 1) (U. Hengstmann and W. Liesack, unpublished). This result indicates that the relative composition of the DNA extracted from the same soil sample can be strongly influenced by the lysis procedure applied.

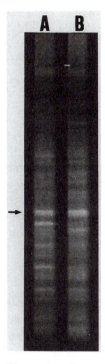

Figure 1. Comparative analysis of a mixed PCR product of partial 16S rDNA fragments using DGGE. The environmental DNA fractions used as the template for the PCR reactions were extracted from the same sample using either (1) only bead-mill homogenization for cell lysis (lane A) or (2) a combination of freeze-thawing, lysozyme, and SDS treatment, followed by bead-mill homogenization (lane B). All steps for the further isolation and purification of the DNA and the PCR-mediated amplification of partial 16S rDNA sequences were the same for both samples. The two different lysis protocols resulted in different banding profiles, with some obvious differences in the 16S rDNA types recovered, but also one common main fragment found with both treatments (arrow). These results indicate that the relative composition of the gene pools extracted varied with the lysis protocol used.

Purification of DNA

The extent of the necessary purification steps after the lysis procedure depends on the soil type under investigation, especially the humic acid content.
In addition, the molecular detection of gene sequences from the indigenous
microbial populations demands a more intensive purification of the environmental DNA than the detection of introduced bacteria (110). Several procedures have been used for the separation of impurities, especially humic acids,
from environmental DNA, which can be applied individually or in combination: 1, treatment with polyvinylpolypyrrolidone (93,111); 2, cesium chloride
density gradient centrifugation (95,97); 3, hydroxyapatite chromatography
(97); 4, precipitation of DNA with potassium acetate, cesium chloride, and
spermine-HCl (101); 5, agarose gel electrophoresis (104,110,112), as well as
6, reversible binding of DNA on glass milk resin (104). Possible inhibition of
PCR-based amplification caused by humic acids still present in the environmental DNA fraction extracted may be overcome by adding bovine serum al-
bumin or T4 gene 32 protein (gp32) to the reaction mixture (113). Jacobsen
(114) developed a magnetic capture-hybridization PCR technique (MCH-
PCR) to eliminate the inhibitory effect of humic acids and other contaminants. This method is based on a single-stranded DNA oligonucleotide probe
complementary to the target sequence and coupled to magnetic beads. After
hybridization in a suspension of crude environmental DNA, magnetic extraction of the beads allows the separation of the target DNA from interfering impurities.

Conclusion

A general protocol for the extraction of total environmental DNA applicable to
all types of soil cannot be recommended. As long as only few data comparing the
different extraction procedures are available, the use of the direct lysis procedure
appears to be more advisable for molecular studies directed toward unravelling
the overall genetic diversity or the microbial community structure. On one hand,
a harsh lysis procedure should probably be applied to facilitate the extraction of
the total microbial gene pool. On the other hand, however, extensive shearing of
the environmental DNA should be prevented as much as possible. The best compromise between these two conflicting goals, and also the amount of purification
necessary, often has to be determined individually for each soil type under study.
For the extraction of environmental DNA from many soil types on a 0.1 to 1
gram scale, the method described by Smalla et al. (101) appears to be a good
compromise to obtain both a largely quantitative lysis of the microorganisms and
a purified DNA fraction with only limited fragmentation. Comprehensive manu-
als with a detailed description of different protocols available for the extraction
and purification of environmental DNA from bulk soil, but also from other habi-

tats, for example, aquatic systems, rhizoplane, rhizosphere, and phyllosphere, have been published (115,116).

PCR-mediated amplification of 16S rDNA

For the analysis of genetic diversity based on comparative 16S rDNA sequence analysis, PCR-mediated amplification of the 16S rDNA fraction present within the total gene pool has to be performed. For this purpose, oligonucleotide primers which target highly conserved regions of the 16S rDNA, and consequently allow the amplification of partial stretches or of the almost complete 16S rRNA genes of members of the domain *Bacteria* (74,75) or of the domain *Archaea* (77), are used. For the study of defined taxa based on PCR-mediated amplification or hybridization assays, specific oligonucleotides have been developed. Listings can be found in the literature (20,23,117). A data base for oligonucleotide probes containing information on 1, design of probes, 2, probe characterization studies, 3, probe specificity, 4, use of probes for in situ hybridization, 5, use of primers for PCR, and 6, specificity of PCR primers is publicly available (118).

Generation and analysis of 16S rDNA clone libraries

Two different techniques have been used for the cloning of mixtures of 16S rDNA fragments amplified by PCR from environmental DNA. The first is a blunt-end cloning strategy, as described by Giovannoni (119). The drawbacks of this approach were the rather low cloning efficiency and the large background of nonrecombinants. For the application of the second technique, a 'forced cloning strategy,' the PCR reaction has to be performed with primers carrying overhangs that contain recognition sites for restriction endonucleases (112,120) (see also Sections 3.1 and 3.3). The recognition sites are incorporated at the 5′ and 3′ termini of the fragments synthesized during PCR. For the subsequent generation of the clone library, the sites are cut with the corresponding restriction endonuclease. This strategy allows directed insertion of PCR products into the vector, but the correct choice of a suitable restriction endonuclease must be made to prevent loss of certain fractions of the genetic diversity as a result of internal cutting of the amplified 16S rDNA fragments (see Sections 3.1 and 3.3).

Commercial kits are available for the cloning of PCR products. The pCR-Script SK(+) cloning kit (Strategene) is based on increased ligation efficiency of blunt-ended DNA fragments by the simultaneous reactions of the *Srf*I restriction enzyme, which cuts only self-ligated vectors without any inserts, and the T4 DNA ligase (121). The most widely used system, the TA cloning kit (Invitrogen), takes advantage of a particular characteristic of the Taq polymerase, namely, to add a single adenosine overhang at both 3′ termini of the double-stranded PCR products (122). The linearized cloning vector carries single 3′ thymidine residues. This eliminates the need for an endonuclease restriction of the PCR product prior to the ligation reaction and results in high cloning efficiency.

The reader should refer to specialist literature for a detailed description of the procedures applied to the analysis of cloned 16S rDNA genes, either by generation of restriction length polymorphism patterns from cloned inserts of plasmid DNA (123) or, more efficiently, via amplified ribosomal DNA restriction analysis (ARDRA) (124,125) or directly by comparative sequence analysis (126), and to the protocols collated by Akkermans et al. (116). A fast computerized identification of microorganisms based on the sequence data obtained is possible online via the Ribosomal Database Project (45).

Phylogenetic analysis

One prerequisite for the phylogenetic analysis of 16S rDNA sequences is the correct alignment of the molecular data. The term 'alignment' means the organization of the sequences in such a way that homologous nucleotides appear in the same column of the alignment so that they can be validly compared to one another. Sequence homology should not be confused with sequence similarity. The first term implies a common ancestry, whereas the second may be the consequence either of common ancestry or of mutational chance. The basic assumption for the inference of phylogenies from molecular data is the hypothesis that nucleotides placed in the same column of the alignment, i.e., homologous nucleotides, are descended from the same base position in a common ancestral gene. The latter is the reason that the 16S rDNA data set used for the construction of phylogenetic trees is restricted to only those positions at which bases in all sequences under study can be unambiguously aligned. Several different methods are available for the inference of the phylogeny, for example, pairwise distance (127), maximum parsimony (128), and maximum likelihood (129).

The most common procedure for phylogenetic inference from 16S rDNA data is the distance matrix analysis. This method tries to reconstruct an evolutionary tree that would be expected to give rise to the differences observed by comparing the present-day sequences. The first step is the calculation of an evolutionary distance value, i.e., the number of nucleotide substitutions per site, for each pair of 16S rDNA sequences under study. The most widely used algorithm is the equation of Jukes and Cantor (130), which assumes that the rate of evolutionary nucleotide substitution is the same for all pairs of the four nucleotides, A, T, C, and G. In a second step, this set of distances is used to construct an inferred phylogenetic tree. Several different treeing programs are in use, which differ in certain principles and algorithms used, such as the neighbor-joining method of Saitou and Nei (131), the least-squares distance method of Fitch and Margioliash (127), or the least-squares algorithm of De

Soete (132). Computational programs for the phylogenetic analysis of molecular data are publicly available, for example, a package of maximum parsimony methods called PAUP (133), PHYLIP (134), MEGA (135), or programs implemented in the Ribosomal Database Project (45). For a more detailed description of the different inference methods used to reconstruct phylogenies, and their meaningful application, the reader is referred to several review articles (31,33,135–137).

3. DIVERSITY OF MICROBIAL 16S rDNA IN SOILS

In the following sections, the results of 16S rDNA-based approaches applied to four geographically different sampling sites are presented. All four studies revealed the power of molecular strategies to detect the presence and genetic diversity of previously unknown and as yet uncultured microbial groups. Of particular interest is the observation that several of these groups were molecularly detected in different habitats, indicating their ubiquitous distribution, for example, planctomycetes and related microbial groups in a forested soil collected in Queensland (Australia), and in a Palouse silt loam soil sampled in Washington (United States) (see Section 3.1). Another example is the TH3 group, a distinct lineage within the actinomycetes, which was found to be present in three completely different sampling sites (see Sections 3.1, 3.2, 3.3). The explanation for the parallel detection of this microbial group might be the acidic nature of the three soil samples investigated. The data presented for a model system of flooded rice cultures show that studies directed to unravelling microbial diversity based on 16S rDNA are not restricted to bulk soil, but can also be applied to soil-related habitats, for example, the rhizoplane and rhizosphere of plants.

Some other 16S rDNA-based studies on microbial diversity in soil (138,139) have been made, but these are difficult to compare with the data from the four studies discussed here because they are mainly derived from different regions of the 16S rDNA. Of the many 16S rDNA sequences from environmental samples, an increasing number comprise either only the first 200–500 nucleotides from the 5′ terminus or the last 400 to the 3′ terminus of the 16S rDNA, so limiting the number of nucleotide positions available for comparison. It would seem that the first 500 nucleotides from the 5′ terminus of the 16S rDNA contain enough information to allow an accurate assignment to one of the main lines of descent, and so this should be recommended as the region to concentrate on in future studies of this nature. More complete sequence information, of course, allows more precise phylogenetic placement, and more stringent comparison with sequences present in the data base or derived from pure cultures isolated in parallel.

3.1   Forested Soil from Mount Coot-tha, Australia

The first soil extensively studied with regard to the phylogenetic analysis of PCR-amplified 16S rDNA clones was a randomly selected sample from the forested southeast slope of Mount Coot-tha, Brisbane, Queensland, Australia. The chemical analysis of a subsurface sample, taken at a depth of between 5 and 10 cm in November 1990, revealed that the soil was acidic (pH 4.2) and contained elevated levels of aluminium and iron. Of the organic carbon, only a small percentage was measured as free aromatic or aliphatic acids. More details of the soil characteristics are given by Stackebrandt et al. (140).

At this early stage of molecular microbial ecology the goal was mainly to develop strategies, rather than to investigate the spatial distribution of taxa within the environment, to measure the function of selected strains, or to determine the location of strains within the soil matrix. Substantial amounts of sample material needed to be processed in order to obtain amplifiable DNA, which made it impossible to determine the microheterogeneity within the microbial community.

In addition, little was known at this time about the methodological pitfalls that make it impossible to estimate the quantitative contribution of populations in a natural microbial community from the frequency of 16S rDNA sequences in a clone library (see Section 4). It was also assumed that the use of a so-called genus-specific PCR primer pair would indeed amplify only DNA of those species against which the primer sequence was designed. A goal of the study with the Mount Coot-tha soil was to investigate the diversity of members of the genus *Streptomyces* in soil. The 16S rDNA was amplified by using a pair of primers that consisted of a universally conserved 5′ oligonucleotide primer and a 3′ oligonucleotide primer that was considered *Streptomyces*-specific. However, an investigation of the influence of primer–template mismatches on the PCR (141) suggested that this 3′ primer allowed the amplification of 16S rDNA from a wider range of *Bacteria* than originally thought (140). In addition, members of the family *Planctomycetaceae*, not considered to be part of the microbial community in soil, exhibit the same probe target sequence as streptomycetes.

Analysis of 133 clones by partial sequence analysis and probing with radioactively labeled taxon-specific oligonucleotides revealed that the majority of clones fell into four main lines of descent: the α-subphylum of the *Proteobacteria*, the planctomycetes, the actinomycetes, and one containing *Verrucomicrobium spinosum*. As compared to the first phylogenetic analysis (112,140,142), the position of several clones can now be determined more precisely by the inclusion of sequences from species analyzed at a later date. The phylogenetic relationships of the clone sequences are discussed in the following paragraphs.

Clones of the α-subphylum of the *Proteobacteria*

Of the clones, 43% are indicative of sequences belonging to members of this subphylum, but information on the phylogenetic position is only available for those 14 clones for which the first 250 nucleotides of the 5' terminus have been analyzed. The majority of the sequences form a moderately related cluster that groups around the two species of *Rhodoplanes* within the α2-subgroup (Fig. 2). As the sequences of these cultivated microorganisms only became available after this study was completed (143), the relationship of the clone sequences to phototrophic purple nonsulfur bacteria could not be determined previously (142). None of the clone sequences is closely related to those of the *Rhodoplanes* spp., but they form four independent lines of significant phylogenetic depth. Consequently, putative phenotypic and physiological characteristics cannot be predicted for the as yet uncultured microorganisms to allow their straightforward isolation. The sequences of clones MC6 and MC23 are almost indistinguishable from those of *Bradyrhizobium japonicum* and close relatives, for example, *Blastobacter denitrificans* and the phototrophic symbionts of the legumes *Aeschynomene* spp. (144) (Fig. 2). Clone MC7 branches between the *Rhodoplanes* and the *Bradyrhizobium* clusters. Two clone sequences, MC74 and MC106, indicate the presence of microorganisms that are related to *Rhodopila globiformis* and members of the *Acidiphilium cryptum/Thiobacillus acidophilus* cluster, all of which are acidophilic. A cultivated relative of clone MC114 has still not been found.

The planctomycetes

Several clones indicated the presence of planctomycetes in the Australian soil sample. This finding was unexpected, as no member of the family *Planctomycetaceae* has been isolated from this kind of environment. The majority of clones were distantly related to *Gemmata obscuriglobus*, while individual clones were found to be neighbors of *Isophaera pallida* and *Planctomyces limnophilus*. Later, the number of representatives of this family isolated from aquatic systems was increased dramatically by modification of the enrichment conditions (145), resulting in the isolation of novel strains that exhibit a closer relationship to the clone sequences than do previously described species (146). In addition, molecular analysis of the microbial community of a Palouse silt loam soil in Washington revealed the presence of a clone, EA73, that was closely related to the clones detected in the Australian soil (A. Ogram, personal communication). In addition, a substantial fraction of the molecularly detected microbial population of marine snow from the Pacific Ocean (147) and the Mediterranean Sea (E. F. DeLong, personal communication) turned out to be members of *Planctomycetaceae*, represented by clone AGG8 in Fig. 3. The presence of planctomycetes in different

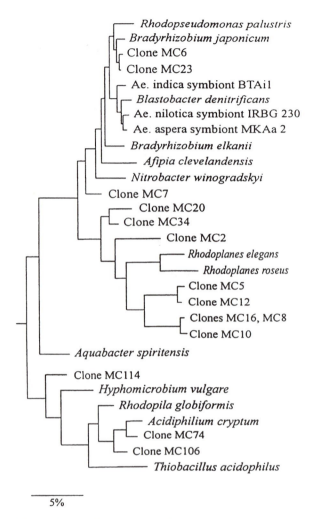

Figure 2. Phylogenetic dendrogram of proteobacterial 16S rRNA gene clones obtained from soil from Mount Coot-tha (MC), indicating the relationship of these sequences to those of cultivated proteobacterial species. The dendrogram was inferred by distance matrix analysis (132) of 250 positions from the 5′ terminus of the molecule. Scale bar represents 5 mutations per 100 nucleotide sequence positions.

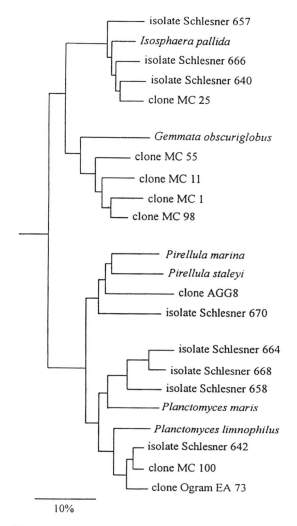

Figure 3. Phylogenetic dendrogram of planctomycete 16S rRNA gene clones obtained from soil from Mount Coot-tha (MC), Palouse Silt Loam Soil in Washington, United States (EA), and marine snow (AGG), illustrating the relationship of these sequences to those of cultivated species of *Planctomycetaceae* and other pure culture isolates. The dendrogram was inferred by distance matrix analysis (132) of 1250 positions from the 5′ terminus of the molecule. Scale bar represents 10 mutations per 100 nucleotide sequence positions.

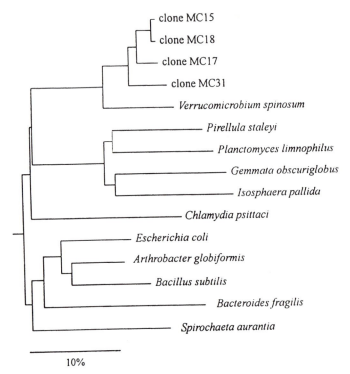

clone MC15
clone MC18
clone MC17
clone MC31
Verrucomicrobium spinosum
Pirellula staleyi
Planctomyces limnophilus
Gemmata obscuriglobus
Isosphaera pallida
Chlamydia psittaci
Escherichia coli
Arthrobacter globiformis
Bacillus subtilis
Bacteroides fragilis
Spirochaeta aurantia

10%

Figure 4. Phylogenetic dendrogram of some 16S rRNA gene clones obtained from soil from Mount Coot-tha (MC), illustrating the relationship of these sequences to those of *Verrucomicrobium spinosum*, planctomycetes, and representatives of some other main lines of descent of the domain *Bacteria*. The dendrogram was inferred by distance matrix analysis (132) of 1250 positions from the 5′ terminus of the molecule. Scale bar represents 10 mutations per 100 nucleotide sequence positions.

habitats and geographically diverse regions clearly indicates that these microorganisms should not be considered a rare group of *Bacteria*. Their ecological role, however, can presently not be assessed because all attempts to cultivate these microorganisms from the sites from which their DNA was isolated have failed.

The *Verrucomicrobium* group

Initial analysis of four clones pointed to the presence of a novel phylum of the domain *Bacteria*, sharing a common but remote ancestry with the planctomycetes and chlamydiae (112). Only after the complete sequence of *Verrucomi-*

crobium spinosum had been determined (148), did the relatedness of the clone sequences and this unusual prosthecate bacterium become obvious (Fig. 4). The analysis of the Palouse silt loam soil in Washington revealed the presence of 16S rDNA (clone EA25), which indicated a close relationship (93% similarity) to clone MC17, identified in the Mount Coot-tha soil (149). The abundance of clone EA25, determined by an additive approach, was estimated to be about 10^7 per gram of soil. However, as the total number of cells was not determined, this value does not allow further speculation about the relative numerical importance of members of the *Verrucomicrobium* lineage or of the related planctomycetes in this environment. Interestingly, new strains belonging to the *Verrucomicrobium* lineage have been isolated in pure culture from the soil of flooded rice (P. H. Janssen and F. A. Rainey, unpublished).

The actinomycetes

Those clone sequences that revealed phylogenetic relatedness to the actino-mycetes (the lineage of gram-positive bacteria with DNA with a high molecular percentage [mol%] G + C content) formed two groups. One group was related to members of the genus *Streptomyces*, the original target group of the study, while the other group constitutes a novel major subline of descent within the *Actino-mycetales*.

The streptomycetes

The presence of streptomycete 16S rDNA was analyzed in two different clone libraries using the same batch of environmental DNA prepared from the Mount Coot-tha soil sample. The methods used for the generation of the clone libraries differed first in the primer pair used for PCR and second in the cloning vector used (150). Significant differences were found in the number and distribution of clones representing different taxa. With respect to clones indicating the presence of *Streptomyces* spp., the number was small in both libraries (Fig. 5). Clone sequences from the two libraries were not identical, but some of them were similar. Interestingly, although more than 40 pure cultures of strepto-mycetes were isolated from the same site from which the molecular analysis was carried out, the 16S rDNA sequences of the cultured strains were in no case identical or even very similar to those of the clone library. This finding lets us speculate that 1, the spores of streptomycetes were not lysed under the conditions used for the isolation of DNA, 2, the number of streptomycete 16S rDNA sequences analyzed was too low to be statistically meaningful, 3, only numeri-cally insignificant representatives of the streptomycetes were isolated in pure culture, or 4, the lack of correlation is due to a combination of these alterna-tives. The dendrogram of relatedness should be considered to be only tentative, bearing in mind that more than 450 *Streptomyces* spp. have been described

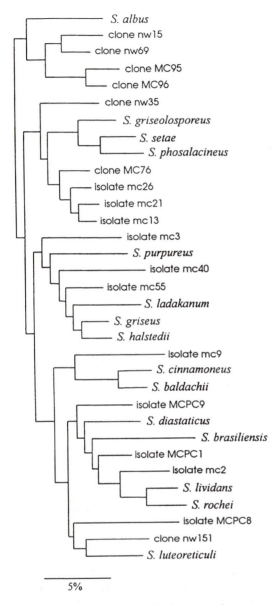

5%

Figure 5. Phylogenetic dendrogram of streptomycete 16S rRNA gene clones obtained from soil from Mount Coot-tha (MC and NW, obtained from two different libraries), illustrating the relationship of these sequences to those of cultivated species of *Streptomyces* and some strains isolated from the same sample from which DNA was isolated. The dendrogram was inferred by distance matrix analysis (132) of 250 positions from the 5′ terminus of the molecule. Scale bar represents 5 mutations per 100 nucleotide sequence positions.

(151). Sequences of the clones and isolates obtained may be found to be closely related to described species once the full range of presently isolated *Streptomyces* spp. has been analyzed by comparative 16S rDNA sequencing.

The TH3 group

The center of this novel, large clone cluster (Fig. 6) is represented by the cultured acidophilic strain TH3 (152,153), characterized by its rod-shaped structure, gram-negative staining type, use of iron as electron donor, and low mol% G + C content of its genomic DNA. Although it forms a coherent taxon, the phylogenetic depth of this cluster is as high as that of the entire order *Actinomycetales*. Their physiological characteristics may be related to the acidic and iron-rich environment from which the sequences were obtained, but whether a suitable isolation strategy can be inferred from these limited data needs to be tested. Surprisingly, the molecular analysis of two other acidic environments revealed the presence of similar clone sequences. Clone libraries of 16S rRNA genes generated from DNA isolated from a peat bog in northern Germany (see Section 3.2) and a thermophilic soil from the center of the North Island of New Zealand (see Section 3.3) contained a substantial number of sequences closely related to those from the Mount Coot-tha soil. As in each case the conditions used for the generation of the clone libraries were different, a constant factor introducing bias toward the selection of these sequences can probably be excluded. Therefore, strain TH3 and its relatives may play a significant role in these kinds of acidic environments, but investigations into the function of these microorganisms in such environments have to await future genetic and biochemical investigations on isolates. Later, a gram-positive filamentous bacterium, morphologically resembling a previously observed morphotype named *Microthrix parvicella*, was isolated from activated sludge by micromanipulation (154). The 16S rDNA of this actinomycete shows about 87% similarity to the sequences of strain TH3 and peat bog clone TM213. Once physiological data are available for *Microthrix parvicella*, the design of novel enrichment and cultivation media for the uncultured strains apparently present in various moderately acidic soils may be possible.

3.2 Peat Bog from Grosses Moor, Germany

Cultivation studies

The presence of gram-positive aerobic species of the domain *Bacteria* in the peat bog Grosses Moor near Gifhorn, Lower Saxony, Germany, was investigated using traditional cultivation techniques. Samples were taken from three different layers (with conventional ^{14}C ages of 3000, 3200, and 6000 years). The majority of the isolated strains were gram-positive bacteria (93%), of which 63% were endospore-formers, 17% were nonsporulating rods, 10% were mycelium-form-

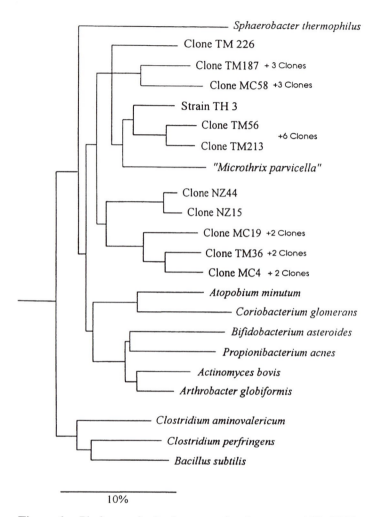

Sphaerobacter thermophilus
Clone TM 226
Clone TM187 + 3 Clones
Clone MC58 +3 Clones
Strain TH 3
Clone TM56
Clone TM213 +6 Clones
"Microthrix parvicella"
Clone NZ44
Clone NZ15
Clone MC19 +2 Clones
Clone TM36 +2 Clones
Clone MC4 + 2 Clones
Atopobium minutum
Coriobacterium glomerans
Bifidobacterium asteroides
Propionibacterium acnes
Actinomyces bovis
Arthrobacter globiformis
Clostridium aminovalericum
Clostridium perfringens
Bacillus subtilis

10%

Figure 6. Phylogenetic dendrogram of actinomycete 16S rRNA gene clones obtained from soil from Mount Coot-tha (MC), acidothermal soil from New Zealand (NZ), and a peat bog in northern Germany (TM), illustrating the relationship of these sequences with those of cultivated species of the order *Actinomycetales* and related taxa. The dendrogram was inferred by distance matrix analysis (132) of 450 positions from the 5′ terminus of the molecule. Scale bar represents 10 mutations per 100 nucleotide sequence positions.

ing bacteria, and 10% others. Only 7% of the isolates were gram-negative bacteria. Sequence analysis of 16S rDNA and subsequent dot blot hybridization with specific oligonucleotide probes revealed that 53 isolates could be assigned to described species. The majority of isolates were identified as members of the species *Bacillus mycoides* (about 20% of all isolated strains) (F. von Wintzingerode, F. A. Rainey, and E. Stackebrandt, unpublished). Also isolated were members of the genera *Mycobacterium*, *Nocardia*, and *Streptomyces*.

Molecular approach

A molecular analysis was performed on the middle layer of the peat bog (155). Based on a conventional cloning approach of PCR-amplified 16S rDNA, using the proofreading DNA polymerase Ultma (Perkin Elmer) and the pCR-Script SK(+) cloning kit (see Section 2.2), the phylogenetic affiliation of 259 cloned 16S rDNA sequences was determined by dot blot hybridization and by sequence analysis. Of these clones, 37 gave a positive signal with a probe specific for gram-positive bacteria, while a probe specific for a group of the α-subphylum of the *Proteobacteria* and another for a wide range of the β-subphylum of the *Proteobacteria* were positive with 110 and 3 clones, respectively. About 80 clones did not react with any of the three probes applied. A probe that was generated on the basis of the sequence of *Bacillus mycoides*, isolated frequently from this environment, did not hybridize with any of the cloned 16S rDNA sequences. Similarly, no sequences were obtained for members of *Mycobacterium*, *Nocardia*, and *Streptomyces*, which had also been isolated from the same sample. It could be argued that the soft lysis protocol applied for the disruption of cells directly in the environmental sample did not lyse the endospores nor the thick gram-positive cell wall of bacilli and actinomycetes.

None of the 92 clones analyzed by comparative sequencing analysis was absolutely identical to that of a described species, or to strains isolated from the same site. The highest similarity was found between two clone sequences and *Burkholderia solanacearum* (about 95% similarity over a stretch of 400 nucleotides from the 5' terminus of the 16S rDNA). All clones that gave a positive signal with the gram-positive probe in the dot-blot analysis were sequenced, and, surprisingly, belonged to a single coherent cluster (clones TM36, 56, 187, etc., in Fig. 6). This group of related clone sequences (88% to 96% similarity) belongs to a large cluster including cloned 16S rDNA sequences that were isolated from Mount Coot-tha, Australia (clones MC4, 58, etc., Fig. 6), and an acidothermal soil in New Zealand (clones NZ15 and NZ44, Fig. 6; see Section 3.3). This group also encompasses strain TH3 and *Microthrix parvicella* (see Section 3.1). Of the 92 clone sequences determined, 35 were not related to any known microorganisms, and the highest homology values with entries in the 16S rDNA data base were between 77% and 85%.

Obviously, there is a discrepancy between the genetic diversity found using purely molecular techniques and that using isolation studies. Whereas the pure cultures obtained were dominated by gram-positive bacteria (more than 90% of isolates), this group made up less than 10% of the clones analyzed. Even then, completely different lineages within the gram-positive bacteria were obtained by the parallel use of these two methods. To what extent each method introduces a bias to the results cannot be ascertained at present.

3.3 Acidothermal Soil from Orakei Karako Thermal Reserve, New Zealand

The biotechnological potential of thermostable enzymes has been the driving force behind numerous isolation studies from thermal environments. A diverse range of thermophilic microorganisms have been isolated and shown to belong to the domains *Bacteria* and *Archaea* (156). The majority of thermophilic isolates has been recovered from thermal environments with pH values around neutral (157), and few strains have been isolated from acidic thermal samples. Those microorganisms with pH optima of less than 5, isolated from such acidothermal environments, have been aerobic *Archaea* (157).

Given the limited diversity of microorganisms recovered by conventional isolation techniques from acidothermal environments, a molecular ecology study was initiated to test the hypothesis that such environments contain only a limited microbial diversity, relative to that found in less extreme environments. The sample selected for investigation was from an environment with a combination of low pH and high temperature. The acidothermal soil sample was collected from the Orakei Korako Thermal Reserve in the central North Island of New Zealand. The pH of the sample was between 3 and 4, and the temperature of the site averaged 75°C.

The environmental DNA was extracted by direct lysis of the microorganisms within the soil matrix using a combined approach of lysozyme, SDS, proteinase K treatment, and sonication for 30 s in a laboratory sonication bath. Humic acids were removed by purifying the DNA by agarose gel electrophoresis (see Section 2.2). In order to recover 16S rRNA gene sequences from both *Archaea* and *Bacteria* in one single PCR reaction, the following primers were used: <u>CGC AGA TCT</u> CAG C(C/A)G CCG CGG TAA T(A/T)C and <u>GGC CGT CGA C</u>AG AAA GGA GGT GAT CCA GCC (targeting positions 519 through 536 and 1525 through 3′ terminus, respectively (*Escherichia coli* numbering [63]). The PCR primers were synthesized with linker regions (underlined) containing recognition sites for the restriction endonucleases *Bgl*II and *Sal*I, respectively, to allow subsequent sticky end cloning of the amplified products. A total of 160 recombinants were selected for further analysis by hybridization with oligonucleotide probes and/or sequence determination.

Archaeal clone diversity

Using a probe specific for members of the domain *Archaea* (158), it was established that 33 of the 160 clones contained archaeal 16S rRNA genes. In order to determine the diversity represented by these archaeal clones, 18 of these 33 were selected for sequence analysis, which showed the clones to comprise two groups. The first was a novel group falling between the *Crenarchaeota* and the *Euryarchaeota* (29), represented by clone NZ115 (Fig. 7). This group of clones originates from a similar branching point to that of the clones pJP89, pJP78, and pJP27 detected by Barns et al. (77), but nevertheless represents a distinct lineage in this intermediate branching position. This group of clones showed no close resemblance to any cultured *Archaea* species for which sequence data are available; the highest similarity is less than 80%. The degree of sequence diversity among the acidothermal soil clones in the group around clone NZ115 was extremely low, with sequence differences of less than 1%.

The second archaeal group, represented by clone NZ26 (Fig. 7), clustered within the *Crenarchaeota* close to members of the genus *Sulfolobus*. Sequence similarities of 90.2% and 92.2% were found to *S. acidocaldarius* and *S. shibatae*, respectively. As with the NZ115 group of clones, limited sequence diversity between clones of the NZ26 group was observed. The NZ26 group possibly represents a species of the genus *Sulfolobus*, for which sequence data have not yet been determined or a strain to date not recovered by isolation studies.

Sequence analysis of the archaeal clones recovered from the acidothermal soil sample demonstrates additional diversity within the domain *Archaea*. The new sequences do not group closely with any of the archaeal clone sequences that have been recovered in other environmental studies, either in thermal environments or in marine environments. Those isolates that have been validly described, but for which no 16S rDNA sequence data are currently available, should be investigated in the future, to allow a better assessment of archaeal diversity. Although many of these strains are physiologically and phenotypically similar, their phylogenetic diversity must be determined for future comparative studies.

Gram-positive bacterial clone diversity

A 16S rDNA oligonucleotide probe specific for gram-positive bacteria (155) was used to determine the number of clones containing 16S rDNA from gram-positive bacterial members of the acidothermal soil community. This probe gave a positive signal with 24 clones. Sequencing of the inserts of these clones indicated that they represented three different groups of the gram-positive bacteria. As was found with the archaeal clones, the degree of sequence diversity within each group was limited.

The majority of these clones grouped within the radiation of the genus *Bacil-*

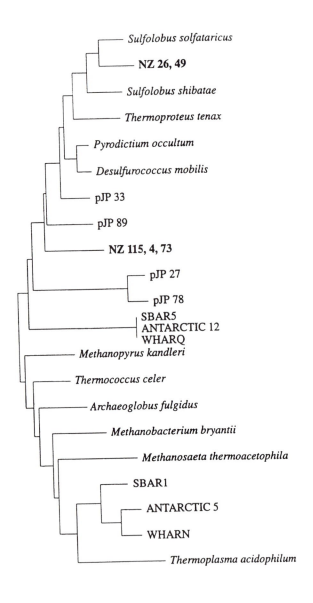

Sulfolobus solfataricus
NZ 26, 49
Sulfolobus shibatae
Thermoproteus tenax
Pyrodictium occultum
Desulfurococcus mobilis
pJP 33
pJP 89
NZ 115, 4, 73
pJP 27
pJP 78
SBAR5
ANTARCTIC 12
WHARQ
Methanopyrus kandleri
Thermococcus celer
Archaeoglobus fulgidus
Methanobacterium bryantii
Methanosaeta thermoacetophila
SBAR1
ANTARCTIC 5
WHARN
Thermoplasma acidophilum

10%

Figure 7. Phylogenetic dendrogram of archaeal 16S rRNA gene clones obtained from acidothermal soil from the Orakei Karako thermal reserve, New Zealand (NZ), illustrating the relationship of these sequences to those of cultivated members of the domain _Archaea_ and 16S rDNA clones retrieved from other habitats: ANTARCTIC (76), SBAR, WHARN (54). The dendrogram was inferred by distance matrix analysis (132) of about 1000 base pairs starting from position 550 of the molecule (_Escherichia coli_ numbering [63]) to the 3' terminus. Scale bar represents 10 mutations per 100 nucleotide sequence positions.

lus, and more specifically, with members of rRNA group 1 (159). The related-ness of clones NZ36, 43, 78, 118, etc., to representatives of the members of the genus *Bacillus* and their relatives is shown in Fig. 8. The cultured thermophilic members of the genus *Bacillus* have been shown to be phylogenetically diverse, but the majority of thermophilic species fall within the group around *B. stearothermophilus* in rRNA group 5 (159,160). The 16S rDNA sequence data have been determined for all thermophilic members of the genus *Bacillus,* and no high similarity was found with these clones. If it is assumed that this group of clones represents an indigenous population in this extreme environment, then it could be considered to represent a new lineage of acidophilic and thermophilic bacteria within the genus *Bacillus.*

A second group of clones, represented by NZ30 and 92 (Fig. 8), was found to cluster with the genus *Alicyclobacillus.* This correlates well with the phenotypic characteristics of members of the genus *Alicyclobacillus* and is what would be expected in terms of microbial composition of an acidothermal environment. These clones would seem to represent a new species of *Alicyclobacillus.* Inter-estingly, an isolate from the same sample, strain OKS1, also grouped with mem-bers of the genus *Alicyclobacillus* but was not identical in sequence composition to any of the clones studied (Fig. 8). This gives an indication of further diversity present in the sample, which was not detected at the sequence level.

As indicated in Section 3.1, clones which represent a deep branch within the *Actinomycetales* affiliated with the iron-oxidizing strain TH3 were also recov-ered.

β-proteobacterial clone diversity

Of the remaining clones not giving positive signals with either the *Archaea* probe or gram-positive probe, 17 were selected and the inserts sequenced. All of these clones were found to comprise two groups within the β-subphylum of the *Proteobacteria* but showed little sequence diversity within their groups. One group, exemplified by clone NZ20, clustered between the *Burkholderia* spp. and *Pseudomonas andropogonis* (Fig. 9). The other group of clones indicative of the β-subphylum of the *Proteobacteria* showed about 98% sequence similarity to *Janthinobacterium lividum* (Fig. 9), a mesophilic, chemoorganotrophic bac-terium.

The presence of these sequences in an acidothermal soil is somewhat surpris-ing, since such microorganisms, although associated with soil or plant material, would normally be isolated from neutral pH environments and would be mesophilic in growth temperature requirements. The environment from which the sample was taken is rich in organic and plant material, and therefore these microorganisms could be present but no longer viable, or their DNA could be present. Since such thermal soil environments are normally unstable, with changing temperature and pH depending on the geothermal conditions prevail-ing at a given time, it is also possible that these microorganisms were present

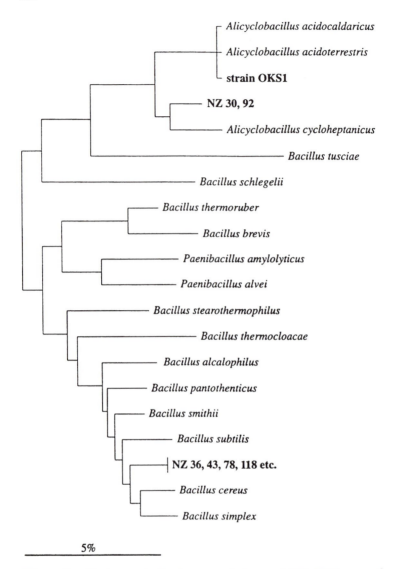

Figure 8. Phylogenetic dendrogram of clones of 16S rRNA genes from gram-positive bacteria obtained from acidothermal soil from the Orakei Karako thermal reserve, New Zealand (NZ), illustrating the relationship of these sequences to those of cultivated members of the *Clostridium/Bacillus* lineage of the gram-positive bacteria. The dendrogram was inferred by distance matrix analysis (132) of about 1000 base pairs starting from position 550 of the molecule (*E. coli* numbering [63]) to the 3′ terminus. Scale bar represents 5 mutations per 100 nucleotide sequence positions.

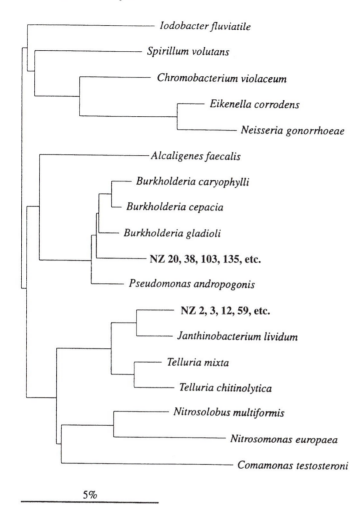

Figure 9. Phylogenetic dendrogram of cloned 16S rRNA genes, belonging to the β-subphylum of the *Proteobacteria*, obtained from acidothermal soil of the Orakei Karako thermal reserve, New Zealand (NZ), illustrating the relationship of these sequences to those of cultivated members of the β-subphylum of the *Proteobacteria*. The dendrogram was inferred by distance matrix analysis (132) of about 1000 base pairs starting from position 550 of the molecule (*E. coli* numbering [63]) to the 3′ terminus. Scale bar represents 5 mutations per 100 nucleotide sequence positions.

and viable in this environment when the conditions were more suited to their growth. The high number of clones shown by oligonucleotide probing to belong to these two groups (81 of 160) may indicate that these microorganisms, or their DNA, were present in a considerable quantity.

Finding the sequences of microorganisms not normally considered to be present and viable in such an extreme environment raises questions about those sequences found in what are considered 'normal' environments. The assumption has been made that the sequences from these normal environments reflect the microbial diversity present, but it is clear from the results of this study on acidothermal soil that other methods need to be applied to determine what fraction of the microbial diversity is present as viable cells, what is present as dead cells, and what is present as naked DNA. The application of in situ probing techniques may indicate whether thermophilic members of the β-subphylum of the *Proteobacteria* are present as intact, perhaps active cells in this soil or whether these sequences are from naked DNA.

This study has demonstrated that the primer pair used amplifies both archaeal and bacterial sequences from the environment. Although this approach gives the advantage of a single PCR and cloning step, it limits analysis to the last 1000 nucleotides of the 16S rRNA molecule.

Although some new lineages have been demonstrated, the degree of diversity recovered from this extreme environment is somewhat limited when compared to the wide range of sequences recovered from the forested soil environment of Mount Coot-tha (see Section 3.1). This lack of diversity may reflect the extreme prevailing conditions. With the exception of the clones related to members of the β-subphylum of the *Proteobacteria*, many of the clone sequences from this site are affiliated with pure culture isolates obtained from similar environments. Many of the clones recovered formed clusters which displayed very little sequence diversity, in contrast to the high 'microheterogeneity' which has been detected in some other habitats (see Section 4.2). These results, therefore, seem to confirm the original hypothesis, that only a limited microbial diversity is present compared to that in less extreme environments.

3.4   Flooded Rice Cultured in a Laboratory Model System

Rice is the world's most important agronomic plant, with globally 143 million ha under cultivation (161). The majority of the rice is cultured under flooded conditions, meaning that the bulk soil is basically anoxic. This is the reason that rice cultures represent one of the major sources for the continual increase in the atmospheric methane concentration (162). It has been assumed that the methanogens present in the completely anoxic bulk soil are responsible for the production of methane. The emission of the methane occurs mainly by diffusion via the aerenchyma of the rice plants, but a substantial amount of the methane originally produced can be oxidized prior to emission by an obligately aerobic methanotrophic population present in the rhizosphere (163–165). Whereas the general scheme of microbial processes involved in the degradation of the polymer substances to monomers and via intermediates finally to methane is well known, only rather limited knowledge is available about the actual microbial populations responsible and their spatial distribution (rhizoplane, rhizosphere, or surrounding bulk soil) in flooded rice soil.

As a first step to unravelling the microbial community of flooded rice cultures, environmental DNA was isolated from intensively washed roots of rice and, in parallel, from bulk soil for a comparative molecular analysis of the microbial populations colonizing these two different habitats of a model rice culture grown in the laboratory. This model system allows some degree of reproducibility and ease of sampling. Rice (*Oryza sativa* var. Roma) was grown in plastic containers in soil collected from Italian rice paddies; the system has been described in detail elsewhere (164). The rate of methane production in this system ($55–94$ mmol $\times$ d^{-1} $\times$ m^{-2}) is similar to that in the paddies from which the soil originates ($79–128$ mmol $\times$ d^{-1} $\times$ m^{-2}) (164,166). The extraction of the environmental DNA from both rice roots and bulk soil, PCR-mediated amplification of 16S rRNA genes with a primer pair considered to be universal for members of the domain *Bacteria*, and the generation of a clone library were done by standard techniques. The comparative phylogenetic analysis of randomly picked clones containing 16S rDNA retrieved either from rice roots or from bulk soil revealed the presence of two phylogenetically different microbial communities in these habitats. Members of the α- and β-subphyla of the *Proteobacteria* dominated the clone library derived from rice roots, suggesting these groups to be the main colonizers of the rhizoplane (D. Rosencrantz and W. Liesack, unpublished). In contrast, these two proteobacterial groups were not retrieved from the bulk soil at all; instead, this clone library was dominated by members of the *Clostridium/Bacillus* lineage of the gram-positive bacteria, related to obligately anaerobic *Clostridium* spp., facultatively anaerobic *Bacillus* spp., and obligately anaerobic gram-positive bacteria with a gram-negative cell wall structure (U. Hengstmann and W. Liesack, unpublished).

Interestingly, the use of a primer pair specifically designed to amplify 16S rRNA genes from members of the archaeal kingdom *Euryarchaeota* resulted in a strong and specific signal from DNA extracted from washed root material. The PCR products were cloned using the TA Cloning Kit (see Section 2.2). The phylogenetic analysis of four completely sequenced clones and another 25 partially analyzed clones indicated the presence of a novel, currently uncultured group of members of the *Euryarchaeota* (clones RRO4, 8, 13, and 20 in Fig. 10); this group constituted the majority of *Archaea* detected by this approach (R. Groβkopf and W. Liesack, unpublished). This phylogenetically coherent

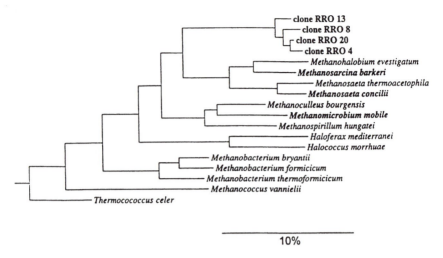

10%

Figure 10.   Phylogenetic dendrogram of archaeal 16S rRNA gene clones (RRO) obtained from the rhizoplane of a model system of flooded rice cultures, indicating the relationship of these sequences to those of cultivated members of the *Euryarchaeota*. The dendrogram was inferred by distance matrix analysis, implemented in the PHYLIP package (134), of about 500 base pairs starting from position 250 of the molecule (*E. coli* numbering [63]). Scale bar represents 10 mutations per 100 nucleotide sequence positions.

group of *Archaea* originates between members of the order *Methanosarcinales*, e.g., the acetoclastic microorganisms *Methanosaeta concilii* and *Methanosarcina barkeri*, and members of the order *Methanomicrobiales*, e.g., the H_2/CO_2- and formate-utilizing *Methanomicrobium mobile*, as shown in Fig. 10. The phylogenetic affiliation of this newly detected group allows one to predict that they are methanogens but gives no clue as to the growth substrate(s) they may be able to utilize and thus makes their selective isolation difficult. That these 16S rDNA sequences are not just present as naked DNA (see Sections 2.2 and 3.3) is indicated by the observation that washed rice root samples from the same system immediately began to produce methane when shifted to anoxic conditions (167).

In contrast to the bulk soil, the rhizoplane offers the potential for a relatively straightforward application of modern technologies such as the use of fluorescently labeled population-specific oligonucleotide probes aimed at the 16S rRNA, coupled with confocal laser scanning microscopy, to visualize the molecularly detected microorganisms in situ by whole-cell hybridization (168) and so to unravel the spatial distribution of the various (aerobic and anaerobic) populations (see Chapter 10).

4. METHODOLOGICAL PROBLEMS OF 16S rDNA-BASED STUDIES

As shown in the previous section, the comparative analysis of the 16S rRNA gene pool is a very sophisticated tool for molecularly based analysis of microbial diversity in soil. However, it should be mentioned that this approach still has some methodological uncertainties, which should be borne in mind when interpreting the data obtained.

4.1 Biased Retrieval of Sequence Information

The 16S rDNA-based strategies to date available for the analysis of genetic diversity in soil facilitate only qualitative rather than quantitative insights into the community structure. The molecular window opened into the microbial world of a soil sample under study is strongly influenced by several parameters, for example, the procedure applied for the extraction of the environmental DNA (see Section 2.2), the oligonucleotide primers and thermal profile used in the PCR, as well as the cloning strategy employed. No detailed comparative studies exist for the extent of bias, but several observations suggest that this is not just a theoretical threat.

Selective priming or higher-order structural elements may cause the different 16S rDNA sequences present in the environmental DNA to be amplified with different efficiencies. This will result in shifts in the relative proportions of the different rDNA types after multiple cycles in the PCR reaction. One important consequence is that the frequency with which a sequence occurs in the 16S rDNA library generated from an environmental sample does not necessarily reflect its relative abundance in the microbial community.

Reysenbach et al. (169) reported a differential amplification of rRNA genes using a mixture of genomic DNAs isolated from a hyperthermophilic archaeon and *Saccharomyces cerevisiae*. In a standard PCR reaction, both 16S rDNA types should ideally be amplified with the same efficiency, but the resulting PCR product was exclusively composed of the yeast 16S rDNA. However, the addition of 5% (w/v) acetamide to the PCR reaction resulted in the amplification of both the archaeal and the yeast 16S rDNAs, indicating that acetamide can alleviate the selective amplification of 16S rDNAs if mixed-target DNA is used.

Suzuki and Giovannoni (170) observed that the extent of bias introduced during the PCR amplification depended on the 16S rDNA primer system used. The studies were performed with defined mixtures of two different 16S rDNA sequence types. One primer pair (519F–1406R) yielded products in the predicted proportions, whereas a second (27F–338R) resulted in a strong bias, tending toward 1:1 mixtures of both sequence types in the final PCR product, regardless of their original proportions in the reaction mixture. The amplification efficiency of the second system was substantially higher than that for the first primer pair tested. In addition, the bias was positively correlated with increased numbers of PCR cycles performed and may be a general phenomenon for PCR reactions with mixed templates and high amplification efficiencies. The results obtained could be fitted to a kinetic model, in which the increasing reassociation of denatured single-stranded PCR product progressively inhibits the priming reaction via annealing of the primer to its template.

Wilson and Blitchington (36) compared the results of culture-based methods and partial sequencing of cloned 16S rRNA genes from a human fecal specimen. Clone libraries from two different PCR products were generated and compared with each other, the first after 9 cycles of PCR and the second after 35 cycles. Interestingly, 50 '9-cycle' clones gave 27 different sequence types, which covered 59% of the diversity present in the library of cloned rDNAs, whereas 39 '35-cycle' clones yielded 13 sequence types, which covered 74% of the clone library analyzed. These data may indicate a strong distortion of the original relative abundance of sequence types, leading to an underestimation of the broad range of diversity present in the environmental gene pool with increasing numbers of PCR cycles performed.

The knowledge of the number of copies of the 16S rRNA gene per species genome is also a prerequisite, if attempts to analyze quantitatively even the simplest community structures based on the relative ratio of the amplified PCR products are to be successful (171). Since this parameter cannot be known for as yet only molecularly detected microorganisms, the exact quantification of microbial population sizes will not be possible from the analysis of 16S rDNA clone libraries.

The products of the PCR reaction are sometimes purified by electrophoresis to prevent cloning of nonspecific amplification products. This means that larger 16S rRNA genes, with intervening sequences such as found in a number of organisms (172–175), may be mistaken for nonspecific products and excluded from the analysis of microbial diversity.

Rainey et al. (150) used the same batch of environmental DNA to evaluate the influence of (1) different 16S rDNA primers and (2) different cloning vectors on the microbial diversity detectable by using molecular approaches. In both cases, the composition of taxon-specific sequence types changed, strongly suggesting that 16S rDNA approaches based on cloning strategies appear not to allow a complete qualitative nor an accurate quantitative determination of the microbial community represented by the gene pool extracted from the habitat under study. The risk of losing detectable genetic diversity, when a sticky-end cloning strategy is used, is discussed elsewhere (see Section 2.2).

4.2 Formation of Chimeric 16S rDNA

Chimeric 16S rDNA molecules are produced in PCR-mediated amplification by the annealing of partial-length fragments of different 16S rDNA types via highly conserved regions and subsequent primer-independent elongation to full-length fragments (176). These fragments are amplified in the next rounds of denaturation, annealing, and elongation, with the same efficiency as nonchimeric fragments. The tendency toward chimera formation is thought to depend on several factors: (1) premature termination of elongation during PCR; (2) availability of partial-length 16S rDNA fragments increasingly present in environmental DNA of low molecular weight, for example, such as may be produced by shearing of DNA during harsh extraction procedures (see Section 2.2); and (3) degree of sequence identity along the primary structure of the rDNA types, and thus, phylogenetic relatedness between the microorganisms under study. Since the chimeras are shuffled together from different rDNA types, the phylogenetic treeing of these sequences will lead to the postulation of novel lines of descent not actually existing in nature. Chimeras were detected in 16S rDNA clone libraries generated from complex environmental samples in the range of 2 of 113 up to 2 of 9 analyzed clones (58,112,177).

A model system with known 16S rDNA sequences has been used to quantify the frequency of chimera formation during PCR amplification in dependency on different parameters (178). Mixtures of two almost identical 16S rDNA sequence types showed a 30% occurrence of chimeras after 30 cycles of coamplification. The frequency of chimera formation decreased to 14.7% and 12.9% for pairs of templates with 86% and 82% similarity, respectively. Furthermore, the frequency of chimera formation was positively correlated with an increase in the number of PCR cycles used but decreased slightly with prolonged elongation periods. The study clearly demonstrated that premature termination during the elongation period is the main reason for the increased occurrence of chimeras. Consequently, for PCR-based studies directed to unravelling microbial diversity via 16S rDNA clone libraries, the use of heat-stable DNA polymerases with high processivity seems to be recommended and may lead to reduced occurrence of premature termination.

Several options are available for checking clone sequences for chimeric structures: (1) reconstruction of separate phylogenetic trees for the 5′ and 3′ regions and (2) checking of the helical regions for complementarity. A study has, however, reported evidence that the latter test is not fail-safe. Two sequences obtained from a hot spring cyanobacterial mat community could only be identified as being chimeric by separate phylogenetic analysis of short sequence regions, as both rDNA sequences exhibited no and one secondary structural abnormality, respectively (177). The Ribosomal Database Project supplies an automated chimera check program through similarity analysis (45). The use of this program is recommended, but an evaluation has revealed that the reliability of this program decreases when the parental sequences which contribute to chimera formation are more than 82% to 84% similar (179).

One general finding of the environmental rRNA approach is the retrieval of 16S rDNA sequences from natural environments which often form closely related clusters with similarities of between 93% and 99%. This raises the question whether this high microdiversity reflects a naturally occurring phenomenon or an artifact due to method-related inaccuracies. In addition to the risk of chimera formation, possible microheterogeneities between different 16S rDNA operons present in the genome of one microbial cell may contribute to an overestimation of the actual microbial diversity. The importance of the latter aspect to the interpretation of closely related sequence data obtained in environmental rRNA approaches cannot be judged with the current limited sequence information. A heterogeneity of approximately 5% between the two 16S rRNA genes of the archaeon *Haloarcula marismortui* was reported (180), but, in contrast, the six 16S rRNA genes present in the genome of *Haemophilus influenzae* are completely identical (181). Amann et al. (182) used simultaneous in situ hybridization with three differently labeled oligonucleotide probes and confocal laser scanning microscopy to examine the microbial complexity of closely related populations within an activated sludge sample. The oligonucleotide probes targeted three different sites in so-called highly variable regions of one of the 16S rDNA clone sequences which had been retrieved from this habitat. Surprisingly, the combined use of the three probes in a 0.7 μm thick optical section of an individual activated sludge floc detected not only the target organisms, but also all of the other six theoretically possible probe-binding patterns. These results clearly demonstrated that at least seven closely related populations, with slightly different 16S rRNA sequence types, were present in the sample under study. These data represent important evidence that the high microdiversity detected by direct 16S rRNA sequence retrieval is indeed realized in nature.

5. DIVERSITY AT THE PHYSIOLOGICAL LEVEL

The assessment of phylogenetic diversity by comparative 16S rRNA sequence analysis is a very important aspect of molecular microbial ecology. It has proven how little is known about the range of evolutionary diversity and highlights the lack of progress in the attempts of the last one hundred years to isolate a representative fraction of prokaryotic species. One major drawback of the 16S rDNA-based approach is the inability to unravel metabolic diversity. Consequently, the functional role of a molecularly detected microorganism in its habitat remains by and large undetermined. The best guess as to the presence of catabolic and anabolic pathways can be made for those strains for which 16S rDNA sequences indicate a close relationship to a cultured strain with known biochemical pathways; examples are those cases where the sequences indicate the position of an as yet uncultured strain within a radiation of well-studied taxa with one basic metabolic type, such as the methanogens, cyanobacteria, spore-forming actinomycetes, and spirochetes. In those lineages in which predictions can be made, i.e., those containing one physiological type, this may simply be a consequence of not having enough reference microorganisms. The ability to make predictions comes from having already studied cultured microorganisms belonging to these lineages. In general, it is almost impossible to predict the metabolic properties of a microorganism whose 16S rDNA sequence forms an individual subline of descent within any of the subphyla of the *Proteobacteria*, the *Clostridium/Bacillus* lineage of gram-positive bacteria, or a deeply rooting line of descent within any of the bacterial and archaeal phyla, not to speak of those uncultured forms that seem to represent novel phyla. But even if the physiological characteristics of a dormant or resting cell can be predicted or are later determined from a cultured strain, it may still be impossible to speak about the role of this microorganism in the environment.

As the direct analysis of mRNA is still in its infancy (86), the presence of potential metabolic activity in a sample is being explored by developing gene probes that target those genes involved in anabolic and catabolic pathways. Understandably, the number of these functional probes is still small; first, the data base of these gene sequences is small and new entries often do not reveal high homologies; second, the degree of conservation of most of these genes is significantly lower than that of ribosomal RNA genes and the range of taxa covered by a probe needs to be determined individually for each probe. A variety of functional genes has been identified for the detection of potential metabolic activity in a broad range of prokaryotes, but only a few probes have been designed for this purpose. For example, among the sulfate-reducing bacteria alone, sequence information is available for APS reductase, bisulfite reductase, cytochrome c_3, ferredoxin, and hydrogenase (183), but only the hydrogenase has been used to identify novel *Desulfovibrio* spp., both in pure culture and in enrichment culture (184). These studies were later extended by the development of a combined approach of PCR assays and DGGE to analyze the genetic diversity of *Desulfovibrio* spp. containing [Ni–Fe] hydrogenase genes in biofilms collected from a bioreactor (185).

Nevertheless, the few probes developed have, in principle, demonstrated their value in screening environmental samples for the presence of certain activities. They complement the available set of polyclonal antibodies which, by Western blotting procedures, are suitable for the detection in whole-cell extracts of homologous proteins against which these antibodies had been raised. Table 1 shows a selection of functional probes together with their size and the probe

Table 1. List of Selected DNA Probes Used in Environmental Studies[a]

Target gene	Size of probe	Organisms
Formyltetrahydrofolate synthetase	1.38 kb	*Clostridium thermoaceticum*
Methane monooxygenase (component B)	2.2 kb	*Methylosinus trichosporium* OB3b
Methane monooxygenase (hydroxylase, component B, reductase)	5.8 kb	*Methylococcus capsulatus* Bath
Methyl monooxygenase (hydrolase and NADH-ferredoxin reductase)	2.35 kb	*Pseudomonas putida* PaW1
Nitrite reductase (copper containing)	1.9 kb	*Pseudomonas* sp. strain 179
Nitrite reductase (heme containing)	0.7 kb	*Pseudomonas stutzeri* JM300
Nitrogenase (*nif*HDK genes)	3.6 kb	*Rhizobium* sp.
Nitrogenase (*nif*H gene)	PCR oligonucleotides	(Gene bank entries of nitrogenases)
Nitrous oxide reductase	1.2 kb	*Pseudomonas stutzeri* ZoBell
Toluene dioxygenase (large and small subunits of oxygenase, ferredoxin, and part of reductase)	3.5 kb	*Pseudomonas putida* F1
Toluene *para*-hydroxylase (monooxygenase and ferredoxin)	3.6 kb	*Pseudomonas mendocina* KR
Toluene *meta*-hydroxylase (α-subunit of monooxygenase)	0.68 kb	*Burkholderia pickettii* PKO1
Toluene *ortho*-hydroxylase	2.2 kb	*Pseudomonas* sp. strain JS-150

[a]The sources of probes are either indicated by Ref. 186 or cited in the text.

source microorganism. This table is based on the information compiled by Fries et al. (186).

As information on the distribution and sequence homology of target genes is sparse, several studies have focused on the hybridization reaction between the probes and microorganisms believed to possess the target gene(s). As an example, DNA from eight acetogens representing six genera was found to hybridize to the formyltetrahydrofolate synthetase (FTHFS) probe. However, the probe also bound, albeit weakly, to DNA from microorganisms that do not have the acetyl-CoA pathway, and it did not bind to DNA from FTHFS-producing purine fermenters. Using the probe on DNA that was isolated from horse manure, several signals of different size were obtained by Southern hybridization of restriction fragments, which probably indicate the presence of different acetogens in this sample (187).

In a similar approach, DNA probes (188,189) and antibodies (190) directed against copper-containing and heme c,d_1–containing nitrite reductase genes and gene products, respectively, were evaluated. Except for a few unexpected signals, results of Southern and Western blot hybridization clearly proved the sensitivity of both probe types. Similarly, the application of probes directed against genes involved in the degradation of toluene to a variety of strains isolated from uncontaminated and toluene-contaminated soil proved the capacity of molecular techniques to identify toluene degraders rapidly (186). These strains, which phylogenetically belong to the genus *Azoarcus*, are widely distributed in nature, even in noncontaminated regions. In order to determine whether those samples from which toluene-degrading *Azoarcus* strains could not be isolated do indeed contain dormant strains of this taxon, the pathway-specific probes should be hybridized to environmental DNA.

The use of a PCR assay, in which the target is first amplified, allows greater sensitivity than probing alone, and in addition, subsequent sequencing of the products, either directly or after cloning, can confirm the specificity of the assay. A PCR assay targeting the gene for a catechol 2,3-dioxygenase, a key enzyme involved in aromatic hydrocarbon degradation, was successfully used to enumerate aromatic hydrocarbon-degrading bacteria in contaminated soil (191). Another example is the detection of methanotrophic bacteria based on the soluble methane monooxygenase (sMMO) in aquatic and terrestrial habitats (192). The disadvantage of this assay is that the presence of the sMMO appears to be restricted to type II methanotrophs such as *Methylosinus trichosporium* OB3b and *Methylosinus sporium* and the type X methanotroph *Methylococcus capsulatus* (Bath) (192). The first sequences of the gene cluster for a particulate methane monooxygenase (pMMO), a version of the MMO which is present in all methanotrophic bacteria including type I such as *Methylomonas* spp., has been published (193). Similarly, gene sequence data for the ammonia monooxygenase (AMO) of *Nitrosomonas europaea*, the key enzyme of ammonia-oxidizing bacteria, have also been published (194,195). Despite their different physiological roles, the pMMO and the AMO share common characteristics, for example, similar substrate ranges and inhibitor profiles (196), which have led to the speculation that both enzyme systems may be homologous. A comparative sequence analysis of a 525-base-pair stretch of *pmo*A and *amo*A from a variety of different methanotrophs and ammonia oxidizers, respectively, showed high sequence identities, supporting the assumption that the pMMO and the AMO are evolutionarily related enzymes. Interestingly, the *amo*A from *Nitrosococcus oceanus*, the only known ammonia oxidizer not belonging to the phylogenetically coherent *Nitrosospira/Nitrosomonas* group (197), showed a higher identity to *pmo*A sequences from methanotrophic members of the γ-subphylum of the *Proteobacteria*, to which *Nitrosococcus oceanus* belongs, than to *amo*A sequences from members of the β-subphylum of the *Proteobacteria* (198). The comparative sequence analysis of partial ammonia monooxygenase genes of the phylogenetically closely related bacteria *Nitrosolobus multiformis* C-71 and *Nitrosospira* sp. AHB1 demonstrated the potential for the use of this functional gene for fine-scale differentiation of ammonia-oxidizing bacteria of the β-subphylum of the *Proteobacteria* in environmental population studies (199). In contrast, the resolution power of the 16S rDNA is very limited for this physiotype (198). On the basis of the *amo*A sequence information obtained, a PCR-based detection system was developed, which allowed the specific detection of ammonia-oxidizing populations in a variety of aquatic and terrestrial environments. The comparative sequence analysis of *amo*A clone sequences revealed the presence of *Nitrosospira* spp. on the rhizoplane of flooded rice cultured in a laboratory model system (see also Section 3.4) (J. Rotthauwe and W. Liesack, unpublished). The detection of *Nitrosospira* spp. on the rice root agrees well with a report in which 16S rRNA genes were targeted that members of this genus are widespread in various environments (200). However, the degeneracy of the genetic code and different evolutionary rates often make it difficult to develop PCR primers that can amplify target sequences of functional genes from a phylogenetically broad range of taxa. Therefore, more comparative information on *amo*A sequences from a variety of nitrifiers will be necessary to judge whether this PCR-based assay facilitates the detection of all *Nitrosospira/Nitrosomonas* lineages present in nature.

    One cluster of genes for which a high degree of conservation is well documented are those involved in nitrogen fixation, for example, *nif*HDK (201,202). The diversity of nitrogenase genes in a natural sample has been demonstrated by amplifying the *nif*H gene from DNA isolated from the rhizosphere of rice (203). None of the 23 *nif*H gene sequences analyzed was identical to those of the 37 different *nif*H gene sequences known from cultured strains. Similarly, most of the sequences isolated from the rice rhizosphere formed distinct clusters which were only remotely related to known nitrogen-fixing bacteria. These results correlate well with those obtained from 16S rDNA studies, indicating that the majority of prokaryotes have indeed not been isolated in pure culture.

A PCR-based assay amplifying a 0.75-kilobase pair region of the gene encoding the α-subunit of the methyl coenzyme M reductase (MCR) was used for the analysis of methanogenic populations in blanket bog peat (204). The primary structure of eight cloned genes analyzed was identical and showed more than 80% identity to the homologous regions of the gene sequences from *Methanosarcina barkeri* and *Methanobacterium thermoautotrophicum*. These high similarities may indicate that genes coding for MCR are useful functional markers for the specific detection of methanogens via PCR-based amplification in environmental samples. A completely new and exciting aspect directed to the analysis of functional diversity and dynamics of microbial communities may be the use of in situ PCR, a method which allows the genetic characterization of individual cells via PCR-based amplification of target sequences. The development of a protocol for prokaryotic in situ PCR (PI-PCR) has been reported (205). The authors demonstrated the use of PI-PCR to detect both DNA and mRNA target sequences in single bacterial cells. Future studies must show whether this new technology will become a reliable and reproducible tool for molecular ecological studies.

A further interesting approach aimed at identifying the metabolic potential of novel as yet only molecularly detected microorganisms was reported by Stein et al. (206). Studies based on 16S rRNA technologies had revealed that previously unknown archaeal lineages are abundant members of the picoplankton of the North Pacific (76). To get more information about one of those lineages, a group rooted deeply within the *Crenarchaeota*, high-molecular-weight DNA was extracted from a marine sample collected at a depth of 200 m. The environmental DNA was used to generate a fosmid DNA library. One 38.5-kilobase pair insert in a recombinant fosmid clone was identified as being derived from a member of the archaeal target group, using the 16S rDNA present in the cloned genomic fragment as a marker gene. Random shotgun sequencing allowed the identification of several functional genes which were also contained within this fragment, e.g., genes coding for the translation elongation factor 2, the RNA helicase, and the glutamate semialdehyde aminotransferase, an enzyme involved in the initial steps of heme biosynthesis and previously not known to be present in *Archaea*. Such molecular approaches may in the future allow the complete genomic potential of as yet uncultivated prokaryotes to be determined and may provide very important information on how to cultivate these microorganisms.

6. STRATEGIES FOR THE ISOLATION OF MICROORGANISMS

As noted in the preceding sections, the use of molecularly based methods in microbial ecology, be it in the study of soil or any other habitat, requires the correlation of molecular phylogenetic–taxonomic data to physiological activities.

With ever-expanding data bases of gene sequences, including 16S rDNA sequences, a lot can be inferred by comparing new sequences with already studied ones, if they are available. However, it is clear that the subtlety of variation in physiological characteristics between closely related organisms is the key to understanding why some are found in certain habitats and others elsewhere. This wide variety of slightly differing physiological types, much more varied than the range of 16S rDNA sequence types, is in fact the great range of microbial diversity, the full extent of which is still only guessed at or inferred. This means that the pure culture approach, far from being a historical relict, is required more than ever to understand this diversity. The microbiologist now has a new challenge to obtain significant habitat-specific organisms or novel newly detected microorganisms in pure culture to allow genetic and physiological studies, which, in turn, facilitate an ecological interpretation of the molecular data.

A wide variety of inventive, often simple and highly effective, but also extremely complicated and sophisticated methods have been used to isolate microorganisms from their environment. Obviously, not all of the strategies can be addressed here, but some points and examples are discussed later. Many strategies for the isolation of specific groups of *Bacteria* and *Archaea* have been collated by Balows et al. (207).

6.1 Selective Enrichment Culture of Microorganisms

The use of selective enrichment in liquid medium is one of the oldest methods for the isolation of a particular microorganism or physiological type out of a mixture of species. The selection of particular culture conditions, such as temperature, pH, redox potential, salt concentration, carbon and energy source, electron acceptor, growth factors, or inhibitors, allows the desired microorganism to grow (relatively) more rapidly and become the numerically dominant species in the culture. The subsequent use of a dilution step combined with a physical separation technique, such as plating, then allows the isolation of the desired microorganism in pure culture. Success depends on the ability of the microbiologist to formulate media which allow better growth of one group and discourage growth of potential, perhaps initially more numerous, competitors. When there are a number of species with similar physiological characteristics, then the microorganism with the highest growth rate, at the high substrate concentrations normally used in enrichment cultures, will be favored. This is not necessarily the microorganism which was dominant or active in the soil, where characteristics such as substrate affinity, stress survival, or others, may be important in determining which will be able to compete successfully.

In an attempt to prevent the selection of only fast-growing strains 'adapted' to high substrate concentrations, continuous culture enrichment has been used to select for microorganisms which are able to grow at low substrate concentrations (208,209). Here the selection is for microorganisms with a high substrate affinity. Continuous cultures result in the establishment of low substrate concentrations by the activity of the microorganisms themselves, ultimately dependent on the microorganism present with the highest substrate affinity. Under correctly selected conditions, the populations in the system will utilize the substrate to a concentration at which microorganisms without a high substrate affinity will no longer be able to maintain a growth rate high enough to prevent their being washed out.

Some microorganisms may not be able to grow in pure culture without a very detailed knowledge of their physiological features and growth requirements. The need for specific growth factors, produced in mutualistic relationships, is an example. Even more complex associations may exist, for example, syntrophic degradation of compounds with the obligate removal of an endproduct, or other metabolic cooperations to degrade compounds. In addition, complex symbioses with eukaryotic organisms in the soil may be important, for example, intracellular or close symbioses with ciliates (210), invertebrates (211), or plants (212).

The problems associated with selective enrichment will be further confounded if the microorganism in question is only known from a gene sequence, and the physiological characteristics can only be guessed at on the basis of related microorganisms. The detection of such microorganisms, for example, by the detection of their 16S rRNA gene in a clone library, may give no clue that there is a specific growth requirement. The use of labeled oligonucleotide probes targeting the 16S rRNA to carry out in situ whole-cell hybridization and observe microscopically the microorganisms in their environment (20,65,78) may in the future offer clues in some cases. This is demonstrated, for example, by the molecular ecological studies combined with microscopy of magnetotactic bacteria in sediment (20,213), where the direct observation of the organisms in nature may give an indication of their growth requirements. Furthermore, labeled oligonucleotide probes can be used to monitor the success of different isolation procedures. For example, enrichment cultures for a sulfate-reducing bacterium were monitored using a fluorescently labeled oligonucleotide hybridization probe specific for a 16S rRNA gene detected in an environmental sample. The cultures in which the target bacterium was present were pursued further, and a pure culture which had the same 16S rRNA gene sequence as originally detected in the sample was isolated (214). Similarly, Huber et al. (215) successfully used a fluorescently labeled oligonucleotide hybridization probe to check enrichment cultures for the presence of a certain hyperthermophilic member of the *Archaea* which had previously been detected as a 16S rDNA sequence in a molecular ecological study (77).

6.2   Isolation of Dominant Microorganisms

The viable count is a commonly used method to determine the numbers of microorganisms and simultaneously obtain representatives in pure culture. Media are selected to allow growth of (ideally) all representatives of certain groups of microorganisms present in a habitat. Brock (216) and Pickup (217) have pointed out the problems associated with the use of the viable count to assess microbial populations. Many typical soil microorganisms form resistant resting stages (218,219) or spores. Dormant cells and survival forms are counted and isolated, together with active cells. This has been found in counts of *Clostridium* spp., where pasteurized and unpasteurized inocula often yield the same cell counts, indicating the majority are present in the sample as spores (220). A multicellular filament or colonial aggregate of cells, such as myxobacteria, will be counted as one individual, as will all the microorganisms adhering to a soil particle. The efficiency of viable counting methods in soil is often reported to be low, often 1% or less, compared with that of direct microscopic counts (221–225). In addition, the use and necessity of special selective or growth-promoting media mean that generally only groups which are searched for, or which can fortuitously grow on these media, can be detected. This leads to a bias, in that the investigator determines which microorganisms will be able to grow through the choice of cultivation methods.

The problem of the low recovery of the total population using viable counting methods may be overcome in part by careful use of growth media, and by choice of appropriate conditions. One important point seems to be the use of low concentrations of appropriate nutrients. Since the bulk (nonrhizosphere) soil seems to be an oligotrophic habitat (16,226), it is likely that a significant proportion of the community will be active, but not growing, as found for sediment environments (227). One of the real problems inherent with cultivation methods is the need to add enough of all nutrients to enable turbidity or colonies to develop, or enough of a product to be detected, so that growth can be recognized. The use of sugar polymers as growth substrates for the recovery of microorganisms from the soil of flooded rice resulted in cell counts up to two orders of magnitude higher than when simpler sugars were used (Fig. 11) (K. J. Chin and P. H. Janssen, unpublished). The organisms isolated in pure culture from the terminal positive tubes of the dilution series with polymers were also able to grow on the simple sugars and so, theoretically, should have been able to grow in those dilution series too. The much lower counts obtained with the simple sugars suggested that growth was inhibited when the organisms were transferred from the soil to the high substrate concentrations in the medium. The sugar polymers have first to be hydrolyzed, and thus the initial substrate concentration is virtually zero. The failure of organisms to grow upon transfer from a low-nutrient environment to growth medium with high substrate concentrations may be attribut-

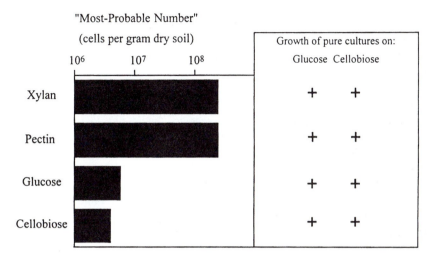

Figure 11. Most probable number counts of microorganisms in the soil of flooded rice showed that sugar polymers (xylan and pectin) were better suited to the recovery of high cell numbers than were simpler sugars (glucose and cellobiose). Different pure cultures isolated from the terminal positive tubes of the dilution series could grow on glucose and cellobiose, but the same organisms apparently could not be isolated from the soil using these growth substrates, since the cell counts were much lower with the simpler sugars.

able to substrate-accelerated death and has been observed in pure cultures (228,229) and in natural microbial communities (230). Other, more provocative explanations have also been suggested for the failure of organisms to grow, especially when small inocula are used (231).

The direct isolation of microorganisms from the most dilute positive tubes of most probable number (MPN) counts or liquid serial dilutions, or direct isolation by dilution onto agar plates or into agar deeps or roll tubes, often results in the isolation of microorganisms which are phylogenetically distinct and physiologically more environmentally relevant than microorganisms obtained by classic enrichment techniques. In MPN counts, the higher (more dilute) dilutions contain microorganisms which were, at the time of sampling, present in higher numbers (assuming quantitative recoveries) than those that grow in the lower (less dilute) dilutions. The lower dilutions of liquid dilution series, by virtue of their nature as enrichment cultures, tend to select for microorganisms which grow faster under the selection pressures imposed by the medium, culture conditions, etc. The length of time the dilution series are incubated is also important (232–234). The dominant microorganisms will not necessarily grow as fast as those obtained in enrichment cultures, since factors other than the specific growth rate in such a medium have determined their ability to survive or thrive in nature. Because the most dilute positive tubes will have to grow up from theoretically one cell an incubation time that is too short will fail to detect these microorganisms. This has been demonstrated in viable counts of methanogenic archaea from a flooded rice field soil, where an incubation time of 1 year was required to ascertain the culturable population size (P. H. Janssen, unpublished).

Microscopic examination of soils shows that most of the microorganisms in soil are smaller than 0.5 μm in diameter (235–237). It has been postulated that unculturable aquatic and marine microorganisms, which often are very much smaller than 'normal' microbial cells, represent dormant forms of nonsporulating bacteria, but evidence has been presented which suggests these microorganisms may be novel taxa which always display a very small cell size (238,239). Some bacteria exhibiting the small cell sizes commonly observed in soils have been isolated from soils (240; P. H. Janssen, unpublished). In contrast, so-called viable but nonculturable bacteria, which may undergo a reduction in size, represent stages which are, at least in some cases, dormant cells of known bacteria. These forms, apparently survival stages during periods of starvation, do not grow on plates when plated out, but will do so after a suitable resuscitation procedure, which is apparently species-dependent (241,242). In some microorganisms, this appears to be more successful in liquid culture than on plates (232,241). Indeed, viable counts from seawater using liquid media were two orders of magnitude higher than those from plating techniques (239). An extensive optimization of liquid dilution techniques and their statistical analysis for a marine ecosystem (234) led to the development of methods which allowed cultivation of up to 60% of the total microbial population. This information facilitated the isolation of numerically significant marine ultramicrobacteria (239), and thus physiological investigations to determine some of the characteristics which allow these bacteria to survive in their habitat (243). Since soil can be regarded as essentially an oligotrophic environment (16,226), these principles should also apply to soil and suggest liquid culture techniques will lead to better recoveries than plating techniques.

6.3   Culture Media and Conditions

Numerous causes may account for an microorganism's being 'unculturable.' Temperature, pH, ionic strength and composition, and oxygen partial pressure must of course all be within the limits allowing growth. Additionally, the mode of energy metabolism must be taken into account: an appropriate carbon source (maybe CO_2), the appropriate energy source (which may not necessarily be an organic compound and may be light), and an appropriate electron acceptor (not necessarily oxygen). The correct oxygen concentration, for example, for microaerophiles, and a suitable redox potential, for example, for anaerobes, have to be considered.

Soil extract agar and similar media have been used to isolate microorganisms from soil (237,244,245). Media containing high concentrations of complex medium additives, such as yeast extract or peptones, certainly do not reflect conditions prevailing in the soil and cannot be expected to yield satisfactory results. Some microorganisms are sensitive to phosphate at concentrations which are often used as the pH buffer in microbiological media (246). Media containing constituents more closely resembling natural concentrations have been formulated (247). By using a range of dilute culture media of varying composition, Schlesner (145) was able to isolate a great number of new budding bacteria belonging to the order *Planctomycetales*, which led to the conclusion that there is a much greater diversity within this group than was previously known (146) (see also Section 3.1). Certain growth factors may also be necessary, in the form of vitamins or anabolic precursors. Metals not normally found in normal trace element mixes may be required as cofactors for special and essential enzyme activities, for example, selenium, tungsten, and molybdenum. However, if the concentration of certain trace elements, such as copper, is too high, growth may be inhibited (248). Temperature may also play an important role in the success in obtaining growth of a microorganism from environmental samples (232,245,249). Standard incubation temperatures of 30°C and 37°C certainly do not apply to many soils. Many of the factors to be considered have been discussed elsewhere (217,250–253).

Diffusion gradient chambers have been used to separate microorganisms on the basis of differing optima or tolerances. Gradients of nutrients, pH, salt, or other parameters are applied to a solid or semisolid growth medium, usually two different parameters at 90° to each other, so that within the range of each of the individual gradients, all possible combinations are present (254–257). On the basis of their different responses (growth or even motility) to the parameters, microorganisms can be separated (258). Swarming motility has, for example, been used to isolate myxobacteria by baiting with a source of growth substrate toward which the bacteria migrate (259).

The gelling agent in solid media may inhibit the growth of some microorganisms. A gel (Gelrite) often applied for the cultivation of thermophilic microorganisms because of its high gel stability at high temperatures (260) also seems useful for isolation of mesophilic microorganisms. Increased viable counts of mesophilic methanogens were obtained, in some cases 1000-fold higher, when it (Gelrite) was used instead of agar (261), leading to the conclusion that agar contains components inhibitory to certain microorganisms. Another useful gelling agent, particularly for the isolation of psychrophilic microorganisms, is pluronic polyol F127. This has been used to cultivate a variety of bacteria from an aquatic system, including denitrifying and sulfate-reducing bacteria and methanogens (262). Silica gel is another alternative to agar, used for example, for nitrifying bacteria (250).

6.4 Physical Methods for the Isolation of Microorganisms

A number of physical methods for the isolation of microorganisms from mixtures have been developed. Some methods are based on physical parameters of the cells, while others are coupled to microscopy and allow a direct observation of the microorganisms being isolated. Filtration can be used to carry out initial separations that are based on the ability of microorganisms to pass through filters or agar gels, for example, spirochetes (263,264), or their retention, as has been used to isolate filamentous *Desulfonema* spp. (265) or *Methanosaeta concilii* (P. H. Janssen, unpublished). The use of micromanipulators in conjunction with microneedles allows specific cells to be removed from a sample and transferred to a suitable growth medium or a range of growth media (154,266). This approach is probably better suited to filamentous microorganisms but has also been successfully applied to those with other morphological characteristics (267,268). Laser trapping (269) of *Bacteria, Archaea*, and protozoa allows the manipulation of single cells by holding the cell in a focused beam of an infrared laser. This method has been used successfully to isolate a previously uncultured *Arthrobacter*-like coccus from rumen fluid after trapping the cells in the laser, then manipulating the microscope stage so that the cells were 'moved' into a sterile glass capillary, followed by transfer to growth medium (270). In this way the bacterium, which made up between 0.01% and 0.1% of the total population, was isolated. A similar method was used to obtain a pure culture of a new member of the *Archaea* which had previously been detected only as a 16S rRNA sequence. A total of 75 enrichment cultures were screened for the presence of cells which contained the target sequence, by using fluorescently labeled gene probes, then the morphotype thus identified was trapped with a laser and 'moved' from the mixed culture into sterile growth medium (215).

Density gradient centrifugation has been used to separate mixtures of microorganisms (271,272) and to separate microorganisms from soil (273,274). The percentage of viable cells recovered from soil was less than 5% of the total extracted cells (273), but this result was certainly due in part to physical damage incurred during the extraction procedure, and in part to the inability of some members of the community to grow on the medium used. Used in conjunction with other techniques, for example, flow cytometry (271) or labeled genetic probes, this method may be useful for isolating members of the soil microbial community.

Flow cytometry coupled with cell sorting can be used to separate microorganisms according to their optical properties. The use of these methods with the soil microflora poses some problems, such as cell adhesion to soil particles and the particulate nature of soils themselves (275). The use of flow-cytometric cell sorting can separate cells and distribute them onto agar plates or microtiter plates or into suitable culture vessels (276). If the optical properties of the target cells can be defined, these cells can be selectively dispersed onto or into various growth media to attempt their isolation. In addition to morphological properties such as size, shape, or intracellular structures, the use of various labels to detect DNA, RNA, or even specific sequences, or the use of labeled antibodies, can add to the selectiveness of the initial sorting parameters. Since four to six parameters may be measured simultaneously, the use of labels may help select nondestructive parameters on the basis of light scattering, which can then be used in subsequent sorting and selection with nonlabelled cells for cultivation. Some staining procedures have been developed which did not greatly affect cell viability (277) and allowed sorting, selection by label detection, and cell recovery in growth media in one run, based directly on selected and not on correlated characteristics.

7. THE NEED FOR A POLYPHASIC APPROACH

Investigations using the 16S rRNA gene as a biological marker have provided some very interesting insights into microbial diversity. They have confirmed that only a minority of the genetic diversity actually existing in nature is at present in culture and have shown the presence of microbial groups in soil which were either not expected in this environment (for example, *Planctomycetes*; see Section 3.1) or previously completely unknown (for example, new lineages of the *Archaea*, see Sections 3.3 and 3.4). Another important outcome is the biogeophylogenetic aspect of these studies: the finding of correlations of 16S rDNA sequence data obtained from different geographic sites, which has revealed the ubiquitous presence of as yet only molecularly detected microbes (see Sections 3.1 and 3.2). An additional advantage of the molecular approach is the possibility of comparing sequence data from studies done in the past with newly retrieved sequences, either directly from the environment or, especially, from newly isolated microorganisms. For example, two clusters of clone sequences detected in the Mount Coot-tha soil (see Section 3.1), originally without known close relatives, were later shown to be related to *Rhodoplanes* spp. (Fig. 2) and *Verrucomicrobium spinosum* (Fig. 4), whose 16S rRNA gene sequences only became available at a later date. Such correlations will become more reliable as the amount of molecular data from known pure cultures accumulates and allows closer matching. This may make the subsequent assignment of a phenotype to the purely molecular data feasible. Currently, one of the problems associated with the molecular biological approach to studying microbial diversity and community structure is the rarity with which matches of detected sequences with known cultivated microorganisms are made (11,57). This may be due to bias problems which do not, to date, allow a quantitative analysis of the microbial community (see Section 4.1), because of the failure to isolate (or impossibility of isolating) all the microorganisms within an environment and lack of coordinated efforts (parallel molecular and cultivation studies) from which a correlation can be reasonably expected. Correlations with known cultivated microorganisms have been found. In a study on a continuously operated bioreactor containing acidic mine runoff and a sulfide ore concentrate, 43 of 57 cloned 16S rRNA genes analyzed were similar or identical to the gene from *Thiobacillus thiooxidans* strain MIM SH12, isolated from the same bioreactor (57,278,279). In another investigation, a 100% 16S rRNA sequence identity between a pure culture and a cloned 16S rRNA gene from the same environment was found in a study of *Synechococcus* spp. in Octopus Spring (280). What is important in these two cases is that the correlating cultivated strains were isolated from the same site in a parallel study. These experiences indicate that investigations on a pure molecular basis are limited in their ecological relevance without accompanying pure culture and physiological studies. One of the special problems in microbiology, with relatively few exceptions, is the general need to be able to cultivate a microorganism to investigate its physiological and genetic characteristics and from these data assemble a more complete picture of its ecological properties. This is true for microorganisms detected by microscopy, by inference from measurements of their activity in a habitat, or by molecular methods. The molecular ecological studies based on 16S rDNA have shown that the need for the isolation of representatives from newly detected lineages is of utmost importance. Molecular studies can indicate the presence of microorganisms but repeatedly fail to tell us what the role of these in the habitat might be. Thus, the use of 16S rDNA sequences to study the microbial ecological characteristics of soils will be greatly strengthened when combined with carefully directed isolation studies.

In recent years, something of a rift between traditional microbiological methods and emerging molecular techniques has begun to develop. As is clearly seen from the examples presented in this chapter, future studies aimed at unravelling microbial community structure and diversity in soil, or any other habitat for that matter, should be a synthesis of two branches of microbial ecology: the molecular approach and cultivation techniques.

ACKNOWLEDGMENTS

We thank our coworkers for providing some research data before publication. Some of these studies were financially supported by grants of the Deutsche Forschungsgemeinschaft (Lie 455/2) and the Bundesministerium für Bildung,

Wissenschaft, Forschung und Technologie (No. 03111221) awarded to Werner Liesack, and by the EC Environment Programme. The mention of specific products in this article does not indicate an endorsement.

REFERENCES

1. M. Alexander, *Introduction to Soil Microbiology, 2nd ed.*, Wiley, New York (1977).
2. T. Hattori and R. Hattori, *Microbial Life in the Soil: An Introduction*, Marcel Dekker, New York (1973).
3. R. C. Foster, *Biol. Fertil. Soils* 6:189 (1988).
4. H. Bolton Jr., J. K. Frederickson, and L. F. Elliot, *Soil Microbial Ecology: Applications in Agricultural and Environmental Management* (F. B. Metting, ed.), Marcel Dekker, New York, p. 27 (1993).
5. M. Begon, J. L. Harper, and C. R. Townsend, *Ecology, 3rd ed.*, Blackwell, Oxford (1996).
6. P. J. Harris, *Beyond the Biomass* (K. Ritz, J. Dighton, and K. E. Giller, eds.), Wiley, Chichester, England, p. 239 (1994).
7. R. K. Peet, *Annu Rev. Ecol. System.* 5:285 (1974).
8. V. Torsvik, J. Goksøyr, and F. L. Daae, *Appl. Environ. Microbiol.* 56:782 (1990).
9. V. Torsvik, J. Goksøyr, F. L. Daae, R. Sørheim, J. Michalsen, and K. Salte, *Beyond the Biomass* (K. Ritz, J. Dighton, and K. E. Giller, eds.), Wiley, Chichester, England, p. 39 (1994).
10. B. S. Griffiths, K. Ritz, and L. A. Glover, *Microb. Ecol.* 31:269 (1996).
11. D. M. Ward, M. J. Ferris, S. C. Nold, M. M. Bateson, E. D. Kopczynski, and A. L. Ruff-Roberts, *Microbial Mats* (L. J. Stal and P. Caumette, eds.), Springer-Verlag, Berlin, p. 33 (1994).
12. F. B. Metting, *Soil Microbial Ecology: Applications in Agricultural and Environmental Management*, (F. B. Metting, ed.), Marcel Dekker, New York, p. 3 (1993).
13. J. A. Shields, E. A. Paul, W. E. Lowe, and D. Parkinson, *Soil Biol. Biochem.* 5:753 (1973).
14. G. Stotzky, *CRC Crit. Rev. Microbiol.* 2:59 (1972).
15. J. W. L. van Vuurde and B. Schippers, *Soil Biol. Biochem.* 12:559 (1980).
16. J. M. Poindexter, *Adv. Microbiol. Ecol.* 5:63 (1981).
17. R. M. Atlas and R. Bartha, *Microbial Ecology: Fundamentals and Applications*, 3rd ed., Benjamin/Cummings, Menlo Park, California (1993).
18. R. L. Tate, *Soil Microbiology*, Wiley, New York (1995).
19. J. F. Darbyshire (ed.), *Soil Protozoa*, CAB International, Wallingford, England (1994).

20. R. Amann, W. Ludwig, and K. H. Schleifer, *Microbiol. Rev.* 59:143 (1995).
21. E. Stackebrandt, *The Prokaryotes, 2nd ed.* (A. Balows, H. G. Trüper, M. Dworkin, W. Harder, and K. H. Schleifer, eds.), Springer Verlag, New York, p. 19 (1992).
22. E. Stackebrandt, *Encyclopedia of Microbiology, Vol. 3* (J. Lederberg, ed.), Academic Press, San Diego, p. 171 (1992).
23. D. M. Ward, M. M. Bateson, R. Weller, and A. L. Ruff-Roberts, *Adv. Microb. Ecol.* 12:219 (1992).
24. C. R. Woese, *Microbiol. Rev.* 58:1 (1994).
25. G. E. Fox, K. J. Pechmann, and C. R. Woese, *Int. J. Syst. Bacteriol.* 27:44 (1977).
26. C. R. Woese, E. Stackebrandt, T. J. Mackie, and G. E. Fox, *Syst. Appl. Microbiol.* 6:143 (1985).
27. E. Stackebrandt, W. Ludwig, and G. E. Fox, *Methods in Microbiology, Vol. 18* (G. Gottschalk, ed.), Academic Press, London, p. 75 (1985).
28. D. J. Lane, B. Pace, G. J. Olsen, D. A. Stahl, M. L. Sogin, and N. R. Pace, *Proc. Natl. Acad. Sci., USA* 82:6955 (1985).
29. C. R. Woese, O. Kandler, and M. L. Wheelis, *Proc. Natl. Acad. Sci., USA* 87:4576 (1990).
30. C. R. Woese, *Microbiol. Rev.* 51:221 (1987).
31. G. J. Olsen and C. R. Woese, *FASEB J.* 7:113 (1993).
32. G. J. Olsen, C. R. Woese, and R. Overbeek, *J. Bacteriol.* 176:1 (1994).
33. W. Ludwig and K. H. Schleifer, *FEMS Microbiol. Rev.* 15:155 (1994).
34. F. A. Rainey, P. H. Janssen, H. W. Morgan, and E. Stackebrandt, *Ant. van Leeuwenkoek* 64:341 (1993).
35. V. Boivin-Jahns, A. Bianchi, R. Ruimy, J. Garcin, S. Daumas, and R. Christen, *Appl. Environ. Microbiol.* 61:3400 (1995).
36. K. H. Wilson and R. B. Blitchington, *Appl. Environ. Microbiol.* 62:2273 (1996).
37. L. G. Wayne, D. J. Brenner, R. R. Colwell, P. A. D. Grimont, O. Kandler, M. I. Krichevsky, L. H. Moore, W. E. C. Moore, R. G. E. Murray, E. Stackebrandt, M. P. Starr, and H. G. Trüper, *Int. J. Syst. Bacteriol.* 37:463 (1987).
38. G. E. Fox, J. D. Wisotzkey, and P. Jurtshuk Jr., *Int. J. Syst. Bacteriol.* 42:166 (1992).
39. E. Stackebrandt and B. M. Goebel, *Int. J. Syst. Bacteriol.* 44:846 (1994).
40. R. A. Clayton, G. Sutton, P. S. Hinkle Jr., C. Butt, and C. Fields, *Int. J. Syst. Bacteriol.* 45:595 (1995).
41. P. Vandamme, B. Pot, M. Gillis, P. de Vos, K. Kersters, and J. Swings, *Microbiol. Rev.* 60:407 (1996).
42. D. J. Lonergan, H. L. Jenter, J. D. Coates, E. J. P. Phillips, T. M. Schmidt, and D. R. Lovely, *J. Bacteriol.* 178:2402 (1996).

43. G. J. Olsen, D. J. Lane, S. J. Giovannoni, and N. R. Pace, *Annu. Rev. Microbiol. 40*:337 (1986).
44. N. R. Pace, D. A. Stahl, D. L. Lane, and G. J. Olsen, *Adv. Microbiol. Ecol. 9*:1 (1986).
45. B. L. Maidak, N. Larsen, M. J. McCaughey, R. Overbeek, G. J. Olsen, K. Fogel, J. Blandy, and C. R. Woese, *Nucl. Acids Res. 21*:3485 (1994).
46. Y. van der Peer, I. Vandenbroek, P. De Rijk, and R. De Wachter, *Nucl. Acids Res. 22*:3488 (1994).
47. V. P. Canhos, G. P. Manfino, and L. D. Blaine, *Ant. van Leeuwenhoek 64*:205 (1993).
48. D. M. Ward, R. Weller, and M. M. Bateson, *Nature 345*:63 (1990).
49. R. Weller and D. M. Ward, *Appl. Environ. Microbiol. 55*:1818 (1989).
50. T. Britschgi and S. J. Giovannoni, *Appl. Environ. Microbiol. 57*:1707 (1991).
51. S. J. Giovannoni, T. B. Britschgi, C. L. Moyer, and K. G. Field, *Nature 345*:60 (1990).
52. J. A. Fuhrman, S. H. Lee, Y. Masuchi, A. A. Davis, and R. M. Wilcox, *Microb. Ecol. 28*:133 (1994).
53. T. M. Schmidt, E. F. DeLong, and N. R. Pace, *J. Bacteriol. 173*:4371 (1991).
54. E. F. DeLong, *Proc. Natl. Acad. Sci. USA 89*:5685 (1992).
55. J. A. Fuhrman, K. McCallum, and A. A. Davis, *Nature 356*:148 (1992).
56. J. A. Fuhrman, K. McCallum, and A. A. Davis, *Appl. Environ. Microbiol. 59*:1294 (1993).
57. N. Ward, F. A. Rainey, B. Goebel, and E. Stackebrandt, *Microbial Diversity and Ecosystem Function*, (D. Allsopp, D. L. Hawksworth, and R. R. Colwell, eds.), CAB International, Egham, England, p. 89 (1995).
58. B. K. Choi, B. J. Paster, F. Dewhirst, and U. B. Göbel, *Infect. Immun. 62*:1889 (1994).
59. E. Stackebrandt and W. Liesack, *Today's Life Sci. 5*:20 (1992).
60. P. A. Rochelle, B. A. Cragg, J. C. Fry, R. J. Parkes, and A. J. Weightman, *FEMS Microbiol. Ecol. 15*:215 (1994).
61. D. M. Ward, R. Weller, and M. M. Bateson, *FEMS Microbiol. Rev. 75*:105 (1990).
62. R. Weller, J. W. Weller, and D. M. Ward, *Appl. Environ. Microbiol. 57*:1146 (1991).
63. J. Brosius, T. J. Dull, D. D. Sleeter, and H. F. Noller, *J. Mol. Biol. 148*:107 (1981).
64. C. Kemmerling, D. Witt, W. Liesack, H. Weyland, and E. Stackebrandt, *Current Topics in Marine Biotechnology*, (S. Miyachi, I. Karube, and Y. Eshida, eds.), Japan Society of Marine Biotechnology, Tokyo, p. 423 (1990).

65. E. F. DeLong, G. S. Wickham, and N. R. Pace, *Science* 243:1360 (1989).
66. P. F. Kemp, S. Lee, and J. LaRoche, *Appl. Environ. Microbiol.* 59:2594 (1993).
67. L. K. Poulsen, G. Ballard, and D. A. Stahl, *Appl. Environ. Microbiol.* 59:1354 (1993).
68. M. Schaechter, O. Maaløe, and N. O. Kjeldgaard, *J. Gen. Microbiol.* 19:592 (1958).
69. K. Flärdh, P. S. Cohen, and S. Kjelleberg, *J. Bacteriol.* 174:6780 (1992).
70. J. Kramer and F. Singleton, *Appl. Environ. Microbiol.* 58:201 (1992).
71. M. Fukui, Y. Suwa, and Y. Urushigawa, *FEMS Microbiol. Ecol.* 19:17 (1996).
72. M. Wagner, G. Rath, R. Amann, H. P. Koops, and K. H. Schleifer, *System. Appl. Microbiol.* 18:251 (1995).
73. R. K. Saiki, D. H. Gelfand, S. Stoffel, S. J. Scharf, R. Higuchi, G. T. Horn, K. B. Mullis, and H. A. Erlich, *Science* 239:487 (1988).
74. D. J. Lane, *Nucleic Acid Techniques in Bacterial Systematics* (E. Stackebrandt and M. Goodfellow, eds.), Wiley, Chichester, England, p. 115 (1991).
75. E. Stackebrandt and W. Liesack, *Non-Radioactive Labelling and Detection of Biomolecules* (C. Kessler, ed.), Springer Verlag, New York, p. 232 (1993).
76. E. F. DeLong, K. Y. Wu, B. B. Prézelin, and R. V. M. Jovine, *Nature* 371:695 (1994).
77. S. M. Barns, R. E. Fundyga, M. W. Jeffries, and N. R. Pace, *Proc. Natl. Acad. Sci. USA* 91:1609 (1994).
78. R. Amann, N. Springer, W. Ludwig, H. D. Görtz, and K. H. Schleifer, *Nature* 351:161 (1991).
79. T. M. Embley, B. J. Finlay, and P. Dyal, *J. Gen. Microbiol.* 138:1479 (1992).
80. B. J. Finley, T. M. Embley, and T. Fenchel, *J. Gen. Microbiol.* 139:371 (1993).
81. N. Dubilier, O. Giere, D. L. Distel, and C. M. Cavanaugh, *Appl. Environ. Microbiol.* 61:2346 (1995).
82. M. Berchtold and H. König, *System. Appl. Microbiol.* 19:66 (1996).
83. B. J. Paster, F. E. Dewhirst, S. M. Cooke, V. Fussing, L. K. Poulsen, and J. A. Breznak, *Appl. Environ. Microbiol.* 62:347 (1996).
84. D. Hahn, R. Kester, M. J. C. Starrenburg, and A. D. L. Akkermans, *Arch. Microbiol.* 154:329 (1990).
85. M. A. Moran, V. L. Torsvik, T. Torsvik, and R. E. Hodson, *Appl. Environ. Microbiol.* 59:915 (1993).
86. A. Ogram, W. Sun, F. J. Brockman, and J. K. Fredrickson, *Appl. Environ. Microbiol.* 61:763 (1995).

87. S. Selenska and W. Klingmüller, *Microb. Releases 1*:41 (1992).
88. H. Christensen and L. K. Poulsen, *Soil. Biol. Biochem. 26*:1093 (1994).
89. D. Hahn, R. I. Amann, W. Ludwig, A. D. L. Akkermans, and K. H. Schleifer, *J. Gen. Microbiol. 138*:879 (1992).
90. K. Fischer, D. Hahn, W. Hönerlage, F. Schönholzer, and J. Zeyer, *System. Appl. Microbiol. 18*:265 (1995).
91. A. Faegri, V. Torsvik, and J. Goksøyr, *Soil Biol. Biochem. 9*:105 (1977).
92. V. L. Torsvik, *Soil Biol. Biochem. 12*:15 (1980).
93. W. E. Holben, J. K. Jansson, B. K. Chelm, and J. M. Tiedje, *Appl. Environ. Microbiol. 54*:703 (1988).
94. S. D. Pillai, K. L. Josephson, R. L. Bailey, C. P. Gerba, and I. L. Pepper, *Appl. Environ. Microbiol. 57*:2283 (1991).
95. R. J. Steffan, J. Goksøyr, A. K. Bej, and R. M. Atlas, *Appl. Environ. Microbiol. 54*:2908 (1988).
96. L. G. Leff, J. R. Dana, J. V. McArthur, and L. J. Shimkets, *Appl. Environ. Microbiol. 61*:1141 (1995).
97. A. Ogram, G. S. Sayler, and T. Barkay, *J. Microbiol. Meth. 7*:57 (1987).
98. G. S. Sayler and A. C. Layton, *Annu. Rev. Microbiol. 44*:625 (1990).
99. C. S. Jacobsen and O. F. Rasmussen, *Appl. Environ. Microbiol. 58*:2458 (1992).
100. P. A. Rochelle, J. C. Fry, R. J. Parkes, and A. J. Weightman, *FEMS Microbiol. Lett. 100*:59 (1992).
101. K. Smalla, N. Cresswell, L. C. Mendonca-Hagler, A. Wolters, and J. D. van Elsas, *J. Appl. Bacteriol. 74*:78 (1993).
102. Y. L. Tsai and B. H. Olson, *Appl. Environ. Microbiol. 57*:1070 (1991).
103. T. Volossiouk, E. J. Robb, and R. N. Nazar, *Appl. Environ. Microbiol. 61*:3972 (1995).
104. J. Zhou, M. A. Bruns, and J. M. Tiedje, *Appl. Environ. Microbiol. 62*:316 (1996).
105. M. I. Moré, J. B. Herrick, M. C. Silva, W. C. Ghiorse, and E. L. Madsen, *Appl. Environ. Microbiol. 60*:1572 (1994).
106. G. Muyzer, E. C. DeWaal, and A. G. Uitterlinden, *Appl. Environ. Microbiol. 59*:695 (1993).
107. S. Rölleke, G. Muyzer, C. Wawer, G. Wanner, and W. Lubitz, *Appl. Environ. Microbiol. 62*:2059 (1996).
108. A. Teske, C. Wawer, G. Muyzer, and N. B. Ramsing, *Appl. Environ. Microbiol. 62*:1405 (1996).
109. A. E. Murray, J. T. Hollibaugh, and C. Orrego, *Appl. Environ. Microbiol. 62*:2676 (1996).
110. J. B. Herrick, E. L. Madsen, C. A. Batt, and W. C. Ghiorse, *Appl. Environ. Microbiol. 59*:687 (1993).

111. M. Berthelet, L. G. Whyte, and C. W. Greer, *FEMS Microbiol. Lett. 138*:17 (1996).
112. W. Liesack and E. Stackebrandt, *J. Bacteriol. 174*:5072 (1992).
113. C. A. Kreader, *Appl. Environ. Microbiol. 62*:1102 (1996).
114. C. S. Jacobsen, *Appl. Environ. Microbiol. 61*:3347 (1995).
115. J. T. Trevors and J. D. van Elsas, *Nucleic Acids in the Environment*, Springer-Verlag, Berlin (1995).
116. A. D. L. Akkermans, J. D. van Elsas, and F. J. de Bruijn, *Molecular Microbial Ecology Manual*, Kluwer, Dordrecht, Netherlands (1995).
117. L. Raskin, J. M. Stromley, B. E. Rittmann, and D. A. Stahl, *Appl. Environ. Microbiol. 60*:1232 (1994).
118. E. W. Alm, D. B. Oerther, N. Larsen, D. A. Stahl, and L. Raskin, *Appl. Environ. Microbiol., 62*:3557 (1996).
119. S. J. Giovannoni, *Nucleic Acid Techniques in Bacterial Systematics* (E. Stackebrandt and M. Goodfellow, eds.), Wiley, Chichester, England, p. 177 (1991).
120. W. G. Weisburg, S. M. Barns, D. A. Pelletier, and D. J. Lane, *J. Bacteriol. 173*:697 (1991).
121. Z. Liu and L. M. Schwartz, *Biotechniques 12*:29 (1992).
122. J. M. Clark, *Nucleic Acids Res. 16*:9677 (1988).
123. C. L. Moyer, F. C. Dobbs, and D. M. Karl, *Appl. Environ. Microbiol. 60*:871 (1994).
124. S. Weidner, W. Arnold, and A. Pühler, *Appl. Environ. Microbiol. 62*:766 (1996).
125. C. L. Moyer, J. M. Tiedje, F. C. Dobbs, and D. M. Karl, *Appl. Environ. Microbiol. 62*:2501 (1996).
126. F. A. Rainey, M. Dorsch, H. W. Morgan, and E. Stackebrandt, *System. Appl. Microbiol. 15*:197 (1992).
127. W. M. Fitch and E. Margoliash, *Science 155*:279 (1967).
128. W. M. Fitch, *Syst. Zool. 20*:406 (1971).
129. J. Felsenstein, *J. Mol. Evol. 17*:368 (1981).
130. T. H. Jukes and C. R. Cantor, *Mammalian Protein Metabolism, Vol. 3* (H. N. Munro, ed.), Academic Press, New York, p. 21 (1969).
131. N. Saitou and M. Nei, *Mol. Biol. and Evol. 4*:406 (1987).
132. C. De Soete, *Pyschometrika 48*:621 (1983).
133. D. L. Swofford, *Phylogenetic Analysis Using Parsimony (PAUP)*, version 3.1, University of Illinois, Champaign (1993).
134. J. Felsenstein, *Cladistics 5*:164 (1989).
135. S. Kumar, K. Tamura, and M. Nei, *Molecular Evolutionary Genetics Analysis*, version 1.0. Pennsylvania State University, University Park (1993).
136. J. Felsenstein, *Annu. Rev. Genet. 22*:521 (1988).

137. M. Nei, *Phylogenetic Analysis of DNA Sequences*, (M. M. Miyamoto and J. Cracraft, eds), Oxford University Press, New York, p. 90 (1991).
138. T. Ueda, Y. Suga, and T. Matsuguchi, *Eur. J. Soil Sci. 46*:415 (1995).
139. J. Bornemann, P. W. Skroch, K. M. O'Sullivan, J. A. Palus, N. G. Rumjanek, J. L. Jansen, J. Nienhuis, and E. W. Triplett, *Appl. Environ. Microbiol. 62*:1935 (1996).
140. E. Stackebrandt, W. Liesack, and B. M. Goebel, *FASEB J. 7*:232 (1993).
141. S. Kwok, D. E. Kellogg, N. McKinney, D. Spasic, L. Goda, C. Levenson, and J. J. Sninsky, *Nucl. Acids Res. 18*:999 (1990).
142. W. Liesack and E. Stackebrandt, *Biodiv. Conserv. 1*:250 (1992).
143. A. Hirashi and Y. Ueda, *Int. J. Syst. Bacteriol. 44*:665 (1994).
144. R. B. So, J. K. Ladha, and J. P. W. Young, *Int. J. Syst. Bacteriol. 44*:392 (1994).
145. H. Schlesner, *System. Appl. Microbiol. 17*:135 (1994).
146. N. Ward, F. A. Rainey, E. Stackebrandt, and H. Schlesner, *Appl. Environ. Microbiol. 61*:2270 (1995).
147. E. F. DeLong, D. G. Franks, and A. L. Alldredge, *Limnol. Oceanogr. 38*:924 (1993).
148. N. Ward-Rainey, F. A. Rainey, H. Schlesner, and E. Stackebrandt, *Microbiology 141*:3247 (1995).
149. S. Y. Lee, J. Bollinger, D. Bezdicek, and A. Ogram, *Appl. Environ. Microbiol. 62*:3787 (1996).
150. F. A. Rainey, N. Ward, L. I. Sly, and E. Stackebrandt, *Experientia 50*:796 (1994).
151. Deutsche Sammlung von Mikroorganismen und Zellkulturen GmbH, *Bacterial Nomenclature Up To Date, April 1996*, Deutsche Sammlung von Mikroorganismen und Zellkulturen GmbH, Braunschweig, Germany (1996).
152. J. A. Brierley, *Appl. Environ. Microbiol. 36*:523 (1978).
153. D. J. Lane, A. P. Harrison Jr, D. Stahl, B. Pace, S. J. Giovannoni, and N. R. Pace, *J. Bacteriol. 174*:269 (1992).
154. L. L. Blackall, E. M. Seviour, M. A. Cunningham, R. J. Seviour, and P. Hugenholtz, *System. Appl. Microbiol. 17*:513 (1994).
155. H. Rheims, C. Spröer, F. A. Rainey, and E. Stackebrandt, *Microbiology 142*:2863 (1996).
156. J. K. Kristjansson, ed., *Thermophilic Bacteria*, CRC Press, Boca Raton, Florida, p. 18 (1992).
157. J. K. Kristjansson and K. O. Stetter, *Thermophilic Bacteria* (J. K. Kristjansson, ed.), CRC Press, Boca Raton, Florida, p. 18 (1992).
158. D. A. Stahl and R. Amann, *Nucleic Acid Techniques in Bacterial Systemat-*

ics (E. Stackebrandt and M. Goodfellow, eds.), Wiley, Chichester, England p. 205 (1991).

159. C. Ash, J. A. E. Farrow, S. Wallbanks, and M. D. Collins, *Lett. Appl. Microbiol. 13*:202 (1991).

160. F. A. Rainey, D. Fritze, and E. Stackebrandt, *FEMS Microbiol. Lett. 155*:205 (1994).

161. P. A. Roger, W. J. Zimmermann, and T. A. Lumpkin, *Soil Microbial Ecology: Applications in Agricultural and Environmental Management* (F. B. Metting, ed.), Marcel Dekker, New York, p. 417 (1992).

162. H. Schütz, W. Seiler, and H. Rennenberg, *Soils and the Greenhouse Effect, Vol. 12* (A. F. Bouwman, ed.), Wiley, Chichester, England, p. 269 (1990).

163. H. Schütz, W. Seiler, and R. Conrad, *Biogeochem. 7*:33 (1989).

164. P. Frenzel, F. Rothfuss, and R. Conrad, *Biol. Fertil. Soils 14*:84 (1992).

165. G. M. King, *Appl. Environ. Microbiol. 60*:3220 (1994).

166. F. Rothfuss and R. Conrad, *Biogeochem 18*:137 (1993).

167. P. Frenzel and U. Bosse, *FEMS Microbiol. Ecol. 26*:25 (1996).

168. B. Aβmus, P. Hutzler, G. Kirchhof, R. Amann, J. R. Lawrence and A. Hartmann, *Appl. Environ. Microbiol. 61*:1013 (1995).

169. A. L. Reysenbach, L. J. Giver, G. S. Wickham, and N. R. Pace, *Appl. Environ. Microbiol. 58*:3417 (1992).

170. M. T. Suzuki and S. J. Giovannoni, *Appl. Environ. Microbiol. 62*:625 (1996).

171. V. Farrelly, F. A. Rainey, and E. Stackebrandt, *Appl. Environ. Microbiol. 61*:2798 (1995).

172. F. A. Rainey, N. L. Ward-Rainey, P. H. Janssen, H. Hippe, and E. Stackebrandt, *Microbiology 142*:2087 (1996).

173. B. K. C. Patel, C. A. Love, and E. Stackebrandt, *Nucl. Acids Res. 20*:5483 (1992).

174. A. Willems and M. D. Collins, *Int. J. Syst. Bacteriol. 43*:305 (1993).

175. N. Springer, W. Ludwig, R. Amann, H. J. Schmidt, H. D. Goertz, and K. H. Schleifer, *Proc. Natl. Acad. Sci. USA 90*:9892 (1993).

176. W. Liesack, H. Weyland, and E. Stackebrandt, *Microb. Ecol. 21*:191 (1991).

177. E. D. Kopczynski, M. M. Bateson, and D. M. Ward, *Appl. Environ. Microbiol. 60*:746 (1994).

178. G. C. Y. Wang and Y. Wang, *Microbiology 142*:1107 (1996).

179. J. F. Robinson-Cox, M. M. Bateson, and D. M. Ward, *Appl. Environ. Microbiol. 61*:1240 (1995).

180. S. Mylvaganam and P. P. Dennis, *Genetics 130*:399 (1992).

181. R. D. Fleischmann, M. D. Adams, O. White, R. A. Clayton, E. F. Kirkness,

A. R. Kerlavage, C. J. Bult, J. F. Tomb, B. A. Dougherty, J. M. Merrick, K. McKenney, G. Sutton, W. FitzHugh, C. Fields, J. D. Gocayne, J. Scott, R. Shirley, L. I. Liu, A. Glodek, J. M. Kelley, J. F. Weidman, C. A. Phillips, T. Spriggs, E. Hedblom, M. D. Cotton, T. R. Utterback, M. C. Hanna, D. T. Nguyen, D. M. Saudek, R. C. Brandon, L. D. Fine, J. L. Fritchman, J. L. Fuhrmann, N. S. M. Geoghagen, C. L. Gnehm, L. A. McDonald, K. V. Small, C. M. Fraser, H. O. Smith, and J. C. Venter, *Science 269*:496 (1995).

182. R. Amann, J. Snaidr, M. Wagner, W. Ludwig, and K. H. Schleifer, *J. Bacteriol. 178*:3496 (1996).

183. E. Stackebrandt, D. A. Stahl, and R. Devereux, *Sulfate-Reducing Bacteria* (L. L. Barton, ed.), Plenum Press, New York, p. 49 (1995).

184. G. Voordouw, V. Niviere, F. G. Ferris, P. M. Fedorak, and D. W. S. Westlake, *Appl. Environ. Microbiol. 56*:3748 (1990).

185. C. Wawer and G. Muyzer, *Appl. Environ. Microbiol. 61*:2203 (1995).

186. M. R. Fries, J. Zhou, J. Chee-Sanford, and J. M. Tiedje, *Appl. Environ. Microbiol. 60*:2802 (1994).

187. C. R. Lovell and Y. Hui, *Appl. Environ. Microbiol. 57*:2602 (1991).

188. G. B. Smith and J. M. Tiedje, *Appl. Environ. Microbiol. 58*:376 (1992).

189. W. Y. Rick, M. R. Fries, S. G. Bezborodnikov, B. A. Averill, and J. M. Tiedje, *Appl. Environ. Microbiol. 59*:250 (1993).

190. M. S. Coyne, A. Arunakumari, B. A. Averill, and J. M. Tiedje, *Appl. Environ. Microbiol. 55*:2924 (1990).

191. S. Hallier-Soulier, V. Ducrocq, N. Mazure, and N. Truffant, *FEMS Microbiol. Ecol. 20*:121 (1996).

192. I. R. McDonald, E. M. Kenna, and J. C. Murrell, *Appl. Environ. Microbiol. 61*:116 (1995).

193. J. D. Semrau, A. Chistoserdov, J. Lebron, A. Costello, J. Davagnino, E. Kenna, A. J. Holmes, R. Finch, J. C. Murrell, and M. E. Lidstrom, *J. Bacteriol. 177*:3071 (1995).

194. H. McTavish, J. A. Fuchs, and A. B. Hooper, *J. Bacteriol. 175*:2436 (1993).

195. D. J. Bergmann and A. H. Hooper, *Biochem. Biophys. Res. Commun. 28*:759 (1994).

196. C. Bédard and R. Knowles, *Microbiol. Rev. 53*:68 (1989).

197. I. M. Head, W. D. Hiorns, T. M. Embley, A. J. McCarthy, and J. R. Saunders, *J. Gen. Microbiol. 139*: 1147 (1993).

198. A. J. Holmes, A. Costello, M. E. Lidstrom, and J. C. Murrell, *FEMS Microbiol. Lett. 132* 203 (1995).

199. J. Rotthauwe, W. de Boer, and W. Liesack, *FEMS Microbiol. Lett. 133*:131 (1995).

200. W. D. Hiorns, R. C. Hastings, I. M. Head, A. J. McCarthy, J. R. Saunders, R. W. Pickup, and G. H. Hall, *Microbiology 141*:2793 (1995).

201. J. D. Kirshtein, H. W. Paerl, and J. Zehr, *Appl. Environ. Microbiol. 57*:2645 (1991).
202. J. P. W. Young, *New Horizons in Nitrogen Fixation*, (R. Palacios, J. Mora, and W. E. Newton, eds.), Kluwer, London, p. 587 (1993).
203. T. Ueda, Y. Suga, N. Nobutaka, and T. Matsuguchi, *J. Bacteriol. 177*:1414 (1995).
204. B. A. Hales, C. Edwards, D. A. Ritchie, G. Hall, R. W. Pickup, and J. R. Saunders, *Appl. Environ. Microbiol. 62*:668 (1996).
205. R. E. Hodson, W. A. Dustman, R. P. Garg, and M. A. Moran, *Appl. Environ. Microbiol. 61*:4074 (1995).
206. J. L. Stein, T. L. Marsh, K. Y. Wu, H. Shizuya, and E. F. DeLong, *J. Bacteriol. 178*:591 (1996).
207. A. Balows, H. G. Trüper, M. Dworkin, W. Harder, and K. H. Schleifer, eds., *The Prokaryotes*, 2nd ed., Springer-Verlag, New York (1992).
208. H. Veldkamp, *Adv. Microb. Ecol. 1*:59 (1977).
209. W. Harder, J. G. Kuenen, and A. Matin, *J. Appl. Bacteriol 43*:1 (1977).
210. J. J. Lee, A. T. Saldo, W. Reisser, M. J. Lee, K. W. Jeon, and H. D. Görtz, *J. Protozool. 32*:391 (1985).
211. J. M. Harris, *Microb. Ecol. 25*:195 (1993).
212. D. Werner, *Symbiosis of Plants and Microbes*, Chapman and Hall, London (1992).
213. S. Spring, R. Amann, W. Ludwig, K. H. Schleifer, H. van Gemerden, and N. Petersen, *Appl. Environ. Microbiol. 59*:2397 (1993).
214. M. D. Kane, L. K. Poulsen, and D. A. Stahl, *Appl. Environ. Microbiol. 59*:682 (1993).
215. R. Huber, S. Burggraf, T. Mayer, S. M. Barns, P. Rossnagel, and K. O. Stetter, *Nature 376*:57 (1995).
216. T. D. Brock, *Ecology of Microbial Communities* (M. Fletcher, T. R. G. Gray, and J. G. Jones, eds.), Cambridge University Press, Cambridge, p. 1 (1987).
217. R. W. Pickup, *Population Genetics of Bacteria* (S. Baumberg, J. W. P. Young, E. M. H. Wellington, and J. R. Saunders, eds.), Cambridge University Press, Cambridge, p. 295 (1995).
218. J. C. Ensign, *Annu. Rev. Microbiol. 32*:185 (1978).
219. H. Voelz and M. Dworkin, *J. Bacteriol. 84*:943 (1962).
220. F. A. Skinner, *Soil Microbiology* (N. Walker, ed.), Butterworths, London, p. 1 (1975).
221. L. R. Bakken and R. Olsen, *Microb. Ecol. 13*:103 (1987).
222. R. Y. Morita, *Can. J. Microbiol. 34*:436 (1988).
223. A. J. Ramsay, *Soil Biol. Biochem. 16*:475 (1984).
224. A. D. Rovira, E. I. Newman, H. J. Bowen, and R. Campbell, *Soil Biol. Biochem. 6*:211 (1974).

225. F. A. Skinner, P. C. T. Jones, and J. E. Mollison, *J. Gen. Microbiol.* 6:261 (1952).
226. F. B. Metting, *Biotechnology: Applications and Research* (P. A. Cheremisinoff and R. P. Ouellette, eds.), Technomic, Lancaster, Pennsylvania, p. 196 (1985).
227. J. A. Novitsky, *Appl. Environ. Microbiol.* 53:2368 (1987).
228. J. R. Postgate, J. R. Hunter, *J. Gen. Microbiol.* 34:459 (1964).
229. W. Reichardt, *Wat. Res.* 13:1149 (1979).
230. V. Straškabová, *J. Appl. Bacteriol.* 54:54 (1983).
231. A. S. Kaprelyants and D. B. Kell, *Trends Microbiol.* 4:237 (1996).
232. H. Jannasch and G. E. Jones, *Limnol. Oceanogr.* 4:128 (1959).
233. G. J. Both and H. J. Laanbroek, *FEMS Microbiol. Ecol.* 85:335 (1991).
234. D. K. Button, F. Schut, P. Quang, R. Martin, and B. R. Robertson, *Appl. Environ. Microbiol.* 59:881 (1993).
235. H. C. Bae, E. H. Cota-Robles, and L. E. Casida, *Appl. Microbiol.* 23:637 (1972).
236. B. Lundgren, *Soil Biol. Biochem.* 16:283 (1984).
237. R. A. Olsen and L. R. Bakken, *Microb. Ecol.* 13:59 (1987).
238. A. S. Kaprelyants, J. C. Gottschal, and D. B. Kell, *FEMS Microbiol. Rev.* 104:271 (1993).
239. F. Schut, E. J. de Vries, J. C. Gottschal, B. R. Robertson, W. Harder, R. A. Prins, and D. K. Button, *Appl. Environ. Microbiol.* 59:271 (1993).
240. L. E. Casida Jr., *Can. J. Microbiol.*, 25:214 (1977).
241. D. B. Roszak, D. J. Grimes, and R. R. Colwell, *Can. J. Microbiol.* 30:334 (1984).
242. L. Nilsson, J. D. Oliver, and S. Kjelleberg, *J. Bacteriol.* 173:5054 (1991).
243. F. Schut, M. Jansen, T. M. Pedro Gomes, J. C. Gottschal, W. Harder, and R. A. Prins, *Microbiology 141*:351 (1995).
244. N. James, *Can. J. Microbiol.* 4:363 (1958).
245. V. Jensen, *The Ecology of Soil Bacteria* (T. R. G. Gray and D. Parkinson, eds.), Liverpool University Press, Liverpool, p. 158 (1968).
246. N. Pfennig, *The Microbe 1984. Part II. Prokaryotes and Eukaryotes* (D. P. Kelly and N. G. Carr, eds.), Cambridge University Press, Cambridge, p. 23 (1984).
247. J. S. Angle, S. P. McGrath, and R. L. Chaney, *Appl. Environ. Microbiol.* 57:3674 (1991).
248. F. Widdel, *Arch. Microbiol.* 148:286 (1987).
249. H. G. Thornton, *Ann. Appl. Bot.* 9:241 (1922).
250. L. E. Casida Jr., *The Ecology of Soil Bacteria* (T. R. G. Gray and D. Parkinson, ed.), Liverpool University Press, Liverpool, p. 97 (1968).
251. P. Gerhardt, R. G. E. Murray, W. A. Wood, and N. R. Krieg, eds., *Methods for General and Molecular Microbiology*, American Society for Microbiology, Washington, D. C. (1994).

252. R. Grigorova and J. R. Norris, eds., *Methods in Microbiology, Vol. 22*, Academic Press, London (1990).

253. J. R. Norris and D. W. Ribbons eds., *Methods in Microbiology, Vol. 3B*, Academic Press, London (1969).

254. D. E. Caldwell, S. H. Lai, and J. M. Tiedje, *Bull. Ecol. Res. Comm. (Stockholm) 17*:151 (1973).

255. J. W. T. Wimpenny and P. Waters, *J. Gen. Microbiol. 130*:2921 (1984).

256. J. W. T. Wimpenny, H. Gest, and J. L. Favinger, *FEMS Microbiol. Lett. 37*:367 (1986).

257. G. M. Wolfaardt, J. R. Lawrence, M. D. Hendry, R. D. Roberts, and D. E. Caldwell, *Appl. Environ. Microbiol. 59*:2388 (1993).

258. D. Emerson, R. M. Worden, and J. A. Breznak, *Appl. Environ. Microbiol. 60*:1269 (1994).

259. H. Reichenbach and M. Dworkin, *The Prokaryotes, 2nd ed.* (A. Balows, H. G. Trüper, M. Dworkin, W. Harder, and K. H. Schleifer, eds.), Springer-Verlag, New York, p. 3416 (1992).

260. C. C. Lin and L. E. Casida Jr., *Appl. Environ. Microbiol. 47*:427 (1984).

261. J. E. Harris, *Appl. Environ. Microbiol. 50*:1107 (1985).

262. S. Gardener and J. G. Jones, *J. Gen. Microbiol. 130*:731 (1984).

263. E. Canale-Parola, *Methods in Microbiology, Vol. 8* (J. R. Norris, and D. W. Ribbons, eds.), Academic Press, New York, p. 61 (1973).

264. R. B. Hespell and E. Canale-Parola, *Arch. Mikrobiol. 74*:1 (1970).

265. F. Widdel, *Arch. Microbiol. 134*:282 (1983).

266. L. A. Hornsby and N. J. Horan, *Wat. Res. 28*:2033 (1994).

267. P. D. Franzman and V. B. D. Skerman, *Ant. van Leeuwenhoek 50*:261 (1984).

268. J. T. Staley, J. A. Fuerst, S. Giovannoni, and H. Schlesner, *The Prokaryotes, 2nd ed.* (A. Balows, H. G. Trüper, M. Dworkin, W. Harder, and K. H. Schleifer, eds.), Springer-Verlag, New York, p. 3710 (1992).

269. A. Ashkin, J. M. Dziedzic, and T. Yamane, *Nature 330*:769 (1987).

270. J. G. Mitchell, R. Weller, M. Beconi, J. Sell, and J. Holland, *Microb. Ecol. 25*:113 (1993).

271. K. P. Putzer, L. A. Buchholz, M. E. Lindstrom, and C. C. Remsen, *Appl. Environ. Microbiol. 57*:3656 (1991).

272. P. Scherer, *J. Appl. Bacteriol. 55*:481 (1983).

273. L. R. Bakken, *Appl. Environ. Microbiol. 49*:1482 (1985).

274. H. Christensen, R. A. Olsen, and L. R. Bakken, *Microb. Ecol. 29*:49 (1995).

275. C. Edwards, J. Diaper, J. Porter, D. Deere, and R. Pickup, *Beyond the Biomass* (K. Ritz, J. Dighton, and K. E. Giller, eds.), Wiley, Chichester, England, p. 57 (1994).

276. H. J. Tanke and M. van der Keur, *TIBTECH 11*:55 (1993).

277. T. Azuma, G. I. Harrison, and A. L. Demain, *Appl. Microbiol. Biotechnol.* *38*:173 (1992).

278. B. M. Goebel and E. Stackebrandt, *Appl. Environ. Microbiol.* *60*:1614 (1994).

279. B. M. Goebel and E. Stackebrandt, *Bacterial Diversity and Systematics*, (F. Priest, A. Ramos-Cormenzana, and B. J. Tindall, eds.), Plenum Press, New York, p. 259 (1994).

280. M. J. Ferris, A. L. Ruff-Roberts, E. D. Kopcyznski, M. M. Bateson, and D. M. Ward, *Appl. Environ. Microbiol.* *62*:1045 (1996).

14

Adaptation of Bacteria to Soil Conditions: Applications of Molecular Physiology in Soil Microbiology

LEO VAN OVERBEEK and JAN DIRK VAN ELSAS   Research Institute for Plant Protection (IPO-DLO), Wageningen, The Netherlands

1. INTRODUCTION

For decades, bacteria that occur naturally in soil have been isolated and characterized in the laboratory with the aim of studying and improving their performance after release into soil as biopesticides (1–3), biofertilizers (4), or bioremediation agents (5,6). Candidates for release are mostly selected by using fast growth of cultures under laboratory conditions and suitability for genetic manipulation as criteria. For introduction into soil, pseudomonads (7–11), rhizobia (12,13), *Flavobacterium* (14), *Escherichia* (15,16), and *Azospirillum* (17) species are commonly used. However, not all species survive well in soil, either because they are not indigenous in soil (like *E. coli*) or because they respond differently to diverse soil types (17). Much effort has further been expended on the isolation and identification of bacteria from rhizosphere soils (18,19), with the purpose of later reintroduction after genetic modifications of selected isolates (20). Copiotrophic (adapted to high nutrient concentrations) gram-negative species like those of the genus *Pseudomonas* (19,20) can be abundant in the rhizosphere and may be well suited as soil inoculants, because of their optimal survival under certain conditions in (rhizosphere) soil, their genetic accessibility, and their possibly high metabolic rates. However, upon introduction, soil-isolated bacteria still have to cope with the harsh conditions often prevailing in soil (21–23). The molecular responses of inoculant bacterial strains to soil conditions are determinative for the prediction and putative manipulation of their behavior in soil. The importance of the colonization of different niches in soil by inoculant bacterial cells to their persistance and activity and hence to the success of their application is summarized in Fig. 1. It is clear that the site of inoculation in soil determines the inoculant's physiological and metabolic status, which in turn influence the persistence and activity of bacterial cells in soil. These two parameters can be considered to represent the two most important factors influencing the success of application of bacterial inoculants in soil.

This chapter will discuss bacterial genetic systems involved in adaptations to soil conditions and the potential utilization of these mechanisms with regard to biocontrol of plant pathogens, containment of genetically modified microorganisms (GMMs), and biomonitoring of toxic agents in soil. In addition to applications of soil inoculants, attention will be paid to measurements of overall metabolic activity in soil. Information about the molecular responses of inoculants to environmental stimuli is not always available from studies with soil bacteria. Data obtained in laboratory experiments and in other ecosystems (like the marine environment) will also be extrapolated to soil.

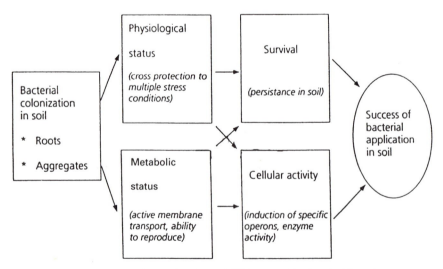

Figure 1.   The effect of colonization of different soil microhabitats by bacterial inoculant cells on the success of application of these cells, e.g., in plant pathogen control or degradation of toxicants in polluted soils.

2. SOIL AND RHIZOSPHERE CONDITIONS AFFECTING INTRODUCED BACTERIA

Bacterial survival and activity in soil are controlled by prevailing (beneficial and deleterious) soil conditions. The effect of stimuli provided by these conditions is related to the site of bacterial application. In this section, these stimuli will be described in general terms before specifying them for different compartments in soil.

Conditions which can be regarded as generally beneficial to soil inoculants are the availability of oxygen, water, and nutrients, and the presence of favorable soil pore spaces, which is related to soil texture. Oxygen is essential for many common soil inoculants, which are mostly obligatorily or facultatively aerobic. In vitro oxygen consumption in rhizosphere soil has been measured by microelectrodes (see Chapter 2; [24]), which may also be applicable for in situ measurements. This technique offers perspectives for on-line monitoring of fluctuations of oxygen supply affecting inoculant fate. The primary effect of soil moisture on introduced bacteria is to provide the water necessary for metabolic activity. However, there are also secondary effects caused by fluctuations or different levels of soil water. These are related to the possible occurrence of anaerobic conditions (water saturation), the release of organic compounds by plant roots (25), and the mineralization and distribution of nutrients through the soil (26). Carbonaceous compounds available to microbes in soil are mostly carbohydrates (27). However, carbonaceous compounds are most abundant in the rhizosphere, and the chemical forms are different; i.e., sugar monoforms, organic acids, and amino acids abound (28,29). For instance, the presence of different aromatic and aliphatic acids can be detected in rhizosphere soil of quackgrass (*Elytrigia repens* L.) by ion exchange chromatography (30). The influence of the organic compounds released from plant roots into the soil has been studied by pulse labeling of plant/soil systems with $^{14}CO_2$ (31,32). The input of nutrients by root exudation supported bacterial activities in soil at a low level, as shown for, e.g., nitrogen fixation (33). Carbonaceous compounds released from plant mucilage, sloughed-off root cells, and root hairs are also important nutrient sources in the rhizosphere. Besides quantitative aspects, the quality of plant root released compounds also influences the bacterial compositon and activity in the rhizosphere, as has been shown by the preference of certain soil bacteria for exudates of different plant roots (34,35). The difference in bacterial activity between cultivars of the same plant species appeared to be related to differences in the exudation spectra, i.e., subtle differences in the compounds released by plant roots of the different cultivars (36).

Whether nutrients are available for bacteria in the rhizosphere depends on the sites in soil where nutrients and bacterial cells are present. Organic compounds tightly bound to the soil matrix are often less available for bacteria (37), and those present in smaller soil pore spaces can be physically protected against mineralization. However, disturbance of the soil will make these nutrients available to soil microorganisms (38). Soil textural aspects also influence bacterial survival, possibly by affecting the level of protection against predation by protozoa. The presence of clay minerals, such as montmorillonite or bentonite, can substantially improve bacterial survival (21,39,40). Some of the pore spaces in soil aggregates serve as protective microhabitats for soil bacteria against predation by protozoa (41); the greater amount of protective pore space in heavier-textured soils, such as clayey soils or sandy soils amended with clay minerals, may allow better protection against predation (42). However, this statement seems contradictory to arguments that available pore space is not the factor limiting bacterial survival in soil (43). Hence, other physical aspects are possibly involved in the protective effect of clay minerals on bacterial survival. Positive correlations can be made between, on the one hand, the availability of oxygen, water, and protective pore spaces in soil and, on the other hand, bacterial survival. However, these conditions cannot be separated; therefore, assessment of single conditions that positively influence bacterial survival and activity is difficult.

Two conditions stressful to bacteria in soil are nutrient (mainly carbon) starvation and low water activity. The availability of organic nutrients in soil largely depends on the soil sites occupied by the bacterial cells or microcolonies. Oligotrophy (defined as nutrient limitation resulting in a restriction of bacterial growth and activity [44]) is probably locally dispersed through most soils. Therefore, soil in general can be regarded as a grossly oligotrophic environment (45). The limited level of available organic nutrients in most soil sites probably leads to local energy depletion in many heterotrophic bacteria, which in turn results in a state of starvation (46,47) and/or nonculturability (48,49). The morphological response of indigenous bacteria (50,51) and of soil bacteria in vitro (22) or in soil (52) to these conditions often is a general shrinkage of cell size.

Conditions triggering the starvation response in soil are local nutrient depletions that are the consequence of previous bacterial growth and activity. Nutritional replenishment is only possible via diffusion through soil water. The reduction of substrate availability appears to be caused by low diffusion during high (>–0.6 MPa) water activity (negative tension of water in soil caused by capillary, osmotic, and adhesion forces) and results in decreased bacterial activity. However, at lower water activity bacterial activity was reduced by dehydration of the cells (53). These observations illustrate the complexity of the soil environment, where different conditions affecting bacterial survival and activity are interrelated. Low water activity in soil is generally related to a low matric potential (retention of free water molecules by water-insoluble soil colloids) but also to low osmotic potential (retention of free water molecules by water-dissolved substances like electrolytes or organic compounds). For instance, the metabolic activity of *P. fluorescens* cells chromosomally marked with *lux* genes was measured by luminescence, dehydrogenase activity, and respiration rate at different matric potentials in soil (54). In this study, the metabolic activities decreased with decreasing matric potentials, whereas survival of the *P. fluorescens* population was not correlated with the matric potential decrease. Furthermore, the stress applied with high salt solutions changes bacterial populations and activities in soil (7,55). Bacterial populations introduced into soil lost viability during water stress, although their deoxyribonucleic acid (DNA) could still be detected (56). To withstand water stress, soil bacteria may adapt intracellularly (57,58) or produce extracellular polysaccharides which tends to exclude the effects associated with the water stress (59).

Both starvation and low (as well as fluctuating) water activity represent typical stress conditions general in most soils. Other stress conditions like extreme temperatures and pH values or toxic compounds may be more specifically related to the climate, geographic location, or soil type and site.

A separation of favorable and adverse conditions in soil is impossible. Soil is to be regarded as a heterogeneous system (60,61), where both favorable and adverse conditions can be present in sites located close to each other. The occurrence of one favorable condition can exclude another favorable condition in the same site, as in the case of water-filled soil pores, which leads to low oxygen diffusion. Individual cells of an introduced population located in different sites in soil will therefore remain in different physiological states, depending on the microhabitats where they are located. However, a rough division of the soil environment into two different compartments, bulk soil and plant rhizosphere, can be made by an evaluation of conditions specifically related to each compartment. Information about these conditions is relevant to the application of soil-controlled genes and the study of the development of stress resistance in soil bacteria. A division of soil into bulk and rhizosphere soil, based on the abundance of nutrient input by crop plants, has long been used (62). Soil closely connected to roots (rhizosphere soil) can be regarded as a favorable environment for copiotrophic bacteria because of the enhanced flow of organic material which serves as substrate. This flow often results in better survival and higher activity of introduced cells in the rhizosphere versus nonrhizosphere (bulk) soil. The conditions typical for the two compartments have been discussed (62). Bacterial responses to these conditions, methods to screen for promoter and structural gene loci which are induced by them, and their potential use will be discussed in the following sections. In addition to the division of soil into rhizosphere and bulk compartments which are useful paradigms for specific effects on bacterial gene expression, other conditions typical for soil can be thought of, such as the low temperature experienced by soil bacteria during wintertime in temperate climate zones and the chemical stress caused by the presence of pollutants.

3. BACTERIAL RESPONSES TO PLANT-RELEASED COMPOUNDS AND OTHER SOIL CONDITIONS

Bacterial cells introduced into soil should be able to rapidly adapt to soil conditions in order to persist and reproduce. Survival strategies depend on the physiological adaptations in the introduced cells, such as adaptation to nutrient-limited conditions and/or other physical/chemical conditions, efficient utilization of root-released compounds, or specific interactions with plants as in symbiosis or pathogenesis. Therefore, bacteria which are well adapted to the soil environment, i.e., that survive and persist, probably have an efficient response to stressful soil conditions, activating molecular mechanisms necessary for their adaptation and survival. Understanding of the physiological and morphological responses in bacteria, which occur upon introduction into soil, is important to predict the inoculant's behavior. Cellular systems, such as membrane-bound proteins (63), that sense signals from the soil and transmit these intracellularly in order to induce operons or regulons, are of fundamental interest, not only for understanding the molecular response to exposure to soil conditions, but also for practical applications, such as differential regulation of heterologous genes in genetically modified soil bacteria.

3.1 Bacterial Colonization of the Rhizosphere and Plant Roots

Efficient colonization and/or physiological adaptation to soil conditions is the only option for soil inoculant bacteria to survive under many adverse soil conditions. Rhizosphere and plant root colonization also is an important initial stage in pathogenic and symbiotic plant–microbe interactions and competition between microbes. Root colonization is defined here as the total of interactions between plant and microbial species in soil which are necessary for successful establishment (growth and abundant cellular activity) on the plant root surface by cells present in soil, in air, at the seed surface, or in animals. Steps in plant root colonization are chemotaxis (movement in response to chemical compounds), anchoring, activation, and expression of genes, as suggested for *Rhizobium* infection (64).

Bacterial movement can be passive via soil water fluxes or active by specific induction of flagellar activity by plant-released compounds (chemotaxis). Some of these compounds, e.g., luteolin (65,66), phenolics (67) for *Rhizobium meliloti*, acetosyringone for *Agrobacterium tumefaciens* (68), and benzoate for *Azospirillum* spp. (69), serve as specific attractants involved in plant–bacterium interactions. Besides specific compounds, substances which are common in root exudates, for instance, sugars for *A. tumefaciens* (70), different amino acids, nucleotides and sugars for *Pseudomones lachrymans* (71), or serine for *P. aerugi-*

nosa (72), can serve as chemoattractants. Plant-induced active movement has also been observed in studies with different rhizobacteria such as *Pseudomomonas fluorescens*, *P. putida* (73), and *Azospirillum brasilense* (74–76). In vitro studies have demonstrated that flagellar movement was coupled with chemotaxis (77), and flagella-minus mutants of *P. fluorescens* were not able to colonize roots (78).

After induction of genes involved in chemotaxis and flagellar activity, a second step in colonization, adsorption to the root, ensues. Adsorption to the root is a step required before anchoring. Bacterial adsorption can be defined as nonspecific and based on electrostatic forces, whereas anchoring can be defined as firm attachment to the root surface. Two distinct mutant types were obtained with *Alcaligenes faecalis*, mutants defective in adsorption but not in anchoring and mutants defective in both adsorption and anchoring (79). Exopolysaccharides and pectinase were shown to play an important role in the attachment. For the attachment of *P. fluorescens* to wheat (*Triticum aestivum*) roots, the production of fimbriae (bacterial fringelike structures) was shown to be essential (80). Agglutination by high-molecular-weight compounds released by different plants (lectins) may be involved in the adherence of soil bacteria to root surfaces; however, these compounds are not the only factors involved in root colonization (81). Another phenomenon involved in root colonization may be the formation of cell aggregates (symplasmata), as shown for a *Pantoea agglomerans* strain on rice roots, the natural host of this strain; however, these symplasmata appeared to be absent on nonhost (wheat) roots (82).

Following adsorption and anchoring, a specific and complex interaction of a bacterium with host plants may ensue, resulting in symbiosis or pathogenesis. For the symbiotic interaction between *Rhizobium* (83–85) or *Bradyrhizobium* (86) and leguminous plants, specific nodulation (*nod*) gene products are necessary to determine host specificity and induction of nodulation in the host meristem (tissue composed of cells capable of diversification). Pathogenic interactions are well known for soil bacteria. In *Agrobacterium tumefaciens*, the virulence (*vir*) gene cluster in the tumor-inducing (Ti) plasmid is responsible for processes leading to crown gall disease in dicotyledonous plants (63), and in *P. syringae*, *P. solanacearum*, and *Erwinia amylovora*, the hypersensitive reaction and pathogenicity (*hrp*) gene cluster is responsible for the hypersensitive reaction (HR), the reaction of the plant to the pathogen (87). Specific plant compounds such as phenolics (88), flavonoids (89,90), and calystegins (91) can elicit bacterial gene expression or serve as metabolic signals in specific plant–bacterium interactions. The specificity of rhizopines influences the outcome of competition between different *Rhizobium* strains in the rhizospheres of leguminous plants (92). Rhizopine catabolism (*moc*) loci of *R. meliloti* isolates of different geographic origins have recently been compared (93). Isolates containing a homologous *moc* locus were not restricted to one geographic origin, and all were isolated from nodules of the same plant species, alfalfa (*Medicago sativa*).

Commensalistic (having an interorganism relationship without mutual effects) bacteria colonizing roots and other plant parts have to compete for nutrients with other species. In a study on bean leaves, different epiphytic species, such as *P. fluorescens*, *P. agglomerans*, *Stenotrophomonas maltophilia*, and *Methylobacterium organophilum*, were able to coexist as a result of differences in nutrient utilization (94). However, common compounds present in root exudates can be utilized by many commensalistic species present in the rhizosphere. Root exudates of maize (95), barley, and wheat (96) have been analyzed; the major compounds are carboxylic acids, sugars, and amino acids. The induction of bacterial genes as a response to exudate compounds can be used for plant-directed expression of beneficial genes. In our laboratory, a transcriptional fusion of *lacZ* with a root exudate–induced promoter of *P. fluorescens* responded specifically to proline present in wheat root exudate but not to 125 other tested substrates (97). The response of this mutant, denoted RIWE8, to nutrients released in the rhizosphere of wheat could be observed in a microcosm system and under field conditions (97,98). The utilization of proline and ornithine as sole carbon and nitrogen sources was also observed with *R. meliloti* (99). Mutations in the proline dehydrogenase gene (*pro*DH) did not result in reduced nitrogen fixation although *pro*DH appeared to be necessary for nodulation and competition on alfalfa roots.

3.2 Bacterial Colonization of Soil and Adaptation to Soil Conditions

Adhesion of bacteria to soil particles may be a first step in the colonization of soil. The *P. fluorescens* mutants deficient in adhesion to surfaces like soil particles were shown to lack flagella and an outer membrane protein (100). The ecological consequences of these deficiencies and their role in adherence remained obscure. However, the physiological condition of cells attached to a solid surface can be different from that of cells in suspension, as shown in a chemostat study with the soil bacterium *P. putida* (101). Furthermore, cell size changes of marine isolates on solid surfaces were studied under starvation (102). More rapid decreases in cell size were observed with hydrophilic isolates attached to surfaces compared to liquid phase cells; however, the opposite behavior was observed for hydrophobic isolates. Adherence of the marine bacterium SW5 to hydrophilic surfaces resulted in sparse colonization with long bacterial cell chains; however, at a hydrophobic substrate, cells colonized tightly packed and at high densities to the substratum (103).

Soil-inoculated bacteria may create their own protective microhabitat by producing exopolysaccharides (EPS) (59), which retain water during desiccation. Nonmucoid (reduced in EPS production) mutants of *Escherichia coli*, *Acinetobacter calcoaceticus*, and *Erwinia stewartii* showed reduced survival during desiccation compared to that of their isogenic parent strains in in vitro studies (104).

Bacterial cells adhered to soil particles and soil pores in microaggregates (41) or on the surface of plant roots may form microcolonies if sufficient nutrients are locally available. Recently, an intercellular communication system based on the autoinducer *N*-(3-oxohexanoyl) homoserine lactone (HSL) was discovered in *Vibrio fischeri* and *V. harveyi* controlling the expression of bioluminescence (*lux*) genes (105). The genes responsible for autoinduction of the *lux* genes in *Vibrio* (*lux*R and *lux*I) have also been found in terrestrial bacteria, e.g., *Erwinia carotovora*, in which antibiotic production was controlled by HSL (106). A *lux*I independently expressed autoinducer (*N*-octanoyl-L-homoserine lactone) also appeared to regulate *lux* gene expression in *V. fischeri* (107). The HSL autoregulatory systems measure cell densities in suspension (105) and may be involved in measuring densities in microcolonies in nature. Application of this mechanism in soil biotechnology may entail the concerted induction of heterologous genes in a cell density–dependent fashion.

As mentioned, introduced bacteria have to adapt metabolically to stress conditions in soil. Carbon starvation is probably the main stress condition in bulk soil, whereas other stressors such as limitations of other nutrients, low water activity, acidity, and low temperature are not restricted to either bulk or rhizosphere soil (62). Adaptation to stress conditions is a general phenomenon occurring in diverse groups of bacteria. Drastic physiological and morphological changes were observed with a *P. fluorescens* strain during different nutrient limitations in a continuous culture system (108). Differences in DNA, ribonucleic acid (RNA), and protein contents were observed during starvation survival (survival under nutrient-limited conditions) at different dilution rates of the marine bacterium ANT-300 (109) with slower-growing cells better adapted to starvation survival. Generally, an increase in cellular protein content, a decrease in cellular RNA, and an accumulation of the signal nucleotide guanosine tetraphosphate (ppGpp) are observed after carbon starvation (110). Normally ppGpp mediates the stringent response (bacterial regulation of operons induced upon amino acid starvation). Both ppGpp and cyclic adenosine monophosphate (cAMP) are involved in metabolic reprocessing after nutrient limitations (110). Other important features of bacterial cells under carbon starvation are a reduction of cell size; the accumulation of storage compounds such as glycogen, polyhydroxyalkanoates, lipids, and polyphosphates, and alterations in the cell envelope (111). A reduction of cellular DNA to one genome equivalent per cell has been suggested, but this has been challenged by recent data (111) which didn't show such a reduction.

During carbon starvation, de novo synthesis of proteins has been observed in *E. coli* (112,113), marine *Vibrio* sp. (114,115), and *P. putida* (116), using two-dimensional gel electrophoresis. Some of these proteins, e.g., heat shock proteins, are involved in carbon-starvation-induced cross-protection to other stress conditions (117). The *E. coli* cells exposed to starvation developed resistance to heat, hydrogen peroxide (H_2O_2) (118), and osmotic (high NaCl concentration) stress (119). However, nutrient-deprived *E. coli* and *Vibrio* cells treated with the protein synthesis inhibitor chloramphenicol during nongrowth still developed protection to heat stress (120). In this study, protein synthesis, as measured by ^{3}H-leucine incorporation, decreased strongly during starvation. The authors therefore concluded that protection to heat stress during prolonged starvation was not mediated by de novo protein synthesis. Starvation-induced cross-protection to heat probably develops during the initial phase after starvation. Those *E. coli* cells exposed to stress conditions (heat, acidic, oxidative, osmotic, and different nutrient stresses) showed enhanced resistance after exposure to seawater, compared to that of controls (121). The *P. putida* cells further showed cross-protection to ethanol, heat, osmotic, and oxidative stress after exposure to carbon limitation (22). Similar observations were made for *P. fluorescens* cells under carbon starvation and in soil (52). Therefore, it can be concluded that macromolecular rearrangements, synthesis of proteins during the initial phase of starvation, and induced cross-protection after exposure to starvation are physiological adaptations which are probably common across the gram-negative bacteria. The observation that cross-protection is also induced during a short period of residence in natural soil clearly indicates that the soil environment generally is to be regarded as a nutrient-limited environment for inoculant bacterial strains.

Important genes expressed during starvation are, for instance, carbon starvation (*cst*) and postexponential response (*pex*) genes. Regulators controlling starvation gene expression are, e.g., the nucleotides cAMP (controlling the *cst* genes), ppGpp, and the alternative sigma factors, σ^{32} (controlling heat shock proteins DnaK, GroEL, and HtpG) and σ^s (122). The genetic locus of σ^s was identified by transcriptional fusion with *lacZ* and insertional inactivation with transposon Tn10. It coincided with the previously described loci *katF* and *appR* (123). Mutants inactivated in the σ^s locus lacked expression of 32 proteins, including some postexponential growth (Pex) proteins (124), as well as enhanced tolerance to different stresses, such as heat shock, hyperosmolarity, and nutrient scarcity (123,125). Sigma factor σ^s also controls the morphogene *bol*A, which is responsible for the morphological change of *E. coli* cells after entrance into the stationary phase (126). Transcription of genes controlled by σ^s is possibly regulated from specific upstream promoter regions called "gearboxes" (127). Expression of the *rpo*S gene (coding for σ^s) but also of other σ^s regulated genes is controlled by a histonelike protein, H-NS (128). However, σ^s is not always involved in the regulation of carbon starvation–induced genes. A DNA sequence analysis of the promoter region of a carbon starvation induced gene in *P. putida* revealed that this gene was controlled by σ^{54} (129). The heat shock protein DnaK also plays an important role in the development of cross-protection to oxidative and thermal stress and in the morphological transition of *E. coli* cells after starvation (130).

Molecular mechanisms involved in the protection of bacteria to osmotic stress are well studied in some enteric bacteria, e.g., *Vibrio, Salmonella*, and *Escherichia* spp. (131). The response to osmotic stress develops in two phases. In the first phase, potassium ions are accumulated by a low-affinity uptake system (Kup and TrkA proteins) and a high-affinity uptake system (Kdp protein). Glutamate is accumulated as a negatively charged counterion. During the second phase, osmoprotectants like glycine betaine, trehalose, and proline are accumulated either by uptake (using the ProU and ProP transport system) or by synthesis. An osmoprotectant commonly found in moderately halophilic bacteria, ectoine, also provided osmoprotection to *E. coli* cells after ProU/ProP mediated uptake (132). Regulation of the *pro*U gene is dependent on DNA supercoiling and two loci involved in DNA supercoiling, *top*A (coding for topoisomerase I) and *osm*Z (coding for a histonelike protein); this influenced expression of *pro*U and other genes (133,134). However, it remains obscure how environmental signals are transmitted to mediate OsmZ-controlled DNA supercoiling as well as the exact mode of action of OsmZ on DNA supercoiling (134). Different mechanisms of osmoprotection have been studied in natural seawater with *E. coli* (135,136) and they were found in bacteria common in soil such as *Streptomyces* spp. (57,58), *Azospirillum* spp., *Acetobacter diazotrophicus*, and unknown halotolerant strains isolated from the rhizosphere of rice (137). Expression of an alginate synthesis gene (*alg*D) in the human pathogen *P. aeruginosa*, which is also commonly observed in soil, is regulated by high osmolarity (138).

Tolerance to low pH is well studied in enteric species such as *Salmonella typhimurium* and *E. coli* (139). A two-step tolerance to low pH was discovered in *S. typhimurium*. Initial adaptation to weakly acidic conditions (preacid shock) was followed by adaptation to a severe acid shock (pH 3.3) (140). The effect of diverse physiological conditions of enteric pathogenic species on acid tolerance was tested (141–143). It appeared that cells incubated for 6 days in seawater, cells in stationary phase (grown in minimal medium with glucose), and C- and N-starved cells showed a pH-dependent acid tolerance response (ATR), although σ^s mediated pH-independent cross-protection to acid stress as part of the general stress resistance response (142). Bacterial responses to low pH are observed in *E. coli* (144) but also in plant pathogens as *A. tumefaciens* (145) and *Erwinia amylovora* (146).

Adaptation to low temperature is a poorly investigated phenomenon. The bacterial response to growth-restricting temperatures was shown to depend on the species (147,148). A temperature downshift to 5°C blocked protein synthesis in the mesophilic species *E. coli*; however, protein synthesis continued at a lower rate with the psychrophilic species *P. fluorescens* (147). Overall protein synthesis of exponential phase *E. coli* cells was studied at 37°C and 10°C; several specific proteins were produced in higher amounts at the lower temperature (149) and may represent cold shock or cold response genes. The functions of some cold shock–induced genes have been characterized in *E. coli* and *Bacillus subtilis*. The *E. coli* cold shock protein CS7.4 maintained the DNA gyrase (DNA super-coiling activity) subunit A at a sustained level during low temperature (150), and the *B. subtilis* cold shock protein *csp*B had an antifreeze function (151). Hence, as expected, in response to low temperature bacteria generate proteins that have protective functions in the cell.

The molecular and physiological basis of bacterial responses to the different soil conditions may vary greatly, and little is understood about the myriad re-sponses seen. However, physiological responses to different stresses also reveal commonalities, as seen in the overlap of stress-induced proteins. Some proteins are expressed during exposure to virtually all types of stresses (generalized stress response), whereas others are restricted to one or a few specific stress conditions. For instance, after exposure to different toxic compounds, *E. coli* showed the presence of several new proteins which were also produced during heat and car-bon starvation stress (152). Besides physiological properties which are common in the response to different conditions, the molecular basis of the transmission of environmental signals for the expression of different genes share common fea-tures with each other. For example, the genes responsible for virulence of *A. tumefaciens* (*vir*) are regulated by a two-component regulatory system, com-posed of VirA and VirG (145). This type of regulation is also present in genes in-volved in the response to nitrogen and phosphorus limitations, and in that to toxic compounds and dicarboxylic acids (153). Generally, one component is in-volved in environmental sensing (VirA), whereas the other component is an in-tracellular transmitter which regulates gene expression (VirG).

Genetic loci found to control gene expression in response to the environment can be used to construct GMMs with differentially controlled genes responding to selected stimuli. Laborious screening of useful promoters (operon-regulating sequences) regulated by the appropriate stimuli can thus be circumvented. How-ever, if such information is not available, other strategies should be explored, such as differential screening of transcriptional fusion (reporter gene expression regulated at messenger RNA [mRNA] level) constructs.

4. DETECTION OF SPECIFIC PROMOTER ACTIVITY IN SOIL AND THE RHIZOSPHERE

The screening for environmentally controlled promoters or genes is facilitated by recent methodological developments. Thus, the random insertion of promoter probes based on transposons followed by screening for differential regulation of the reporter gene is an approach to identify, characterize, and, in a later stage, isolate environmentally induced regulatory loci. Suitable loci are selected by testing for expression of the reporter gene under selected inducing (promoter induction and transcription) as well as noninducing conditions. A technique based on the synthesis of mRNA during differential screening (differential display) has been devised. Differential display has initially been developed for eukaryotic species (154,155). It can be adopted for prokaryotic species to isolate differentially expressed mRNA templates (156). Using this method, suitably induced genes can be isolated from the bacterial genome.

To isolate differentially induced promoters, a bank of operon fusions with promoterless reporter genes should be prepared and screened for reporter gene expression under inducing versus noninducing conditions. Transcriptional fusion vehicles (promoter probes) have been developed which are based on transposons or phages that insert randomly into the genome, e.g., Mu, Tn*10*, or Tn*5*. Various elements constructed by different groups are available (157–171). Promoter probe elements based on transposon Tn*5*, which contain *trp'-lacZ*, *lacZ*, *luc*, *npt*II, *pho*A, *lux*AB, or *xyl*E as reporter genes and kanamycin, tetracycline, or gentamycin resistances as selectable markers, have been constructed (172,173). A schematic representation of the promoter probe construct Tn*5*-B20 (173) inserted into the genome of *P. fluorescens* strain R2f (resulting in the proline-responsive mutant strain RIWE8; 97) is shown in Fig. 2. Construct Tn*5*-B20 contains a promoterless *lacZ* gene at the beginning of the IS50L site of transposon Tn*5* and is controlled by the intrinsic *Pseudomonas* promoter upstream of the inserted Tn*5*-B20 element.

The screening for stationary phase–induced promoters has been facilitated by

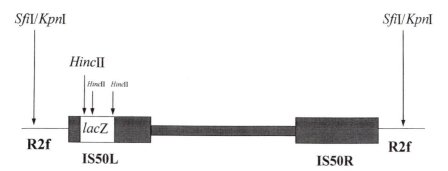

Figure 2. Insertion of the Tn5-B20 (*lacZ* as reporter gene) promoter probe construct in the chromosome of *P. fluorescens* R2f, resulting in a proline-responsive mutant, RIWE8. R2f, chromosomal DNA of *P. fluorescens* R2f; IS50L/R, left and right insertion elements of Tn5; *Sfi*I, *Kpn*I, *Hinc*II, restriction sites in Tn5-B20 and chromosomal *P. fluorescens* RIWE8 DNA.

the use of a bicistronic reporter system (157). A novel reporter gene system has been based on the LamB surface protein equipped with a corona virus epitope (174). Upon expression of the reporter, the epitope is exposed at the bacterial surface and can be detected with a specific antibody. An advantage of this system over other reporter genes is that induction of gene expression can be easily detected in situ without prior extraction or addition of substrates.

Promoter probe inserts have been used for diverse objectives. Acid-induced promoters have been characterized in *E. coli* (144) and in *Lactococcus lactis* (159); a phosphate-regulated promoter in *P. fluorescens* (175); a carbon- and nitrogen-starvation–induced promoter in *R. meliloti* (176); nitrogen-, phosphorus-, (177) and carbon-starvation-induced promoters (178) in *P. fluorescens*; and wheat root exudate–induced promoters in *P. fluorescens* (98,179). In addition, carbon starvation, cold shock, and wheat root–induced promoters have been identified in *Burkholderia cepacia*. The screening procedure and the differential responses of the promoters found in *B. cepacia* are discussed later.

The *B. cepacia* strain P2 has been isolated from the grass (*Lolium perenne*) rhizosphere (20). It showed superior root colonization capability with gramineous species as well as survival in bulk soil. This isolate is a promising candidate to serve as a host (functional delivery vector) for the expression of beneficial genes in the rhizosphere. A pool of random promoter probe insertion mutants was prepared by using the Tn5-B20 construct (173), and these were screened for reporter gene expression as a response to either carbon starvation, cold shock, or root exudates. Mutants were transferred (in duplicate) to minimal medium containing the chromogenic substrate for β-galactosidase, X-gal(5-bromo-4-chloro-3-indolyl-β-D-galactopyranoside) and incubated at 27°C (noninducing condition) and 4°C (inducing condition). Colonies turning blue after 5–7 days of incubation under inducing conditions but remaining white under noninducing conditions were isolated and further screened in a standard β-galactosidase assay (180). The mutant obtained, clone PD19, was shown to respond to cold shock (Fig. 3a). To select for a response to carbon starvation, a bank of P2 Tn5-B20 mutants was screened for differential expression of *lacZ* on minimal medium with X-gal amended with 0.01% and 0.025% glucose (carbon limitation–inducing condition) versus 0.1% glucose (noninducing condition). The P2 mutant PC3 responded specifically to carbon starvation (Fig. 3b) but to a significantly lower extent to nitrogen and phosphorus starvation (not shown). Mutants responsive to wheat root exudate were selected by screening for a differential response in minimal medium with X-gal and wheat root exudates (collected as described by van Overbeek and van Elsas, 1995 [97]) (inducing condition) and 0.1% glucose (noninducing condition). Two mutants showing the desired differential responses were obtained. Further testing of the muants was necessary because the compound in the root exudate that triggers the reporter gene expression was still unknown. The response to 95 different substrates was tested in (BioLog

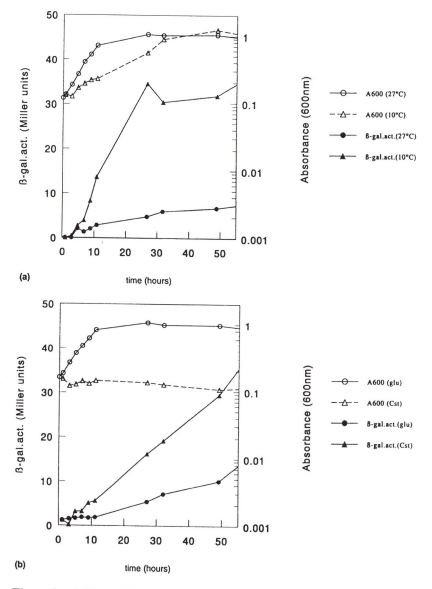

Figure 3. Differential reporter gene expression of *Burkholderia cepacia* P2 mutants; PD19 at 10°C (a), PC3 by carbon starvation (b), and PEWE2 by wheat root exudate and leucine (c). The optical density of exponentially growing and carbon-starved cells was measured by absorbance at 600 nm (A600), whereas reporter gene activity was measured by a β-galactosidase assay according to (180). This assay measures the hydrolysis of the lactose analog *o*-nitrophenyl-β-D-galactopyranoside (ONPG) to *o*-nitrophenol, which can be determined by spectrophotometry at a wavelength of 420 nm. The specific activity (β-gal. act.), in Miller units, is the measured absorbance at 420 nm corrected for the bacterial biomass (A_{600}); glu, glucose; Cst, carbon starvation; RE, wheat root exudate; leu, leucine.

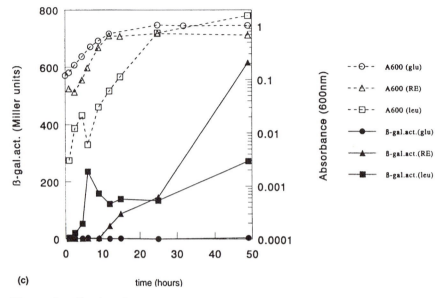

Figure 3. Continued

GN) plates, developed for determination of the substrate use of bacterial isolates. These consist of 96-well microtiter plates, in which the wells contain different carbon sources. Mutant cells were suspended in mineral medium plus X-gal and pipetted into the different wells. One mutant, clone PEWE2, responded specifically to a single compound, leucine (Fig. 3c), whereas the other responded to a broad spectrum of substrates.

Promoter probe reporter systems are also commonly used to investigate the expression of genes involved in plant–bacterium interactions. Reporter gene expression was used in transcriptional fusions of *hrp-lacZ* in *P. solanacearum* (181,182); *hrp-gusA* (β-glucuronidase) in *Erwinia amylovora* (146); *virB-lacZ* in *Agrobacterium tumefaciens* (183); *avrD-lux* and *gus* in *P. syringae* (184); different pectinase genes with *gusA* in *Erwinia chrysanthemi* (185); *nodA-*, *nodC-*, and *nodD-lacZ* in *Rhizobium* species (186,187); *nifA-lacZ* (188,189); *nifB-lacZ* (189); *nifH-gusA* (190); *nifH-lacZ* (189) in *Azospirillum brasilense*; *nifA-*, *nodG-*, and *ntrC*-promoters of *A. brasilense*, coupled to *lacZ*, in *A. lipoferum* (188), and *aggA-xylE* in *P. putida* (191). A new sensitive reporter gene, the ice nucleation gene *inaZ*, has been tested under the control of promoters from the β-lactamase-, *hrpR-*, and pyruvate decarboxylase genes in *Zymomonas mobilis* (192). This reporter gene has also been used under the control of the phenazine production promoter in *P. aureofaciens* on radish seeds and sugar beet in soil under different soil matric potentials (193). The *inaZ* reporter gene offers great opportunities for promoter activity studies in soil environments, since it is highly sensitive and specific in this environment. Reporter genes are also used to screen for and test promoters which are not involved in plant–bacterium interactions. Genes responsible for biphenyl degradation (*bph*) were transcriptionally fused with *lacZ* in *P. fluorescens*, and expression was detected on sugar beet seed and on germinating seeds in sterile soil (194). Promoters of two heat shock genes (*dnaK* and *grpE*) were fused with *luxCDABE* and reporter gene activities as a result of the application of different toxic compounds in *E. coli* measured (195). This heat shock–regulated luminescence system can be used as a biosensor for environmental pollution.

So far, suitable mutants have been mostly tested in in vitro experiments; however, for the evaluation of differential gene expression under soil conditions, soil microcosm studies should be performed. To study differential expression in rhizosphere soils, specific soil microcosm systems have been developed which allow one to follow reporter gene expression either in situ or via extraction from soil.

For studies on in situ root exudate–controlled reporter gene expression in the rhizosphere or on the rhizoplane, plants should be preferentially grown in natural (unsterilized) soil. For destructive analysis at any point in time, plant roots can be isolated from soil and reporter gene expression measured by extractive means or in situ. Different reporter genes can be used to assess induced gene expression by this approach. In situ root exudate–controlled reporter gene activity under natural conditions has been reported (190,196) (see Table 1). However, in situ root colonization studies under gnotobiotic conditions have been performed with various different markers and bacterial species. De Weger et al. (197) used naphthalene-inducible *luxCDABE* and constitutively induced *luxCDABE* and *luxAB* gene constructs for in situ root colonization studies with *P. fluorescens*. Reporter gene activity was measurable at a colonization level of 10^3–10^4 colony forming units (CFUs)/cm root. In situ reporter gene (glucuronidase [Gus]) expression in *A. brasilense* (190) and *Azoarcus* sp. (198) was determined in gnotobiotic root systems of wheat and rice, respectively. Specific *nifH-gusA* (fusion of the structural nitrogen fixation gene *nifH* and the glucuronidase reporter) induction in *A. brasilense* was observed on gnotobiotic wheat roots.

To follow in situ reporter gene expression during root development in soil, the roots can be grown alongside a glass plate or on a microscope slide. For studying root colonization, specific incubation chambers (root boxes) have been developed (199,200). Reporter gene activity should ideally be measured without disturbance of the system. Therefore, light-emitting reporter genes (*lux* and *luc*) are best suited for in situ determinations. However, the addition of exogenous factors (substrate) may be necessary (see Chapter 7) and the effects of this should be considered. Ideally, any disturbance of the system should be avoided. Emission of light can be measured by autoradiography (light emission sensed by X-ray

Table 1. Examples of Environmentally Controlled Reporter Genes Tested in Natural Systems[a]

Inducing compound or condition	Bacterial species	Regulating locus	Reporter gene	Reference
Naphthalene and salicylate	P. fluorescens	nahG Promoter (naphthalene)	luxCDABE	254
Wheat root association	Azospirillum brasilense	nifH Promoter (N fixation)	gus	190
Phosphorus starvation under gnotobiotic conditions in rhizosphere of potato and bulk soil	P. putida	Unidentified P-starvation-induced promoter	lacZ	175
Bean rhizosphere	P. fluorescens	rrnB Promoter (ribosomal)	luxCDABE	196
Iron restriction in bean rhizosphere and phylosphere	P. fluorescens P. syringae	pvd Promoter (pyoverdine)	inaZ	255
Carbon starvation in bulk soil	P. fluorescens	Unidentified C-starvation-induced promoter	lacZ	178
Wheat rhizosphere/proline	P. fluorescens	Unidentified wheat exudate/proline induced-promoter	lacZ	97

[a]luxCDABE, bioluminescence genes; gus, β-glucuronidase gene; lacZ, β-galactosidase gene; inaZ, ice nucleation active gene.

films), by optical fiber light measurements (197), or by use of extremely light-sensitive instruments such as a charged coupled device (CCD-photomultiplier) camera. For microscopic detection of in situ root exudate–induced reporter gene activity, specific fluorescent markers can be used. A new and promising marker is the green fluorescent pigment (GFP) gene (*gfp*), which recently has been cloned from jellyfish (201). Using promoterless *gfp* as a promoter probe reporter, specifically induced expression can be observed by ultraviolet (UV) microscopy in which intracellular GFP will emit green light.

For induced reporter gene activity measurements by extractive means, soil suspensions from rhizosphere and bulk samples can be prepared and analyzed using in vitro activity assays. Reporter gene (*lacZ*) activity controlled by a root exudate–inducible promoter of *P. fluorescens* R2f (mutant RIWE8) has thus been detected in the rhizosphere of wheat plants under field conditions (98). A soil microcosm setup (202) was used to study the activity of this mutant at different distances from wheat roots (97) in soil. The β-galactosidase activity in RIWE8-inoculated soil decreased with increasing distance from the roots, indicating that the compound triggering reporter gene expression was exuded by the wheat roots into the soil.

Information about promoters that are induced under environmental conditions is scarce. Examples of activities of potential environmentally controlled promoters under natural or gnotobiotic conditions are shown in Table 1. The promoter–reporter gene combinations can be used in the natural environment, e.g., as a biosensor system (the P_{nahG}-*lux*CDABE system) or in determination of expression of genes of interest in situ (the P_{nifH}-*gus*, the P_{rib}-*lux*CDABE and the P_{pvd}-*inaZ* systems). If the promoter is still unidentified, it can be tested for in situ induction before eventual cloning of a gene of interest (P_P-starvation-, P_C-starvation-, and $P_{root-exudate}$-*lacZ* systems). To our knowledge, environmentally controlled promoters have not yet been applied in soil inoculants. Two possible applications of soil environment–regulated promoters are discussed in the next section.

5. PROSPECTS FOR THE USE OF ENVIRONMENTALLY INDUCED PROMOTERS IN SOIL INOCULANTS

Promoter probe mutants responding to soil signals are prime candidates for cloning or gene replacement experiments. Practical applications of these constructs are the biosensing of toxic waste compounds or other stress conditions in soil, the biological control of plant diseases, and the biological containment of released GMMs. In particular, rhizosphere-induced promoters can be used for the biological control of soil-borne plant pathogens such as fungi, nematodes, or insect larvae.

Bacterial antagonism toward phytopathogenic microorganisms in the rhizosphere can be the result of competition for nutrients (203), excretion of antagonistic compounds like siderophores (204), or antibiotics (1), or a combination of both actions (205). Genes in rhizosphere bacteria involved in the antagonism toward plant-pathogenic fungi are of special interest and may be used as targets for cloning experiments. When these genes are brought under transcriptional regulation of a rhizosphere-induced promoter, the timing and extent of gene expression can be controlled, depending on the promoter of choice. This new strategy in the utilization of GMMs as biocontrol agents is promising, since it can enhance and concentrate the beneficial action of the control agent to the site and time where/when it is most needed, i.e., at the onset or peak of pathogen activity. Bacterial genes involved in the synthesis of antifungal antibiotics have been identified and cloned. Production of pyrrolnitrin has been observed in *Burkholderia cepacia* (206,207) and *P. fluorescens* (208) and that of phenazine-1-carboxylic acid, 2,4-diacetylphloroglucinol (209–211) and pyoverdine (205) in fluorescent *Pseudomonas* spp. The genetic loci expressing pyrrolnitrin (212) and pyoluteolin (213) in *P. fluorescens* and the locus of an antibiotic-like compound from *P. aureofaciens* (214) which inhibits the growth of the fungus *Aphanomyces euteiches* have been cloned. Antibiotic compounds are also known to be produced by bacilli, such as *B. subtilis*. Furthermore, in *B. cereus*, the production of two antifungal compounds has been described (215).

To protect agricultural crops against attack by soil-borne insect larvae, GMMs that contain bioinsecticidal genes can also be applied in the rhizosphere. For instance, the *Bacillus thuringiensis* crystal (Cry) proteins can be used as crop-protecting agents, as has been shown by their inclusion and expression in plants. Soil pseudomonads have been equipped with chromosomal inserts of the *cry*IVB gene, targeting the larvae of leatherjackets (*Tipula* spp.), which attack roots of gramineous plants. Studies with the (P_{tac}-controlled) *cry*IVB gene have been performed in *P. fluorescens* in vitro (2) and in soil (216). These GMMs, containing a constitutively expressed *cry*IVB gene, showed a slightly reduced survival rate in soil compared to that of their isogenic parent strain (216). This was most likely due to the increased metabolic load experienced by the GMM. Hence, developing strains with induced expression of *cry* or antifungal genes only at the target sites when required (i.e., in the rhizosphere in concert with the pathogen activity peak) will be advantageous for inoculant survival, since no extra metabolic load would occur under conditions of absence of root-released nutrients.

The efficacy and in particular the safety of GMMs in the field are current concerns. Major topics are the putative uncontrolled spread of the GMM and ecological disturbances caused by it. To reduce potential hazards after accidental or deliberate releases of GMMs, a biological containment system, which allows the introduced strains to show activity and survival only at the site of application, can be constructed. Ideally, the GMM is eliminated under other conditions, preferentially by an intrinsic factor such as the expression of a containment (cell-killing) gene. Soil-controlled promoters are prime candidates to serve as regulatory systems controlling this containment gene. Van Elsas and van Overbeek (62) opted for carbon starvation–or cold shock–induced promoters for this goal. A carbon starvation–induced promoter will preferentially be switched on in bulk soil and may be repressed in most rhizosphere sites where bacterial activity is high, whereas a cold-inducible promoter will be active during wintertime in the absence of crop plants. Possible pitfalls in the application of stress-induced promoters as regulators of containment genes may be low promoter activity (62) or expression of the containment genes during stress conditions. Host cells may be metabolically arrested when passing through a nonculturable state. During metabolic arrest, containment gene expression may stop and hence the host cell may survive the stress until resuscitation by a more favorable condition. It is thus possible that the action of the containment construct is counteracted during nonculturability of the bacterial host.

The potential problems with starvation- or cold-induced promoters can be overcome by using root exudate–controlled promoters as regulators of containment genes. For containment, the promoter would have to act in reverse order, since the containment gene will have to be repressed at the root site. Therefore, a "negative genetic loop" should be constructed, which consists in root exudate–controlled repressor protein expression and a repressor-controlled promoter directing expression of the containment gene. This system has been described for *m*-methylbenzoate induction (217) and tested in soil (218); the addition of *m*-methylbenzoate in combination with the constitutively expressed XylS regulator controls LacI repressor expression, which in turn controls containment gene expression from the lactose promoter P_{lac}. The disadvantage of this system was the need for an exogenously added compound in soil. With root exudate–controlled containment gene expression, such additions are not needed.

Genes proposed for biological containment are the *E. coli gef* and homologous *hok* (host killing) and *rel*F genes (219) and the *Serratia marcescens nuc* (nuclease) gene (220). Induced overexpression of *gef*, *hok*, and *rel*F genes causes a membrane potential collapse resulting in cell death (221,222), whereas *nuc* gene expression results in intracellular degradation of DNA (220). An improved containment system has been developed by the insertion of replicate copies of *rel*F, which decreases the possibility of genetic inactivation in single *rel*F copies by spontaneous mutations (223). An alternative type of biocontainment gene regulation is stochastic (random) induction using the type I fimbriae promoter (P_{fim}) from *E. coli* (224), which is an invertible promoter responsible for phase variation in the expression of fimbriae in *E. coli*. Using this promoter, stochastic *gef* expression in *E. coli* could be accomplished.

Maximal safety of released GMMs can only be offered if 100% killing efficiency of the GMM is achieved. However, escape from killing can be the result of inadequate promoter induction, (partial) resistance to the host killing proteins, or mutations in the host killing gene. Although mechanisms can be developed to enhance the killing rate after induction (223), a waterproof containment system will probably never be obtained. Hence, one should always take into account the possibility of escape from host killing in a few cells. The question then is whether the putative survival of a small population of GMMs in soil poses an ecological risk and whether the forces of ecological selection can enhance the magnitude of this potential.

6. PHYSIOLOGICAL AND METABOLIC ASPECTS OF THE SURVIVAL OF BACTERIA INTRODUCED INTO SOIL

The initial physiological state of cells is important in the adaptation and metabolic activity of soil inoculants. Different approaches to determine the physiological and metabolic conditions of cells in soil have been taken into account. In this section, adaptation to the soil environment, assessments of metabolic activity, and influence of specific mutations on survival are discussed.

The response of gram-negative bacteria to soil stress (like carbon starvation) can be investigated by assessing the development of protection to other stresses. The development of such cross-protection in soil bacterial species has recently been reported for fluorescent *Pseudomonas* cells under carbon starvation (22,52) and in two different soils (52). The adaptation can be seen to represent a bacterial strategy to cope with stresses encountered in soil. It is likely that a major trigger, under the conditions of favorable temperature and moisture used, is the limited availability of easily degradable C sources found in soil. As discussed before, prolonged residence of bacterial cells in soil may even result in a metabolic arrest, depending on the microhabitat colonized. Indirect evidence comes from the assessment of an introduced bacterial population in soil by CFU and direct microscopic (immunofluorescence [IF]) cell counts, which often show a discrepancy. IF cell counts which significantly outnumbered CFU counts indicated the presence of nonculturable fractions in the introduced bacterial populations (23,48,225). It is of obvious interest to assess whether such nonculturable cells are alive (viable) and metabolizing, alive but in a metabolically arrested state, or dead (see Chapter 3).

Methods for determining metabolic activity (direct viable cell counts [DVCs], not to be confounded with CFU counts) are based on the ability of cells to increase in size after application of nutrients together with a cell division inhibiting antibiotic, nalidixic acid. Alternatively, the presence of a membrane potential can be assessed using redox dyes. For determination of the metabolic activity of a specific population in the complex bacterial community in soil, new approaches based on hybridization with fluorescently labeled ribosomal (rRNA) oligonucleotide probes, in combination with viable staining procedures, have been developed. Stained cells can be studied by microscopy or by flow cytometry.

The DVC is a direct approach to detect and assess viable cells from natural environments by microscopy (226). Cells extracted are treated with nutrients and nalidixic acid and transferred to microscope slides for analysis. The rationale of this technique is that the division of cells which are still capable of taking up nutrients after residence in soil is inhibited by nalidixic acid, which results in the formation of elongated cells. For cells which are nalidixic acid–resistant, another cell division–inhibiting antibiotic, cephalexin (227), can be used. Using fluorescein isothiocyanate– (FITC)-labeled antibodies, the introduced strain can be distinguished from indigenous cells. Cells that are significantly larger than control cells (e.g., cells which are killed after extraction by sodium azide) in the introduced population can be quantified and their percentage calculated. This fraction represents the share of viable cells in the introduced population. Viability determined by the DVC is based on the cellular response to high nutrient concentrations, much as in CFU counts. A drawback of the DVC is that active cells adapted to oligotrophic conditions may show nongrowth, similar to what was observed with CFU determinations (228), and will thus be classified as metabolically inactive. Recently, the DVC has been used in combination with other techniques to determine cellular activity in aquatic systems (121,227,229–231). The DVC was also applied in combination with CFU counts to assess epiphytic *P. syringae* populations (232), as well as in a field microplot, to describe the activity of a *P. fluorescens* R2f mutant, RIWE8 (containing a root exudate–inducible reporter gene), in combination with the determination of cross-protection to stress factors (98).

Metabolic activity of individual cells (extracted or in situ) as well as of populations of cells (using flow cytometry) recovered from natural environments can be determined using specific dyes. Differentiation between active and inactive cells is based on the presence of a membrane potential, evidenced by redox dyes. Commonly used dyes are 2-(*p*-iodophenyl)-3-(*p*-nitrophenyl)-5-phenyl tetrazolium chloride (INT) (233) and 5-cyano-2,2-ditolyl tetrazolium chloride (CTC) (234). Respiring cells actively transport the colorless and soluble dyes across the plasma membrane, after which colored insoluble deposits are formed inside the cell, which can be visualized by microscopy. Metabolically active cells from environmental samples can be counterstained with a fluorescent DNA-intercalating stain like 4′-6-diamino-2-phenylindole (DAPI) (235). Viability studies with CTC have been performed with *E. coli* in an aquatic system (236), and with soil indigenous species (237) as well as a *Flavobacterium* strain in soil (238).

The fluorescent reagent rhodamine 123, which accumulates intracellularly after passing an active plasma membrane, was used for studies by flow cytometry with starved *Aeromonas salmonicida* (239), *Micrococcus luteus* (240), and *E.*

coli, and *S. typhimurium* cells (241). The oxonol stain bis-(1,3-dibutylbarbituric acid) trimethine oxonol (DiBAC$_4$) has a high binding affinity to depolarized membranes and was used to detect dead *E. coli, P. aeruginosa, Staphylococcus aureus,* and *Salmonella typhimurium* cells (241,242). Another viability stain which is used in flow cytometry to detect viability of *Saccharomyces cerevisiae* (243) and different gram-negative bacterial species (242) is carboxyfluorescein diacetate (CFDA). This compound is hydrolysed to the fluorescent stain carboxyfluorescein (CF) after passing the plasma membrane; only active membranes can prevent the diffusion of intracellularly accumulated CF, which results in an increased fluorescent signal in living cells. Besides viability measurements of bacterial populations, physiological parameters such as DNA content and biovolume of soil bacteria under different growth conditions have been determined in vitro (244). Biovolumes and frequencies of dividing cells (a parameter for the potentially active bacterial fraction in soil) were determined in bacteria indigenous in soil by using confocal laser scanning microscopy combined with image analysis in soil smears (245).

The detection of metabolic activity of specific bacteria in natural environments should be based on the detection of the target cells in connection with a measurement of activity, e.g., viable staining. Classically, the detection of specific strains in a natural environment like soil is based on IF cell counts (246), for which the availability of specific antibodies that recognize the target strain is an absolute prerequisite. New detection techniques have been developed that are based on the hybridization of cells with oligonucleotide probes targeted to the ribosomal RNA (rRNA). Probes labeled with fluorescent dyes have been developed to assay bacterial cells in situ, thereby circumventing cultivation steps (247–249). A combination of specific ribosomal (16S or 23S) RNA-directed oligonucleotide probes (250,251) and charged-coupled-device-enhanced microscopy was used to determine the RNA content of specific pseudomonads at different growth rates and in complex aquatic environments. Moreover, the in situ colonization of *Azospirillum brasilense* on wheat roots was studied using fluorescently labeled rRNA targeted oligonucleotides and scanning confocal laser microscopy (252). This combination of species-specific staining (rRNA- or IF-based detection), viable staining techniques, and sophisticated quantification methods is a promising development which is hoped to enable one to determine the metabolic activity of introduced populations in situ, e.g., at specific root or soil sites.

The investigation of genes in soil bacteria that are involved in adaptation to and survival in soil is still in its infancy. Bacterial strains with reporters inserted in genes showing expression upon exposure to stress conditions like carbon (178) or phosphorus starvation (175) have been studied in vitro and in soil. A *P. fluorescens* mutant strain interrupted in a carbon starvation–induced gene (mutant RA92) showed a survival rate similar to that of its parent strain under carbon limitation in vitro and in rhizosphere soil. However, it showed reduced survival in bulk soil (178). These data suggest that the physiological characteristics observed in vitro as to the expression of the reporter gene and survival under carbon starvation may appear pleiotropic in soil. Mutations in genes responsible for production of antifungal compounds in *P. fluorescens* also revealed complex pleiotropic effects in vitro (208) as well as under natural soil conditions (253). Both the *gac*A mutant strain (253) and strain RA92 (178) showed more complex behavior in soil than might be expected from in vitro observations. Therefore, the effects on bacterial behavior in natural ecosystems of, at first sight, straightforward molecular modifications are often still poorly understood.

7. CONCLUSIONS

Recent developments in molecular microbiology shed new light on the autecology of bacteria in soil and offer good opportunities to improve insights and strategies in the fundamentals of strain applications. In the forthcoming decade, molecular biology will certainly gain a permanent position in soil microbial ecology. However, the role of molecular biology in soil ecology is still too reductionistic to address all aspects of the complex soil environment. To date, molecular biology has only been able to supply fragmentary information on the network of conditions affecting bacterial behavior in soil. Hence, an integrative approach, including other relevant disciplines and approaches, is the key to successful application of molecular biology in ecology.

One field of application for molecular techniques is the ecology of phytopathological processes. Attempts to control plant-pathogenic fungi by antifungal antibiotic-producing bacteria have met with variable success in the past (1). Current knowledge of the sequence of genes involved in antibiotic production and loci responding to environmental stimuli will permit the antagonistic action to be tailored to the ecological needs of the application. However, doubts remain about the activity of antibiotics in soil. Factors such as competition for nutrients between antagonists and pathogens have been suggested to play a more dominant role than antibiotic production per se (203). Obviously, competition for nutrients can play an important role in the action of antagonistic agents toward plant pathogens. However, it is unlikely that the capacity to produce antibiotics has evolved only for the benefit of a minor competitive advantage to the host. Therefore, a strategy for antagonism by bacteria against plant pathogens may be the production of antibiotics at the first stages of plant colonization. At this stage, bacterial antagonists will have a selective advantage from antibiotic production, since the growth of competitors for nutrients (including pathogens) can be suppressed. Genes involved in antibiotic production may be repressed after colonization to prevent a reduction of growth rate or loss of viability caused by the metabolic expense of antibiotic production. The combination of effects of antibiotic production and other ecological factors is essential for pathogen growth inhibition, as shown with fluorescent pseudomonads antagonistic to *Phytophtora parasitica* (205). In this study, in vitro growth of the fungus was only inhibited by successful colonization of hyphae and pyoverdin production. Therefore, both molecular and ecological approaches are necessary to unravel the regulation of antibiotic-synthetic genes in situ. For the successful use of a root exudate–controlled antibiotic production gene cluster, the ecological behavior of the host strain plays an important role. Antibiotic production can only be successfully used for antagonistic action toward plant pathogens when the producing organism inhabits the same microhabitat at the same time as the pathogen. Studying the behavior of novel plant pathogen–controlling strains in vitro as well as under natural conditions will offer fundamental insight into the ecological properties of antagonist–pathogen interactions and may indicate strategies based on ecological and molecular principles for improved biological control.

ACKNOWLEDGMENTS

This work was supported by grants of the EU BRIDGE and BIOTECH programs. We thank J. T. Trevors for critical comments on the manuscript.

REFERENCES

1. D. M. Weller and R. J. Cook, *Phytopathology 73*: 463 (1983).
2. C. Waalwijk, A. Dullemans, and C. Maat, *FEMS Microbiol. Lett. 77*: 257 (1991).
3. R. S. Bora, M. G. Murty, R. Shenbagarathai, and V. Sekar, *Appl. Environ. Microbiol. 60*: 214 (1994).
4. A. H. Bosworth, M. K. Williams, K. A. Albrecht, R. Kwiatkowski, J. Beynon, T. R. Hankinson, C. W. Ronson, F. Cannon, T. J. Wacek, and E. W. Triplett, *Appl. Environ. Microbiol. 60*: 3815 (1994).
5. R. N. Pertsova, F. Kunc, and L. A. Golovleva, *Folia Microbiol. 29*: 242 (1984).
6. M. Briglia, P. J. M. Middeldorp, and M. S. Salkinoja-Salonen, *Soil Biol. Biochem. 26*: 377 (1994).
7. M. Dupler and R. Baker, *Phytopathology 74*: 195 (1984).
8. J. L. Morel, G. Bitton, G. R. Chaudry, and J. Awong, *Curr. Microbiol. 18*: 355 (1989).
9. J. D. van Elsas, A. F. Dijkstra, J. M. Govaert, and J. A. van Veen, *FEMS Microbiol. Ecol. 38*: 151 (1986).

10. G. Compeau, B. J. Al-Achi, E. Platsouka, and S. B. Levy, *Appl. Environ. Microbiol. 54*: 2432 (1988).
11. D. J. Drahos, G. F. Barry, B. C. Hemming, E. J. Brandt, H. D. Skipper, E. L. Kline, D. A. Kluepfel, and D. T. Gooden, *Pre-Release Testing Procedures: US Field Test of a lacZY-Engineered Soil Bacterium* (M. Sussman, C. H. Collins, F. A. Skinner, and D. E. Stewart-Tull, eds.), Academic Press, London, p. 181 (1988).
12. B. Brunel, J. C. Cleyet-Marel, P. Normand, and R. Bardin, *Appl. Environ. Microbiol. 54*: 2636 (1988).
13. J. Postma, J. D. van Elsas, J. M. Govaert, and J. A. van Veen, *FEMS Microbiol. Ecol. 53*: 251 (1988).
14. I. P. Thompson, K. A., Cook, G. Lethbridge, and R. G. Burns, *Soil. Biol. Biochem. 22*: 1029 (1990).
15. M. A. Devanas and G. Stotzky, *Curr. Microbiol. 13*: 279 (1986).
16. G. Recorbet, C. Steinberg, and G. Faurie, *FEMS Microbiol. Ecol. 101*: 251 (1992).
17. Y. Bashan, M. E. Puente, M. N. Rodriguez-Mendoza, G. Toledo, G. Holguin, R. Ferrera-Cerrato, and S. Pedrin, *Appl. Environ. Microbiol. 61*: 1938 (1995).
18. H. J. Miller, G. Henken, and J. A. van Veen, *Can. J. Microbiol. 35*: 656 (1989).
19. P. Lemanceau, T. Corberand, L. Gardan, X. Latour, G. Laguerre, J. Boeufgras, and C. Alabouvette, *Appl. Environ. Microbiol. 61*: 1004 (1995).
20. E. H. Nijhuis, M. J. Maat, I. Zeegers, C. Waalwijk, and J. A. van Veen, *Soil Biol. Biochem. 25*: 885 (1992).
21. J. D. van Elsas, *The Release of Genetically Modified Microorganisms* (D. E. S. Steward-Tull and M. Sussman, eds.), Plenum Press, New York, p. 1 (1992).
22. M. Givskov, L. Eberl, S. Møller, L. K. Poulsen, and S. Molin, *J. Bacteriol. 176*: 7 (1994).
23. L. S. van Overbeek, J. D. van Elsas, J. T. Trevors, and M-E. Starodub, *Microb. Ecol. 19*: 239 (1990).
24. O. Højberg and J. Sørensen, *Appl. Environ. Microbiol. 59*: 431 (1993).
25. J. K. Martin, *Soil Biol. Biochem. 9*: 303 (1977).
26. J. Cortez, *Biol. Fertil. Soils 7*: 142 (1989).
27. L. E. Lowe, *Developments in Soil Science. Vol. 8. Soil Organic Matter* (M. Schnitzer and S. U. Khan, eds.), Elseviers, Amsterdam, p. 65 (1978).
28. J. M. Lynch and J. M. Whipps, *Plant Soil 129*: 1 (1990).
29. E. A. Curl and B. Truelove, *The Rhizosphere* (D. F. R. Bommer, B. R. Sabey, G. W. Thomas, Y. Vaadia, and L. D. Van Vleck, eds.), Springer Verlag, Berlin, Heidelberg (1986).

30. R. Baziramakenga, R. R. Simard, and G. D. Leroux, *Soil Biol. Biochem.* 27: 349 (1995).
31. H. Keith and J. M. Oades, *Soil Biol. Biochem. 18*: 445 (1986).
32. P. J. Gregory and B. J. Atwell, *Plant Soil, 136*: 205 (1991).
33. S. M. Beck and C. M. Gilmour, *Soil Biol. Biochem. 15*: 33 (1983).
34. E. C. S. Chan, H. Katznelson, and J. W. Rouatt, *Can. J. Microbiol. 9*: 187 (1962).
35. A. D. Rovira, *Annu. Rev. Microbiol. 19*: 241 (1965).
36. C. Christiansen-Weniger, A. F. Groneman, and J. A. van Veen, *Plant Soil 139*: 167 (1992).
37. D. B. Knaebel, T. W. Federle D. C., McAvoy, and J. R. Vestal, *Appl. Environ. Microbiol. 60*: 4500 (1994).
38. J. Hassink, *Biol. Fertil. Soils 14*: 126 (1992).
39. J. A. van Veen and J. D. van Elsas, *Perspectives in Microbial Ecology, Slovena Society for Microbiology* (F. Megusar and M. Gantar, eds.), Ljubljana: p. 481 (1986).
40. C. E. Heijnen, C. H. Hok-a-Hin, and J. D. van Elsas, *Soil. Biol. Biochem.* 25: 239 (1993).
41. T. Hattori, *Rep. Inst. Agr. Tohoku Univ. 37*: 23 (1988).
42. J. D. van Elsas, C. E. Heijnen, and J. A. van Veen, *Biological Monitoring of Genetically Engineered Plants and Microbes* (D. R. MacKenzie and S. C. Henry, eds.), Proc. Kiawah Island Conference, Agricultural Research Institute, Bethesda, Maryland, p. 67 (1990).
43. J. Postma and J. A. van Veen, *Microb. Ecol. 19*: 149 (1990).
44. J. S. Poindexter, *Adv. Microb. Ecol. 5*: 63 (1981).
45. R. Y. Morita and C. L. Moyer, *Recent Advances in Microbial Ecology* (L. Hatton, Y. Ishida, Y. Maruyama, R. Y. Morita, and A. Udida, eds.), Proc. 5th Int. Symp. Microbial Ecol., Japan Scientific Societies Press, p. 75 (1989).
46. R. Y. Morita, *Can. J. Microbiol. 34*: 436 (1988).
47. J. Postgate, *New Scientist 1665*: 43 (1989).
48. R. R. Colwell, P. R. Brayton, D. J. Grimes, D. B. Roszak, S. A. Huq, and L. M. Palmer, *Bio/Technology 3*: 817 (1985).
49. D. B. Roszak and R. R. Colwell, *Microbiol. Rev. 51*: 365 (1987).
50. H. C. Bae, E. H. Cota-Robles, and L. E. Casida Jr, *Appl. Microbiol. 23*: 637 (1972).
51. J. C. Gottschal, *J. Appl. Bacteriol. Symp. Suppl. 73*: 39s (1992).
52. L. S. van Overbeek, L. Eberl, M. Givskov, S. Molin, and J. D. van Elsas, *Appl. Environ. Microbiol. 61*: 4208 (1995).
53. J. M. Stark and M. K. Firestone, *Appl. Environ. Microbiol. 61*: 218 (1995).
54. A. Meikle, S. Amin-Hanjani, L. A. Glover, K. Killham, and J. I. Prosser, *Soil Biol. Biochem. 27*: 881 (1995).

55. D. R. Polonenko, C. I. Mayfield, and E. B. Dumbroff, *Plant Soil 59*: 269 (1981).

56. J. C. Pedersen and C. S. Jacobson, *Appl. Environ. Microbiol. 59*: 1560 (1993).

57. K. Killham and M. K. Firestone, *Appl. Environ. Microbiol. 47*: 301 (1984a).

58. K. Killham and M. K. Firestone, *Appl. Environ. Microbiol. 48*: 239 (1984b).

59. E. B. Roberson and M. K. Firestone, *Appl. Environ. Microbiol. 58*: 1284 (1992).

60. R. C. Foster, *Biol. Fertil. Soils 6*: 189 (1988).

61. D. E. Smiles, *Biol. Fertil. Soils 6*: 204 (1988).

62. J. D. van Elsas and L. S. van Overbeek, *Starvation in Bacteria* (S. Kjelleberg, ed.), New York, Plenum Press, p. 55 (1993).

63. S. C. Winans, *Microbiol. Rev. 56*: 12 (1992).

64. B. J. J. Lugtenberg and R. A. de Maagd, *Cell to Cell Signals in Plant and Animals* (V. Neuhoff and J. Friend, eds.), Springer Verlag, Berlin, Heidelberg, p. 15 (1991).

65. A. J. Dharmatilake and W. D. Bauer, *Appl. Environ. Microbiol. 58*: 1153 (1992).

66. G. Caetano-Anollés, D. K. Christ-Estes, and W. D. Bauer, *J. Bacteriol. 170*: 3164 (1988).

67. W. D. Bauer and G. Caetano-Anollés, *The Rhizosphere and Plant Growth* (D. L. Keister and P. B. Cregan, eds.) Kluwer Academic, The Netherlands, p. 155 (1991).

68. A. M. Ashby, M. D. Watson, and C. H. Shaw, *FEMS Microbiol. Lett. 41*: 189 (1987).

69. G. Lopez-de-Victoria and C. R. Lovell, *Appl. Environ. Microbiol. 59*: 2951 (1993).

70. G. J. Loake, A. M. Ashby, and C. H. Shaw, *J. Gen. Microbiol. 134*: 1427 (1988).

71. Chet, I., Zilberstein, Y., and Henis, Y., 1973, *Physiol. Plant Pathol. 3*: 473–479 (1973).

72. T. Nikata, K. Sumida, J. Kato, and H. Ohtake, *Appl. Environ. Microbiol. 58*: 2250 (1992).

73. M. F. T. Begonia and R. J. Kremer, *FEMS Microbiol. Ecol. 15*: 227 (1994).

74. P. van Rhijn, M. VanStockem, J. vanderLeyden, and R. de Mot, *Appl. Environ. Microbiol. 56*: 990 (1990).

75. Y. Bashan and G. Holguin, *Appl. Environ. Microbiol. 60*: 2120 (1994).

76. Y. Bashan and G. Holguin, *Microb. Ecol. 29*: 269 (1995).

77. M. C. Hawes, L. Y. Smith, and A. J. Howarth, *Mol. Plant-Microbe Interact. 1*: 182 (1988).

78. L. A. de Weger, C. I. M. van der Vlugt, A. H. M. Wijfjes, P. A. H. M. Bakker, B. Schippers, and B. Lugtenberg, *J. Bacteriol., 169*: 2769 (1987).

79. C. B. You, M. Lin, X. J. Fang, and W. Song, *Soil Biol. Biochem. 27*: 463 (1995).

80. S. J. Vesper, *Appl. Environ. Microbiol. 53*: 1397 (1987).

81. D. C. M. Glandorf, I. van der Sluis, A. J. Anderson, P. A. H. M. Bakker, and B. Schippers, *Appl. Environ. Microbiol. 60*: 1726 (1994).

82. W. Achouak, T. Heulin, G. Villemin, and J. Balandreau, *FEMS Microbiol. Ecol. 13*: 287 (1994).

83. H. P. Spaink, D. M. Sheeley, A. A. N. van Brussel, J. Glushka, W. S. York, T. Tak, O. Geiger, E. P. Kennedy, V. N. Reinhold, and B. J. J. Lugtenberg, *Nature 354*: 125 (1991).

84. H. P. Spaink, A. Aarts, G. Stacey, G. V. Bloemberg, B. J. J. Lugtenberg, and E. P. Kennedy, *Mol. Plant-Microb. Interact. 5*: 72 (1992).

85. R. F. Fisher and S. R. Long, *Nature 357*: 655 (1992).

86. G. Stacey, J. Sanjuan, S. Luka, T. Dockendorff, and R. W. Carlson, *Soil Biol. Biochem. 27*: 473 (1995).

87. D. K. Willis, J. J. Rich, and E. M. Hrabak, *Mol. Plant-Microb. Interact. 4*: 132 (1991).

88. N. K. Peters and D. P. S. Verma, *Mol. Plant-Microbe Interact. 3*: 4 (1990).

89. D. A. Phillips, C. A. Maxwell, U. A. Hartwig, C. M. Joseph, and J. Wery, *The Rhizosphere and Plant Growth* (D. L. Kleister and P. B. Cregan, eds.), Kluwer Academic, The Netherlands, p. 149 (1991).

90. U. A. Hartwig, C. M. Joseph, and D. A. Phillips, *Plant Physiol. 95*: 797 (1991).

91. A. Goldmann, B. Message, P. Lecoeur, L. Delarue, M. Maille, and D. Tepfer, *Phytochemistry and Agriculture* (T. A. Van Beek and H. Breteler, eds.), Clarendon Press, Oxford, p. 76 (1993).

92. P. J. Murphy, W. Wexler, W. Grzemski, J. P. Rao, and D. Gordon, *Soil Biol. Biochem., 27*: 525 (1995).

93. S. Rossbach, G. Rasul, M. Schneider, B. Eardly, and F. J. De Bruijn, *Mol. Plant-Microb. Interact. 8*: 549 (1995).

94. M. Wilson and S. Lindow, *Appl. Environ. Microbiol. 60*: 4468 (1994).

95. I. Kraffczyck, G. Trolldenier, and H. Beringer, *Soil Biol. Biochem. 16*: 315 (1984).

96. V. Vančura, *Plant Soil 21*: 231 (1964).

97. L. S. van Overbeek and J. D. van Elsas, *Appl. Environ. Microbiol. 61*: 890 (1995).

98. L. S. van Overbeek, J. D. van Elsas, and J. A. van Veen, *Appl. Environ. Microbiol.* (in press).

99. J. I. Jiménez-Zurdo, P. Van Dillewijn, M. J. Soto, M. R. De Felipe, J. Olivares, and N. Toro, *Mol. Plant-Microb. Interact.* 8: 492 (1995).

100. M. F. DeFlaun, B. M. Marshall, E.-P Kulle, and S. Levy, *Appl. Environ. Microbiol.* 60: 2637 (1994).

101. S. Møller, C. S. Kristensen, L. K. Poulsen, J. M. Carstensen, and S. Molin, *Appl. Environ. Microbiol.* 61: 741 (1995).

102. B. Humphrey, S. Kjelleberg, and K. C. Marshall, *Appl. Environ. Microbiol.* 45: 43 (1983).

103. H. M. Dalton, L. K. Poulsen, P. Halasz, M. L. Angles, A. E. Goodman, and K. C. Marshall, *J. Bacteriol.* 176: 6900 (1994).

104. T. Ophir and D. L. Gutnick, *Appl. Environ. Microbiol.* 60: 740 (1994).

105. W. C. Fuqua, S. C. Winans, and E. P. Greenberg, *J. Bacteriol.* 176: 269 (1994).

106. P. Williams, N. J. Bainton, S. Swift, S. R. Chhabra, M. K. Winson, G. S. A. B. Stewart, G. P. C. Salmond, and B. W. Bycroft, *FEMS Microbiol. Lett.* 100: 161 (1992).

107. A. Kuo, N. V. Blough, and P. V. Dumlap, *J. Bacteriol.* 176: 7558 (1994).

108. A. Persson, G. Molin, and C. Weibull, *Appl. Environ. Microbiol.* 56: 686 (1990).

109. C. L. Moyer and R. Y. Morita, *Appl. Environ. Microbiol.* 55: 2710 (1989).

110. A. Matin, E. A. Auger, P. H. Blum, and J. E. Schultz, *Annu. Rev. Microbiol.* 43: 293 (1989).

111. C. A. Mason. and T. Egli, *Starvation in Bacteria* (S. Kjelleberg, ed.), Plenum Press, New York/London, p. 81 (1993).

112. R. G. Groat, J. E. Schultz, E. Zychlinsky, A. Bockman and A. Matin, *J. Bacteriol.* 168: 486 (1986).

113. R. G. Groat and A. Matin, *J. Industrial Microbiol.* 1: 69 (1986).

114. T. Nyström, R. M. Olsson, and S. Kjelleberg, *Appl. Environ. Microbiol.* 58: 55 (1992).

115. D. S. Morton and J. D. Oliver, *Appl. Environ. Microbiol.* 60: 3653 (1994).

116. M. Givskov, L. Eberl, and S. Molin, *J. Bacteriol.* 176: 4816 (1994).

117. A. Matin, *Mol. Microbiol.* 5: 3 (1991).

118. D. E. Jenkins, J. E. Schultz, and A. Matin, *J. Bacteriol.* 170: 3910 (1988).

119. D. E. Jenkins, S. A. Chaisson, and A. Matin, *J. Bacteriol.* 172: 2779 (1990).

120. Å. Jouper-Jaan, A. E. Goodman, and S. Kjelleberg, *FEMS Microbiol. Ecol.* 101: 229 (1992).

121. P. M. Munro, R. L. Clément, G. N. Flatau, and M. J. Gauthier, *Microb. Ecol.* 27: 57 (1994).

122. A. Matin, *J. Appl. Bacteriol. Symp. Suppl.* 73: 49s (1992).

123. R. Lange and R. Hengge-Aronis, *Mol. Microbiol.* 5: 49 (1991).

124. M. P. McCann, J. P. Kidwell, and A. Matin, *J. Bacteriol. 173*: 4188 (1991).
125. P. M. Munro, G. N. Flatau, R. L. Clément, and M. J. Gauthier, *Appl. Environ. Microbiol. 61*: 1853 (1995).
126. R. Lange and R. Hengge-Aronis, *J. Bacteriol. 173*: 4474 (1991).
127. M. Vicente, S. R. Kushner, T. Garrido, and M. Aldea, *Mol. Microbiol. 5*: 2085 (1991).
128. M. Barth, C. Marschall, A. Muffler, D. Fischer, and R. Hengge-Aronis, *J. Bacteriol. 177*: 3455 (1995).
129. Y. Kim, L. S. Watrud, and A. Matin, *J. Bacteriol. 177*: 1850 (1995).
130. D. Rockabrand, T. Arthur, G. Korinek, K. Livers, and P. Blum, *J. Bacteriol. 177*: 3695 (1995).
131. I. R. Booth and C. F. Higgins, *FEMS Microbiol. Reviews 75*: 239 (1990).
132. M. Jebbar, R. Talibart, K. Gloux, T. Bernard, and C. Blanco, *J. Bacteriol. 174*: 5027 (1992).
133. C. F. Higgins, C. J. Dorman, D. A. Stirling, L. Wadell, I. R. Booth, G. May, and E. Bremer, *Cell 52*: 569 (1988).
134. C. S. J. Hulton, A. Seirafi, J. C. D. Hinton, J. M. Sidebotham, L. Wadell, G. D. Pavitt, T. Owen-Hughes, A. Spassky, H. Buc, and C. F. Higgins, *Cell 63*: 631 (1990).
135. P. M. Munro, M. J. Gauthier, V. A. Breittmayer, and J. Bongiovanni, *Appl. Environ. Microbiol. 55*: 2017 (1989).
136. M. J. Gauthier, G. N. Flatau, D. Le Rudulier, R. L. Clément, and M. Combarro Combarro, *Appl. Environ. Microbiol. 57*: 272 (1991).
137. A. Hartmann, S. R. Prabhu, and E. A. Galinski, *Plant Soil 137*: 105 (1991).
138. A. Berry, J. D. DeVault, and A. M. Chakrabarty, *J. Bacteriol. 171*: 2312 (1989).
139. J. L. Slonczewski, *ASM News 58*: 140 (1992).
140. J. W. Foster and B. Bearson, *J. Bacteriol. 176*: 2596 (1994).
141. M. J. Gauthier and R. L. Clément, *FEMS Microbiol. Ecol. 21*: 275 (1994).
142. I. S. Lee, J. L. Slonczewski, and J. W. Foster, *J. Bacteriol. 176*: 1422 (1994).
143. K. W. Arnold and C. W. Kaspar, *Appl. Environ. Microbiol. 61*: 2037 (1995).
144. J. L. Slonczewski, T. N. Gonzalez, F. M. Bartholomew, and N. J. Holt, *J. Bacteriol. 169*: 3001 (1987).
145. S. C. Winans, *Mol. Microbiol. 5*: 2345 (1991).
146. Z.-M. Wei, B. J. Sneath, and S. V. Beer, *J. Bacteriol. 174*: 1875 (1992).
147. R. J. Broeze, C. J. Solomon, and D. H. Pope, *J. Bacteriol. 134*: 861 (1978).
148. J. Cloutier, D. Prévost, P. Nadeau, and H. Antoun, *Appl. Environ. Microbiol. 58*: 2846 (1992).
149. P. G. Jones, R. A. VanBogelen, and F. C. Neidhardt, *J. Bacteriol. 169*: 2092 (1987).

150. P. G. Jones, R. Krah, S. R. Tafuri, and A. P. Wolffe, *J. Bacteriol. 174*: 5798 (1992).

151. G. Willimsky, H. Bang, G. Fischer, and M. A. Marahiel, *J. Bacteriol. 174*: 6326 (1992).

152. A. Blom, W. Harder, and A. Matin, *Appl. Environ. Microbiol. 58*: 331 (1992).

153. C. W. Ronson, B. T. Nixon, and F. M. Ausubel, *Cell 49*: 579 (1987).

154. P. Liang and A. B. Pardee, *Science 257*: 967 (1992).

155. D. Bauer, H. Müller, J. Reich, H. Riedel, V. Ahrenkiel, P. Warthoe, and M. Strauss, *Nucl. Acids Res. 21*: 4272 (1993).

156. A. A. Bhagwat and D. L. Keister, *Appl. Environ. Microbiol. 58*: 1490 (1992).

157. V. De Lorenzo, I. Cases, M. Herrero, and K. Timmis, *J. Bacteriol. 175*: 6902 (1993).

158. E. A. Groisman and M. Casadaban, *J. Bacteriol. 168*: 357 (1986).

159. H. Israelsen, S. M. Madsen, A. Vrang, E. B. Hansen, and E. Johansen, *Appl. Environ. Microbiol. 61*: 2540 (1995).

160. E. Joseph-Liauzun, R. Fellay, and M. Chandler, *Gene 85*: 83 (1989).

161. B. Kessler, V. De Lorenzo, and K. N. Timmis, *Mol. Gen. Genet. 233*: 293 (1992).

162. W. Kokotek and W. Lotz, *Gene 84*: 467 (1989).

163. M. P. Krebs and W. S. Reznikoff, *Gene 63*: 277 (1988).

164. L. Kroos and D. Kaiser, *Proc. Natl. Acad. Sci. USA 81*: 5816 (1984).

165. C. Manoil, *J. Bacteriol. 172*: 1035 (1990).

166. D. J. O'Sullivan and F. O'Gara, *Appl. Environ. Microbiol. 54*: 2877 (1988).

167. P. Ratet, J. Schell, and F. De Bruijn, *Gene 63*: 41 (1988).

168. J. J. Shaw, L. G. Settles, and C. I. Kado, *Mol. Plant-Microbe Interact. 1*: 39 (1987).

169. H. Shen, S. E. Gold, S. J. Tamaki, and N. T. Keen, *Gene 122*: 27 (1992).

170. S. E. Stachel, G. An, C. Flores, and E. W. Nester, *EMBO J. 4*: 891 (1985).

171. J. C. Way, M. A. Davis, D. Morisato, D. E. Roberts, and N. Kleckner, *Gene 32*: 369 (1984).

172. V. De Lorenzo, M. Herrero, U. Jacubzik, and K. N. Timmis, *J. Bacteriol. 172*: 6568 (1990).

173. R. Simon, J. Quandt, and W. Klipp, *Gene 80*: 161 (1989).

174. A. Cebolla, C. Guzman, and V. De Lorenzo, *Appl. Environ. Microbiol. 62*: 214 (1995).

175. L. A. de Weger, L. C. Dekkers, A. van der Bij, and B. J. J. Lugtenberg, *Mol. Plant-Microb. Interact. 7*: 32 (1993).

176. P. O. Lim, D. Ragatz, M. Renner, and F. J. De Bruijn, *Trends in Microbial Ecology* (R. Guerrero and C. Pedrós-Alió, eds.), Proc. 6th Int. Symp. Microb. Ecol., Spanish Society for Microbiology, p. 97 (1993).

177. L. Kragelund, B. Christoffersen, O. Nybroe, and F. J. De Bruijn, *FEMS Microbiol. Ecol., 17*: 95 (1995).
178. L. S. van Overbeek, unpublished results.
179. S. T. Lam, D. M. Ellis, and J. M. Ligon, *The Rhizosphere and Plant Growth* (D. L. Kleister and P. B. Cregan, eds.), Kluwer Academic, The Netherlands, p. 43 (1991).
180. J. H. Miller, *Experiments in Molecular Genetics.* Cold Spring Harbor Laboratory, Cold Spring Harbor, New York (1972).
181. M. Arlat, C. L. Gough, C. Zischek, P. A. Barberis, A. Trigalet, and C. A. Boucher, *Mol. Plant-Microbe Interact. 5*: 187 (1992).
182. M. Arlat, P. Barberis, A. Trigalet, and C. Boucher, *Plant Pathogenic Bacteria* (Z. Klement, ed.), Proc. 7th Int. Conf. Plant. Path., Budapest, Hungary, p. 419 (1989).
183. L. S. Melchers, A. J. G. Regensburg-Tuïnk, R. A. Schilperoort, and P. J. J. Hooykaas, *Mol. Microbiol. 3*: 969 (1989).
184. H. Shen and N. T. Keen, *J. Bacteriol. 175*: 5916 (1993).
185. N. Hugouvieux-Cotte-Pattat, H. Dominiguez, and J. Robert-Baudouy, *J. Bacteriol. 174*: 7807 (1992).
186. K. K. Le Strange, G. L. Bender, M. A. Djordjevic, B. G. Rolfe, and J. W. Redmond, *Mol. Plant-Microbe Interact. 4*: 214 (1990).
187. M. Hungria, A. W. B. Johnston, and D. A. Phillips, *Mol. Plant-Microbe Interact. 5*: 199 (1992).
188. S. Katupitiya, J. Millet, M. Vesk, L. Viccars, A. Zeman, Z. Lidong, C. Elmerich, and I. R. Kennedy, *Appl. Environ. Microbiol. 61*: 1987 (1995).
189. F. Arsène, S. Katupitiya, I. R. Kennedy, and C. Elmerich, *Mol. Plant-Microb. Interact. 7*: 748 (1994).
190. A. van de Broek, J. Michiels, A. van Gool, and J. van der Leyden, *Mol. Plant-Microbe Interact. 6*: 592 (1993).
191. C. R. Buell and A. J. Anderson, *Mol. Plant-Microbe Interact. 6*: 331 (1993).
192. C. Drainas, G. Vartholomatos, and N. J. Panopoulos, *Appl. Environ. Microbiol. 61*: 273 (1995).
193. D. G. Georgakopoulos, M. Hendson, N. J. Panopoulos, and M. N. Schroth, *Appl. Environ. Microbiol. 60*: 4573 (1994).
194. G. M. Brazil, L. Kenefick, K. M. Callanan, A. Haro, V. De Lorenzo, D. N. Dowling, and F. O'Gara, *Appl. Environ. Microbiol. 61*: 1946 (1995).
195. T. K. van Dyk, W. R. Majarian, K. B. Konstantinov, R. M. Young, P. S. Dhurjati, and R. A. LaRossa, *Appl. Environ. Microbiol. 60*: 1414 (1994).
196. M. V. Brennerova and D. E. Crowley, *FEMS Microbiol. Ecol. 14*: 319 (1994).
197. L. A. de Weger, P. Dunbar, W. F. Mahafee, B. J. J. Lugtenberg, and G. S. Sayler, *Appl. Environ. Microbiol. 57*: 3641 (1991).

198. T. Hurek, B. Reinhold-Hurek, M. Van Montagu, and E. Kellenberger, *J. Bacteriol. 176*: 1913 (1994).

199. D. R. Polonenko and C. I. Mayfield, *Plant Soil 51*: 405 (1979).

200. J. W. L. van Vuurde and B. Schippers, *Soil Biol. Biochem. 12*: 559 (1980).

201. M. Chalfie, Y. Tu, G. Euskirchen, W. W. Ward, and D. C. Prasher, *Science 263*: 802 (1994).

202. A. F. Dijkstra, J. M. Govaert, G. H. N. Scholten, and J. D. van Elsas, *Soil Biol. Biochem. 19*: 351 (1987).

203. J. W. Deacon, *Biocontrol Sci. Technol. 1*: 5 (1991).

204. J. Leong, *Annu. Rev. Phytopathol. 24*: 187 (1986).

205. C. Yang, J. A. Menge, and D. A. Cooksey, *Appl. Environ. Microbiol. 60*: 473 (1994).

206. Y. Homma, Z. Sato, F. Hirayama, K. Konno, H. Shirahama, and T. Suzui, *Soil Biol. Biochem. 21*: 723 (1989).

207. K. D. Burkhead, D. A. Schisler, and P. J. Slininger, *Appl. Environ. Microbiol. 60*: 2031 (1994).

208. T. D. Gaffney, S. T. Lam, J. Ligon, K. Gates, A. Frazelle, J. Di Maio, S. Hill, S. Goodwin, N. Torkewitz, A. M. Allshouse, H.-J. Kempf, and J. O. Becker, *Mol. Plant-Microb. Interact. 7*: 455 (1994).

209. D. J. O'Sullivan and F. O'Gara, *Microbiol. Rev. 56*: 662 (1992).

210. M. Mazzola, R. J. Cook, L. S. Thomashow, D. M. Weller, and L. S. Pierson, *Appl. Environ. Microbiol. 58*: 2616 (1992).

211. M. Mazzola, D. K. Fujimoto, L. S. Thomashow, and R. J. Cook, *Appl. Environ. Microbiol. 61*: 2554 (1995).

212. D. S. Hill, J. I. Stein, N. R. Torkewitz, A. M. Morse, C. R. Howell, J. P. Pachlatko, J. O. Becker, and J. M. Ligon, *Appl. Environ. Microbiol. 60*: 78 (1994).

213. J. Kraus and J. E. Loper, *Appl. Environ. Microbiol. 61*: 849 (1995).

214. F. L. Carruthers, A. J. Conner, and H. K. Mahanty, *Appl. Environ. Microbiol. 60*: 71 (1994).

215. L. A. Silo-Suh, B. J. Lethbridge, S. J. Raffel, H. He, J. Clardy, and J. Handelsman, *Appl. Environ. Microbiol. 60*: 2023 (1994).

216. J. D. van Elsas, L. S. van Overbeek, A. M. Feldman, A. M. Dullemans, and O. de Leeuw, *FEMS Microbiol. Ecol. 85*: 53 (1991).

217. A. Contreras, S. Molin, and J. Ramos, *Appl. Environ. Microbiol. 57*: 1504 (1991).

218. M. C. Ronchel, C. Ramos, L. B. Jensen, S. Molin, and J. L. Ramos, *Appl. Environ. Microbiol. 61*: 2990 (1995).

219. L. K. Poulsen, N. W. Larsen, S. Molin, and P. Andersson, *Mol. Microbiol. 3*: 1463 (1989).

220. I. Ahrenholtz, M. G. Lorenz, and W. Wackernagel, *Appl. Environ. Microbiol. 60*: 3746 (1994).

221. K. Gerdes, F. W. Bech, S. Troels, A. Jørgensen, A. Løbner-Olesen, P. B. Rasmussen, T. Atlung, L. Boe, O. Karlstrom, S. Molin, and K. von Mayenburg, *EMBO J. 5*: 2023 (1986).
222. L. K. Poulsen, A. Refn, S. Molin, and P. Andersson, *Mol. Microbiol. 5*: 1627 (1991).
223. S. Knudsen, P. Saadbye, L. H. Hansen, A. Collier, B. L. Jacobsen, J. Schlundt, and O. H. Karlström, *Appl. Environ. Microbiol. 61*: 985 (1995).
224. P. Klemm, L. B. Jensen, and S. Molin, *Appl. Environ. Microbiol. 61*: 481 (1995).
225. P. E. Turpin, K. A. Maycroft, C. L. Rowlands, and E. M. H. Wellington, *J. Appl. Bacteriol. 74*: 421 (1993).
226. K. Kogure, U. Simidu, and N. Taga, *Can. J. Microbiol. 2*: 415 (1979).
227. I. Rahman, M. Shahamat, P. A. Kirchman, E. Russek-Cohen, and R. R. Colwell, *Appl. Environ. Microbiol. 60*: 3573 (1994).
228. T. Shiba, R. T. Hill, W. L. Straube, and R. R. Colwell, *Appl. Environ. Microbiol. 61*: 2583 (1995).
229. S. Duncan, L. A. Glover, K. Killham, and J. I. Prosser, *Appl. Environ. Microbiol. 58*: 1308 (1994).
230. J. D. Oliver and R. Bockian, *Appl. Environ. Microbiol. 61*: 2620 (1995).
231. J. D. Oliver, F. Hite, D. McDougald, N. L. Andon, and L. M. Simpson, *Appl. Environ. Microbiol. 61*: 2624 (1995).
232. M. Wilson and S. E. Lindow, *Appl. Environ. Microbiol. 58*: 3908 (1992).
233. R. Zimmermann, R. Iturriaga, and J. Becker-Birck, *Appl. Environ. Microbiol. 36*: 926 (1978).
234. G. G. Rodriguez, D. Phipps, K. Ishiguro, and H. F. Ridgway, *Appl. Environ. Microbiol. 58*: 1801 (1992).
235. O. Nybroe, *FEMS Microbiol. Ecol. 17*: 77 (1995).
236. B. H. Pyle, S. C. Broadaway, and G. A. McFeters, *Appl. Environ. Microbiol. 61*: 2614 (1995).
237. A. Winding, S. J. Binnerup, and J. Sørensen, *Appl. Environ. Microbiol. 60*: 2869 (1994).
238. C. E. Heijnen, S. Page, and J. D. van Elsas, *FEMS Microb. Ecol. 18*: 129 (1995).
239. J. A. W. Morgan, G. Rhodes, and R. W. Pickup, *Appl. Environ. Microbiol. 59*: 874 (1993).
240. T. V. Votyakova, A. S. Kaprelyants, and D. B. Kell, *Appl. Environ. Microbiol. 60*: 3284 (1994).
241. R. López-Amorós, J. Comas, and J. Vives-Rego, *Appl. Environ. Microbiol. 61*: 2521 (1995).
242. R. I. Jepras, J. Carter, S. C. Pearson, F. E. Paul, and M. J. Wilkinson, *Appl. Environ. Microbiol. 61*: 2696 (1995).

243. P. Breeuwer, J. Drocourt, F. M. Rombouts, and T. Abee, *Appl. Environ. Microbiol.* *60*: 1467 (1994).
244. H. Christensen, L. R. Bakken, and R. A. Olsen, *FEMS Microbiol. Ecol.* *102*: 129 (1993).
245. J. Bloem, M. Veninga, and J. Shepherd, *Appl. Environ. Microbiol.* *61*: 926 (1995).
246. B. B. Bohlool and E. L. Schmidt, *Adv. Microb. Ecol.* *4*: 203 (1980).
247. R. I. Amann, L. Krumholz, and D. A. Stahl, *J. Bacteriol.* *172*: 762 (1990).
248. R. I. Amann, N. Springer, W. Ludwig, H.-D. Görtz, and K.-H. Schleifer, *Nature 351*: 161 (1991).
249. K. Trebesius, R. Amann, W. Ludwig, K. Mühlegger, and K. Schleifer, *Appl. Environ. Microbiol.* *60*: 3228 (1994).
250. T. D. Leser, M. Boye, and N. Hendriksen, *Appl. Environ. Microbiol.* *61*: 1201 (1995).
251. M. Boye, T. Ahl, and S. Molin, *Appl. Environ. Microbiol.* *60*: 1384 (1995).
252. B. Assmus, P. Hutzler, G. Kirchhof, R. Amann, J. R. Lawrence, and A. Hartmann, *Appl. Environ. Microbiol.* *61*: 1013 (1995).
253. A. Natsch, C. Keel, H. A. Pfirter, D. Haas, and G. Défago, *Appl. Environ. Microbiol.* *60*: 2553 (1994).
254. A. Heitzer, K. Malachowsky, J. E. Thonnard, P. R. Bienkowski, D. C. White, and G. S. Sayler, *Appl. Environ. Microbiol.* *60*: 1487 (1994).
255. J. E. Loper and S. E. Lindow, *Appl. Environ. Microbiol.* *60*: 1934 (1994).

15

Microbial Ecology, Inoculant Distribution, and Gene Flux Within Populations of Bacteria Colonizing the Surface of Plants: Case Study of a GMM Field Release in the United Kingdom

MARK J. BAILEY, ANDREW K. LILLEY, RICHARD J. ELLIS, PENNY A. BRAMWELL, and IAN P. THOMPSON   Institute of Virology and Environmental Microbiology, Oxford, United Kingdom

In the following chapter we describe the rationale used to design investigations for the release and monitoring of a genetically modified microorganism (GMM) in the environment. We have used a GMM to investigate specifically the microbial ecological properties of plant surfaces in an attempt to understand community succession, adaptation of individuals to changing environments through genetic reassortment and phenotypic variation, and the nature and extent of active gene transfer in the natural environment. To cover all the associated disciplines and expertise necessary to undertake a field release of a GMM we have presented a case history describing the microbiological characteristics of the target habitat, the strategy for the modification of the candidate bacteria, the prerelease evaluation and development of precise monitoring assays for the GMM in the field, and an assessment of gene transfer potential in the natural environment. In order to accommodate this chapter within the available space we recommend that it be read in conjunction with current reviews of the environmental release of GMMs to soils and crop plants (1–4) and the methods available for the monitoring of genetically modified bacteria in the natural environment (5).

1. INTRODUCTION

An understanding of the underlying principles of microbial ecology that influence how communities or individuals survive, persist, and colonize particular habitats is fundamental to the selection, design, and introduction of inocula with potential for environmental management. The plant surface has been the target of a number of releases of genetically modified microorganisms (GMMs) which have been introduced as mediators of plant protection or plant growth (4). However, few biocontrol programs have been effective, and the unpredictable outcome of these programs, when compared to successes with chemical-based control regimens, remains a major restriction to the commercial and widespread use of bacterial inocula in agriculture and related industries (see Chapter 17). Although the reasons for failure are undoubtedly multifactorial, the choice of candidate strains and a more intimate understanding of their interactions with the environment and other microbes are essential.

2. MICROBIOLOGY OF SOIL AND PLANT SURFACES

A number of investigations have revealed that plant surfaces (phytosphere) and soils influenced by plant root growth (rhizosphere) are consistently colonized by fluorescent pseudomonads (6,7). Because of the intimate association between pseudomonads and the plant, many of these bacteria affect plant health and yield either as phytopathogens or as promoters of plant growth by the exclusion of pathogens through aggressive colonization or the production of antimicrobial compounds. Other pseudomonads, which represent a large proportion of the community, have no apparent effect on the plant, where they exist as saprophytes and possible reservoirs of bacteria pathogenic to susceptible plant species (8). The spectrum of beneficial environmental traits, the ease with which they can be genetically manipulated, and their ability to persist and actively colonize the phytosphere of a variety of plants have made fluorescent pseudomonads prime candidates for the carriage of foreign genes suitable for biocontrol and other functions.

A common concern associated with the potential use and widespread release of GMMs relates to the impact they may have on the target and wider environment by perturbing established populations or by exchanging genetic material. Therefore, one of the basic considerations in our approach for the development of inocula, with which to study the autecological properties of introduced bacterial strains and to assess the potential for gene transfer, was that the candidate bacteria should be selected from indigenous populations adapted to the target habitat. However, studies of the ecological characteristics of microbial communities are limited by the difficulties associated with the extraction, characterization, and representative sampling of isolates. Traditional methods based on morphological characteristics and the ability to utilize specific substrates provide the foundations of bacterial systematics, but in the last decade more precise molecular methodologies which describe component biomolecules have been developed. These include comparative analyses of genomes, the structure and sequence of specific operons, e.g., ribosomal ribonucleic acid (RNA), or the random amplification of nucleic acids to produce cellular or community fingerprints. With the exception of ribosomal RNA (rRNA) sequence information, the available data bases are limited to the identification of isolates. However, one commercial system provides rapid identification of bacteria by the comparison of profiles generated after gas/liquid chromatography of extracted cellular fatty acid methyl ester (FAME profile) using the Microbial Identification System (MIS-MIDI, Newark). We have made extensive use of this apparatus to describe the microbiological properties of sugar beet plants grown at the same field site proposed for the release of a GMM (6,9–11). These investigations, which qualitatively and quantitatively compared bacterial populations isolated from sugar beet (Fig. 1a), demonstrated the abundance of a group of related fluorescent pseudomonads (Fig. 1b) that were isolated on every sampling occasion and proliferated over consecutive seasons. These analyses indicated the niche adaptation of specific isolates (9) with a concomitant rapid turnover of predominant genotypes (clones) as the habitat changed with plant maturation (11). As identical bacterial genotypes have been isolated over consecutive seasons, colonizers must have developed survival strategies appropriate to the habitat. If succession can be predicted, then epiphytic bacterial colonizers on leaf surfaces must consist of specialists. We speculate that pseudomonads are represented by populations of individual genotypes each adapted to local conditions, where intense selection from plant development and changing environmental conditions results in the succession of populations within the community by the proliferation of better adapted genomic groups. As a result of these studies we were able to select a candidate bacterium to allow detailed investigations into the ecological properties of microorganisms in the plant environment.

3. SELECTION AND GENETIC MODIFICATION OF CANDIDATE BACTERIA SUITABLE FOR FIELD RELEASE

3.1 The Candidate Microorganism

A plasmid-free, nonpathogenic ribosomal RNA group 1 fluorescent pseudomonad, *Pseudomonas fluorescens* SBW25, was selected for marking from the natural microflora of the sugar beet plant surface (Fig. 1) to generate a GMM which survives and naturally colonizes plants after seed inoculation, can be easily and accurately monitored in the natural environment, and permits direct assessment of genetic stability and gene transfer.

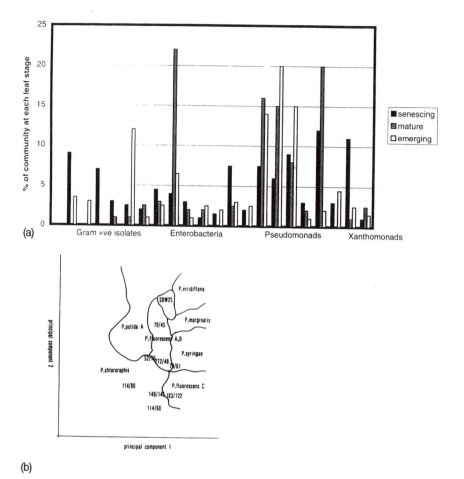

(a)

(b)

Figure 1. Microbiology of sugar beet: (a) Community profile of bacteria isolated from the leaves of sugar beet grown at the field site proposed for the release of the genetically modified microorganism (GMM). Sampled leaves were homogenized, diluted, and spread-plated onto tryptic soy broth agar (TSBA). Isolated colonies were identified by FAME-MIS analysis. The relative proportion of each bacterial group within the community is presented (6). Enterobacteriaceae, Pseudomonads, Xanthomonads. (b) Taxonomic relatedness of phytosphere fluorescent pseudomonads and SBW25. The pseudomonads collected throughout the growing seasons 1990, 1991, and 1992 were clustered by two-dimensional analysis of the principal fatty acid components of each isolate using the Microbial Identification System software. In total 502 isolates were compared with SBW25. The figure delineates boundaries by species named according to the MIS data base (version 3.8). *P. chlororaphis* ($n = 165$), *P. fluorescens* A, B ($n = 179$), *P. putida* A ($n = 111$), *P. syringae* ($n = 20$), *P. viridiflava* ($n = 12$), *P. marginalis* ($n = 8$), *P. fluorescens* C ($n = 7$), SBW25 ($n = 24$). *P. fluorescens* SBW25, originally isolated from the phyllosphere of mature sugar beet plants, was confirmed as a typical representative of the indigenous pseudomonad community.

3.2 Taxonomic Position of SBW25

The original isolate, *Pseudomonas fluorescens* SBW25 (formally *P. aureofaciens,* [12]), was identified by the Microbial Identification System (Aerobe-TSBA library, version 3.8). SBW25 was not able to grow at 37°C and was non-pathogenic to plants, such as sugar beet, pea, tomato, bean, and tobacco.

3.3 Genetic Modification of SBW25

Two marker gene cassettes, *kan*^{r}-*xyl*E (kanamycin resistance and catechol 2,3 dioxygenase activity) and *lac*ZY (β-galactosidase and lactose permease), which allow modified pseudomonads to utilize lactose uniquely as a sole carbon source (13), were inserted, each into separate sites on the bacterial chromosome (14). The marker genes were chosen, as they had been shown to be well expressed in pseudomonads and because they provided phenotypes that could be detected by simple substrate utilization or colorimetric assays (Fig. 2). In addition, the *lac*ZY genes had already been used in the United States for the monitoring of a number of GM pseudomonads released into the environment (1,2). The marked sites were approximately 1 Mbp distant on the physically mapped 6.6 Mbp bacterial chromosome (15). Sites were marked by site-directed homologous recombination after the transformation of recipient SBW25 bacteria with modified deoxyribonucleic acid (DNA). This approach was taken so that retention of vector sequences or other motifs often used in genetic modification was avoided, e.g., the use of transposons or disarmed transposons, where the flanking IS repeats are retained. The chromosomal location of the markers was preferred to the marking of a plasmid to minimize any possibility of environmental transfer of a marked plasmid to other hosts, any growth disadvantage to inocula carrying a plasmid, or the loss of the plasmid resulting in an inability to detect the introduced bacteria.

3.4 Detection

The GMM (*P. fluorescens* SBW25EeZY6KX) was marked in the chromosome with unique constitutively expressed genes that allowed the sensitive detection of a single GMM cell by using (1) selective agars from a background of at least 1 × 10^{12} colony-forming units of bacteria, (2) a most probable number method using a gram of soil (16), or (3) polymerase chain reaction (PCR) amplification using specific oligonucleotides directed to the marker genes or sequences that flank the site of insertion (17).

3.5 Marker Gene Transfer

Separate chromosomal sites were selected to provide a method for the direct assessment of marker gene transfer. All GMMs were isolated as blue colonies on

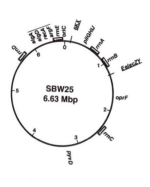

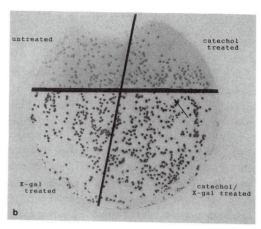

Figure 2. Physical map and modified phenotype of GMM *P. fluorescens* SBW25EeZY6KX: (a) Physical map of the SBW25 chromosome (15) showing the location of the introduced marker genes, *6KX* and *EelacZY* (14). The position and direction of transcription on the circular chromosome of the five ribosomal RNA operons (rrn) and other genes have been included. The marker gene cassettes were under the control of constitutive promoters and provided additional methods for the selection and detection of the modified bacteria. Separate sites were chosen to allow the insertion of multiple markers and allow methods for the sensitive detection of gene transfer or loss of genotype to be developed. KX (kan^r, *xyl*E), kanamycin resistance and 2,3-catechol dioxygenase activity, *lac*ZY, lactose utilization and β-galactosidase activity. (b) Phenotypic detection of the GMM in environmental samples by simple isolation on indicator bacteriological agar. The GMM was introduced as a seed dressing to sugar beet. Forty-nine days after planting leaves were collected and washed with sterile saline solution and the bacterial suspension spread onto petri dishes containing TSB agar. After 24 h incubation at 28°C colonies were transferred to filter paper and each quarter treated with chromogenic substrate diagnostic for the GMM. After treatment with 1% catechol colonies turn *yellow* (*xyl*E gene product 2,3-catechol dioxygenase), colonies treated with 0.025% X-gal turn *blue* (*lac*Z gene product, β-galactosidase), or colonies treated with both substrates turn *green*. Note arrow pointing at indigenous blue colonies identified as Enterobacteriaceae.

Pseudomonas selection agar supplemented with X-gal and kanamycin. There-fore, any white colonies would either represent indigenous kanamycin-resistant pseudomonads, GMM mutants that had lost the ability to express *lacZ*, or an in-digenous population of pseudomonads to which the *kan*ʳ-*xyl*E cassette had been transferred. The latter phenotype was confirmed by the application of catechol, which in the presence of the *xyl*E gene product is reduced to a yellow semialde-hyde. The design of the system to assess chromosomal gene transfer was based on two assumptions: (1) that the kanamycin-catechol 2,3-dioxygenase pheno-type was unique in the study habitat and (2) that transfer, if it were to occur, would be between closely related organisms. To bias such an outcome, chromo-somal sites were selected for marking as their disruption did not affect the be-havior or growth of the marked bacteria and on the basis of their relative genetic conservation in the indigenous pseudomonad community. By cross-hybridiza-tion analysis of isolated pseudomonads (Fig. 1) the site to be marked with the *lac*ZY cassette (site Ee) was detected in only 6% of the population, whereas the other site, to be marked with *kan*ʳ-*xyl*E cassette (site 6), was ubiquitous (100% of isolates). It was considered unlikely that both regions (ca. 15% apart, Fig. 2) would be transferred together, but if chromosomal mobilization did occur then the most likely outcome would be the stable transfer of the marked fragment which had the highest probability of recombining with another pseudomonad re-cipient, the fragment marked with the *kan*ʳ-*xyl*E cassette. Hence, assays for marker gene transfer concentrated on the detection of this novel phenotype in *lac*ZY⁻ pseudomonads.

4. PRERELEASE SAFETY EVALUATION AND GREENHOUSE-BASED MICROCOSM STUDIES OF THE FATE OF THE GMM IN THE PHYTOSPHERE

4.1 Regulatory Considerations

A number of interrelated issues concern the release of GMMs have been raised repeatably; they relate to the risks or potential environmental harm that may arise as a result of the persistence, replication, and dispersal of the GMM; the ex-change of genetic information; and the displacement of indigenous communities after field release. Within the United Kingdom, consent for the small-scale re-lease of GMMs must be obtained from the secretary of state for the environment. The granting of consent is based on an assessment of the potential for hazard to human health or the environment, which may follow the release of a GMM. Such factors as toxicity, pathogenicity, and the parental organism's known eco-logical properties are considered, alongside any change in the behavior of the or-ganism resulting from the genetic modification. In order to be able to make predictions of the outcome of the field trials, to develop monitoring assays, determine survival characteristics, and assess genetic stability, microcosm experiments in normal field soils were undertaken. These investigations were done in collaboration with Professor J. M. Lynch and Dr. J. M. Whipps at Horticultural Research International, United Kingdom, where the behavior and impact of SBW25 in the phytosphere of wheat were also studied (18,19).

4.2   Autecological Characteristics of the GMM in the Phytosphere of Greenhouse-Grown Sugar Beet

Inocula of the GMM were prepared from washed overnight cultures and introduced as a seed dressing to sugar beet in greenhouse investigations. Seed inoculation was preferred to spray application or seedling inoculation so that natural competition and establishment of the GMM could be determined. In greenhouse studies on sugar beet, the GMM survived for the duration of the 531 day investigation and established populations throughout the phytosphere (20), preferentially colonizing immature leaves (21). From this evidence it was predicted that the GMM would establish on a field-grown crop and persist on both the leaves and roots throughout the season. After successful contained greenhouse investigations on wheat and sugar beet, consent for small-scale field releases of the GMM to sugar beet (IVEM) and wheat (HRI) were granted in April 1993. A second release was undertaken on sugar beet in 1994. These were the first releases in the United Kingdom of a genetically modified free-living bacterium.

5.   STATISTICAL CONSIDERATIONS FOR THE DESIGN OF FIELD EXPERIMENTS

We recommend that the reader refer to more detailed text pertinent to the description of the application and use of statistics in microbial ecology (see Chapter 16) (22–24). A simplified outline of the considerations included in our investigations is provided.

5.1   Experimental Considerations

The central aims of the study were to demonstrate the ability of the GMM to survive and colonize a crop after seed inoculation and to determine whether any impact on the microbial community associated with a sugar beet crop could be detected. Three treatments (uninoculated, wild type–inoculated, and GMM-inoculated) were compared. Studies intended to calculate population parameters (e.g., mean values and variances), to test hypotheses (e.g., comparison of means between treatments), or to test the fit of mathematical models require careful design. The design must take into account some knowledge of the habitat and presumed performance of the inocula and integrate careful planning with realistic and practical sampling methods which provide data suitable for statistical analyses. This approach facilitates the distinction of treatment effects from natural variation. Factors which were taken into account included field design, demands of sampling, monitoring and detection methods, need for controls, replicates, reproducibility, and introduction of the bacterial inoculum.

5.2 Design of the Release Site

The design must provide randomization, replication, and the use of 'treatment blocks' to account for natural gradients or subdivisions of the material. We applied a nine plot *randomized* Latin square arrangement, one plot in each row and one plot in each column assigned to each of the three treatments (Fig. 3). The three main considerations in choosing this arrangement were the following:

1. The field had a gentle slope which may influence experimental conditions (to prevent bias, the Latin square was laid out perpendicularly to the slope so that all three treatments were applied at all three positions on the gradient).
2. 'Blocking' facilitated the use of statistical methods which increase the *precision* of the comparison of treatment effects by estimating the differences between blocks.
3. The use of more than one plot per treatment is necessary to improve the statistical *power* of the tests applied.

As part of the practical limitations imposed by the number of samples that could be collected, processed, and accurately analyzed with the resources available, we had to resolve whether it was more valid to increase the numbers of replicates (plots) at the expense of the number of individual samples (plants) or vice versa. Many statistics texts discuss this problem and some offer formulae for estimating the number of plots and samples needed to achieve specific levels of precision. In the first release the distribution of the GMM was investigated in six phytosphere habitats (senescing, mature, and emerging leaves, and rhizosphere soil, rhizoplane, and the cortex of the storage root), and hence each sampling occasion required the weighing, extraction, dilution, and plating of 54 (3 treatments × 3 plots per treatment × 6 phytosphere habitats per plot) samples. This was the maximum that could be processed in a single day without the detrimental effect which might result from prolonged storage of samples or the extraction of the sample on subsequent days after chilling at 4°C. For meaningful comparison, all samples were collected at the same time of day, early morning, on every sampling occasion to prevent adverse diurnal influences (10). In the first season, tissues from three individual plants from each plot were bulked (composited) to provide material for each of six habitats from each plot ($n = 3$).

(a)

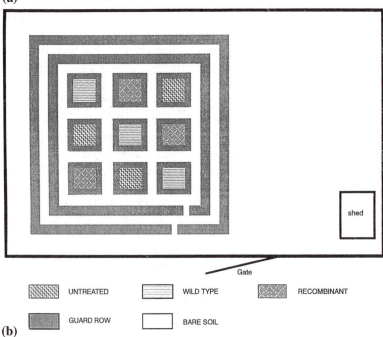

▨ UNTREATED	▦ WILD TYPE	▨ RECOMBINANT
▨ GUARD ROW	▢ BARE SOIL	

(b)

Figure 3. Experimental field plot used for the GMM field release, 1993–1994: (a) Release site 90 days after release (treatment and planting) of sugar beet seeds. (b) Schematic of field plot design. Triplicate plots of 100 plants (10 × 10 rows) were planted. Each plot was enclosed by three guard rows arranged within two further guard row blocks. The area was surrounded by a 2 m deer and hare fence to exclude herbivores. Bare soil was maintained within the enclosure by herbicide treatment, between blocks of sugar beet plants, walk boards were positioned to facilitate access and minimize contamination.

Although bulking plants reduces variance, it can result in an overestimate of mean values (25) (e.g., GMM counts and other bacterial counts) and reduce the power of the statistical tests that can be performed. For the second release (1994), three tissue types (mature leaf, immature leaf, and rhizoplane) were sampled from three individual plants taken from each of the three replicate treatment plots, i.e., $n = 9$ for each treatment. This regimen increased the power of the tests and allowed plant-to-plant variation to be determined.

Statistical analysis of the data was applied to estimate population parameters, to calculate the confidence of these estimates, and to distinguish significant from nonsignificant differences. Populations with normal distributions are most commonly described by a measure of central tendency (mean) and a measure of variability such as standard deviation, standard error, or coefficient of variation (Chapter 16). Because estimates of the population size of epiphytic bacteria are often strongly skewed, where counts vary by several orders of magnitude, data were normalized by logarithmic transformation, where the mean of $\log_{10}$ data is expressed as the arithmetical mean. To standardize the sample data, they were to be expressed as colony-forming units per gram of dry weight of tissue; this minimized plant wet weight variation.

6. SURVIVAL, DISPERSAL, AND PERSISTENCE OF GMM AFTER FIELD RELEASE

6.1 Colonization and Establishment of GMM Populations in the Phytosphere of Field-Grown Sugar Beet

The autecological properties of the GMM were followed over two separate releases to sugar beet in 1993 and 1994 (Fig. 4). In 1993 the GMM reached a maximum population density of 1.2×10^5 CFU g^{-1} in the rhizosphere soil 56 days after planting (30 days after first true leaf emergence) and remained above detectable limits (20 CFU/g) in the phytosphere for 270 days after introduction (Fig. 4). In the second release experiment (1994) estimates of the phytosphere population density of the GMM were approximately 10,000 times greater than those recorded in 1993 (Fig. 4). In 1994, the GMM represented between 25% and 81% of the isolated pseudomonad population taken from the surface of growing plants sampled up to day 102 after introduction (60 days after first true leaf emergence). By comparison, the highest relative density of the GMM within the 1993 pseudomonad community was 6% (recorded 28 days after introduction on emerging plants). In both seasons, GMM population densities declined as the plants matured and differentiated, whereas total pseudomonad counts did not differ significantly between samples or between seasons.

It has not been possible to define the environmental or plant conditions which favored the better establishment of the GMM in the 1994 trial than in the previ-

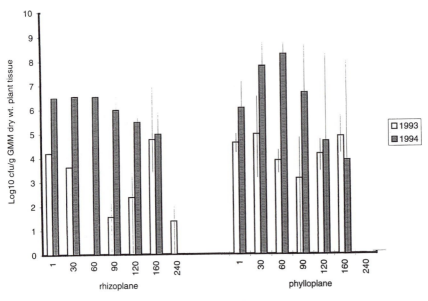

Figure 4. GMM survival in the phytosphere of seed-inoculated sugar beet crop over two consecutive releases, 1993 and 1994: Estimates of bacterial density were made by counting blue colonies on GMM selection agar (PSA/CFC, Difco UK, supplemented with 75 μg ml^{-1} kanamycin and 0.025% X-gal). Mean, maximum, and minimum counts of the GMM for 1993 [☐] (n = 3, pool of three plants taken from each of the three plots) and 1994 [▨] (n = 9, individual plants, three from each plot) are given. The mean values for the total pseudomonad count in the rhizoplane and emerging leaf phyllosphere, calculated from each sample over the 2 year study period, were 8.9 × 10^7 CFU/g dry weight and 6.3 × 10^9 CFU/g dry weight, respectively. Data have been adjusted to permit comparison of the two releases with respect to the time in days taken for the emergence of first true leaf set after the introduction of treated seeds (i.e., day 28, 1993; day 41, 1994). Washed inocula were applied to commercially pelleted sugar beet seed (var. Amethyst, EB3, Germains UK Ltd, Kings Lynn, United Kingdom), ca. 1 × 10^7 CFU/seed. Less than 10^{10} CFU of wild type and GMM were released over the 2-year period. Strains were subcultured less than 10 times from the original isolation of SBW25 wild type from the field to minimize the selection of mutants adapted to laboratory conditions.

ous year. Possibly, the initial slow growth of the plants, after germination (the first true leaf set developed within 28 days in 1993, compared with 41 days in 1994), provided an opportunity for the GMM inocula to colonize the plant preemptively. A sustained population established and colonized each new leaf pair as they developed (approximately every 16 days in maturing sugar beet). This advantage was maintained for a further 60 days of plant growth. On certain plants the displacement of the GMM population was extremely aggressive, resulting in the disappearance of the GMM from isolation plates when some plants were sampled. The GMMs were not detected in the phyllosphere (limits of detection 20 CFU g^{-1}) in two of nine individual plants sampled on both day 120 and day 160 (1994). It was assumed that these plants had been colonized by the GMM earlier in the season as the variation between replicates was small within the first three sample sets (coefficient of variation [CV] range, 5% to 10%) with the GMM detected on every plant sampled. The reason for the elimination of the GMM population in some plants was unresolved, but it contributed to the large variations observed in these later samples. The GMM population densities could have declined as a result of competition by better adapted pseudomonads that actively excluded the GMM from its adopted niche. This exclusion reflects changes in the local environment at the plant surface to which the GMM was less well adapted. An alternative explanation for the loss of the GMM may be predation by bacteriophage. Plaques, produced by lytic bacteriophage infection, were frequently observed in bacterial lawns after spread plating the least diluted samples taken from the field (Lilley and Bailey, unpublished observations). However, the importance of lytic bacteriophage as bacterial predators in terrestrial habitats has yet to be determined, although phage infection is considered to play an important role in the control of bacterial population densities in aquatic environments (26) and transfer of phage between bacteria has been modeled in the phyllosphere (27). The GMM survived well at the root surface, where it was detected on every individual plant sampled in the second release. These observations support the view that strong selective pressures are exerted on microflora at the leaf surface (e.g., temperature, rain, wind, ultraviolet [UV] radiation, and variation in nutrient and moisture availability), factors which are more constant in the soil. The microbiological properties at the plant surface are clearly influenced by plant activity. Typically, success or decline in the population dynamics of the GMM in the leaf was also paralleled in the root; note, for instance, the biphasic curve for the 1993 data (Fig. 4). By contrast, total pseudomonad numbers varied little on each sampling occasion throughout or between seasons. Without a selectable indicator population (the GMM), it would not have been possible to make any assumptions as to whether plant responses have any direct or indirect influence on microbial population dynamics.

6.2 Persistence and Dispersal of the GMM in the Soil Environment

The GMM could not be detected on the leaves or roots of overwintering plant tissue or in the soil. When sugar beet was resown the following season, colonies of GMM bacteria were detected by selective plating in only one of nine samples collected on a single occasion, 43 days after seed introduction. These observations indicate that the GMM can survive at low levels in the soil. Lateral dispersal through the soil could not be demonstrated. The detection of GMM bacteria that had dispersed to the foliar parts of indigenous weed species and to uninoculated sugar beet leaves in the guard rows was rare and at the limits of detection. This sporadic transfer of GMM was thought to result only after prolonged direct physical contact between plant leaves during wet conditions (20).

In a separate study we investigated the potential role of invertebrates in the physical dispersal of the GMM through the phyllosphere (Lilley et al., in preparation). Briefly, 20 third instar *Mamestra brassicae* (Lepidoptera: *Noctuidae*) larvae were placed on the leaves of mature sugar beet plants that had been seed-inoculated with the GMM. These naturally colonized plants, which supported established GMM populations similar to those observed in Fig. 4, were surrounded by a ring of untreated mature sugar beet. Plants were kept in a netted enclosure to contain the caterpillars, and the dispersal of the GMM from the treated to the untreated plants was monitored. The GMMs were transferred by the insects and survived and colonized 'trap plants.' Forty-two days after insect introduction, approximately 20 days after pupation of the larvae, a GMM population was established on the immature leaves of trap plants similar to those determined for seed inoculated plants of the same age. Hence, under appropriate conditions of insect density, GMM density, and plant age, phylloplane bacteria can be dispersed by invertebrates and become established on new plants.

6.3 Statistical Testing of the Distribution and Impact of Inocula on Indigenous Communities

To distinguish significant from nonsignificant differences, analysis of variance (ANOVA) was performed (see Chapter 16), (23,24,28). For example, the ANOVA test was applied to data collected on each sampling date to determine whether any of the mean bacterial counts estimated from each of the plots differed significantly and whether differences could be related to either the treatment or the position of each plot in the field.

Data for this ANOVA were initially tested to ensure that they complied with the assumptions implicit in the test: (1) unbiased collection of data (fulfilled by the random assigning of treatments to plots in the Latin square and the randomized selection of plants for sampling), (2) normally distributed collected data, (3) similar variances within samples. Conditions 2 and 3 were tested by using standard procedures after $\log_{10}$ transformation of the data.

The ANOVA tested a null hypothesis (H_0), that there were no significant differences in sample means from the three treatments. Essentially, this is a two-tailed test because if assumptions made in H_0 prove incorrect, then the alternative hypothesis can only be that there are differences and not that one treatment has a greater effect than another. One tailed tests are less stringent, but require *prior* predictions of 'greater than' and are not commonly applied to field data. The ANOVA compares two or more sample means by dividing or partitioning the variability of the samples into components. This variability is calculated and partitioned as the sums of squares (SS) using standard formulae and is depicted here as

$$SS_{total} = SS_{treatment} + SS_{plots\ across\ slope} + SS_{plots\ up\ the\ slope} + SS_{residual}$$

The ANOVA was applied to examine whether any of the total bacterial counts for a given plot was significantly different from that of other plots and whether the influence of inoculum treatment was greater than the influence of the position of each plot in the field. When plot position and treatment were found to have no significant effect, a second ANOVA was applied to test treatment effect. The analysis generated probability values of the null hypothesis with respect to the sources of variation. A result with a probability less than 0.05 ($P < 0.05$) is generally considered to be statistically significant. Probability values of less than 0.01 are accepted as highly significant. By the analysis of variance it was determined that the release of the GMM or wild type had no significant effect on the total numbers of bacteria colonizing the immature leaves of sugar beet plants.

6.4   Assessment of Environmental Impact

An important aspect of the investigations was to ascertain whether any impact could be detected and whether methods of assessing impact in the field could be developed. The apparent competitiveness of the GMM under appropriate conditions, as seen in the 1994 release or in the greenhouse studies, resulted in the proliferation of the inoculant and the possible exclusion of indigenous populations. We attempted to measure such effects by assessing the metabolic potential of the sampled community (21), measuring the rate of development of pseudomonad colonies on agars (19), and evaluating community diversity by the MIS-FAME identification of isolates (20). Effects were recorded in the greenhouse, where the exclusion of Enterobacteriaceae species and changes in community diversity in the phyllosphere were observed in wild-type *P. fluorescens* SBW25 and GMM-treated plots. These effects were transient and had no lasting impact on the development of the microbial community in maturing tissue. Another interesting observation was made during the assessment of whether inocula underwent adaptive change as a strategy for survival in leaf and root associated habitats (the phytosphere). On isolation from colonized plants, it was often observed that the GMM produced at least two colony variants on agar, and FAME profiles of GMM isolates collected over a 220 day period differed from those of the original laboratory-maintained inoculum (Fig. 5). These adaptations underline the need to monitor the genotype and phenotype of the progeny of GMMs released in the field closely since they may have important implications for risk assessment and prediction of behavior.

7. GENE STABILITY AND ASSESSMENT OF GENE TRANSFER IN THE PHYTOSPHERE

The dissemination of genes within and between bacterial communities in the natural environment has been considered an important evolutionary and adaptive survival mechanism (29). Although conjugation has been described as the most probable method of gene transfer in the terrestrial environment (30), studies which describe plasmid mobilization between laboratory strains in simulated environments only demonstrate the potential for gene mobilization and fail to provide the data necessary to assess whether transfer occurs at ecologically significant frequencies in natural habitats. Because of the concerns expressed that the release of GMMs to the environment might result in the transmission of undesirable traits, an unequivocal demonstration of the movement of genetic material to or from GMMs in the natural environment is required to make valid impact assessments.

7.1 Transfer Proficiency of Phytosphere Microbial Communities

Laboratory and field-based investigations with *P. fluorescens* SBW25EeZY6KX confirmed the genetic stability of this GMM, and under all conditions tested no loss of phenotype or transfer of the marker genes could be detected. Therefore, additional investigations were undertaken during the field release to determine whether plasmid-mediated gene transfer could be detected in the leaf and root habitats of sugar beet. The sugar beet phytosphere supports microbial communities containing cryptic transfer-proficient plasmids (31) and conjugative mercury resistance plasmids which transfer, between simultaneously introduced donor and recipient strains, at higher frequencies in the rhizosphere than in soil or on bacteriological agars in the laboratory (32). Mercury-resistant bacteria have been reported in all natural environments studied and mercury-resistance determinants are frequently located on plasmids or within transposons (33). Mercury resistance was therefore selected as a suitable natural phenotype with which to monitor gene transfer to the released GMM.

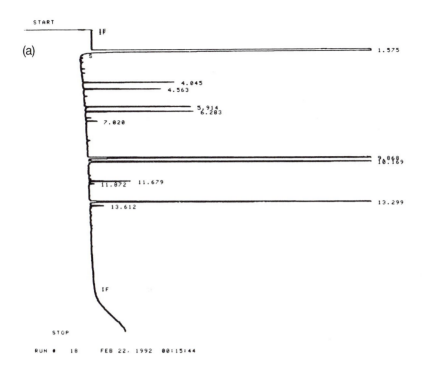

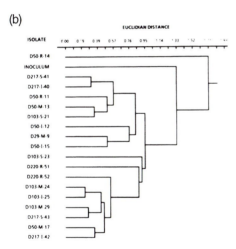

Figure 5. Fatty acid methyl ester-Microbial Identification System (FAME-MIS) analysis of GMM-SBW25 in the phyllosphere of sugar beet: (a) Chromatogram showing the FAME profile of *P. fluorescens* SBW25. (b) Dendrogram based on FAME analysis of the GMM isolates taken from the field demonstrating the changing phenotypic relationship between 18 colonies isolated up to 220 days after seed inoculation. The day (D) of isolation, assigned culture number, and origin—I, immature; M, mature; S, senescent leaves; R, root—are shown in the codes alongside the original inoculum. (Data from Ref. 20.)

7.2   Natural Transfer of Genetic Information from the Indigenous Microflora to the GMM

Selective agars specific for the GMM were supplemented with $HgCl_2$ to determine whether colonies of mercury-resistant SBW25EeZY6KX bacteria could be isolated directly from leaf and root homogenates of field-grown sugar beet plants. Mercury-resistant recombinant SBW25EeZY6KX colonies were isolated from plants collected midseason and represented a significant proportion of the GMM population isolated in 1993 (Bailey and Lilley, unpublished). Molecular analysis of the plasmid DNA extracted from the GMM isolates confirmed that they were closely related to other large mercury plasmids isolated by more traditional methods from the microflora of sugar beet grown at this site (34). There was no evidence to indicate that the chromosomally located marker genes in the GMM could be mobilized to the indigenous pseudomonad community sampled directly from the field or after the transfer of these plasmids in the laboratory. No GMM isolate lost the ability to express all the markers. It was concluded that, even under the test environmental conditions, the chromosomally located marker genes in the GMM were stably integrated.

8.   CONCLUSIONS

A number of field releases with genetically modified bacteria have been undertaken to assess the behavior of inocula and risks associated with such releases. Although representatives of several bacterial genera have been studied, the majority of investigations have focused on fluorescent pseudomonads in the rhizosphere of cereal crops, predominately wheat. These studies revealed that only limited lateral dispersal of inocula occurred and that vertical dispersal correlated with the depth of penetration of the root system. Aerial dispersal of inocula has been less well investigated. Typically, such inocula were applied as aerosols to introduce the high population densities necessary to compete with the indigenous microflora (35). Our investigations into the ecological properties of the released GMM to the phytosphere demonstrated that after seed inoculation it was able to colonize the roots and leaves of sugar beet and persist throughout the growing season of the crop. The detection of the GMM against high background counts of indigenous microorganisms was facilitated by the sensitivity of the monitoring methods developed through the choice of marker genes. Further, we believe that the successful colonization and persistence of the GMM in the natural environment are testimony to the careful design of the strategy for genetic modification and the selection of a candidate organism from the target habitat.

In vitro investigations demonstrated that the maintenance and expression of transferred marker genes were only possible after their integration into the genome of recipient pseudomonads that shared considerable homology with SBW25 (14). Therefore, if mobile genetic elements with the capacity to transfer chromosomal regions (e.g., transducing bacteriophage, conjugative plasmids with chromosome-mobilizing activity, or conjugative transposons) were prevalent and active in the natural environment, then the design of the GMM should have detected such events. No transfer of the marker genes was observed. This indicated that the mobilization of chromosomal DNA from bacteria in the phytosphere does not occur at detectable frequencies with the methods applied. However, we have been able to demonstrate uniquely the active transfer of conjugative plasmids to bacteria-colonizing leaves and roots of field-grown plants after seed inoculation with recipient bacteria. The transfer of plasmids to these colonizers results from active interaction between the inocula and the indigenous microflora which normally maintain mercury-resistance plasmids. It is also apparent that distinct genetic groups of mercury-resistance plasmids exist which transfer between populations of bacteria colonizing sugar beet plants. Because of the different plasmid types isolated on different sampling occasions, from root and leaf habitats, it was concluded that a large number of individual transfer events had occurred.

Agricultural crops have been commercially important targets for the application of unmodified inocula and remain potential targets for GMMs introduced as mediators of plant protection or growth promotion (see also Chapter 17) (1,2,4,7). The GMMs, containing novel phenotypes, also provide a unique opportunity to investigate the ecological characteristics of bacterial inocula and their genes. Important issues with respect to the impact GMMs may have on target and other environments are their potential perturbation of natural microbial populations and their ability to exchange genetic material, although little evidence exists to describe the natural frequency of such events. Knowledge of the frequency and probability of gene exchange and reassortment is essential for understanding the dynamics of bacterial population genetics and bacterial adaptation and for making predictions of the fate and impact of released GMMs. The uptake and exchange of DNA between bacterial populations may also be a cause for concern when predicting the outcome of the activity of GMMs in the environment as the exchange of genetic material between populations may be common and provide an effective strategy for survival in extreme and fluctuating environments (such as the phyllosphere and rhizosphere). We are currently undertaking long-term monitoring of the released inoculant to assess whether phenotypic changes, gene uptake, or genomic reassortment can be detected and whether such changes alter the ability of the original inocula to persist or adapt to colonize alternate plant species. To further our interests in natural gene flux in the phytosphere, the GMM described here has also been included in experiments to determine the contribution of plasmids to host survival and persistence and to provide biological assays necessary to identify the plant stimuli that promote gene transfer between colonizing populations of bacteria. It is hoped that these approaches will lead to the development of inocula better able to survive and remain biologically active once released by exploiting the potential for indigenous populations to exchange genetic material. By engineering required traits, e.g., genes with biopesticide, bioinsecticide, or bioremediation activity or genes that enhance environmental competence, into conjugative plasmids, bioactivity could be retained in the environment as the plasmid transfers to succeeding populations of bacteria, each better adapted to survival. Before this scenario is realized we must continue to develop a better understanding of the consequences of and factors that control genetic transfer to make the necessary estimates of risk associated with the release of GMMs.

ACKNOWLEDGMENTS

We would like to thank the Natural Environment Research Council (United Kingdom) for its continuing support of the Molecular Microbial Ecology Group at the Institute. Support for the first field release in 1993 was provided for IPT, RJE, and PAB by U.K. DOE funded projects (PECD 6/8/143 and PECD 7/8/161) held in collaboration with Horticulture Research International (United Kingdom). Support for the environmental plasmid transfer studies and the second release field release, conducted by Andrew Lilley, was through an EU Biotechnology program grant (Bio2-CT92-0491).

REFERENCES

1. D. J. Drahos, *Biotechnology 9*:157 (1991).
2. D. J. Drahos, G. F. Barry, B. C. Hemmin, E. J. Brandt, E. L. Kline, H. D. Skipper, D. A. Kluepfel, D. T. Gooden, and T. A. Hughes, *Release of Genetically Engineered and Other Microorganisms* (J. C. Fry and M. J. Day eds.), Cambridge University Press, pp. 147–159 (1992).
3. J. C. Fry and M. J. Day, *Release of Genetically Engineered and Other Microorganisms*, Cambridge University Press p. 178 (1992).
4. M. Wilson and S. E. Lindow, *Annu. Rev. Microbiol. 47*:913 (1993).
5. J. Prosser, *Microbiology 140*:5 (1994).
6. I. P. Thompson, M. J. Bailey, J. S. Fenlon, T. R. Fermor, A. K. Lilley, J. M. Lynch, P. J. McCormack, M. McQuilken, K. J. Purdy, P. B. Rainey, and J. M. Whipps, *Plant Soil 150*:177 (1993).
7. D. B. O'Sullivan and F. O'Gara, *Microbiol. Rev. 56*:662 (1992).
8. F. Dane and J. J. Shaw, *Microbial Releases 2*:223 (1994).

9. I. P. Thompson, M. J. Bailey, R. J. Ellis, and K. J. Purdy, *FEMS Microbiol. Ecol. 102*:75 (1993).

10. I. P. Thompson, M. J. Bailey, R. J. Ellis, A. K. Lilley, P. J. McCormack, K. J. Purdy, and P. B. Rainey, *FEMS Microbiol. Ecol. 16*:205 (1995).

11. P. B. Rainey, M. J. Bailey, and I. P. Thompson, *Microbiology 140*:2315 (1994).

12. M. J. Bailey and I. P. Thompson, *Genetic Interactions Between Microorganisms in the Microenvironment* (E. M. Wellington and J. D. van Elsas, eds.), Pergamon Press, Oxford, p. 126 (1992).

13. G. F. Barry, *Gene 71*:75 (1988).

14. M. J. Bailey, A. K. Lilley, I. P. Thompson, P. B. Rainey, and R. J. Ellis, *Mol. Ecol.* (1995).

15. P. B. Rainey and M. J. Bailey, *Mol. Microbiol. 19*:521 (1996).

16. F. A. A. M. De Leij, M. J. Bailey, and J. M. Lynch, and J. M., Whipps, *Lett. Appl. Microbiol. 16*:307 (1993).

17. P. A. Bramwell, R. V. Barallon, H. J. Rogers, and M. J. Bailey, *Molecular Microbial Ecology Manual* (A. D. L. Akkermans, J. D. Van Elsas, F. J. De Bruijn, eds.), Kluwer Academic Publishers, pp 36–55 (1995).

18. F. A. A. M. De Leij, M. J. Bailey, J. M. Whipps, and J. M. Lynch, *FEMS Microbiol. Ecol. 13*:249 (1994).

19. F. A. A. M. De Leij, M. J. Bailey, J. M. Whipps, and J. M. Lynch, *App. Soil Ecol. 1*:207 (1994).

20. I. P., Thompson, R. J. Ellis, and M. J. Bailey, *FEMS Microbiol. Ecol. 17*:1 (1995).

21. R. J. Ellis, I. P. Thompson, and M. J. Bailey, *FEMS Microbiol. Ecol. 16*:9 (1995).

22. M. MacIntosh, *Risk Assessment in Genetic Engineering* (M.A. Levin, and H. Strauss eds.), McGraw-Hill, New York, pp. 219–239, (1991).

23. J. C. Fry, *Biological Data Analysis.* IRL Press, Oxford, pp. 1–39 (1993).

24. R. R. Sokal and F. J. Rohlf, *Biometry*, Freeman, New York (1981).

25. S. S. Hirano, E. V. Nordheim, D. C. Arny, and C. D. Upper, *Appl. Environ. Microbiol. 44*, 695 (1982).

26. G. Bratbak, M. Heldal, S. Norland, S, and T. F. Thingstad, *Appl. Environ. Microbiol. 56*:352 (1990).

27. S. P. Kidambi, S. Ripp, and R. V. Miller, *Appl. Environ. Microbiol. 60*, 496 (1994).

28. D. S. Brown, *Statistical Testing in Biology*, Plenum Press, Oxford (1989).

29. D. L. Coplin, *Annu. Rev. Phytopath. 27*:187 (1989).

30. S. B. Levy and R. V. Miller, *Gene Transfer in the Environment*, McGraw-Hill, New York, p. 434 (1989).

31. B. J. Powell, K. J. Purdy, I. P. Thompson, and M. J. Bailey, *FEMS Microbiol. Ecol. 12*:195 (1993).

32. A. K. Lilley, J. C. Fry, M. J. Day, and M. J. Bailey, *Microbiology 140*:27 (1994).

33. S. Silver and M. Walderhaug, *Microbiol. Rev. 56*:195 (1992).

34. A. K. Lilley, Transfer of natural plasmids between bacteria in the rhizosphere of sugar beet. Ph.D. Thesis, University of Wales, Cardiff (1994).

35. J. J. Shaw, F. Dane, D. Geiger, and J. W. Kloepper, *Appl. Environ. Microbiol. 58*:267 (1992).

16

Application of Common Statistical Tools

BRIAN B. McSPADDEN GARDENER   Michigan State University, East Lansing, Michigan

ANDREW K. LILLEY   Institute of Virology and Environmental Microbiology, Oxford, United Kingdom

1. INTRODUCTION

The soil is a dynamic and complex environment for its resident microorganisms. The activity of macro- and mesofauna can have drastic consequences on microbial activities and population structures. Most often these changes occur on spatial and temporal scales much smaller (and in some instances, much greater) than scientists can conveniently study. The dynamics of individual occurrences become obscured by these differences in scale, as people study bulk effects on easily handled samples much larger and on time scales much longer than those of the distinct phenomenon. Given this natural constraint, the question arises as to how macroorganisms such as ourselves can study the ecological characteristics of microorganisms in a meaningful way.

A number of answers to that question are discussed in several chapters of this book. Some techniques allow for the direct observation of microorganisms in situ, while others require their extraction, in whole or in part, from the environment. Data from all of these techniques may be amenable to statistical analysis, provided that the preconditions of the use of the chosen statistical tools are met in the experimental design. Thus, statistics can be used to investigate further the natural phenomena that have been detected by various techniques.

Statistical tools can be applied most productively when the patterns of experimental data are not readily discernible or when replicates give conflicting results. The first case indicates that natural variation is high, relative to the magnitude of the variable being studied. The second case arises as a consequence of the first when sample sizes are too small. The effective application of statistics, though, does not take place as an afterthought attempt to make sense of confusing data. Rather, it is integral to the design of the experiment, just as any other analytical tool is. In this regard, one will most often use statistics in experiments after initial observations of high natural variation.

This chapter will begin by discussing the nature of experimental data and how they can be handled mathematically. Next, the theoretical basis of statistical analysis will be outlined with emphasis on the practical implementation of statistical tools. The most common methods of comparison and recognition of functional relationships will then be briefly explained. Some recommendations for planning and presenting statistical analyses will be made. At the end of this chapter, an example of the proper application of statistics to microbial data will be given. It is hoped that this introduction will promote the expanded application of various microbiological and molecular techniques to questions of soil microbiology once thought to be unanswerable because of the high degree of variation noted in such data.

The step-by-step procedures for performing the statistical tests presented in this chapter can be found in a number of texts (e.g., 10,11,19). The necessary tables for evaluating the significance of the calculated test statistics can be found in those same texts or in separate volumes (e.g., 16). It should be noted, however, that not all tests are described in all books, and algebraic conventions vary from text to text. Furthermore, few of the published texts reviewed by us give a critical and comprehensive discussion of all the statistical methods described in this chapter. Two of the best, in our opinion, are those by Fisher (2) and Fry (3).

2. EXPERIMENTAL DATA

Data are classified according to their relative quantitativeness. They may be cardinal, ordinal, or nominal. Cardinal data are numerical, with the relationships between one value and another being countable on a given measurement scale. They are the most common type of data in the empirical sciences. Ordinal data are also numerical, but the difference between values is one of inconsistent magnitude. Ordinal data are commonly used in plant pathology and epidemiology, where disease incidence is scaled to arbitrarily defined levels that are clearly distinct. Nominal data are purely descriptive and are used when distinctions are being made between categorical groups (e.g., bacteria and fungi).

Data can also be classified according to the distinctiveness of the variable being measured with regard to the population from which the measurement is drawn. It may be discrete or continuous. Discrete variables are whole numbers, while continuous variables may have any value on a measurement scale. Depending on their numerical significance within a sample, data may be considered discrete or continuous. For example, the number of viable pseudomonads in a mixed microcolony would yield discrete data, say, 17 out of 30 bacteria, not some fraction between 17 and 18. However, counting viable pseudomonads in a large or bulked sample (e.g., from a gram of soil) would yield essentially continuous data, say, 6.54×10^7 CFU/g. As it is possible to observe a value near to but not exactly 6.54×10^7, the count data are essentially continuous at this level of scale. Recognizing this aspect of a measured variable is critical in choosing an appropriate statistical test. For example, the Kolmogorov–Smirnov test for comparing two samples requires that the data belong to continuous distributions, and it would be inappropriate for comparing the number of pseudomonads in different microcolonies. However, if one wanted to compare the number of viable pseudomonads from 5 g soil samples of different fields, then the Kolmogorov–Smirnov test would be appropriate.

One should recognize that wherever there is variation in a measured variable, a population of measurements will conform to a particular distribution. Thus, distributions are simply mathematical descriptions of populations of data points. Distributions may arise from the natural variation of a measured character within a population and/or the spatial/temporal structuring of the measured variable in the environment. The distributions in these two cases are not generally known a priori and may themselves be the basis of scientific investigation (e.g., 9). Distributions of data also result from imprecise measurement and/or from sampling error. In these instances, the distributions are most often assumed to be normal (i.e., to follow a Gaussian curve). This is generally a fair assumption, for example, when making dilutions of soil samples. However, when applying new methodologies one should still be cautious, because the distribution of the data may depend on the precision of the technique and the sampling strategy (7,15).

In general, the types of statistical tests applied to a given data set will depend on whether or not the underlying distribution of the measured variable is known. The distribution of a measured variable may be assessed graphically and numerically. Histograms provide a quick visual assessment of the data, while statistical analyses give more precise assessments of the distribution. Proper assessment of the distribution requires a relatively large sample size (e.g., ≥ 30 independent samples). This comprehensive sampling may or may not be convenient, depending on the demands of the empirical procedure. If the distribution of the data can be established, parametric tests derived from the mathematical properties of that distribution can be used in analyzing data. Nonparametric, or distribution-free, tests can be used instead when the underlying distribution of the measured variable is not known. The choice of when to use parametric vs. nonparametric tests is discussed in more detail later (see Section 3.1, *Hypothesis testing*).

It is possible to change the numerical values of a data set systematically so as to facilitate its analysis. These transformations are generally used to make the underlying distribution of the data approach normality and improve the homogeneity of the sample variances. This facilitates the use of parametric tests, which are more powerful than nonparametric analogs in cases where data sets are small. The logarithmic transformation has been commonly applied to data taken on soil systems (14). The justification for such a manipulation is that the distribution of such data (e.g., colony counts, denitrification rates) is log-normally distributed in most cases. However, caution must be used, because exceptions to this rule have been noted (6,7). In addition, as the logarithmic transformation reduces the ratio of mean and standard deviation, data may appear more precise than they really are. Alternatively, cardinal data may be transformed to ordinal or nominal scales in nonparametric tests. This type of transformation is most useful when the underlying distribution of the data is unknown, or is known not to conform to an easily defined distribution, such as the normal or log-normal. Ultimately, if one chooses to transform a data set, some justification in terms of the underlying distribution should accompany the presentation of the data.

3. STATISTICAL MEASURES

Data sets can be characterized mathematically in several ways. The most common descriptors are those of central tendency and scatter about that central tendency. Additionally, limits can be assigned to a range of values which will encompass a specified percentage (e.g., 95%) of the measured values. These three types of measurements are the essence of basic statistical analyses. The measurements have analogous whole population and sample population definitions.

Central tendency can be defined by the mean, median, or mode (Table 1). The choice of measure largely depends on the distribution of the data. As a general rule, the mean is the best choice if the distribution is symmetrical. If it is asymmetrical, the median often provides a better gauge of the central value. In experiments where there are few replicates, the mean is the best central measure regardless of the underlying distribution of the measured variable. The mode is not commonly used, except in noting patterns of repetitive values.

Scatter about the central tendency can also be measured in a variety of ways (Table 2). Variance and standard deviations are the most commonly reported

Table 1. Measures of Central Tendency

Mean		$\bar{x} = \left(\sum\limits_{i=1}^{n} x_i \right) / n$
Median	for odd n	value of observation numbered: $(n + 1)/2$
	for even n	mean of the values numbered: $n/2$ and $(n + 2)/2$
Mode		most common value
	where	x = sample value
		n = number of measurements in a data set

Table 2. Measures of Data Scatter

Deviation	$d_x = (x - \bar{x})$
Range	$R_x = x_{max} - x_{min}$
Variance	$s^2 = \sum\limits_{i=1}^{n} (x_i - \bar{x})^2 / (n - 1)$
Standard deviation	$s = \sqrt{s^2}$
Standard error of the mean	$SEM = s / \sqrt{n}$
Coefficient of variation	$V = s / \bar{x}$

measures, and they are most meaningful in situations where the data are from symmetrical or unimodal distributions. The range can be informative when the relative importance of the data scatter is being assessed. Note that the standard measures of scatter decrease with increasing sample or replicate size. This reflects the fact that repeated measurements increase one's confidence in the central measure as the true value for the population as a whole.

Confidence intervals arise from mathematical descriptions of probability. They indicate the range of values over which a measurement may be expected to fall with a specified certainty. The size of the intervals depends on the degrees of freedom within an experiment and the amount of scatter in the data (Table 3). The degrees of freedom in an experiment equal the number of ways the value of an estimated population parameter could have been calculated. Thus, it is a measure of the strength of the approximation of the calculated sample measure. The greater the number of degrees of freedom, the lower the value of a test statistic required to detect a significant difference. Generally speaking, the degrees of freedom equal one less than the number of units used in making a comparison. Confidence limits for the mean and variance of normally distributed data are given by the well-established t and χ^2 distributions, respectively. In the case of

Table 3. Confidence Limits of Sample Measures[a]

Parametric[b]	$CL_{mean} = \bar{x} \pm t_{(\alpha/2, n-1)} \, (s/\sqrt{n})$
	$CL_{variance} = (\chi^2_{(\alpha/2, n-1)})^{-1} (s^2)(n-1)$ and
	$(\chi^2_{(1-\alpha/2, n-1)})^{-1}(s^2)(n-1)$
Nonparametric[c]	$^{d}CL_{mean} = \bar{x} \pm ks$ where $1/k^2 = \alpha$

[a]The level of significance, α, is chosen by the investigator.
[b]Parametric statistics require that the sample come from a normally distributed population.
[c]Nonparametric statistics make no assumption about the distribution of the sampled population.
[d]Approximation using Chebyshev's theorem.

data which do not come from normally distributed populations, confidence limits of the mean are broadly specified by Chebyshev's theorem. The theorem specifies the theoretically largest range of values within which the true population mean exists. Alternatively, one can calculate confidence intervals for the median based on the sign test (10).

3.1 Hypothesis Testing

Like all empirical methods, statistics is used to test hypotheses. Mathematically, the choice is between the null hypothesis (H_0) and an alternative hypothesis (H_1). The null hypothesis is generally stated as "H_0: all quantities are equivalent," with the corresponding alternative "H_1: not all of the quantities are equivalent." Still, it is possible to define the hypotheses more precisely. Comparisons using inequalities, coefficients, and/or constants can be formulated for statistical testing. For example, one may wish to test whether or not the rate of ammonification of soil type A is five times greater than that of soil type B. The null hypothesis could be stated as H_0: rate of A < 5 × rate of B, H_1: rate of A ≥ 5 × rate of B. To test this hypothesis, the data from soil B must be multiplied by 5 before performing the statistical test.

For any given test, a value, called a test statistic, is calculated according to a particular equation using the experimental data. Thus, the test statistic is a summation of all the experimental data in a single number. The value which is calculated is meaningful only in relation to the degrees of freedom of the experiment (e.g., the number of treatments and replicates thereof). In general, if the test statistic has a high calculated value, the null hypothesis is rejected and the alternative hypothesis is accepted as the better explanation. However, the opposite is not true; failure to disprove the null hypothesis is *not* equivalent to proof of it.

Now, what is "high" for a particular statistic depends on the test being performed. Values specific to different tests (e.g., Student's t, Wilcoxon T) are set out in tables. Collections of these can be found as appendices in various books (e.g., 2,10) or as separate volumes (16). For each value of a test statistic, the level of significance, α, can be found. The level of significance is equal to the probability (P) that the value of the test statistic occurred by chance. These values, then, are used to make quantitative statements about the experimental data.

In performing tests of the null hypothesis, there are two types of errors that can occur. Type I is an unfair rejection of the H_0, and type II is an unfair acceptance of H_0 when it is, in fact, false. The probability of making a type I error is designated by α, and the probability of making a type II error is designated by β. α and β are referred to as "levels of significance" for making type I and type II errors, respectively. For every distribution there is a specific relationship between α and β depending on the scatter of the data. If the distribution is complex or undetermined, it is not possible to calculate β for a given α, or vice versa. Because of this, investigators should designate a low α (e.g., 0.05 or 0.01), take the largest practical sample size, and refrain from declaring proof of the null hypothesis (20).

While performing hypothesis tests, one can choose between one-tailed and two-tailed comparisons. One-tailed comparisons are used when there is a predicted directional difference between the two data sets (i.e., one has values greater than the other). Two-tailed comparisons are used when there is no such a priori prediction. For example, if one expected the addition of a metabolizable antibiotic to stimulate the population size of a microbial inoculant in the soil, one would compare treatments with and without the added compound using a one-tailed comparison. However, if the study also involved tracking the impact of the inoculant on denitrification, comparison of denitrification rates should be done using a two-tailed comparison because one cannot predict ahead of time in which direction the values will change. For any specified α, a two-tailed test is more conservative and is less likely to disprove the null hypothesis than the corresponding one-tailed test. In fact, a two-tailed comparison at any given α is equivalent to a one-tailed comparison at $(0.5)\alpha$ if the distributions are symmetrical. Because of this mathematical equivalency, it is important to specify whether a one- or two-tailed comparison was made when presenting the results of statistical tests.

The choice of statistical test depends on the question asked and the nature of the data that can be obtained experimentally (see Table 4). Statistics are used to assess comparisons and apparent functional relationships within and between data sets. The relative quantitativeness of that assessment depends on the test used. In this regard, there is a tradeoff: nonparametric tests have few assump-

Table 4. Common Statistical Tools Discussed in the Text[a]

	One or two samples		Multiple samples	
Purpose	Parametric	Nonparametric	Parametric	Nonparametric
Making comparisons between samples	Student's t Fisher's LSD	Wilcoxon T Mann– Whitney U Kolmogorov– Smirnov D Sign test Wald– Wolfowitz	One-way ANOVA Two-way ANOVA Scheffé's multiple range Tukey's multiple range	Kruskall– Wallace H Friedman's S Dunn's multiple range
Describing functional relationships between variables	Linear regression Pearson's r	Monotonic regression Spearman's r Kendall's tau	Cluster analysis[b] Principal components analysis Factor analysis	Cluster analysis[b]

[a]Analogous tests are placed side by side.
[b]The description of parametric or nonparametric will depend on the coefficient and algorithm used.

tions but are less likely to detect and quantitate differences, while parametric tests have more assumptions but greater power in establishing differences for any given data set.

All of the statistical tests discussed here assume that the measured variables are randomly sampled and that the variance of the individual sample groups is roughly the same. Failure to validate the first assumption indicates experimental bias in the sampling procedure. Homoscedasticity is the condition that sample variances are similar. If the sample variances are not similar, then some aspect of the samples is different (e.g., the underlying distribution of the variable may differ between samples). This may confound proper statistical analysis, or it may lead to additional discovery of relevant variables impacting the experimental system being investigated. Preliminary analysis of the data can indicate whether or not these assumptions can be safely made. Some statistical tests for these assumptions are listed in Table 5.

Parametric tests have the additional assumption that the measured variable comes from a population with an established (generally normal) distribution. If the distribution of the measured variable can be ascertained, then it is possible to

Table 5. Some Test Statistics for the Key Assumptions
Required for Performing Statistical Tests

Randomness[a]	(a) Wald–Wolfowitz runs test[c]
Homoscedasticity[b]	(a) Hartley's F^d_{max}
	(b) Cochran's C^d
Normality[c]	(a) Shapiro–Wilks W^e

[a]Samples must be taken randomly from a larger population.
[b]Sample groups should have similar variances.
[c]Parametric tests require that samples be taken from populations
whose values follow a normal/Gaussian distribution.
[d]See Ref. 10.
[e]See Ref. 2.
[f]See Ref. 13.

use a parametric test. However, a rigorous assessment of the distribution of a measured variable requires a relatively large data set (e.g., ≥ 30 independent samplings), which may be impractical for certain experimental systems. Then, an appropriate statistical test must be applied to verify the approximation of the data to the distribution. The commonly used parametric statistics presented in this chapter all require that the sampled variable come from a normally distributed population. If these parametric tests are applied to data sets whose distributions are, in fact, not normal, then their validity is undermined. Thus, preliminary testing of one's data is recommended, particularly when parametric tests are to be performed (see Table 5).

The alternative to using a parametric test is using a nonparametric or distribution-free analogue (see Table 4). While such tests are less likely to detect significant differences or recognize functional relationships in small data sets, the difference is not always substantial. In general, the successful application of such tests requires only a slightly larger sample size to distinguish equally significant differences (e.g., 8). The benefit of using such tests is that no assumption is made about the distribution of the data. Thus, nonparametric statistics are used if the distribution of the sample variable is not, or cannot be, ascertained with a desired certainty.

Last, a note on experimental bias: As noted, all statistical tests assume random sampling of variables (unbiased measurements). The elimination of bias from sampling is as critical to the proper statistical analysis as it is to any experimental analysis. While much has been published on how to design experiments to randomize samples and treatments, the criteria by which bias is eliminated are the same for all empirical analysis. Thus, homogenizing the control and experimental groups with respect to all variables except those being tested is the ideal.

Maximizing the conditions toward that ideal will yield data of greater scientific value.

3.2 Making Comparisons

One of the first questions asked in a study is, Do the control and experimental data differ? There are a variety of nonparametric and parametric tests for detecting significant differences between two or more groups of data. "Groups" may be different treatments or replicates, or blocks of the same treatment. Generally speaking, the null hypothesis for comparisons is that there is no difference between groups. The alternative hypothesis is that there is a difference. While a statistical test will indicate the mathematical significance of such a comparison, it is left to the researcher to justify its phenomenological relevance.

Parametric comparisons

In making parametric comparisons, the whole data set is summarized by one or a few numbers (parameters) which specify the distribution of the sampled variable. The most commonly used parameters for comparison are the mean and the variance. The following tests are all derived from the normal distribution, which is specified by the mean and variance, and as such require that the measured variables come from a normally distributed population. Transformations of raw data can be used to make the underlying distribution approximate normality, but deviations from absolute normality will weaken the validity of the parametric test.

To test the difference between two means, Student's t tests are used. If comparisons are made between two treatments per sample location, and many sample locations are used to compare the two treatments, a paired t test is used. The test statistic assumes that the number of samples is the same between treatments and that the variance of the two sample groups is the same. A more general comparison can be made with the two-sample t test. Fisher's least significant difference (LSD) is a useful and algebraically equivalent alternative to the t test. In using Fisher's LSD an interval is calculated to assess whether or not two sample values are significantly different. When more than two groups are to be compared with one another, multiple t tests should not be performed, as the probability of making a type I error increases. Although such applications can be found in the literature, the practice should be discouraged. This is because it is not always clear whether the stated α represents the value for individual tests or the sum total of comparisons. Instead, multiple range tests should be used, as outlined later.

The parametric analysis of variance (ANOVA) is based on the observation that variables controlled by similar phenomena give rise to homoscedastic sample populations. In other words, the variance of similar populations will be similar. The ANOVA will indicate whether two or more sets of data are significantly different for a given α, but it does not indicate the magnitude nor direction of that difference. For a single nominal variable, two sample groups can be tested for distinctiveness using an F test. The F statistic is simply the ratio of sample variances of each group. In more complex data sets, comparisons use the sum of squares (SS) and the mean of squares (MS), which are the component elements of the variance used to calculate an F statistic. The SS is equal to the sum of the square of the deviations of a group, and the MS is simply the SS divided by the degrees of freedom of the comparison.

The equations used to calculate these statistics will vary according to the evenness of sample sizes, the number of independent variables in an experiment, and the expected interaction between them. The simplest form, a one-way ANOVA, states that the total variability is equal to the sum of the between group variability and the within group/residual variability. In two-way ANOVA the effects of two independent variables, A and B, on the dependent variable can be assessed. Again, the total SS is partitioned into the SS of each independent variable, SS_A and SS_B, and the $SS_{residual}$. If the independent variables interact, an additional source of variability, $SS_{interaction}$, can be calculated. Each partial SS is converted to a variance (the MS) by division of the appropriate number of degrees of freedom and compared to the total variance using an F test. Each F test will indicate the relative significance of each variable's contribution to the total variance. In presenting the results of such analyses, it is most informative to include the type of ANOVA performed and the significance level of the F statistic obtained in the comparisons.

After an analysis of variance indicates that there is a significant difference within the analyzed data, an effort can be made to identify which subsets of the data are significantly different from the others. Multiple range tests control for this increased uncertainty in making multiple comparisons in a variety of ways. Some of the common tests found in the literature include the Scheffé, Tukey, Student–Neuman–Keulls, and Duncan tests. The first two of these are recommended because they are less likely to allow for a type I error. For a complete and succinct discussion of the different methods of making multiple comparisons, see Chew (1).

Nonparametric comparisons

The simplest tests for distinguishing two sets of data rely on binary coding of the data with regard to the null hypothesis. That is to say, an H_0 is chosen such that all the data points can be designated as (+) or (–). This may be the presence or absence of a particular scorable trait or measurement above a certain threshold. Using the sign test one can calculate the probability that a particular set of data will occur given the null hypothesis that the two alternatives are equally likely. The Wald–Wolfowitz test is based on the premise that if two groups are similar, then the order of their values will be randomly distributed when combined in a sequence. The null hypothesis is rejected if there are too few or too many runs of values coming from the same sample. Thus, it will detect most nonrandom trends in the data. The runs test can be used to confirm quickly that samplings are random.

Rank sum tests take into account the magnitude of the differences between samples and are more powerful than the preceding tests. They are most commonly used to compare the central measures of two or more data sets. The null hypothesis is that there is no difference between the central points of the compared data sets. They require that the data be ordered from lowest to highest value and that an ordinal number be assigned to each value in order. Thus, the lowest value is ranked as 1, the next higher value as 2, and so on. In cases of tied values, an arithmetic mean rank is calculated and assigned to all tied values.

The Wilcoxon T and Mann–Whitney U tests compare differences in location between two central measures. It can be shown that the statistical functions T and U are linearly related and, thus, interchangeable. A more general comparison of two groups uses the Kolmogorov–Smirnov D statistic. The D test detects all distribution differences (location, dispersion, and skewness), though the singular value calculated gives only an indication of the sum total of those differences. In instances where the primary difference is in the location of the distribution center, the Kolmogorov–Smirnov test is less likely to detect significant differences than the Wilcoxon and Mann–Whitney tests. The D test assumes that the measured variable comes from a continuous distribution and, therefore, can only be applied to cardinal data sets.

When more than two groups are being compared, one can use either the Kruskall–Wallis H or the Friedman S test. Respectively, these two tests are the nonparametric equivalents of parametric one-way and two-way ANOVA described. Significant values of H or S indicate that one or more of the groups are different from the rest. Identifying which group is different requires the use of Dunn's multiple range test. The procedure generates a range for every possible pairwise comparison between groups. Groups whose mean ranks differ by more than the critical range are considered significantly different.

Other tests specifically gauge the similarity (as opposed to the difference) between two data sets, or between a data set and a theoretical expectation. Collectively they are referred to as "goodness of fit" tests. Again, the null hypothesis is that there is no difference. Best known of these is the χ^2 test, which measures the difference between observed and expected values. A single comparison between an experimental data set and a theoretical model can be made. In cases where there are multiple groups to be compared, χ^2 contingency tables are used. Data are organized in rows (r) and columns (c) representing independent treatments, samplings, or variables. Calculations from such tables result in a χ^2 statistic with $(r - 1)(c - 1)$ degrees of freedom. Valid tests using contingency tables require uniform sampling because bias due to nonuniform sampling will be detected as a significant difference.

3.3 Functional Relationships

The second empirical question that can be addressed statistically is, How are the tested variables related? Observed relationships may be causal or simply correlative, and statistics alone will not distinguish between the two. The following methods merely establish the mathematical relationships between tested variables. The test statistics which use cardinal data will yield information regarding the nature (linear or curvilinear) and direction (positive or negative) of the functional relationship. Alternative forms that use ordinal and nominal data will only indicate the direction of that relationship.

It is probably worthwhile here to distinguish regression from correlation. The difference stems from the type of experiment performed. Regression analysis is performed on data from experiments where the independent variable is under direct control of the investigator. Correlation analysis is applied to experiments where there is no direct control of the independent variable by the investigator. While the two terms are commonly interchanged, the literature should keep the definitions distinct.

Simple parametric methods

The simplest functional relationship between two cardinal variables is a linear one. Linear regression is calculated as the change of one variable with respect to another. Thus, for any two variables there are two possible regression lines, designated by the coefficients b_{xy} and b_{yx}. The latter is more generally used, as x corresponds to the independent variable (i.e., under direct control or hypothesized to be causative) and as such is not impacted by y in the experiment. Fitting a regression line to a set of observed values entails positioning the line such that the sum of the deviations (vertical distances from the points to the line) is minimized. This is done by the least squares method, which uses the square of the deviations to prevent problems with negative values.

The commonly calculated correlation coefficient, Pearson's r, is also based on the least squares method. This correlation coefficient is mathematically related to the linear regression coefficient by the factor of the ratio of the variances of the two variables, but it does not reflect the slope of the linear relationship. Values of r range from -1 to 1, indicating how tightly the data adhere to a linear relationship. High absolute values indicate a strict adherence to the linear model. Low absolute values indicate a greater degree of dispersion in the data.

It is important to note that a regression line or a correlation coefficient is simply a description of the functional relationship present in the data. To evaluate the significance of either requires an ANOVA test. For instance, the variability of the data with respect to the regression line can be partitioned between the $SS_{regression}$ arising from the fitted values and the $SS_{residual}$ arising from the difference between the observed and fitted values. After dividing by their degrees of freedom, an F test of the two variances indicates the significance of a calculated regression curve. The same can be done for a correlation coefficient. As might be expected, the level of significance will depend on the number of data points and the degree of variation in the data. Such an analysis is important in determining the meaningfulness of a functional relationship.

Simple nonparametric methods

For noncardinal data, rank correlation methods exist which involve managing data as noted for rank sums tests. These methods establish the degree of monotonic relationship between two variables (i.e., how they increase or decrease together). These methods can also be applied to cardinal data when the variance of such data is large and/or the data values are not normally distributed. They are particularly useful when the mathematical relationship between the variables is not easily defined.

Iman and Conover have described a simple graphical procedure for graphing the monotonic regression curve (10). This distribution-free method results in a jagged curve which can be used as the basis for more precisely defining the relationship between the two variables.

There are two easily calculated nonparametric correlation coefficients. Spearman's correlation coefficient provides a monotonic description of the relationship between variables and is the nonparametric analogue to Pearson's r. Kendall's correlation coefficient, τ, is calculated in a somewhat different way but yields similar results. Both give values from -1 to 1, indicating the degree of monotonic association. The level of significance of these coefficients can be determined from prepared tables which can be found in many statistical texts.

More complex mathematical methods

There are more complex statistical methods for identifying and analyzing functional relationships in multivariate data. Among these are cluster, principal component, and factor analyses. The final results of these methodologies are presented in tables and/or in some graphical form (e.g., dendrograms or three-dimensional graphs). Because of the amount of data and computation required, such analyses are almost always performed by using computers. A comprehensive explanation of these methods is beyond the scope of this chapter, and only a brief discussion is given. For a more complete description of such methods, the reader is referred elsewhere (4,5,8,17,18).

Multivariate data arise when two or more individuals are compared by measuring many different characters or attributes for each individual unit. Characters may be the biochemical tests applied to strains in classification of bacterial species, or the physical and chemical measurements from soil samples. Individuals may be bacterial strains, collections of bacteria sampled from different parts of a root, or samples of soil. Character data may be cardinal (continuous or discontinuous), ordinal, or nominal.

Cluster analysis is an extension of correlation analysis and has been widely used in taxonomic studies and sequence analysis. A variety of methods can be used, but all follow the same general scheme: identify units by a number of characters, evaluate the relationship among them by use of a calculated coefficient, and sort the units accordingly by some algorithm.

Data are clustered according to similarities or dissimilarities in sample units, and numerical coefficients are most often calculated between individuals taken in pairs. Similarity coefficients seek to give expression to the proportion of shared characters (see Table 6). The simple matching coefficient calculates the proportion of matches (positive or negative) between two sample units, while the Jaccard and Dice coefficients are concerned only with the proportion of positive shared traits. Conversely, distance coefficients express some proportion of dissimilar characters. The most common forms are Euclidean measures, derived from the Pythagorean theorem applied in multiple dimensions. Euclidean distances start at zero (identical samples) and have no limit. The actual range of coefficient values in an analysis will be a function of the amount of character data and their scatter. The calculated coefficients (of either type) are often presented

Table 6. Similarity Coefficients[a]

Simple matching coefficient	$S_{SM} = (w + z)/(w + x + y + z)$
Jaccard's coefficient	$S_J = w/(w + x + y)$
Dice coefficient	$S_D = 2w/(2w + x + y)$

where

w = number of characters positive for both samples

z = number of characters negative for both samples

x = number of characters positive for sample A but not sample B

y = number of characters positive for sample B but not sample A

[a]See Ref. 5.

in a pairwise matrix, with column and rows specifying distinct sample units (e.g., bacterial strains in a taxonomic study). The information provided by the matrix may also be displayed graphically (e.g., as a dendrogram).

Different methods of clustering are available in software packages because clustering entails a compromise for which no ideal solution is available. The problem arises from attempting to take distances measured in many dimensions and express them in just two or three dimensions. While the literature may suggest preferred clustering methods for particular applications, researchers can validate their conclusions by checking to see whether alternative methods give generally similar results.

Principal component and factor analyses go beyond correlation to describe the elements responsible for clustering. Both methodologies assume that the principal-components or factors identified are standardized normal variables and are uncorrelated with one another. This assumption may restrict their applicability to in situ data from soil systems, because most of the commonly measured variables relevant to microbiological activity (e.g., biological activity, moisture, temperature, and chemical activity of various ions) are known to be functionally related and may follow nonnormal distributions. Nonetheless, the methods are outlined because applicable situations may still be found.

A multivariate data set can be condensed into a smaller collection of linear relationships, called principal components, which describe most of the variation in the data. The principal components can then be used to identify which combinations of variables are associated with each other and the degree to which they impact the system under study. Factor analysis is used to explain the relationships between measured variables in terms of unidentified variables called factors. The methodology assumes that a few unobserved factors control those relationships. Thus, factor analysis tests for the existence of independent controlling factors without identifying them. This aspect of indeterminancy facilitates the association of measurable variables that would not otherwise be thought to be related.

4. PLANNING AND PRESENTATION OF STATISTICAL ANALYSES

When designing any experiment, care should be taken so that the data acquired can be appropriately analyzed. With regard to statistical analyses, specific assumptions are made in the use of particular tests. These assumptions are essentially no different from those applied to experiments that are not analyzed with statistics. The major exception is that parametric tests require that the analyzed data be normally distributed.

In planning experiments where statistical analyses will be used, the first questions one would like to answer are, How large a sample? and How many replicates? The confidence limits defined earlier can be used to answer those questions. As can be seen from Table 3, the standard deviation and sample size affect the certainty of the measured mean. Thus, one can solve algebraically for the required sample size to detect a significant difference. The results of such an exercise are tabulated in Table 7. Because Chebyshev's theorem gives the maximum error, it yields an upper limit to the confidence interval for nonparametric data sets, given a specified α. In practice, most distribution-free tests require a minimal sample size of 5 to 10 data points to attain the standard levels of significance ($\alpha = 0.05$ or 0.01) if the sample values from different groups overlap. The precise sample size required depends on the test and the degree of relatedness of compared data sets, as well as the scatter of the data. Perusing statistical tables relevant to the nonparametric tests one expects to perform in an experimental study will readily reveal the sample sizes required for specific tests.

As for presentation, clarity should be the overriding concern in determining how to present statistical data. The choice between graphical presentation and tabulation of calculated values will depend on the question being addressed. Graphical analyses alone may not be sufficient, as they are more easily viewed subjectively. By including the appropriate mathematical measures, one can solidify relationships presented graphically. In all cases, however, the statistical methods used and the key determinants of their validity should be presented (21).

With regard to the methods described in this chapter, the following guidelines should be adhered to strictly. The sample size and the number of replicates in a given experiment should be clearly indicated. The measure of sample variation presented (e.g., error bars on graphs, MS in ANOVA) should also be specified. In presenting ANOVA results, the actual α level attained by the calculated F statistic should be used. In making multiple comparisons, the specified α level should

Table 7. Minimum Sample Size Required for Detecting Significant Differences Between Two Means Taken from Normal Populations

Difference in means[a]	$\alpha = 0.05$	$\alpha = 0.01$
$10s$	2	3
$5s$	3	4
$2s$	4	6
s	7	11
$0.5s$	18	30

[a]Calculated as a multiple of the standard deviation, s, of the sample.

represent the error rate for the whole experiment, not that of individual comparisons. And all such information should be placed in the legend of the figure or table used to present the results. Where simple functional relationships are reported, calculated coefficients should be presented along with the graphs of the primary data. If complex mathematical analyses are performed, a complete description should be presented in the methods section. Justification of the chosen analytical technique in phenomenologically relevant terms should also be given in instances where the reader is unlikely to be familiar with the applicability of a chosen method. Following these guidelines will facilitate meaningful understanding and judicious review of published experimental results.

5. EXAMPLE OF STATISTICAL ANALYSIS

The data in Table 8 were taken from a field release experiment of genetically modified bacteria. The purpose of the experiment was to test the effect of plasmid carriage on the colonization of sugar beet seedlings by *Pseudomonas fluorescens* strain LN.

Four seed inoculation treatments were analyzed; the host strain carried either (1) no plasmid, (2) pQ195, (3) pQ196, or (4) pQ197. Because of the high degree of variation noted in earlier experiments using this strain, 9 replicate trays were sown with seeds from each treatment. The 36 trays were then arranged randomly in a greenhouse. After 22 days, one seedling from each tray was removed for analysis. The phyllosphere bacteria were dispersed in excess distilled water by a bead beater, diluted, and plated. Viable counts were taken from selective and

Table 8. CFU per Gram Fresh Weight Recovered on Selective Media Containing Hg

No plasmid	pQ195	pQ196	pQ197
556000	110000	145000	104000[a]
317000	165000	605000	71900
431000	150000[a]	128000	160000
104000	93200	405000[a]	40000
538000	88300	496000	276000
665000	152000	574000	153000
357000[a]	191000	226000	63400
223000	67500	609000	155000
196000	324000	80600	64200

[a]Median value for each treatment.

nonselective plates. Only the data from the selective plates containing Hg are analyzed here. Significant probability (P) values are highlighted in bold in the accompanying tables (Tables 9, 10 and 11).

The statistical analysis involves comparing plate count data from four different treatments. To determine whether to use parametric or nonparametric tests, the raw data were subjected to evaluations listed in Table 5. The results for this example are presented in Table 9.

The design of the experiment suggests that the data would follow a random pattern for each treatment. Thus, the counts for each treatment should be randomly distributed above and below its median measurement. The Wald–Wolfowitz runs test indicated a highly probable number of runs for each treatment. Therefore, no evidence for experimental bias exists, and the data meet the requirement of random sampling. To test for homogeneity of the variance, the count data were analyzed using Hartley's F_{max} test. The test statistic was relatively high; that means that the probability the data are from distributions with similar variance is slight. Thus, the data set probably does not meet the criterion of homoscedasticity. To address the question of whether the sample data come from normally distributed populations, the Shapiro–Wilks test was performed for each of the treatments. The relatively low test statistics indicate that the data from three of the four treatments may not have been taken from normally distributed populations.

At this point a choice has to be made by the data analyst. Either the data must be transformed to yield a more homoscedastic and normally distributed data set,

Table 9. Summary of Test Statistics of Raw and Transformed Data

	Test statistic[a]	Probability (P)[b]
Wald–Wolfowitz runs test	6, 6, 7, 7	0.4, 0.4, 0.3, 0.3
Hartley's F_{max}		
raw data	8.81	**<0.05**
log transformed data	2.69	>0.05
ranked data	2.36	>0.05
Shapiro–Wilks test		
raw data	0.97, 0.86, 0.87, 0.88	0.8, **0.1**, **0.1**, **0.2**
log transformed data	0.89, 0.97, 0.87, 0.95	0.3, 0.9, **0.1**, 0.7

[a]Tests were performed on the whole data set or the four individual treatments, as required.
[b]P values indicate the probability that the null hypothesis is true given the observed data set. Significant P values, which likely disprove the null hypothesis, are highlighted in bold.

or parametric tests must be abandoned in favor of analogous nonparametric methods (see Table 4). While the logarithmic transformation is commonly applied to bacterial counts from the phyllosphere, it does not guarantee that the data will meet the requirements for performing parametric tests (7). However, in this case, the logarithmic transformation of the count data *does* yield a set that can be further analyzed by parametric ANOVA. As one can still perform nonparametric ANOVA on the untransformed data and generally obtain very similar results, we present the results of both methods for completeness. It should be noted that when using nonparametric ANOVA, such as the Kruskall–Wallis test, the homoscedasticity of the ranked data should be tested. In this example, the condition is met satisfactorily (see Table 9).

The analyses of variance are summarized in Table 10. The null hypothesis is that there is no difference in colony-forming unit (CFU) counts between treatments. Both the nonparametric and the parametric methods suggest the same conclusion. The probability that the null hypothesis is true is extremely low; therefore, one or more of the treatments is different from the others. To find out which treatments are significantly different, multiple comparison tests were performed. Again, both methodologies give similar results. Treatments 2 and 4 differed most significantly from the control, treatment 1. Treatment 3 resembles the control and is markedly different from treatment 4. The difference between 3 and 2 is not statistically significant, but the *P* values are relatively low.

The statistical analysis concludes by stating that carriage of plasmids pQ195 and pQ197 results in significantly lower numbers of recoverable *P. fluorescens* LN containing these plasmids. The biological interpretation of these results is somewhat less clear. What is responsible for the roughly twofold decrease in ob-

Table 10.   Summary of ANOVA Results[a]

Parametric ANOVA:					
Source of variation:	SS	df	MS	F	*P*
Between treatments	1.65	3	0.551	7.54	**0.001**
Within treatments (error)	2.34	32	0.073		
NP ANOVA:					
Source of variation:		df		H	*P*
Between treatments		3		13.5	**0.004**

[a]*P* values indicate the probability that the null hypothesis is true given the observed data set. Significant *P* values, which likely disprove the null hypothesis, are highlighted in bold. SS, sum of square; df, degrees of freedom; MS, mean of squares/variance; Fisher's F and Kruskall–Wallis H, test statistics; P, probability that the null hypothesis is true given the observed data set.

Table 11. Matrices of P Values for Multiple Comparison Tests[a]

Parametric: Tukey's test			
treatment	1	2	3
2	**0.023**		
3	0.97	0.061	
4	**0.002**	0.80	**0.007**
Nonparametric: Dunn's test			
treatment	1	2	3
2	**0.10**		
3	>0.99	0.26	
4	**0.013**	>0.99	**0.047**

[a]P values indicate the probability that the null hypothesis is true given the observed data set. Significant P values, which likely disprove the null hypothesis, are highlighted in bold.

served CFU? The reduced numbers may be due to reduced culturability of the strains carrying these plasmids, or the plasmids may not stably replicate with the bacteria in the phyllosphere environment. In contrast, plasmid pQ196 had no measurable effect on recovery. This may be due to differences that allow for more stable replication in the host, or there may be sequence differences which confer a selective advantage on the host carrying pQ196. All of these testable hypotheses arise from the careful statistical analysis of the data.

6. CONCLUSION

Statistical tests of hypotheses offer quantifiable analyses of otherwise confusing data. However, one should recognize that application of statistics post experiment is unlikely to reveal meaningful patterns unless the requirements for particular tests are integrated into the experimental design. This is best done by planning experiments with a clear idea of which statistical tools will later be applied to the data.

The nature of empirical data directly impacts the choice and validity of statistical tests. The common parametric tests outlined require that the measured variable come from an approximately normal distribution. Deviations from normality negatively impact the validity of parametric tests. Nonparametric tests can be used when the assumption of the corresponding parametric test cannot be met.

The most common statistical tests have been briefly described. The procedures for actually calculating the various test statistics can be found elsewhere (e.g., 2,3,10). Particular effort should be made to understand the conditions under which chosen tests can be most fruitfully applied. Failure to do so may allow inappropriate conclusions to be drawn from otherwise valid analyses.

Empirical analyses that fail to acknowledge or control unmeasured variables will be biased. Statistics will not eliminate nor obscure that bias. As long as the methods of analysis are presented fully, a fair evaluation of an experimental protocol is possible.

REFERENCES

1. V. Chew, *HortScience II*: 348–357 (1976).
2. L. D. Fisher and G. van Belle, *Biostatistics: A Methodology for the Health Sciences*, Wiley & Sons, New York (1993).
3. J. C. Fry, ed., *Biological Data Analysis: A Practical Approach*, Oxford University Press, Oxford (1993).
4. J. E. Jackson, *A User's Guide to Principal Components*. Wiley & Sons, New York (1991).
5. L. Kaufman and P. J. Rousseeauw. *Finding Groups in Data: An Introduction to Cluster Analysis*, Wiley & Sons, New York (1990).
6. C. van Kessel, D. J. Pennock, and R. E. Farrell, *Soil Sci. Soc. Am. J. 57*: 988–995 (1993).
7. Kinkel, M. Wilson, and S. E. Lindow, *Microbial Ecol. 29*: 283–297 (1994).
8. P. Kline, *An Easy Guide to Factor Analysis*, Routledge, New York (1994).
9. Loper, T. V. Suslow, and M. N. Schroth. *Phytopathology 74*: 1454–1460 (1984).
10. Neave and P. L. Worthington, *Distribution-Free Tests*. Unwin Hyman, London (1988).
11. G. E. Noether, *Introduction to Statistics the Nonparametric Way*, Springer-Verlag, New York (1991).
12. T. B. Parkin, *Agronomy J. 85*: 747–753 (1993).
13. Parkin and J. A. Robinson, Analysis of lognormal data. In *Advances in Soil Sciences*, Vol. 20 (B. A. Stewart, ed.), Springer Verlag: New York (1992).
14. Parkin and J. A. Robinson, Statistical treatment of microbial data. In *Methods of Soil Analysis*, Part 2. *Microbiological and Chemical Properties* Book Series No. 5 Soil Science Society of America, Madison, Wisconsin (1994).
15. Parkin, J. L. Starr, and J. J. Meisinger. *Soil Sci. Soc. Am. J. 51*: 1492–1501 (1987).

16. F. C. Powell. *Statistical Tables for the Social, Biological, and Physical Sciences.* Cambridge University Press, Cambridge (1982).
17. A. C. Rencher, *Methods of Multivariate Analysis*, Wiley & Sons, New York (1995).
18. H. C. Romesburg, *Cluster Analysis for Researchers.* Robert E. Krieger: Malabar, Florida (1990).
19. B. A. Rosner, *Fundamentals of Biostatistics*, 3rd ed, PWS-Kent, Boston (1990).
20. F. J. Wall, *Statistical Data Analysis Handbook*, McGraw-Hill, New York (1986).
21. W. G. Warren, *Can. J. For. Res. 16*: 1185–1191 (1986).

17

Ecological Considerations Involved in Commercial Development of Biological Control Agents for Soil-Borne Diseases

JOHN M. WHIPPS   Horticulture Research International, Wellesbourne, United Kingdom

1. INTRODUCTION

Biological control, the use of one organism to eliminate or control another, has received considerable attention over the last decade since the publication of the seminal book by Cook and Baker (1). It has stemmed from increasing public concern over the use of chemicals in the environment in general, as well as a reduction in the availability of previously widely used, effective fungicides. However, despite the huge amount of work carried out, reflected by an increasing number of books or reviews on the subject (2–12) the number of commercially available disease biocontrol agents for soil-borne pathogens is still small (Table 1). This raises the question as to why, with all the research effort directed to achieving commercially acceptable biocontrol agents, so few have made it to the marketplace. Numerous factors may be involved. For example, biocontrol agents may be less effective than existing chemical controls; they may be more expensive and difficult to produce, store, and apply; they may require changes in current agronomic practices to be used successfully. These are often commercially perceived problems or requirements (13) rather than problems with the biocon-

Table 1. Examples of Antagonists Available or in the Process of Registration for Use in Commercial Disease Biocontrol Preparations for Soil-Borne Plant Pathogens

Antagonist	Target pathogen	Disease/host	Product name and source
Bacteria and actinomycetes			
Agrobacterium radiobacter	*Agrobacterium tumefaciens*	Crown gall of crucifers, roses, and fruit trees	Diegall (Fruit Growers Chemical Co., NZ); Galltrol (AgBio Chem, Inc., CA, USA); NoGall (Root nodule Pty Ltd, Bio-Care, Australia); Norbac 84-C (New BioProducts Inc., CA, USA)
Bacillus subtilis	*Pythium ultimum* *Rhizoctonia solani*	Damping off of cotton and legumes	Kodiak and Epic (Gustafson, Inc. TX, USA)
Pseudomonas fluorescens	*Fusarium oxysporum* f. sp. *raphani*	Fusarium wilt of radish	BioCoat (S&G Seeds, BV, The Netherlands)
Streptomyces griseoviridis	*Alternaria brassicicola* *Fusarium oxysporum* f. sp. *dianthi*	Damping off of crucifers Carnation wilt	Mycostop (Kemira Agro Oy, Finland)
Fungi			
Coniothyrium minitans	*Sclerotinia sclerotiorum*	Sunflower disease	Coniothyrin (Russian government)

Biological control agent	Pathogen	Disease	Product
Fusarium oxysporum (nonpathogenic)	*Fusarium oxysporum* f. sp. *batatas*	Fusarium wilt of sweet potato	Japanese government
	Fusarium oxysporum	Fusarium wilt of tomato and carnation	Fo47 (Natural Plant Protection, France)
		Fusarium wilt of basil, carnation, and tomato	Biofox-C (S.I.A.P.A., Bologna, Italy)
Gliocladium catenulatum	*Pythium* spp.	Damping off	Primastop (Kemira Agro Oy, Finland)
Gliocladium virens	*Pythium ultimum* and *Rhizoctonia solani*	Damping off of bedding plants	GlioGard and SoilGard (W. R. Grace & Co., USA)
Peniophora (Phlebia) gigantea	*Heterobasidion annosum*	Stem and root rot of pine	P.g. suspension (Ecological Laboratories Ltd, UK)
Pythium oligandrum	*Pythium ultimum*	Damping off of sugar beet	Polygandron (Vyzkummy ustov, Czechoslovakia)
Trichoderma spp.	*Botrytis, Pythium, Sclerotinia* and *Verticillium* spp.	Fruit and vegetables	Trichodermin (Bulgarian and Soviet governments)
	Armillaria mellea	Honey fungus of trees	BINAB-T (Bio-Innovation AB, Töreboda, Sweden)
Trichoderma harzianum	*Heterobasidion annosum*	Stem and root rot of pine	
	Pythium sp.	Damping off of pea	F-stop (TGT, Inc/Cornell University, USA)

trol agents per se. Nevertheless, the need to understand the ecological behavior of any antagonist may be the key to furthering the development of biocontrol agents in general (14,15).

Manipulation of the soil environment by management processes such as the use of soil amendments (16), crop rotations, use of fumigants (17), or soil solarization (18) may be practical ways to control soil-borne pathogens, frequently stimulating activity of the natural soil microbiota and providing a general soil suppressiveness to disease. Indeed, many soils gradually become specifically suppressive to a single disease if the same crop is grown repeatedly in the same location, resulting in disease decline soils. This is related to the activity of individual or select groups of antagonistic microorganisms which develop during some stage of the life cycle of the pathogen. Examples include wheat (*Triticum* spp.), sugar beet (*Beta vulgaris*), and potato (*Solanum tuberosum*) (19–21). Nevertheless, despite the obvious value of these soil management approaches for disease control, it must be assumed that commercial development of biocontrol will focus on the production of specific biocontrol inocula for repeated, possibly inundative use, much as chemicals are used currently. Consequently, to allow successful development of any biocontrol agent, its interactions with the pathogen, the crop, the natural resident microflora, and the environment must be examined (Fig. 1). Such ecological studies are time-consuming; the spray and

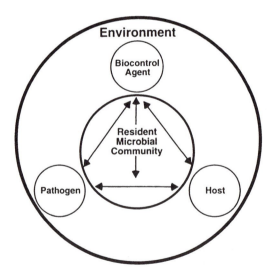

Figure 1. Schematic diagram of the potential interactions among a biocontrol agent, a pathogen, and the host plant and the additional interrelationships with the resident microflora and environment.

slay techniques devised for developing pesticides do not lend themselves readily to this type of work and so new conceptual approaches are required.

This chapter will identify areas where ecological studies have helped elucidate biocontrol activity and demonstrate how the knowledge gained in this way can be applied to the development of successful biocontrol agents in the future.

2. NATURAL BIOCONTROL

2.1 Suppressive Soils

Various soils from around the world are known to be naturally suppressive to a range of different soil-borne pathogens (22). In these soils, the pathogen may be present but does not cause disease on a susceptible host. This suppressiveness is related to both physicochemical characteristics and microbial components in the soil and is most encouraging in that it demonstrates that biocontrol can be achieved in the field. This is the ultimate commercial target. Understandably, therefore, much effort has gone into obtaining detailed analysis of the physicochemical environments and microbial components present in these soils.

The physicochemical characteristics of a soil have often been shown to be involved in soil suppressiveness, but specific characteristics apparently responsible for suppression of one pathogen in one soil may not always be involved in suppression of the same pathogen in another soil. Also, a soil which is suppressive to one pathogen is not necessarily suppressive to another. For example, soil suppressiveness to fusarium wilt (a disease in which the pathogen is almost entirely confined to the vascular system, resulting in loss of turgidity, usually in leaves, and, consequently, leaf collapse and fall) of banana (*Musa* spp.) caused by *Fusarium oxysporum* f. sp. *cubense* in Central America was correlated with the presence of montmorillonite clay (23), whereas soil suppressiveness to black root rot (a disease in which root infection by the pathogen is accompanied by cell death and root blackening) of tobacco (*Nicotiana tabacum*) caused by *Thielaviopsis basicola* in Switzerland was related to the presence of vermiculitic clays (24). In contrast, in the United States, suppressiveness to black root rot of tobacco was dependent on the interactions among soil pH, base saturation, and exchangeable aluminium rather than clay content (25). Such complex interactions are known to be involved in suppressiveness in other soils. For example, low pH values and high levels of organic matter, calcium, potassium, and magnesium were associated with suppression of *Pythium aphanidermatum* in a Mexican Chinampa agricultural system (26), but high calcium content was the single most important physicochemical characteristic related to suppressiveness to *P. splendens* in Hawaii, although several other factors with some influence were identified (27,28).

In many of these suppressive soils, biological components of suppression were also identified. In both the Mexican Chinampa system and the Hawaiian soils suppressive to *Pythium*, a very active soil microbial population was present. Indeed, suppressiveness to fusarium wilt in the famous Châteaurenard region of France is related to the combined effect of a competition for nutrients between the general soil microbiota and the entire *Fusarium* population as well as competition between pathogenic and nonpathogenic *Fusarium* species (29).

Numerous other specific microorganisms or groups of organisms have been identified with soil suppressiveness. For example, Swiss soils suppressive to black root rot of tobacco favor development of fluorescent pseudomonads, and *Pseudomonas fluorescens* CHA0 has been strongly suggested as the microbe responsible for suppressiveness in this soil (24,30). Both fluorescent pseudomonads and nonpathogenic fusaria have been implicated in suppression of fusarium wilt (31), various pseudomonads in suppression of take-all in wheat (a root disease caused by *Gaeumannomyces graminis* var. *tritici* in which the pathogen initially causes black lesions on seminal and crown roots and generally results in plant death some time after heading) (32), *Pythium nunn*, and *P. oligandrum* in suppression of other *Pythium* spp. (33,34), *Trichoderma* spp. in suppression of *Rhizoctonia solani* (35), *Phialophora graminicola* in suppression of take-all in grass pasture or turf (a disease similar to take-all in wheat but caused by *G. graminis* var. *avenae*) (36), and *Coniothyrium minitans* and *Sporidesmium sclerotivorum* in suppression of *Sclerotinia sclerotiorum* (10,37,38). There is also evidence that mycorrhizal fungi can be involved in natural disease suppression. The ectomycorrhizal fungus *Pisolithus tinctorius* can protect *Eucalyptus* spp. from *Phytophthora cinnamomi* (39), and arbuscular mycorrhizas may decrease severity of citrus root rot (a root disease of *Citrus* spp.) caused by *Phytophthora parasitica*, verticillium wilt of cotton (*Gossypium* spp.), and *Thielaviopsis basicola* root rot of cotton (40–42).

The suppressiveness identified in many soils has often been related to microbial competition for nutrients. Although competition for carbon and nitrogen has been shown to be involved in suppression of *Fusarium* or *Pythium* in some soils (43–46), it is competition for ferric iron (Fe^{3+}) that has generated most interest. Numerous studies have shown that fluorescent *Pseudomonas* spp. are efficient competitors for ferric iron, mediated by the production of iron-chelating compounds termed siderophores (47,48). Suppression of disease development of several soil-borne pathogens by fluorescent pseudomonads has repeatedly been suggested to occur through siderophore action (43,49–54). However, production and activity of siderophores are pH-dependent, with ferric iron availability increasing with decreasing pH. Consequently, this implies that fluorescent pseudomonads causing disease suppression through siderophore production would be relatively inactive in soils of low pH (55). Further, siderophore –Fe^{3+} complexes can be taken up by other microorganisms in the soil and possibly the pathogen itself, making this mode of action very difficult to identify with certainty. Soil pH may also influence nutrient availability and microbial activity in general and thereby have a direct effect on disease suppression (56,57).

Suppression may also be related to production of antibiotics by soil microbiota. Until recently, direct evidence for this hypothesis was lacking as a result of difficulties in identifying small quantities of specific metabolites in soil. Now, through improved analytical techniques, this has been confirmed and extended by molecular studies of mode of action. Several researchers have produced transposon mutants lacking antibiotic production in vitro which have been accompanied by decreased biocontrol activity in in vivo bioassays. By complementing the mutant gene from a gene bank, antibiotic production and biocontrol activity could be restored (58–61). Nevertheless, the fact that an organism is capable of producing an antibiotic or a siderophore in vitro is still no guarantee of a direct role for that moiety in disease suppression in the field (14,62,63).

The numerous complex interactions between physicochemical characteristics and microbial populations may well explain why specific suppressions cannot always be transferred easily to other soils and why there are so many different types of suppressive soils. Nevertheless, these soils have provided excellent sources of antagonists for future development. Indeed, the commercially available biocontrol product Mycostop (Kemira Agro Oy, Helsinki, Finland) contains a *Streptomyces griseoviridis* isolate which originated from Finnish sphagnum peat, suppressive to many soil-borne pathogens (64). Similarly, a *Fusarium* sp. (Fo 47) undergoing registration for control of fusarium wilt in France was isolated from fusarium-suppressive soil (C. Alabouvette, INRA, Dijon, France, personal communication). Several pseudomonads from suppressive soils are also undergoing intensive study with potential for commercialization (24,65,66).

Importantly, the environmental limits of activity and modes of action of many of these potential biocontrol agents have been identified during the ecological phase of study. This should help any subsequent commercial developmental stage by establishing the soil–crop–pathogen combination within which biocontrol activity should be expected. It is unrealistic to expect that a broad-spectrum biological control agent that is active in a range of soils will be found. Nevertheless, even if one should be discovered, it is an arguable point whether such an agent should be used unless extensive environmental safety testing is first completed.

2.2 Container Media and Composts

Other areas in which natural biological disease control has been found to occur include container media and composts (67). Composted softwood and hardwood barks have been widely examined as materials for incorporation into soilless potting mixes and have been shown to be naturally suppressive to damping off (collapse and death of seedlings which result from a lesion caused by a pathogen at approximately soil level) caused by *Pythium ultimum* in cucumber, damping off in bedding plants and radish due to *Rhizoctonia solani*, and also *Phytophthora* root rot of lupins (*Lupinus angustifolius*) and woody ornamentals such as *Aucuba japonica* (68–73). In composted hardwood bark from Ohio, United States, this effect appears to be due to high levels of an antagonistic isolate of *Trichoderma harzianum* although bacterial antagonists such as *Bacillus cereus*, *Enterobacter cloacae*, *Pseudomonas* spp., and *Xanthomonas maltophilia* may also play a role (74). Recent research in Japan has suggested the application of bark compost to farmland to control fusarium wilts may be due to the presence of antifungal lipids such as sitosterol (24R-ethylcholest-5-en-3β-o1) of low polarity and sterols in the composts (75,76). Recently, complex amendments based on pine bark with inorganic additives such as $Al_2(SO_4)_3$, $CaCl_2$, KCl, NH_4NO_3, and $(NH_4)_2SO_4$ have been devised to give control of *R. solani* and *Pythium aphanidermatum* in soil and appear to be related to increased populations of *T. harzianum* and *Penicillium oxalicum* (77,78). This has been extended in Taiwan, where addition of a complex mixture of materials (4.4% bagasse, 8.4% rice husks, 4.25% oyster shell powder, 8.25% urea, 1.0% KNO_3, 13.2% calcium superphosphate, and 60.5% siliceous slag) termed S-H mixture (79) to soils has resulted in the control of an extensive range of diseases, apparently related to different components and properties of the mixture. Thus, with a pH of 8, S-H mix inhibits development of *Plasmodiophora brassicae*, the cause of clubroot of brassicas, and spore germination of some fusaria is decreased; the high calcium levels inhibit *P. brassicae*, *Pythium* spp., and fusarium wilt; urea inhibits *Pythium* spp., *Fusarium* spp., *Sclerotium rolfsii*, and *Phytophthora* spp.; oyster shell stimulates antagonistic chitinolytic microorganisms (capable of degrading chitin, which is a polymer of *N*-acetyl-D-glucosamine); slag provides high levels of calcium, aluminium, and silica which inhibit germination of chlamydospores (asexual spores, frequently resting stages, primarily for perennation) of *Fusarium*; and, overall there may be a stimulation of plant growth from the nutrients alone. It remains to be seen whether this empirical approach to disease suppression is adopted outside Taiwan.

Other useful composted materials incorporated into potting media include liquorice roots, which suppressed damping off in cucumber caused by *Pythium aphanidermatum* (80) and grape marc (residue left after juice extraction from grape, *Vitis* spp.), which decreased *R. solani* disease on radish (*Raphanus sativus*) and *Sclerotium rolfsii* infection on chickpea (*Cicer arietinum*) and bean (*Phaseolus vulgaris*) (81).

Another potentially useful source of organic amendment involves composted municipal waste. Composted sewage sludge incorporated into soil has been shown significantly to reduce aphanomyces root rot of peas (*Pisum sativum*) caused by *Aphanomyces euteiches*; rhizoctonia root rot of bean, cotton, and radish caused by *Rhizoctonia solani*; sclerotinia drop (a basal rot) of lettuce

(*Lactuca sativa*) caused by *Sclerotinia minor*; fusarium wilt of cucumber (*Cucumis sativus*); and phytophthora crown rot (an infection of the crown of the roots at soil level and the stem due to *Phytophthora capsici*) of pepper (*Capsicum frutescens*) (16). The control was correlated with a general increase in microbial activity, and there was no evidence that specific organisms were involved. Similarly, composted organic household waste incorporated into sand gave reduction of root rot of peas, beans, and beetroot (*Beta vulgaris*) caused by *Pythium ultimum* and *Rhizoctonia solani* and *Mycosphaerella pinodes* on peas (82,83). Nevertheless, some care should be taken with such materials as they may cause other problems by introducing heavy metals into soils and there may be a risk that some human and plant pathogens are introduced if the composting process is not carefully controlled.

3. INOCULANTS FOR BIOLOGICAL DISEASE CONTROL

The concepts and difficulties associated with obtaining commercially successful disease biocontrol agents have been considered before (84–91). The development of a biocontrol agent generally follows a stepwise progression from selection, screening, inoculum production, and formulation to application and field testing. At any stage, the biocontrol agent may fail and it may be necessary to return to a previous step and reevaluate the situation. However, if ecological principles are applied throughout the procedure, it must enhance the chances of obtaining a successful agent in the end. These principles involve understanding the pathogen, crop, and methods of cultivation; the microbiological characteristics and environment where biocontrol must take place; and finally the ecological properties of the antagonist itself. The antagonist must be targeted to work at the correct stage or stages of the life cycle of the pathogen, where it will have maximum impact and, ideally, thus provide a level of reproducible control which companies and scientists will consider acceptable enough to warrant commercialization.

3.1 Selection

Soil environments where diseases should occur but fail to materialize are obvious sites for initial selection of antagonists. As mentioned earlier (Section 2.1), disease-suppressive soils have already provided a range of potential antagonists which are undergoing development. Another option is to select microorganisms from the environment where biocontrol is required to take place. This could include the seed, root, or hypocotyl (the region of the stem beneath the cotyledons [seed leaves] and directly above the young root of a seedling) of a host plant, ideally growing in the soil or medium where the plant would be grown subsequently (91–96). Alternatively, the soil and the growing medium itself where the crop would be grown could be sampled, and any antagonists isolated should, theoretically, be adapted to that environment. However, if the pathogen is active in the soil of choice and disease is not in decline, then suitable antagonists may never be obtained from that site.

An alternative approach is to bait soils with pathogen(s) and isolate antagonists directly from pathogen structures. Each isolate would have the potential to attack the pathogen and be ecologically adapted to the environment of the pathogen. This procedure has been widely used to obtain antagonists from sclerotia (long-lived, firm masses of hyphal tissue involved in perennation) of a range of pathogens, including *Sclerotinia minor, S. sclerotiorum, Sclerotium cepivorum*, and *Phymatotrichum omnivorum* (97–100). Mycelial mats and spores of pathogens as well as pathogen-infected plant material have also been incorporated into soil as baits for antagonists (101–105).

Selection of antagonists can also be made from soils enriched with materials such as chitin. Populations of chitinolytic microorganisms such as actinomycetes (*Micromonospora, Nocardia*, and *Streptomyces*), and some bacteria including *Serratia marcescens* have been shown to increase under such conditions (106,107) and to provide effective biocontrol of a range of pathogens. Soil enrichment with laminarin-containing material has also recently been used to isolate a β-1,3-glucanase-producing *Pseudomonas cepacia* with ability to decrease disease incidence caused by *Rhizoctonia solani, Sclerotium rolfsii*, and *Pythium ultimum* (108). This reflects a prior assumption that biocontrol in the environment can take place through the activity of fungal cell wall–degrading enzymes produced by antagonists.

3.2 Screening

When potential antagonists have been collected, they have to be assessed for reproducible biocontrol activity in comparison with some kind of standard strain or strains. It is often at this stage that ecological principles are abandoned in favor of some kind of arbitrary in vitro screening procedure involving agar plate tests. The literature is full of such examples and in many cases these are taken no further. The reasons for this are obvious: such tests are quick and easy and large numbers of antagonists can be assessed rapidly. However, they bear little or no resemblance to the environmental and microbiological factors facing the antagonist where biocontrol is expected to act, and so frequently organisms viewed as successful in in vitro screens go on to fail in larger container, glasshouse, or field trials. Nevertheless, they can provide information on the limits to growth and help with subsequent inoculum production, formulation, and physiological studies and thus have some basic value. Information on potential modes of action such as production of antibiotics, siderophores, and lytic enzymes can also be determined (and some screens are positively designed for this), but other modes of action, such as competition, cannot. Biocontrol mechanisms involving the plant, including induced resistance or niche exclusion, cannot even be considered. Similarly, rhizosphere competence and plant growth promotion, useful features exhibited by several known biocontrol agents, are also missed in in vitro agar plate studies. Consequently, screens should logically be designed to mimic as closely as possible the natural situation where biocontrol is required and yet still allow a rapid throughput of isolates to optimize the identification of useful ones. A tall order indeed, and one which may require a different, potentially time-consuming approach for each host–pathogen combination considered. A selection of plant-based bioassays of varying ecological relevance is described in the next section.

In vivo bioassays

In vivo bioassays can take many forms and can include simple excised plant tissue tests, direct tests on fungal propagules (structures involved in perennation and dissemination such as spores or sclerotia), and assays involving seedlings or even more mature plants. They are generally carried out under standardized controlled environmental conditions and are designed to cope with large numbers of potential control agents. Consequently, they are also easily adapted to allow investigation of the effect of any number of variables influencing the possible tripartite interactions of biocontrol agent, pathogen, and host plant and often these are of great value in later steps in development of a biocontrol agent. Environmental conditions, nutrient availability, and inoculum form have all been investigated using such assays.

Several bioassays involving excised plant tissues are known. For example, potato seed pieces were used to assess fluorescent pseudomonads for suppression of the potato soft rot pathogen *Erwinia carotovora* (109); an excised cotton hypocotyl assay has been used to assess biocontrol of post-emergence damping off caused by *Rhizoctonia solani* (110); and an excised tomato stem, celery (*Apium graveolens*), or lettuce petiole (stalk that attaches a leaf blade to the stem) assay has been used to assess biocontrol potential against *Sclerotinia sclerotiorum* (111), all designed to mimic ecologically realistic situations. Use of plant tissue in this way has been extended in Australia, where sclerotia of *S. sclerotiorum* were placed within stem tissue of bean and buried in soil to simulate the natural occurrence of sclerotia (99). Any potential biocontrol agent must first penetrate the plant tissue to attack and destroy the sclerotia within.

Inoculation of sclerotia with antagonists or burial of sclerotia or other pathogen propagules in soil or other substrates accompanied by treatment with antagonists has been widely used as a first step in assessing for biocontrol potential. After set periods, the pathogen propagules can be recovered, either observed directly or surface-sterilized with, for instance, sodium hypochlorite or ethanol, and their viability assessed. These procedures have been used for sclerotia of *Phymatotrichum omnivorum*, *Sclerotinia minor*, *S. sclerotiorum*, *Sclerotium cepivorum*, *Sclerotium rolfsii*, and *Verticillium dahliae* (97–99,112–117), and spores of *Pythium* and *Phytophthora* (118,119). These assays provide direct quantitative estimates of ability to reduce inoculum potential of the pathogen.

Although a number of excised tissue or pathogen propagule bioassays have been used to assess biocontrol potential, the majority of in vivo bioassays have involved use of seed or seedling tests in which control is estimated in terms of seed or seedling infection or seedling emergence or survival after a set time. Essentially there are three basic formats: (1) the putative antagonist is applied to the seed directly before planting into pathogen-infested soil or growing medium, (2) the antagonist is applied to pathogen-infested soil or growing medium after seeds or seedlings are sown, or (3) the antagonist is incorporated into the soil or growing medium containing the pathogen before seeds or seedlings are planted. Depending on whether bacteria or fungi are the antagonists under study, they may be applied simply as cell or spore suspensions or solid-substrate inocula or after some kind of formulation (120). Nevertheless, in a primary in vivo bioassay screen, procedures are generally simple and rapid and bear less resemblance to the natural situation than more complex secondary screens involving greater numbers of plants, cultivars, and strains of pathogen. For example, tissue culture plate assays have been devised to select antagonists against *Pythium* spp. (94,121) and a petri dish assay has been designed to identify bacteria active against a range of pathogens (122). These types of assay, lacking soil or normal growing media, have been extended to include tests for antagonists against *Phytophthora megasperma* f. sp. *medicaginis* in culture tubes containing vermiculite (123), for *Gaeumannomyces graminis* causing take-all of wheat in pots of sand infected with the pathogen (91), and for *Pythium ultimum* and *Rhizoctonia solani* in styrofoam cups containing a mixture of sand and vermiculite (124).

Slightly more complicated assays may contain mixtures of soil and inert material. For instance, bacteria suppressive to take-all of wheat were screened in tubes containing soil infested with the pathogen placed on a column of vermiculite (85) and antagonists against *Aphanomyces euteiches* f. sp. *pisi* in cones containing soil placed on vermiculite which was then watered with a suspension of zoospores of the pathogen (125).

In spite of the speed of tests involving artificial substrata, most screening tests involve assessments carried out in soil, a soil mix, or peat-based potting mixes. Ideally, these materials would be naturally infested with the pathogen. But, generally, in order to standardize conditions and achieve reproducible levels of disease, the soil or potting mixture is artificially infested, and frequently it is steamed or sterilized before use to remove unwanted or variable microbial ef-

Table 2. Examples of Studies Screening for Biocontrol Activity Using in vivo Bioassays in Soil and Peat-Based Potting Mixes

Pathogen	Host	Reference
Aphanomyces cochlioides	Sugar beet	138
Erwinia carotovora	Potato	109
Gaeumannomyces graminis	Wheat	92, 103, 130, 132
Phytophthora megasperma	Soybean	139
Pythium aphanidermatum	Cucumber	44
P. ultimum	Cotton	140
	Pepper	131
	Snapdragon	84, 129
	Sugar beet	96
Rhizoctonia solani	Cotton	141
	Lettuce	142
	Pepper	84, 129
	Radish	105, 141
	Wheat	129, 141
Sclerotium cepivorum	Onion	143
S. rolfsii	Bean	113

fects. A huge number of screens of this type, varying markedly in scale and complexity, have been devised, examples are given in Table 2. They have also been used to screen for plant growth–promoting rhizobacteria or fungi which may act either by control of minor pathogens on the roots or by direct growth stimulation perhaps mediated by hormone production (126–128). Indeed, plant growth promotion has been found quite frequently in pathogen-minus controls in biocontrol screens (e.g., 129–132).

The same basic procedures have also been adapted to screen for rhizosphere competence in potential biocontrol or plant growth–promoting microorganisms (133–135), growth in the rhizosphere is viewed as an ecologically desirable attribute of any control agent of root-infecting pathogens (3).

Subsequently, such in vivo bioassays are generally repeated, scaled-up, and varied to gather as much information as possible to help achieve reproducible control in the field. Environmental limits on antagonist growth and methods of inoculum production, formulation, and application are all examined concomitantly. Eventually, field trials of the most promising biocontrol agents are carried out, and frequently it is at this stage that even these isolates fail to give the control required or desired.

3.3 Field Trials

Numerous biocontrol agents which have been selected through various types of screening procedures have now gone on to large-scale greenhouse or field testing. Some examples are given in Table 3. Significantly, many more have been tested than have reached the stage of commercialization (see Table 1 for comparison). Weller (3) discussed the problems found with many results of tests using bacteria in the field and these arguments hold equally for many later trials carried out with both bacteria and fungi. Essentially, inconsistent levels of disease control were obtained from test to test and from place to place, despite all the careful screening and preparatory work leading up to the field test. This result emphasizes the influence of the environment on biological control. In addition, some agents gave reproducible control but levels were too low to be commercially acceptable; others were uneconomic because of the quantities of inoculum required (4), and, in some instances, antagonists were less effective than existing chemical controls. Consequently, it is worthwhile considering the attributes of some current commercially available biocontrol agents for soil-borne plant pathogens.

4. COMMERCIALLY SUCCESSFUL BIOCONTROL AGENTS

Deacon (14) reviewed the attributes of the two major disease biocontrol agents (*Phlebia gigantea* and *Agrobacterium radiobacter*) used routinely on an international scale. These will not be referred to again; suffice it to say that their selection and mode of use showed remarkable parallels and were based on sound ecological principles. Perhaps most importantly, both antagonists control specific disease problems in specific environmental niches and do not resemble broad-spectrum fungicides or bactericides.

Recently, new products based on *Streptomyces griseoviridis* (Mycostop [Kemira Agro Oy, Helsinki, Finland]) and *Gliocladium virens* (GlioGard and SoilGard [Grace-Sierra, Milpitas, California]) have entered the marketplace after a long and intensive period of screening and developmental work (64,136,137). Again, both are aimed at relatively small and specialized markets, the former for control of damping off of crucifers and carnation wilt, the latter for control of damping off of bedding plants. Significantly, control of carnation wilt and damping off of bedding plants takes place under glass where the environment is controlled; control of damping off in general requires protection for a short time only, and control of damping off in bedding plants involves the use of partially sterilized potting mix. All these factors serve to reduce environmental and microbial variability and the period when control is required and must indicate the way forward for biological control provided the market can give a satisfactory return on the cost of development.

Numerous other biocontrol agents are now approaching commercialization

Table 3. Examples of Biocontrol Agents Tested in Large-Scale Trials in Soil in the Greenhouse or the Field

Antagonist	Pathogen	Crop	Reference
Bacteria			
Bacillus subtilis A.13	*Rhizoctonia solani*	Wheat	129
Bacillus pumilus	*Gaeumannomyces graminis*	Wheat	144
Enterobacter aerogenes	*Phytophthora cactorum*	Apple	145
E. cloacae	*Sclerotinia homeocarpa*	Turfgrass	146
Pseudomonas spp.	*Aphanomyces euteiches*	Pea	125
	Erwinia carotovora	Potato	147
	Pythium ultimum	Cotton	148
	Rhizoctonia solani	Cotton	148
P. fluorescens	*Gaeumannomyces graminis*	Wheat	149
P. fluorescens 2-79	*G. graminis*	Wheat	32
P. fluorescens Q29z-80	*Pythium ultimum*	Chickpea	150
Fungi			
Coniothyrium minitans	*Sclerotinia sclerotiorum*	Sunflower Lettuce	151 152
Fusarium oxysporum (nonpathogenic)	*Fusarium oxysporum*	Tomato, muskmelon	31
Laetisaria arvalis	*Pythium ultimum*	Table beet	153
Penicillium oxalicum	*Pythium ultimum*	Chickpea	150
Pythium oligandrum	*Aphanomyces cochlioides*	Chickpea	150
	Pythium ultimum	Sugar beet	McQuilken and Whipps (unpublished)
Sporidesmium sclerotivorum	*Sclerotinia minor*	Lettuce	154
Talaromyces flavus	*Verticillium dahliae*	Aubergine	155
Trichoderma harzianum	*Rhizoctonia solani* *Sclerotium rolfsii*	Sugar beet Peanut	156 157
Verticillium biguttatum	*Rhizoctonia solani*	Potato	158

(R. D. Lumsden, USDA, Beltsville, Maryland, personal communication), and it will be interesting to see whether these also have specific ecological niches for activity.

5. CONCLUSIONS

Clearly, to achieve commercial success, any biocontrol agent must be ecologically adapted to its environment of use. Without this attribute it will never work reproducibly. Activity may be enhanced by reducing environmental variability either through working under glass or through partially sterilized soil or artificial potting mixes. Optimizing inoculum quality, formulation, and application technologies may also help. However, any biocontrol agent must also meet the exacting criteria of the marketplace. Control equivalent to currently available pesticides, appropriate shelf life, and ease of use, giving an acceptable cost–benefit balance to a company, are required. Unfortunately, although in many cases reproducible biocontrol has been achieved in the field or in large-scale greenhouse trials already, development of these potential biocontrol agents cannot go forward because of such commercial considerations. Nevertheless, if pesticides continue to be withdrawn and there is a public demand for biological disease control, there are many candidate biocontrol agents ready for further development for which the ecological groundwork has already been carried out.

ACKNOWLEDGMENTS

This work was supported by the U.K. Ministry of Agriculture, Fisheries and Food (MAFF) and the Biotechnology and Biological Sciences Research Council (BBSRC).

REFERENCES

1. R. J. Cook and K. F. Baker, *The Nature and Practice of Biological Control of Plant Pathogens*, American Phytopathological Society, St. Paul, Minnesota, pp. 539 (1983).
2. R. J. Cook, *Phil. Trans. R. Soc. Lond. B 318*: 171 (1988).
3. D. M. Weller, *Annu. Rev. Phytopathol. 26*: 379 (1988).
4. P. B. Adams, *Phytopathology 28*: 59 (1990).
5. D. Hornby, ed., *Biological Control of Soil-Borne Plant Pathogens*, CAB International, Wallingford, England, p. 479 (1990).
6. R. Baker, *Crop Prot. 10*: 85 (1991).

7. A. B. R. Beemster, G. J. Bollen, M. Gerlagh, M. A. Ruissen, B. Schippers, and A. Tempel, eds., *Biotic Interactions and Soil Borne Disease: Developments in Agriculture and Managed Forest Ecology*, Vol. 23, Elsevier, Amsterdam, p. 428 (1991).

8. J. A. Lewis and G. C. Papavizas, *Crop Prot. 10*: 95 (1991).

9. E. B. Nelson, *Handbook of Applied Mycology*. Vol. 1, *Soil and Plants* (D. K. Arora, B. Rai, K. G. Mukerji, and G. R. Knudsen, eds.), Marcel Dekker, New York, Basel, Hong Kong, p. 327 (1991).

10. R. D. Lumsden, *The Fungal Community: Its Organization and Role in the Ecosystem* 2nd ed. (G. C. Carroll and D. T. Wicklow, eds.), Marcel Dekker, New York, p. 275 (1992).

11. I. Chet, ed., *Biotechnology in Plant Disease Control*, Wiley-Liss, New York, p. 373 (1993).

12. J. M. Whipps and M. P. McQuilken, *Exploitation of Microorganisms* (D. Gareth Jones, ed.) Chapman and Hall, London, p. 45 (1993).

13. K. A. Powell and J. L. Faull, *Biotechnology of Fungi for Improving Plant Growth* (J. M. Whipps and R. D. Lumsden, eds.), Cambridge University Press, Cambridge, p. 259 (1989).

14. J. W. Deacon, *Biocontrol Sci. Technol. 1*: 5 (1991).

15. H. C. Huang, *Can. J. Plant Pathol. 14*: 86 (1992).

16. R. D. Lumsden, J. A. Lewis, and G. C. Papavizas, *Environmentally Sound Agriculture* (W. Lockeretz, ed.), Praeger Press, New York, p. 51 (1983).

17. D. E. Munnecke and S. D. van Gundy, *Annu. Rev. Phytopathol. 17*: 405 (1979).

18. A. J. L. Phillips, *Plant Pathol. 39*: 38 (1990).

19. D. Hornby, *Soil-Borne Plant Pathogens* (B. Schippers and W. Gams, eds.), Academic Press, London, p. 133 (1979).

20. M. Hyakumachi, K. Kanzawa, and T. Ui, *Biological Control of Soil-Borne Plant Pathogens* (D. Hornby, ed.), CAB International, Wallingford, England, p. 227 (1990).

21. A. Tanii, T. Takeuchi, and H. Horita, *Biological Control of Soil-Borne Plant Pathogens* (D. Hornby, ed.), CAB International, Wallingford, England, p. 143 (1990).

22. R. W. Schneider, ed., *Suppressive Soils*, American Phytopathological Society, St. Paul, Minnesota, p. 88 (1982).

23. G. Stotzky and R. T. Martin, *Plant Soil 18*: 317 (1963).

24. G. Défago, C. H. Berling, U. Burger, D. Haas, G. Kahr, C. Keel, C. Voisard, P. Wirthner, and B. Wüthrich, *Biological Control of Soil-Borne Plant Pathogens* (D. Hornby, ed.), CAB International, Wallingford, England, p. 93 (1990).

25. J. R. Meyer and H. D. Shew, *Phytopathology 81*: 946 (1991).

26. R. D. Lumsden, R. Garcia-E., J. A. Lewis, and G. A. Frias-T, *Soil Biol. Biochem. 19*: 501 (1987).
27. C. W. Kao and W. H. Ko, *Phytopathology 76*: 215 (1986a).
28. C. W. Kao and W. H. Ko, *Phytopathology 76*: 221 (1986b).
29. C. Alabouvette, *Agronomie 6*: 273 (1986).
30. E. W. Stutz, G. Défago, and H. Kern, *Phytopathology 76*: 181 (1986).
31. C. Alabouvette, P. Lemanceau, and C. Steinberg, *Pestic. Sci. 37*: 365 (1993).
32. D. M. Weller and R. J. Cook, *Phytopathology 73*: 463 (1983).
33. R. Lifshitz, B. Sneh, and R. Baker, *Phytopathology 74*: 1054 (1984).
34. F. N. Martin and J. G. Hancock, *Phytopathology 76*: 1221 (1986).
35. S. Liu and R. Baker, *Phytopathology 70*: 404 (1980).
36. J. W. Deacon, *Plant Pathol. 22*: 88 (1973).
37. H. C. Huang and G. C. Kozub, *Bot. Bull. Academia Sinica 32*: 163 (1991).
38. J. M. Whipps, S. P. Budge, and S. J. Mitchell, *Mycol. Res. 97*: 697 (1993).
39. D. H. Marx and S. V. Krupa, *Interactions Between Nonpathogenic Soil Microorganisms and Plants* (Y. R. Dommergues and S. V. Krupa, eds.), Elsevier Scientific, Amsterdam, p. 373 (1978).
40. F. Schoenbeck and H. W. Dehne, *Plant Dis. Reporter 61*: 266 (1977).
41. R. M. Davis, J. A. Menge, and D. C. Erwin, *Phytopathology 69*: 453 (1979).
42. R. M. Davis and J. A. Menge, *Phytopathology 70*: 447 (1980).
43. Y. Elad and R. Baker, *Phytopathology 75*: 1053 (1985).
44. Y. Elad and I. Chet, *Phytopathology 77*: 190 (1987).
45. A. Sivan and I. Chet, *Phytopathology 79*: 198 (1989).
46. C. Alabouvette, *Biological Control of Soil-Borne Plant Pathogens* (D. Hornby, ed.), CAB International, Wallingford, England, p. 27 (1990).
47. J. Leong, *Annu. Rev. Phytopathol. 24*: 187 (1986).
48. P. Demange, S. Wendenbaum, A. Bateman, A. Dell, and M. A. Abdallah, *Iron Transport in Animals, Plants and Microorganisms*, (G. Winkleman, D. Van der Helm and J. B. Neilands, eds.), VCH Chemie, Weinheim, Germany, p. 167 (1987).
49. F. M. Scher and R. Baker, *Phytopathology 72*: 1567 (1982).
50. J. W. Kloepper, J. Leong, M. Teintze, and M. N. Schroth, *Curr. Microbiol. 4*: 317 (1980).
51. P. A. H. M. Bakker, J. G. Lamers, A. W. Bakker, J. D. Marugg, P. J. Weisbeek, and B. Schippers, *Neth. J. Plant Pathol. 92*: 249 (1986).
52. P. A. H. M. Bakker, A. W. Bakker, J. D. Marugg, P. J. Weisbeek, and B. Schippers, *Soil Biol. Biochem. 19*: 443 (1987).
53. J. O. Becker and R. J. Cook, *Phytopathology 78*: 778 (1988).
54. J. E. Loper, *Phytopathology 78*: 166 (1988).

55. J. M. Lynch, *The Rhizosphere* (J. M. Lynch, ed.), Wiley, Chichester, England, p. 177 (1990).

56. A. Simon and K. Sivasithamparam, *Biological Control of Soil-Borne Plant Pathogens* (D. Hornby, ed.), CAB International, Wallingford, England, p. 215 (1990).

57. B. H. Ownley, D. M. Weller, and L. S. Thomashow, *Phytopathology 82*: 178 (1992).

58. L. S. Thomashow and D. M. Weller, *J. Bacteriol. 170*: 3499 (1988).

59. A. R. Poplawsky and A. H. Ellingboe, *Phytopathology 79*: 143 (1989).

60. L. S. Thomashow, D. M. Weller, R. F. Bonsall, and L. S. Pierson III, *Appl. Environ. Microbiol. 56*: 908 (1990).

61. C. Keel, U. Schnider, M. Maurhofer, C. Voisard, J. Laville, U. Burger, P. Wirthner, D. Haas, and G. Défago, *Mol. Plant–Microbe Interact. 5*: 4 (1992).

62. H. Hamdan, D. M. Weller, and L. S. Thomashow, *Appl. Environ. Microbiol. 57*: 3270 (1991).

63. J. Kraus and J. E. Loper, *Phytopathology 82*: 264 (1992).

64. M.-L. Lahdenperä, E. Simon, and J. Uoti, *Biotic Interactions and Soil-Borne Diseases* (A. B. R. Beemster, G. J. Bollen, M. Gerlagh, M. A. Ruissen, B. Schippers, and A. Tempel, eds.), Elsevier, Amsterdam, p. 258 (1991).

65. M. H. Ryder, P. G. Brisbane, and A. D. Rovira, *Biological Control of Soil-Borne Plant Pathogens* (D. Hornby, ed.), CAB International, Wallingford, England, p. 123 (1990).

66. L. S. Thomashow and D. M. Weller, *Biological Control of Soil-Borne Plant Pathogens* (D. Hornby, ed.), CAB International, Wallingford, England, p. 109 (1990).

67. H. A. J. Hoitink, Y. Inbar, and M. J. Boehm, *Plant Disease 75*: 869 (1991).

68. H. A. J. Hoitink, D. M. Van Doren Jr., and A. F. Schmitthenner, *Phytopathology 67*: 561 (1977).

69. S. Spencer and D. M. Benson, *Plant Dis. 65*: 918 (1981).

70. C. T. Stephens, L. J. Herr, H. A. J. Hoitink, and A. F. Schmitthenner, *Plant Dis. 65*: 796 (1981).

71. G. A. Kuter, E. B. Nelson, H. A. J. Hoitink, and L. V. Madden, *Phytopathology 73*: 1450 (1983).

72. C. T. Stephens and T. C. Stebbins, *Plant Dis. 69*: 494 (1985).

73. W. Chen, H. A. J. Hoitink, and L. V. Madden, *Phytopathology 78*: 1447 (1988).

74. O. C. H. Kwok, P. C. Fahy, H. A. J. Hoitink, and G. A. Kuter, *Phytopathology 77*: 1206 (1987).

75. H. Kai, T. Veda, and M. Sakaguchi, *Soil Biol. Biochem. 22*: 983 (1990).

76. T. Ueda, H. Kai, and E. Taniguchi, *Soil Biol. Biochem. 22*: 987 (1990).

77. J. W. Huang and E. G. Kuhlman, *Phytopathology 81*: 163 (1991a).

78. J. W. Huang and E. G. Kuhlman, *Phytopathology 81*: 171 (1991b).

79. Y. S. Lin, S. K. Sun, S. T. Hsu, and W. H. Hsieh, *Biological Control of Soil-Borne Plant Pathogens* (D. Hornby, ed.), CAB International, Wallingford, England, p. 249 (1990).

80. Y. Hadar and R. Mandelbaum, *Crop Protection 5*: 88 (1986).

81. B. Gorodecki and Y. Hadar, *Crop Protection 9*: 271 (1990).

82. C. Schüler, J. Biala, C. Bruns, R. Gottschall, S. Ahlers, and H. Vogtmann, *J. Phytopathology 127*: 227 (1989).

83. C. Schüler, J. Pikny, M. Nasir, and H. Vogtmann, *Biol. Agric. Hort. 9*: 353 (1993).

84. P. Broadbent, K. F. Baker, and Y. Waterworth, *Aust. J. Biol. Sci. 24*: 925 (1971).

85. D. M. Weller, B.-X. Zhang, and R. J. Cook, *Plant Dis. 69*: 710 (1985).

86. J. M. Whipps, Crop Protection in Vegetables *Aspects Appl. Biol. 12*: 75, (1986).

87. R. Campbell, *Biol. Agric. Hort. 3*: 317 (1986).

88. R. Campbell, *Crop Protection 13*: 4 (1994).

89. R. D. Lumsden and J. A. Lewis, *Biotechnology of Fungi for Improving Plant Growth*, (J. M. Whipps and R. D. Lumsden, eds.), Cambridge University Press, Cambridge, p. 171 (1989).

90. P. Merriman and K. Russell, *Biological Control of Soil-Borne Plant Pathogens* (D. Hornby, ed.), CAB International, Wallingford, England, p. 427 (1990).

91. A. Renwick, R. Campbell, and S. Coe, *Plant Pathol. 40*: 524 (1991).

92. M. M. Dewan and K. Sivasithamparam, *Plant Soil 109*: 93 (1988).

93. K. P. Hebbar, A. G. Davey, and P. J. Dart, *Soil Biol. Biochem. 24*: 979 (1992).

94. E. B. Nelson and C. M. Craft, *Phytopathology 82*: 206 (1992).

95. J. P. Clarkson and J. A. Lucas, *Plant Pathol. 42*: 543 (1993).

96. P. M. Stephens, J. J. Crowley, and C. O'Connell, *Soil Biol. Biochem. 25*: 1283 (1993).

97. P. B. Adams, *Phytopathology 77*: 575 (1987).

98. C. M. Kenerley and J. P. Stack, *Can. J. Microbiol. 33*: 632 (1987).

99. P. Trutmann and P. J. Keane, *Soil Biol. Biochem. 22*: 43 (1990).

100. D. C. Sandys-Winsch, J. M. Whipps, J. S. Fenlon, and J. M. Lynch, *Biocontrol Sci. Technol. 4*: 269 (1994).

101. B. Sneh, S. J. Humble, and J. L. Lockwood, *Phytopathology 67*: 622 (1977).

102. A. R. Wynn and H. A. S. Epton, *Trans. Br. Mycol. Soc. 73*: 255 (1979).

103. P. T. W. Wong and R. Baker, *Soil Biol. Biochem. 16*: 397 (1984).

104. A. Fradkin and Z. A. Patrick, *Can. J. Microbiol. 31*: 411 (1985).

105. L. Mihuta-Grimm and R. C. Rowe, *Phytopathology 76*: 306 (1986).

106. R. Mitchell and M. Alexander, *Soil Sci. Soc. Am. Proc. 1962 26*: 556 (1962).

107. A. Ordentlich, Y. Elad, and I. Chet, *Phytopathology 78*: 84 (1988).

108. M. Fridlender, J. Inbar, and I. Chet, *Soil Biol. Biochem. 25*: 1211 (1993).

109. G.-W. Xu and D. C. Gross, *Phytopathology 76*: 414 (1986a).

110. J. W. Kloepper, *Phytopathology 81*: 1006 (1991).

111. J. M. Whipps, *J. Gen. Microbiol. 133*: 1495 (1987).

112. M. Artigues and P. Davet, *Soil Biol. Biochem. 16*: 413 (1984).

113. Y. Henis, J. A. Lewis, and G. C. Papavizas, *Soil Biol. Biochem. 16*: 391 (1984).

114. J. D. Mueller, M. N. Cline, J. B. Sinclair, and B. J. Jacobsen, *Plant Dis. 69*: 584 (1985).

115. P. Davet, *Agronomie 6*: 863 (1986).

116. W. C. Wong and I. K. Hughes, *J. Appl. Bacteriol. 60*: 57 (1986).

117. J. M. Whipps and S. P. Budge, *Mycol. Res. 94*: 607 (1990).

118. R. D. Lumsden, *Phytopathology 71*: 282 (1981).

119. E. D. Sutherland and G. C. Papavizas, *J. Phytopathol. 131*: 33 (1991).

120. G. E. Harman and R. D. Lumsden, *The Rhizosphere* (J. M. Lynch, ed.), Wiley, Chichester, England, p. 259 (1990).

121. C. N. Rosendahl and L. W. Olson, *J. Phytopathol. 134*: 324 (1992).

122. P. S. Randhawa and N. W. Schaad, *Phytopathology 75*: 254 (1985).

123. J. Handelsman, S. Raffel, E. H. Mester, L. Wunderlich, and C. R. Grau, *Appl. Environ. Microbiol. 56*: 713 (1990).

124. C. Hagedorn, W. D. Gould, and T. R. Bardinelli, *Appl. Environ. Microbiol. 55*, 2793 (1989).

125. J. L. Parke, R. E. Rand, A. E. Joy, and E. B. King, *Plant Dis. 75*: 987 (1991).

126. P. Broadbent, K. F. Baker, N. Franks, and J. Holland, *Phytopathology 67*: 1027 (1977).

127. Y. Elad, I. Chet, and R. Baker, *Plant Soil 98*: 325 (1987).

128. T. V. Suslow and M. N. Schroth, *Phytopathology 72*: 199 (1982).

129. P. R. Merriman, R. D. Price, K. F. Baker, J. F. Kollmorgan, T. Piggot, and E. H. Ridge, *Biology and Control of Soil-Borne Plant Pathogens* (G. W. Bruehl, ed.), American Phytopathological Society, St. Paul, Minnesota, p. 130 (1975).

130. M. M. Dewan and K. Sivasithamparam, *Mycol. Res. 93*: 156 (1989).

131. A. R. Harris, D. A. Schisler, and M. H. Ryder, *Soil Biol. Biochem. 25*: 909 (1993).

132. M. H. Ryder and A. D. Rovira, *Soil Biol. Biochem. 25*: 311 (1993).

133. J. E. Loper, C. Haack, and M. N. Schroth, *Appl. Environ. Microbiol. 49*: 416 (1985).

134. J. W. Kloepper, F. M. Scher, M. Laliberté, and I. Zaleska, *Can. J. Microbiol. 31*: 926 (1985).
135. J. S. Ahmad and R. Baker, *Phytopathology 77*: 182 (1987).
136. R. D. Lumsden, J. C. Locke, J. A. Lewis, S. A. Johnston, J. L. Peterson, and J. B. Ristaino, *Biol. Cultural Tests 5*: 90 (1990).
137. R. D. Lumsden, J. C. Locke, S. T. Adkins, J. F. Walter, and C. J. Ridout, *Phytopathology 82*: 230 (1992).
138. M. P. McQuilken, J. M. Whipps, and R. C. Cooke, *Plant Pathol. 39*: 452 (1990).
139. A. B. Filonow and J. L. Lockwood, *Plant Dis. 69*: 1033 (1985).
140. E. B. Nelson, *Plant Dis. 72*: 140 (1988).
141. M. Ichielevich-Auster, B. Sneh, Y. Koltin, and I. Barash, *Phytopathology 75*: 1080 (1985).
142. P. A. Maplestone, J. M. Whipps, and J. M. Lynch, *Plant Soil 136*: 257 (1991).
143. S. J. Kay and A. Stewart, *Plant Pathol. 43*: 371 (1994).
144. A. L. Capper and R. Campbell, *J. Appl. Bacteriol. 60*: 155 (1986).
145. R. S. Utkhede and E. M. Smith, *Soil Biol. Biochem. 25*: 383 (1993).
146. E. B. Nelson and C. M. Craft, *Plant Dis. 75*: 510 (1991).
147. G.-W. Xu and D. C. Gross, *Phytopathology 76*: 423 (1986b).
148. C. Hagedorn, W. D. Gould, and T. R. Bardinelli, *Plant Dis. 77*: 278 (1993).
149. A. L. Capper and K. P. Higgins, *Plant Pathol. 42*: 560 (1993).
150. A. Trapero-Casas, W. J. Kaiser, and D. M. Ingram, *Plant Dis. 74*: 563 (1990).
151. H. C. Huang, *Can. J. Plant Pathol. 7*: 26 (1980).
152. S. P. Budge and J. M. Whipps, *Plant Pathol. 40*: 59 (1991).
153. H. C. Hoch and G. S. Abawi, *Phytopathology 69*: 417 (1979).
154. P. B. Adams and D. R. Fravel, *Phytopathology 80*: 1120 (1990).
155. J. J. Marois, S. A. Johnston, M. T. Dunn, and G. C. Papavizas, *Plant Dis. 66*: 1166 (1982).
156. E. G. Ruppel, R. Baker, G. E. Harman, J. P. Hubbard, R. J. Hecker, and I. Chet, *Crop Prot. 2*: 399 (1983).
157. P. A. Backman and R. Rodrigues-Kabana, *Phytopathology 65*: 819 (1975).
158. G. Jager and H. Velvis, *Neth. J. Plant Pathol. 91*: 49 (1985).

18

Pesticides: Microbial Degradation and Effects on Microorganisms

EDWARD TOPP   Agriculture and Agri-Food Canada, London, Ontario, Canada

TATIANA VALLAEYS and GUY SOULAS   National Institute of Agricultural Research (INRA), Dijon, France

1. INTRODUCTION

The agricultural environment can support a variety of unwanted organisms that reduce the yield and value of crops. Modern agriculture must use measures to control the population density of these pests to remain economically viable. Thus herbicides, insecticides, fungicides, nematocides, and biocides, collectively called pesticides, are used to kill weeds, insects, fungi, nematodes, and everything, respectively. Unfortunately, leaching into the soil, run-off, and volatilization move these chemicals into ground and surface waters, and into the atmosphere. Detection of trace levels of pesticides throughout the environment has prompted concerns about their impact on humans and other nontarget organisms. Pesticides and other organic chemicals in soil and water can be degraded by photolytic, chemical, and biological mechanisms. Photolytic degradation can occur when a pesticide molecule is irradiated by sunlight. Chemical degradation occurs when the molecule is chemically unstable in the conditions of its environment. The term "biodegradation" refers to the transformation of pesticides by living microorganisms. A wealth of data comparing the persistence of pesticides in sterile and nonsterile soils shows that biodegradation is the primary mechanism of pesticide destruction in soils. Soil microorganisms thus perform an invaluable function in attenuating the environmental impact of these chemicals. However, excessively rapid biodegradation is problematic because it can hinder the efficacy of some pesticides by reducing the concentration below that required to control the targeted organisms.

This chapter will present aspects of the biodegradation of organic pesticides by soil microorganisms. The possible effects of pesticide use on soil microorganisms will be briefly discussed. Other aspects of pest control pertinent to the soil microbiologist such as integrated pest management and biocontrol will not be addressed. The literature concerning pesticide biodegradation in soils and in laboratory culture is voluminous, and many comprehensive reviews are available for more detailed information (1–11).

2. FACTORS INFLUENCING PESTICIDE BIODEGRADATION IN SOILS

Much of what is known about the biochemical, physiological and genetic characteristics of pesticide-degrading bacteria comes from studies with pure or mixed microbial cultures. However, the rates and pathways of biodegradation of a pesticide in soils are determined by factors which cannot be duplicated in nonsoil systems such as broth or slurries. These factors are specific to the pesticide, the soil, and pesticide-degrading microorganisms and are in many cases interactive (Fig. 1). The role that each has in determining the rate and pathway of degradation will be considered in turn.

2.1 Pesticides

Structure

A glance at just a few of the several hundred different pesticides which are or have been used commercially reveals the remarkable variety and complexity of these chemicals (Table 1). The first pesticides to be extensively used were the insecticide 1,1,1-trichloro-bis-(4-chlorophenyl)ethane (DDT) and the herbicide 2,4-dichlorophenoxyacetic acid (2,4-D) during the Second World War. During the 1950s and 1960s new classes of chemicals, the organophosphate and carbamate insecticides, and the triazine and substituted urea herbicides came into use. More recently, the pyrethroid insecticides and sulfonylurea herbicides have become available. The trend has been to replace nonselective environmentally recalcitrant (i.e., long-lived) pesticides which must be used at high concentrations, with chemicals that are effective at much lower application rates and are less threatening to nontarget organisms. Pesticides should also persist in soils or on

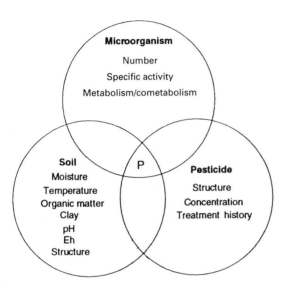

Figure 1. Factors influencing the persistence (P) of a pesticide in soil. Factors specific to the pesticide, soil, and degrading microorganisms are frequently interactive. For example, soil temperature and moisture have a tremendous impact on microbial activity. Soils sorb and retain pesticides, reducing their bioavailability and extending their persistence. The chemical structure of the pesticide will inherently influence its biodegradability.

plant surfaces sufficiently long to kill the target but not so long that there is a high probability that they will migrate from the point of application into, for example, groundwater. Thus the persistence of the chemical in the environment is very important.

The structure of the pesticide molecule determines its physical and chemical properties and inherent biodegradability. At one extreme of environmental recalcitrance lie the chlorinated hydrocarbon insecticides such as DDT, 1, 1a, 2, 2, 3, 3a, 4, 5, 5, 5a, 5b, 6-dodecachlorooctahydro-1,3,4-metheno-1H-cyclobuta-[cd]pentalene (mirex), and (1aα, 2β, 2aα, 3β, 6β, 6aα, 7β, 7aα)-3,4,5,6,9,9-hexachloro-1a,2,2a,3,6,6a,7,7a-octahydro-2,7:3,6-dimethanonaphth[2,3-b]oxirene (dieldrin). These chemicals are now notorious for their ubiquity in the environment and have been largely removed from the marketplace. They are insoluble in water, sorb tightly to soil, and are thus relatively unavailable for biodegradation. Furthermore, they are heavily substituted with bulky chlorine atoms which hinder an enzymatic attack on the carbon skeleton. They dissolve readily in fat (i.e., are lipophilic) and become increasingly concentrated in organisms (i.e., are subject to biomagnification) as they move up the trophic chain.

Table 1. Selected Pesticides Representing Some Major Classes of Herbicides, Insecticides, and Fungicides and Examples of Microorganisms Which Degrade Them

Class	Example	Structure	Microorganisms
HERBICIDES			
Triazine	Atrazine		*Rhodococcus*
Phenoxyalkanoic	2,4-D		*Alcaligenes eutrophus*
Phenylurea	Linuron		*Bacillus sphaericus*
Sulfonylurea	Chlorsulfuron		*Streptomyces*
Phosphinic acid	Glyphosate		*Arthrobacter*

INSECTICIDES

Class	Pesticide	Structure	Organism
Organophosphate	Parathion		*Flavobacterium* sp.
N-Methylcarbamate	Carbofuran		*Arthrobacter*
Pyrethroid	Deltamethrin		*Pseudomonas fluorescens*
Organochlorine	DDT		*Aerobacter aerogenes* *Trichoderma viride* Algae

FUNGICIDES

Class	Pesticide	Structure	Organism
Dicarboximide	Iprodione		*Pseudomonas putida*

Atrazine: 2-Chloro-4-ethylamino-6-isopropylamino-1,3,5-triazine; *2,4-D*: 2,4-Dichlorophenoxyacetic acid; *Linuron*: 3-(3,4-Dichlorophenyl)-1-methoxy-1-methylurea; *Chlorsulfuron*: 2-Chloro-*N*-[[(4-methoxy-6-methyl-1,3,5-triazin-2-yl)amino]carbonyl]benzenesulfonamide; *Glyphosate*: *N*-(phosphonomethyl)glycine; *Parathion*: *O*,*O*-diethyl *O*-(4-nitrophenyl)phosphorothioate; *Carbofuran*: 2,3-Dihydro-2,2-dimethylbenzofuran-7-yl methylcarbamate; *deltamethrin*: 3-(2,2-dibromoethenyl)-2,2-dimethylcyclopropanecarboxylic acid; *DDT*: 1,1,1-Trichloro-bis-(4-chlorophenyl)ethane; *Iprodione*: 3-(3,5-Dichlorophenyl)-*N*-(1-methylethyl)-2,4-dioxo-1-imidazolidinecarboxamide.

Sometimes toxic concentrations are reached, as in the case of laying birds which had absorbed sufficient concentrations of DDT to be infertile as a result of eggshell fragility. At the other extreme of environmental recalcitrance are molecules which, for example, have labile linkages that are broken quite readily. The insecticide 2,3-dihydro-2,2-dimethylbenzofuran-7-yl methylcarbamate (carbofuran) and the herbicide 2,4-D can be degraded in a matter of a few days in field soils which have been treated repeatedly and have developed a very active degrading microbial population. Minor differences in the position or nature of substituents in pesticides of the same class can influence the rate of degradation.

Concentration

Pesticide concentration is an important parameter in determining the rate of biodegradation. The degradation kinetics of many pesticides approaches first-order; the rate of degradation decreases roughly in proportion with the residual pesticide concentration. A plot of the logarithm of the concentration of pesticide remaining over time will yield a straight line. After prolonged periods, the rate of degradation of very low concentrations of residual pesticide can become unexpectedly slow. It may be that over extended periods soil retention limits the bioavailability of pesticide residues and thus prolongs their persistence. At very low pesticide concentrations, there may be insufficient nutrients available to support metabolizing microorganisms and cometabolic transformation may dominate. The differences and significance of metabolic versus cometabolic transformations will be discussed in Section 4.

Pesticides enter the soil directly through incorporation, through spraying, or in seed coatings, or indirectly, adhering to plant residues or washed off the surface of crops after foliar applications. Granular, emulsified, aqueous, or sometimes microencapsulated pesticide formulations are used. The formulation and method of application will influence the concentration, availability, and distribution of the chemical in the soil. Usually almost all of the applied pesticide is found in the top 5 cm of the soil profile. Typical bulk concentrations immediately following application range from a few micrograms to a few milligrams per kilogram soil. The concentration can, however, be very heterogenous, particularly with granular formulations.

Treatment history

Repeated application of a pesticide, or a structurally similar member of the same chemical class, can result in "enhanced" or accelerated degradation. The half-life of the pesticide can be shortened from a few weeks to a few days, resulting in significantly reduced efficacy (12). Enhanced degradation appears to be due to an enrichment of microbial populations capable of metabolizing the chemical. Classes of chemicals for which this phenomenon has been observed include the phenoxyalkanoic and thiocarbamate herbicides and N-methylcarbamate insecticides. Chemicals with similar structure degraded by the same pathways can coinduce the enhanced degradation of each other, a phenomenon called cross-enhancement. Once this problem has developed it can last for variable periods of up to several years during which different pesticides must be used. Rotating different types of chemicals from season to season may prevent the enrichment of the degrading flora.

2.2 Soil

The role of the soil in pesticide biodegradation is critical because it provides the environment for degradative microorganisms. Soils can also sorb pesticides, regulating their bioavailability and influencing their persistence.

Effects of soil properties and conditions on biodegradative microorganisms

A number of soil and climatic parameters influence the activity of pesticide-degrading microorganisms. These are soil moisture and temperature, pH, and E_h (oxidation–reduction potential). The response of pesticide-degrading microorganisms to extremes of pH is as that expected for other soil microorganisms. Under conditions of low E_h, reductive transformation reactions are likely to occur. Soils rich in organic matter tend to have a larger, more active biomass. Soil moisture and temperature are the most important factors regulating activity.

Water is a solvent for pesticide movement and diffusion and a reagent in hydrolytic reactions and is essential for microbial functioning. Pesticide degradation is slow in dry soils. The rate of pesticide transformation generally increases with water content. In very wet soils, such as rice paddies, the rate of diffusion of atmospheric oxygen into the soil is limited and anaerobic pesticide transformation mechanisms can become important. Poor oxygen transfer at high moisture content can frequently increase the persistence of pesticides. However, compounds that are heavily chlorinated and subject to reductive dechlorination are an exception. For example, DDT is relatively stable in aerobic soils, but under anaerobic conditions it is rapidly converted to the metabolite 1,1-dichloro-2,2-bis(4-chlorophenyl)ethane (DDD) (13).

Pesticide biodegradation is optimal at mesophilic temperatures (around 25°C to 40°C) and occurs slowly or not at all at very cold or very hot temperatures. Seasonal variations in temperature and moisture must be taken into consideration when pesticides are used in various climates; pesticides degrade less rapidly in drier or colder climates than in more temperate areas.

Soils influence the bioavailability and persistence of pesticides

Organic matter and clay content and composition play a critical role in the availability of pesticides. Sorptive processes increase pesticide persistence by removing the chemical from the aqueous phase, in which it is available to microorganisms (14). Likewise, pesticide molecules may diffuse into tiny spaces in clay or humus that are too small to be penetrated by bacteria, a phenomenon called molecular entrapment (15,16). Retention of pesticides by field soils is subject to hysteresis; retained residues are released more slowly than the rate with which they are formed. It appears that the longer the contact time of a pesticide with the soil, the longer the time subsequently required for the molecule to be released and be available for biodegradation. The persistence of "aged" residues that have been in contact for several years with soil in the field is much longer than that of freshly applied pesticide (16). Clay surfaces can catalyze pesticide transformations (17). For example, parathion sorbed to clay can be hydrolyzed abiologically (17). Soil pH can influence adsorption of pesticides and hence their bioavailability. For instance, more of the herbicide 2,4-bis(isopropylamino)-6-methylthio-1,3,5-triazine (prometryn) sorbed to clay montmorillonite at pH 3 than at pH 7 (17).

2.3 Microorganisms

Examples of pesticide-degrading microorganisms can be found throughout the microbial world in the domains eubacteria, archaea, and eukarya. Every physiological type is represented in the pesticide-degrading microflora: aerobic, anaerobic (fermentative, methanogenic, sulfate-reducing), chemolithotrophic, and photosynthetic organisms. Pesticide residues can be mineralized by single microorganisms or by assemblages of microorganisms. Many different soil microorganisms catalyzing the same or different types of transformation reactions may simultaneously attack a pesticide. The complete degradation of a pesticide may require a community of organisms that sequentially exchange and then transform excreted metabolites as the molecule is gradually broken down. Transformation reactions that occur under reducing and oxidizing conditions can sometimes be combined to accelerate degradation. For example, methoxyclor is mineralized more rapidly in soil that is first incubated under anaerobic conditions, and then under aerobic conditions, than it is in soil that is maintained under only anaerobic or only aerobic conditions (18). Transformation reactions catalyzed under anaerobic conditions produced a product which was readily mineralizable in the presence of oxygen.

Soil heterogeneity results in the variable distribution and activity of pesticide-degrading microorganisms from the micrometer to the regional (thousands of kilometers) scale (19). Most of the microorganisms in soil are located in the thin layers of water surrounding soil particles. Rapid and significant changes in E_h due to microbial respiration and gas diffusion limitations occur at this scale (20). The extremely heterogeneous and rapidly changing conditions will be reflected in the distribution and activity of aerobic pesticide-degrading bacteria. At the field scale, pesticide degradation can be slower in exposed drier soils on hillocks than in sheltered moist areas. At the regional scale, climatic differences of temperature and moisture make their presence felt. Pesticide degradation is slower in cold climates than in more temperate or tropical zones. Application rates and timing have to be adjusted accordingly. Pesticide degradation rates generally decrease with depth through the soil profile (21), and in aquifers chemical transformation mechanisms may dominate.

3. MICROBIAL PESTICIDE TRANSFORMATION REACTIONS

Given the great chemical diversity of pesticides and pesticide-degrading microorganisms it is to be expected that a wide variety of transformation reactions catalyzed by microorganisms should be involved in their degradation. A given pesticide is usually biodegraded by a limited number of consistent reactions in soils. Microbial pesticide transformation reactions can be grouped into those that are oxidative, reductive, hydrolytic, or synthetic in nature (2). A brief discussion of the major pesticide transformation reactions follows. Various aspects of pesticide enzymology have been reviewed in detail elsewhere (2,4,5,7,8,10).

3.1 Oxidative Reactions

The major oxidative reactions are hydroxylation, dealkylation, epoxidation, and sulfoxidation. Oxidation reactions are among the most important pesticide-transforming reactions in aerobic environments. They are catalyzed by monooxygenase, dioxygenase, laccase, and peroxidase enzymes. Aerobic conditions are required for these reactions because in addition to functioning as a terminal electron acceptor for respiration, oxygen is an obligate reagent in the transformation reaction. The products of oxidative reactions frequently have anionic hydroxyl or carboxyl substituents and are more polar and water-soluble than the parent pesticide. They are thus more biodegradable and more likely to be immobilized by covalent attachment to humic materials. Examples of each of the major transformation reactions follow.

Hydroxylation reactions

Hydroxylation reactions are those in which hydroxyl groups, the oxygen atoms of which originate from dioxygen gas, are inserted into target molecules. In contrast, in hydrolytic reactions the oxygen atom originates from water. Hydroxylation is a very frequent mechanism of initial attack on pesticide molecules under aerobic conditions. Both alkyl and aryl (aromatic) groups can be hydroxylated. For example, an actinomycete hydroxylates the alkyl groups of the herbicide metolachlor (Fig. 2) (22). A *Rhodococcus* strain hydroxylates the benzene nu-

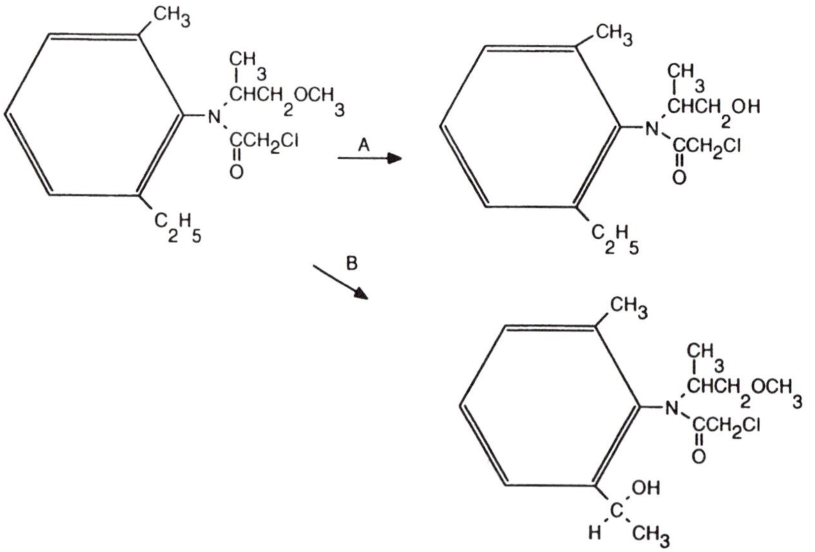

Figure 2. Hydroxylation of the herbicide metolachlor by an actinomycete. Hydroxylation of the alkyl chain in pathway A results in demethylation.

cleus of the insecticide carbofuran, producing 5-hydroxycarbofuran as an end-product (Fig. 3 pathway B) (23). A *Flavobacterium* strain hydroxylates the biocide pentachlorophenol (PCP), producing tetrachloro-*para*-hydroquinone (Fig. 4 step A) (24). The ligninolytic white-rot fungus *Phanerochaete chrysosporium* oxidizes and ultimately mineralizes numerous pesticides, including the herbicides 2,4-D and 2,4,5-trichlorophenoxyacetic acid (2,4,5-T) (25). The substrate specificity for these reactions is, in some cases, narrow and in others broad. For example, the *Flavobacterium* strain will hydroxylate only at the *para* position of a rather narrow range of halogenated phenols, whereas *Phanerochaete chrysosporium* will hydroxylate many aromatic and aliphatic organic pollutants that are extremely varied in structure.

Hydroxylation reactions are doubly important because they are essential in the aerobic metabolism of all pesticides containing aromatic nuclei (see Section 4.1).

Dealkylation reactions

Many pesticides possess alkyl groups linked to the rest of the molecule via nitrogen or oxygen atoms. The biodegradation of these pesticides is frequently initiated by oxidative removal of these substituents. Dealkylation reactions catalyzed

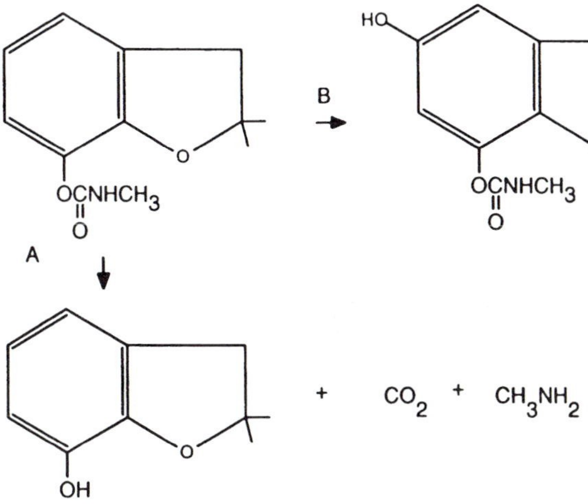

Figure 3. A, Metabolism of the insecticide carbofuran by a gram-negative methylotrophic bacterium, strain ER2. Hydrolysis of the alkyl group by a carbofuran hydrolase produces methylamine, carbon dioxide, and 7-phenolcarbofuran, which is excreted as an endproduct. Methylamine supports the growth of the organism as the sole source of carbon and nitrogen. B, Cometabolic ring hydroxylation of carbofuran to form the endproduct 5-hydroxycarbofuran by *Rhodococcus* TE1.

by mono- and dioxygenases require oxygen and reductant in the form of reduced nicotinamide-adenine dinucleotide phosphate (NADPH). N-dealkylation is the primary route of degradation of the *s*-triazine herbicides in soil and by pure cultures (Fig. 5) (26). The first reaction in the degradation of 2,4-D by *Alcaligenes eutrophus* JMP134(pJP4) is removal of the alkylether group by means of an α-ketoglutarate-dependent dioxygenase, formerly thought to be a monooxygenase (Fig. 8) (27). Hydroxylation of the methoxy group on metolachlor by an actinomycete was accompanied by removal of the methyl group, i.e., demethylation (Fig. 2) (22).

Epoxidation

Examples of microbial transformation reactions converting a pesticide to a metabolite which is more of a health concern than the parent compound are rather rare. However, the production of epoxides (compounds which contain saturated three-membered cyclic ethers) by reactions which insert an oxygen atom into carbon–carbon double bonds is one such case. For example, heptachlor (1, 4, 5, 6, 7, 8, 8-heptachloro-3α, 4, 7, 7α-tetrahydro-4,7-methano-1*H*-indene) is

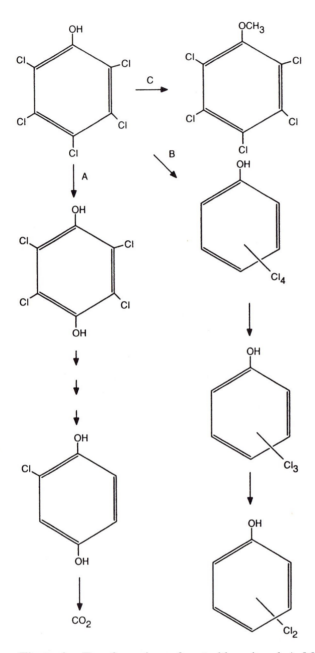

Figure 4. Transformations of pentachlorophenol: A, Metabolism by a *Flavobacterium* sp. An initial hydroxylation is followed by sequential glutathione-dependent reductive dehalogenation steps. The intermediate chlorohydroquinone is mineralized by as yet unknown reactions. B, Sequential reductive dechlorination in anaerobic sludges. Various isomers of tetra-, tri-, and dichlorophenol accumulate sequentially and transiently. C, Cometabolic methylation to produce volatile tetrachloroanisole.

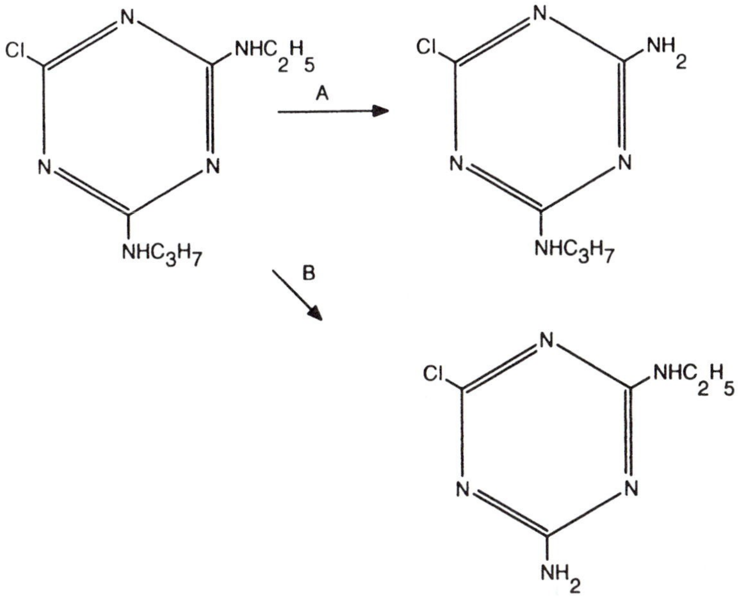

Figure 5. Dealkylation of atrazine by *Rhodococcus* TE1. Both the ethyl (pathway A) and the isopropyl (pathway B) groups can be removed.

epoxidized to the suspected carcinogen heptachlor epoxide by a number of microorganisms including the fungi *Rhizopus* and *Fusarium* and the bacteria *Bacillus* and *Streptomyces* (28).

Sulfoxidation

Sulfoxidation reactions oxidize a divalent sulfur atom to the sulfoxide, and sometimes to the sulfone. Numerous pesticides such as the insecticide 2-methyl-2-(methylthio)propanal *O*-[(methylamino)carbonyl]oxime (aldicarb) are sulfides having the structure RSR′. Sulfoxidation is a frequent mechanism of degradation of these compounds in soils (2). Sulfoxidation of some sulfides in soils probably occurs chemically as well as biologically because this reaction can also occur in sterile soils (5).

3.2 Reduction Reactions

Reactions which add electrons to pesticide molecules are common, particularly in anaerobic flooded soils. The saturation of double bonds is important in the anaerobic degradation of aromatic molecules (29). Nitro groups can be reduced to the amino, as in the biodegradation of the insecticide parathion in lake sediments (30). Reductive dehalogenation reactions are extremely important in the transformation of halogenated pesticides in anaerobic environments. A number of mixed or pure bacterial cultures reductively dehalogenate the organochlorine pesticides DDT, dieldrin, heptachlor, the γ-isomer of 1, 2, 3, 4, 5, 6-hexachloro-cyclohexane (lindane), and 2,2-bis(4-methoxyphenyl)-1,1,1-trichloroethane (methoxychlor) (7). The heterocyclic herbicide 5-bromo-6-methyl-3-(1-methyl-propyl)-2,4(1H, 3H)-pyrimidinedione (bromacil) is dehalogenated in anaerobic slurries of aquifer material (31). Reductive dehalogenation is favored under extremely reducing methanogenic conditions and is frequently inhibited by oxygen, nitrate, or sulfate (7). However, reductive dehalogenation reactions can occur at higher redox potentials. For example, several reductive dehalogenation steps are involved in the aerobic mineralization of PCP by a *Flavobacterium* (Fig. 4) (32).

3.3 Hydrolytic Reactions

Hydrolytic reactions are very important in the degradation of numerous pesticides. The hydroxyl group (OH) originates from water, and therefore hydrolytic reactions can occur in the presence or absence of oxygen. Since no cofactors are required, hydrolytic reactions can be catalyzed by intracellular enzymes, or extracellularly by hydrolases excreted by living microorganisms, or released into the surrounding soil upon their death. Molecules which possess chemical linkages containing electron-deficient atoms, good targets for nucleophilic attack, are frequently hydrolyzed by soil microorganisms. Examples of such pesticides include the organophosphorus and methylcarbamate insecticides such as parathion and carbofuran, and the phenylurea herbicides such as linuron. A carbofuran-degrading methylotroph, strain ER2, initiates the attack on carbofuran by hydrolyzing the carbamate linkage, producing 7-phenolcarbofuran, carbon dioxide, and methylamine (Fig. 3, pathway A) (33).

3.4 Synthetic Reactions and the Formation of Immobilized Residues

Synthetic reactions covalently attach pesticides, or pesticide transformation products, to other organic molecules. Molecules which contain reactive nucleophilic groups, amino (-NH$_2$), hydroxyl (-OH), or carboxyl (-COOH), for example, can participate in these reactions. Likewise, extremely reactive radicals consisting of aromatic molecules which have had an electron removed by laccases or other enzymes readily bind to each other or other organic molecules. Although all products of synthetic reactions are larger than the parent compounds, the mobility and bioavailability of the reaction products are variable, depending on the size of the molecule to which the residue is attached. For example, the methylation of the hydroxyl group in PCP by fungi produces a volatile methoxy derivative, tetrachloroanisole (Fig. 4) (34). This is a rare example of a reaction product's being much more volatile, and therefore more mobile in the environment, than the parent compound. Activated transformation products can react together to form polymers. For example, hydrolysis of the phenylurea herbicides produces chlorinated anilines that then readily dimerize to form azobenzene and other condensation products (35).

No doubt the most significant outcome of synthetic transformation reactions is the covalent attachment of pesticide residues to soil humus. This effectively immobilizes the pesticide residues to the soil matrix and greatly reduces their bioavailability and movement through the soil profile. A significant fraction of some soil-applied pesticides can form bound residues. Bound pesticide residues can perhaps be slowly released during turnover of organic matter, but the long-term fate of these bound residues is at present unclear.

4. METABOLISM AND COMETABOLISM OF PESTICIDES

Pesticide transformation by microorganisms which derive some nutritional benefit from the process, using the chemical as a source of carbon and energy, or another nutrient, is called metabolism. Metabolism frequently results in the complete mineralization of a pesticide, in other words, in its conversion to carbon dioxide, water, and inorganic ions. In contrast, the microbial transformation of a pesticide without apparent nutritional benefit to the microorganism is called cometabolism. Cometabolism is characterized by a usually subtle transformation of the pesticide molecule. Cometabolic reactions do not support the growth of the degrading biomass but may sometimes be used as a detoxification mechanism. Metabolism and cometabolism are distinguished by differences in the specificity and regulation of the enzymes involved, and the ability of the pesticide to support the growth of the degrading biomass, differences which have important consequences for the kinetics of pesticide biodegradation in the field.

4.1 Pesticide Metabolism

Metabolic pathways consist of a series of sequential transformation reactions, whose purpose is to convert the xenobiotic pesticide, or components thereof, into "natural" molecules which can be processed by intermediary and central metabolism and support growth. The first peripheral reactions are usually catalyzed by enzymes which are induced by the pesticide substrate and which have a fairly narrow substrate specificity. The genes encoding the enzymes catalyzing these initial reactions frequently reside on plasmids (autonomous self-replicating extrachromosomal genetic elements). The persistence of some pesticides which are metabolized in field soils can be greatly reduced by repeated applications. Rapid degradation is preceded by a lag, presumably during which the degradative biomass increases in size until the activity is high enough to be detected.

Most pesticides contain aromatic components (Table 1). Complete metabolism of an aromatic nucleus requires that the ring be opened, a reaction which cannot be accomplished easily because the benzene is inherently very stable. Aromatic nuclei are prepared for ring fission by first hydroxylating, making the ring more reactive. The basic principles of aerobic aromatic metabolism have been elucidated with nonpesticide molecules, and experience suggests that they are also universally applicable to the aromatic pesticides (36). The first principle is that ring fission is only accomplished on phenyl nuclei containing at least two hydroxyl groups, usually adjacent to each other (i.e., catechol derivatives). The second principle is that aromatic substrates differing substantially in the number, location, and nature of substituents can frequently be converted to common hydroxylated intermediates. For example, 4-chloro-2-nitrophenol, 4-chlorophenol, and 3-chlorobenzoate can be converted to 4-chlorocatechol (Fig. 6). The common intermediate 4-chlorocatechol is a substrate for a single ring-cleaving dioxygenase. Only the initial transformation reaction is different in the metabolic pathway for each of these substrates. This strategy of funnelling substrates of diverse structure into common intermediates results in a significant economy for the microorganism since completely independent suites of enzymes need not be synthesized for each substrate. The third principle is that the mechanism of catechol formation will vary according to the location and nature of substituents. For example, in the case of 4-chloro-2-nitrophenol and 4-chlorophenol, both of which already contain a hydroxyl group, a monooxygenase inserts a second one. On the other hand, a dioxygenase is required to insert two hydroxyl groups into the 3-chlorobenzoate nucleus. Substituent groups are frequently removed to generate the desired catechol. The hydroxylation of 4-chloro-2-nitrophenol is accompanied by loss of the nitro group, and that of 3-chlorobenzoate by decarboxylation. The fourth and final principle is that the catechol ring cleavage substrate is opened by a dioxygenase which cleaves between (intradiol, or *ortho* cleavage) or adjacent to (extradiol, or *meta* cleavage) the two adjacent hydroxyl groups. The routes by which the ring carbons enter central metabolism are consequently different (Fig. 7).

Three examples of metabolic pathways will be considered: 2,4-D, carbofuran, and pentachlorophenol. The most thoroughly studied pesticide biodegradation pathway is that of 2,4-D in *Alcaligenes eutrophus* JMP134(pJP4). The pathway illustrates how several different types of reactions are combined to convert the herbicide into the central metabolites succinate and acetyl-CoA. The 2,4-D degradation pathway has been elucidated and the genes encoding the enzymes involved have been cloned (Fig. 8). The initial dealkylation reaction is achieved

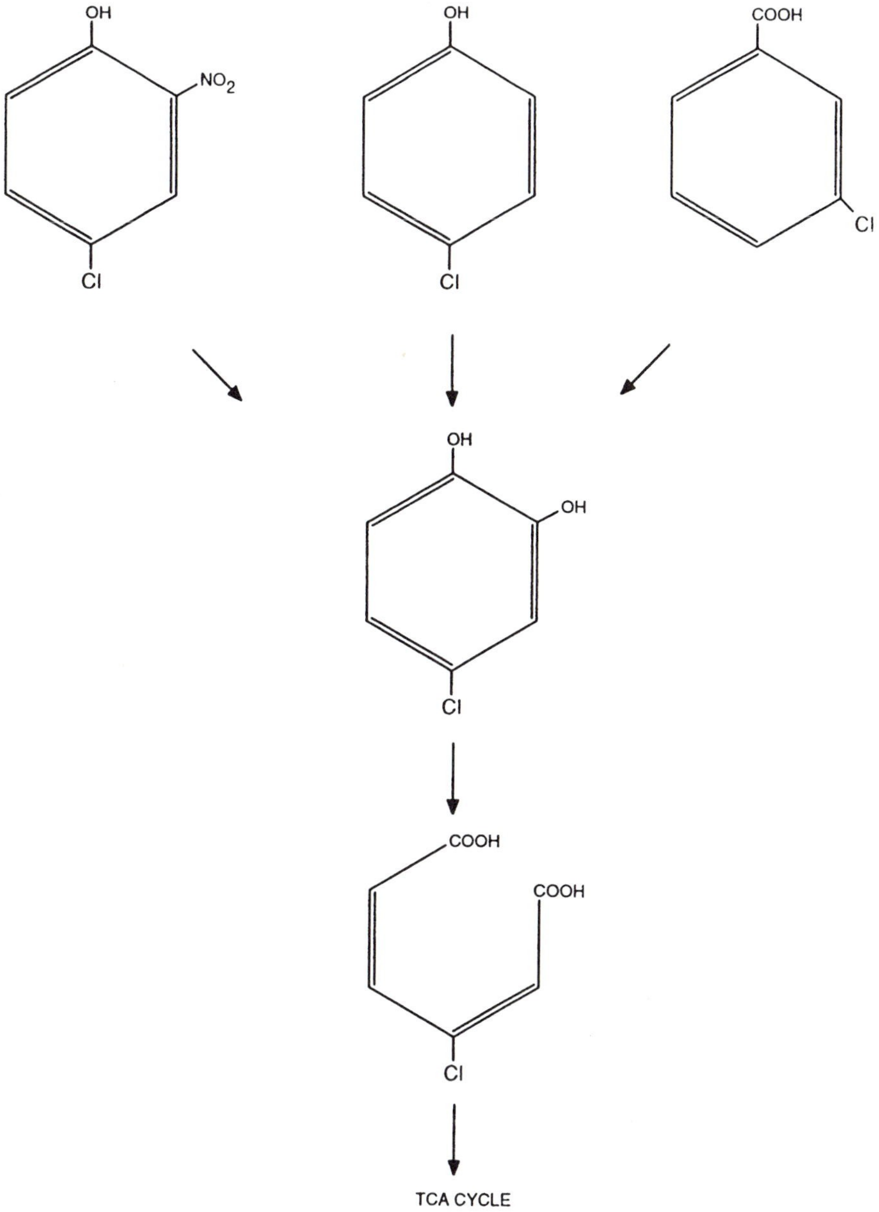

Figure 6. Examples of how aromatic substrates with different substituent groups, 2-nitro-4-chlorophenol, 4-chlorophenol, and 3-chlorobenzoate, can be converted to a common ring cleavage substrate, 4-chlorocatechol. In this example, the 4-chlorocatechol is attacked by an ortho-cleaving dioxygenase to form 3-chloro-*cis*, *cis*-muconate.

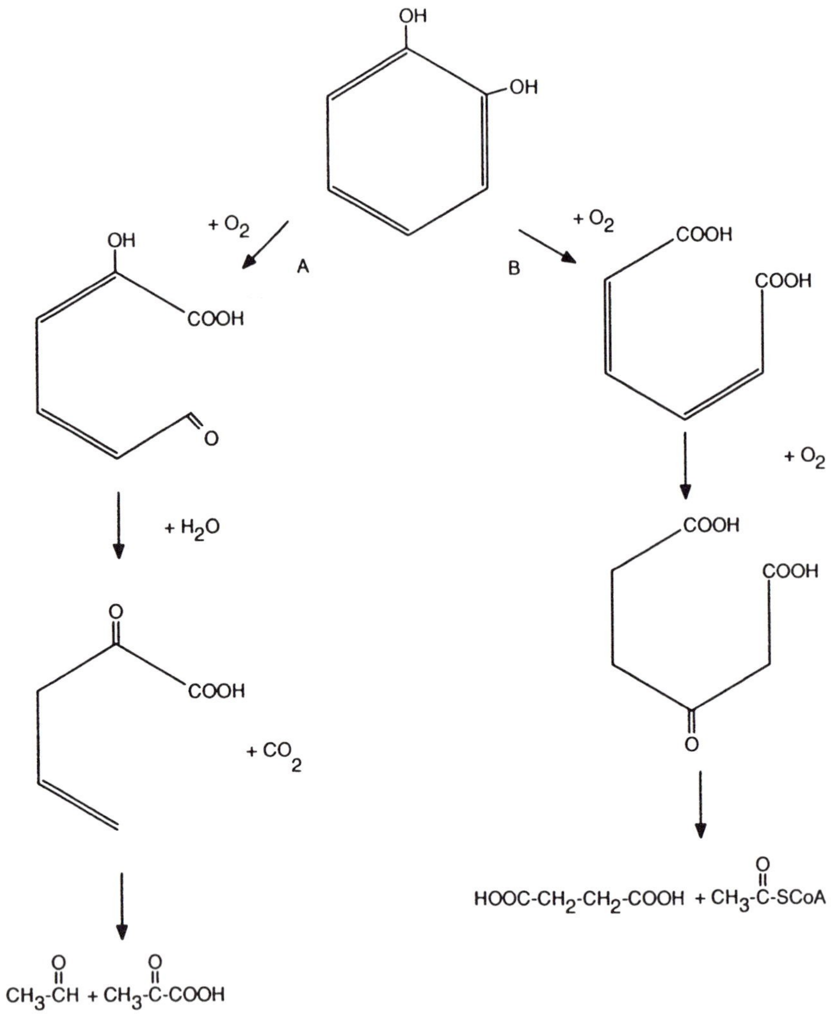

Figure 7. The meta pathway (A) and ortho pathway (B) of catechol degradation. Entry into central metabolism is via different ring cleavage products. In the meta pathway the ring carbon atoms are converted to acetaldehyde and pyruvic acid via the intermediates 2-hydroxymuconic semialdehyde and 2-ketopent-4-enoate. In the ortho pathway the ring carbon atoms are converted to succinate and acetyl-CoA via the intermediates *cis, cis*-muconate, and 3-ketoadipate.

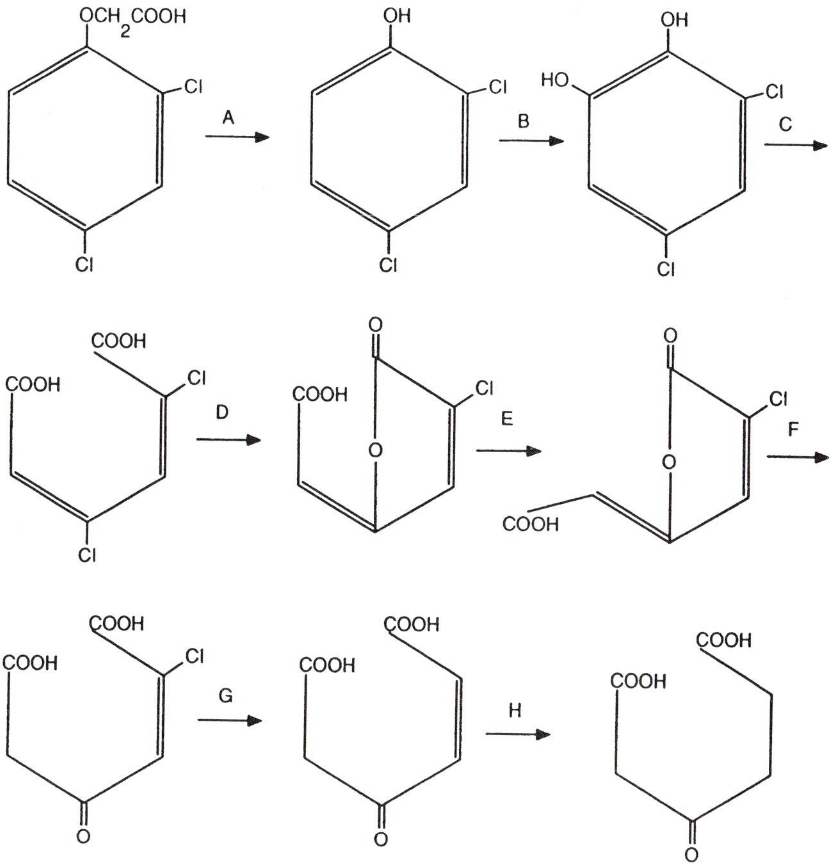

Figure 8. The pathway of 2,4-D degradation by *Alcaligenes eutrophus* JMP134(pJP4). Enzymes in the pathway: A, dioxygenase; B, monooxygenase; C, ring-cleaving dioxygenase; D, chloromuconate cycloisomerase; E, lactone isomerase; F, hydrolase; G, H, chloromaleylacetic acid reductase. The 3-ketoadipate produced by the pathway is converted to succinate and acetyl-CoA.

by cleaving the ether bond with an α-ketoglutarate-dependent dioxygenase. The alkyl chain is cleaved off in the form of glyoxylic acid, which is then mineralized by the organism. The aromatic ring is converted to the intermediate 2,4-dichlorophenol, which is then hydroxylated by means of a monooxygenase to form 3,5-dichlorocatechol, a ring cleavage substrate. An ortho-fission dioxygenase opens the ring, producing 2,4-dichloromuconic acid. The chlorine atoms must be removed before the ring carbon can enter the tricarboxylic acid cycle.

The molecule is cyclized to form a lactone (the cyclic ester of the hydroxy carboxylic acid), displacing one of the chlorine atoms. The lactone is hydrolyzed to yield 2-chloromaleylacetic acid, which is then dechlorinated and converted to 3-oxoadipic acid by two sequential reduced nicotinamide-adenin dinucleotide (NADH) consuming reductive steps, each catalyzed by chloromaleylacetate reductase (37). The molecule is converted to succinate and acetyl-CoA for further metabolism by the tricarboxylic acid (TCA) cycle. Genes encoding the enzymes which catalyze the dioxygenation, cycloisomerization, and lactone hydrolysis reactions are located on the broad host range P1 incompatibility group plasmid pJP4, whereas the gene encoding chloromaleylacetate reductase is found on the chromosome (38).

Carbofuran is one of a number of *N*-methylcarbamate insecticides (Fig. 3). A gram-negative soil bacterium, strain ER2, hydrolyzes the molecule at the carbamate linkage to form 7-phenolcarbofuran, carbon dioxide, and methylamine. The aromatic moiety of the molecule is not metabolized by this bacterium; in soils it will readily sorb or covalently attach to organic matter or be further degraded by other microorganisms. Strain ER2 grows at the expense of the methylamine originating from the side chain, using enzymes of the serine pathway of carbon assimilation common to many other methylotrophic bacteria. The full complement of enzymes required for growth on carbofuran, the hydrolase, and the C-1 assimilation pathway, are induced upon exposure to the insecticide. The gene encoding the hydrolase resides upon a 130 kb plasmid. Other insecticides which differ greatly in the structure of the aromatic moiety but retain the *N*-methylcarbamate linkage are also degraded by the same pathway (Fig. 9).

Pentachlorophenol is an uncoupler of oxidative phosphorylation and is used as a biocide in a number of applications. This molecule can serve as the sole carbon source for growth by a *Flavobacterium* isolate (Fig. 4). The initial attack on the molecule is catalyzed by an inducible monooxygenase enzyme which hydroxylates the carbon atom at the 4 position, simultaneously releasing chloride. The hydroxyl group originates from dioxygen gas and NADPH is an obligate cofactor in the reaction; consequently the reaction will not occur under anaerobic conditions or extracellularly. The tetrachlorohydroquinone intermediate is subjected to three sequential reductive dechlorination reactions, ultimately producing chlorohydroquinone. Reduced glutathione is an obligate cofactor in these reductive dehalogenation reactions which appear to be catalyzed by an inducible enzyme system. The chloroquinone is mineralized by as yet unknown reactions. Di-, tri-, and tetrachlorophenols and the herbicide bromoxynil (3,5-dibromo-4-hydroxybenzonitrile) are also metabolized by the *Flavobacterium* using the same suite of enzymes. In anaerobic sludges and undefined anaerobic enrichment cultures the molecule is reductively dechlorinated to tetra-, tri-, di-, and monochlorinated products.

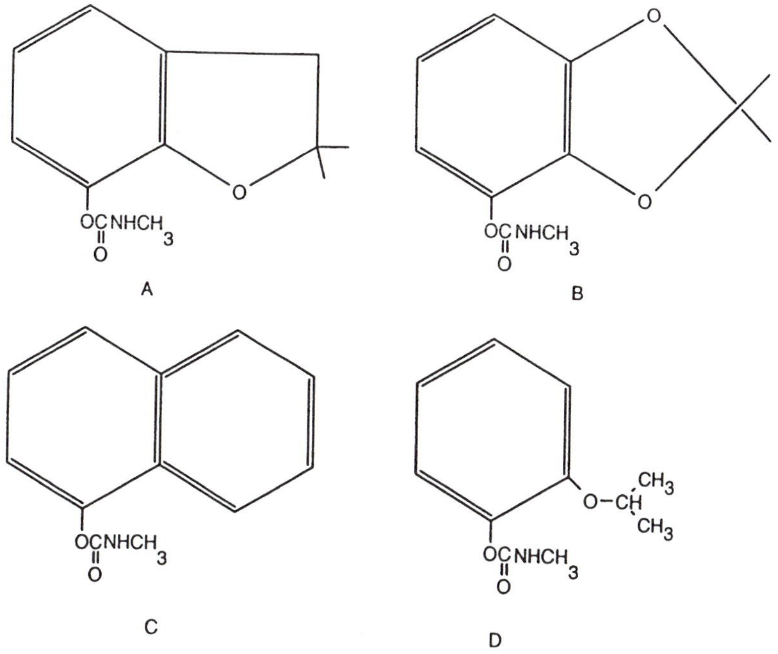

Figure 9. Different chemicals in the same class of pesticides can be metabolized by a common pathway. *N*-methylcarbamate insecticides differing in the aromatic moiety are all metabolized by carbofuran by a gram-negative methylotrophic bacterium, strain ER2: A, carbofuran; B, bendiocarb; C, carbaryl; D, propoxur.

4.2 Cometabolism of Pesticides

The fortuitous transformation of substrates by enzymes with low substrate specificity is probably the dominant form of degradation of many pesticides, especially when they are at very low concentration. Enzymes that catalyze cometabolic reactions are usually not induced by the substrate. Cometabolism of an applied pesticide is not preceded by a lag phase and does not accelerate with repeated application since there is no growth of the degrading biomass. Examples of cometabolic reactions include the hydroxylation of (2-chloro-*N*-(2-ethyl-6-methylphenyl)-*N*-(2-methoxy-1-methylethyl)acetamide (metolachlor) by an actinomycete (Fig. 2) (22) and of carbofuran by a *Rhodococcus* sp. (Fig. 3) (23). Pentachlorophenol and other chlorinated phenols can be methylated by various bacteria and fungi to form volatile anisoles, perhaps as a detoxification mechanism (Fig. 4) (34).

5. ADAPTATION AND THE DEVELOPMENT OF NEW DEGRADATIVE CAPABILITIES

How is it that microorganisms possess the enzymes required to degrade xenobiotic molecules whose structures are apparently foreign to anything previously seen in nature? The degree of "foreignness" is actually variable from pesticide to pesticide. There appear to be no natural counterparts for many pesticide structures, for example, the chlorinated hydrocarbons DDT, mirex, and dieldrin. However, some seemingly unusual structures found in pesticide molecules are also in fact found in nature. Soil fungi can, for example, produce large amounts of various halogenated aromatic compounds (39). Enzymes to degrade these structures undoubtedly predate the age of the organic chemist. Enzymes involved in the metabolism of natural chemicals may have sufficiently low substrate specificity that they also attack xenobiotic analogs. Pesticides can also be degraded cometabolically by enzymes with low substrate specificities. For example, the oxidative lignin-degrading system of *Phanerochaete chrysosporium* is remarkably nonspecific and can also degrade a very wide variety of pollutants (25). Remember that lignin is the material that gives wood much of its strength and consists of a highly complex three-dimensional heterogeneous benzylic polymer.

All natural compounds can be catabolized, given suitable environmental conditions, because the evolution of biosynthesis of natural chemicals was sufficiently slow to permit the parallel evolution of new catabolic functions required for their degradation (2). Microorganisms can occupy an infinite variety of niches in the environment because of their rapid growth rate, large numbers, and small size. The breadth of selective pressures experienced gave them the opportunity to develop tremendous biochemical diversity. There is an important selective advantage in the ability to utilize a new substrate in otherwise carbon-limited soils.

Microorganisms growing at the expense of a xenobiotic (i.e., foreign to nature) pesticide can frequently be isolated from soil only a few years after the introduction of the chemical. Some soils degrade a pesticide much more rapidly after repeated applications, suggesting that some kind of adaptation and change in the properties of the degrading microflora may have taken place. This observation has prompted speculation that the biodegradation of newly introduced pesticides becomes possible because of the rapid evolution and selection of catabolic phenotypes (38).

There is as yet no direct evidence proving that the degradation of a given pesticide in the environment is the result of selection of a new genotype by the introduction of that pesticide. However, studies with other organic pollutants show that catabolic phenotypes do evolve in the environment quite rapidly. There is no reason to believe that this should not also be the case with pesticides.

A number of genetic mechanisms may be involved in the evolution of new pesticide-degrading capabilities in soils. Microorganisms can in principle acquire new catabolic capabilities through the recruitment of genes encoding pesticide-degrading enzymes and by modification of the substrate specificity and regulation of preexisting enzymes which have other functions (38). Genes which encode enzymes involved in the degradation of organic pollutants frequently reside on plasmids and sometimes in transposons, deoxyribonucleic acid (DNA) elements able to replicate and insert new copies in the genome. These mobile genetic elements can be exchanged between microorganisms in soil and water (40). Moreover, genes involved in a catabolic pathway are frequently clustered together, facilitating coregulation and transfer of the pathway. Plasmids are involved in the degradation of a large number of pesticides, including the phenoxyalkanoic and thiocarbamate herbicides and the methylcarbamate and organophosphorus insecticides (41). The transfer of these catabolic plasmids may be involved in the adaptation of the soil microflora and development of enhanced degradation. The 2,4-D-degradation encoding plasmid pJP4 can be transferred into a wide variety of bacteria, including *Escherichia coli, Rhodopseudomonas sphaeroides, Agrobacterium tumefaciens, Rhizobium* sp., *Pseudomonas putida, Pseudomonas fluorescens*, and *Acinetobacter calcoaceticus* (42). Small modifications to a catabolic gene sequence can alter the properties of the encoded enzymes. The enzyme substrate, or gene effector specificity, can be altered by substitutions of single base pairs, point mutations (43). The movement and rearrangement of sections of DNA through genetic recombination or transposition of DNA accompanying transposon replication can modify the regulation and expression of catabolic genes (44). Duplicated genes may facilitate rapid evolution since one gene copy can accumulate mutations and yield enzymes with altered properties, while the other copy can continue its normal function.

6. APPLIED ASPECTS OF PESTICIDE BIODEGRADATION

6.1 Prediction of Pesticide Fate in Soils

Pesticide fate simulation models are useful for environmental hazard analyses and for development of agricultural management practices which minimize the environmental risk and threat to human health of pesticide use. Simulation models try to describe and integrate the chemical, biological, and physical processes that determine pesticide fate and movement in soils (45). Models of varying complexity have been developed for research, management, regulatory, and screening purposes. Pesticide persistence models have subroutines which describe biodegradation rates in different soils under field conditions and can usually also describe vertical or horizontal movement of pesticide residues in the field, plant uptake, equilibrium sorption to soil particles, and volatilization.

A frequently employed biodegradation subroutine incorporates the relationships proposed by Walker (46–48) for simulating herbicide persistence in surface soils under different conditions of soil moisture and soil temperature. The model combines the first-order reaction rate law with equations describing the dependence of the rate constant, or the related half-life, on temperature and moisture. The rate constants and relationship to soil conditions will vary for each pesticide. Rate data are usually obtained in the laboratory with soils incubated at various temperatures and moisture contents. The degradation rate is estimated by quantifying residual pesticide concentrations in soil extracts sampled over time. The predicted residual concentration of the herbicide in the field is then calculated with the model for actual field soil temperature and moisture values estimated from meteorological data.

Most models assume that the degradation rate decreases exponentially with the pesticide concentration, in other words, first-order degradation kinetics. However, pesticide retention to the soil and multiple biological and chemical degradation pathways frequently cause this assumption to be incorrect (49).

The development and improvement of simulation models provide an area of active research. Phenomena that are not well understood include macropore and preferential flow, formation of "bound residues" by nonequilibrium sorption, and degradation kinetics which deviate from first-order, spatial and temporal variability, and pesticide fate in subsoils.

6.2   Applications of Pesticide-Degrading Bacteria, Enzymes, and Genes

The disposal of pesticide wastes and cleanup of pesticide spills can be problematic. The addition of microorganisms or pesticide-degrading enzymes has been proposed as a method of decontaminating polluted water or soil (6,50). Pesticide residues could either be mineralized by the introduced agents or immobilized to soil organic material via "oxidative coupling." In oxidative coupling the parent compound is enzymatically transformed by oxidation to a reactive intermediate which rapidly reacts with soil organic matter. For example, laccase and peroxidase can catalyze the oxidative coupling of 2,4-dichlorophenol to fulvic and humic acids (50). It has been suggested that the covalently attached residues are not bioavailable or mobile and therefore are effectively detoxified.

In situ pesticide degradation rates can be manipulated by modifying the soil environment. For example, the plant rhizosphere can accelerate pesticide degradation, presumably through enhancement of microbial activity via the provision of carbon in the form of root exudate or modification of oxygen concentrations. The organophosphorus insecticides O,O-diethyl O-[6-methyl-2-(1-methylethyl)-4-pyrimidinyl] phosphorothioate (diazinon) and O,O-diethyl O-(4-nitrophenyl) phosphorothioate (parathion) were mineralized about twice as fast in soil containing a bush bean (*Phaseolus vulgaris*) plant as in soil without a plant (51).

A number of bacterial genes have been used to modify plants genetically and make them resistant to specific herbicides. For example, the herbicide bromoxynil inhibits photosynthesis and uncouples oxidative phosphorylation. A gene originating from *Klebsiella pneumoniae* subsp. *ozaenae* encoding a bromoxynil-modifying nitrilase was used to generate bromoxynil-resistant transgenic plants (52).

7. INHIBITORY EFFECTS OF PESTICIDES ON SOIL MICROORGANISMS

The literature on the effects of pesticides on soil microorganisms is enormous. Each pesticide's toxicity is examined individually in a variety of soils under a variety of conditions. It is difficult to make meaningful generalizations. A potential inhibitory effect of a pesticide can be assessed at many different levels: total microbial biomass, enumerations of total bacterial or fungal populations, measurement of functional (e.g., respiration rate, nitrification rate, incorporation of radiolabeled substrate, nodulation of legumes) or specific soil enzyme (e.g., phosphatase, dehydrogenase, urease) activity, energy charge, or enumeration of specific microorganisms. New molecular techniques now make it possible to study the effects of pesticides on microbial diversity. Toxicity tests with pure microbial cultures, or in soils contaminated with unrealistically high pesticide concentrations, are of little value. There are a number of problems with assessing the significance of the side-effect tests. It can be difficult to discriminate a real effect from background variation in the measured activity. Consideration should be given to how the results of a microbial toxicity test can be interpreted in terms of a real effect on soil fertility. How long must an effect last to be of concern, days, months, or years? Finally, what threshold values which, if exceeded, are cause for concern, should be used?

Suffice it to say that at field concentrations and under field conditions, some pesticides can sometimes have a short-lived inhibitory effect, but there is little solid scientific evidence to suggest that pesticides generally have a measurable long-term effect on the soil microbial population (53). Biocides are toxic to microorganisms, but as the chemical is degraded microbial populations recover. Fungicides and soil fumigants seem to be more inhibitory than insecticides or herbicides. A fungicide may inhibit soil fungi, and a herbicide photosynthetic microorganisms. There is no reason to expect that an insecticide will be toxic to microorganisms because they do not have sensitive targets. Generally, there seems to be little effect of pesticides on soil enzymes or cyanobacteria (54,55). Herbicides can have a positive or a negative effect on the incidence of plant diseases by modifying the populations of fungi in soil or by increasing the sensitivity of plants to opportunistic pathogens (56). In a field study examining the effects on the microflora of 35 consecutive years of 2,4-D use at Indian Head, Saskatchewan, Canada, there was no statistically significant difference in total microbial biomass, respiration, nitrogen mineralization, or nitrification rates, or activity of urease, phosphatase, or dehydrogenase enzymes, in herbicide-treated and untreated control plots (57). It was concluded that there was no significant ecological effect or decline in soil fertility associated with 2,4-D use.

8. CONCLUSIONS

Pesticide breakdown in soils is largely mediated by microorganisms. Much is known about the enzymological characteristics of pesticide transformations reactions and the ways pesticides are metabolized to support growth. The rate and pathway of pesticide biodegradation in soils are heavily influenced by the soil environment. Pesticide-degrading microorganisms and enzymes have applications in bioremediation and production of herbicide-resistant transgenic plants. Numerous aspects of pesticide microbiological properties are areas of active research. Interactions among the soil, pesticide, and pesticide-degrading microorganisms are complex and variable. Adaptation of pesticide-degrading microorganisms and acceleration of pesticide degradation following repeated treatments are poorly understood phenomena. Molecular techniques now available will be used for gaining insights into adaptation, evolution, and diversity and the effects of pesticides on soil microorganisms (see Section 2).

REFERENCES

1. M. Alexander, *Adv. Appl. Microbiol 7*: 35 (1965).
2. J. M. Bollag and S.-Y. Liu, *Pesticides in the Soil Environment Processes, Impacts, and Modelling* (H. H. Cheng, ed.), Soil Science Society America, Madison, Wisconsin, p. 169 (1990).
3. H. H. Cheng ed., *Pesticides in the Soil Environment: Processes, Impacts, and Modelling*, Soil Science Society America, Madison, Wisconsin (1990).
4. D. J. Hardman, *Crit. Rev. Biotechnol. 11*: 1 (1991).
5. I. R. Hill, *Pesticide Microbiology* (I. R. Hill and S. J. L. Wright ed.), Academic Press London, p. 137 (1978).

6. J. S. Karns, M. T. Muldoon, W. M. Mulbry, M. K. Derbyshire, and P. C. Kearney, *Biotechnology in Agricultural Chemistry* (H. LeBaron ed.), American Chemical Society, Washington DC (1987).
7. W. W. Mohn, and J. M. Tiedje, *Microbiol. Rev. 56*: 482 (1992).
8. D. M. Munnecke, L. M. Johnson, H. W. Talbot and S. Barik, *Biodegradation and Detoxification of Environmental Pollutants* (A.M. Chakrabarty ed.), CRC Press, Boca Raton, (1982).
9. B. L. Sawhney and K. Brown (ed.), *Reactions and Movement of Organic Chemicals in Soils*, Soil Science Society America, Madison, Wisconsin (1989).
10. A. E. Smith, *Environmental Chemistry of Herbicides* (R. Grover ed.), Vol. 1, CRC Press, Boca Raton, (1988).
11. A. E. Smith, *Rev. Weed Sci. 4*: 1 (1989).
12. R. A. Chapman and C. R. Harris, *Enhanced Biodegradation of Pesticides in the Environment* (K. D. Racke and J. R. Coats ed.), American Chemical Society, Washington DC, p. 83 (1990).
13. W. D. Guenzi and W. E. Beard, *Proc. Soil Sci. Soc. Am. 32*: 522 (1968).
14. J. J. Pignatello, *Reactions and Movement of Organic Chemicals in Soils* (B. L. Sawhney and K. Brown ed.), Soil Science Society America, Madison, Wisconsin, p. 45 (1989).
15. K. M. Scow and M. Alexander, *Soil Sci. Soc. Am. J. 56*: 128 (1992).
16. S. M. Steinberg, J. J. Pignatello, and B. L. Sawhney, *Environ. Sci. Technol. 21*: 1201 (1987).
17. P. M. Huang, *Soil Biochemistry* (J.-M. Bollag and G. Stotzky eds.), Vol. 6, Marcel Dekker, New York, p. 29 (1990).
18. S. Fogel, R. L. Lancione, and A. E. Sewal, *Appl. Environ. Microbiol. 44*: 113 (1982).
19. T. B. Parkin, *J. Environ. Qual. 22*: 409 (1993).
20. A. J. Sextone, N. P. Revsbech, T. B. Parkin, and J. M. Tiedje, *Soil Sci. Soc. Am. J. 49*: 645 (1985).
21. W. G. Johnson and T. L. Lavy, *J. Environ. Qual. 23*: 556 (1994).
22. A. Krause, W. G. Hancock, R. D. Minard, A. J. Freyer, R. C. Honeycutt, H. M. LeBaron, D. L. Paulson, S.-Y. Liu, and J.-M. Bollag, *J. Agric. Food Chem. 33*: 584 (1985).
23. R. M. Behki, E. Topp, and B. Blackwell, *J. Agric. Food Chem. 42*: 1375 (1994).
24. L. Xun, E. Topp, and C. S. Orser, *J. Bacteriol. 174*: 5745 (1992).
25. J. S. Yadav and C. A. Reddy, *Appl. Environ. Microbiol. 59*: 2904 (1993).
26. R. Behki, E. Topp, W. Dick, and P. Germon, *Appl. Environ. Microbiol. 59*: 1955 (1993).
27. F. Fukumori and R. P. Hausinger, *J. Bacteriol. 175*: 2083 (1993).

28. J. R. W. Miles, C. M. Tu, and C. R. Harris, *J. Econ. Entomol. 62*: 1334 (1969).
29. D. Grbic-Galic, *Soil Biochemistry*, Vol. 6 (J.-M. Bollag and G. Stotzky ed.), Marcel Dekker, New York, p. 117 (1990).
30. D. A. Graetz, G. Chesters, T. C. Daniel, L. W. Newland, and C. B. Lee, *J. Wat. Poll. Cont. Fed. 12*: 76 (1970).
31. N. R. Adrian and J. M. Suflita, *Appl. Environ. Microbiol. 56*: 292 (1990).
32. L. Xun, E. Topp, and C. S. Orser, *J. Bacteriol. 174*: 8003 (1992).
33. E. Topp, R. S. Hanson, D. B. Ringelberg, D. C. White, and R. Wheatcroft, *Appl. Environ. Microbiol. 59*: 3339 (1993).
34. A. J. Cserjsei and E. L. Johnson, *Can. J. Microbiol. 10*: 45 (1972).
35. R. Bartha and D. Pramer, *Adv. Appl. Microbiol. 13*: 317 (1970).
36. S. Dagley, *Essays Biochem. 11*: 81 (1975).
37. V. Seibert, K. Stadler-Fritzsche, and M. Schlömann, *J. Bacteriol. 175*: 6745 (1993).
38. J. R. van der Meer, W. M. de Vos, S. Harayama, and A. J. B. Zehnder, *Microbiol. Rev. 56*: 677 (1992).
39. E. De Jong, J. A. Field, H.-E. Spinnler, J. B. P. A. Wijnberg, and J. A. M, *Appl. Environ. Microbiol. 60*: 264 (1994).
40. R. R. Fulthorpe and R. C. Wyndham, *Appl. Environ. Microbiol. 58*: 314 (1992).
41. J. S. Karns, W. W. Mulbry, and P. H. Tomasec, *Recent Advances in Microbial Ecology* (Hattori, T. Y. Ishida, Y. Maruyama, R. Y. Morita, and A. Uchida eds.), Japan Scientific Societies Press, Tokyo, p. 595 (1989).
42. R. H. Don and J. M. Pemberton, *J. Bacteriol. 145*: 681 (1981).
43. P. H. Clarke, *Microbial Degradation of Organic Compounds* (D. T. Gibson ed.), Marcel Dekker, New York, p. 11 (1984).
44. L. N. Ornston, J. Houghton, E. L. Neidle, and L. A. Gregg, *Pseudomonas: Biotransformations Pathogenesis, and Evolving Biotechnology* (Silver, S., A. M. Chakrabarty, B. Iglewski, and S. Kaplan eds.), American Society for Microbiology, Washington, D.C., p. 207 (1990).
45. R. J. Wagenet and P. S. C. Rao, *Reactions and Movement of Organic Chemicals in Soils* (Cheng, H. H. ed.), Soil Science Society America, Madison, Wisconsin, p. 351 (1990).
46. A. Walker, *Pestic. Sci. 1*: 237 (1970).
47. A. Walker, *Pestic. Sci. 7*: 41–49 (1976).
48. A. Walker, *Pestic. Sci. 7*: 50–58 (1976).
49. J. W. Hamaker, *Organic Chemicals in the Soil Environment* (C. A. I. Goring and J. W. Hamaker eds.), Marcek Dekker, New York, p. 253 (1972).
50. P. Nannipieri and J.-M. Bollag, *J. Environ. Qual. 20*: 510 (1991).
51. T.-S. Hsu and R. Bartha, *Appl. Environ. Microbiol. 37*: 36 (1979).

52. D. M. Stalker, K. E. McBride, and L. D. Malyj, *Science 242*: 419 (1988).
53. L. Somerville and M. P. Greaves, *Pesticide Effects on the Soil Microflora*, Taylor & Francis, London (1987).
54. A. Schäffer, *Soil Biochemistry*, Vol. 8 (J.-M. Bollag and G. Stotzky ed.), Marcel Dekker, New York, p. 273 (1993).
55. K. Venkateswarlu, *Soil Biochemistry*, Vol. 8 (J.-M. Bollag and G. Stotzky ed.), Marcel Dekker, New York, p. 137 (1993).
56. C. A. Lévesque and J. E. Rahe, *Annu. Rev. Phytopathol. 31*: 579 (1992).
57. V. O. Biederbeck, A. C. Campbell, and A. E. Smith, *J. Environ. Qual. 16*: 257 (1987).

19

A Case Study of Bioremediation of Polluted Soil: Biodegradation and Toxicity of Chlorophenols in Soil

KAM TIN LEUNG,* DEENA ERRAMPALLI, MIKE CASSIDY HUNG LEE, and JACK T. TREVORS University of Guelph, Guelph, Ontario, Canada

HIDEO OKAMURA Okayama University, Kurashiki, Japan

HANS-JÜRGEN BACH GSF-National Research Center for Environment and Health, Neuherberg, Germany

B. HALL University College Dublin, Dublin, Ireland

1. INTRODUCTION

For about 60 years, chlorophenols, mainly pentachlorophenol (PCP), have been used worldwide as wide-spectrum biocides to control bacteria, fungi, algae, mollusks, and insects (1). In addition, lower chlorinated phenols have been used as precursors of other pesticides. For instance, 2,4-dichlorophenol and 2,4,5-trichlorophenol are used to synthesize 2,4-dichlorophenoxyacetic acid (2,4-D) and 2,4,5-trichlorophenoxyacetic acid (2,4,5-T), respectively; 2,4-D and 2,4,5-T are two of the most commonly used herbicides in the world. In the 1970s and early 1980s, the worldwide production of chlorophenols was estimated to be 200,000 tons/year, when PCP was about 50% of the total production (2). More than 80% of the PCP production has been consumed by the lumbering industry as wood preservatives (2). Because of the long-term usage and recalcitrant nature of the highly chlorinated phenols, the chemicals have been detected in soils, sediments, natural waters, and higher animals. Because of the toxicity of these

---
*Current affiliation: University of Tennessee, Knoxville, Tennessee

compounds, five chlorinated phenols, including PCP, have been listed as priority pollutants by the U.S. Environmental Protection Agency (3). In Japan, PCP was used as a herbicide for paddy and upland rice with production levels of about 14,000 tons/year. However, use of PCP as herbicide in Japan was banned in the 1970s as a result of its accumulation in humans and other higher animals. In Finland, PCP production ceased in 1984. Despite the banning or restricted usage of chlorophenols in some countries, such as Finland, Germany, and Japan, chlorinated phenols are still employed as wood preservatives in many other countries (2,4).

2. TOXICITY OF CHLOROPHENOLS

The toxicity of chlorophenols has been investigated by bioassays with ecologically relevant organisms such as bacteria, zooplanktons, shrimps, fish, and higher plants (Table 1). Higher chlorinated phenols tend to be more toxic than the less chlorinated phenols in most biological systems. Bioavailabilities of chlorophenols in soils depend on the pKa [–log (acid dissociation constant)] and Koc (organic carbon sorption coefficient, which is equal to the ratio of the amount of substrate adsorbed per unit of organic carbon in soil to the concentration of the chemical in soil solution) of the chlorinated contaminants. Therefore, toxicity of the pollutants is affected by both the pH and the organic carbon content of the contaminated samples.

PCP has an oral lethal dosage for 50% mortality (LD$_{50}$) ranging from 27 to 210 mg/kg in rats, and an estimated minimum lethal dose in humans of 29 mg/kg (5). Acute exposure of laboratory animals to PCP results in vomiting, hyperpyrexia (elevated body temperature), elevated blood pressure and respiration rate, and cardiovascular stress (6). PCP is particularly toxic to aquatic lives and has a lethal concentration for 50% mortality (LC$_{50}$) of about 0.2 mg/L for salmon (7). It is also known to be an uncoupling agent of oxidative phosphorylation in mitochondria and membranes of microorganisms. For instance, the minimum inhibitory concentration (MIC) of PCP for *Clostridium perfringens* is 0.062 mM (8).

3. BIODEGRADATION AND BIOTRANSFORMATION OF CHLOROPHENOLS IN SOIL

3.1 Aerobic and Anaerobic Degradation of Chlorophenols

Despite the toxicity of chlorophenols, these compounds can be degraded by some soil microorganisms under either aerobic or anaerobic conditions (Table 2). In general, the rate of aerobic decomposition of chlorophenols in soil decreases as the number of chlorine substituents increases, whereas the opposite is observed for anaerobic decomposition of chlorinated phenols (9).

Under aerobic conditions, biodegradation of chlorinated phenols depends on the number and position of the chlorine substituent(s). In general, those with fewer chlorine substituents and *ortho*-and/or *para*-substituted chlorine(s) have shorter half-lives in the environment. In a soil microcosm study, 2-chlorophenol, 4-chlorophenol, 2,4-dichlorophenol, and 2,4,6-trichlorophenol required only 5 to 14 days for complete degradation. Other chlorophenols, such as 3-chlorophenol, 2,4,5-trichlorophenol, 2,3,4,6-tetrachlorophenol, and PCP, persisted for more than 75 days in soil (10). In nature PCP is very recalcitrant. Although it can be degraded by photolysis, it is primarily decomposed by specific microorganisms in soil. At some contaminated areas, PCP can persist for 5 years (11).

The half-life of PCP in a flooded paddy field containing an initial PCP concentration of 100 ppm was reported to be 40 days (12). However, under aerobic conditions, only a small amount was degraded in 2 months (12). Although complete degradation of PCP under anaerobic conditions has been documented, anaerobic degradation of PCP often results in accumulation of fewer chlorinated phenols, indicating preferential dechlorination of higher chlorinated phenols in anaerobic environments.

Biodegradation of chlorophenols can be greatly affected by environmental factors. Therefore, both the efficiency and the degradative strategy of the chlorophenol-degrading microorganisms may vary under different conditions. Some environmental parameters that influence chlorophenol degradation will be discussed later in this chapter.

3.2 Biotransformation of Chlorophenols

Some microbial strains, such as *Acinetobacter, Mycobacterium, Pseudomonas,* and *Rhodococus* spp., can O-methylate chlorophenols. This involves methylation of the hydroxy group of PCP. Although chloroanisoles are usually less toxic to bacteria, these compounds may be more harmful to other organisms than their precursors. Chloroanisoles are more lipophilic than chlorophenols. Therefore, O-methylation of chlorophenols may increase the potential of bioaccumulation of the pollutants (9).

Chlorophenols can be incorporated into soil organic matter through oxidative coupling, which is catalyzed by phenoloxidases and peroxidases. These enzymes are produced by bacteria, fungi, and plants. They catalyze the polymerization of chlorophenols with humic substances in soil. It has been demonstrated that a laccase, an oxidase involved in lignin formation, can couple chlorophenols with model phenolic humus constituents forming dimers, trimers, and higher polymers (13). After being covalently bound to soil humic substances, chlorophenols are immobilized and stabilized against biodegradation.

Table 1. Chemical Properties and Toxicity Data of Chlorophenols[a]

Substrate[b]	Chemical property[c]		Toxicity					Acute oral toxicity[c]		
	pKa	log K_{ow}	Microtox[d]	Shrimp[e]	Bluegill[f]	Trout[g]	Millet root[h]	Male mouse	Female mouse	Rat
Phenol	9.8	1.5	380.00	79.40	—[j]	4.47	120	—	—	—
2-CP	8.3	2.2	263.00	40.74	56.20	—	58	—	—	570
3-CP	8.8	2.5	109.60	—	—	—	—	521	530	670
4-CP	9.1	2.4	64.50	35.48	30.90	3.39	—	1373	1422	—
2,3-DCP	7.7	3.2	30.20	—	—	—	—	2585	2376	—
2,4-DCP	7.9	3.1	33.88	—	28.80	3.98	25	1276	1352	—
2,5-DCP	7.5	3.2	57.54	—	—	—	—	1600	946	—
2,6-DCP	6.8	2.9	81.30	117.50	—	8.91	—	2198	2120	—
3,4-DCP	8.6	3.4	10.00	—	—	—	—	1685	2046	—
3,5-DCP	8.2	3.5	16.98	9.12	—	—	—	2643	2389	—
2,3,4-TriCP	6.5	4.1	6.31	10.00	—	—	—	—	—	—
2,3,5-TriCP	6.5	4.2	5.62	—	—	—	—	—	—	—
2,3,6-TriCP	6.5	3.9	64.56	13.80	—	—	—	—	—	—
2,4,6-TriCP	6.1	3.7	38.90	—	3.63	—	12	—	—	—
3,4,5-TriCP	6.5	4.4	1.82	—	—	—	—	—	—	—
2,3,4,5-TetCP	5.3	5.0	0.76	—	—	—	—	—	—	—

2,3,4,6-TetCP	5.3	4.1	5.50	51.30	0.81	—	—	—	—	—
2,3,5,6-TetCP	5.3	4.9	9.55	—	1.74	—	—	—	177	—
PCP	4.8	5.0	1.95	12.30	—	0.14	—	177	177	27–175

[a]More information can be obtained from Ref. 72.

[b]CP, chlorophenol; DCP, dichlorophenol; TriCP, trichlorophenol; TetCP, tetrachlorophenol; PCP, pentachlorophenol.

[c]pKa and log K_{ow} represent -log (acid dissociation constant) and log (octanol–water partition coefficient), respectively.

[d]30-Minute effective concentration (30EC$_{50}$, concentration of chemical which reduces 50% of the bacterial activity in 30 min) of the toxicity (Microtox) values, are presented in micromoles.

[e]Lethal thresholds to shrimp in 96 hours (96LT) were measured in micromoles.

[f]50% Lethal concentrations to bluegill in 24 hours (24LC$_{50}$) were measured in micromoles.

[g]50% Intraperitoneal injection lethal dosages to trout (IPLD$_{50}$) were measured in micromoles.

[h]Concentrations of chemicals which reduced elongation of millet root by 50% in 120 hours (120EC$_{50}$) were measured in milligrams per liter.

[i]50% Acute oral lethal dosages (LD$_{50}$) to mice and rats were measured in milligrams per kilogram.

[j]—Data not available.

Table 2. Microorganisms Capable of Either Partial or Complete Degradation of
Chlorophenols[a]

Substrates	Microorganisms	Respiratory pathway[c]
	Bacteria	
Chlorophenols	*Pseudomonas* sp. strain B13	A
2-Chlorophenol	*Desulfitobacterium dehalogenans* strain 2 CP-1	AN
4-Chlorophenol	*Azotobacter chroococcum* strain MSB1	A
	Arthrobacter sp.	A
	Alcaligenes sp. strain A7-2	A
3-Chlorophenol, 4-chlorophenol	*Rhodococcus* sp. strains An117 and An213	A
4-Chlorophenol, 2, 4-dichlorophenol	*Rhodococcus erythropolis* 1cp	A
2,4-Dichlorophenol[b]	*Alcaligenes eutrophus* JMP134	A
	Pseudomonas sp. strain NC1B 9340	A
	Pseudomonas cepacia BRI6001	A
	Pseudomonas cepacia	A
	Acinetobacter sp.	A
	Xanthobacter sp. strain CP	A
	Flavobacterium sp. strain 50001	A
	Arthrobacter sp.	A
	Flavobacterium sp. strain MH	A
2,4,-Dichlorophenol	*Desulfitobacterium dehalogenans* JW/IU-DC1	AN
2,5-Dichlorophenol	*Pseudomonas* sp. strain JS6	A
Dichlorophenols, tetrachlorophenols	*Desulfomonile tiedjei* strain DCB-1	AN
2,4,5-Trichlorophenol	*Pseudomonas cepacia* strain AC1100	A
2,4,5-Trichlorophenol	*Pseudomonas pickettii*	A
	Azotobacter sp. strain GP1	A
2,4,6-Trichlorophenol, dichlorophenols	*Desulfomonile tiedjei* strain DCB-2	AN
2,4,6-Trichlorophenol, tetrachlorophenols	*Streptomyces rochei* 303	A
Trichlorophenols, tetrachlorophenols	*Mycobacterium chlorophenolicum* strains PCP-1, CP-2, CG-1	A
	M. fortuitum strain CG-2	A
	Flavobacterium sp. strain ATCC 39723	A
2,4,6-Trichlorophenol,- 2,3,4,6- tetrachlorophenol	*Arthrobacter* sp. strain KC-3	A

Table 2. Continued

Substrates	Microorganisms	Respiratory pathway[c]
2,4,6-Trichlorophenol, 2,3,4,6-tetrachlorophenol	*Pseudomonas saccharophila* strains KF1, NKF1	A
Pentachlorophenols	*Arthrobacter* sp. strain ATCC 33790	A
	Arthrobacter sp. strain KC-3	A
	Arthrobacter sp. strain NC	A
	Desulfomonile tiedjei strain DCB-1	AN
	Flavobacterium sp. strain ATCC 39723	A
	Mycobacterium chlorophenolicum strains PCP-1, CP-2, CG-1	A
	M. fortuitum strain CG-2	A
	Pseudomonas sp.	A
	Pseudomonas spp.	A
	Pseudomonas sp. strain RA2	A
	Pseudomonas sp. strain SR3	A
	Pseudomonas aeruginosa	A
	Pseudomonas cepacia strain AC1100	A
	Sphingomonas sp. strain RA2	A
	Streptomyces rochei 303	A
	Fungi	
3-Chlorophenol, 4-chlorophenol	*Candida tropicalis*	A
	Candida maltosa	A
4-Chlorophenol, chlorophenols	*Penicillium frequentens* Bi 7/2	A
2,4,5-Trichlorophenol, 2,4-dichlorophenol	*Phanerochaete chrysosporium*	A
2,4,5-Trichlorophenol, 2,4,6-Trichlorophenol	*Pleurotus cornucopiae*	A
Pentachlorophenol	*Ceriporiopsis subvermispora*	A
	Mycena avenacea	A
	Phanerochaete chrysosporium	A
	Phanerochaete sordida	A
	Phoma glomerata	A
	Tramatis versicolor	A
	Zygomycetes	A

[a]For more details on chlorophenol-degrading microorganisms, see Ref. 2 and 4.
[b]2,4-Dichlorophenol is an intermediate in the 2,4-dichlorophenoxyacetic acid (2,4-D) degradation pathway.
[c]A, aerobic; AN, anaerobic.

4. CHLOROPHENOL-DEGRADING MICROORGANISMS AND THEIR DEGRADATIVE PATHWAYS

4.1 Aerobic Degradation of PCP

Bacteria are mainly responsible for the degradation of chlorinated phenols in the environment. Mono- and dichlorophenol are degraded by an *ortho*-ring cleavage pathway (Fig. 1). However, degradative mechanisms of tri-, tetra-, and pentachlorophenol involve the formation of chloro-*p*-hydroquinones in the first step of dechlorination (Fig. 2).

Some fungi can also degrade chlorophenols (Table 2). Most studies on PCP degradation were conducted with members of the white rot Basidiomycetes. In liquid culture, *Phanerochaete chrysosporium* and *P. sordida* have been shown to mineralize 23% and 11.64% of 1.1 ppm and 25 ppm, PCP, respectively (14,15). However, in a soil study, only 2% of the soil PCP was mineralized by *P. chrysosporium*. Both *P. chrysosporium* and *P. sordida* can O-methylate PCP. In a field study, inoculation of *Phanerochaete* spp. showed success in decreasing the PCP concentration in soil from 300–400 mg/kg to less than 100 mg/kg of soil. However, mineralization of PCP was not analyzed in this study (16).

Degradation of mono- and dichlorophenols

Some aerobic bacterial strains isolated from environmental samples can degrade mono- and dichlorophenols (Table 2). However, they are traditionally isolated from 2,4-D or chlorobenzoate enrichment cultures. *Pseudomonas* strain B13 (17) and *Alcaligenes eutrophus* JMP134 (18) have been used to study the degradative pathways of mono- and dichlorophenols (Fig. 1). Mono- or dichlorophenols are first hydroxylated by phenol hydroxylases forming chlorocatechols and the reactions are reduced nicotinamide-adenine dinucleotide- (NADH-) or reduced nicotinamide-adenine dinucleotide phosphate- (NADPH-) dependent. In addition, molecular oxygen is required for the reactions. Chlorocatechols are then cleaved at the *ortho* position by chlorocatechol 1,2-dioxygenases, which are analogous to catechol 1,2-dioxygenase. One chlorine substituent is eliminated spontaneously after the ring cleavage. The second chlorine of dichlorophenols is removed by a maleylacetate reductase. Both mono- and dichlorophenols are broken down to β-ketoadipate, which will eventually be metabolized through the tricarboxylic acid (TCA) cycle.

Degradation of polychlorinated phenols

Polychlorinated phenols (defined as chlorophenols with three or more chlorine substituents in this chapter) can be resistant to microbial degradation and are degraded differently from the mono- and dichlorophenols. Polychlorophenol degrading strains share similar strategies in decomposing polychlorinated phenols.

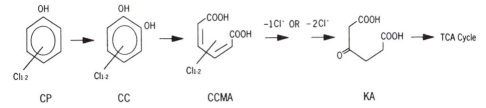

Figure 1. Biodegradation of mono- and dichlorophenol by the *ortho* ring cleavage pathway. CP, CC, and CCMA are substituted with either one or two chlorine substituent(s). CP, chlorophenol; CC, chlorocatechol; CCMA, chlorinated *cis*, *cis*-muconic acid; KA, β-ketoadipic acid; TCA cycle, tricarboxylic acid cycle.

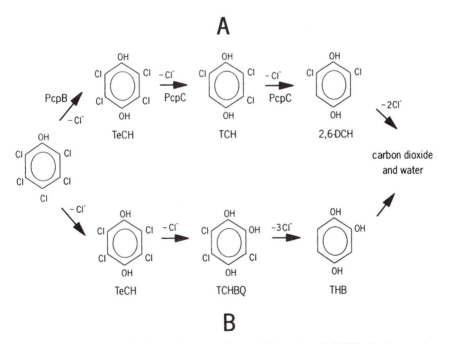

Figure 2. Biodegradation pathways of pentachlorophenol (PCP). Pathways A and B are based on the PCP-degradative mechanism of *Flavobacterium* sp. ATCC 39723 and *Rhodococcus chlorophenolicus* strain-PCP-1, respectively. TeCH, tetrachloro-*p*-hydroquinone; TCH, trichloro-*p*-hydroquinone; 2,6-DCH, 2,6-dichloro-p-hydroquinone; TCHBQ, trichlorohydroxybenzoquinone; 1,2,4-THB, 1,2,4-trihydroxybenzene; PcpB, *Flavobacterium* PCP-4-monooxygenase, PcpC TeCH, reductive dehalogenase.

First, the degradative bacteria transform polychlorophenols to chloro-p-hydro-quinones. More chlorine substituents will then be removed prior to the ring cleavage reaction (Fig. 2).

Several microbial strains can degrade trichlorophenols (Table 2). *Pseudomonas cepacia* strain AC1100 is one of the most studied trichlorophenol-degrading strains (19). The initial step for trichlorophenol degradation involves the formation of a dichloro-p-hydroquinone. Another chlorine will be removed, forming a monochlorinated trihydroxybenzene, which will eventually be miner-alized by the trichlorophenol-degrading bacteria.

Bacteria capable of degrading PCP were first isolated in the 1970s. They be-longed to the genera *Pseudomonas* and *Arthrobacter*. Since then, more PCP-de-grading strains have been isolated and classified under the *Flavobacterium, Sphingomonas*, and *Mycobacterium* spp. (Table 2). Gram-positive *Mycobac-terium chlorophenolicum* strain PCP-1 has a concentration that inhibits growth by 50% (IC_{50}) of 5.3 mg/L PCP in liquid culture (20). Other PCP-degrading bac-teria can mineralize PCP at concentrations between 160 and 300 mg/L in liquid culture. Despite the taxonomic diversity of the PCP-degrading bacteria, gram-negative degradative strains may share a similar PCP degradative pathway. The gram-positive *M. chlorophenolicum* strains have a different PCP degradation pathway (Fig. 2) (4).

Despite being classified into different bacterial genera, PCP-mineralizing *Arthrobacter* sp. American Type and Culture Collection (ATCC) 33790; *Flavobacterium* sp. ATCC 39723; *Pseudomonas* spp. SP3, UG25, and UG30; and *Sphingomonas* sp. RA2 are similar in many biochemical, genetic, and mor-phological aspects (21–23). Recent data, based on the presence of sphingogly-colipids, similarity of 16S ribosomal DNA (rDNA) sequences, and serological relatedness, have indicated that these PCP-degrading bacteria should be re-grouped under the genus *Sphingomonas* (21,22). Members of this genus are gram-negative, rod-shaped, aerobic, and non-spore-forming. They also possess unusual sphingoglycolipids which contain a sphingosine (a long-chain aliphatic amine) backbone. In addition, the sphingosine backbone is linked to a fatty acid and sugar substituents. The genus *Sphingomonas* is characterized by the absence of O antigens. Therefore, purified lipid fractions of *Sphingomonas* strains are weak immunogens (21). Furthermore, both *pcpB* and *pcpC* dechlorination genes are found in these bacteria, indicating similar PCP dechlorination pathways (21,23).

Detailed biochemical and genetic information on a PCP-degrading strain, *Flavobacterium* sp. ATCC 39723, has been generated (24). First PCP is con-verted to tetrachloro-p-hydroquinone (TeCH) in the presence of O_2 and NADPH by a PCP-4-monooxygenase. Degradation of the oxygenolytic product, TeCH, proceeds by a tetrachlorohydroquinone-reductive dehalogenase, forming first 3,5,6-trichlorohydroquinone (TCH) and then 2,6-dichlorohydroquinone (DCH).

The reductive dehalogenase requires reduced glutathione as a cofactor. Although it has been shown that the dichloro-*p*-hydroquinone intermediate will be broken down to water, carbon dioxide, and chloride ions, the biochemical mechanism of this part of the PCP-degrading pathway has not yet been elucidated (Fig. 2).

The PCP-4-monooxygenase of *Flavobacterium* sp. ATCC 39723 has been purified. The enzyme is approximately 63 kilodaltons (kDa) and inducible by PCP. It exhibits a broad substrate range and is active toward various halogenated phenols, including 2,3,5,6-tetrachlorophenol, 2,4,6-trichlorophenol, pentafluorophenol, 2,4,6-triiodophenol, and tribromophenol. It also catalyzed the release of amino, nitro, and cyano groups from the *para* position of 4-amino-2,6-dichlorophenol, 2,6-dibromo-4-nitrophenol, and 3,5-dibromo-4-hydroxybenzonitrile, respectively (25). The monooxygenase can also convert the herbicide bromoxynil (3,5-dibromo-4-hydroxybenzonitrile) to dibromohydroquinone (26). The *Flavobacterium* sp. ATCC 39723 TeCH reductive dehalogenase expresses constitutively and uses reduced glutathione as a cofactor to dechlorinate TeCH and TCH. The enzyme is approximately 30 kDa and oxygen-sensitive. In terms of the requirement for reduced glutathione by the enzyme, TeCH reductive dehalogenase may be classified as a bacterial glutathione-*S*-transferase.

The PCP-4-monooxygenase is encoded by a *pcpB* gene which has been cloned and sequenced; it has an open reading frame of 1614 base pairs (bp) of nucleotides that translates to a product of 538 amino acids with a predicted molecular weight (MW) of 59,932 daltons (27). The *pcpB* gene also exhibits 56% nucleotide similarity to the *tfdB* gene encoding 2,4-dichlorophenol hydroxylase (27). The *Flavobacterium pcpC* gene, encoding the TeCH reductive dehalogenase, is encoded in an open reading frame of 747 bp which translates to an enzyme of 248 amino acids with a molecular weight of 28,263 daltons. Sequence comparison analysis has revealed some similarity between the *pcpC* gene and the glutathione-*S*-transferase gene of carnations and maize (28).

Mycobacterium chlorophenolicum PCP-1 (formerly classified as *Rhodococcus chlorophenolicus* PCP-1) also transforms PCP to TeCH. However, the *pcpB* gene has not been found in *M. chlorophenolicum* PCP-1. A membrane-associated P_{450} monooxygenase in strain PCP-1 appeared to displace chlorine from the *para* position of PCP hydrolytically. The TeCH is further degraded through a hydrolytic dechlorination reaction and three consecutive reductive dechlorination steps, forming 1,2,4-trihydroxybenzene (29). The intermediate will eventually be mineralized by the gram-positive bacterium (Fig. 2).

4.2 Anaerobic Degradation of PCP

Unlike aerobic degradation of chlorophenols, anaerobic degradation of chlorinated phenols requires a consortium of bacterial strains. Therefore, most studies have been conducted in mixed cultures enriched from sewage sludges, sediments, or soils. To date only a few anaerobic chlorophenol-dechlorinating strains have been isolated (Table 2). *Desulfomonile tiedjei* DCB-1 is the most studied chlorophenol-dechlorinating strain. The strain was originally isolated from sewage sludge enriched by 3-chlorobenzoate. It is able to dechlorinate PCP and some other chlorophenols (30).

Anaerobic degradation of chlorophenols is mostly found in either methanogenic or sulfate-reducing environments. Under methanogenic conditions, reductive dechlorination is initiated. Chlorinated phenols serve as electron acceptors in the methanogenic community. After complete dechlorination, the phenol intermediate will eventually be transformed to methane and carbon dioxide by the methanogenic bacteria (31).

Degradation of chlorophenols has also been demonstrated in sulfate-reducing environments. In one study, oxidation of chlorophenol to CO_2 was stoichiometrically proportional to the reduction of sulfate. In addition, inhibition of sulfate reducers by a molybdate supplement also stopped the dechlorination process (32). These observations are consistent with the fact that *D. tiedjei* DCB-1 is a sulfate-reducing bacterium. Some early studies showed that sulfate supplement inhibited reductive dechlorination of methanogenic microbial communities (33). This can be explained by the competition for reducing equivalents between the dechlorinating microorganisms and the nondechlorinating sulfate reducers. In addition, the inhibitory effect of sulfate may be caused by the competition for reducing equivalents between the dechlorination and sulfate reduction reaction within the dechlorinating bacterium. The occurrence of reductive dechlorination of chlorophenols in methanogenic, sulfate-reducing, or iron-reducing environments indicates the diversity of the anaerobic dechlorinating microorganisms.

Reductive dechlorination of PCP is a stepwise replacement of chlorine substituents with hydrogen. In most cases, chlorine substituents are preferentially removed in the order *ortho*, *para*, and *meta* substitutions, forming 3,4,5-trichlorophenol, 3,5-dichlorophenol, and phenol, respectively. However, initial dechlorination at the *meta-* or *para*-position of PCP has been observed. Although complete mineralization of PCP and other chlorophenols has been well documented, incomplete degradation of PCP resulting in the accumulation of the less chlorinated phenols has also been reported.

5. ENVIRONMENTAL FACTORS AFFECTING THE BIODEGRADATION OF CHLOROPHENOLS

A soil environment is a complex and dynamic system which involves interactions of various biotic and abiotic components (Fig. 3). Understanding the effect of environmental conditions is important because the growth, survival, and activity of microorganisms in soil depend on various environmental factors, such

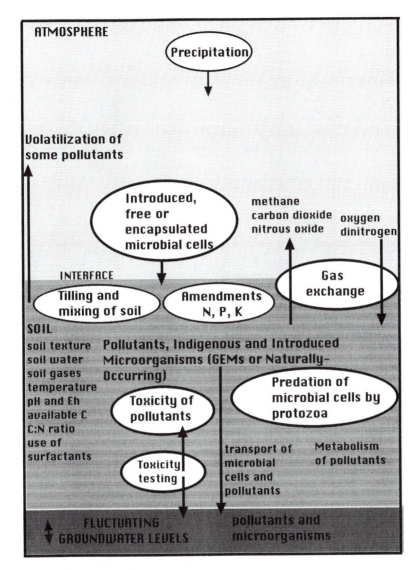

Figure 3.   Interactions of various biotic and abiotic factors in a soil environment during bioremediation.

as soil temperature, pH, moisture content, soil texture, organic matter content, availability of nutrients, and soil redox potential. In addition, composition of the indigenous microbial community, bioavailability of the chlorophenol contaminants, and presence of a suitable electron acceptor can also affect the survival and activity of chlorophenol-degrading microorganisms. Adverse environmental conditions can decrease or inhibit the growth of the microorganisms and expression of genes that produce pollutant-degrading enzymes. Information gained on the effects of environmental factors has direct relevance to the application of bioremediation technologies in chlorophenol-contaminated soils. Environmental conditions are often different at soil sites, so it is important to optimize conditions to achieve the best possible bioremediation of the pollutants.

5.1 Soil Temperature

Soil temperature affects biodegradation of chlorophenols by having a direct effect on reaction rates and exerting selective pressures on microbial populations. Temperature can also affect contaminant solubility, sorption, viscosity, and volatilization. Most chlorophenol-degrading microorganisms degrade chlorophenol contaminants optimally at temperatures between 24°C and 35°C. For instance, *Flavobacterium* sp. ATCC 39723 mineralized 40% to 50% of PCP at an initial concentration of 100 µg/g of dry soil between 24°C and 35°C (34). Biodegradation of chlorophenols is adversely affected by low temperatures in contaminated sites because of the higher temperature optima required for most of the chlorophenol-degrading microorganisms. However, in rare cases, microorganisms degrade chlorophenol pollutants at low temperatures (8°–16°C) while some others, such as *M. chlorophenolicum*, degrade PCP up to 41°C (4).

5.2 Soil pH

Soil pH affects the activities of chlorophenol-degrading microorganisms as well as the bioavailability of the chlorophenol contaminants. Biodegradation of PCP occurs at pH 5.6–8.0, whereas most PCP microorganisms function optimally at or near pH 7. In a nutrient medium, *Arthrobacter* sp. ATCC 33790 degraded PCP optimally at pH 7.1 and no degradation was observed at pH 6.2 (35). A *Mycobacterium* strain methylated PCP in nutrient broth at pH 6.5–7.0. However, hydroxylation of PCP by the *Mycobacterium* strain was dominant at pH 6.0 (36).

5.3 Soil Moisture Content

In a sandy soil, *Flavobacterium* sp. ATCC 39723 mineralized 60% of PCP at an initial concentration of 100 µg/g of dry soil at soil moisture content between 15% and 20% (34). In the same study, no mineralization was observed at 50%

soil water content. At 30 mg PCP/kg soil, *M. chlorophenolicum* strain PCP-1 mineralized 25%–30% of PCP at moisture content between 58% and 79% of the water-holding capacity of a sandy soil (37). In another study, *Flavobacterium* sp. ATCC 39723 mineralized 66% of the PCP at 60% soil water content, as compared to 55% mineralization at 30% soil water content in a silt loam soil containing 175 mg PCP/kg soil (38). Low oxygen content in saturated soils may adversely affect microbial populations in polluted soils.

5.4 Soil Texture

Texture of a given soil is defined as the proportion of clay, silt, and sand of the soil. It dictates both physical and chemical properties of soils and affects the transport and biodegradation processes of pollutants in soils. Sandy soil contains at least 70% sand (i.e., mineral soil particles with size between 0.02 and 2 mm in diameter) and 15% or less clay (<0.002 mm). It has low water-holding capacity and porosity. Therefore, pollutants tend to leach through the soil readily. Clay has about 10,000 times as much surface area as the same weight of sand. The specific surface (area per unit weight) of colloidal clay ranges from 10 to 1000 m^2/g. Therefore, clay soils have high adsorption and water-holding capacity. Clay colloid particles usually carry negative charges on their surfaces. Consequently, positively charged ions are bound on the surfaces of the clay particles. The high surface activity of clay particles affects the mobility and bioavailability of pollutants in soil. In a study by Zielke et al. (39), 3-chlorophenol, 3,5-dichlorophenol, and 3,4,5-trichlorophenol were adsorbed on a clay mineral at 2.0, 3.8, and 4.7 mmol/kg of clay at pH 7.4, respectively. In addition, adsorption equilibrium of the sorbates was achieved in less than 5 hours.

5.5 Sorption of Chlorophenols in Soil

Bioavailability and mobility of chlorophenols are dependent on sorption of the contaminants in soil. Presence of the ionizable -OH group on chlorophenols determines the partitioning of the compounds between the organic and aqueous phases in soil–water systems. At lower soil pH, chlorophenols will predominantly be in nonionic forms and will be sorbed to soil organic matter. At higher soil pH, chlorophenols will be ionized and become more soluble in the aqueous phase of the soil–water system. In a study by Schellenberg et al. (40), sorption of the nonionized chlorinated phenols was found to behave in a manner that was consistent with the principle of solute partitioning. Boyd et al. (41) also observed that the sorption of chlorophenols was proportional to the aqueous concentration of the contaminants as well as the amount of organic matter in the soil. Because of the relatively high water solubilities of the lower chlorinated phenols in neutral and alkaline soils, there is evidence that the chlorophenol contaminants are being leached from contaminated soils (42).

At lower soil pH, chlorophenols (especially the highly chlorinated phenols) are tightly sorbed to the organic phase of soil and may not be available for biodegradation. Synthetic or natural emulsifiers can be used to increase the bioavailabilities of the pollutants. In some cases, biosurfactants/bioemulsifiers exhibit low toxicity and perhaps assist in biodegradation.

5.6 Heavy Metals

In the wood preservation industry, heavy metals are used in conjunction with chlorophenols. Most of the heavy metals, such as cadmium, chromium, cobalt, copper, mercury, nickel, and zinc, either alone or in certain compounds, exert detrimental effects on soil microorganisms. Accordingly, the presence of heavy metals in contaminated sites may directly affect microorganisms involved in biodegradation. For instance, in both nutrient medium and soil extract, pentachlorophenol degradation by *Flavobacterium* sp. ATCC 53874 was inhibited by a combination of chromate, copper, and arsenate at concentration of 2, 2, and 10 mg/L, respectively (43,44). In a study by Said and Lewis (45), Cu^{2+} levels as low as 0.06 µg/ml could significantly inhibit biodegradation of 2,4-dichlorophenoxyacetic acid methyl ester.

5.7 Nutrients

Organic carbon content in soil can influence the extent and rate of bioremediation. Under aerobic conditions, addition of nutrients, such as glutamate (a readily metabolizable carbon source), nitrogen, phosphorus, and potassium, to chlorophenol contaminated soils enhances the rate of biodegradation of chlorophenols and prevents the loss of viability of the chlorophenol-degrading microorganisms. The rate of PCP degradation by *Flavobacterium* sp. ATCC 39723 was increased in liquid culture after the addition of glutamate at a concentration of 3 g/L (46). Glucose, at 20 mg/L, also enhanced degradation of 50 mg/L by *Flavobacterium* sp. (46); however, it did not stimulate PCP degradation by *Sphingomonas* sp. strain RA2 (47).

Information on the effect of carbon sources on the anaerobic degradation of chlorophenols is limited. The degradation of 2-chlorophenol stopped completely when yeast extract and peptone were absent in the growth medium. Dechlorination of 2-chlorophenol required the presence of *n*-butyrate or another fatty acid as an electron donor (48). In a study by Hendriksen et al. (49), the researchers showed that amendment of reactors with a carbon source such as glucose increased PCP degradation, decreased lag phases, increased stability of the degradative process, and caused concomitant increase in biomass. In addition to the readily metabolizable carbon sources such as glucose, some anaerobic microorganisms require specific reductive equivalents. For instance, *Desulfomonile tiedjei* DCB-1 requires pyruvate and formate for growth. The bacterium also needs acclimation (i.e., preexposure) by 3-chlorobenzoate to allow PCP dechlorination to proceed (30).

5.8 Electron Acceptors

Electron acceptors are a dominant factor in determining the composition of a microbial population. Thus, they can also affect the presence of dechlorinating microorganisms in a microbial community. Reduction potential of an electron acceptor is the potential of the chemical to accept electrons. For example, oxygen has a higher reduction potential than sulfate. Chlorophenols can be used as electron acceptors in dechlorination reactions. In a study by Mohn and Tiedje (50), *D. tiedjei* DCB-1 was able to conserve energy for growth from reductive dechlorination coupled to formate oxidation. However, in the presence of other electron acceptors with higher reduction potential, such as sulfate and nitrate, the latter acceptors may channel electron flow away from the reductive dechlorination process. Madsen and Aamand (51) showed an inhibitory effect of sulfate on PCP dechlorination. In the presence of 10 mM sulfate, the PCP dechlorination rate of a methanogenic culture was significantly reduced in 30 hours. Addition of molybdate, an inhibitor of sulfate reduction, reversed the effect of the sulfate on PCP dechlorination.

5.9 Microbial Community

A natural soil microbial community is usually well adapted to its habitat. The presence of chlorophenols may enrich for a subpopulation of chlorophenol-degrading microorganisms. A 1000-fold enrichment of PCP-degrading bacteria has been demonstrated during annual applications of PCP to an agricultural soil over a period of 3 years (52). However, without proper treatment, chlorophenol contaminants may persist for an extended period despite the presence of chlorophenol-degrading microorganisms in a contaminated soil.

Bioaugmentation, an application of degradative microorganisms for bioremediation, has been used in chlorophenol-contaminated soils where the indigenous soil microbial community is unable to degrade the contaminants. Chlorophenol-degrading microbial inoculum, either naturally occurring or genetically engineered, may not be competitive against the indigenous microbial population. Briglia et al. (53) showed that PCP-degrading *Flavobacterium* sp. ATCC 39723 cells declined nine orders of magnitude in 60 days. However, in the same study, the authors demonstrated that *M. chlorophenolicum* PCP-1 survived in soil at a cell density of 1×10^8 cells/g of soil for 200 days. Therefore, caution must be taken to ensure that the inoculum can survive in the environment and degrade the target pollutants. In addition, the inoculum in use should not have any adverse effects on the target environment.

6. CHLOROPHENOL BIOREMEDIATION: TECHNOLOGY AND FIELD APPLICATION

6.1 Soil Bioremediation Methods

Various field-scale bioremediation technologies have been developed. The criteria that determine the treatments of a contaminated soil are (1) biodegradability of the contaminant(s); (2) environmental conditions of the contaminated site; (3) size, geological characteristics, and hydrodynamics of the contaminated area; (4) land use of the site after bioremediation; and (5) cost of the remediation treatments.

In situ bioremediation technology is based on the stimulation of microbial activities at the contaminated areas. This method is suitable for large-scale and non-point-source pollutions. Electron donors and/or acceptors can be added to a contaminated site to stimulate catabolic activities of the indigenous pollutant-degrading microorganisms. For instance, oxygen can be introduced into the contaminated area by injection of aerated water, air sparging, venting, and injection with hydrogen peroxide. In addition, microbial inoculum (natural isolates or enriched indigenous bacteria) can also be introduced.

If conditions are not favorable for in situ treatments, contaminated soil may have to be excavated and treated by ex situ treatments, which include bioslurry, composting, and land farming. Bioslurry treatment involves application of bioreactors which are operated and maintained under optimal conditions for biodegradation. Composting and land-farming treatments are both based on stimulations of microbial catabolic activities by spreading the contaminated soil on specially constructed areas which are also maintained under optimal conditions for bioremediation. Because of the high operation cost of the ex situ treatments, they are usually used for remediating contaminated sites with low soil volume and high concentrations of contaminants.

Bioslurry (bioreactor) treatment has been shown to be successful in treating PCP-contaminated soils. In a combination of a PCP-degrading mixed culture and nutrient supplements (at C:N:P ratio of 25:8:1), more than 99% of the soil PCP contaminants were degraded in 2 weeks (54). In another ex situ bioremediation treatment, PCP-contaminated soil was excavated and incubated as a soil slurry with a consortium of PCP-degrading microorganisms in a bioreactor. The results from this study showed that PCP at an initial concentration of 370 mg/kg dry soil was degraded to below 0.5 mg/kg in 14–21 days. A bench-scale study confirmed that the soil slurry treatment was more efficient in degrading PCP from PCP-contaminated materials than the land-farming remediation process (55).

Composting is usually conducted in piles, where the excavated soil is actively aerated and mixed with highly degradable bulking material such as straw. To prevent leaching by rain, the composting area is usually lined with plastic which is covered with a layer of gravel or sand to allow drainage. Composting treatment, with or without addition of microbial inoculum, is currently being implemented at chlorophenol-contaminated sawmill sites in Finland. Results of the treatment were dependent on the soil type of the samples. In one compost treatment (without microbial inoculum), a soil containing an initial concentration of chlorophenols of 8520 mg/kg dry soil was remediated. After a 32-month period of treatment, the chlorophenol concentration decreased to 18 mg/kg dry soil (56).

For land-farming treatment, contaminated soil is spread out in a layer and tilled on a regular basis to promote homogenization and oxygen availability. Nutrients and/or degradative microbial inoculum can be added to optimize the efficiency of the remediation process. This method has been used for treating PCP and polycyclic aromatic hydrocarbons. In a study by Compeau et al. (54), 50% reduction of PCP at initial concentrations between 400 and 1000 mg/kg dry soil was observed in 8–12 weeks.

6.2 Use of Immobilized Cells in Chlorophenol Degradation

"Immobilization of cells" is a general term that describes many different forms of cell attachment, including flocculation, adsorption on surfaces, covalent bonding to carriers, cross-linking of cells, and cell encapsulation in a polymer matrix (57,58). The use of immobilized and encapsulated cells in the biodegradation of chlorophenols has been investigated. A wide range of microorganisms have been immobilized on various carriers and have been shown to enhance chlorophenol degradation (Table 3). Hu et al. (59) demonstrated the ability of polyurethane foam–immobilized *Flavobacterium sp.* ATCC 39723 cells to degrade 700 mg/L PCP in liquid culture.

Coimmobilization is defined as the use of microorganisms with known degradative capabilities, immobilized together with an adsorbent within a spherical permeable gel matrix or membrane (60). Because of the strong adsorption nature of adsorbents toward contaminants, coimmobilization offers some advantages over cell immobilization: the presence of adsorbents quickly removes the contaminant, allowing the use of high flow rates in reactor systems; adsorbents dampen variations in contaminant concentrations; they help lower concentrations when the contaminant is toxic; and they help retain any extracellular enzymes produced by bacterial cells. Lin et al. (61) demonstrated the effectiveness of coimmobilizing cells and activated carbon in alginate in degrading PCP in so-

Table 3. Examples of Solid Carriers Used to Immobilize Microbial Cells for Use in Biodegradation

Contaminants	Microorganisms	Carriers
Chlorophenols	Mixed culture	Alginate
Chlorophenols	*Rhodococcus* spp.	Polyurethane
Chlorophenols	Activated sludge	Celite R-633 microcarriers
Chlorophenols	Mixed culture	Glass, cellulose, or chitin
4-Chlorophenol	*Alcaligenes* sp. A7-2	Alginate, granular clay
Pentachlorophenol	*Flavobacterium* sp. ATCC 39723	Alginate or polyurethane
Pentachlorophenol	*Arthrobacter* sp. ATCC 33790	Alginate and/or activated carbon
Pentachlorophenol	*Phanerochaete chrysosporium*	Alginate
Pentachlorophenol	Mixed culture	Anaerobic granules

From Ref. 73.

lution, soil extract, and sand. Coimmobilized *Arthrobacter* cells were exposed to 117 μM PCP (30.7 mg/L). Complete PCP removal was observed in 30 hours and 50% of the PCP was mineralized. Degradation of PCP by coimmobilized cells was dependent on the type and concentration of adsorbent used. It was reported that the larger the quantity of adsorbent used, the lower the PCP concentration, and the lower the mineralization rate by bacteria.

Investigations using immobilization of microbial cells for biodegradation of contaminated soils are few in number, but promising in their results. Encapsulation of microbial cells for soil applications provides a range of advantages over free cell soil applications. These advantages include ease of application to the soil, reduced possibility of off-site drifting, and protection of the cells from environmental stress (58). For example, Salkinoja-Salonen et al. (62) observed that alginate-immobilized *M. chlorophenolicum* inoculum retained its PCP mineralizing activity in soil more effectively than the free cell inoculum. In our laboratory, a soil from a wood treatment site contaminated with 350 ppm PCP as well as polycyclic aromatic hydrocarbons was used to determine the effectiveness of PCP-degrading microorganisms encapsulated in a k-carrageenan matrix. Degradation of PCP was stimulated with the addition of nutrients (a combination of carbon, nitrogen, potassium, and phosphorus), regardless of the presence of an unencapsulated PCP-degrading microbal inoculum. After 30 weeks of optimal conditions, approximately 50% of the added ^{14}C-PCP was recovered as $^{14}CO_2$. In comparison, the encapsulated cells, added at a level of 10^8 colony forming units (CFUs)/g dry soil, mineralized over 70% of the ^{14}C-PCP within 12 weeks.

More information on immobilization and encapsulation of microorganisms can be obtained from the Bioencapsulation Research Group at ESAIA, 2, Avenue de la Foret de Haye, B.P. 172, 54505 Vandoeuvre, France, or on the World Wide Web at http://ensaia.u-nancy.fr/BRG/BRG.html.

7. BIOLOGICAL ASSESSMENT OF CHLOROPHENOL BIOREMEDIATION

Traditionally, success of bioremediation is assessed by monitoring the disappearance of contaminants and appearance of the degradative products. However, a complete assessment of the performance of the degradative microorganisms also requires detailed analysis of the degradative and nontarget microbial populations as well as the reduction of toxicity of the contaminated areas (Fig. 4). In this section, some approaches in studying the dynamics of PCP-degrading bacteria and toxicity bioassays will be discussed.

7.1 Detection of PCP-Degrading Microorganisms in Soil

Several methods have been used for monitoring microorganisms in the environment. These include antibiotic resistance selective plating, most probable number (MPN), immunofluorescence, deoxyribonucleic acid (DNA)/DNA hybridization, polymerase chain reaction (PCR), and genetic markers, such as the *lacZY* and *luxAB* reporter genes. Because of the widespread antibiotic resistance gene pool in the environment and the apparent instability of spontaneous antibiotic resistance mutants, antibiotic selective plating has not been successful in studying PCP-degrading bacteria in soil. Because of the lack of O antigen on the surface of most PCP degraders, immunofluorescence used to study these bacteria may suffer from a lack of specificity.

Traditionally, the only method used for quantifying PCP-degrading bacteria was a MPN/[^{14}C]PCP mineralization assay. This method is based on the measurement of [^{14}C]PCP mineralization in a dilution series of a soil sample. Release of $^{14}CO_2$ from the dilution samples indicated the presence of PCP-degrading bacteria. With this method, Briglia et al. (53) showed that *Mycobacterium chlorophenolicum* PCP-1 survived in soil for as long as 200 days. In the same study, the cell density of *Flavobacterium* sp. ATCC 39723 decreased by nine orders of magnitude in 60 days. The study also showed a positive correlation between the efficiency of PCP degradation and persistence of the degradative bacteria. However, efficiency of the MPN/[^{14}C]PCP mineralization assay has not been calibrated.

Recently, an MPN/PCR protocol has been developed in our laboratory to

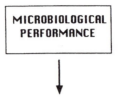

- Chemical analyses- determine if environmental conditions are suitable for degradation of pollutant(s)

- Gene probing- determine if microorganisms present in soil samples have genetic information required for coding of necessary catabolic enzymes

- Soil microcosms- incubate contaminated soil samples with pollutants(s) to determine rate and extent of ^{14}C-labelled pollutant(s) being mineralized to CO_2 and H_2O

- Additional measurements- soil respiration (O_2 consumption and CO_2 evolution), colony forming unit counts (CFU), direct microscopic counts

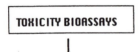

- Determine if bioremediation treatment decreased the toxicity of the soil

- Examples of soil toxicity tests are seedling emergence test, earthworm test, Microtox solid-phase test, TOXI-Chromotest, Tradescantia-micronucleus (TRAD-MCN) test

Figure 4.   Microbiological performance and toxicity bioassays used in soil bioremediation.

quantify a PCP-degrading bacterium, *Pseudomonas* sp. UG30, in soil. The 744-bp fragment of the *pcpC* dechlorination gene was targeted for the MPN/PCR assay. The detection limit of this protocol was 4 cells/g of dry soil. This method showed a good correlation to the UG30 inoculum density, ranging from 36 to 2×10^9 CFU/g of soil. However, the MPN/[^{14}C]PCP mineralization assay underestimated the inoculum density approximately 100-fold. An MPN/PCR assay targeting a specific ribosomal DNA (rDNA) sequence of *M. chlorophenolicum* PCP-1 has also been developed to monitor the cell density of *M. chlorophenolicum*

PCP-1 in different soils (J. D. van Elsas, IPO-DLO, Wageningen, Netherlands, personal communication).

7.2 Application of Toxicity Tests for Contaminated Soils

Since chemical waste sites are seldom contaminated with single compounds, chemical analyses can be expensive and, in fact, uninformative regarding environmental hazards. The use of bioassays (assays which are based on biological effects) to evaluate soils, soil elutriates, and surface and subsurface water from hazardous chemical waste sites may provide a more direct, integrated estimate of environmental toxicity. On the basis of bioassay data, chemical waste sites can be ranked according to their toxic potential or mapped for cleanup operations. Some aspects of the toxicity of field soils contaminated by hazardous chemicals including wood-preserving chemicals (63), pentachlorophenol (64), polyaromatic hydrocarbons (65), creosote (66), and fuel spills (67) have been examined.

Soil toxicity tests

Seedling emergence/root elongation test

Higher plants are selected because vegetation is the dominant biological component of terrestrial ecosystems and may reflect the toxicity of hazardous wastes located at the contaminated site. Lettuce and radish are recommended as the preferred test species (68). Methodology guidelines also indicate valid test results for cucumber, red clover, and wheat.

The 5-day seedling emergence test exposes seeds to total available toxic constituents in dilutions of soil, while the root elongation test exposes seeds to water-soluble constituents eluted from the soil. The seedling emergence test is, therefore, likely to demonstrate greater sensitivity than the root elongation test if non-water-soluble toxic constituents are present in the soil. The root elongation test could be conducted if toxic constituent mobility, caused by precipitation events, were of specific interest.

Earthworm survival test

Earthworms have been selected as indicator species because they are representative of the terrestrial environment and are of considerable importance in improving soil aeration, drainage, and fertility. Earthworms differ from aquatic organisms in that they may be exposed to toxic chemicals in the aqueous phase via soil/sediment moisture, in the vapor phase, or through direct contact with particulate matter on the surface of soil/sediment constituents. The earthworm test currently recommended is the 14 day-test using *Eisenia andrei* or *E. foetida*, proposed by the U.S. EPA (68); the test with *E. andrei* may be preferred because it is standardized and reproducible.

Other tests

The solid-phase test (Microtox, Microbics Inc., Carlsbad, CA, USA), and the chromotest (TOXI-chromotest, EBPI, 14 Abacus Rd., Brampton, Ontario, Canada) are bacterial test kits used as toxicity evaluation tools for contaminated soil/sediment samples. The Microtox test utilizes a luminescent marine bacterium (*Photobacterium phosphorium*) which decreases its level of luminescence in a toxic environment. The assay is rapid and the use of elutriates or leachates for assessing soil toxicity has been shown to be variably sensitive (69).

An application of TOXI-chromotest (color change test for toxicity) has been developed as a soil/sediment chromotest (70). The bacterium is a mutant *Escherichia coli*, which has permeable cell walls to increase sensitivity to toxicants. The bacterial cells are grown in medium in direct contact with the soil/sediment sample for 2 hours. A drop of the soil–bacteria suspension is put on a filter paper containing the substrate X-gal. β-Galactosidase activity is measured indirectly as a change in blue color intensity (result of cleavage of X-gal).

The heterogeneity of soil and contaminants at any site may necessitate taking a large number of samples to develop an accurate toxicity map. Screening soil toxicity can be done by using seedling emergence and earthworm survival tests. Ideally, it is recommended that the set of screening tests be augmented to include tests such as the algal growth inhibition test, bacterial test, seedling emergence test, arthropod test using spiders or mites, earthworm survival/reproduction test, and other soil-dependent organism (e.g., nematodes) reproduction tests. Results of a literature review of organisms used in soil toxicity tests have been summarized by Keddy et al. (71).

8. CONCLUSIONS

Biodegradation of chlorophenols has been relatively well studied and information has been obtained from both laboratory- and pilot-scale studies. However, the success of chlorophenol bioremediation is often hampered by the lack of understanding of the composition and dynamics of the chlorophenol-degrading microbial populations in soil. For instance, several gram-negative aerobic PCP-degrading bacterial strains have been isolated. Despite their biochemical and physiological similarities, they have been classified under different bacterial genera, not to mention that little is known about their distribution and ecological properties in the environment. Recently, studies have shown that these gram-negative PCP-degrading bacteria possess homologous *pcpB* and *pcpC* dechlorination genes and should be reclassified under the genus *Sphingomonas*. With the application of various genetic analyses, the dechlorination genes may be used in tracking PCP-degrading microbial communities in given soil and water samples. In addition, the dechlorination genes may also prove useful as genetic indicators to predict the PCP-degrading potential of contaminated environments.

Effectiveness of chlorophenol biodegradation is dependent on the survival of the degradative microorganisms and the environmental conditions of the contaminated areas. On the basis of the prevalent environmental conditions of the contaminated sites, either in situ or ex situ treatments can be used to bioremediate the contaminants. Evidence has shown that indigenous microbial communities can be stimulated to degrade chlorophenols by various chemical/physical treatments. If microbial inoculum is needed to increase the efficiency of the bioremediation process, immobilization of the inoculum has been proved to be effective. Encapsulation of microbial cells can improve both the survival rate and activity of the inoculants in soil. More information on bioremediation and bioencapsulation technologies can be obtained from the World Wide Web, as listed in Table 4.

Risk assessment of bioremediation is important in determining the success of the operation. First, complete mineralization of chlorophenols has to be confirmed by chemical analyses. Second, microbiological assays such as MPN, gene probing, and immunofluorescence can be used to assess the persistence of the degradative microorganisms and/or genes as well as the effect of the bioremediation treatment on the indigenous microbial community of the environment. Finally, a successful bioremediation treatment has to eliminate the toxicity of the contaminated site. This requires a combination of toxicity bioassays involving various ecologically relevant organisms such as microorganisms, higher animals, and plants.

Table 4. Some Selected Internet Information Resources Related to Microbiology, Bioremediation, Toxicology, and Their Uniform Resource Locators (URLs)

URL1.	Enzyme metabolic pathways at Argonne National Laboratory http://www.mcs.anl.gov/home/towell/metabhome.html
URL2.	The National Center for Biotechnology Information http://www.ncbi.nlm.nih.gov/
URL3.	Brookhaven National Laboratory/Protein Database http://www.pdb.bnl.gov/
URL4.	University of Minnesota Biocatalysis/Biodegradation Database http://dragon.labmed.umn.edu/~lynda/index.html
URL5.	American Type Culture Collection (ATCC) http://www.atcc.org/atcc.html
URL6.	Bioencapsulation Research Group (BRG) http://ensaia.u-nancy.fr/BRG/BRG.html
URL7.	Extension Toxicology Network http://ace.ace.orst.edu/info/extoxnet

ACKNOWLEDGMENT

This research was supported by the Natural Sciences and Engineering Research Council of Canada (NSERC) Operating and Strategic Grants Program to J. T. T. and H. L.

REFERENCES

1. P.-Y. Lu, R. L. Metcalf, and L. K. Cole, *Pentachlorophenol: Chemistry, Pharmacology, and Environmental Toxicology* (K. R. Rao, ed.), Plenum Press, New York, p. 53 (1978).
2. M. M. Haggblom and R. J. Valo, *Microbial Transformation and Degradation of Toxic Chemicals* (L. Y. Young and C. E. Cerniglia, eds.), Wiley & Sons, New York, p. 389 (1995).
3. L. H. Keith and W. A. Telliard, *Environ. Sci. Technol. 13*: 416 (1979).
4. K. A. McAllister, H. Lee, and J. T. Trevors, *Biodegradation 7*: 1 (1996).
5. U. G. Ahlborg and T. M. Thunberg, *CRC Crit. Rev. Toxicol. 7*: 1 (1980).
6. D. G. Crosby, *Pure Appl. Chem. 53*: 1051 (1981).
7. G. A. Chapman and D. L. Shumway, *Pentachlorophenol: Chemistry, Pharmacology, and Environmental Toxicology* (K. R. Rao, ed.), Plenum Press, New York, p. 285 (1978).
8. G. Ruckdeschel, G. Renner, and K. Schwarz, *Appl. Environ. Microbiol. 53*: 2689 (1987).
9. D. D. Hale, W. Reineke, and J. Wiegel, *Biological Degradation and Bioremediation of Toxic Chemicals* (G. R . Chaudhry, ed.), Dioscorides Press, Portland, Oregon, p. 74 (1994).
10. M. Alexander and M. I. H. Aleem, *J. Agric. Food Chem. 9*: 44 (1961).
11. D. P. Cirelli, *Fed. Reg. 43*: 48446 (1978).
12. S. Kuwatsuka, *Environmental Toxicology of Pesticides* (F. Matsumura, G. M. Boush, and T. Misato, eds.), Academic Press, New York, p. 385 (1972).
13. J.-M. Bollag, S.-Y. Liu, and R. Minard, *Soil Sci. Soc. Am. J. 44*: 52 (1980).
14. R. T. Lamar, M. J. Larsen, and T. K. Kirk, *Appl. Environ. Microbiol. 56*: 3519 (1990).
15. G. J. Mileski, J. A. Bumpus, M. A. Jurek, and S. D. Aust, *Appl. Environ. Microbiol. 54*: 2885 (1988).
16. R. T. Lamar and D. M. Dietrich, *Appl. Environ. Microbiol. 56*: 3093 (1990).
17. H.-J. Knackmuss and M. Hellwig, *Arch. Microbiol. 117*: 1 (1978).
18. R. H. Don, A. J. Weightman, H.-J. Knackmuss, and K. N. Timmis, *J. Bacteriol. 161*: 85 (1985).
19. U. M. X. Sangodkar, P. J. Chapman, and A. M. Chakrabarty, *Gene 71*: 267 (1988).
20. J. H. A. Apajalahti and M. S. Salkinoja-Salonen, *Appl. Microbiol. Biotechnol. 25*: 62 (1986).

21. U. Karlson, F. Rojo, J. D. Van Elsas, and E. Moore, *System. Appl. Microbiol. 18*: 539 (1995).

22. L. J. Nohynek, E. L. Suhonen, E.-L. Nurmiaho-Lassila, J. Hantula, and M. Salkinoja-Salonen, *System. Appl. Microbiol. 18*: 527 (1995).

23. K. T. Leung, M. B. Cassidy, K. W. Shaw, H. Lee, J. T. Trevors, E. M. Lohmeier-Vogel, and H. J. Vogel, *World J. Microbiol. Biotechnol.*, in press. (1997).

24. C. S. Orser, and C. C. Lange, *Biodegradation 5*: 277 (1994).

25. L. Xun, E. Topp, and C. S. Orser, *J. Bacteriol. 174*: 2898 (1992).

26. E. Topp, L. Xun, and C. S. Orser, *Appl. Environ. Microbiol. 58*: 502 (1992).

27. C. S. Orser, C. C. Lange, L. Xun, T. C. Zaht, and B. J. Schneider, *J. Bacteriol. 175*: 411 (1993).

28. C. S. Orser, J. Dutton, C. Lange, P. Jablonski, L. Xun, and M. Hargis, *J. Bacteriol. 175*: 2640 (1993).

29. J. H. A. Apajalahti and M. S. Salkinoja-Salonen, *J. Bacteriol. 169*: 675 (1987).

30. W. W. Mohn and K. J. Kennedy, *Appl. Environ. Microbiol. 58*: 1367 (1992).

31. W. W. Mohn and J. M. Tiedje, *Microbiol. Rev. 56*: 482 (1992).

32. M. M. Haggblom and L. Y. Young, *Appl. Environ. Microbiol. 56*: 3255 (1990).

33. G.-W. Kohring, X. Zhang, and J. Wiegel, *Appl. Environ. Microbiol. 55*: 2735 (1989).

34. R. L. Crawford and W. W. Mohn, *Enzyme Microb. Technol. 7*: 617 (1985).

35. G. J. Stanlake and R. K. Finn, *Appl. Environ. Microbiol. 44*: 1421 (1982).

36. T. Suzuki, *J. Pesticide Sci. 8*: 419 (1983).

37. M. Briglia, P. J. M. Middeldorp, and M. S. Salkinoja-Salonen, *Soil Biol. Biochem. 26*: 377 (1994).

38. A. G. Seech, J. T. Trevors, and T. L. Bulman, *Can. J. Microbiol. 37*: 440 (1991).

39. R. C. Zielke, T. J. Pinnavaia, and M. M. Mortland, *Reactions and Movement of Organic Chemicals in Soils* (B. L. Sawhney and K. Brown, eds.), Soil Science Society of America, Madison, Wisconsin, p. 81 (1989).

40. K. Schellenberg, C. Leuenberger, and R. Schwarzenbach, *Environ. Sci. Technol. 18*: 652 (1984).

41. S. A. Boyd, M. D. Mikesell, and J.-F. Lee, *Reactions and Movement of Organic Chemicals in Soils* (B. L. Sawhney and K. Brown, eds.), Soil Science Society of America, Madison, Wisconsin, p. 209 (1989).

42. V. H. Kitunen, R. Valo, and M. Salkinoja-Salonen, *Environ. Sci. Technol. 21*: 96 (1987).

43. A. J. Wall and G. W. Stratton, *Can. J. Microbiol. 40*: 388 (1994).

44. E. Topp and R. S. Hanson, *Can. J. Soil Sci. 70*: 83 (1990).

45. W. A. Said and D. L. Lewis, *Appl. Environ. Microbiol. 57*: 1498 (1991).

46. E. Topp, R. L. Crawford, and R. S. Hanson, *Appl. Environ. Microbiol. 54*: 2452 (1988).

47. P. M. Radehaus and S. K. Schmidt, *Appl. Environ. Microbiol. 58*: 2879 (1992).

48. G. Dietrich and J. Winter, *Appl. Microbiol. Biotechnol. 34*: 253 (1990).

49. H. V. Hendriksen, S. Larsen, and B. K. Ahring, *Appl. Environ. Microbiol. 58*: 365 (1992).

50. W. W. Mohn and J. M. Tiedje, *Arch. Microbiol. 153*: 267 (1990).

51. T. Madsen and J. Aamand, *Appl. Environ. Microbiol. 57*: 2453 (1991).

52. I. Watanabe, *Soil Biol. Biochem. 9*: 99 (1977).

53. M. Briglia, E.-L. Nurmiaho-Lassila, G. Vallini, and M. S. Salkinoja-Salonen, *Biodegradation 1*: 273 (1990).

54. G. C. Compeau, W. D. MaHaffey, and L. Patras, *Environmental Biotechnology for Waste Treatment* (G. S. Sayler, R. Fox, and J. W. Blackburn, eds.), Plenum Press, New York, p. 91. (1991).

55. J. G. Mueller, S. E. Lantz, B. O. Blattmann, and P. J. Chapman, *Environ. Sci. Technol. 25*: 1055 (1991).

56. R. J. Valo and M. Salkinoja-Salonen, *Appl. Microbiol. Biotechnol. 25*: 68 (1986).

57. M. B. Cassidy, K. T. Leung, H. Lee, and J. T. Trevors, *J. Microbiol. Meth. 23*: 281 (1995).

58. J. T. Trevors, J. D. van Elsas, H. Lee, and L. S. van Overbeek, *Microbial. Rel. 1*: 61 (1992).

59. Z. C. Hu, R. A. Korus, W. E. Levinson, and R. L. Crawford, *Environ. Sci. Technol. 28*: 491 (1994).

60. A. R. Siahpush, J. E. Lin, and H. Y. Wang, *Biotechnol. Bioeng. 39*: 619 (1992).

61. J. E. Lin, H. Y. Wang, and R. F. Hickey, *Biotechnol. Bioeng. 38*: 273 (1991).

62. M. S. Salkinoja-Salonen, P. J. M. Middeldorp, M. Briglia, R. Valo, M. Haggblom, and A. McBain, *Advances in Applied Biotechnology*, Vol. 4 (D. Kamely, A. Chakrabarty, and G. Omenn, eds.), Gulf Publishing, Houston, p. 344 (1989).

63. J. E. Matthews and L. Hastings, *Toxicity Assessment 2*: 265 (1987).

64. D. P. Middaugh, S. M. Resnick, E. E. Lantz, C. S. Heard, and J. G. Mueller, *Arch. Environ. Contam. Toxicol. 24*: 165 (1993).

65. M. P. Eisman, S. Landonarnold, and C. M. Swindoll, *Bull. Environ. Contam. Toxicol. 47*: 811 (1991).

66. F. Baud-Grasset, J. M. Bifulco, J. R. Meier, and T. H. Ma, *Mutation Res. 303*: 77 (1993).

67. X. Wang and R. Bartha, *Soil Biol. Biochem. 22*: 501 (1990).

68. J. C. Greene, C. L. Bartels, W. J. Warren-Hicks, B. R. Parkhurst, G. L. Lin-

der, S. A. Perterson, and W. E. Miller, *Protocol for Short Term Toxicity Screening of Hazardous Waste Sites*, U. S. EPA 3-88-029, p. 102 (1989).

69. J. M. Thomas, J. R. Skalski, J. F. Cline, M. C. McShane, J. C. Simpson, W. E. Miller, S. A. Peterson, C. A. Callahan, and J. C. Greene, *Environ. Toxicol. Chem. 5*: 487 (1986).

70. K. K. Kwan, *Environ. Toxicol. Water Qual. 8*: 223 (1993).

71. C. J. Keddy, J. C. Greene, and M. A. Bonnell, *Ecotoxicol. Environ. Safety 30*: 221 (1995).

72. K. L. E. Kaiser, D. G. Dixon, and P. V. Hodson, *QSAR In Environmental Toxicology* (K. L. E. Kaiser, ed.), D. Reidel Publishing, Dordrecht, Holland, p. 189 (1984).

73. M. B. Cassidy, H. Lee, and J. T. Trevors, *J. Ind. Microbiol. 16*: 79 (1996).

20

The Impact of Heavy Metals on Soil Microbial Communities and Their Activities

STEFAN WUERTZ* and MAX MERGEAY Flemish Institute for Technological Research (VITO), Mol, Belgium

1. INTRODUCTION

When attempting to assess the impact of any element, chemical, or mixture of chemicals on microorganisms, the first impulse is always to look at the cellular, subcellular, or molecular level of interaction. It is here that the greatest body of evidence has been collected over the last three decades, in carefully conducted laboratory studies involving strains well defined in both genetic and physiological terms. Yet, the greatest challenge arises when these data are extrapolated to mixed microbial populations while encountering ecosystem variables beyond the immediate control of the investigator. In this chapter we will attempt to summarize the effects of heavy metal pollution on microbial populations and the environmental processes mediated by them. Such an endeavor would be unthinkable without attention to the mechanisms which have evolved to protect certain microorganisms against heavy metal stress. Several excellent recent reviews are

---

*Current affiliation: Technical University of Munich, Garching, Germany

available which deal with the specific mechanisms of resistance to heavy metals in bacteria (see, for example, 1–9). This chapter will be limited to a discussion of their ecological implications and occurrence in specific biotopes such as the polluted areas around copper mines and nonferro factories or the natural, nickel-rich, ecosystems of New Caledonia in Oceania.

The group of heavy metals is a loose classification of about 65 elements, which are defined as having a density greater than 5 g cm^{-3} (10). A variety of units are used to report metal concentrations. In keeping with the recommendations made by Duxbury (11) and Cooney and Wuertz (12), we have converted units to molar concentrations wherever possible. For the purpose of this review we will follow the classification system of Duxbury (11) and Wood (13), which is based on the potential toxicity of heavy metals (Table 1). For a critical view on the definition of heavy metals, the reader is referred to a review by Gadd (10). Several metals are listed by a European Commission directive as hazardous, and their maximum allowed levels in soils are compared with average concentrations found in Table 2. Some heavy metals are essential for cell functions; they include Co, Cr, Cu, Ni, Zn, and Mo. Others have no known essential biological function, e.g., Al, Ag, Cd, Sn, Au, Sr, Hg, Tl, and Pb. Metal toxicity is an area subject to intense research activity. It is generally agreed that most metals can be toxic if present at high enough concentrations. The difficulty in assessing toxicity lies in the many other environmental factors which can determine metal effects. In both laboratory and field studies, the effects of metal speciation and the physicochemical environment can make interpretation of available data difficult. Even with an axenic microbial culture, these parameters may change as a result of the interaction between cells and metals. Changes in pH and the synthesis of intra- and extracellular products and metabolites affect metal speciation. The latter, although crucial to metal toxicity, is usually not known (14) partly because of technical limitations or unrealistic efforts required.

In addition to these problems, it is frequently not possible to distinguish the effects of metals from those of other environmental factors which may impact natural microbial populations. This does not take away from efforts, for example, to define metal resistance in a bacterial isolate by the ability to grow in defined media containing high levels of metals and use this phenotypic observation in genetic studies. It simply puts these into perspective and avoids the pitfalls of trying to make global assumptions about metal resistance when studying a highly adapted microorganism.

In broad terms, microorganisms can adapt to heavy metals in two ways: They can survive by means of intrinsic properties including those related to their cell wall structure, extracellular polymeric substances (EPS), and binding/precipitation of metals inside or outside the cell (10). Alternatively, they can develop specific mechanisms of detoxification as a result of being challenged by a metal. These latter mechanisms include active efflux mechanisms as well as reduction/oxidation,

Table 1.   Classification of Heavy Metals According to Their Potential Toxicity

Type 1: noncritical	Type 2: potentially toxic and relatively accessible	Type 3: potentially toxic but insoluble or rare
Fe	Co, Ag, Sb, Ni, Cd, Bi, Cu, Pt, Cr,Zn, Au, Mn, Sn, Hg, Mo, Pd, Pb, Tl	Hf, Ta, Os, Zr, Re, Rh, W, Ga, Ir, Nb, La, Ru

Modified From Ref. 11 and Ref. 13.

alkylation/dealkylation, and intracellular compartmentalization/sequestration (production of metallothioneins and phytochelatins) (10). Metallothioneins are small (molecular weight [MW] 6000–10000) cysteine-rich proteins which do not contain aromatic amino acids and which can fix heavy cations; their synthesis is inducible. Phytochelatins are metal-binding peptides formed by γ-glutamyl-cysteine synthetase; their formula is (γ glu-cys) n-gly, where $n = 3$ to 8. The mechanisms frequently overlap, as in extracellular precipitation and crystallization due to initial metal efflux, which change the physicochemical environment of the cell (10). Consequently, microbial adaptation to heavy metal stress must be viewed as a dynamic process involving a general cell response (for example, development of slime in biofilms), specific resistance mechanisms, and changes in metal speciation and toxicity which are influenced by both microbial and environmental components.

The total metal concentrations in soil can be difficult to determine and may vary, depending on the extraction techniques used (11). Comparing metal exposures at different environmental soil sites (usually expressed as parts per million [ppm] or milligrams per kilogram [mg kg^{-1}] dry weight) can be erroneous because the bulk density of soils containing different amounts of organic and mineral matter varies. Comparisons should therefore be based on organic matter (OM) content (humus, litter)—at least in forest soils with a clay content <10% (15). All too often important soil parameters are not reported. Similarly, the influence of abiotic factors on heavy metal toxicity to microorganisms cannot be overemphasized. Their individual effects have often been thoroughly studied under laboratory conditions (e.g., 16,17). The cation exchange capacity (CEC), the capacity to replace cations with heavy metal ions, of a soil depends on the organic matter, clay minerals, and hydrous metal oxides present. In general, the higher the CEC (organic and clay soils), the lower the toxicity of heavy metals to microorganisms. The pH is another critical parameter. It affects both the speciation of a metal (metal toxicity) and the degree of complexation by ligands (metal availability). For example, hydroxylated cations and metal hydroxides which form at high pH may bind to the cell surface, thereby altering the net charge of

Table 2. Maximum Metal Concentrations in Soils Permitted Under European Commission Regulations and the Average Concentrations Found on Earth

Metal	Maximum allowed soil concentration ($\mu mol\ g^{-1}$)	Earth's crust ($\mu mol\ g^{-1}$)	Average concentrations found in[a]		
			Rocks with highest concentration	Soil ($\mu mol\ g^{-1}$)	Soils[b] ($mol\ ha^{-1}$)
Zn	4.59	1.14	Shales and clays	0.15–4.59	300–9177
Cu	2.20	0.79	Basic	0.03–3.93	63–7868
Ni	1.28	1.36	Ultrabasic	0.03–17.04	$68–3.4 \times 10^4$
Cd	0.027	8.9×10^{-4}	Shales and clays	$8.9 \times 10^{-5}–0.02$	0.18–43
Pb	1.45	0.068	Granite	$9.65 \times 10^{-3}–1.45$	19–2890
Hg	0.007	2.5×10^{-4}	Sandstones	$5.0 \times 10^{-5}–1.50 \times 10^{-3}$	0.1–3

[a]Adapted from Ref. 85.
[b]Amount of metal per hectare calculated for a soil depth of 15 cm and a bulk density of 1.3 (approximate mass 2000 t).

the cell (18). This in turn can affect cell physiological properties and interactions with the surrounding cells and particulates. In addition, metabolic processes *inside* microorganisms as well as soil processes in general are affected by the external pH (15). Other environmental factors include the redox potential, temperature, buffering capacity, and aeration status. Finding out how toxic a metal is in situ represents a difficult task. It is not surprising, therefore, that there is considerable variability across all studies on metal effects. It is in this context that we will review the effects of heavy metals on different levels of microbial communities in the soil and the processes mediated by them.

2. EFFECTS OF HEAVY METALS ON MICROBIAL COMMUNITIES

2.1 Abundance and Biomass

The numbers of microbes in soils can be affected by metal contamination (11,19). For instance, the biomass produced when fumigated low- and high-metal soil was inoculated with unfumigated soil and a glucose solution was about 50% lower in the high-metal soil (20). This was due to less efficient conversion of substrate C to biomass C in the presence of metals since more CO_2-C was produced in the high-metal soil, whereas the amount of adenosine triphosphate (ATP) produced per gram biomass C was not affected.

Yet, microbial abundance is a less sensitive indicator of heavy metal effects than microbial processes, in part because of the rather crude and limited use of colony-forming units (CFU) as the main measure. It is estimated that only 0.1% to 10% of the bacterial soil population can be readily cultivated on low-nutrient growth media. Likewise, fungal numbers are underestimated because the method favors spores over growing hyphae (21). Fungal biomass, however, can be more indicative of pollution than fungal CFU along a heavy metal gradient (22). It appears that fungi are more tolerant to heavy metals than bacteria, and among the latter, gram-negative and high-GC gram-positive bacteria adapt better to metal stress than low-GC gram-positive ones (21). Most studies have focused on the characterization of metal-tolerant bacterial and fungal isolates in an attempt to demonstrate selective pressures (for example, 23–26). It is known that metals in soils affect microbial diversity to the point that certain genera are found less frequently, while other genera seem to adapt well. This was the case for the fungal genera *Penicillium* and *Oidiodendron*, which were isolated less frequently from Cu-contaminated soils, whereas *Paecilomyces farinosus* was only found in soils contaminated with 15.7 μmol Cu g^{-1} (27). It is, however, misleading to generalize from isolated studies since other species of *Penicillium* have been found in abundance in polluted soils (23,28).

Several studies have attempted to correlate the presence of a metal in soils or sediments with the emergence of metal-resistant communities (25,26,29–31). The general trend which emerges from these studies is that a positive correlation between the metal concentration in soil and the frequency of bacterial tolerance could only be demonstrated when metal levels in growth media were highly elevated (without knowledge of chemical speciation) and when the proportion of tolerant microorganisms was low (11). This is illustrated in Fig. 1 for Cd. Doelman et al. (31) described a sensitivity-resistance (S/R) index, which they derived by comparing the growth of isolated bacteria and fungi in the presence of an arbitrarily chosen concentration of heavy metals. The threshold for resistance was set at 142 μM Cd. Bacteria unable to grow in the presence of 4.5 μM Cd were defined as sensitive. For Zn, the thresholds were 765 and 77 μM, respectively. Bacteria were isolated from four sites representing polluted and unpolluted sandy and clay soils (Table 3).

The S/R index was 0.53 ± 0.24 in the unpolluted clay soil (containing 5.1 nmol Cd g⁻¹ and 2.14 μmol Zn g⁻¹) and 0.24 ± 0.07 in the polluted clay soil (containing 54 nmol Cd g⁻¹ and 10.3 μmol Zn g⁻¹). In sandy soil, the S/R index was 1.50 ± 0.28 in the unpolluted sample (containing 0.53 nmol Cd g⁻¹ and 180 nmol Zn g⁻¹) and 0.19 ± 0.03 in the polluted sample (containing 20 nmol Cd g⁻¹ and

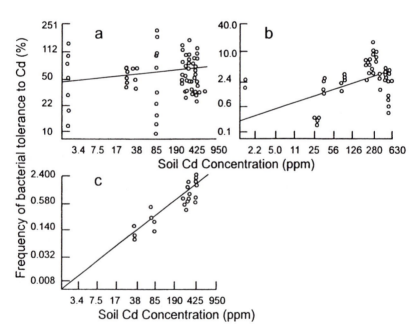

Figure 1. Percentage of bacteria from soils able to grow on media supplemented with (a) 0.09 μmol g⁻¹; (b) 0.9 μmol g⁻¹; or (c) 1.8 μmol g⁻¹. (Data from Ref. 11 and Ref. 25.)

Table 3. Physicochemical Characteristics of Sandy and Clay Soils in The Netherlands

Soil origin	Moisture content (%)	pH	Organic matter (%)	Clay (%)	Silt (%)	Sand (%)	Total heavy metals (μmol g^{-1} dry soil)			
							Cd	Pb	Cu	Zn
Tungelerwallen	11.2	3.6	3.6	3.3	4.5	88.6	0.0005	0.09	0.05	0.18
Weerterberger	9.4	2.8	35.1	2.1	2.6	60.2	0.02	0.92	0.30	3.85
Nude	13	5.6	2.9	32.7	49.9	14.5	0.005	0.41	0.79	2.14
Uiterwaarden	23	7.2	9.4	30.7	39.8	12.5	0.05	0.84	1.65	10.3

Adapted from Ref. 31.

3.85 µmol Zn g^{-1}). A multiple comparison analysis was not performed. Yet, it is obvious that for sandy soils there was a significant drop in the S/R index in the polluted soil sample (Weerterbergen). It is unclear from the data whether the difference in S/R indices in the clay soils is statistically significant. A higher toxicity—and therefore an increased degree of resistance among surviving bacteria—might be expected in sandy soil compared to clay soil on the basis on soil chemical characteristics and decreased metal complexation properties due to the absence of surfaces such as hydrated iron, manganese, and aluminium oxides, which provide hydroxyl groups for metal binding. In general, minerals of sand and silt have little effect on the chemical properties of soil. The S/R indices of the polluted clay (Uiterwaarden) and sandy soils (Weerterbergen), however, were similar, a finding which argues against increased selection pressure in a sandy environment. It follows that the S/R index is useful only when comparing similar soil types. This example illustrates that there is no universal means by which to measure heavy metal pollution.

The question arises whether the concentrations of metals used in growth media (and the chemical speciation therein) are still ecologically relevant and whether the ability of bacteria to persist in the presence of such elevated metal levels is an inherent property rather than a selected trait. The first question must remain unanswered until specific information on the speciation and toxicity of heavy metals in soil environments versus those in growth media becomes available. In response to the second question, it is known that plasmids conferring metal resistance are found in the gram-negative genera *Alcaligenes* and *Pseudomonas*, among others (32). We will later discuss the ecological significance of these microorganisms, which are colonizers of contaminated soil.

Another approach to the determination of metal effects at the community level uses the [^{3}H] thymidine incorporation rate, which is proportional to the growth rate of bacteria. Bååth (33) contaminated a sandy loam soil with 0 to 6.3 µmol Cu as CuSO$_4$ g^{-1} soil and, after a 2-year storage period, extracted soil bacteria and added various metals to measure community tolerance. No attempt was made to quantitate the amount of metal present in the soil samples which were extracted or in the actual soil extract. Both the ecological dose (ED$_{50}$ the dose responsible for an effect which reduces control values by 50%) values relating to effects of Cu addition to the soil (Table 4) and 50% inhibition of bacterial suspensions (IC$_{50}$) after addition of Cu (IC$_{50}$) were measured. The ED$_{50}$ values determined on intact soil and soil extract were identical after 2 years of storage. Hence, the soil bacterial activity was reduced by 50% by a single addition of 2.0 µmol Cu g^{-1} soil. This correlated well with ATP measurements, which also included fungal activity. Interestingly, viable counts were not affected. When Cu was added to bacterial suspensions extracted from stored soil, there was a clear difference between unpolluted and polluted soil in the way thymidine incorporation rates were affected (Fig. 2). Bacterial communities in soil previously conta-

Table 4. ED$_{50}$ for Different Measurements in Artificially Cu-Polluted Soil After Different Storage Times

	Storage time (yr)	ED$_{50}$[a]
Thymidine incorporation	0.5	2.8
(whole soil)	1	2.4
	2	2.1
Thymidine incorporation	1.75	2.1
(bacterial suspension)		
Adenosine triphosphate	1	2.6
Total bacteria	2	NE
Viable counts	1	NE

Data from Ref. 35.
[a]Expressed as log microgram added Cu (as $CuSO_4$)g^{-1} soil. NE: no ED$_{50}$ value could be calculated, since no effect of the Cu pollution was found.

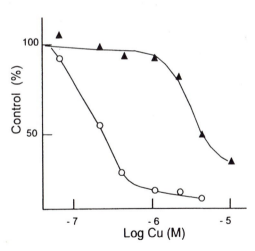

Figure 2. The effect of addition of Cu on thymidine incorporation into macromolecules of bacteria extracted from soil by homogenization-centrifugation. $\bigcirc$, bacterial suspension from non-polluted soil; $\blacktriangle$, bacterial suspension from Cu-polluted soil (6.3 μmol g^{-1}) (Data from Ref. 33.)

minated with 2.5 µmol Cu g^{-1} soil or more showed increased resistance to Cu. The order of metal toxicity was Ag > Cu > Cd > Zn > Pb (33). It can be concluded that the thymidine incorporation rate is a useful tool to monitor effects of metals on bacterial activity/growth rate. Metal concentrations should be measured once samples are withdrawn to establish how much of the metal is present in soil or bacterial suspension before adding further metals. It would be particularly useful to have an idea of the bioavailable fraction. A Cu biosensor suitable for this purpose has been developed (34), and other metal biosensors should become available soon. Measurements can be made at the same pH as in the field. It must be remembered, however, that extraction from soil may yield only 10% to 15% of the total number of bacteria present. Hence, our definition of a microbial community is an operational one.

Interestingly, the addition of Cu to soil also increased the bacterial community tolerance to other metals. This observation was further developed in a follow-up study (35). The same soil type (sandy loam, 4.4% organic matter) was contaminated with one of five different metals: Cd, Cu, Zn, Ni, or Pb. With the exception of Pb, all amendments resulted in an increase in tolerance to the metal added, as measured by [^{3}H] thymidine incorporation. All amendments led to increased tolerance to some or all of the metals tested. This was expressed as the difference in 50% inhibitory concentrations between contaminated and uncontaminated soils ($\Delta IC_{50} = IC_{50}$ in polluted soil – IC_{50} in unpolluted soil). Since IC_{50}, expressed as log [metal concentration], was generally proportional to the logarithm of the heavy metal added to the soil, the threshold concentration above which a particular metal showed an effect on the microbial communities could be estimated by extrapolation (Table 5). The actual thresholds are likely to be lower since there is no evidence that the linear relationship holds at low levels of metal contamination. The study nonetheless offers a clue about the effects of low levels of metals. Threshold concentrations for single metal additions were 2, 3, 1, 1, and 3 µmol of Cu, Cd, Zn, Ni, and Pb, respectively, per gram dry soil. Using the same soil, Frostegaard et al. (36) determined that changes in phospholipid fatty acid composition (PLFA) were statistically significant at 2, 1, 8, 2 and 8 µmol g^{-1} soil for Cu, Cd, Zn, Ni, and Pb, respectively (see next section). These determinations are in good agreement, and any differences may be attributable to the fact that the PLFA method was based on a whole soil sample, whereas the thymidine technique relied on extraction of bacterial cells.

2.2 Diversity

Bacterial community changes at low metal concentrations are subtle, apart from a shift to gram-negative genera. Since fungi exhibit variations even within single genera, diversity changes must be studied at the species level (21). Physiological groups of bacteria were the most sensitive index of metal pollution in a study of soil

Table 5. Threshold Soil Metal Concentrations for Changes in Levels
of Bacterial Community Tolerance Estimated by Using the
Thymidine Incorporation Technique

Soil pollutant	Threshold concentration for tolerance ($\mu mol\ g^{-1}$ [dry wt])[a] to				
	Cu	Cd	Zn	Ni	Pb
Cu	6	12	10	17	NT[b]
Cd	2	3	1	6	3
Zn	9	6	4	23	10
Ni	NT	20	6	1	13
Pb	24	16	13	19	NT

[a]The data were calculated by assuming that there was a linear relationship between Δ
IC_{50} values (on a log scale) and log metal addition.
[b]NT, no tolerance detected.
Data from Ref. 35.

transects taken from a smelter in northern Sweden (27). Using multivariate analysis,
the impact of heavy metals and pH on numbers of bacteria capable of producing
acid from various substrates such as maltose, arabinose, cellobiose, and xylose was
shown. Hydrolysis of chitin, starch, cellulose, and xylan was also strongly affected
by metals, but here the pH effect was less evident. Total mycelial length (of fungi)
or phosphatase activity did not correlate with heavy metal content. In general, a sig-
moidal dose–response curve was obtained when the abundance of bacteria belonging to a functional group was plotted against soil metal concentrations (27). Logistic
response curves have been described for effects on the rate of thymidine incorporation (35) and enzymatic activities, too, but on a logarithmic scale (37). While both
metal concentration and pH affected parameters like functional bacterial group, the
pH variation itself was independent of pollution but related to Cd and Mn levels
(27). This study exemplifies the need for proper statistical analytical tools (multivariate analysis), on the one hand (see Chapter 16), and an appropriate number of
environmental and biological variables, on the other hand.

Only recently have modern methodologies become available for the study of
microbial community structure (see Chapters 12a, 12b, and 13). A larger toolbox
is needed in order to distinguish among effects on physiological capacities and
changes in species diversity. In a very thorough study, the phospholipid fatty
acid composition (PLFA) in a forest humus and an arable soil experimentally
polluted with Cd, Cu, Ni, Pb, or Zn was measured by extracting the PLFAs directly (36). The composition was compared with other determinations of biomass (adenosine triphosphate [ATP] content, total PLFA, and lipid phosphate
content) and activity (respiration rate). This study was particularly interesting

because it involved two soils with different pH and organic matter content. Metal concentrations reported were nominal; i.e., the actual amount of metal in the sample which was withdrawn was not determined. Changes in PLFA patterns were demonstrated for both soil types at levels of metal contamination frequently lower than those at which effects on ATP content, soil respiration, or total PLFA had been found (Tables 6 and 7). Using principal component analysis, the authors showed that most of the variation in PLFA patterns was due to the presence of added metals irrespective of which heavy metal had been added, except for changes due to Cu contamination in the arable soil. Some groups of PLFA responded similarly to metals; e.g., levels of the isobranched i15:0 and i17:0 always decreased. These changes did not necessarily reflect changes in the same organism groups in the two soil types (36).

A general shift toward gram-negative bacteria could not be demonstrated, despite an increase in cy17:0 (a species of cyclopropyl fatty acid), because other fatty acids showed variable responses. Levels of 15:0 and 17:0 PLFA, indicative of bacterial genera, were elevated in all metal-contaminated samples taken from the arable soil but not in those from the forest soil. Actinomycete-specific PLFA (10 Me 16:0, 10 Me 17:0, and 10 Me 18:0, where 10 indicates the position of a methyl group at C position from the carboxylic end and 0 the absence of a double bond) increased in the forest soil but not in the arable soil, indicating that different types of actinomycetes were present. Fungi increased in the arable soil (measured as 18:2ω6) except in the presence of Cu, which is known to be a fungal inhibitor. This did not occur in the forest soils; because of the presence of plant material rich in 18:2ω6, which would mask any metal-induced changes. The order of metal toxicity in the forest soil was Cd = Ni > Zn = Cu > Pb, which is in line with

Table 6. Levels of Heavy Metal Contamination of Forest Soil at Which Changes in PLFA Patterns Were Found[a]

| Metal | PLFA pattern | ATP | | Respiration | | totPLFA |
		ED_{50}	ED_{10}	ED_{50}	ED_{10}	ED_{10}
Cd	1	25	1.2	14	0.2	3.9
Cu	8	68	5.3	>128	12	15
Ni	2	15	1.6	6	1.5	1.5
Pb	2	15	3.8	33	11	11
Zn	4	>256	>256	24	2.3	>256

[a]Calculated by SIMCA analysis and levels at which an ED_{50} or ED_{10} in ATP content, respiration, and total PLFA were found, expressed as micromoles per gram (dw) of soil. PLFA, phospholipid fatty acid; tot PLFA, total PLFA
From Ref. 36.

Table 7. Levels of Heavy Metal Contamination of a Sandy Loam at Which Changes in PLFA Patterns Were Found[a]

Metal	PLFA pattern	ATP		totPLFA
		Ed_{50}	Ed_{10}	ED_{10}
Cd	1	6.3	1.2	3.2
Cu	2	14	3.1	>64
Ni	2	23	9.7	4.3
Pb	8	28	11	7.3
Zn	8	17	1.9	2.9

[a]Calculated by SIMCA analysis and levels at which an ED_{50} or ED_{10} in ATP content and totPLFA were found, expressed as micromoles per gram (dw) of soil.
From Ref. 36.

other reports. In addition, Cd was the most toxic metal in arable soil. To summarize, the PLFA pattern represents a sensitive indicator of changing microbial communities and is more representative than the colony-forming unit (CFU). The use of PLFA has allowed a first glimpse at the complexity of community responses to heavy metals. Among other emerging new technologies, it can be expected that the application of fluorescently labeled, ribosomal ribonucleic acid (rRNA)–targeted oligonucleotide probes for in situ monitoring of bacteria (38,39) in combination with confocal laser scanning microscopy (40) should greatly advance the study of diversity changes imparted by pollutants including heavy metals.

3. EFFECTS OF HEAVY METALS ON MICROBIAL PROCESSES

Even more important than the effect of metals on microbial communities and species diversity is the impact they have on microbial processes relevant to the whole soil ecosystem. It could be argued that only when such process parameters are affected can one talk of a significant ecological perturbation. The following section addresses these effects as they apply to the soil environment. Tables listing literature reports on the effects of heavy metal pollution have been published elsewhere and are recommended to the interested reader (21).

3.1 Litter Decomposition

Several studies have shown that litter decomposition is decreased in the presence of certain heavy metals (41–44). On the other hand, Friedland et al. (45) reported no inhibition of litter decomposition on a forest floor by 0.48–0.97 µmol Pb g^{-1} soil, 1.53 µmol Zn g^{-1}, 0.16–0.31 µmol Cu g^{-1}, 0.17 µmol Ni g^{-1}, and 0.018 µmol Cd g^{-1}. Table 2 reveals that these concentrations are all below the allowable levels under European Commission regulations. Importantly, if inhibition does occur, it happens at a later stage, when more recalcitrant molecules such as lignin are attacked (41,42,46). For Cu, 0.61–22 µmol g^{-1} caused an inhibition (21). In forest soils, *Pinus* needles were affected at higher concentrations (3.62 µmol g^{-1}, 80% organic matter) (21) than *Quercus* and *Smilacina* leaves (0.61 and 0.66 µmol g^{-1}, respectively, organic matter content not reported) (46). For Zn, the effective metal levels were 4.31 µmol g^{-1} (*Smilacina* leaves) to 10.71 µmol g^{-1} (*Pinus* leaves) in the same two studies. However, comparisons are not straightforward since Cu was one of several metal contaminants and one study was a point estimate (few samples were taken) and the other a gradient study (many samples taken). The difference found in the comparison is interesting, since point estimates often lead to higher toxicity estimates when samples containing high levels of metals are included. In the preceding comparison, however, the gradient study led to a higher estimated toxicity level than the point study. It can be concluded from this observation that not only soil contamination but also litter quality plays a role in determining decomposition rates.

3.2 Carbon Mineralization

Soil respiration rate can be measured in different ways, i.e., by using the organic substrates already in the soil or by adding different substrates. Duxbury (11) pointed out that the latter approach could hardly be justified in an environment where nutrients are normally low. Yet it was shown that degradation of large molecules like cellulose and starch is more readily inhibited than degradation of glucose (48). It is also useful to add a substrate that is readily available because even then exponential CO_2 production is preceded by a lag period. The duration of this phase is a function of metal toxicity (49) and displays a linear relationship with the log of metal concentration (50). This is useful as a screening tool to analyze contaminated soils but does not reveal much about the state of carbon mineralization in the unamended soil.

Another important parameter is soil moisture content, which some workers choose to standardize. This has been called a biased sample (51) and may alter some of the existing matrix interactions between microorganisms and soil components. However, it has been used successfully to minimize variability among samples (52). Evolution of CO_2 was affected by 10% in organic soil (64%) at pH 6.2 in the absence of added substrates by 4.98 µmol Cd, 5.00 µmol Cu, 5.00 µmol Zn, or 5.00 µmol Ni (53). In loamy sand (pH 4.9, 2.1% C) only 0.09 µmol Cd g^{-1} soil was required to inhibit CO_2 formation by 17%; 1.57 µmol Cu or 0.48 µmol Pb g^{-1} soil was needed for a 25% inhibition, 0.15 µmol Zn for a 21% inhibition, and 1.70 µmol Ni for a 28% inhibition. These results confirm the importance of soil type to metal toxicity. Hence, C mineralization is a sensitive indicator for metal pollution. In field studies where soils contaminated with several metals are used, effects are detectable at even lower metal concentrations, presumably because of their synergistic toxicity. For example, in a forest humus (pH 3.3–3.5), a 30% reduction of CO_2 evolution occurred in the presence of 0.014 µmol Cd, 0.55 µmol Cu, 0.38 µmol Pb, 1.09 µmol Zn, and 0.71 µmol As g^{-1} soil. The corresponding metal values in nonpolluted soil were 7.47 nmol Cd, 0.35 µmol Cu, 0.19 µmol Pb, 0.78 µmol Zn, and 0.43 µmol As g^{-1}, respectively.

3.3 Nitrogen Transformations

Of the major steps in the nitrogen cycle, nitrification is potentially most sensitive to heavy metal pollution. Field studies have suggested little effect of heavy metals on N_2 fixation (54). This may be explained, in part, by the fact that metal pollution in this study was chronic, although the concentrations of Pb and Zn present were rather high at 38.6 and 398 µmol g^{-1} soil, respectively. Long-term pollution of soils with metals, combined with a suitable pH, can reduce the amount of bioavailable metal (21). The same holds, of course, for pollution with organic contaminants like polycyclic aromatic hydrocarbons. It has been shown in sludge-amended soil that free N_2 fixation is extremely sensitive to Zn (0.46 µmol g^{-1} soil), Cu (0.24 µmol g^{-1}), Ni (0.034 µmol g^{-1}), and Cd (0.018 µmol g^{-1}) as ethylenediaminetetra acetic acid– (EDTA) extractable fractions (55). In laboratory studies, free N_2 fixation was 10 times more sensitive than soil respiration rate, urease activity, or rate of nitrate formation (56). The major stumbling blocks for the routine use of N_2 fixation assays are thus natural variability (what constitutes a control soil?) and low activity (21).

Symbiotic N_2 fixation has been studied with a view to the whole plant response, as summarized by Duxbury (11). Interestingly, heavy metals can have both adverse and stimulatory effects as a function of concentration. For example, Cd inhibited nitrogenase in root nodules of *Almus rubra* (red alder) at concentrations greater than 0.03 µmol Cd g^{-1} soil (57). At lower concentrations (0.09–0.9 µmol Cd g^{-1} soil) nitrogenase activity was still affected, yet nodulation was stimulated. This was explained as a compensating effect by the plant to make up for reduced enzyme activity. Unlike Cd, which is not an essential metal, Mo and Zn are required for enzyme activity in symbiotic N_2 fixation. Addition of Mo and Zn stimulated N_2 fixation in field experiments involving *Vigna unguiculata* using sandy loam with optimal fixation rates at 153 mole Zn/ha and 91 mole Mn/ha (58). The N_2 fixation by *Rhizobium–Trifolium repens* (white clover) was not affected in various metal-contaminated soils when the host plant was indigenous to the sward (59). However, soils which did not contain indigenous populations of *T. repens* failed to support nodulation in agar-grown host plants. This was interpreted as a metal-induced reduction of effective rhizobial population size relative to control soils. Other studies report negative effects of Cd, Cu, Co, Ni, and Zn on both N_2 fixation and nodulation of plants (60–63).

Nitrogen mineralization has been studied in the field (64–67) and in the laboratory (66,68–74). As stated by Bååth (21), both stimulatory and inhibitory effects are observed, with increased N mineralization at low metal concentrations, e.g., when 1.70 µmole Ni g^{-1} soil was added to a forest organic layer containing 6% C, and decreased N mineralization at higher metal concentrations, when the same soil received 8.52 µmole Ni g^{-1} soil (70). Similarly, the addition of 4.45 µmole Cd, 7.87 µmole Cu, 2.41 µmole Pb, or 7.67 µmole Zn g^{-1} sandy loam soil had a positive effect, whereas twice the amount of any of the metals inhibited N mineralization (72).

Likewise, soil nitrification showed an opposing response to different metal concentrations after addition of 1.70 µmol Ni g^{-1} versus 8.52 µmol Ni g^{-1} (70) and can be affected more than N mineralization. In denitrification, only the reduction of NO_2^- to NO was sensitive to Cd added at 0.48 (sandy soil) or 0.55 (clay soil) µmol g^{-1} soil (75). The same process was affected in silt loam (pH 6.75, 1.8%C) by addition of 0.09 µmol Cd, 0.79 µmol Cu, or 1.53 µmol Zn g^{-1} soil.

To summarize, several processes of the nitrogen cycle are affected by metals at rather low concentrations, although not always in a negative and predictable way. Best suited for the study of heavy metal effects in soil are probably free nitrogen fixation, nitrification, and denitrification. The problem lies in field tests, which do not always support evidence from laboratory investigations.

3.4 Enzyme Activity

Enzyme activities in soil, both aspecific and specific, are affected by heavy metals (21). Of the former, dehydrogenase and hydrolytic enzymes, with the exception of β-glucosidase, are good indicators of metal pollution. Among the more specific enzymes rank urease and phosphatase. The effective concentrations ranged from 0.55 to 22 µmol Cu g^{-1} soil in field studies. The lowest concentration of Cu alone which had an effect on phosphatase was 2.52 µmol g^{-1} in histosol, with ca. 40% C and a pH of 5.4 (76). Most field studies involve more than one metal, making comparison difficult. There are a multitude of laboratory studies which show effects of Cu at concentrations as low as 0.16 µmol g^{-1} (sandy soil, pH 6.9, 1.1% C) on dehydrogenase activity (77). In the same study, using alluvial soil at pH 7.1 and with a carbon content of 1.8%, dehydrogenase activity was only inhibited half as much as in sandy soil at 7.87 µmol Cu g^{-1}.

A decrease in enzyme activity can be caused by a drop in enzyme synthesis, which, in the case of more specific enzyme activities, relates to an effect on a physiological group of organisms (27,78). In laboratory amendments, 0.63–2.36

μmol Cu g⁻¹ was the lowest effective concentration which inhibited urease activity by 10% in sandy soil at pH 7.0 with an organic matter content of 1.6% and a CEC of 1–2 mEq/100 g soil (79). The same degree of inhibition in sandy peat (pH 4.4, 12.8% organic matter, CEC 50–55 mEq/100 g soil) required 31.2 μmol Cu g⁻¹ soil. Nickel inhibited urease activity at 0.51 to 2.04 μmol g⁻¹ in sandy soil and at 9.2 to 187 μmol g⁻¹ in sandy peat soil (79).

All these studies refer to the addition of a single metal. When several metals are present, as is usually the case in polluted environments, the amount of a specific metal required to inhibit enzyme activity may be significantly lower because of synergistic action. This was evident in the field study by Tyler and Westman (80). Phosphatase and urease activities in forest humus (pH 3.3–3.5) were inhibited by 28% and 29%, respectively, in the presence of 0.01 μmol Cd, 0.55 μmol Cu, 0.38 μmol Pb, 1.09 μmol Zn, and 0.71 μmol As g⁻¹ soil. The corresponding metal values in unpolluted control soil were 0.007, 0.35, 0.19, 0.78, and 0.43 μmol g⁻¹ soil, respectively. Hence, in all cases less than a doubling of the baseline concentration led to a significant inhibition of enzyme activity. This study used soil from Rönnskär in northern Sweden, which together with another site at Gusum represents an area polluted by metal works that has been under intense study for over 20 years (15). Gusum contains the secondary smelter and brass foundry and is impacted mostly by Cu, Zn, and some Pb; Rönnskär houses the primary smelter. The surrounding forests have been exposed to Cu, Pb, Zn, and As, as well as large SO₂ emissions. This unique environment offers the opportunity for gradient studies not usually available to the investigator. At Rönnskär, the pH has remained low throughout the gradient, compared to the site at Gusum (21). Table 8 summarizes the relative concentration ratios for metals, which significantly inhibited microbial communities and their activities along the gradient. Effects are observed at ratios as little as two times those of baseline samples in the case of urease and phosphatase activity at Rönnskär.

3.5 Impact of Heavy Metals on Degradation of Organic Pollutants

There are reports indicating that heavy metals can impair not only organic matter (OM) decomposition but also the biodegradation of xenobiotics, for example, the herbicide 2,4-dichlorophenoxyacetic acid methyl ester (2,4-DME) (81). This study involved the addition of 2,4-DME and metal solutions to lake sediments and Aufwuchs (floating mats of filamentous algae), collected in the field or from laboratory microcosms. Because experiments were performed on slurries and metal concentrations were expressed in terms of molarity, comparison with soil studies is difficult. The concentrations required to inhibit biodegradation rates significantly as estimated by the minimum inhibitory concentration (MIC) (the intercept of the plot of percentage reduction in rate coefficient k, versus log [mo-

Table 8. Effects on the Biological Characteristics of Coniferous Forest Litter and Mor Soils, Obtained in Field Studies at Gusum (Secondary Smelter, Brass Foundry) and Rönnskär (Primary Smelter of Sulfide Ore) in Sweden[a]

Variable/function considered	Percentage depression	Elements in pollution and concentration ratio	Reference
Gusum		Cu + Zn	
Litter decomposition	Decrease	5	(21)
Soil respiration	18%	4	(48)
	25%	6	(52)
	30%	7	(47)
	40%	10	(22)
N mineralization	20%	3	(64)
Urease activity	35%	6	(52)
Phosphatase activity	30%	6	(52)
Amylase activity	25%	7	(47)
Community structure	Altered	4	Nordgren et al. (1985)
Microbial biomass	20%	7	(21)
Fungal length	35–40%	10	(22)
Rönnskär		Cu + Pb + Zn + As	
Litter decomposition	Decrease	7	(21)
Soil respiration	25%	2	(80)
	35%	4	(27)
Urease activity	25%	2	(80)
	35%	8	(27)
Phosphatase activity	30%	2	(80)
Colony-forming units of hydrolytic bacteria	45%	8	(27)

[a]Metal concentrations are calculated as the ratio between those of the studied materials and those of the same materials from sites not influenced by local pollution.
From Ref. 21.

lar metal concentration]) were low for Cu (1.2 μM), Zn (0.097 μM), and Cd (0.89 μM). The actual concentration of free metal ion was not known in these experiments.

In general, bacterial strains capable of metabolizing xenobiotics including naphtalene, phenanthrene, house fuel oil, and toluene tend to be sensitive to heavy metals (31,82). The ability of soil bacterial isolates to utilize 20 aromatic substances as C sources was related to the contamination of soils by heavy metals (83). A negative correlation was found for the ratio of the 10 lowest-scoring to the 10 highest-scoring characteristics tested but, significantly, not for the average number of substrates used per isolate. This would suggest a loss of rare biochemical traits in bacterial populations due to the presence of heavy metals. Attempts to improve the heavy metal resistance of valuable degrader strains have been successful for the genus *Alcaligenes* (82,84). The actual performance of these strains in bioreactors or in situ on contaminated lands must be evaluated before conclusions can be drawn about their rate of survival and possible applications in soil bioremediation projects.

3.6 Mycorrhizae

The most likely way for metals to enter the food chain is via plants. For this reason, it is important to evaluate effects of metals on plants and to study uptake routes. In nature, most plants are mycorrhizal: that is, the uptake of nutrients, including essential metals such as zinc and copper, is enhanced through an extended network of mycelia (86–91). However, the presence of mycorrhizae does not necessarily imply that plants survive differently in soils polluted with heavy metals. For example, vesicular–arbuscular (VA) mycorrhizae decreased zinc toxicity to grasses growing in soil polluted with zinc (92) but were responsible for increased shoot uptake of copper and zinc in perennial bunch grass (93). Ericaceous mycorrhizae, on the other hand, seem to aid plant survival in metal-polluted areas, leading to lower metal concentrations in the shoots of plants (94,95). Similarly, ectomycorrhizae can increase metal tolerance of plants (96–99).

Mycorrhizae reduce toxicity to their host plant by binding heavy metals. The efficiency of protection differs among fungal isolates and different heavy metals (100). Binding can occur in the mycelium (101–103), either in the hyphal cell walls or in the surrounding polysaccharide slime. Metals are known to adsorb to fungal cell wall components, including cellulose, chitin, and melanin (10,100). Binding sites within cells of ectomycorrhizal fungi are metallothioneinlike proteins, which have a high cysteine content (104), and perhaps polyphosphate granules (105), although the presence of phosphorus-containing proteins in mycorrhizal fungi has not been proved (100).

To summarize, it is clear that mycorrhizal fungi play an important role in the effect heavy metals have on their host plants. In general, they either increase the uptake of metals at low concentrations or reduce their uptake at toxic concentrations. This suggests a form of regulation at the level of mycelial cells, which may involve metallothioneinlike proteins.

4. MICROORGANISMS ADAPTED TO METAL-CONTAINING BIOTOPES

Until now we have mainly reviewed the acute or chronic toxic effects of heavy metals on the microbial communities and the metabolism of their constituents as well as the variety of bacterial responses to toxic effects. In the present section, attention will focus on biotopes with a high content in heavy metals and where the toxicity is permanent and on bacteria which specifically colonize such biotopes.

4.1 Bacteria from Zinc- and Copper-Polluted Areas

Some bacteria with plasmid-borne multiple resistance to heavy metals have exclusively been found in industrial biotopes with a high content of heavy metals. This is the case for strains clustered around *Alcaligenes eutrophus* CH34. Strain CH34 was first isolated from sediments of a decantation basin in a nonferro metallurgical factory in Belgium (106). It was classified as *A. eutrophus* because of its facultative chemolithotrophic metabolism supported by the presence of two hydrogenases, one soluble and one membrane-bound. It harbors two large plasmids, pMOL28 and pMOL30. Plasmid pMOL30 carries an operon, *czc*, involved in resistance to cadmium, zinc, and cobalt. The *czc* operon was used as a probe to detect the presence of similar genotypes in other environmental samples (107).

Bacteria hybridizing with the *czc* probe and generally closely related to *A. eutrophus* CH34 were present in a variety of soil samples highly contaminated by heavy metals. This observation was reported for a zinc desertified area in Belgium (107), where the sandy soil was covered by aerial deposits of zinc oxide released from a zinc factory built at the end of the nineteenth century and dismantled 60 years later. Bacteria with a *czc*+ phenotype were also found in soils loaded with tailings from metallurgical factories in Zaïre and in Belgium. These factories processed mainly copper, cobalt, and zinc leachates.

Figure 3 shows one site in Zaïre where *czc*+ bacteria have been detected in viable counts using *czc* as a probe (1,107). Table 9 reports corresponding chemical and viable count data. The picture shows a canal conveying the tailings at a distance from the production site of a copper/cobalt processing plant. The banks of this canal are regularly flooded by the tailings and never give rise to any vegetation. Samples of these highly contaminated soils which sheltered metal-resistant

Figure 3. Picture of a river bank flooded with tailings from a copper factory in Zaïre (Likasi, Shaba Province). In the bottom of the picture, the sampling place is visible. Table 11 shows chemical analysis and viable counts (CFU) from this sample. czc^+ *A. eutrophus* strain VA1 (1) is representative of this biotope.

bacteria related to *A. eutrophus* CH34 display a total heavy metal content of up to 20%. The presence of bioavailable metal is determined by the fraction of EDTA extractable metal. Bacteria resistant to metals in the range 10 to 600 ppm of bioavailable zinc, copper, or cobalt were found and constituted an appreciable fraction of the viable counts estimated in nonselective conditions. On the contrary, when the metal-containing soil sample did not release metal after EDTA, Mg^{2+}, or hot water treatment, only less resistant bacteria were found. This situation applied to the extraction sites of the ores (malachite, blende). These barely

Table 9. Chemical and Microbiological Analysis of a Polluted Soil Sample[a]

| | Chemical analysis | | |
Metal	Total extraction	EDTA extraction	Hot water extraction
Cd	110	6	1
Cu	69,760	4,817	17
Co	12,845	125	43
Pb	858	6	<1
Ni	116	1	<1
Fe	163,310	<1	<1
Zn	9,325	350	55

Bacterial viable counts (CFU/g)	
Rich medium	2.10^4
Rich medium + Zn^{2+} 10 mM	7.10^3
Min. medium (106)	$2.5.10^3$
Min. medium + Cu^{2+} 0.8 mM	3.10^3

[a]The chemical data are expressed in milligrams per kilogram. EDTA and hot water data give an approximation of biologically available quantities. Viable counts (expressed as colony-forming units (CFUs)/per gram of soil) are of the same order of magnitude on media provided or not provided with heavy metals. Also see Fig. 3.

weathered sites do not contain much sediment with bioavailable metals. Searches for metal-resistant heterotrophic bacteria in copper/cobalt extraction sites were disappointing. It should be stated that the ores tested were mainly accompanied by a dolomitic (basic) matrix. This is not a favorable condition for the development of acidophilic leaching bacteria. Through the leaching processes, bacteria like *Thiobacillus* would make metals available to heterotrophic bacteria, like metal-resistant *A. eutrophus* which would occupy ecological habitats different from those of the acidophilic bacteria or participate at a different stage in the chemical/biological processing of ores that mining operations bring to the air.

Although metal-resistant *czc*[+] *A. eutrophus* and related bacteria seem to account for a high fraction of the CFU in biotopes highly contaminated by heavy metals, other bacteria of interest were also encountered.

Special mention should be given to *Pseudomonas aeruginosa* isolates which exhibited resistances to cadmium (up to 1 mM) and to zinc (up to 5 mM). Such soil isolates have been found in the zinc-desertified area mentioned near the rhizosphere of dwarfed elders. The genetic basis of metal resistance of environmental isolates of *Pseudomonas aeruginosa* was studied recently in strains isolated from polluted rivers in Pakistan (M-e-Talat, Karachi, personal communication) and will be briefly discussed later. Other isolates which have recently been found at zinc-rich industrial sites together with *A. eutrophus* strains include *Sphingomonas yanokuyae* and *Arthrobacter ilicis* isolates (M. Mergeay and L. Diels, unpublished).

4.2 Nickel-Resistant Bacteria

Besides bacteria which are czc^+, a systematic search for nickel-resistant bacteria was carried out. This search was of special interest because it led to a comparison between nickel-resistant bacteria isolated from anthropogenically polluted and naturally nickel-rich ecosystems. The first category of ecosystems included galvanization tanks, sewage plants, and metal-working fluids. The nickel-resistant bacteria obtained from these samples were isolated after special enrichment of nickel-resistant bacteria or after enrichment of a specific group (enterics, denitrifiers) and further testing for nickel resistance.

Naturally nickel-rich ecosystems were mainly represented by serpentine soils originating from weathering of serpentinite rocks (108,109). The flora of these soils is characterized by many endemic species; some of these are not only nickel-resistant but also nickel-accumulating.

Another remarkable naturally nickel-rich ecosystem is provided by the *Sebertia acuminata* in New Caledonia. This tree contains more than 25% nickel weight/dry weight (wt/dry wt) in its latex and more than 1% in its leaves. Stoppel and Schlegel (110) suggest the existence of a "nickel cycle" driven by the decay of falling leaves of the tree, the liberation of nickel from the leaves, the percolation of nickel in the groundwater, and the reaccumulation of nickel in the plant root system. The viable counts of nickel-resistant (20 mM) bacteria in the humus and soil layers underneath the canopy of these nickel-hyperaccumulating trees were almost as high as the numbers observed in the absence of nickel-selective pressure. As mentioned previously, this kind of observation was also reported for soil samples highly contaminated by zinc (107). In both cases, it will be of interest to analyze further the microbial content of these samples and to see the relationships of the CFU fraction with (1) the nonculturable bacteria and (2) the culturable bacteria nonselectable under aerobic heterotrophic conditions (i.e., some obligate chemolithotrophs). This information is probably essential to understanding the role of metal-resistant culturable bacteria in the processing of metals in any metal-rich biotope. The PCR-based methods are promising tools to approach such questions.

The nickel-resistant bacteria from anthropogenically polluted biotopes (111) and from naturally nickel-rich soils (112) have been hybridized with various probes (110) carrying, e.g., *cnr* from plasmid pMOL28 (*A. eutrophus* CH34)

(113), *ncc* from pTOM8 (*A. xylosoxidans* 31A) (114), *nre* from pTOM8 (*A. xylosoxidans* 31A) (114), and *nre* from *Klebsiella oxytoca* 15788 (115). Four major hybridization patterns could be distinguished (110):

1. *cnr/ncc* type—strong homologies with the *cnr* or the *ncc* operons, respectively. Various Belgian and Zaïran *czc* isolates also exhibit this pattern. However, it should be remarked that *czc* and *cnr* were never carried by the same plasmid. In *A. eutrophus* CH34, *czc* is carried by pMOL30 and *cnr* by pMOL28 (106). It is not known whether the nickel-resistant strains displaying the pattern *cnr/ncc* are also *czc*$^+$.

2. *cnr/ncc-nre* type—strong homologies with both the *ncc/cnr* operons and the *nre* operon. In *A. xylosoxidans* 31A, which is characteristic of this pattern, the gene clusters *ncc* and *nre* are strongly linked on plasmid pTOM8 and are carried by the same 14.5 kb fragment. The *ncc-nre* is well expressed in *Alcaligenes* strains, while *nre* is also well expressed in enterics. Recently, a taxonomic study based on 16S ribosomal RNA has shown that strain 31A is much more closely related to other metal-resistant strains than to the type strain of *A. xylosoxidans* (H. Brim, personal communication).

3. *Klebsiella oxytoca* type—strong homologies with nickel resistance determinants of *K. oxytoca* 15788 and slightly positive hybridization signals with the *nre* operon of strain 31A. *K. oxytoca* were found in a metal factory in Göteborg, Sweden (111,115). Some New Caledonian strains belong to this type and were selected through enrichment for enteric bacteria.

4. Others, showing no hybridization reaction with any of the DNA probes used in the reported study—strains isolated from naturally nickel-rich biotopes from Scotland and California and six New Caledonian isolates belong to this group.

In conclusion, the study suggested that *cnr/ncc* strains may be found in both anthropogenically polluted areas and in natural metal-rich (at least nickel) areas. In polluted areas, strains which are *cnr*$^+$/*ncc*$^+$ are also often *czc*$^+$. This may indicate that the origin of *czc*$^+$/*cnr*$^+$/*ncc*$^+$ strains in industrially polluted soils is to be found in natural soils which neighbor (weathered) surface ores.

Natural areas rich in copper/cobalt also contain a diversified endemic flora which is specific of the African copper belt stretching from southern Zaïre to Zimbabwe (116–118). However, the much desired microbiological analysis of the rhizosphere of cuprophytes is still waiting. These natural areas are located not far from the industrial sites where metal-tolerant *Alcaligenes* have been isolated.

As far as Europe is concerned, a study of samples from old mines, now closed for exploitation, would be of interest to examine the dispersion of strains related to *A. eutrophus* CH34 and its relatives such as *A. xylosoxidans* 31A and likely most of the New Caledonian isolates.

The study of Stoppel and Schlegel (110), which is mainly based on DNA hybridization data, also points to the diversity (or the evolutionary divergence at the DNA level) of the nickel resistance mechanism, as a substantial class of test strains did not hybridize with the tested probes. This is supported by the recent discovery of gram-positive nickel-resistant bacteria found in a metal factory in Tirol, Austria. These strains displayed a resistance up to 30 mM Ni^{2+} and have been classified as *Arthrobacter ilicis* (147). Some high-GC gram-positive bacteria (*Thallobacteria*) display plasmid-borne resistance to high levels of heavy metals.

Pseudomonas aeruginosa is often encountered in areas heavily polluted by heavy metals (107). Various plant-growth-promoting bacteria also belong to this species. These observations are of importance because this species is an opportunistic pathogen. It may raise the fear that polluted areas would be enriched in putative pathogens. As far as *P. aeruginosa* is concerned, a profound study to compare clinical and environmental isolates would surely be of major interest. Some observations of interest regarding the behavior of *P. aeruginosa* toward heavy metals have been recorded. In *P. aeruginosa* CMG103 isolated from a Pakistan river polluted by heavy metals, a chromosomal determinant for high resistance to zinc and cadmium, *czr*, was identified. The DNA sequence determination of *czr* showed a significant homology with all the components of the efflux mechanism mediated by *czc*CBA (*cnr*CBA, *ncc*CBA). The *czr* probe of *P. aeruginosa* CMG103 hybridized with DNA of other *P. aeruginosa* strains, including the reference strain PAO1. *czr* was mapped on the *P. aeruginosa* PAO chromosome (U. Roming, personal communication; M-e-Talat Hassan, personal communication). Other observations made with other soil isolates of *P. aeruginosa* show that metals such as zinc, cadmium, and nickel may induce the production of pyoverdin, the major siderophores of *P. aeruginosa* (119). The genetic determinant responsible for this effect was identified (120). This observation raises the question of the still obscure relationship between iron metabolism and transport and response to heavy metals.

4.3 Mechanisms of Resistance to Heavy Metals in Soil Bacteria

Genes conferring resistance to toxic concentrations of heavy metals are often encountered on plasmids, albeit not exclusively so. Figure 4 describes the interactions between microorganisms and metals as a function of the external concentration of heavy metals. Concentrations range from those in which all microorganisms have to ensure homeostasis for essential metals (nanomolar/micromolar level) up to the very high concentrations at the decimolar/molar level where acidophilic chemolithotrophs (*Thiobacillus ferrooxidans* and *Archaea*) can survive. However, the substantial difference in growth conditions between acidophilic chemolithotrophs and heterotrophs has to be taken into consideration.

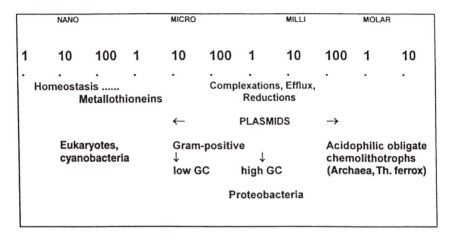

Figure 4. Concentration ranges of toxic heavy metals are presented in a logarithmic way. Microorganisms are ranged according to maximum tolerated concentrations. Arrows delimit the concentration range where plasmid-borne resistance to heavy metals is generally observed.

Plasmids allow heterotrophic bacteria to cope with concentrations at the millimolar level while metallothioneins offer a possibility to eukaryotic cells and to some bacteria such as the cyanobacteria (121–123) to cope with concentrations which are toxic at the micromolar level. Chemical elements which are involved in plasmid-borne resistance are indicated in the periodic table (Fig. 5). Besides, Table 10 reports on the main mechanisms described up to now for resistance to heavy metals. Resistances are known for the cations of Co (106); Cu (3,124–127); Ni (106,114); Zn (106); Cd (3); Ag (3); Hg (3,128,129); Pb, Tl (thallium), and Bi (bismuth); the oxyanions of Cr (chrome) (130–132); As (arsenic) (4); B (boron) (133); Sb (antimony) (3,4); and Te (tellurium) (134–136); and organometallic compounds of Hg, Se (selenium), and Sn (tin) (12). Information related to Pb and Tl is still scarce although mapping data or insertional mutants are available (1). Resistance markers associated with *Alcaligenes eutrophus* CH34 and related strains (1) are indicated in Figs. 5 and 6. The best known *A. eutrophus* metal resistance genes are those involved in multiple cation resistances. The *czc* genes present on plasmid pMOL30 encode resistance to Cd^{2+}, Co^{2+}, and Zn^{2+} ions, *cnr* genes on plasmid pMOL28 code for resistance to Co^{2+} and Ni^{2+} (106). In other related strains, *ncc* genes code for resistance to nickel, cadmium, and cobalt (114). Table 11 reports the minimum inhibitory concentrations of heavy metals which are associated with these operons. All of them share much homology at the DNA level and a common organization of the structural

PERIODIC TABLE OF THE ELEMENTS

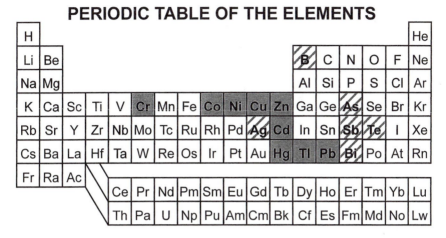

Figure 5. Boxes with elements in bold characters refer to plasmid-borne genetic determinants for bacterial resistance. Full gray boxes refer to those genetic determinants found in *Alcaligenes eutrophus* CH34.

Table 10. Main Mechanisms for Heavy Metal Resistance

Genotype	Ion	Mechanisms
mer	Hg^{2+}	Detoxification of mercuric salts (operon *mer* RTPAD) or organomercurials (*mer* RTPBAD) by reduction of Hg^{2+} into Hg^0
*cad*A	Cd^{2+}	ATPase-mediated efflux of cations
ars	AsO^-_2	ATPase-mediated efflux of anions
	AsO^{3-}_4	Reduction of arsenate followed by efflux of arsenite
cop	Cu^{2+}	Sequestration of Cu^{2+} outside the cytoplasmic membrane
pco	Cu^{2+}	Efflux
czc	$Cd^{2+}, Zn^{2+}, Co^{2+}$	Cation/proton antiporter chemiosmotic efflux
cnr	Co^{2+}, Ni^{2+}	Cation/proton antiporter chemiosmotic efflux

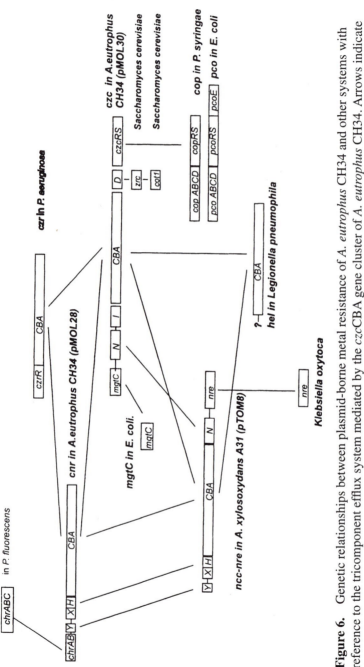

Figure 6. Genetic relationships between plasmid-borne metal resistance of *A. eutrophus* CH34 and other systems with reference to the tricomponent efflux system mediated by the *czc*CBA gene cluster of *A. eutrophus* CH34. Arrows indicate the relationships as revealed by sequences of DNA and the corresponding putative proteins. *czc* resistance to Cd^{2+}, Co^{2+}, and Zn^{2+} (*czc*NICBADRS); *cnr* resistance to Co^{2+} and Ni^{2+} (*cnr*YXHCBA); *ncc*, resistance to Ni^{2+}, Cd^{2+} and Co^{2+} (*ncc*YXHCBAN); *czr*, resistance to Cd^{2+} and Zn^{2+} (*czr*RCBA); *hel*, genetic determinant involved in virulence in *L. pneumophila*; *chr*, resistance to chromate; *pco/cop*, resistance to copper; *nre*, resistance to Ni^{2+} *mgt*, transport of Mg^{2+}, *zrc*, zinc/cadmium tolerance in yeast; *cot*, cobalt tolerance in yeast.

Table 11. Minimum Inhibitory Concentrations of Metals in *Alcaligenes eutrophus* CH34[a]

Genotype	Metal				Ref.
	Cd^{2+}	Zn^{2+}	Co^{2+}	Ni^{2+}	
Plasmid-free derivative	0.6	0.2	0.2	0.6	(106)
czc[+]	2.5	12	20	0.6	(106)
cnr[+]	0.6	0.2–0.4	5	2.5	(106)
cnr Y968	0.6	1.5	8	8	(141)
ncc	2	0.4	12	40	(114)
nre	NR	NR	NR	3	(114)

[a]Minimum inhibitory concentrations (millimoles) were measured on minimal medium as described in Ref. 106.

genes which are involved in the resistance mechanisms: *czc*CBA, *ncc*CBA, and *cnr*CBA. These genes encode an efflux pathway via a cation proton antiporter chemiosmotic system in order to move out the toxic metal which accumulated in the uptake of essential divalent cations (e.g., Mg^{2+}, Mn^{2+}) (137–139).

Sequence data suggest that the efflux proceeds via a tricomponent system made of a pump (cation/proton antiporter system: two protons are taken up for every exported cation) and two accessory proteins (1,140). The pump is clearly an inner membrane–associated protein (*czc*A, *ncc*A, and *cnr*A). The accessory proteins include membrane fusion proteins (*czc*B, *ncc*B, and *cnr*B) and an outer membrane protein (*czc*C, *ncc*C, and *cnr*C). The tricomponent system belongs to a family of export systems which ensure the export of a great variety of molecules including cations, siderophores, virulence factors, proteins, and antibiotics. Direct physiological evidence of efflux was first reported for nickel resistance (*cnr* operon) (141) and is now reported for the *czc* operon (137). Since then, regulatory genes for these operons have been identified and mapped, but the mechanism of the regulation looks very complex and is far from being understood.

Regulatory genes in *cnr* and *ncc* operons seem to be very similar. *cnr*YXH and *ncc*YXH are located directly upstream of the structural genes. Both *cnr*Y and *ncc*Y are believed to be repressor genes (113,114,142), while *cnr*H and *ncc*H show extensive homology with a new family of genes encoding σ factors, called σ_{70} EFP (114,142). The functions which are controlled by these σ factors mainly are responses to environmental signals (143).

No equivalent of *cnr*YXH/*ncc*YXH has been found for the regulation of the *czc* operon. The organization of the *czc* operon differs in many aspects from its *cnr* and *ncc* counterparts.

Figure 6 represents the organization of genes of various heavy metal resistance operons and their genetic relationships. The *czc*, *ncc*, and *cnr* CBA genes have an equivalent in *czr* which encodes chromosomal resistance to zinc and cadmium in *P. aeruginosa* (M-e-Talat, personal communication) and in *hel* which encodes the export of a virulence factor in *Legionella pneumophila* (1,144). In Fig. 6, the *czc* region of pMOL30 is shown as the reference material. Upstream of *czc*CBA lies *czc*N, for which no phenotype is known at the moment, but which has a counterpart in the last gene of the *ncc*YXHCBAN gene cluster (1,114). Downstream of *czc*CBA lie *czc*D, which exhibits homology with yeast resistance genes, and *czc*RS (D. van der Lelie et al., in preparation), which have counterparts in the copper resistance operons *cop* (*P. syringae*) (125,127) and *pco* (*Escherichia coli*) (125) Figure 6 strongly suggests a modular organization of metal resistance genes by successive association of genes, operons, and gene clusters.

Although not a typical soil inhabitant, *P. syringae* is of interest for its association with crop plants and the potential of gene exchange with soil bacteria. Copper-resistant *P. syringae* strains have been isolated from leaves of crop vegetables sprayed with copper-containing fungicides. Copper resistance has not only been reported in *P. syringae* (on a plasmid and the chromosome) but also in *E. coli* plasmids (124–128). The genetic organization of both systems is similar and includes at least four gene products. Proteins CopA and CopC are periplasmic and fix 1 and 11 copper atoms, respectively. CopB is an outer membrane protein involved in the efflux of Cu^{2+} ions to the extracellular medium, and CopD seems to ensure Cu^{2+} transport to the cytoplasm. The role of *cop*D is far from being clear. It is possible that *cop*D ensures that essential Cu^{2+} requirements for the cell are met in the case of maximal expression of the efflux-mediated resistance operon. However, *P. syringae* *cop*+ displays a phenotype which is not observed in *E. coli* *pco*+ (125). Cu^{2+} is sequestered from the medium and accumulates in and around the outer cell membrane. Indeed, *cop*+ colonies become dark blue while the medium color substantially fades. It is not yet clear whether the *cop* operon alone is required to ensure this sequestering or whether additional functions are required. This observation of Cu sequestration points to a crucial point in efflux-mediated metal resistances of how to prevent or minimize the reentry of extruded toxic metal. Does post-efflux processing of metal which would prevent or decrease the reentry occur? In *Alcaligenes eutrophus* CH34, the supernatant of cultures grown in the presence of Cd^{2+} or Zn^{2+} was almost fully depleted of these cations during the stationary phase (1,145,146). This depletion goes hand-in-hand with a pH increase proportional to the initial concentration of Cd^{2+} or Zn^{2+}, with the observation of precipitates of carbonates and hydroxydes of both cations. This removal of cations was observed with a variety of carbon sources including acetate, lactate, and gluconate (146), but also under chemolithotrophic conditions (Cools and Diels, unpublished). Precipitation of Cd and Zn in the cell envelopes was visualized by electron microscopy (146).

The pH increase favors the formation of insoluble carbonates and hydroxydes and may be a direct consequence of the cation/proton antiporter efflux. The bioprecipitation of Cd^{2+} or Zn^{2+} may be associated with a strategy to prevent reentry of toxic cations, as suggested by another observation. Cadmium and Zn seem to induce the production of one protein (25 kD) which is associated with the outer membrane and which can be found in the extracellular medium. The 25 kD protein was found to be associated with the metal precipitates which were harvested from membrane-based bioreactors designed to remove heavy metals from effluents (Diels, personal communication). The genetic basis of this inducible protein is currently under examination. Any strategy to prevent reentry of toxic metals via bioprecipitation may also change the metal speciation. This is of special relevance in soils highly contaminated by heavy metals and may even have some (although probably limited) consequences at the geochemical level. For example, it could allow the partial revegetation of such polluted soils or create the conditions for further colonization by endemic metal-associated plants.

REFERENCES

1. L. Diels, Q. Dong, W. Baeyens, D. van der Lelie, and M. Mergeay, *J. Industrial Microbiol. 14*: 142 (1995).
2. G. Ji and S. Silver, *J. Industrial Microbiol. 14*: (1995).
3. S. Silver and M. Walderhaug, *Microbiol. Rev. 56*: 195 (1991).
4. C. Cervantes, G. Ji, J. L. Ramirez, and S. Silver, FEMS *Microbiol. Rev. 15*: 355 (1994).
5. C. Cervantes and S. Silver, *Plasmid 27*: 65 (1992).
6. N. L. Brown, O. A. Rouch, and B. T. O. Lee, *Plasmid, 27*: 41 (1992).
7. T. K. Misra, *Plasmid 27*: 4 (1992).
8. E. G. Walter and D. E. Taylor, *Plasmid 27*: 52 (1992).
9. R. M. Slawson, M. I. Van Dyke, H. Lee, and J. T. Trevors, *Plasmid 27*: 72 (1992).
10. G. M. Gadd, FEMS *Microbiol. Lett. 100*: 197 (1992).
11. T. Duxbury, *Adv. Microb. Ecol. 185*: (1985).
12. J. J. Cooney and S. Wuertz, *J. Industrial Microbiol. 14*: (1995).
13. J. M. Wood, *Science 183*: 1049 (1974).
14. M. N. Hughes and R. K. Poole, *J. Gen. Microbiol. 137*: 725 (1991).
15. G. Tyler, A.-M. Balsberg Pahlsson, G. Bengtsson, E. Bååth, and L. Tranvik, *Water Air Soil Pollut. 47*: 159 (1989).
16. H. Babich and G. Stotzky, *CRC Crit. Rev. Microbiol. 8*: 99 (1980).
17. H. Babich and G. Stotzky, *Environ. Res. 36*: 111 (1985).
18. Y. E. Collins and G. Stotzky, *Appl. Environ. Microbiol. 58*: 1592 (1992).
19. K. Chandler and P. C. Brooks, *Soil Biol. Biochem. 23*: 1169 (1991).

20. K. Chandler and P. C. Brooks, *Soil Biol. Biochem. 613*: (1992).
21. E. Bååth, *Water Air Soil Pollut. 47*: 335 (1989).
22. A. Nordgren, E. Bååth, and B. Söderström, *Appl. Environ. Microbiol. 46*: 1829 (1983).
23. K. Arnebrandt, E. Barth, and A. Nordgren. *Mycologia 79*: 890 (1987).
24. H. Yamamoto, K. Tatsuyama, and T. Uchiwa, *Soil Biol. Biochem. 17*: 785 (1985).
25. B. H. Olson and I. Thornton, *J. Soil Sci. 33*: 271 (1982).
26. P. Doelman and L. Haanstra, *Soil Biol. Biochem. 11*: 487 (1979).
27. A. Nordgren, T. Kawri, E. Bååth and B. Söderström, *Environ. Poll. 41*: 89 (1986).
28. P. Carter, Master of Science Thesis, University of Ontario, Toronto (1978).
29. L. E. Hallas and J. J. Cooney, *Appl. Environ. Microbiol. 41*: 446 (1991).
30. S. Wuertz, C. M. Miller, R. M. Pfister, and J. J. Cooney, *Appl. Environ. Microbiol. 57*: 2783 (1991).
31. P. Doelman, E. Jansen, M. Michels, and M. van Till, *Biol. Fertil. Soils 17*: 177 (1994).
32. M. Mergeay, *Trends Biotech. 9*: 17 (1991).
33. E. Bååth, *Soil Biol. Biochem 24*: 1167 (1992).
34. P. Corbisier, E. Thiry, and L. Diels, *Environ. Tox. Water Qual. 11*: 171 (1996).
35. M. Diaz-Ravina, E. Bååth, and A. Frostegård, *Appl. Environ. Microbiol. 60*: 2238 (1994).
36. A. Frostegård, A. Tunlid, and E. Bååth, *Appl. Environ. Microbiol. 59*: 3605 (1993).
37. L. Haanstra, J. H. Oude Voshaar and P. Doelman, *Plant Soil 84*: 293 (1985).
38. E. F. De Long, G. S. Wickham, and N. R. Pace, *Science 243*: 1360 (1989).
39. R. I. Amann, W. Ludwig, and K. H. Schleifer, *Microbiol. Rev. 59*: 143 (1995).
40. D. E. Caldwell, D. R. Korber, and J. R. Laurence, *Adv. Microb. Ecol. 12*: 1 (1992).
41. C. B. Strojan, *Oecologia* (Berlo) *32*: 203 (1978).
42. P. J. Coughtey, C. H. Hjones, M. H. Martin, and S. W. Shales *Oecologia* (Berlo) *39*: 51 (1979).
43. B. Freedman and T. C. Hutchinson *Can. J. Bot. 58*: 2123 (1980).
44. S. T. Williams, T. McNeilly, and E. M. H. Wellington, *Soil Biol. Biochem. 9*: 271 (1977).
45. A. J. Friedland, A. H. Johnson, and T. G. Siccama, *Can. J. Bot. 64*: 1349 (1986).
46. J. C. Inman and G. R. Parker, *Environ. Pollut. 17*: 39 (1978).
47. A. Ebregt and J. M. A. M. Boldewijn, *Pl. Soil 47*: 137 (1997).

48. G. Tyler, *Intern. Conf. Heavy Metals Environ.* Toronto, Canada 217 (1975).
49. A. Nordgren, E. Bååth, and B. Söderström, *Soil Biol. Biochem.* 20: 949 (1988).
50. L. Haanstra and P. Doelman, *Soil Biol. Biochem.* 161: 595 (1984).
51. D. Parkinson, T. R. G. Gray, and S. T. Williams, *Methods for Studying the Ecology of Soil Microorganisms,* Blackwell Scientific, Oxford (1971).
52. G. Tyler, Swedish Natural Environmental Protection Board PM542 (1974).
53. B. Lighthart, J. Baham, and V. V. Volk, *J. Environ. Qual.* 10: 543 (1983).
54. J. A. Rother, J. W. Millbank, and J. Thornton, *J. Soil Sci.* 33: 101 (1987).
55. P. C. Brookes, S. P. McGrath, and C. Heijnen, *Soil Biol. Biochem.* 18: 345 (1986).
56. J. Skugins, H. Ö. Nohrstedt, and S. Oden, *Swedish J. Agric. Res.* 16: 81 (1986).
57. C. Wickliff, H. J. Evans, K. R. Carter, and S. A. Russell, *J. Environ. Qual.* 9: 180 (1980).
58. E. G. Rhoden and J. R. Allen, *Commun. Soil Sci. Plant Anal.* 18: 243 (1982).
59. J. P. Obbard and K. C. Jones, *Environ. Pollut.* 79: 105 (1993).
60. J. R. Porter and R. P. Seridan, *Plant Physiol.* 68: 143 (1983).
61. N. D. Mellveen and H. J. Cole, *Phytopathol.* 64: 583 (1974).
62. S. J. Vesper and T. C. Weidensaul, *Water Air Soil Pollut.* 9: 413 (1978).
63. S. P. McGrath, P. C. Brookes, and K. E. Giller, *Soil Biol. Biochem.* 20: 415 (1988).
64. G. Tyler, *Nature 255:* 701 (1975).
65. M. H. Minnich and M. B. McBride, *Plant Soil 91:* 231 (1986).
66. U. Necker and C. Kunze, *Angew. Bot.* 60: 81 (1986).
67. A. Rüling and G. Tyler, *Swedish Natural Environ. Protection Board,* PM 1150.
68. Bhuiya and A. H. Cornfield, *Environ. Pollut.* 7: 161 (1974).
69. F. H. Chang and F. E. Broadbent, *J. Environ. Qual.* 11: 1 (1982).
70. J. B. De Catanzaro and T. C. Hutchinson, *Water Air Soil Pollut.* 24: 153 (1985).
71. M. S. I. Quraishi and A. H. Cornfield, *Environ. Pollut.* 4: 159 (1973).
72. R. F. Morrissey, Ph.D. Thesis, University of Connecticut (1975).
73. C. N. Liang and M. A. Tabatabou, *Environ. Pollut.* 12: 141 (1977).
74. C. N. Liang and M. A. Tabatabou, *J. Environ. Qual.* 7: 291 (1978).
75. D. J. McKenny and J. R. Vriesacker, *Environ. Pollut.* 38: 221 (1985).
76. S. P. Mathur and R. B. Sanderson, *Can. J. Sci.* 58: 125 (1978).
77. W. Maliszewska, S. Dec, H. Wierzbicka, and A. Wozniakowska *Environ. Pollut.* 37: 195 (1985).
78. M. A. Cole, *Appl. Environ. Microbiol.* 33: 262 (1977).
79. P. Doelman and L. Haanstra *Biol. Fertil. Soils 2:* 231 (1986).

80. G. Tyler and L. Westman, Swedish Natural Environment Protection Board, PM 1203 (1979).
81. W. A. Said and D. L. Lewis, *Appl. Environ. Microbiol. 57*: 1498 (1991).
82. D. Springael, L. Diels, and M. Mergeay, *Biodegradation 5*: 343 (1994).
83. H. H. Reber, *Biol. Fertil. Soils 13*: 181 (1992).
84. D. Springael, L. Diels, L. Hooyberghs, S. Kreps, and M. Mergeay, *Appl. Environ. Microbiol. 59*: 334 (1993).
85. A. Wild, *Soils and the Environment: An Introduction*, Cambridge University Press, Cambridge (1993).
86. A. E. Gilmore, *J. Am. Soc. Hortic. Sci. 96*: 35 (1971).
87. G. D. Bowen, M. F. Skinner, and D. I. Bevege, *Soil Biol. Biochem. 6*: 141 (1974).
88. N. R. Benson and R. P. Covey, *Hortic Sci. 11*: 252 (1976).
89. K. Swaminathan and B. C. Verna, *New Phytol. 82*: 481 (1979).
90. L. W. Timmer and R. F. Leyden, *New Phytol. 85*: 481 (1980).
91. A. Gildon and P. B. Tinker, *New Phytol. 95*: 263 (1983).
92. T. A. Dueck, P. Visser, W. H. O. Ernst, and H. S. Chat, *Soil Biol. Biochem. 18*: 331 (1986).
93. D. S. Hayman and M. Tavares, *New Phytol. 100*: 367 (1985).
94. R. Bradley, A. J. Burt, and D. J. Read, *Nature 292*: 335 (1981).
95. R. Bradley, A. J. Burt, and D. J. Read, *New Phytol. 91*: 197 (1982).
96. D. H. Marx and J. D. Artman, *Reclam. Rev. 2*: 23 (1979).
97. R. E. Preve, J. A. Burger, and R. E. Kreh, *J. Environ. Qual. 13*: 387 (1984).
98. M. T. Brown and D. A. Wilkins, *New Phytol. 99*: 101 (1985).
99. K. F. Buglio and H. F. Wilcox, *Can. J. Bot. 66*: 55 (1988).
100. U. Galli, H. Schüepp, and C. Brunold, *Physiol. Plant 92*: 364 (1964).
101. J. Denny and D. A. Wilkins, *New Phytol. 106*: 545 (1987).
102. M. D. Jones and T. C. Hutchinson, *New Phytol. 108*: 461 (1988).
103. J. V. Colpaert and J. A. van Assche, *New Phytol. 123*: 325 (1993).
104. A. F. W. Morselt, W. T. M. Smits, and T. Limonard, *Plant Soil 96*: 417 (1986).
105. M. Ling-Lee, G. A. Chilvers, and A. E. Ashford, *New Phytol. 15*: 551 (1975).
106. M. Mergeay, D. Nies, H. G. Schlegel, J. Gerits, and F. Van Gysegem, *J. Bacteriol. 162*: 328 (1985).
107. L. Diels and M. Mergeay, *Appl. Environ. Microbiol. 56*: 1485 (1990).
108. R. R. Brooks, *Serpentine and Its Vegetation: A Multidisciplinary Approach*, Groom Helm, London (1987).
109. A. J. M. Baker and R. R. Brooks, *Biorecovery 1*: 81 (1989).
110. R. D. Stoppel and H. G. Schlegel, *Appl. Environ. Microbiol. 61*: 2276 (1995).
111. T. Schmidt and H. G. Schlegel, *FEMS Microbiol. Ecol. 62*: 315 (1989).

112. H. G. Schlegel, J. P. Cosson, and A. M. Baker, *Bot. Acta 104*: 18 (1991).

113. H. Liesegang, K. Klemke, R. A. Siddiqui, and H. G. Schlegel, *J. Bacteriol. 175*: 767 (1993).

114. T. Schmidt and H. G. Schlegel, *J. Bacteriol. 176*: 7045 (1994).

115. A. D. Stoppel, M. Meyer, and H. G. Schlegel, *Bio. Metals 8*: 70 (1995).

116. F. Malaisse, Grégoire, J., Brooks R. R., Morrison R. S., and Reeves R. D., *Science 199*: 887 (1978).

117. A. Baker and R. Brooks, *Biorecovery 1*: 81 (1989).

118. F. Malaisse, R. Brooks, and A. Baker, *Belg. J. Bot. 127*: 3 (1994).

119. M. Höfte, S. Buysens, N. Koedam, and P. Cornelis, *Bio Metals 6*: 85 (1993).

120. M. Höfte, Q. Dong, S. Kourambas, V. Krishnapillai, D. Sherratt, and M. Mergeay, *Mol. Microbiol. 14*: 1011 (1994).

121. R. W. Olafson, W. D. McCubbin, and C. M. Kay, *Biochem. J. 251*: 691 (1988).

122. N. J. Robinson, A. Gupta, A. P. Fordham-Skelton, R. R. D. Croy, B. A. Whitton, and J. W. Huckle, *Proc. R. Soc. London 242*: 241 (1990).

123. J. W. Huckle, A. P. Morby, J. S. Turner, and N. J. Robinson, *Mol. Microbiol. 7*: 177 (1993).

124. N. L. Brown, D. A. Rouch, and B. T. O . Lee, *Plasmid 27*: 41 (1992).

125. D. A. Cooksey, *Mol. Microbiol. 7*: 1 (1993).

126. N. L. Brown, B. T. O. Lee, and S. Silver. In *Metal Ions in Biological Systems*, Vol. 30 (Sigel H. and H. Sigel, eds.) Marcel Dekker, New York, p. 405 (1994).

127. M. A. Mellano and D. A. Cooksey, *J. Bacteriol. 170*: 2879 (1988).

128. A. O. Summers and S. Silver, *J. Bacteriol. 112*: 1228 (1972).

129. T. K. Misra, *Plasmid 27*: 4 (1992).

130. C. Cervantes, H. Ohtake, L. Chu, T. K. Misra, and S. Silver, *J. Bacteriol. 172*: 287 (1990).

131. A. Nies, D. Nies, and S. Silver, *J. Biol. Chem. 265*: 5648 (1990).

132. H. Ohtake, C. Cervantes, and S. Silver, *J. Bacteriol. 169*: 3853 (1987).

133. A. O. Summers and Jacoby, *Antimicrob. Agents Chemother. 13*: 637 (1978).

134. G. Lloyd-Jones, D. A. Ritchie, and P. Strike, *FEMS Microbiol. Lett. 81*: 19 (1991).

135. E. G. Walter, J. H. Weiner, and D. E. Taylor, *Gene 101*: 1 (1990).

136. E. G. Walter, C. M. Thomas, J. P. Ibbobson, and D. E. Taylor, *J. Bacteriol. 170*: 1111 (1991).

137. D. H. Nies, *J. Bacteriol. 177*: 2707 (1995).

138. D. H. Nies, A. Nies, L. Chu, and S. Silver, *Proc. Natl. Acad. Sci. USA 88*: 7351 (1989).

139. D. H. Nies and S. Silver, *J. Bacteriol. 171*: 896 (1989).

140. Q. Dong and M. Mergeay, *Mol. Microbiol. 14*: 185 (1994).
141. C. Sensfus and H. G. Schlegel, *FEMS Microbiol. Lett. 55*: 295 (1988).
142. J. M. Collard, A. Provoost, S. Taghavi, and M. Mergeay, *J. Bacteriol. 175*: 779 (1993).
143. M. A. Lonetto, K. L. Brown, K. E. Rudd, and M. J. Buttner, *Proc. Nat. Acad. Sci. USA 91*: 7573 (1992).
144. M. McClain, M. C. Hurley, J. Arroyo, M. Wolf, and N. C. Engleberg, *Infect. Immun. 62*: 4075 (1994).
145. J. M. Collard, P. Corbisier, L. Diels, Q. Dong, C. Jeanthon, M. Mergeay, S. Taghavi, D van der Lelie, A. Wilmotte, and S. Wuertz, *FEMS Microbiol. Rev. 14*: 405 (1994).
146. L. Diels, S. Van Roy, K. Somers, I. Willems, W. Doyen, M. Mergeay, D. Springael, R. Leysen, *J. Membr. Sci. 100*: 249 (1995).
147. R. Margesin and F. Schinner, *J. Basic Microbiol. 38*: 269 (1996).

21

Agricultural Soil Manipulation: The Use of Bacteria, Manuring, and Plowing

ERIC G. BEAUCHAMP and DAVID J. HUME   University of Guelph, Guelph, Ontario, Canada

1. INTRODUCTION

Life in soils, especially those used for agricultural purposes, continues to be a matter of concern. It is important that soil productivity be sustained or even enhanced for economical reasons and food production for an ever-increasing population. The use or manipulation of soils for agricultural purposes normally involves soil disturbance and addition of various substances to enhance crop production. The objective of this chapter is to examine some of the effects of these agronomic practices on microbial populations and activities in the soil. The addition of some selected microbes to soil will also be examined for their importance in enhancing crop productivity. Thus the importance of N_2 fixers and soil phosphorus solubilizers will be discussed. A major focus will be placed on the effects of tillage practices and the additions of manures and fertilizers to microbial communities.

Along with the various agronomic practices and substances applied to agricultural soils, the soil environment, including temperature, oxygen supply, water content, and nutrient availability, has a large effect on microbial communities and their activities. Moreover, these conditions constantly change in field soils in contrast to soil in most laboratory incubation setups. Soil structure and compaction variation usually result in very different physical conditions. Water content may be extremely variable both spatially and temporally as a result of evapotranspiration and precipitation events. Soil temperatures vary from well below freezing to higher than 45°C. Crop growth cycles contribute residues or C substrate and nutrients for microorganisms in a generally predictable time sequence.

2. TILLAGE

Tillage of soils is usually done to prepare a seedbed, although other purposes may include weed control and incorporation of residues, manures, fertilizers, or herbicides. More recently, many farmers have been changing to reduced or mulch tillage practices. Although soil disturbance may be greatly reduced, some normally occurs during the planting operation. For the purposes of this chapter, the two extremes of tillage will be emphasized, namely, conventional and no-till practices. Conventional tillage involves breaking up of the soil with a moldboard plow turning over the upper 10–20 cm, usually 15 cm, followed by disking and/or harrowing to prepare a seedbed. The no-till practice involves direct planting of seeds with minimal soil disturbance caused by the planting equipment. Thus, conventional tillage results in the incorporation plus some mixing of aboveground crop residues, whereas no-till results in most of the residues remaining on the soil surface. The benefits of no-till are usually considered to be less time input, lower energy consumption, and reduced soil erosion, all resulting in greater profitability.

Tillage of a soil for the first time from a virgin to a cultivated state resulted in an increase in the numbers of microorganisms of 1.5 times in one study (1). The percentages of bacteria and actinomyctes were relatively similar, but there was a general decline in fungi. More recent comparisons deal with the kind of tillage practice, e.g., conventional tillage vs. no-till. For example, Doran (2) showed that the microbial numbers with these tillage treatments depended on the depth of sampling (Table 1). This is not surprising inasmuch as the no-till practice results in crop residue concentration at the soil surface that is concomitant with a greater microbial activity. The relatively lower availability of crop residues in the 7.5–15 cm zone with no-till treatment obviously resulted in a relative decrease in numbers of aerobic microorganisms. Anaerobic bacteria increased in both layers probably as a result of greater soil compaction and lower air-filled porosity, resulting in decreased availability of oxygen.

Soil conditions for microorganisms are influenced by water content, soil compaction, structure, and texture. The influence of water content is shown in Fig. 1.

Table 1. Ratio of Microbial Groups Populations in
No-Till to Conventional Till Soils at Two Depths[a]

| | NT:CT ratio | |
Microbial group	0–7.5 cm	7.5–15 cm
Fungi	1.57	0.76
Actinomycees	1.14	0.98
Aerobic bacteria	1.41	0.68
NH_4^+ oxidizers	1.25	0.55
NO_2^- oxidizers	1.58	0.75
Facultative anaerobes	1.57	1.23
Denitrifiers	7.31	1.77

[a]NT, no-till; CT, conventional till.
From Ref. 2.

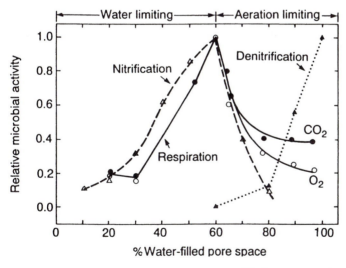

Figure 1. The relationship between water-filled pore space in soil and relative
microbial activity with respect to nitrification, dentrification, and respiration (O_2
uptake and CO_2 production). (From Ref. 3.)

Soil porosity or pore volume as a proportion of the total volume is determined by
the bulk density (BD) and soil particle density (PD) according to the following
equation. The densities of various soil mineral particles average about 2.65 Mg
(megagrams) m^{-3}.

$$\text{Pore volume} = \left(1 - \frac{BD}{PD}\right)$$

The water content of soil determines the water-filled porosity (WFP), the proportion of soil pore volume filled with water, and consequently the air-filled
porosity. The preceding equation indicates that soil bulk density or soil compaction will affect soil porosity. The rate of diffusion of air into the soil is affected by the continuity of air-filled pores, the rate of gas molecule movement
(diffusion), and the microbial demand for oxygen. The diffusion of air through
water-filled pores is about 10,000 times slower than that through air-filled pores
with similar pore continuities. It is believed that, in addition to diffusion rates, O_2
availability in localized zones in the soil is partly dependent on the rate of O_2
consumption by microorganisms. No-till soils may have higher water content
and bulk density, thus having a higher water-filled porosity than conventionally
tilled soil (Table 2). Thus one may expect the relative activities of microbial
groups to differ with tillage practices.

According to Fig. 1, aerobic microbial activity increases with water content
until air diffusion is restricted (3). Maximum rates of respiration, nitrification,
and mineralization occur at the highest water content at which soil aeration remains nonlimiting. At about 60% WFP, there is a sharp transition to O_2-limiting
conditions. Respiration begins to decrease and denitrification begins and increases exponentially, reaching a maximum when soil approaches saturation. At

Table 2. Bulk Density, Soil Water Content, and Water-Filled Porosity for
Blount Silt Loam That Received Conventional or No-Tillage Treatments for 9
Years with Continuous Maize Cropping

Tillage treatment	Depth (cm)	Bulk density (Mg m^{-3})	Water content (weight basis) (%)	Water-filled pore space (%)
No-till	0–7.5	1.46	20.2	65.4
	7.5–15	1.50	19.6	69.2
Conventional[a]	0–7.5	1.35	12.3	36.5
	7.5–15	1.39	17.3	50.6

[a]Spring plowing depth = 20 cm; spring disking depth = 10 cm
From Ref. 3.

the same time, nitrification decreases rapidly because of limited O_2 supply above 60% WFP.

Some recent studies have focused on the effects of tillage practices on the microbial biomass carbon (MBC). Microbial biomass C is relatively easy to measure and may allow an assessment of tillage effects on microbial numbers and, indirectly, activities. The MBC measurement essentially involves the fumigation of soil with chloroform to kill microorganisms. The quantity of C extracted is then compared with that from nonfumigated soil. The difference between these two measurements is considered to be microbial C, taking into account the proportion of microbial C that is not extractable (4). The microbial biomass nitrogen (MBN) content can be estimated in the same manner.

It has been observed that tillage of a virgin soil for annual crops like maize (*Zea mays*) or cereals results in a gradual decrease in organic matter that is rapid at first but slows down with time. After many years of cultivation (e.g., 50–70), the organic matter content in the soil levels off, indicating that annual C additions from crop residues or manures are similar to C losses from microbial decomposition. The early, rapid release of organic C after initial soil tillage of a virgin soil is thought to be due to an increase in the accessibility of organic matter to microorganisms by mixing and greater aeration. Soil erosion losses also contribute to the organic matter decrease.

The size of the soil microbial biomass is dependent on the cropping regime. An example of this effect with continuous wheat (*Triticum aestivum*), wheat–fallow, and cereal–grass rotations on a solonetzic (saline) soil is shown in Table 3 (5). It is noteworthy that the wheat–fallow rotation with a crop every second year had the lowest MBC and MBN values in both horizons. The cereal–grass rotation, with greatest return of crop residues to the soil, resulted in the greatest MBC and MBN values.

Table 3. Microbial Biomass C and N Content in the A and B_{nt} Horizons of Solonetzic Soil on Which Continuous Wheat, Wheat–Fallow, and Cereal–Grass Rotations Were Grown[a]

Soil horizon	Continous wheat		Wheat–fallow		Cereal–grass	
	C	N	C	N	C	N
			(kg ha^{-1})			
Ap	171 b	30 b	69 c	11 c	288 a	48 a
B_{nt}	74 b	14 b	74 b	13 b	182 a	43 a

[a]Data for microbial biomass C or N within soil horizons followed by different letters are significantly different ($P = 0.05$).
From Ref. 5.

In another study on a Prince Edward Island (Canada) soil, the effects of conventional tillage and no-till practices on the size and activity of microbial populations were measured (Table 4). The microbial activity (respired CO_2-C production) as well as MBC and MBN were generally greater in the 0–5 cm depth with the zero tillage practice. Alternatively, with the conventional tillage practice, these parameters were generally greater in the 5–10 cm depth probably because of incorporation of residues to that depth during plowing. This study showed that tillage-induced changes in crop residue incorporation, root activity, and soil moisture regimen affect microbial sizes and activities in soil.

Angers et al. (6) studied soil organic C and MBC levels after moldboard plowing (to 15 cm) of a 20-year-old timothy crop (*Phleum pratense*) and then grew maize for 3 successive years. Plowing to a depth of 15 cm resulted in a redistribution of the organic C (Table 5). There was a decrease in total soil organic C down to 24 cm shortly after plowing, probably due to decomposition of easily decomposable residues. Thereafter, during the 3-year period with maize, there was little relative change in total organic C. Thus, it appeared that crop residue C returned to the soil was sufficiently great to maintain soil organic C even though the tops of maize plants were harvested. It has been estimated that about 6000 kg ha^{-1} of aboveground residues are needed annually to maintain the organic C level in a soil in maize-growing areas of the temperate regions (7).

It was evident that the MBC, as a percentage of the total organic C, decreased somewhat, especially in the third year. This change has been suggested to portend a decrease in the organic C level of the soil (8). If MBC provides a prediction of such a change, this would be a distinct advantage because measurements of changes in total soil organic C during short periods are quite insensitive for this purpose. Thus the effects of management practices such as tillage may be predictable on a relatively short-term basis by calculation of an index of organic

Table 4. Microbial Biomass C and N and Respiration (CO_2-C Production) at Two Depths in a Prince Edward Island Soil with Conventional Tillage and Zero Tillage Practices for Cereal Grain Crop

Soil depth (cm)	Conventional tillage			Zero tillage		
	Respiration	MBC	MBN	Respiration	MBC	MBN
			(kg ha^{-1})			
0–5	35	111	23	68	182	35
5–10	41	247	46	20	164	32

Differences between tillage treatment data for each parameter within each depth are significant ($P = 0.05$). MBC, microbial biomass carbon; MBN, microbial biomass nitrogen. From Ref. 5.

Table 5. Average Contents of Microbial Biomass Carbon at Different Depths and Total Organic C and MBC/Organic C Ratio of Two Loam Soils with Timothy (*Phleum pratense*) Meadow and After Plowing and 3 Years of Maize

Depth (cm)	Meadow	After plowing	Year 1 (kg C ha⁻¹)	Year 2	Year 3
			Maize		
0–6	484	245	307	246	204
6–12	279	414	262	303	266
12–18	177	275	345	268	217
18–24	212	133	129	187	190
0–24	1152	1067	1043	1004	877
Total organic C (0–24 cm)	59,200	53,300	51,700	51,800	53,200
MBC/organic C (%)[a]	1.9	2.0	2.0	1.9	1.6

[a]MBC, microbial biomass carbon.
From Ref. 6.

matter dynamics, i.e., the proportion of the soil organic matter C in the microbial biomass (9). One adverse effect of plowing the timothy crop and growing another crop (maize) was a reduction of the mean weight diameter of water-stable aggregates. This was attributed to a reduction of fungal hyphae and the absence of fine roots normally present with a timothy crop that act as mechanical binding agents and are disrupted by tillage practices.

Precautions need to be taken with respect to time and location of soil sampling in relation to location of crop plants in the field for MBC and MBN determinations. It has been shown that the distribution of MBC and MBN values differs at different depths (Tables 3 and 4). It has also been shown that distance from maize plants, for example, may affect these values (10). This is probably related to the distribution of roots with their associated microbial rhizosphere populations.

Microbial biomass C levels change during the growing season. An example with averaged values for winter wheat and spring barley grown (*Hordeum vulgare*) in rotation with peas (*Pisum sativum*) on a silt loam soil in the northwest United States is shown in Fig. 2 (11). The changes over the season were particularly great in the 0–5 cm layer with the no-till practice. Substantial changes in the MBC during the growing season also have been observed with wheat in the United Kingdom (12) and maize in Ontario, Canada (10).

In summary, it is evident that the size of the microbial population in soil tilled in different ways is dependent on soil physical conditions, location with respect

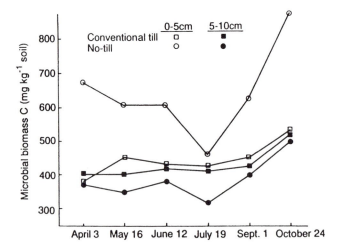

Figure 2. Changes in microbial biomass carbon in two soil depths with conventional and no-tillage practices. (From Ref. 11.)

to depth and distance to plants, and time or stage of crop growth during the growing season. These factors need to be taken into account when interpreting soil manipulation effects.

3. TILLAGE AND PHOSPHORUS AVAILABILITY

Phosphorus absorption by maize seedlings is greater in undisturbed than in disturbed soil. Soil disturbance reduces mycorrhizal hyphae colonization of soils, which, in turn, reduces the extent of exploration for nutrients in the soil. The hyphae of the mycorrhizae extend from the interior of roots of many crops into the soil and effectively increase the surface area contact with soil while deriving nutrients from the host plant in a symbiotic relationship (13). It is noteworthy, however, that some crop species such as oilseed rape (*Brassica napus*) and spinach (*Spinacea oleracea*) do not have mycorrhizal relationships.

Soil disturbance through tillage may affect the phosphorus status of maize seedlings, as shown in the following example (14). Four soil treatments in a field trial were imposed as follows: (1) SD = severe disturbance by passing soil through a 5 mm sieve, (2) CT = conventional tillage, (3) NT = no tillage, and (4) NTHP = no tillage with hand planting by removing a soil core with a cork borer, placing a seed in the hole, and replacing the soil core. A mechanical planter was used for treatments 1, 2, and 3. It was evident that the greater the tillage (soil disturbance), the lower the P concentration and dry biomass accumulation by the corn seedlings (Table 6). Arbuscular (finely branched hyphae in root cortex),

Table 6. Dry Biomass and P Concentrations in Maize Seedling Shoots at the Five- to Six-Leaf and Seven- to Nine-Leaf Stages in Response to Tillage Treatments on a P-Deficient Soil.

Tillage treatment[a]	Five- to six-leaf stage		Seven- to nine-leaf stage	
	DM (g plant^{-1})	P conc (mg g^{-1})	DM (g plant^{-1})	P conc (mg g^{-1})
SD	0.21	2.1	0.9	2.4
CT	0.26	2.7	1.4	2.5
NT	0.23	3.0	1.8	3.0
NTHP	0.33	3.3	2.2	2.8
LSD ($P = 0.05$)	0.10	0.5	0.9	0.5

[a]SD, severe disturbance; CT, conventional tillage; NT, no-tillage; NTHP, no tillage and hand planting; LSD, least significant difference; DM, dry matter; P conc., phosphorus concentration.
From Ref. 14.

vesicular (expanded hyphae between root cells), and other hyphae colonization of the roots increased from the five- to six-leaf to the seven- to nine-leaf stage but was only slightly affected by the tillage treatment. Increased P absorption was due to greater numbers of extraradicle hyphae in contact with soil. Soil disturbance resulted in hyphae disruption. In undisturbed soil, the network of hyphae in the soil apparently remains more intact, consequently increasing root colonization and enhancing P supply for the plant.

4. PHOSPHORUS-SOLUBILIZING MICROORGANISMS

Phosphorus-solubilizing (PS) microorganisms are diverse and are represented by bacterial, actinomycete, and fungal species in soils. The PS bacteria constitute 0.5% of the microbial population, whereas PS fungi represent 0.1% in Canadian prairie soils (15). The PS bacteria outnumber PS fungi 2-fold to 150-fold in soils. Interest in the ability of these microorganisms to solubilize P minerals is based on the enhancement of P availability to plants and reduction of fertilizer P requirements.

Mineral P compounds are solubilized by organic acids excreted by PS microorganisms. The organic acids dissolve phosphate-containing minerals and/or chelate with cationic partners of the phosphate ions. Lactic, citric, malic, oxalic, and succinic acids are among those released by these organisms and all have chelation properties. These microorganisms may also have root growth–enhancing properties as well as capabilities to solubilize other plant nutrients such as Cu and Zn.

Table 7. Yields of Wheat Grain and Straw and P Uptake with *Penicillium bilaji* and Phosphate Fertilizer Additions to a P-Deficient Soil

Treatment	Yields		P uptake
	Grain	Straw	
		(g 0.9m^{-2})	
P. bilaji	207	237	0.95
Rock phosphate	167	204	0.73
Rock phosphate + *P. bilaji*	198	223	0.82
Monoammonium phosphate	238	296	1.12
Monoammonium phosphate + *P. bilaji*	231	285	1.09
Control	163	202	0.71

From Ref. 16.

Phosphorus-solubilizing organisms must be able to demonstrate significant solubilization of rock phosphate or increased availability of phosphatic fertilizers. Identification of these organisms is obtained by the clearing of an agar medium containing precipitated calcium phosphate in their vicinity.

An example of enhanced P availability to wheat plants by *Penicillium bilaji* was demonstrated by Kucy (16). Some data from a field trial on a P-deficient soil are shown in Table 7. The recommended rate of 20 kg P ha^{-1} was applied in a band with the wheat seeds at planting time. The data indicate that *P. bilaji* increased wheat grain and straw yields as well as P uptake. The addition of rock phosphate (low P solubility) had little effect while monoammonium phosphate (commercial fertilizer with high soluble P content) resulted in the highest yields and P uptake. The addition of *P. bilaji* to these phosphate sources did not increase P availability but increased release of P from soil.

Many soils in the world have a very low P status so the addition of a PS organism and a modest application of rock phosphate may result in enhanced P availability. If this is so, then the processing of phosphate rock into phosphate fertilizers may not be necessary; a source of P for crops at lower cost and requiring less technological sophistication would result.

5. MICROORGANISMS TO ENHANCE NITROGEN FIXATION

5.1 Introduction

The most important dinitrogen (N_2) fixers are the bacterial genera *Rhizobium* (fast-growing, with doubling times of less than 6 hours in broth culture) and *Bradyrhizobium* (slow-growing, with doubling times of more than 6 hours), which will be collectively referred to here as rhizobia. These bacteria are involved in symbiotic relationships with legume plants and some nonlegumes. In symbiosis, the host plant provides energy, carbon skeletons, and nodules which restrict the availability of oxygen. Rhizobia convert atmospheric N_2 to forms of N the plant can use.

Crop plants that fix N_2 have fairly specific requirements in terms of symbiotic rhizobia. Some effective crop–rhizobia combinations are shown in Table 8. When a crop capable of N_2 fixation is planted, the compatible rhizobia are usually added to the seed or soil in a process called inoculation. The material used to introduce the bacteria is called an inoculant. Most inoculants are made by introducing rhizobia into or onto a peat-based carrier. Sedge peat, which is usually found in ancient peat deposits under more recent sphagnum peat, is the most common carrier. Other carriers which have been used include ground rice husks, compost made from bagasse (residue after sugar extraction from sugarcane), clay granules, and vermiculite (expanded-layer clay mineral). Various kinds of inoculants and their performance were reviewed by Smith (17). Table 9 contains a list of the major types of inoculants currently in use. The sterile-carrier, powdered peat inoculants have been outperforming the other types of inoculants. Response of crops to inoculation depends on two major factors: (1) whether the required rhizobia are present in the soil and (2) available N level in the soil.

The rhizobia for many crops had to be introduced when the crop was first grown. For example, populations of *Bradyrhizobium japonicum*, the main symbiont for soybean, did not naturally exist in soils of Europe or the Americas, and inoculation is imperative for good nitrogen fixation and yield when the crop is first grown. After initial inoculation, the rhizobia may persist indefinitely or die out, depending on the conditions. With *B. japonicum*, persistence is favored by soils of medium texture (loams), near-neutral pH, and moderate temperatures. Disappearance of soybean rhizobia from soils is most frequent in sandy soils subjected to hot, dry conditions (18); in locations with extremely cold winters; with waterlogging; and in acidic soils (with pH levels less than 6). Other rhizobia, such as those which nodulate the forage legume birdsfoot trefoil (*Lotus corniculatus*), survive well in acid soils but do not persist in neutral soils when birdsfoot trefoil is not being grown. Rhizobia multiply in the rhizosphere and nodules of the host plant when it is grown and then survive saprophytically until the host crop is grown again in the rotation. Other nonlegume crop plants may support increases in populations of rhizobia. Inoculation of wheat, oat, lupin, and pea seeds with *B. japonicum* improved nodulation in following soybean crops (19). Weaver et al. (20), in Iowa, found about 80% of fields previously cropped to soybean had adequate *B. japonicum* populations of more than 10^4 cells g^{-1} soil. In soils which had not grown a soybean crop for more than 12 years, at least 5% still retained populations greater than 10^5 cells g^{-1}, indicating long survival times under these conditions.

Table 8. Major Nitrogen–Fixing Crops, Their Rhizobial Symbionts, and Their Relative Ability to Fix Atmospheric Nitrogen

Crop	Genus and species	Rhizobia	Relative amounts of N_2 fixed[a]
Alfalfa	*Medicago sativa*	*Rhizobium meliloti*	High
Birdsfoot trefoil	*Lotus corniculatus*	*Rhizobium loti*	Intermediate
Black gram	*Vigna mungo*	*Rhizobium* species	High
Chickpea	*Cicer arientinum*	*Rhizobium* species	Variable
Cowpea	*Vigna unguiculata*	*Rhizobium* species	High
Faba bean	*Vicia faba*	*Rhizobium-leguminosarum* biovar *vicae*	High
Field bean	*Phaseolus vulgaris*	*Rhizobium leguminosarum* biovar *phaseoli* R. etli biovar *phaseoli* R. tropici	Low
Green gram	*Vigna radiata*	*Rhizobium* species	High
Lentil	*Lens culinaris*	*Rhizobium leguminosarum* biovar *viceae*	Variable
Lima bean	*Phaseolus lunatus*	*Rhizobium* species	Low
Lupin	*Lupinus spp.*	*Bradyrhizobium* species	High
Pea	*Pisum sativum*	*Rhizobium leguminosarum* biovar *vicae*	High
Peanut (groundnut)	*Arachis hypogaea*	*Rhizobium* species	High
Pigeon pea	*Cajanus cajan*	*Bradyrhizobium* species	Intermediate
Red clover	*Trifolium repens*	*Rhizobium leguminosarum* biovar *trifolii*	High
Soybean	*Glycine max*	*Bradyrhizobium japonicum* B. elkanii *Rhizobium fredii*	Intermediate

[a]Actual amounts of N_2 fixed will vary with available soil N and environmental conditions. Under average soil conditions and favorable temperature and moisture conditions, high denotes approximately 150–200 kg N_2 fixed per year; intermediate denotes 50–100 and low denotes 25–50 kg N_2 fixed per year. Variable denotes a broad range from 25–200 kg N_2 fixed per year, with variability mainly resulting from the compatibility of the host legume with the specific rhizobial strains in the soil.

The response to inoculation of legumes in rhizobia-free soil depends considerably on the level of available N in the soil. Merrien et al. (21) reported that in France, N_2 fixation supplied 10% to 80% of the plant N in fields, with the greatest fixation occurring with low amounts of soil N assimilation. In soils with high levels of available N, nodule formation is decreased or even inhibited at very high N levels. Lack of nodulation, therefore, does not necessarily mean the inoculation was unsuccessful. When a pale green color in the plant, symptomatic of N deficiency, accompanies a lack of nodulation, then the inoculation should be considered unsuccessful.

5.2 Factors Affecting the Performance of Inoculants

Effective nodulation results when efficient strains of the appropriate rhizobia are provided in sufficient numbers and in a vigorous state. Rhizobial species include many strains, which may vary in ability to nodulate and to fix N_2 efficiently with the host plant. Commercial inoculants commonly contain effective strains that have been tested in both indoor and field trials prior to their use as inoculant strains. Good strains should be effective with all the crop species and cultivars that they are likely to encounter.

In areas of the world where crops or their related species have been grown for centuries, the rhizobial strains nodulating the crop are unselected and nitrogen fixation often is poor. In northern China, considered the center of origin for soybean, the crop nodulates well but the effectiveness of individual nodules ranges from high to totally ineffective. In this situation, N_2 fixation is so poor that fertilizer N is added in order to achieve high soybean yields. Techniques are needed to replace ineffective strains with effective ones, but reliable methods to do so are not available for the situations in which rhizobia persist in soils. Often N_2 fixation is much higher where initial inoculation of crops was necessary.

When a rhizobial inoculant is introduced into a field soil free of the same rhizobia, crop responses generally increase with the number of viable rhizobia applied per seed. The response in seed yield to different numbers of soybean rhizobia applied with the seed is shown in Fig. 3. The original inoculant, which provided about 10^6 rhizobia per seed, was diluted with sterile peat to achieve 10^5, 10^4, and 10^3 rhizobia per seed (22). At 10^3 rhizobia per seed, yields were not different from the uninoculated controls. At higher numbers of rhizobia, yields increased sharply. At some level of inoculation beyond 10^6 per seed, other factors would be expected to become limiting and the response would level off.

When various commercial inoculants are used, their performance differs even at the same number of viable cells per seed. Vigor and nutritional status of the rhizobia also influence inoculant performance. Effective inoculant carriers should provide adequate nutrition, pH, oxygen, and moisture for the rhizobia so they are multiplying and vigorous when inoculation occurs (17).

Table 9. Major Types of Rhizobial Inoculants and Their Effects on Grain Yields of Soybeans Grown in Fields Containing No *Bradyrhizobium japonicum* Populations

Type of inoculant	Product name	1988	1989	1990	1991	1992	1993	6 Yr average	Increase over control (%)
			(grain yield, t/ha^{-1} at 14% moisture)						
Sterilized powdered peat	Grip	3.17	2.30	2.82	3.09	2.06	2.75	2.70	26.7
	Hi-Stick	3.09	2.16	3.01	301	2.27	2.73	2.71	27.4
	Pulse R					1.80	2.56		
	Sow-fast						2.62		
Conventional powdered peat	Nitragen Powdered Peat	3.20	1.91	3.16	2.74	1.79	2.81	2.60	22.1
	Urbana Powdered Peat	2.92	1.77	2.98	2.67	1.83	2.60	2.46	15.4
	Urbana Rhizostick						2.58		

Granular peat	Nitragen Soil Inoculant Granular	3.03	1.94	2.92	3.04	2.05	2.25	2.58	20.9
	Urbana Granular	3.13	1.97	2.63	3.04	2.26	2.56	2.56	20.2
Liquid	Cell Tech			3.04	2.48	1.99	2.74	2.56	16.9[a]
	Liqui-Prep			3.17	2.07	1.66	2.37	2.32	5.8[a]
Preinoculated coating	Hi-Coat			2.98	2.43	1.60	—		
Uninoculated control		2.61	1.42	2.65	2.01	1.63	2.45	2.13	—

[a] These values are based on 4-year averages. All trials were conducted with four replications at each of two locations in each year, except in 1992, when only one location was used.
From D. J. Hume, unpublished data.

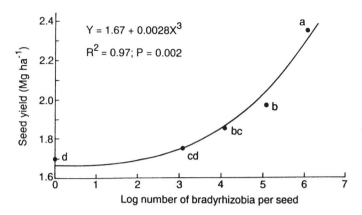

Figure 3. Relationship between soybean seed yield and log number of *Bradyrhizobium japonicum* applied per seed as full-strength inoculated and successive 10-fold dilutions with sterile powdered peat in 1989. Values with the same letter are not different at $P < 0.05$ by Duncan's New Multiple Range Test. (From Ref. 22.)

5.3 Rhizobium Nodulation of Nonlegumes

Rhizobium bacteria cause effective nodulation on the nonlegume *Parasponia* (23). Recently, techniques to remove the cell walls of root hairs have been developed. This allows entry of rhizobia, which has induced nodulelike structures in oilseed rape, rice (*Oryza sativa*), and wheat (24). These results may lead to eventual N_2 fixation by nonlegume crops, but such fixation is not possible at present.

5.4 N₂ Fixation by Other Organisms

There are over 170 species of woody dicotyledonous plants which are nodulated by the bacterium *Frankia* (25), including alder (*Alnus* species), a forest tree important in recolonization of forests because of its N_2-fixation capability. The ability to isolate *Frankia* occurred quite recently (26,27) and microbiological understanding of it is much less developed than for rhizobia.

    The cyanobacterium *Anabaena azollae* has an important symbiotic relationship with azolla, an aquatic fern. In southeastern China and northern Vietnam, azolla has been used as a green manure crop to provide N for rice (28). Inoculation usually involves broadcasting azolla plants from a nursery onto a prepared field.

    There are a number of free-living, N_2-fixing bacteria, called diazotrophs, which live in loose association with roots of legumes and nonlegumes. In the tropics, *Azospirillum* species carry on associative N_2 fixation with maize, sugarcane, and other grasses. Associative N_2-fixing bacteria are mainly free-living organisms which proliferate near or within the roots because carbon is available for growth and fix N_2 in the presence of reduced oxygen concentrations (29). Other diazotrophic genera include *Azotobacter, Beijerinckia, Bacillus, Klebsiella, Enterobacter*, and *Clostridium*.

6. CHEMICAL FERTILIZERS AND SOIL MICROORGANISMS

Additions of chemical fertilizers or soil amendments, if needed, usually increase crop dry matter production. Normally, the greater the harvested yield, the greater the quantity of crop residues returned to the soil. Decomposition of these residues normally promotes microbial populations and activity in the soil. Thus, chemical fertilizers have a beneficial effect on the heterotrophic microbial community, which, in turn, will exert a positive effect on soil structure, improve nutrient availability, and enhance soil humus content.

It is noteworthy that transformations of some fertilizers in soil are dependent on soil microorganisms. Several examples include the nitrification of ammoniacal fertilizers, the production of phosphatase that catalyzes the hydrolysis of polyphosphate fertilizers, and the production of urease that catalyzes urea hydrolysis to produce ammonium carbonate.

Chemical fertilizers can create hostile environments for soil microorganisms, especially in the zone close to the fertilizer particle or granule. This hostile environment could be caused by increased salt concentration in the soil solution, imbalance of ionic nutrients, low pH, high pH, or a high concentration of nitrite. Fertilizer granules are normally 1 to 3 mm in diameter. When such granules of potassium chloride dissolve, the local soil solution concentration of salt may be high enough to affect microbial activity adversely (30). Also urea granules hydrolyse, often causing soil pH to exceed 9 and consequently produce toxic levels of NH_3. The nitrification of ammoniacal fertilizer sometimes results in a buildup of nitrite (NO_2^-) that may inhibit nitrite oxidizers. Other fertilizers such as triple superphosphate (46% P_2O_5) may lower the soil pH to about 1.5 close to the granule, resulting in the increase of high concentrations of Al and Mn in the local soil solution. Other phosphate fertilizers have a lesser effect on the local soil pH (e.g., monoammonium phosphate, pH 4.7, or diammonium phosphate, pH 8). One of the most potent soil microcides is anhydrous ammonia, a common N fertilizer (31). The soil zone affected may extend several centimeters from the point of injection and adversely affect local roots and microorganisms.

In spite of the foregoing there are no reports of permanent damage to soil microbial populations by chemical fertilizers. The reasons may be that relatively small quantities are normally applied and the effects are localized close to the fertilizer granules. The localized hostile environment gradually dissipates through chemical or biochemical reactions, transformations, or diffusion of ions from the granule site. Microorganisms at the periphery of the concentrated zone may carry out transformations that convert $NH_3(NH_4^+)$ to NO_2^- and hence NO_3^- (32). There is some concern that H^+ ion production during this transformation process may exceed the effects of acid rain on acidification of agricultural soils. Acid soils may be amended by limestone or other liming materials to restore productivity. Also, there is considerable interest in using nitrification inhibitors such as dicyandiamide to inhibit nitrifier activity and delay the transformation of NH_4^+ to NO_3^- and thereby lessen the potential for loss through leaching or denitrification during the crop growth period.

7. LIVESTOCK MANURES AND SOIL MICROORGANISMS

7.1 Unprocessed Manures

Microorganisms are added with manure to soil. Some species may survive whereas others disappear as the manure decomposes. Morrison and Martin (33) provided information on the survival of pathogenic microorganisms applied in manures. In this section the microbial processes stimulated by animal manures will be discussed.

Manures usually provide a rich organic substrate for soil microflora and fauna. The decomposition of manures in soil is influenced by soil conditions such as moisture, aeration, and temperature as well as manure characteristics such as C:N ratio and lignin and "bedding" (used for the animals in the barn) contents. The proportion of the total N in the ammoniacal form also influences the rate of decomposition as well as the kind of bedding (e.g., straw vs. wood sawdust). Wood sawdust normally has a much higher C:N ratio (>200/1) than straw (typically about 60/1). The high C:N ratio usually results in greater immobilization of NH_4^+ or NH_3 by microorganisms and consequently a slowing of the decomposition process. This process is enhanced by higher ammoniacal N content in manure but slowed by higher lignin content.

A conceptionalization of several important microbial processes involving C and N is shown in Fig. 4. These processes affect the fate of the N in the manure and in the soil, regulating the availability of mineral N (NH_4^+ and NO_3^-) for plants. Depending on factors such as the C:N ratio, bedding kind and content, and ammonium content of different manures, these processes proceed at varying rates. Figure 5 shows the rate of N gas production through denitrification in a silt loam soil treated with liquid and solid cattle manures during a 49-day period in the early growing season. It is evident that the manures, particularly the straw-bedded solid cattle manure, resulted in the highest denitrification rates. This was due to the greater organic substrate availability to denitrifiers in this manure. Second, it is noteworthy that the rate was highly dependent on the soil water

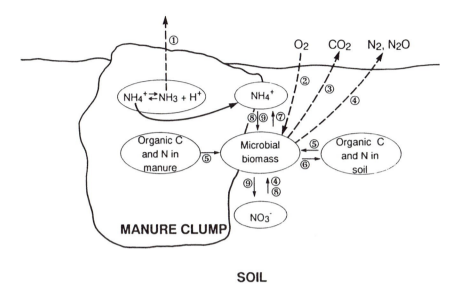

Figure 4. Conceptualization of important processes occurring during microbial decomposition of a partially exposed manure clump in soil. 1, ammonia volatilization; 2, oxygen consumption and diffusion to microorganisms; 3, carbon dioxide production and diffusion from microorganisms; 4, denitrification; 5, consumption of organic C and N substrates by microorganisms; 6, return of organic C and N in dead microbial cells to soil organic matter; 7, mineralization of organic N to ammonium; 8, microbial immobilization of ammonium and nitrate; 9, nitrification.

content (rainfall) and consequently was quite variable. The periodic increase in soil moisture decreases air diffusion to facultatively heterotrophic bacteria, resulting in an O_2 deficiency which, in the presence of large quantities of organic substrate and NO_3^-, stimulates the denitrification process.

It should be noted that manures contain other mineral nutrients for plants such as phosphorus, potassium, magnesium, and zinc. The organically bound phosphorus is released into the soil solution through a microbially mediated mineralization process. Once in the soil solution it may be absorbed by plants or be precipitated as various phosphate minerals.

7.2 Manure Composting

Composting of manures and other organic materials involves an aerobic microbial process (34). The procedure usually begins by the stacking of manures in small piles or windrows. Aeration of the moist manure is necessary and is usually accomplished by turning or pumping air through the composting mass.

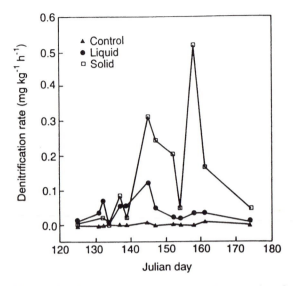

Figure 5. Denitrification rates in soil cores taken from the 2.5–7.5 cm layer of silt soil during a 49-day period following application and incorporation of liquid cattle manure and solid (straw bedded) cattle manure. (From D. W. Bergstrom, M. Tenuta, and E. G. Beauchamp, unpublished data.)

A desirable C:N ratio is between 20/1 and 25/1. Many manures have a C:N ratio lower than this range but may be adjusted by adding other organic substances such as straw or wood shavings or chips as additional bedding. If the C:N ratio is too low, excessive N is lost. If the C:N is too high, the composting process is slowed because of inadequate N availability to the microorganisms.

Shortly after the composting process begins, the temperature of the manure mass increases to over 60°C. Mesophilic microbial activity diminishes and is replaced by thermophilic microbial activity. After 1 or 2 weeks, microbial activity gradually subsides as the availability of easily decomposable substrate diminishes, and over several months the manure returns to ambient temperatures.

The benefits of composting include reduced viable weed seed and pathogen survival due to the heat buildup. The lower volume of manure reduces spreading time and energy expenditure during handling. Also odor is reduced during the composting process.

The composting procedure requires additional labor input. The process results in the loss of C as CO_2 and N from volatilization or leaching if the composting procedure is carried out on bare soil. The result of the loss of C is lower availability of C for soil microbes when the composted material is applied. The decomposition of raw manure in the soil enhances microbial effects on soil structure improvement. Moreover, the prior loss of N during the composting process results in lower N availability for plants. The N remaining in the composted manure occurs mainly in slowly degradable forms and is only slowly available to plants (35).

8. CONCLUSIONS

Soil microorganisms carry out vital functions in conjunction with the various agricultural practices we follow to grow crops efficiently. Along with environmental conditions such as temperature and moisture, different practices such as tillage methods or manure application affect the size and activities of microbial populations. The microbial processes involved in the cycling of soil nitrogen and plant residue decomposition are probably the most important.

Some microorganisms are involved in special functions. An example is the enhancement of soil phosphorus availability to plants by mycorrhizal fungi. Another well-known example is the microbial fixation of atmospheric nitrogen. Superior symbiotic N_2-fixing bacterial strains have been developed and are used to inoculate the seeds of cultivated leguminous plants in a particularly successful commercial endeavor.

REFERENCES

1. E. N. Mishustin, *Soils Fertilizers 19*: 385 (1956).
2. J. W. Doran, *Soil Sci. Soc. Am. J. 44*: 765 (1980).
3. D. M. Linn and J. W. Doran, *Soil Sci. Soc. Am. J. 48*: 1267 (1984).
4. E. D. Vance, P. C. Brookes, and D. S. Jenkinson, *Soil Biol. Biochem. 19*: 703 (1987).
5. M. R. Carter, *Soil Tillage Res. 7*: 29 (1986).
6. D. A. Angers, A. Pesant, and J. Vigneux, *Soil Sci. Soc. Am. J. 56*: 115 (1992).
7. W. E. Larson, R. F. Holt, and C. W. Carlson, *Crop Residue Management Systems*. (W. R. Oschwald, ed.), Am. Soc. Agron., Madison, Wisconsin (1978).
8. D. A. Angers, N. Bissonnette, A. Légére, and N. Samson, *Can. J. Soil Sci. 73*: 39 (1993).
9. M. R. Carter, *Biol. Fertil. Soils 11*: 135 (1991).
10. R. Chang, MSc Thesis, University of Guelph (1989).
11. D. M. Granatstein, D. F. Bezdicek, V. L. Cochran, L. F. Elliott, and J. Hammel, *Biol. Fertil. Soils 5*: 265 (1987).
12. J. M. Lynch and L. M. Panting, *Soil Biol. Biochem. 12*: 29 (1980).
13. M. H. Miller and T. P. McGonigle, *Mycorrhizas in Ecosystems* (D. J. Read, D. H. Lewis, A. H. Fitter, and I. J. Alexander, eds.), C.A.B. International, Oxon, England (1992).

14. T. P. McGonigle, D. G. Evans, and M. H. Miller, *New Phytol. 116*: 629 (1990).
15. R. M. N. Kucey, H. H. Janzen, and M. E. Leggett, *Adv. Agron. 42*: 199 (1989).
16. R. M. N. Kucey, *Appl. Environ. Microbiol. 53*: 2699 (1987).
17. R. S. Smith, *Can. J. Microbiol. 38*: 485 (1992).
18. C. Vidor, *World Soybean Research Conf. IV Proceedings.* (A. J. Pascale, ed.), Orientacion Grafica Editora, Buenos Aires, Argentina, p. 438 (1989).
19. A. Diatloff, *Aust. J. Exp. Agric. Anim. Husb. 9*: 357 (1969).
20. R. W. Weaver, L. R. Frederick, and L. C. Dumenil, *Soil Sci. 114*: 137 (1972).
21. A. Merrien, B. Dechamps, M. J. Durouchoux, C. Maisonneuve, and B. Roux, *World Soybean Research Conf. IV Proceedings.* (A. J. Pascale, ed.), Orientacion Grafica Editora, Buenos Aires, Argentina, p. 485 (1989).
22. D. J. Hume and D. H. Blair, *Can. J. Microbiol. 38*: 588 (1992).
23. M. J. Trinick and J. Galbraith, *New Phytol. 86*: 17 (1980).
24. R. W. Ridge, B. G. Rolfe, Y. Jing, and E. C. Cocking, *Symbiosis 14*: 345 (1992).
25. M. D. Stower, *Symbiotic Nitrogen Fixation Technology* (G. H. Elkan, ed.) Marcel Dekker, New York, p. 29 (1987).
26. D. Callanham, P. Del Trediei, and J. G. Torrey, *Science 199*: 899 (1978).
27. A. Quispel and T. Tak, *New Phytol. 81*: 587 (1978).
28. T. A. Lumpkin and D. P. Bartholomew, *Crop Sci. 26*: 107 (1986).
29. D. A. Zuberer, *Symbiotic Nitrogen Fixation Technology* (G. H. Elkan, ed.) Marcel Dekker, New York, p. 95 (1987).
30. R. F. Harris. *Water Potential Relations in Soil Microbiology.* (J. F. Parr, W. R. Gardner and L. F. Elliott, eds.), Soil Sci. Soc. Am., Madison, Wisconsin, p. 23 (1981).
31. C. F. Eno, W. G. Blue, and J. M. Good Jr., *Soil Sci. Soc. Am. Proc. 19*: 55 (1955).
32. J. F. Parr, *Soil Sci. 107*: 94 (1969).
33. S. M. Morrison and K. L. Martin, *Land as a Waste Management Alternative*, Proc. 1976 Cornell Agricultural Waste Management Conference, Ann Arbor Science Publishers, Ann Arbor, Michigan, p. 371 (1977).
34. J. K. R. Gasser, *Composting of Agricultural and Other Wastes*. Elsevier Applied Science Publishers, New York (1985).
35. J. W. Paul and E. G. Beauchamp, *Can. J. Soil Sci. 73*: 253 (1993).

Index